Current Topics in
Developmental Biology
Volume 81

Multiscale Modeling of Developmental Systems

Series Editor

Gerald P. Schatten
Director, PITTSBURGH DEVELOPMENTAL CENTER
Deputy Director, Magee-Women's Research Institute
Professor and Vice-Chair of Ob-Gyn-Reproductive Sci. & Cell Biol.-Physiology
University of Pittsburgh School of Medicine
Pittsburgh, Pennsylvania 15213

Editorial Board

Peter Grüss
Max-Planck-Institute of Biophysical Chemistry
Göttingen, Germany

Philip Ingham
University of Sheffield, United Kingdom

Mary Lou King
University of Miami, Florida

Story C. Landis
National Institutes of Health
National Institute of Neurological Disorders and Stroke Bethesda, Maryland

David R. McClay
Duke University, Durham, North Carolina

Yoshitaka Nagahama
National Institute for Basic Biology, Okazaki, Japan

Susan Strome
Indiana University, Bloomington, Indiana

Virginia Walbot
Stanford University, Palo Alto, California

Founding Editors

**A. A. Moscona
Alberto Monroy**

Current Topics in Developmental Biology
Volume 81

Multiscale Modeling of Developmental Systems

Edited by

Santiago Schnell
Indiana University School of Informatics and
Biocomplexity Institute

Philip K. Maini
Centre for Mathematical Biology and
Oxford Centre for Integrative Systems Biology
University of Oxford

Stuart A. Newman
Department of Cell Biology and Anatomy
New York Medical College

Timothy J. Newman
Department of Physics and School of Life Sciences
Arizona State University

Published in Affiliation with the Society for Developmental Biology

ELSEVIER

AMSTERDAM • BOSTON • HEIDELBERG • LONDON
NEW YORK • OXFORD • PARIS • SAN DIEGO
SAN FRANCISCO • SINGAPORE • SYDNEY • TOKYO
Academic Press is an imprint of Elsevier

Cover Photo credit: Published in Affiliation with the Society for Developmental Biology

Academic Press is an imprint of Elsevier
84 Theobald's Road, London WC1X 8RR, UK
Radarweg 29, PO Box 211, 1000 AE Amsterdam, The Netherlands
Linacre House, Jordan Hill, Oxford OX2 8DP, UK
30 Corporate Drive, Suite 400, Burlington, MA 01803, USA
525 B Street, Suite 1900, San Diego, CA 92101-4495, USA

First edition 2008

Copyright © 2008 Elsevier Inc. All rights reserved

No part of this publication may be reproduced, stored in a retrieval system or transmitted in any form or by any means electronic, mechanical, photocopying, recording or otherwise without the prior written permission of the publisher

Permissions may be sought directly from Elsevier's Science & Technology Rights Department in Oxford, UK: phone (+44) (0) 1865 843830; fax (+44) (0) 1865 853333; email: permissions@elsevier.com. Alternatively you can submit your request online by visiting the Elsevier web site at http://elsevier.com/locate/permissions, and selecting *Obtaining permission to use Elsevier material*

Notice
No responsibility is assumed by the publisher for any injury and/or damage to persons or property as a matter of products liability, negligence or otherwise, or from any use or operation of any methods, products, instructions or ideas contained in the material herein. Because of rapid advances in the medical sciences, in particular, independent verification of diagnoses and drug dosages should be made

ISBN: 978-0-12-374253-7

ISSN: 0070-2153

For information on all Academic Press publications
visit our website at books.elsevier.com

Printed and bound in USA

08 09 10 11 12   10 9 8 7 6 5 4 3 2 1

Working together to grow libraries in developing countries

www.elsevier.com | www.bookaid.org | www.sabre.org

ELSEVIER   BOOK AID International   Sabre Foundation

Contents

Contributors xiii
Introduction xvii

1

Models of Biological Pattern Formation: From Elementary Steps to the Organization of Embryonic Axes
Hans Meinhardt

    I. Introduction   2
   II. Primary Pattern Formation by Local Self-Enhancement and Long-Ranging Inhibition   6
  III. The Two Main Body Axes   21
  IV. Subpatterns   33
   V. Conclusion   53
       Acknowledgments   54
       References   54

2

Robustness of Embryonic Spatial Patterning in *Drosophila melanogaster*
David Umulis, Michael B. O'Connor, and Hans G. Othmer

    I. Introduction   65
   II. Robustness in the Developmental Context   69
  III. Scaling of AP Patterning in *Drosophila*   77
  IV. Models of the Segment Polarity Network   80
   V. Dorsal–Ventral Patterning in *Drosophila*   87
  VI. Conclusions   105
       Acknowledgments   108
       Note added in proof   108
       References   108

3

Integrating Morphogenesis with Underlying Mechanics and Cell Biology
Lance A. Davidson

    I. Introduction 113
   II. *Xenopus laevis* as a Model System 114
  III. Distinct and Separable Tissue-Scale Processes 115
  IV. Complex Trajectories through Dynamic Microenvironments 121
   V. At the Cell-Scale: Act Locally Move Globally 124
  VI. Molecular-Scale: Mechanics, Adhesion, and Traction 127
 VII. Modeling Morphogenesis: A Grand Challenge 128
      Acknowledgments 129
      References 129

4

The Mechanisms Underlying Primitive Streak Formation in the Chick Embryo
Manli Chuai and Cornelis J. Weijer

     I. Introduction 136
    II. Structure of the Early Embryo 136
   III. Experimental Observations of Streak Formation 137
   IV. Mesoderm Induction 140
    V. Cellular Mechanisms of Streak Formation 143
   VI. Mechanisms of Movement 150
  VII. Challenges for Modeling and Computational Approaches 151
 VIII. Outlook 152
      References 153

5

Grid-Free Models of Multicellular Systems, with an Application to Large-Scale Vortices Accompanying Primitive Streak Formation
Timothy J. Newman

    I. Introduction 158
   II. Grid-Free Models of Multicellular Systems 158
  III. Recent Experimental Results Concerning Primitive Streak Formation 164
  IV. Components of the Planar Cell Polarity Mechanism 167
   V. Phenomenological Cell-Based Model of the Chick Epiblast 170

Contents

    VI. Computer Simulations: Formation and Maintenance of Vortices during Streak Formation 170
   VII. Discussion and Conclusions 174
       Acknowledgments 179
       References 181

6

Mathematical Models for Somite Formation

Ruth E. Baker, Santiago Schnell, and Philip K. Maini

     I. Introduction 183
    II. Models for Somite Formation 186
   III. Discussion 198
   IV. Perspective 200
       Acknowledgments 200
       References 200

7

Coordinated Action of N-CAM, N-cadherin, EphA4, and ephrinB2 Translates Genetic Prepatterns into Structure during Somitogenesis in Chick

James A. Glazier, Ying Zhang, Maciej Swat, Benjamin Zaitlen, and Santiago Schnell

     I. Introduction 206
    II. Patterns of Gene Expression and Protein Distribution during Somitogenesis 208
   III. From Genetic Oscillators to Adhesion/Repulsion-Protein Patterns 211
   IV. From Adhesion-Protein Patterns to Segmentation 214
    V. Computer Simulation of Segmentation 215
   VI. Results and Discussion 222
   VII. Conclusion 229
       Acknowledgments 230
       Introduction to Appendices 230
       Appendix A. Python Code to Execute Somitogenesis Simulations (somite.py) 231
       Appendix B. CC3D ML Code to Execute Somitogenesis Simulations (somite.xml) 233

Appendix C. Python Steppables for Somitogenesis Simulations (somiteSteppables.py) 236
References 244

8

Branched Organs: Mechanics of Morphogenesis by Multiple Mechanisms
Sharon R. Lubkin

    I. Introduction 249
    II. Background 251
    III. Candidate Physical Mechanisms 253
    IV. Models of Branching 258
    V. Discussion 263
        Acknowledgments 265
        References 265

9

Multicellular Sprouting during Vasculogenesis
Andras Czirok, Evan A. Zamir, Andras Szabo, and Charles D. Little

    I. Introduction 270
    II. Empirical Data, *in vivo* 272
    III. Elongated Structures, *in vitro* 277
    IV. Mathematical Model of Sprout Formation 281
    V. Conclusions 287
        Acknowledgments 287
        References 287

10

Modeling Lung Branching Morphogenesis
Takashi Miura

    I. Introduction 291
    II. Modeling *in vitro* Lung Branching Morphogenesis 296
    III. Functional Modeling—Structure and Air Flow 300
    IV. Future Directions 300
    V. Numerical Simulations of Branching Morphogenesis Models 301
        References 306

11

Multiscale Models for Vertebrate Limb Development

Stuart A. Newman, Scott Christley, Tilmann Glimm, H. G. E. Hentschel, Bogdan Kazmierczak, Yong-Tao Zhang, Jianfeng Zhu, and Mark Alber

    I. Introduction 312
   II. Tissue Interactions and Gene Networks of Limb Development 313
  III. Models for Chondrogenic Pattern Formation 316
  IV. Simulations of Chondrogenic Pattern Formation 323
   V. Discussion and Future Directions 332
       Acknowledgments 336
       References 336

12

Tooth Morphogenesis *in vivo*, *in vitro*, and *in silico*

Isaac Salazar-Ciudad

     I. Introduction 342
    II. The Use of Mammalian Tooth for Developmental and Evolutionary Biology 343
   III. Morphological Changes During Tooth Development 344
   IV. Gene Networks in Tooth Development 347
    V. The Formation of the Cusps 348
   VI. Spacing Between Cusps 349
  VII. Morphodynamic Model 1 350
 VIII. Model 1 and Tooth Dynamics 353
   IX. Morphodynamic Model 2 355
    X. What Do Model Dynamics Reveal About Developmental Dynamics 358
   XI. Tooth Model in Comparison to Other Models of Organ Development 366
  XII. Concluding Remarks 367
       Acknowledgments 368
       References 368

13

Delaunay-Object-Dynamics: Cell Mechanics with a 3D Kinetic and Dynamic Weighted Delaunay-Triangulation
Michael Meyer-Hermann

    I. Overview of Methods in Theoretical Biology 374
    II. Delaunay-Based Interaction 378
    III. Voronoi-Cells Approximate Real Cells 380
    IV. Delaunay-Dynamics 382
    V. Equation of Motion for Vertices 384
    VI. Mechanics Matters 392
    VII. Conclusion 396
        Acknowledgments 396
        References 397

14

Cellular Automata as Microscopic Models of Cell Migration in Heterogeneous Environments
Haralambos Hatzikirou and Andreas Deutsch

    I. Introduction 402
    II. Idea of the LGCA Modeling Approach 406
    III. LGCA Models of Cell Motion in a Static Environment 408
    IV. Analysis of the LGCA Models 414
    V. Results and Discussion 420
        Acknowledgments 424
        References 432

15

Multiscale Modeling of Biological Pattern Formation
Ramon Grima

    I. Introduction 436
    II. Quantitative Modeling 437
    III. Building Cellular and Tissue-Level Models for a Simple Biological System 439
    IV. Mean-Field Theory and the Interrelationship of Models at Different Spatial Scales 444

V. Multiple Scale Analysis 447
VI. Discussion 457
   References 458

16

Relating Biophysical Properties Across Scales

Elijah Flenner, Francoise Marga, Adrian Neagu, Ioan Kosztin, and Gabor Forgacs

I. Introduction 462
II. Theory and Computer Modeling 463
III. Results 470
IV. Conclusions 480
   Acknowledgments 482
   References 482

17

Complex Multicellular Systems and Immune Competition: New Paradigms Looking for a Mathematical Theory

Nicola Bellomo and Guido Forni

I. Introduction 485
II. Conceptual Lines Towards a Mathematical Biological Theory 486
III. From Hartwell's Theory of Modules to Mathematical Structures 488
IV. A Simple Application and Perspectives 491
V. What Is Still Missing for a Biological Mathematical Theory 496
   References 500

Index 503
Contents of Previous Volumes 515

Contributors

Numbers in parentheses indicate the pages on which the authors' contributions begin.

Mark Alber (311), Interdisciplinary Center for the Study of Biocomplexity, University of Notre Dame, Notre Dame, Indiana 46556; Department of Mathematics, University of Notre Dame, Notre Dame, Indiana 46556

Ruth E. Baker (183), Centre for Mathematical Biology, Mathematical Institute, University of Oxford, 24-29 St. Giles,' Oxford OX1 3LB, UK

Nicola Bellomo (485), Department of Mathematics, Politecnico, Turin, Italy

Scott Christley (311), Department of Computer Science and Engineering, University of Notre Dame, Notre Dame, Indiana 46556; Interdisciplinary Center for the Study of Biocomplexity, University of Notre Dame, Notre Dame, Indiana 46556

Manli Chuai (135), Division of Cell and Developmental Biology, Wellcome Trust Biocentre, College of Life Sciences, University of Dundee, Dundee DD1 5EH, United Kingdom

Andras Czirok (269), Department of Anatomy and Cell Biology, University of Kansas Medical Center, Kansas City, Kansas 66160; Department of Biological Physics, Eotvos University, Budapest 1117, Hungary

Lance A. Davidson (113), Department of Bioengineering, University of Pittsburgh, Pittsburgh, Pensylvania 15260

Andreas Deutsch (401), Center for Information Services and High-Performance Computing, Technische Universität Dresden, Nöthnitzerstr. 46, 01069 Dresden, Germany

Elijah Flenner (461), Department of Physics and Astronomy, University of Missouri-Columbia, Columbia, Missouri 65211

Gabor Forgacs (461), Department of Physics and Astronomy, University of Missouri-Columbia, Columbia, Missouri 65211; Department of Biological Sciences, University of Missouri-Columbia, Columbia, Missouri 65211

Guido Forni (485), Department of Clinical and Biological Sciences, Molecular Biotechnology Center, University of Turin, Turin, Italy

James A. Glazier (205), Biocomplexity Institute and Department of Physics, 727 East Third Street, Indiana University, Bloomington, Indiana 47405

Tilmann Glimm (311), Department of Mathematics, Western Washington University, Bellingham, Washington 98225

Ramon Grima (435), Institute for Mathematical Sciences, Imperial College, London SW7 2PG, United Kingdom

Haralambos Hatzikirou (401), Center for Information Services and High-Performance Computing, Technische Universität Dresden, Nöthnitzerstr. 46, 01069 Dresden, Germany

H. G. E. Hentschel (311), Department of Physics, Emory University, Atlanta, Georgia 30322

Bogdan Kazmierczak (311), Polish Academy of Sciences, Institute of Fundamental Technological Research, 00-049 Warszawa, Poland

Ioan Kosztin (461), Department of Physics and Astronomy, University of Missouri-Columbia, Columbia, Missouri 65211

Charles D. Little (269), Department of Anatomy and Cell Biology, University of Kansas Medical Center, Kansas City, Kansas 66160

Sharon R. Lubkin (249), Department of Mathematics, North Carolina State University, Raleigh, North Carolina 27695-8205; Department of Biomedical Engineering, North Carolina State University, Raleigh, North Carolina 27695-8205

Philip K. Maini (xvii, 183), Centre for Mathematical Biology, Mathematical Institute, University of Oxford, 24-29 St. Giles,' Oxford OX1 3LB, UK; Oxford Centre for Integrative Systems Biology, Department of Biochemistry, University of Oxford, South Parks Road, Oxford OX1 3QU, UK

Francoise Marga (461), Department of Physics and Astronomy, University of Missouri-Columbia, Columbia, Missouri 65211

Hans Meinhardt (1), Max-Planck-Institut für Entwicklungsbiologie, Spemannstr. 35, D-72076 Tübingen, Germany

Michael Meyer-Hermann (373), Frankfurt Institute for Advanced Studies (FIAS), Ruth-Moufang Str. 1, 60438 Frankfurt am Main, Germany

Takashi Miura (291), Department of Anatomy and Developmental Biology, Kyoto University Graduate School of Medicine, Yoshida Konoe-chou, Sakyo-Ku 606-8501, Japan; JST PRESTO

Adrian Neagu (461), Department of Physics and Astronomy, University of Missouri-Columbia, Columbia, Missouri 65211; University of Medicine and Pharmacy Timisoara, 300041 Timisoara, Romania

Stuart A. Newman (xvii, 311), Department of Cell Biology and Anatomy, New York Medical College, Valhalla, New York 10595

Timothy J. Newman (xvii, 152), Department of Physics and School of Life Sciences, Arizona State University, Tempe, Arizona 85287

Contributors

Michael B. O'Connor (65), Department of Genetics, Cell Biology and Development and Howard Hughes Medical Institute, University of Minnesota, Minneapolis, Minnesota 55455

Hans G. Othmer (65), School of Mathematics and Digital Technology Center, University of Minnesota, Minneapolis, Minnesota 55455

Isaac Salazar-Ciudad (341), Developmental Biology Program, Institute of Biotechnology, P.O. Box 56, FIN-00014, University of Helsinki, Helsinki, Finland

Santiago Schnell (xvii, 183, 205), School of Informatics and Biocomplexity Institute, 1900 East Tenth Street, Indiana University, Bloomington, Indiana 47406; Complex Systems Group, Indiana University School of Informatics, 1900 East 10th Street, Eigenmann Hall 906, Bloomington, Indiana 47406

Maciej Swat (205), Biocomplexity Institute and Department of Physics, 727 East Third Street, Indiana University, Bloomington, Indiana 47405

Andras Szabo (269), Department of Biological Physics, Eotvos University, Budapest, 1117 Hungary

David Umulis (65), Department of Chemical Engineering and Materials Science, University of Minnesota, Minneapolis, Minnesota 55455

Cornelis J. Weijer (135), Division of Cell and Developmental Biology, Wellcome Trust Biocentre, College of Life Sciences, University of Dundee, Dundee DD1 5EH, United Kingdom

Benjamin Zaitlen (205), Biocomplexity Institute and Department of Physics, 727 East Third Street, Indiana University, Bloomington, Indiana 47405

Evan A. Zamir (269), Department of Anatomy and Cell Biology, University of Kansas Medical Center, Kansas City, Kansas 66160

Ying Zhang (205), Biocomplexity Institute and Department of Physics, 727 East Third Street, Indiana University, Bloomington, Indiana 47405

Yong-Tao Zhang (311), Interdisciplinary Center for the Study of Biocomplexity, University of Notre Dame, Notre Dame, Indiana 46556; Department of Mathematics, University of Notre Dame, Notre Dame, Indiana 46556

Jianfeng Zhu (311), Department of Mathematics, University of Notre Dame, Notre Dame, Indiana 46556

Introduction

Santiago Schnell,* Philip K. Maini,[†,‡] Stuart A. Newman,[§] and Timothy J. Newman[¶]
*Indiana University School of Informatics and Biocomplexity Institute, 1900 East Tenth Street, Eigenmann Hall 906, Bloomington, Indiana 47406
[†]Centre for Mathematical Biology, Mathematical Institute, University of Oxford, 24-29 St. Giles,' Oxford OX1 3LB, United Kingdom
[‡]Oxford Centre for Integrative Systems Biology, Department of Biochemistry, University of Oxford, South Parks Road, Oxford OX1 3QU, United Kingdom
[§]Department of Cell Biology and Anatomy, New York Medical College, Valhalla, New York 10595
[¶]Department of Physics and School of Life Sciences, Arizona State University, Tempe, Arizona 85287

Organisms are composed of cells, and cells are in turn made up of organelles, macromolecular complexes, proteins, polysaccharides, lipids and small molecules. The living world thus has a hierarchical organization as far as its composition and architecture are concerned (Harris, 1999). Causal interactions in biological systems, in contrast, move both up and down these scales. On the level of the individual organism, molecules are synthesized in a spatiotemporal fashion as result of, and resulting in, changes in cell number and state, and organismal geometry and topology. On the level of the organism type, or species, boundary conditions and compositional details are propagated across generations largely through the continuity of the genes. So while organisms are self-organizing systems, they are not entirely so. The multicomponent, multiscale, hybrid nature of living systems continues to provide novel challenges to theoretical and experimental biology.

Occupying an intermediate level of the organizational hierarchy, the cell provides a focal plane from which one can scale upward and downward. Because the cell is also the minimal unit of life, cooperation among such units is the driving principle of multicellular organization. The cell's central role was already recognized by those scientists who first came to understand that they constitute the bodies of all living things. This conviction is encapsulated in a legendary phrase of Raspail: "*Donnez-moi une vésicule organique douée de vitalité, et je vous rendrai le monde organisé*"[1] (Raspail, 1833).

Almost two centuries later Raspail's assertion remains a still-distant goal, appearing to confirm Kant's doubts that there would ever be a "Newton of the grass blade" (Kant,

---

[1] This phrase can be translated as: "Give me an organic vesicle endowed with life and I will give you back the whole of the organized world."

1790). Multicellular systems have proven to be very complex. Whether cells are members of a community or components of a tissue, they are interacting continuously with one another and with their local environment. Cells live in an aqueous porous medium, where signaling molecules are subject to diffusion, transport and decay along the flow patterns generated by the movements of the cells and other structures. Cells undergo switches in type and thus in biosynthetic capability and behavior. Experimental work over the past three decades has established connections among these phenomena, and their molecular underpinnings and correlates. Some progress has been made in understanding multicellular developing systems at the theoretical level, that is, in the form of mathematical or computational models encompassing the dynamical behavior of cells and molecules at realistic spatiotemporal scales. This approach, though currently less developed than the experimental ones, is nonetheless essential for an understanding of morphogenesis and cellular pattern formation. This volume provides descriptions of recent research in this area using a wide range of systems and modeling strategies.

Constructing models is something of an art. Several models might be consistent with the data at hand; they might even yield the same mathematical or computational representation. The first role of a model is to test a verbal description arising from a biological hypothesis, using the language of mathematics which, unlike verbal reasoning, allows the outcome of complex nonlinear interactions to be precisely computed. If the model produces the incorrect predictions, then the biological hypothesis is incorrect. At a minimum, modeling can refine intuition. But a rigorously developed model coupled with experiments has the potential to accomplish far more.

The most realistic models of multicellular systems require the use of subcellular models, which can take into account biochemical kinetics and cytoskeletal mechanics. Such models are multiscale. When the subcellular and supercellular regimes are coupled these models can be used to investigate the large-scale, often visible, patterns seen in tissues.

Whether on one scale or many, then, modeling can serve two purposes. When the significant details of a developmental process are known, a mathematical model can be used as a surrogate for the living system, providing a way to carry out virtual experiments. These have the potential advantage of being more comprehensive and more rapidly performed than corresponding *in vivo* or *in vitro* tests. In such cases the model can provide a proof-of-principle that the known components are indeed sufficient to replicate important behaviors and that they interact as expected. Where important details of the developmental process are not known, modeling can serve as a tool for testing hypotheses and generating predictions. Increased understanding can arise from both approaches. Therefore we need a suite of models, each designed to address specific questions (Schnell *et al.*, 2007).

Introduction

The Volume

This volume arises from the Ninth Biocomplexity Workshop held at Lake Monroe, Bloomington, Indiana in May 2006. This event was part of the Biocomplexity Institute workshop series organized by the Biocomplexity Institute and Indiana University School of Informatics. Biocomplexity 9 was titled "Multiscale modeling of multicellular systems: An interdisciplinary workshop." Participants discussed current and future theoretical and experimental problems in the study of multicellular systems. Researchers were brought together from many disciplines, including experimental and theoretical developmental biology, applied mathematics, biophysics, engineering and computer science. In addition to containing contributions from most participants of Biocomplexity 9, the present volume has chapters contributed by several leaders in the field who were not in attendance.

We have organized the volume by starting with the chapters dealing with general concepts of pattern formation, a focus of much mathematical and theoretical biology over the last three decades (Chapters 1 and 2). Chapters 3–5 deal with the process of gastrulation, the period in embryogenesis during which cell patterns and morphological complexity are first established from relatively simple multicellular systems. After gastrulation in vertebrates, the mesoderm lying along the dorsal side of the embryo to either side of the notochord gives rise to the serially repeated somites. In Chapters 6 and 7, the reader will find two contributions dealing with models of somitogenesis, an area that has experienced a productive confluence of theoretical and experimental work. In the five chapters that follow (Chapters 8–12), the development of specific structures and organs crucial to the formation of fully functional organisms, such as lungs, glands, limbs and teeth, are explored. This volume on theoretical approaches would be incomplete without the introduction of new methodologies. Chapters 13–17 illustrate important advances in theoretical approaches to develop more realistic models of multicellular systems.

In the middle of the last century, Alan Turing showed, using a simple mathematical model, that a system of chemical reactions with stable spatially uniform dynamics in the absence of diffusion, could be destabilized by diffusion so as to assume a stable spatially nonuniform configuration (Turing, 1952). The result was highly counterintuitive at the time, since diffusion generally evens out spatial heterogeneities. Turing suggested that the chemical pattern set up by the instability could serve as a prepattern for a cellular response. The plausibility of Turing's "reaction–diffusion" model as a biological mechanism was reinforced when Gierer and Meinhardt presented a realistic reaction–diffusion system that undergoes the Turing instability and produces a pattern. This system comprises a pair of reacting chemicals labeled activator and inhibitor (Gierer and Meinhard, 1972). The first activates the production of itself and the second chemical; the inhibitor in turn inhibits the growth of the autocatalytic activator. Pattern formation is possible if the activator in this system diffuses much more slowly than the inhibitor, and has a shorter half-life. This led to an important principle of pattern formation: activation at short range coupled with inhibition at long range.

This principle has proved to have general utility in developing systems even when the morphogenetic signals and means of their propagation are not literally soluble molecules and free diffusion. In Chapter 1, Hans Meinhardt shows in an elegant fashion how simple activator–inhibitor systems can produce cell patterns and morphogenetic changes reminiscent of those observed in many areas of development and how evolution may have employed similar dynamics in different ways to generate different forms in different taxonomic groups.

The paper by David Umulis and coworkers (Chapter 2) shows that patterns can arise by different mechanisms. One of the major problems of pattern formation is discovering the mechanisms of localized production and transport that generate positional information. During the last 30 years, molecular biologists have focused on studying molecular components involved in signal transduction and gene expression in a number of model systems in developmental biology. Umulis *et al.* focus their attention on two patterns in the *Drosophila melanogaster* embryo which have been studied extensively by molecular developmental biologists: these are the anterior–posterior and dorsal–ventral patterning of the embryo. They show how molecular components are integrated in networks, and how these networks transduce the inputs they receive and produce the desired patterns of gene expression. Umulis *et al.* discuss a number of different aspects of robustness in *Drosophila* embryonic patterning and show how the models lead to new insights concerning scale-invariance in anterior–posterior patterning, the role of network topology and signature in the switching network used for control of the segment polarity genes, and the role of signaling via heterodimers in dorsal–ventral patterning.

The stunning successes of molecular biology in recent decades have mainly provided the ingredients for the complex mechanisms of morphogenesis. The challenge now facing us is to understand how these entities integrate in the correct manner so that the whole is greater than merely the sum of the parts. Biochemical dynamics and tissue mechanics must play key roles in this process. In Chapter 3, Lance Davidson shows that there is no single molecular mechanism that controls morphogenesis, but rather there is a collection of cellular processes that work together to generate the architecture and modulate the forces responsible for changes in tissue form. Davidson summarizes the early development of the frog *Xenopus laevis* from a biomechanical perspective. He describes the cells, their behaviors and the unique microenvironments they traverse during gastrulation, demonstrating the important role of tissue mechanics in development.

Manli Chuai and Cornelis Weijer discuss the current understanding of the mechanisms underlying the initial phases of gastrulation, in particular the formation of the chick primitive streak (Chapter 4). The genetic basis of anterior–posterior axis development, germ layer and streak formation has been studied extensively. However, because of the small size of the embryo at these stages, little attention has been paid to the cell movement patterns associated with gastrulation. Chuai and Weijer review current experimental evidence of gastrulation movements and the possible cellular mechanisms underlying these processes. Cellular mechanisms involved in gastrulation

Introduction

may include oriented cell division, cell–cell intercalation, chemotactic cell movement in response to attractive and repulsive signals and a combination of chemotaxis and "contact following." Chuai and Weijer critically examine the experimental evidence in favor for and against these different mechanisms and outline open questions in gastrulation research. An important conclusion of their work is that mathematical models and computer simulations have a fundamental role to play in furthering our understanding of gastrulation, since many of the interactions between cell signaling and movement are dynamic and nonlinear.

In Chapter 5, Timothy Newman explores a mechanism based on planar cell polarity to explain coordinated cell movement lateral to the primitive streak during its formation. These complex cell movements were recently observed by the Weijer group (Cui et al., 2005). Newman shows via computer simulations that planar cell polarity can generate large-scale cell movement resulting in two counter-rotating vortices, similar to those observed experimentally. The complexity of coordinated cell motion is modeled with a new computational method for studying multicellular systems, known as the Subcellular Element Model. This new methodology is a powerful tool for modeling intracellular mechanisms and adaptive cell shape changes. Newman also provides a brief review of grid-free modeling approaches. In this volume, the readers will find other methods for modeling morphogenesis at the cell level, such as the Cellular Potts Model (Chapters 7 and 11), finite element methods (Chapter 11), agent-based methods (Chapter 13) cellular automata (Chapter 14), and Monte Carlo simulations (Chapter 16). All these methods have helped us in understanding the important role of individual cells and their interactions in modeling morphogenesis.

In vertebrate embryos, the anterior–posterior axis segments into similar morphological units, known as somites, after the formation of the primitive streak. These segments constitute a prepattern for the formation of the vertebrae, ribs and other associated repetitive features of the body axis. The formation of the repeated somites is one of the areas of developmental biology in which an interplay between experimental and theoretical investigations has met with great success. The idea that temporal oscillations may underlie the spatial periodicity of somite organization was anticipated by the evolutionary morphologist William Bateson in the late 19th century (Bateson, 1894), first made part of a specific model by the experimentalist Jonathan Cooke and the mathematician Christopher Zeeman more than 80 years later (Cooke and Zeeman, 1976) and confirmed experimentally by the group led by Olivier Pourquié two decades after that (Palmeirim et al., 1997). In Chapter 6, Baker et al. review and discuss a series of mathematical models motivated by newer experimental findings which account for different stages for somite formation. These models range from the creation of a genetic prepattern to the mechanisms involved in generating morphological somites. In his contribution, mentioned above, Hans Meinhardt proposes a reaction–diffusion model for somite formation. The paper by Baker et al. shows that the segmentation pattern can also arise from other mechanisms. As noted previously, this is a commonplace of mathematical biology—several models can produce the same results and make similar predictions. The challenge for theoreticians is to suggest carefully designed experiments to distinguish between models, and the challenge for the experimentalists is to design ways of doing these experiments, which may help decide between alternative mechanisms.

In Chapter 7, James Glazier and coworkers propose a model accounting for how cell determination and subsequent differentiation may translate into somite morphology. The model starts from an established prepattern of adhesive and repulsive molecules, which gives rise to the patterns of cell movement and morphological changes leading to segmentation. The simulations of Glazier *et al.* are implemented using the extended Cellular Potts Model. In this model cells are extended domains of pixels on a lattice. Cell interactions are described by an effective energy and fields of local concentrations of chemicals. The effective energy combines true energies, like cell–cell adhesion, and terms that mimic energies, e.g., the response of a cell to a chemotactic gradient.

The chapter by Sharon Lubkin (Chapter 8) describes the physical forces responsible for the morphogenesis of branched ducts such as those found in glands and in the lung. Developmental biologists have experimentally uncovered a great deal of information concerning the genes and signaling pathways of branching morphogenesis. Lubkin illustrates how development must also take into account the physical aspects of change in tissue shape and form. Indeed, physics can be seen as the primary means by which alterations in molecular expression bring about such tissue changes (Forgacs and Newman, 2005). In her contribution, Lubkin reviews a collection of relatively simply and physically justifiable models for branching morphogenesis. The models presented have potentially measurable parameters which can be used to quantify the relative contributions of different mechanisms to morphogenesis. A challenge for experimentalists is to develop contexts and methods within which these novel models can be tested.

In Chapter 9, Andras Czirok and coworkers review the patterning of the primary vascular plexus of warm blooded vertebrates. This is a process operating on various length scales. They show that the formation and rapid expansion of multicellular sprouts is a key mechanism by which endothelial cell clusters join to form an interconnected network. The work of Czirok *et al.* employs sophisticated microscopic methods to track cells and extracellular matrix fibers over an extended area of tissue. On the basis of these experimental observations, they propose a mathematical model of preferential attraction to elongated structures that can explain multicellular sprouting during vasculogenesis. This paper is another example of the benefits of theoretical frameworks for the comprehension of complex experimental results and *in vivo* reality.

Takashi Miura shows how mathematical models can help us understand the mechanism of branching morphogenesis, with emphasis on the lung (Chapter 10). With close attention to experimental findings, including those from his own laboratory, he reviews several models which make predictions concerning the *in vitro* systems. One of these models generates tree-like branching patterns by applying a set of simple rules iteratively. Models have been useful for understanding some aspects of branching morphogenesis, and they can be helpful for understanding the functional aspects of the bronchial tree. Miura explains how simple multiscale models can help in understanding both morphological and functional aspects of the bronchial tree. In addition, he shows how mathematical models can help developmental biologists with little experience in modeling gain new insights into the dynamics of pattern formation.

The development of the vertebrate limb has similarities to body axis segmentation in that a series of repetitive elements are formed, but also resembles branching morphogenesis in that more than one spatial dimension is needed to formally characterize the pattern. It is an apt developmental system for mathematical and computational modeling since it has been the subject of extensive experimental studies at the molecular and cellular level. In Chapter 11, Stuart Newman and coworkers describe features of the developing limb itself, as well as a planar culture system that utilizes isolated mesenchymal cells of the embryonic limb to provide a simplified *in vitro* model for chondrogenic pattern formation. They present several different kinds of models for the various patterning processes, including a Turing-type continuum "reactor–diffusion" model that generates the well-known proximodistal order of appearance of skeletal elements, as well as a multiscale discrete stochastic model that reproduces several quantitative aspects of pattern formation *in vitro*. Since the full *in vitro* developmental process has both continuous and discrete aspects, they suggest that the most satisfactory model will have a hybrid nature.

In Chapter 12, Isaac Salazar-Ciudad reviews still another developmental system that produces repeated elements—the dentition, or teeth, of vertebrates. As the author shows, moreover, this system brings into focus an important but neglected question in developmental pattern formation—the interplay between the released chemical signals termed morphogens and changing tissue geometry. It is commonly assumed that pattern formation proceeds in a "morphostatic" fashion, i.e., by the generation of spatiotemporal patterns of morphogens, followed by shape-changing tissue responses to these gradients. Salazar-Ciudad shows that this is not always the case: morphogen patterns can be dramatically affected when generated in concert with tissue rearrangements in three-dimensional-space in the form of complex developmental mechanisms that the author terms "morphodynamic." Using computational models he shows that morphodynamic mechanisms can predict important topographic properties of tooth cusp formation. Equally important, such mechanisms must dictate more complex genotype–phenotype relationships over the course of evolution than simpler morphostatic mechanisms.

The chapter by Michael Meyer-Hermann (Chapter 13) starts with an overview of mathematical methods in biology to model multicellular systems, paying special attention to the current agent-based methods, their strengths and limitations. By developing a new agent-based method, he shows that the construction of a physically well-defined modeling architecture for dynamic cellular systems is essential in order to gain predictive power. Only when the parameters of the model are observable quantities does the model acquire the potential to be falsified, which is a prerequisite of any scientific approach. The novel method developed by Meyer-Herman is an agent-based model for cell mechanics based on geometrical representations known as Delaunay triangulations and Voronoi tessellations. The methodology combines physically realistic cell mechanics with a reasonable computational load. He illustrates the power of the new method with two examples, avascular tumor growth and genesis of lymphoid tissue in cell-flow equilibrium.

Cells can be modeled as discrete entities distributed on a two-dimensional artificial grid. This type of simulation is called a cellular automaton, completed by assigning a set of rules governing the behavior of cells. In Chapter 14, Haralambos Hatzikirou and Andreas Deustch use cellular automata to understand the interplay of moving cells in the typical heterogeneous environment of multicellular systems. This is of great importance as cells move in a complex extracellular matrix composed of fibrillar structures, collagen matrices and other cells, which can affect the cell response to external signals. They introduce a special subtype of automaton, known as a lattice-gas cellular automaton, which has been widely used as a discrete model of fluid dynamics (Wolf-Gladrow, 2000). The extension of this automaton for investigation of cell–cell interactions and cell–environment interactions enables the observation of the macroscopic evolution of the cell population and estimation of cell dispersion speed under different environments.

Modeling approaches have strengths and weaknesses. Cellular automata are computationally efficient, and allow a wide range of cell behaviors to be implemented; however, they are also strictly defined on a grid, which may lead to artifacts, and do not easily lend themselves to analytic calculation. In Chapter 15, Ramon Grima addresses this and other issues. Multicellular systems are complex and can be studied at different scales. Grima discusses how mathematical models can be constructed at different spatial scales so as to provide insight into the fundamental biological processes central to cellular pattern formation. He concludes that the simultaneous theoretical and numerical analysis of models of the same biological system at different spatial scales provides a better understanding than a single-scale model.

In Chapter 16, Elijah Flenner and coworkers use a combination of experiment, theory and modeling to relate measured tissue-level biophysical quantities to subcellular parameters. Their work concentrates on the morphogenetic process of tissue fragment fusion, a phenomenon seen in many episodes of organogenesis, by following the coalescence of two contiguous multicellular aggregates. The time evolution of this process can be described accurately by the theory of viscous liquids. They study fusion by Monte Carlo simulations and a Cellular Particle Dynamics model equivalent to the Subcellular Element Model described in T. Newman's chapter. The multidisciplinary approach of combining experiment, theory and modeling provides a general and versatile way to study multiscale problems in living systems.

Nicola Bellomo and Guido Forni look at the problem of developing a general mathematical theory to model multicellular systems (Chapter 17). They use the mathematical kinetic theory for living particles to describe complex multicellular systems dealing with cell expansions, cell death and immune surveillance. Kinetic modeling describes the statistical evolution of large systems of interacting particles (e.g., cells) whose microscopic state includes *activity*, a variable related to the expression of biological function. The modeling is developed at the cellular scale, as an intermediate between the subcellular and macroscopic scales. Bellomo and Forni apply their new theory to investigate competition between neoplastic and immune cells. This work is important in that it illustrates that while theoreticians strive to include more realistic biology in their models, they must also not fail to neglect the development of mathematical theory to underpin and justify their modeling and computational approaches.

The contributions collected in this volume show how major questions in developmental systems can be addressed through a multidisciplinary effort, with particular focus on the importance of mathematical and computational biology. The biology of multicellular systems is now among the most active areas in all of science, relating not only to embryonic development, the focus of most of these contributions, but also reparative medicine, cancer biology and immunology. The unprecedented growth of these fields and the complexity of their experimental findings require new and innovative theoretical frameworks for their successful comprehension and application. This book is intended to serve as a comprehensive review of the current state-of-the-art in the subject.

Acknowledgments

We are very grateful to James A. Glazier, Director of the Biocomplexity Institute of Indiana University, for his encouragement and financial support of Biocomplexity Workshop 9. Biocomplexity 9 was also supported by the Multidisciplinary Ventures and Seminars Fund of the Office of the Vice Chancellor for Academic Affairs and Dean of the Faculties, Indiana University, the National Science Foundation, and Indiana University School of Informatics.

References

Bateson, W. (1894). "Materials for the Study of Variation." Macmillan, London.
Cooke, J., and Zeeman, E. C. (1976). A clock and wavefront model for control of the number of repeated structures during animal morphogenesis. *J. Theor. Biol.* **58**, 455–476.
Cui, C., Yang, X. S., Chuai, M. L., Glazier, J. A., and Weijer, C. J. (2005). *Dev. Biol.* **284**, 37–47.
Forgacs, G., and Newman, S. A. (2005). "Biological Physics of the Developing Embryo." Cambridge Univ. Press, Cambridge, UK.
Gierer, A., and Meinhard, H. (1972). *Kybernetik* **12**, 30–39.
Harris, H. (1999). "The Birth of the Cell." Yale University Press, Manchester, UK.
Kant, I. (1790). "Critique of Judgement (trans. J.H. Bernard, 1966)." Hafner, New York.
Palmeirim, I., Henrique, D., Ish-Horowicz, D., and Pourquié, O. (1997). Avian hairy gene expression identifies a molecular clock linked to vertebrate segmentation and somitogenesis. *Cell* **91**, 639–648.
Raspail, R. V. (1833). "Nouveau système de chimie organique, fondé sur des méthodes nouvelles d'observation." Ballière, Paris.
Schnell, S., Grima, R., and Maini, P. K. (2007). *Am. Sci.* **95**, 134–142.
Turing, A. M. (1952). *Philos. Trans. R. Soc. London Ser. B* **237**, 37–72.
Wolf-Gladrow, D. A. (2000). Lattice-gas cellular automata and lattice Boltzmann models: An introduction. *In* "Lecture Notes in Mathematics." Springer-Verlag, Berlin.

Models of Biological Pattern Formation: From Elementary Steps to the Organization of Embryonic Axes

Hans Meinhardt
Max-Planck-Institut für Entwicklungsbiologie, Spemannstr. 35, D-72076 Tübingen, Germany

I. Introduction
   A. The Body Pattern of Hydra-Like Ancestral Organisms Evolved into the Brain of Higher Organisms
II. Primary Pattern Formation by Local Self-Enhancement and Long-Ranging Inhibition
   A. A Mathematical Description
   B. Polar Patterns, Gradients, and Organizing Regions
   C. Formation of Periodic Patterns
   D. Stripe-Like Patterns and the Role of Saturation in the Self-Enhancement
   E. The Antagonistic Reaction Can Result from a Depletion of a Substrate or Co-Factor
   F. Pattern Formation within a Cell
   G. Oscillations and Traveling Waves
   H. How to Avoid Supernumerary Organizers: Feedback on the Competence
   I. A Graded Competence Allows Small Organizing Regions
   J. An Inhibition in Space *and* in Time: The Generation of Highly Dynamic Patterns
III. The Two Main Body Axes
   A. The Blastopore (Marginal Zone) as Organizer for the AP Axis
   B. The Spemann-Type Organizer Induces Midline Formation
   C. The Orthogonal Orientation of the Main Body Axes in Vertebrates
   D. The Hydra-Type Organizer (Marginal Zone) Provides the Prerequisites to Generate the Spemann-Type Organizer
   E. The Spemann-Type Organizer: The *Chordin/BMP/ADMP* System as a Pattern-Forming Reaction
   F. The Role of Maternal Determinants
   G. Pattern Regulation and Unspecific Induction: How Dead Tissue Can Induce a Second Embryonic Axis
   H. The AP Patterning of the Trunk: A Time-Based Sequential Posterior Transformation and a Ring-to-Rod Conversion
   I. The Left–Right Polarity: A Second Pattern Is Squeezed to the Side
IV. Subpatterns
   A. Switch-Like Gene Activation Requires Autocatalytic Genes
   B. Activation of Several Genes by a Morphogen Gradient: A Step-Wise Promotion
   C. Mutual Induction
   D. Hydra Tentacles as Example
   E. Sequences of Structures and their Dynamic Control: Planarians as Examples
   F. Segmentation: A Superposition of a Periodic and a Sequential Pattern
   G. Formation of a Precise Number of Different Segments during Terminal Outgrowth
   H. Somite Formation: The Conversion of a Periodic Pattern in Time into a Periodic Pattern in Space

I. Borders and Intersections of Borders Become the New Organizing Region for Secondary
      Structures such as Legs and Wings
   J. Filamentous Branching Structures: Traces behind Moving Signals
V. Conclusion
   Acknowledgments
   References

   An inroad into an understanding of the complex molecular interactions on which development is based can be achieved by uncovering the minimum requirements that describe elementary steps and their linkage. Organizing regions and other signaling centers can be generated by reactions that involve local self-enhancement coupled to antagonistic reactions of longer range. More complex patterns result from a chaining of such reactions in which one pattern generates the prerequisites for the next. Patterning along the single axis of radial symmetric animals including the small freshwater polyp hydra can be explained in this way. The body pattern of such ancestral organisms evolved into the brain of higher organisms, while trunk and midline formation are later evolutionary additions. The equivalent of the hydra organizer is the blastopore, for instance, the marginal zone in amphibians. It organizes the anteroposterior axis. The Spemann organizer, located on this primary organizer, initiates and elongates the midline, which is responsible for the dorsoventral pattern. In contrast, midline formation in insects is achieved by an inhibitory signal from a dorsal organizer that restricts the midline to the ventral side. Thus, different modes of midline formation are proposed to be the points of no return in the separation of phyla. The conversion of the transient patterns of morphogenetic signaling into patterns of stable gene activation can be achieved by genes whose gene products have a positive feedback on the activity of their own gene. If several such autoregulatory genes mutually exclude each other, a cell has to make an unequivocal decision to take a particular pathway. Under the influence of a gradient, sharply confined regions with particular determinations can emerge. Borders between regions of different gene activities, and the areas of intersection of two such borders, become the new signaling centers that initiate secondary embryonic fields. As required for leg and wing formation, these new fields emerge in pairs at defined positions, with defined orientation and left–right handedness. Recent molecular-genetic results provide strong support for theoretically predicted interactions. By computer simulations it is shown that the regulatory properties of these models correspond closely to experimental observations (animated simulations are available at www.eb.tuebingen.mpg.de/meinhardt).
© 2008, Elsevier Inc.

I. Introduction

The formation of a higher organism within each life cycle is a most fascinating process. With modern molecular-genetic techniques it is possible to monitor simultaneously the mutual interference of hundreds of genes. However, it is notoriously

1. Models for the Organization of the Embryonic Body Axes

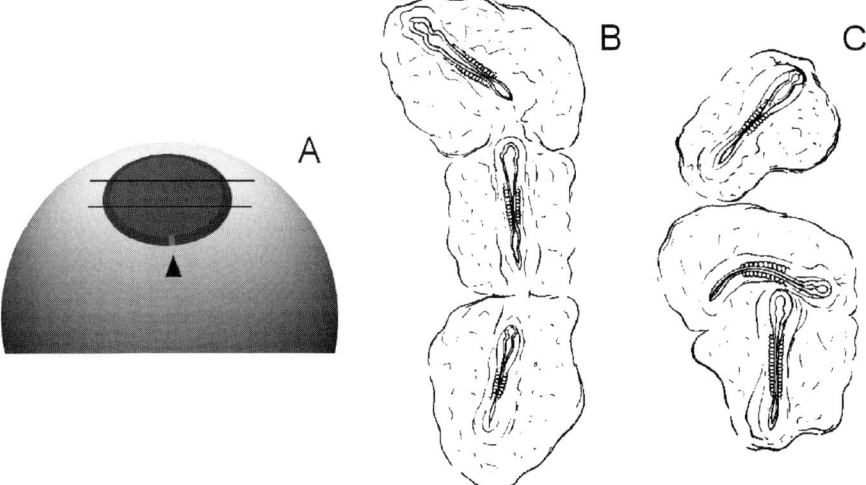

Figure 1 An example of pattern regulation. (A) The early chick embryo has the shape of a disk that is located on the huge yolk. The triangle marks the normal position of the organizing region, Koller's sickle. (B–C) Fragmentation of such a disk at an early stage can lead to complete embryo formation in each fragment (Lutz, 1949; Spratt and Haas, 1960). The fragment that contains the organizer forms an embryo in the normal orientation. In the other fragments the orientation is more or less arbitrary. Nevertheless, in all embryos the correct mutual orientation of the anteroposterior and the mediolateral axes (as indicated by the paired somites) reestablished (B, C after Lutz, 1949).

difficult to deduce from such a plethora of data the functioning of underlying complex networks. Long before the molecular-genetic methods became available, we followed a different approach by asking what type of molecular machinery would be required at least to account for the observed patterns, including pattern regulation after experimental interference. It turned out that interactions employing relatively few components are able to describe elementary steps in surprising detail. In order to find the appropriate hypothetical interactions a mathematical formulation of the reactions was mandatory since our intuition is often unreliable to predict the behavior of systems that are based on strong positive and negative feedback loops.

The final complexity of an organism does not already exist in a mosaic-like fashion in the egg. For instance, each cell of an eight-cell mouse embryo can give rise to a complete embryo. Likewise, after an early partition of the disk-shaped chick embryo, complete embryos can emerge in each fragment (Fig. 1). Obviously, communication between the cells is essential to achieve this spatial organization, and an interruption of this communication can lead to a dramatic rearrangement of the main body axes. It follows that axes formation has a strong self-organizing aspect. Most surprisingly, even after such a severe perturbation, the two main body axes, anteroposterior (AP) and dorsoventral (DV), still have the correct orientation relative to each other, indicating a strong coupling between the system that patterns these two axes.

Simple radial-symmetric animals including the freshwater polyp hydra or the small sea anemone *Nematostella* are evolutionary ancestral organisms, close to the branch point where bilaterality was invented. Since mechanisms in development are so well preserved during evolution (de Jong *et al.*, 2006), it is reasonable to assume that these animals provide a key to understanding the patterning along a single axis and provide information about the steps that occurred towards more evolved bilateral-symmetric body plans.

Hydra tissue is famous for its almost unlimited capability for regeneration (Trembley, 1744; von Rosenhof, 1755; see also Gierer, 1977; Bode, 2003). Even more dramatic, hydra tissue can be dissociated into individual cells and, after reaggregation, these clumps of cells again form viable organisms (Gierer *et al.*, 1972; Fig. 2). Obviously, pattern formation does not require any initiating asymmetry. The small cone-shaped region around the gastric opening, the so-called hypostome, has organizing capabilities (Browne, 1909). A small tissue fragment from this region transplanted into the body column of another animal can induce the formation of a secondary body axes. Although Ethel Browne did not use explicitly the term 'organizer,' she discovered a phenomenon that became of central interest 15 years later with the discovery of the amphibian organizer (Spemann and Mangold, 1924; see Lenhoff, 1991). Thus, hydra can be used as a guide to find the corresponding interactions that allow *de novo* organizer formation and its regeneration.

Hydra is also a convenient model organism to study more complex patterning steps. In many developmental systems particular structures are formed with a precise spatial relation. In hydra the primary organizer, the hypostome, is surrounded by a necklace of tentacles. Since the tentacles resemble a periodic pattern, hypostome and tentacle formation is governed by two separate but coupled pattern-forming systems, providing an inroad into the question of how to induce two structures next to each other. Why do tentacles appear close to each other around a narrow ring, but do not form with a similar spacing along the body column?

Hydra is under the control of two antipodal organizing regions, the head and the foot. Both appear at maximum distance from each other. Again, this is a frequent occurrence; shoot and root in plants or head and tail in planarians are other examples. Which interaction enforces a maximum distance from each other but allows, nevertheless, both terminal structures to be formed close to each other at early stages or during regeneration of small fragments?

In the first part of this paper such elementary steps in pattern formation will be discussed and compared with more recent molecular-genetic observations.

A. The Body Pattern of Hydra-Like Ancestral Organisms Evolved into the Brain of Higher Organisms

A step of primary importance in the development of higher organisms is the generation of the main body axes, anteroposterior (AP), dorsoventral (DV) and, in ver-

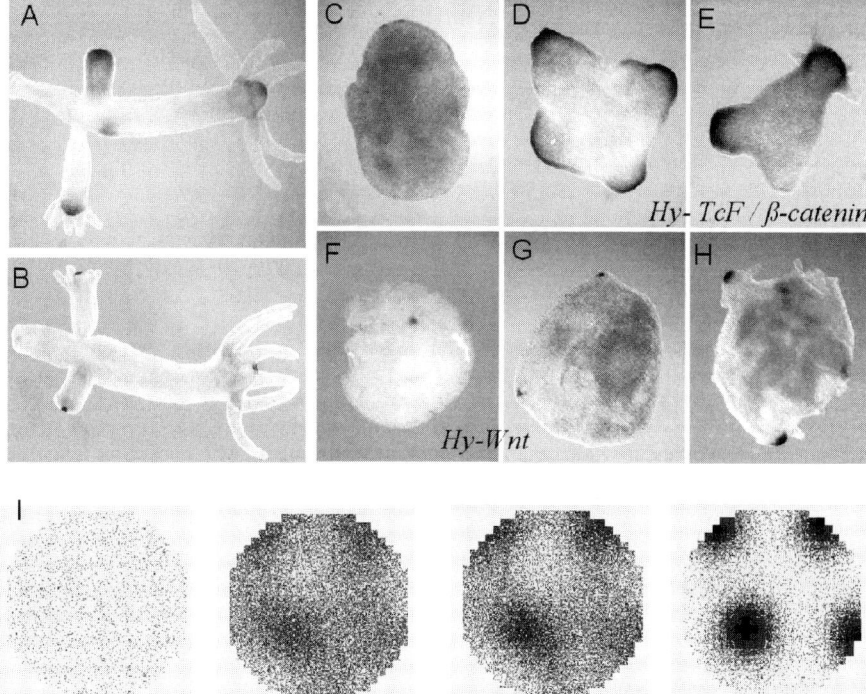

Figure 2 The canonical *Wnt* pathway is involved in organizer formation in the freshwater polyp hydra. From classical experiments it was known that the region around the opening of the gastric column, the cone-shaped hypostome, has organizer activity. (A) *TCF* (and β-catenin) expression occurs in a graded fashion at the tip and also precedes the formation of a new axis during bud formation (Hobmayer *et al.*, 2000). (B) *Hy-Wnt* expression is more sharply confined to the tip. (C–E) In reaggregating cells, *Hy-TCF* (and β-catenin) appears first uniformly distributed and become subsequently more restricted to regions that eventually form the new heads. (F–H) In contrast, *Hy-Wnt* appears directly in sharp spots that form the future oral opening. (I) This nested pattern formation can be accounted for by the assumption that both *Hy-β*-catenin/*Hy-Tcf* (gray) and *Hy-Wnt* (black) are pattern forming systems and that a high *Hy-β*-catenin concentration is the precondition to trigger *Hy-Wnt*. Thus, *Wnt* peaks require a high *Hy-β*-catenin/*Hy-Tcf* level and appear as sharp spots at the highest level of the more graded *Hy-β*-catenin/*Hy-Tcf* distributions. Such a superimposed patterning allows the specification of a large region for head formation and provides nevertheless a sharp signal, e.g., for the opening of the gastric column (Photographs by courtesy of B. Hobmayer, T. Holstein and colleagues; see Hobmayer *et al.*, 2000.)

tebrates, left–right (LR). Radially-symmetric organisms themselves provide a key to understanding the essential inventions required for the transition from radial- to bilateral-symmetric body plans. For long it was unclear whether the single axis of hydra corresponds to any of the main body axes in higher organisms, and, if so, to which axis and in which orientation. Almost all components involved in higher organisms to pattern the AP as well as the DV axes are already present in hydra. However, systems that control the orthogonal axes in higher organisms, e.g., *WNT* for the AP axis and *Chordin/BMP* for the DV axis, are expressed in hydra along the only existing axis. Thus, bilaterality is proposed to be achieved by a realignment of at least two already existing, originally parallel axial systems and not by the invention of a new signaling system (Meinhardt, 2004a). Some coelenterates already show pronounced deviations from radial symmetry (Martindale, 2005).

As the expression patterns of more and more genes became available, the situation became more and more difficult to interpret. Around the gastric opening genes are expressed that are characteristic for both head *and* tail formation, *Goosecoid* and *Brachyury* (Broun *et al.*, 1999; Technau and Bode, 1999). This apparent discrepancy can be resolved by the assumption that the body pattern of a hydra-like ancestor evolved into the most anterior (and most important) part of higher organisms, the brain and the heart (Meinhardt, 2002). The *Otx* gene, in vertebrates, characteristic for the fore- and midbrain, is expressed in the hydra all over the polyp with the exception of the most terminal regions (Fig. 3). This suggests that in an ancestral radial-symmetric organism the posterior end was at a position that corresponds in vertebrates roughly to the midbrain/hindbrain border. Thus, although in hydra the region around the gastric opening with the tentacles is commonly called 'the head,' it represents the most posterior part as indicated by *Wnt* and *Brachyury* expression. The recently observed highly conserved patterning in the brain of such distantly related organisms as insects and vertebrates (Hirth *et al.*, 2003; Lowe *et al.*, 2003; Sprecher and Reichert, 2003) is proposed to have its origin in the preserved body pattern of a common radially-symmetric ancestor.

In later parts of this paper it will be shown that this relation is a key to understanding the different modes in the generation of bilateral-symmetric body plans. In the final part mechanisms will be discussed that lead to insertion of new structures within the frame of the body axes, such as legs and wings or of branching structures such as blood vessels and tracheae.

II. Primary Pattern Formation by Local Self-Enhancement and Long-Ranging Inhibition

The observation that patterns can emerge in an initially more or less uniform assembly of cells (Fig. 2) raises the question of what type of molecular interaction would be able to generate local concentration maxima. In a pioneering paper, Alan Turing (Turing, 1952) showed that pattern formation is possible by an interaction of two components with different diffusion rates, now collectively called reaction–diffusion systems. However, most reactions in which two substances interact have no pattern-forming capability whatsoever, even if they spread with different rates. We have shown that pattern formation is possible if, and only if, a locally restricted self-enhancing reaction is coupled with an antagonistic reaction that acts on a longer range (Gierer and Meinhardt, 1972; Gierer, 1977; Meinhardt, 1982;

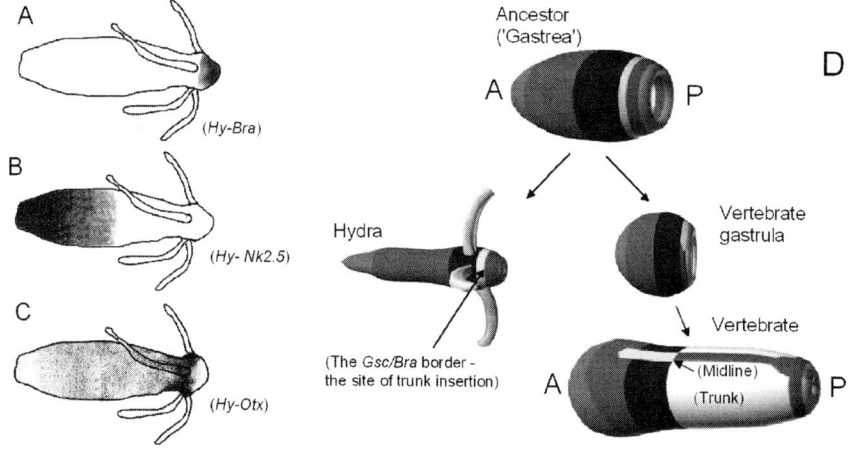

Figure 3 Gene expression of well-known genes in hydra suggests that the body pattern of radial-symmetric ancestors evolved into the head pattern of higher organisms. (A) *Brachyury* is expressed in hydra around the gastric opening (Technau and Bode, 1999). In higher organisms, *Brachyury* marks the blastoporal opening, which is always most posterior. (B) In the antipodal foot region *Nkx2.5* is expressed (Grens *et al.*, 1996). In higher organisms *Nkx2.5* is expressed at very anterior positions (e.g., Tonissen *et al.*, 1994) and is responsible for heart formation. The hydra foot and the vertebrate heart, both involved in pumping, have a common ancestry (Shimizu and Fujisawa, 2003). (C) *Otx*, a gene typical for the fore- and midbrain in higher organisms, is expressed throughout the hydra except in the terminal parts (Smith *et al.*, 1999). The posterior *Otx* border marks in vertebrates the midbrain–hindbrain border (reviewed in Rhinn and Brand, 2001), in hydra the border between the tentacle zone and the hypostome. (D) These expressions suggest that the body pattern of an ancestral organism—the *gastrea* in terms of Haeckel (Haeckel, 1874)—evolved into the brain of higher organisms. The so-called hydra head is in fact the posterior end. The vertebrate gastrula can be regarded as a remnant of this ancestral organism. The midline and the trunk (see Figs. 12–14) are two major later evolutionary inventions. For instance, the trunk-typical Hox gene clusters are absent in Cnidarians (Kamm *et al.*, 2006).

Meinhardt and Gierer, 2000). This condition, not inherent in Turing's paper, is essential for a homogeneous distribution to become unstable. We have derived a general criterion for which interactions lead to a stable pattern and which do not. Pattern formation from more or less homogeneous initial situations is also common in the inorganic world. Examples are the formation of sand dunes, cumulus clouds, stars or lightning. Pattern formation is also common in social interactions. These processes are based on the same principle (Meinhardt, 1982).

A prototype of such a pattern-forming reaction consists of a short-ranging substance, to be called the activator, which promotes its own production directly or indirectly. It also regulates the synthesis of its rapidly diffusing antagonist, the inhibitor. The latter slows down the autocatalytic activator production (Fig. 4; Gierer and Meinhardt, 1972) or catalyzes the activator decay. A homogeneous distribution is unstable since, for example, a small local elevation of the activator will increase fur-

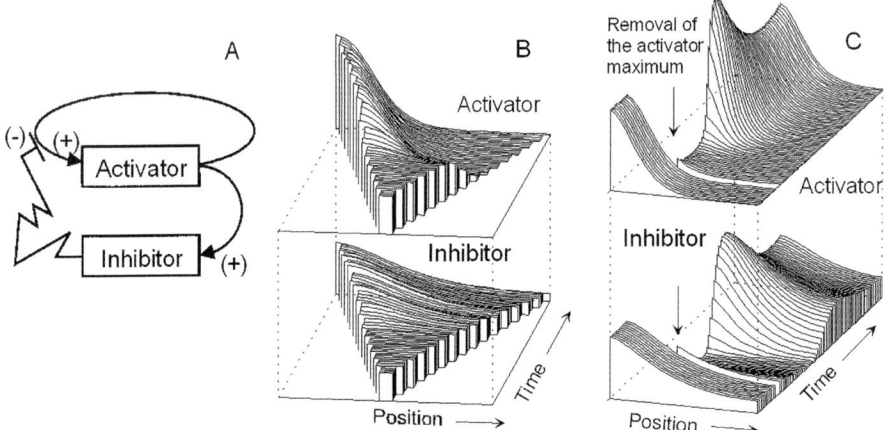

Figure 4 Pattern formation by an activator–inhibitor interaction. (A) Reaction scheme: The activator catalyses its own production. The production of its rapidly spreading antagonist, the inhibitor, is also under activator control (Eqs. (1a) and (1b); Gierer and Meinhardt, 1972). In such a reaction, the homogeneous distribution of both substances is unstable. (B) The simulation illustrates pattern formation in a growing chain of cells as a function of time. Whenever a certain size is exceeded, random fluctuations are sufficient to initiate patterning. A high concentration appears at a marginal position. Thus, although the genetic information is the same in all cells, such a system is able to generate a reproducible polar pattern, appropriate to accomplish space-dependent cell differentiation (see Figs. 19 and 20). (C) Regeneration. After removal of the activated region, the inhibitor is no longer produced. After decay of the remnant inhibitor, a new activation is triggered. The graded profiles are restored as long as the remaining fragment is still large enough (see also Fig. 10).

ther due to autocatalysis despite the fact that a surplus of the inhibitor is also produced at the same position. The latter, however, dilutes rapidly by a fast diffusion into the surroundings of this incipient maximum, slowing down the autocatalysis there. Therefore, a local rise is intimately connected with a down-regulation at larger distances. A new patterned steady state is reached when a local high activator concentration is in a dynamic equilibrium with the surrounding cloud of the inhibitor (Fig. 4). Both the more localized activator and the more smoothly distributed inhibitor can be used as a morphogenetic signal. Thus, pattern formation depends critically on the spatial distribution of signals. Although diffusion is a good approximation, the real process is usually much more complex requiring a chain of several molecules: secreted ligands, receptors and components that transmit the signal from the cell surface to the nucleus. The term 'diffusion' is only used as shorthand for a long-ranging signaling. Other modes of redistribution of molecules are conceivable as well. In plants, for instance, active transport of auxin plays a major role.

At the time the theory was proposed (1972), activator–inhibitor systems were completely hypothetical. Since then several systems have been found that correspond to this scheme. For instance, *Nodal* is a secreted factor that has a positive feedback on its own production; *Lefty2* is under the same control as *Nodal* and acts as an antagonist. *Lefty2* cannot dimerize and blocks in this way the *Nodal* receptor. This system is involved in mesoderm and midline formation as well as in left–right patterning (Chen and Schier, 2002; Nakamura et al., 2006). Another example is the specification of heterocyst cells in the blue-green alga *Anabaena* (see Fig. 7). In hydra, *Wnt* and β-catenin are expressed in the hypostome, suggesting that the canonical *Wnt* pathway is involved in the formation of the hydra organizer (Fig. 2; Hobmayer et al., 2000; Broun et al., 2005). However, the molecular basis of the self-enhancement and of the long-ranging inhibition is not yet clear.

A. A Mathematical Description

Since all biological processes are assumed to be accomplished by the interaction of molecules, a theory of biological pattern formation has to describe the changes of concentrations in space and time as function of the local concentration of the relevant substances involved. The following set of equations describes the local change of the activator $a(x)$ and inhibitor concentration $h(x)$ per time unit (Gierer and Meinhardt, 1972), for simplicity written here for a one-dimensional array of cells:

$$\frac{\partial a}{\partial t} = \frac{\rho a^2}{h} - \mu a + D_a \frac{\partial^2 a}{\partial x^2}, \tag{1a}$$

$$\frac{\partial h}{\partial t} = \rho a^2 - \nu h + D_h \frac{\partial^2 h}{\partial x^2}. \tag{1b}$$

Such equations are easy to read. Equation (1a) states that the concentration change of the activator a per unit time $(\partial a/\partial t)$ is proportional to a nonlinear autocatalytic production term (a^2). The autocatalysis is slowed down by the action of the inhibitor $1/h$. The second term, $-\mu a$, describes the degradation. The number of activator molecules that disappear per time unit is proportional to the number of activator molecules present (like the number of people dying per year in a city is on average proportional to the number of inhabitants). The autocatalysis must be nonlinear (a^2) since it must overcome disappearance by linear decay $(-\mu a)$. This condition is satisfied if the active component is not the activator itself but a dimer of two activator molecules. An example is the dimerization of the activator that leads to heterocyst formation in *Anabaena* (see Fig. 7). The factor ρ, the *source density*, describes the general ability of the cells to perform the autocatalytic reaction. Its function is close to what is described as 'competence' in the biological literature. Slight asymmetries in the source density can have a strong influence on the *orientation* of the emergent pattern. The concentration change of a and h also depends on the exchange of molecules with neighboring cells. This exchange is assumed to occur by simple diffusion but other mechanisms are conceivable as well.

Equation (1b) can be read in an analogous manner. A necessary condition to enable spatial pattern formation is that the inhibitor spreads more rapidly than the activator, i.e., the condition $D_h \gg D_a$ must be satisfied. In addition, the inhibitor must have a more rapid turnover ($\nu > \mu$), otherwise the system will have the tendency to oscillate (see Fig. 9).

For many simulations Eqs. (1a) and (1b) are used with a few extensions:

$$\frac{\partial a}{\partial t} = \frac{\rho a^2}{h(1 + \kappa a^2)} - \mu a + D_a \frac{\partial^2 a}{\partial x^2} + \rho_a, \qquad (2a)$$

$$\frac{\partial h}{\partial t} = \rho a^2 - \nu h + D_h \frac{\partial^2 h}{\partial x^2} + \rho_h. \qquad (2b)$$

The term κ leads to a saturation of the self-enhancing reaction at higher activator concentrations and thus to an upper limit of activator production. This allows, for instance, the formation of stripes (see Fig. 6). The last term in Eq. (2a) is a small activator-independent activator production. This term ensures that the concentration of the activator never sinks to zero and enables the reformation of an activator maximum after removal of an established maximum. It is important for the initiation of the autocatalytic reaction at low activator concentrations as required for regeneration (Fig. 4) and for oscillations (see Fig. 9). The last term in Eq. (2b) is a small activator-independent inhibitor production. It has the consequence that a uniform low concentration of activator can be a semistable situation. This baseline inhibitor level can suppress a spontaneous trigger. In this way the pattern-forming system can be "asleep" until a trigger occurs that raises the activator concentration above a threshold from which the patterning proceeds further due to the self-enhancement. Such a trigger can be supplied, for example, by adjacent activated cells during the spread of traveling waves (see Fig. 9C).

For simulations, the concentration changes are calculated for small but discrete time steps and the space is subdivided into discrete units or 'cells.' Starting with initial distributions of both substances, these equations allow the computation of their changes over a short time interval. Adding these changes to the given concentrations leads to new concentrations. By repeating this computation the total time course can be calculated in an iterative way. The exchange of molecules between adjacent cells requires special conditions at the boundaries of the field of cells. Usually it is assumed that the boundaries are impermeable. Simple and well commented programs that can be compiled and executed on a PC are available on our website.

B. Polar Patterns, Gradients, and Organizing Regions

For the generation of a primary body axis, it is essential that one side of the developing organism becomes different from the other. Pattern formation requires a certain field size such that the different diffusion rates can come into play. If an activator–inhibitor mechanism is involved, a high concentration emerges at one and a low concentration at the opposite side whenever a certain size of tissue is exceeded (Fig. 4). The generation of such a polar pattern is a most important step: Although the genetic information is the same in all the cells, different genetic information can be activated in a position-dependent manner.

1. Models for the Organization of the Embryonic Body Axes

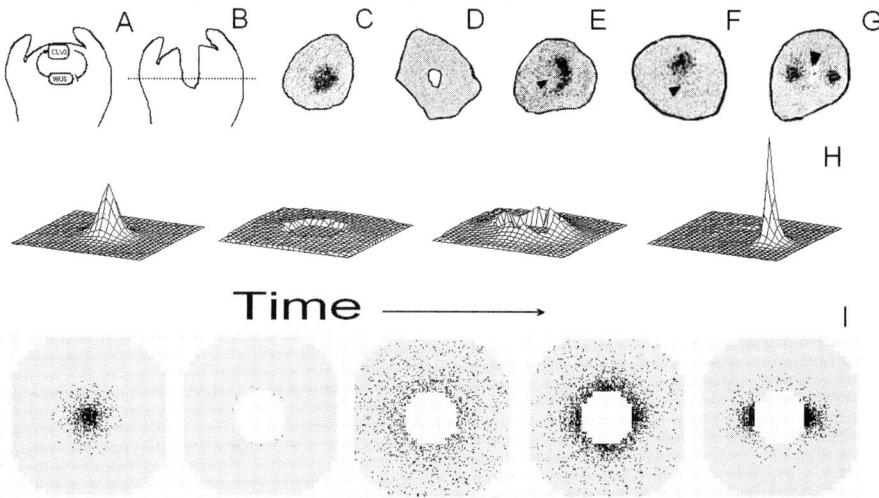

Figure 5 Regeneration of the organizer in the shoot apical meristem. (A) *Wuschel (Wus)* is a crucial component in the maintenance of the shoot apical meristem (Mayer *et al.*, 1998). The size of *Wus* expression is controlled by a negative feedback via *Clavata3 (Cl3)* (reviewed in Clark, 2001; Tsiantis and Hay, 2003). The autocatalytic component expected from the model is not yet known. (B–G) Restoration of *Wus* activity after laser-ablation of the *Wus* expressing cells (Reinhardt *et al.*, 2003; hole in B). *Wus* activity in the plane indicated in B before (C) and after (D) ablation. *Wus* expression reappears in cells that were previously not expressing *Wus*; first in a rather diffuse way. After two days the expression has a ring- or crescent-like distribution (E). Eventually one (F) or two maxima (G) emerge. The latter case leads to a split of the meristem. (H, I). Simulations: killing the cells that produce the activator leads to a decline of the inhibitor and to a new trigger of the activator in the surrounding competent cells. The activation is first more diffuse. The concomitantly produced inhibitor (not shown) leads to a competition and peak sharpening. Either one (H) or two (I) maxima survive. In (I) the activator concentrations are plotted as a pixel density. To localize the new activations near the center of the meristem, a graded competence that decreases towards the periphery has to be assumed (see Fig. 10). (B–G is drawn after Reinhardt *et al.*, 2003.)

The model accounts not only for the generation of a pattern but also for pattern regulation (Figs. 4 and 5). With the removal of an area of high activator concentration, the area of inhibitor production is also removed. After the decay of the remnant inhibitor the formation of a new activator maximum is triggered in the remaining cells, starting from a low level activator production [ρ_a in Eq. (2a)]. The pattern becomes restored in a self-regulatory way.

C. Formation of Periodic Patterns

In many developmental situations, periodic structures are formed. Examples are the spacing of the leaves on a growing plant, the bristles on insects or feathers of birds. In terms of the model, periodic structures can be formed if the field is or becomes larger

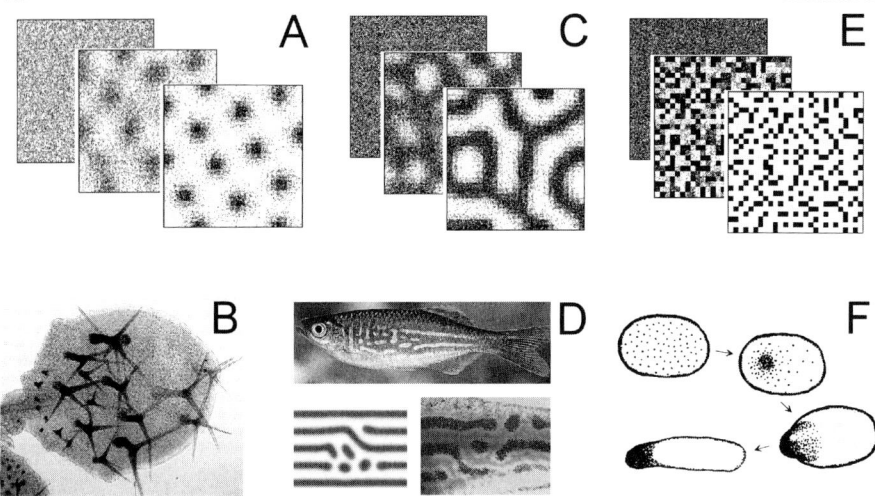

Figure 6 Generation of elementary periodic patterns in a cell sheet. Assumed is an activator–inhibitor system (Eq. (2a) and (2b); only the activator distributions are shown). (A) Several maxima emerge if the size of the field is larger than the range of the inhibitor. When initiated by random fluctuations, the spacing is somewhat irregular but a maximum and minimum distance is maintained. (B) Biological example: Trichome formation in *Arabidopsis* (reviewed in Pesch and Hülskamp, 2004). The three spines of each trichome are produced by a single cell. The gene *GL2* (dark) is involved in trichome activation (see Szymanski et al., 2000). The genes *TRY* and *CPC* (not shown) are involved in the inhibition. Although *TRY* and *CPC* inhibit trichome formation, they are only expressed in the future trichome cells, as expected from our model. (C) Stripe-like distributions emerge if the activator production cannot surpass a certain level, e.g., due to a saturation [$\kappa > 0$ in Eq. (2a)]. Stripe formation requires some spread of the activator. (D) Examples: The stripes of pigment cells in zebrafish. Although the molecular details are not yet clear (reviewed in Parichy, 2006), the reappearance of the stripes after laser ablation occurs as expected by our theory (Yamaguchi et al., 2007). (E) Without spread of the activator, a segregation into two different cell types occurs. Activated and nonactivated cells appear in a certain ratio in a salt-and-pepper distribution. (F) Such a pattern is characteristic for the early prestalk/prestalk patterning in *Dictyostelium discoideum* (Maeda and Maeda, 1974; see Zhukovskaya et al., 2006; for modeling see Meinhardt, 1983cc). (B kindly supplied by Martina Pesch and Martin Hülskamp; D—by Shigeru Kondo.)

than the range of the inhibitor (the range is the mean distance a molecule can travel between its production and degradation). When the pattern is formed in a field that has already a substantial extension, the resulting pattern will be somewhat irregular; only a maximum and minimum distance will be maintained (Fig. 6). The cilia on the surface of *Xenopus* embryo (Deblandre et al., 1998, 1999) or the trichomes of leaves (Hülskamp, 2004; Fig. 6B) are biological examples of this type of pattern. In contrast, in growing fields new maxima will be formed whenever the distance to existing maxima becomes too large. Then the inhibitor concentration between the maxima can become so low that autocatalysis is no longer repressed. New maxima are inserted at the maximum distance of existing peaks. The resulting pattern will be more regular.

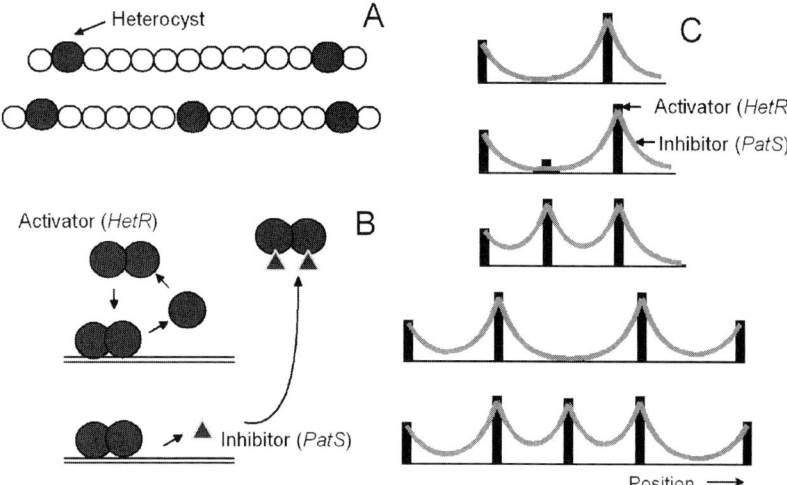

Figure 7 Insertion of new maxima during growth. (A) Biological example: the insertion of new nitrogen-fixating cells, so-called heterocysts, in the blue green algae *Anabaena*. Whenever the distance between two heterocysts (dark circles) in the linear chain of cells becomes larger then ca. 12–14 cell, a normal cell differentiates into a larger, nondividing heterocyst. It is the cell that has the largest distance from the existing heterocysts. (B) Heterocyst formation is under control of the transcription factor *HetR*. *HetR* form dimers that directly activate *HetR* transcription (Huang *et al.*, 2004), Dimerization satisfies our prediction that the autocatalysis must be nonlinear [Eqs. (2a), (2b)]. *HetR* also activates the formation of a small peptide, *PatS* (triangles) that can spread through intercellular junctions (Yoon and Golden, 1998) and that can bind to *HetR*. If *PatS* is bound to *HetR* DNA-binding of *HetR* is no longer possible. Thus, *PatS* inhibits the activator autocatalysis, as predicted. (C) Simulation: only the inhibitor is diffusible across the cells. Therefore, activation occur in isolated cells. Whenever the inhibitor drops below a threshold level, a new autocatalysis of the activator is triggered from a baseline activation [ρ_a in Eq. (2a)]. Since the inhibitor distribution around a minimum is shallow, initially more than one cell can start this activation process. Due to competition only one isolated cell eventually becomes activated. In agreement with the expectation from our theory, if *HetR* is mutated, no heterocysts are formed. In contrast, if *PatS* is mutated, most cells form heterocysts (Buikema and Haselkorn, 2001).

Examples are the insertion of new leaves at a growing shoot, new heterocyst cells in *Anabaena* (Fig. 7), new trichomes (Fig. 6B) or new bristles in insects (Wigglesworth, 1940).

D. Stripe-Like Patterns and the Role of Saturation in the Self-Enhancement

Stripe-like patterns, i.e., structures with a long extension in one dimension and a short extension in the other, are formed in many instances during embryogenesis. Proverbial are the stripes of zebras. Stripe-like distributions can emerge if activator production has an upper bound (Meinhardt, 1989, 1995). If activator autocatalysis saturates at a relatively low concentration [$\kappa > 0$ in Eq. (2a)] the inhibitor production is limited too and the mutual competition between neighboring cells is reduced. Due to the sat-

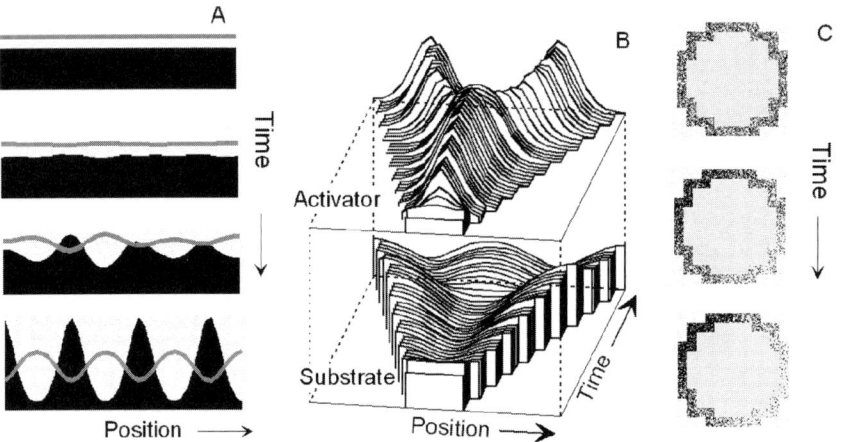

Figure 8 Pattern formation by an activator-depletion mechanism. (A) Autocatalysis proceeds at the expense of a rapidly spreading substrate or co-factor (Eqs. (3a), (3b); Gierer and Meinhardt, 1972). The concentration of the antagonist is lowest in regions of high activator concentration, in contrast to the situation in an activator–inhibitor system (Figs. 4 and 7). (B) During growth, activator maxima have the tendency to split due to the inherent saturation of the activator autocatalysis in this system. Since the substrate supply is higher at the flanks than in the center, activator production can become higher in the flanks, which leads to a deactivation in the center and a shift of the maxima towards higher substrate levels. This is in contrast to the behavior of activator–inhibitor systems without saturation in self-enhancement (Fig. 7). Splitting activator maxima are crucial for dichotomous branching during filament elongation (Fig. 27). (C) Such a system is appropriate for *intracellular* pattern formation. In this simulation the self-enhancing reaction is assumed to proceed by a cooperative aggregation of molecules at the membrane. This aggregation proceeds at the expense of freely diffusible monomers that can spread rapidly in the cytoplasm (not shown). Local high concentrations emerge at a particular part of the cell membrane. Corresponding mechanisms are discussed for the *Dictyostelium discoideum* (Charest and Firtel, 2006; see also Fig. 11).

uration more cells remain activated although at a lower level. Thus, activated cells tolerate activated cells in their neighborhood, independent of the range of inhibition. In addition to saturation, a further condition for stripe formation is a modest diffusion of the activator. Due to this diffusion, activated regions tend to occur in large coherent patches since activated cells tend to activate adjacent cells. On the other hand, pattern formation requires that activated cells are close to nonactivated cells into which the inhibitor can diffuse. These two seemingly contradictory features, coherent patches and proximity of nonactivated cells, are characteristic for stripe-like patterns (Fig. 6C). If initiated by random fluctuations, the stripes have random orientations too. It is a feature of this mechanism that the width of the stripe and the interstripe-region is of the same order. Therefore, the formation of a single straight, nonbranching stripe as necessary for midline formation in higher organisms requires additional constraints. As shown further below, different mechanisms evolved in different phyla to solve this intricate patterning problem.

In growing tissues that are patterned by systems with saturating activator production [$\kappa > 0$ in Eq. (2a)] periodic structures can emerge by splitting of existing maxima, in contrast to the insertion of new maxima in the absence of saturation (see Fig. 7). Saturation leads to a plateau-like widening of the maxima. If the area into which the inhibitor can escape enlarges in the course of growth, the plateau-like activation enlarges too, i.e., the pattern is size-regulated. From a certain extension onwards, however, the activator production at the center of a maximum can be lower than that at the flanks due to the rising inhibitor level at the center. This leads to a deactivation in the center (see also Fig. 8B). Splitting of existing maxima is the basis for branching in the lung and in tracheae (see Fig. 27).

E. The Antagonistic Reaction Can Result from a Depletion of a Substrate or Co-Factor

Instead of an inhibition that is produced at an activator maximum, the antagonistic effect can also result from the depletion of a substrate $s(x)$ which is a prerequisite for the self-enhancing reaction and which is consumed during the autocatalytic activator production (Fig. 8; Gierer and Meinhardt, 1972):

$$\frac{\partial a}{\partial t} = \rho s a^2 - \mu a + D_a \frac{\partial^2 a}{\partial x^2} + \rho_a, \tag{3a}$$

$$\frac{\partial s}{\partial t} = \delta - \rho s a^2 - \nu s + D_s \frac{\partial^2 s}{\partial x^2}. \tag{3b}$$

According to Eq. (3b), the factor s is produced everywhere with constant rate δ; s is removed by the autocatalytic reaction at the same rate as the activator is produced. The activator-depletion mechanism has an inherent upper bound of the activator production since the production comes to a halt if most of the substrate is consumed. As mentioned above, such saturation can lead in growing systems to new maxima that result from the splitting of existing maxima (Fig. 8B). Thus, an activator-depletion mechanism is unsuitable for the formation of organizing regions, i.e., of isolated signaling centers that are surrounded by large regions that are devoid of signaling centers.

F. Pattern Formation within a Cell

Pattern formation does not only occur between cells but also within a cell. Intracellular pattern formation is often the first pattern that is generated in development. Pattern formation in the egg of the brown alga *Fucus* is an example of an unstable system where almost any external asymmetry can orient the emerging pattern (Jaffe, 1968; Leonetti et al., 2004). In the absence of such asymmetries, a polar pattern will nevertheless arise, although with a random orientation. The pattern consists of a localized influx and efflux of calcium ions.

To satisfy our general conditions for pattern formation within a cell, the self-enhancing reaction is expected to be restricted to parts of the cell cortex while the antagonistic reaction spreads more rapidly within the entire cytoplasm. Activator-depleted substrate mechanisms appear as especially suitable for such intercellular patterning. Activation can occur by a cooperative aggregation of molecules at the cell cortex, i.e., aggregation proceeds more rapidly at positions where some of these molecules are already present. This aggregation is antagonized by the depletion of unbound molecules diffusing freely in the cytoplasm. In such a system the condition for different ranges of the autocatalytic and antagonistic reactions is satisfied in a straightforward manner. In intracellular patterning 'long range' denotes a communication over the entire cell while 'short range' indicates a cooperative process that covers only a part of the cell cortex.

G. Oscillations and Traveling Waves

The discussion so far has considered only patterns that are stable at least for a certain period of time. Stable patterns can result if the antagonistic reaction has a shorter time constant than the activator. Under this condition any deviation from the steady state concentrations is rapidly back-regulated. In contrast, if in an activator–inhibitor system, the inhibitor has a longer half-life than the activator (i.e., if in Eq. (1) $\nu < \mu$), oscillations will occur (Fig. 9A). Since the inhibitor follows too slowly, activation proceeds in a burst-like manner. In the course of time, however, more and more inhibitor

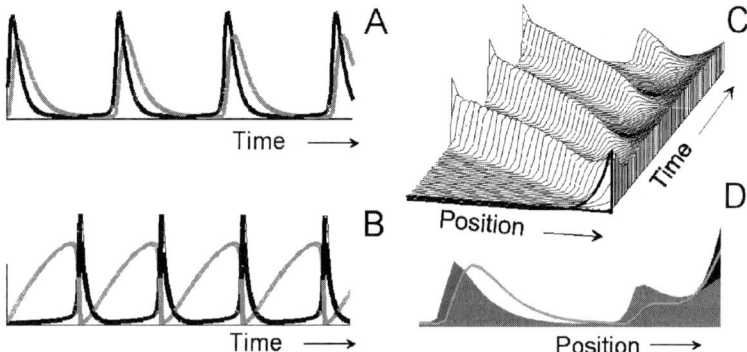

Figure 9 Oscillations and traveling waves. Oscillations can occur in single cells if the antagonist reacts too slowly to a change in the activator concentration. (A) In an activator–inhibitor system oscillations occur if the inhibitor half-life is longer than that of the activator. (B) In an activator–substrate system, oscillations occur if the production rate of the substrate is lower than the removal rate of the activator. (C) Traveling waves can occur if the activator but not the antagonist spreads. An activated cell 'infects' its neighbor. Such wave formation needs an initiation site, a pacemaker region. In this simulation a stable pattern is assumed (black) that causes a high baseline activator production in the oscillating system. Waves spread into the surrounding cells at regular time intervals. (D) A snapshot of a distribution as shown in C. A new wave is just detaching from the pacemaker region (see also somite formation, Fig. 24).

accumulates until the activator production breaks down suddenly. The slow decay of the inhibitor leads to a refractory period until the next trigger occurs that starts from a baseline activator production ρ_a. In the activator–substrate mechanism, oscillations occur if substrate production is insufficient to maintain a steady state [$\delta < \mu$ in Eq. (3)]. Substrate concentration increases until a threshold is reached. The burst-like activation leads to a collapse in substrate concentration and thus to a switching off of activator production. Substrate can accumulate again until the next activation is triggered, and so on (Fig. 9B). Depending on the spread of the components, global oscillations or traveling waves can emerge (Fig. 9C). Thus, the same reaction can generate patterns in space or in time, depending on the spread and the time constants of the components involved. Oscillations and traveling waves will play an important role in the discussion of somite formation (see Fig. 24).

H. How to Avoid Supernumerary Organizers: Feedback on the Competence

The generation of more and more signaling centers during growth is characteristic of the formation of periodic patterns (Fig. 7). For the generation of embryonic axes, however, a single organizer has to be maintained despite growth, otherwise supernumerary and possibly partially fused embryos will result. As mentioned, a fragment of the chicken blastodisc can regenerate a complete embryo even if it does not contain the original organizing region (Fig. 1). This capability for pattern regulation, however, is lost at later stages. Obviously, cells distant to the organizer lose their competence to form an organizer and become unable to trigger a secondary organizer even when the inhibitor drops to very low levels due to growth. Such fading of the competence to form an organizing region is a process of primary importance in making development reproducible. One way to suppress the trigger of new organizing regions is an elevated baseline inhibitor production [ρ_h in Eq. (2b)]. Under such a regime, a nonactivated fragment cannot regenerate a new organizer.

Experiments in hydra, however, suggest a different mechanism since fragments from all positions remain able to regenerate a new polyp. In fragments the regeneration of a head always occurs at the side pointing towards the original head. Thus, the tissue has a systematic polarity; the competence is graded. It is the *relative* position of a group of cells within the fragment that is decisive as to whether they will form a head, a foot, or something in between (Fig. 10). Dissociation and reaggregation experiments have shown that this polarity is based on a systematic change of the tissue composition and not on the orientation of the individual cells (Gierer *et al.*, 1972). This polarity is a very stable tissue property. After head removal, it takes about 1 hour to reform a β-catenin and *Wnt* signal, indicative for a regenerating organizer (Hobmayer *et al.*, 2000). In contrast, as revealed by grafting experiments, reversal of polarity requires about two days (Wilby and Webster, 1970a, 1970b).

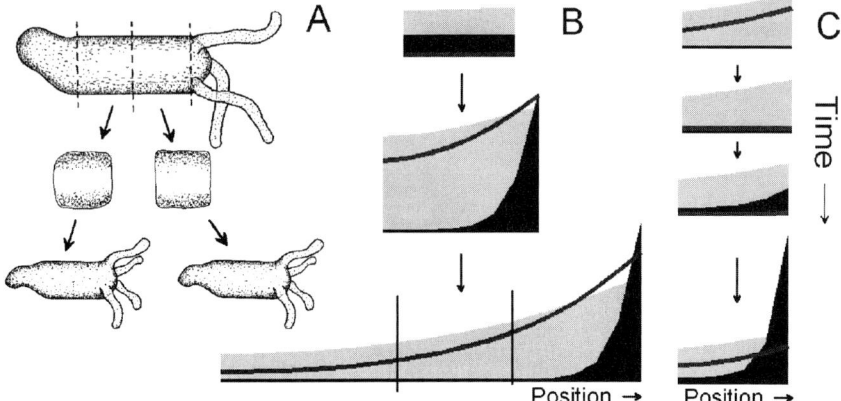

Figure 10 Feedback of the organizer on the competence to avoid secondary organizing regions. (A) Evidence for a graded competence: fragments of a hydra regenerate a new head always at the side pointing towards the original head. (B) Model: if the activator (black) has a long-ranging and long-lasting feedback on the ability of the cells to perform the autocatalysis (source density; gray distribution, ρ in Eqs. (2a), (2b) and (4); the black line is the inhibitor), cells distant to the organizer become unable to compete with the primary organizer for activation. Despite substantial growth, a single organizer and thus a monotonic gradient is maintained, although the range of the activator can be very small (compare with the periodic pattern formed in Fig. 7). (C) Due to the graded competence, regeneration can be a rapid process within the competent region since no time-consuming competition is required for one region to win. Since the source density (gray area) changes only slowly, it remains nearly unchanged during regeneration. The new activation occurs at a predictable position as suggested by the experiment (A). Due to the graded competence, the activator maximum appears at a marginal position. Due to the short range of the activator relative to the field size, regeneration can occur in small fragments.

In the model the competence corresponds to the ability of the cells to perform the autocatalytic reaction, a property we have called 'source density' [ρ in Eq. (1)]. Activator production is assumed to depend not only on the presence of activator molecules itself but also on the presence of other factors necessary to accomplish this autocatalysis. If these other factors are missing, an activation would be impossible even at low inhibitor levels. If the organizer exerts a positive feedback on these prerequisites, the ability to perform the self-enhancing reaction is preferentially maintained in the region closer to the organizer. In contrast, regions distant to a once established organizer become unable to generate secondary maxima. In this way, a single maximum can be maintained even in fields that grow substantially. Equation (4) gives a possible interaction:

$$\frac{\partial \rho}{\partial t} = \gamma a - \mu_\rho \rho + D_\rho \frac{\partial^2 \rho}{\partial x^2}. \tag{4}$$

A simulation using Eqs. (2a), (2b), and (4) is given in Fig. 10. This model is in agreement with the observation in *Xenopus*, zebrafish and chick that development is found to proceed normally after removal of the proper organizer as long as cells next to the organizer remain present (Steward and Gerhart, 1990; Yuan and Schoenwolf, 1998; Saùde *et al.*, 2000). As discussed further below, maternally provided factors may restrict from the beginning the competence to a small part of the developing embryo, counteracting in this way the formation of secondary organizing regions. Such a head start also shortens the time needed for one region to win the competition with all other regions to become the organizing region.

An organizing region thus exerts two seemingly conflicting effects. On the one hand, it inhibits the formation of other organizing regions. On the other hand, it promotes organizer formation in the first place. Why do both effects not cancel each other? This is because inhibition and promotion have different time constants. To allow stable patterns and pattern regulation, the inhibitor must have a rapid turnover such that a new organizer can reappear shortly after removal of the original organizer. In contrast, the competence has a long time constant such that within the time scale required for pattern regulation it remains almost unchanged (Fig. 10C).

While organizer formation is a local event, the graded competence extends over a much larger region. This suggests that some agent spreads from the organizer that, in turn controls directly or indirectly, the competence. The molecular basis of this graded source density (or head activation gradient as it is frequently called in the experimental literature on hydra regeneration) is not yet clear. After treatment of hydra with Alsterpaullone, a drug that stabilizes β-catenin, the whole polyp obtains properties that are normally restricted to the tissue near the genuine organizer (Broun *et al.*, 2005). In terms of the model, the source density becomes high everywhere. This effect, however, does not allow the conclusion that β-catenin as such acts as source density. β-Catenin reappears about 1 hour after head removal (Hobmayer *et al.*, 2000), indicating that it has a short time constant, suggesting that β-catenin belongs to the activator loop. This 1 hour is short compared with the 2 days required to change the intrinsic polarity. An ectopic elevation of the activator concentration would also lead to an overall increase of the source density [Eq. (4)].

Since the complementary influence of organizing regions is crucial for explaining many biological observations, it is worth illustrating the situation with an anthropomorphic analogy. A king, president or any other figure in power usually has a strong tendency to suppress others from taking over—a long-range inhibition. On the other hand, he promotes individuals among his courtiers to obtain a higher ranking, to become ministers, etc. In this way, the center of power generates a hierarchy. Inhibition and promotion are two closely interwoven processes. If the top position becomes vacant, due to this nonuniformity, a fight will set in only between the few who have high ranking in the hierarchy. Usually proximity to the former center is an advantage. In the short time interval until a new hero is selected, the ranking in the hierarchical pyramid remains essential unchanged. This analogy also illustrates what can happen if the whole hierarchy is eliminated, for instance, in a revolutionary situation. Many rivaling centers and civil-war like situations could emerge, with all their unpredictable consequences.

I. A Graded Competence Allows Small Organizing Regions

As shown earlier (Fig. 4), if the competence is homogeneously distributed, the range of the activator must be comparable to the size of the field if a polar pattern with a terminal maximum should emerge. In such a situation a small fragment cannot regenerate since its size would be smaller than the range of the activator. However, hydra fragments of 1/10 of the body length can regenerate perfectly. Again, this problem disappears if the competence is graded since the maximum will appear at the highest level of the graded source density, i.e., at a terminal position even if the range of the activator is small (Fig. 10).

J. An Inhibition in Space *and* in Time: The Generation of Highly Dynamic Patterns

In the preceding section it has been shown that a positive feedback of an activator maximum on its own sources, i.e., on the ability of the system to perform autocatalysis, can stabilize an existing maximum and can enhance its dominance over more distant regions. The opposite interference, a *destabilization* of established maxima by a second local-acting antagonistic reaction, is also a frequently used strategy in development. It can lead to highly dynamic systems that never reach a stable state. Imagine a conventional two-component system as described above. On its own it would produce a stable pattern. Imagine further a second antagonist that has the opposite properties to the normal antagonist: a short range and a long time constant. Shortly after the generation of a maximum, this maximum will be 'poisoned' by the second antagonistic reaction and thus locally quenched. Depending on the parameters, the system can respond in two ways. (i) The maximum shifts into an adjacent position, only to become quenched there too: traveling waves result. These waves can have unusual properties, e.g., they can emerge without a pacemaker region and can penetrate each other. (ii) The maxima disappear and reappear somewhat later at a displaced position, only to become quenched there again. Either regular out-of-phase oscillations between adjacent regions occur or maxima appear and disappear at somewhat irregular positions. Examples of both modes are given in Fig. 11. We came across this reaction type by searching for mechanisms that account for the pigmentation pattern on some tropical sea shells (Meinhardt and Klingler, 1987; Meinhardt, 2003). Subsequently it has turned out that such three-component systems—an activator–inhibitor system coupled to a quenching component—are appropriate to describe a wide range of biological phenomena. Examples are the pole-to-pole oscillations in the bacterium *E. coli* for the determination of the division plane (Meinhardt and de Boer, 2001), the separation of the barbs of an avian feather (Figs. 11A–11E; Harris *et al.*, 2005), the highly sensitive orientation of chemotactic cells and growth cones by minute external cues (Figs. 11F–11K; Meinhardt, 1999), and the initiation of new leaves around a growing shoot with a displacement of the golden angle (Meinhardt, 2004b). A detailed discussion of these systems is beyond the scope of the present article.

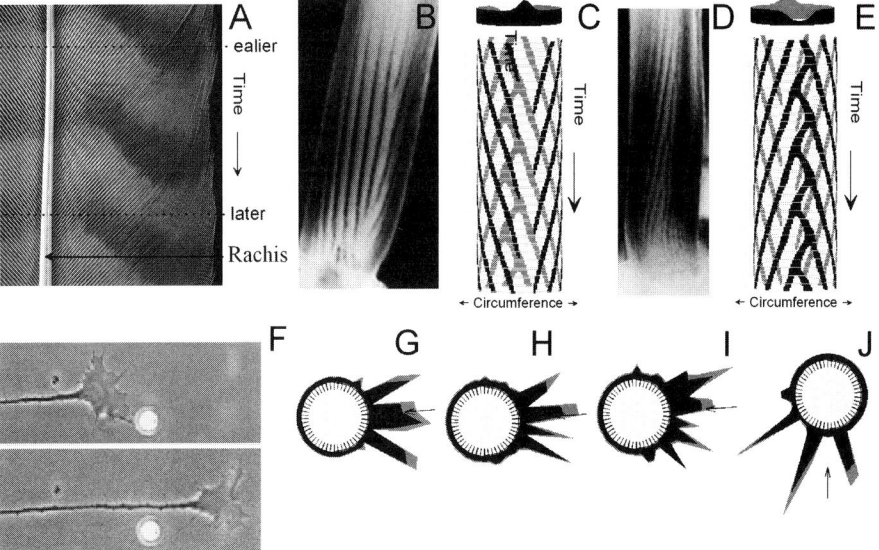

Figure 11 Patterning in avian feathers and orientation of chemotactic cells as examples for the role of destabilization by a second antagonist. (A–E) Formation of feather filaments (barbs) by traveling waves. Feathers are formed by proliferation of stem cells at the base of feather buds. Therefore, the tip of a feather is the oldest part (in contrast, for example, to the situation of a tree!). Also the cells forming the tip of the barbs are born earlier than those forming the connection to the rachis. A permanent regional cell death along the ventral side (the side opposite to the rachis; dark branching line in D and E) allows an opening of the cylindrical sheet into a plane. The signal is formed by *Shh* (B and D)/*BMP2* that acts as an activator–inhibitor system (Harris *et al.*, 2005). Like cutting with scissors, traveling waves of high *Shh* expression (dark lines) separate the individual barbs (white oblique stripes). These waves run from the "cut-open" region on the ventral side towards the rachis at the dorsal side. For the simulation C and E a second, short-ranging but long lasting inhibitor was assumed that locally quenches the maxima, enforcing traveling waves by a permanent shift of the maxima. This cutting comes to rest near the future rachis (wave-free region in B and C), otherwise the filaments would detach from the feather. (F–K) Orientation of growth cones (F) and other chemotactic sensitive cells by minute external asymmetries. The problem is to maintain sensitivity for the minute external asymmetries although a strong internal amplification is involved to generate a pronounced cell-internal pattern. In the model (Meinhardt, 1999), isolated signals for filopods (black) are generated by a saturating self-enhancing reaction together with an inhibition that covers the whole cell (not shown). They appear preferentially at the side where the guiding signal (arrow) is slightly higher. A second antagonistic component (gray) quenches locally the signal after a certain time interval. Thus, signals for filopods appear and disappear permanently at the cell surface (G–I). After a change in the direction of the guiding signal, the internal signals emerge at the new side (J), although the guiding asymmetries are minute (2% across the cell plus 1% random fluctuation between cell surface elements).

III. The Two Main Body Axes

In vertebrates the famous Spemann organizer and its relatives such as Hensen's node play a crucial role in axes formation. Many of the molecular components involved are known (reviewed in Harland and Gerhart, 1997; De Robertis and Kuroda, 2004; Stern, 2001; Boettger *et al.*, 2001; Niehrs, 2004; Schier and Talbot, 2005). However, which axes the organizer controls—AP, DV or both—remained remarkably fuzzy. How can a single organizer organize two axes that are oriented perpendicular to each other? In amphibians even the orientation of the main body axes in the early embryo relative to the animal–vegetal axis of the egg is controversial (Gerhart, 2002). What is the relation of the hydra-type organizer of ancestral organisms and the Spemann-type organizer? Is there a hidden organizer for the second axis? How it is achieved that the two axes are so rigidly coupled (Fig. 1)?

The finding that β-catenin and *Wnt* are expressed in the hydra organizer (Fig. 2A; Hobmayer *et al.*, 2000) was very exciting since it was well known that the same pathway also plays a crucial role in the formation of the Spemann-type organizer. Indeed, hydra-derived β-catenin mRNA injected into an early amphibian embryo can induce a second embryonic axis, as would a graft of the Spemann organizer. However, as shown below in more detail, in vertebrates, control for the AP axis does not reside in a Spemann-type organizer but in the equivalent of the hydra-type organizer, the blastoporal ring. The Spemann-type organizer, located on this ring, initiates the formation of a stripe-shaped midline organizer. The DV (or better mediolateral) specification of a cell depends on its distance to this midline rather than on the distance to the original organizer. Midline formation is realized with different structures in the brain and in the trunk, with the prechordal plate and the notochord, respectively. Thus, in the early gastrula there are two separate organizers, one for the primary AP axis and one for the secondary DV axes. Both have a stripe-like extension with orthogonal orientation, convenient for generating a near-Cartesian coordinate system (Figs. 12A and 12B). The actual positional information for the mediolateral axis is presumably a reversed gradient since the midline acts as sink for BMP (Dosch *et al.*, 1997).

A. The Blastopore (Marginal Zone) as Organizer for the AP Axis

As mentioned above, the small hydra organizer located around the gastric opening became in vertebrates a large ring (Figs. 3 and 12A). The canonical *Wnt* pathway is a crucial component of the hydra organizer (Hobmayer *et al.*, 2000). Likewise in the vertebrates, *Wnt-8* expression is high in the marginal zone/germ ring (Christian and Moon, 1993) and evidence has accumulated that a gradient of *Wnt* controls the AP pattern of the brain (Kiecker and Niehrs, 2001a; Nordström *et al.*, 2002; Dorsky *et al.*, 2003). This suggests that the signaling center required for the AP patterning is the blastopore itself, i.e., the ancestral organizing region of radially-symmetric organisms. Cells distant to the blastopore are exposed to low *Wnt* levels and form the forebrain; cells closer to the blastopore form the midbrain. In agreement with recent observations this early AP specification is essentially independent of the Spemann-type organizer. For instance, by removal of maternal determinants from the fish embryo it was possible to suppress the formation of the organizer completely. By

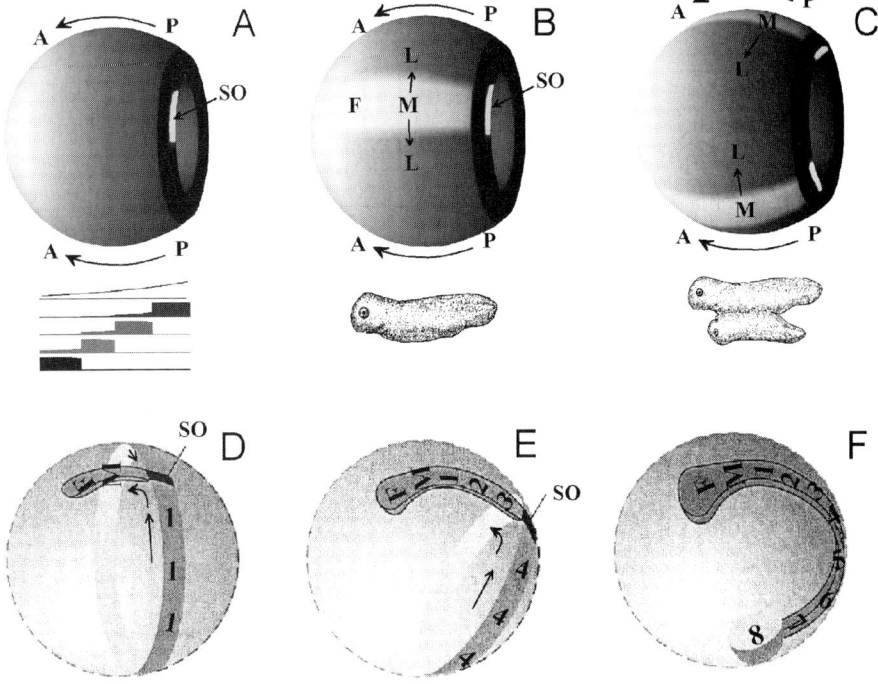

Figure 12 Model for the generation of a near-Cartesian coordinate system in two steps—the amphibian embryo as example. (A) AP-patterning of the early gastrula is assumed to be accomplished by an ancestral system (see Fig. 3). The marginal zone (black) is assumed to be equivalent to the hydra organizer and controls the posterior-to-anterior pattern in a gradient-based manner (fading gray), a process that does not require an organizer. (B) The Spemann organizer (SO, white) forms on the blastoporal ring. The ingressing organizer-derived mesodermal cells form the prechordal plate (light gray), which acts as the midline organizer for the mediolateral (L ← M → L) pattern of the brain and induces neuronal development in the overlying ectoderm. The distance from this midline determines the mediolateral specification. Both signaling sources have a stripe-like extension and provide a near-Cartesian positional information system determining the pattern of the fore- (F) and midbrain (M). (C) Induction of a second organizer (see Fig. 17) leads to two embryos that are fused at the ventral side. (D–F) AP patterning of the trunk. The cells at the blastopore obtain in a time-autonomous process more and more posterior determinations (1, 2, 3 . . .). The pace of this process is given by an oscillation that leads also to the periodic patterning of the somites (see Fig. 24). Cells near the blastoporal ring move towards the organizer and the incipient midline, forming the mediolateral pattern along the emerging AP axis. When cells obtained a certain distance from the organizer, the somitic oscillation stops (see Fig. 24), somites are formed and the cells obtain their final AP determination. Cells antipodal to the organizer have to move further to reach the region near the organizer/midline and are later integrated into the axial structures. Due to the prolonged posteriorization these cells form more posterior structures. Thus, cells antipodal to the dorsal organizer form posterior and not primarily ventral structures. In this schematic drawing, the animal–vegetal axis is fixed and the shape changes due to the conversion–extension mechanism are ignored (partially after Meinhardt, 2006).

a simultaneous suppression of *BMP* signaling due to the *swirl* (*ZBmp-2b*) mutation, neuronal development was enabled all around the circumference of the early embryo (Ober and Schulte-Merker, 1999). Most remarkable, the expression of neuronal markers such as *Otx* and *Krox 20* appeared in the normal order although no organizing region was present, supporting the view that the Spemann-type organizer does not play a decisive role in AP patterning. Similar results were observed in *Xenopus* (Reversade *et al.*, 2005). Also during later development, newly expressed Hox genes, which are required to specify more posterior structures, become activated exclusively near and around the marginal zone but not in the organizer itself (Wacker *et al.*, 2004a, 2004b, see below).

B. The Spemann-Type Organizer Induces Midline Formation

In principle, any second signaling source with an-off-axis localization would lead to symmetry breaking and bilaterality. However, to pattern the DV or mediolateral axis of an animal with a long-extended AP axis, a spot-like signaling center is inappropriate. A reasonable, i.e., a near-Cartesian combinatorial patterning requires for the DV axis an organizing region with a stripe-like geometry. Only then, for instance, can floor plate, notochord, somites, neural crest cells or particular nerve cells be formed at the midline or at a certain distance from it, independent of the AP level.

The generation of a solitary straight stripe-like organizing region is an intricate patterning problem. As mentioned above (Fig. 6), under certain conditions—saturation of autocatalysis—stripe-like patterns can be generated. Stripe formation requires a reduced lateral inhibition because otherwise the stripes would disintegrate into isolated patches. The reduced lateral inhibition, however, has the consequence that an existing stripe can suppress other stripes only up to a moderate distance. The width of the stripe and the width of the interstripe regions are of the same order. Nevertheless, the formation of a single straight organizing region is possible by an appropriate interaction of a spot-forming and a stripe-forming system whereby the spot-forming system makes sure that only one stripe can emerge. A surprising diversity of such midline-generating mechanisms has evolved (Meinhardt, 2004a). In vertebrates midline formation occurs by sequential elongation under the driving force of the organizer. The midline appears on the *same* dorsal side as the organizer (Fig. 12). In insects and spiders, a dorsal organizer repels the midline (Moussian and Roth, 2005; Akiyama-Oda and Oda, 2006). The midline and thus the central nervous system appear on the *opposite* side, i.e., ventrally (Fig. 13A). The well-known DV–VD inversion between insects and vertebrates is proposed to have this origin. While in vertebrates the midline becomes sequentially elongated, in insects the midline was predicted to have from the beginning the full AP extension but becomes narrower in the course of time (Meinhardt, 1989, 2004a), in agreement with recent observations (Fig. 13B) (Chen *et al.*, 2000). Thus, neither a moving node nor a moving prechordal plate is involved in midline formation in insects. For planarians the situation is very different. A primary DV confrontation is the precondition for generating the secondary AP axis (Chandebois, 1979;

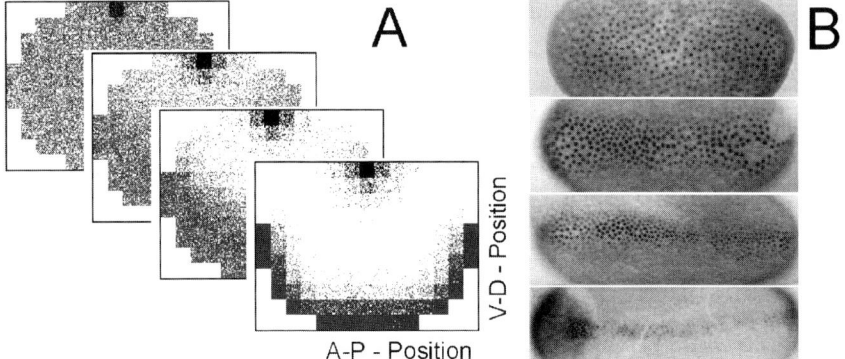

Figure 13 Model for midline formation in insects by an inhibition from a dorsal organizer. (A) Predicted mechanism: A patch-like dorsal organizer repels the midline; the midline forms at the ventral side (Meinhardt, 1989, 2004a). Two pattern-forming systems are assumed, one with a patch-forming (black) and one with a stripe-forming characteristic (gray). The first has an inhibitory influence on the second. A high activation of the stripe-forming system is possible only at a distance from the patch-shaped organizer. The midline appears simultaneously along the whole AP axis but sharpens in the course of time to a narrow ventral midline, in contrast to the sequential elongation in vertebrates (Fig. 12). (B) Such sharpening has been found in *Tribolium*. Shown is the *Dorsal* expression on the ventral side at successive stages (Chen et al., 2000). (Figure kindly supplied by Siegfried Roth.)

Kato et al., 2001). This suggests that the DV pattern is primary and that a DV border is the precondition for forming the terminal structures of the AP axis. The DV border represents an organizing line that circumvents the organisms (Fig. 14).

The temporary expression of *Noggin* at this DV-border (Ogawa et al., 2002) suggests that what in vertebrates is the single dorsal midline corresponds in planarians to *two* lateral organizing stripes that extend from the anterior to the posterior pole. This corresponds well with the *two* lateral nerve cords that are typical for planarians, one at each side. The assumption of a repulsion from the circumventing organizing borders explains the formation of a single dorsal midline (Fig. 14), realized by a high *BMP* expression (Orii et al., 1998). These very different modes of generating a bilateral body plan suggest that different ancestral bilateral organisms evolved in a different manner from radially-symmetric ancestors without a common urbilaterian.

C. The Orthogonal Orientation of the Main Body Axes in Vertebrates

To obtain a near-Cartesian coordinate system in vertebrates, it is crucial that the midline has an orientation parallel to the AP axis, i.e., perpendicular to the blastopore. In the early gastrula the blastopore and thus the organizer is posterior to the brain anlage while the trunk does not yet exist. Therefore, from the initial position of the organizer,

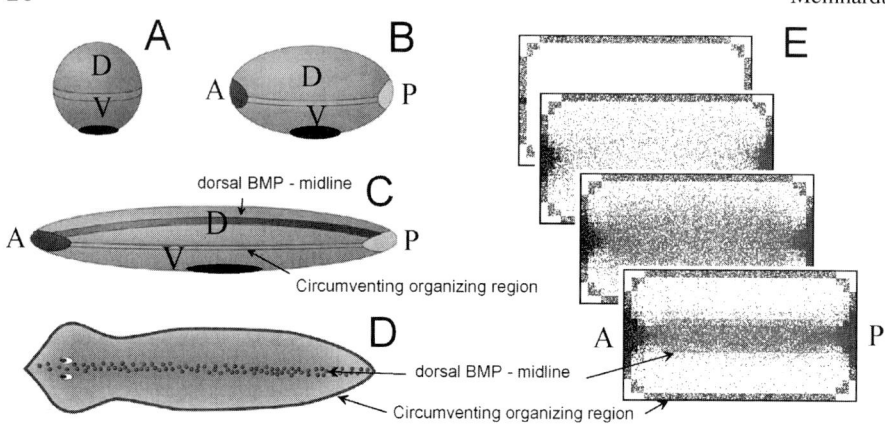

Figure 14 Model for axes formation in planarians. (A) Primary is a DV subdivision. The DV border obtains organizing function. The single opening is ventrally located, in contrast to the posterior localization in vertebrates and insects. (B) The formation of the terminal A and P regions depend on this DV border. (C, D) The expression of *Noggin* (Ogawa *et al.*, 2002) suggests that these two lateral organizing lines are equivalent to the single dorsal midline in vertebrates. The inhibitory influence of these borders allows *BMP* expression only at the central dorsal midline as observed (Orii *et al.*, 1998). (E) Simulation: A stripe forming system (*BMP*) is repelled from the two circumventing lateral borders and forms the midline. The terminal regions A and P are assumed to have an activating influence of midline formation in order that the midline can reach the anterior and posterior pole (Meinhardt, 2004a).

the midline has to be extended both towards the anterior to organize the brain and towards the posterior together with the sequential elongation of the trunk.

Mesodermal cells all around the marginal zone ingress. During early ingression, these cells move necessarily away from the blastopore, i.e., towards the anterior. Cells of the organizer, being a small part of the large mesodermal ring, populate with this ingression a narrow stripe, the prechordal plate (Kiecker and Niehrs, 2001b; Gritsman *et al.*, 2000). This stripe has necessarily an orientation perpendicular to the blastopore and parallel to the AP axis as required (Fig. 12). Most interestingly, although this polarized cell movement occurs normally in a cell sheet, this ingression can be accomplished by single cells. Single wildtype cells transplanted into a mutant in which ingression is no longer possible move in the correct way (Carmany-Rampey and Schier, 2001), suggesting that some chemotaxis-like orientation of the cell movement away from the blastopore is involved (see also Fig. 11F–11J).

D. The Hydra-Type Organizer (Marginal Zone) Provides the Prerequisites to Generate the Spemann-Type Organizer

Based on the considerations above, it is expected that an ancestral molecular system is involved in blastopore formation. To localize the organizer on the blastopore, it is expected that this ancestral system provides the precondition for a second pattern-forming system that generates the proper Spemann-type organizer. The canonical *Wnt*-pathway including β-catenin is a good candidate to provide this precondition since it marks the blastopore already in cnidarians and in *Amphioxus* (see Fig. 2; Hobmayer *et al.*, 2000; Holland *et al.*, 2000, 2005; Yu *et al.*, 2007). Indeed, during early stages the *Wnt* pathway is crucial for the formation of dorsal structures in all vertebrates (see Fagotto *et al.*, 1997; De Robertis *et al.*, 2000).

E. The Spemann-Type Organizer: The *Chordin/BMP/ADMP* System as a Pattern-Forming Reaction

The Spemann-type organizer displays many regulatory features typical of a genuine pattern-forming process. For instance, after organizer ablation a more or less normal development can follow in many systems (see Harland and Gerhart, 1997; Yuan and Schoenwolf, 1998; Saùde *et al.*, 2000). As mentioned, complete chicken embryos can be formed after early fragmentation of the blastodisc (Fig. 1). In *Xenopus*, co-culture of dissociated animal and vegetal cells leads not only to mesoderm induction but also to the formation of axial structures such as notochord and neural tube (Nieuwkoop, 1992), although such a procedure certainly wipes out any maternally imposed asymmetries. Thus, it is expected that the formation of the Spemann-type organizer is based on a pattern-forming process in which both self-amplification and long-range inhibition play an important role. Evidence for this mechanism has accumulated recently (Chen and Schier, 2002; Lee *et al.*, 2006; Reversade and De Robertis, 2005).

Central in organizer formation is the mutual inhibition of *BMP* and *Chordin* (reviewed in Harland and Gerhart, 1997; Niehrs, 2004; De Robertis and Kuroda, 2004; Stern, 2001; Boettger *et al.*, 2001; Schier and Talbot, 2005). Two components that mutually inhibit each other produce as a system a positive autoregulation. For instance, an increase of the first component leads to an enforced repression of the second, which, in turn, leads to a further increase of the first as if this substance would be directly autoregulating. To obtain a balanced activation of *BMP* and *Chordin* at opposite positions, a third component is anticipated that acts antagonistically on one of these indirectly self-enhancing reactions. A candidate is the *Anti-Dorsalizing Morphogenetic Protein* (*ADMP*) (Moos *et al.*, 1995). Its properties have been frequently regarded as counterintuitive: being expressed in the organizer, but functioning by reducing organizer activity. However, as its acts over a longer range (Lele *et al.*, 2001; Willot *et al.*, 2002; Reversade and De Robertis, 2005), it satisfies the theoretical expectations: being produced in the organizer region, and yet antagonizing a self-enhancing reaction. Equations (5a)–(5c) describe this simplified *Chordin* (c)–*BMP* (b)–*ADMP* (a) interaction and Fig. 13 shows a simulation of pattern formation based on this model system.

$$\frac{\partial c}{\partial t} = \frac{\rho}{\gamma_c + b^2/a^2} - \mu_c c + D_c \frac{\partial^2 c}{\partial x^2} + \rho_c, \tag{5a}$$

$$\frac{\partial b}{\partial t} = \frac{\rho}{(\gamma_b + c^2)} - \mu_b b + D_b \frac{\partial^2 a}{\partial x^2} + \rho_b, \tag{5b}$$

$$\frac{\partial a}{\partial t} = \gamma_a c - \mu_a a + D_a \frac{\partial^2 a}{\partial x^2}. \tag{5c}$$

In this model equation c (*Chordin*) is inhibited by b (*BMP*) and *vice versa*; a (*ADMP*), under control of the organizer c, undermines the c-inhibition by b. This model is certainly a simplification. The mutual interference also depends on a competitive binding to receptors. Further, *Tolloid* plays an important role in degradation and transport. The side antipodal to the organizer might harbor a complete pattern-forming system in itself since *BMP* autoregulation (Hild *et al.*, 1999) and long-ranging antagonistic component (*swirl* and *sizzled*) have been found (Martyn and Schulte-Merker, 2003; Lee *et al.*, 2006). Thus, more complex interactions are to be envisaged. Nevertheless, the simplified scheme given in Eqs. (5a)–(5c) accounts for many observations, such as broadening and shrinking of expression regions if *BMP*, *Chordin* or *ADMP* are misexpressed or for the regeneration of an organizer after ablation. The more complex system encompasses the fact that residual pattern formation can take place even if some components of are nonfunctional.

F. The Role of Maternal Determinants

Frequently in the literature the formation of the Spemann-type organizer is regarded more as a chain of inductions rather than as a self-organizing process since organizer formation depends on the correct placement of maternally derived determinants. This is the case in amphibians and in fish (Harland and Gerhart, 1997; Ober and Schulte-Merker, 1999). After separation of the blastomeres of a frog embryo at the two-cell state, the cell not containing the future Spemann organizer is unable to form dorsal structures. Only unstructured tissue, 'Bauchstücke' in terms of Spemann, results (see Sander and Faessler, 2001).

As shown above, if pattern formation starts from a nearly homogeneous initial distribution, a single organizing region can be generated only in a field small enough that the range of the antagonistic reaction is sufficient to suppress secondary organizing regions (Fig. 4). In contrast, for instance, the early amphibian gastrula is very large. Even if organizer formation is restricted to the marginal zone, the inhibitor would be further diluted by spreading into the whole embryo, making inhibition of a secondary activation at the antipodal side difficult (Fig. 16). A way to ensure the formation of only a single organizing region in a large embryo is to make only a small portion of the marginal zone competent. Since the *Wnt* pathway provides the precondition for organizer formation, an asymmetric distribution of components of the *Wnt* pathway can provide a strong bias for the positioning of the organizer up to the point that the disadvantaged side becomes incompetent to form the organizer. Indeed, by cortical rotation, components of the *Wnt* pathways are displaced from the vegetal pole to the marginal zone, making the cells competent for organizer formation. The model also provides a rationale for why strong initial asymmetries are especially common in large embryos such as amphibians but less important for small ones such as the mouse.

Wnt signaling is not required to maintain the organizer. Axis formation can be induced by injection of *Chordin* RNA even if the appropriate localization of components of the *Wnt* pathway was suppressed by UV irradiation (Smith and Harland, 1992). Moreover, although required for organizer initiation, *Wnt* synthesis is switched off in the organizer as soon as the organizer is formed (Christian and Moon, 1993). In terms of the model proposed, this is required since *Wnt* is the signal for posteriorization. When organizer-derived cells move anteriorly to form the prechordal plate, this posterior signaling has to be switched off; otherwise no anterior brain structures could be formed. *Goosecoid*, a downstream target in the organizer, directly suppresses *Wnt8* (Yao and Kessler, 2001).

G. Pattern Regulation and Unspecific Induction: How Dead Tissue Can Induce a Second Embryonic Axis

Many classical transplantation experiments have shown that an existing organizer has a strong inhibiting effect on the formation of a secondary organizer. This has been shown for the hydra (Wilby and Webster, 1970a, 1970b) and for the chick organizer (Khaner and Eyal-Giladi, 1989). The simulations in Fig. 15 show that our model describes these regulations correctly.

One of the problems in the early search for molecules involved in organizer formation was that very unspecific manipulations, such as implantation of denatured tissue or injury, can trigger the formation of a secondary embryonic axis. Waddington *et al.* (1936) proposed that this nonspecificity results from the removal of an inhibitor. The tendency for unspecific induction is species-dependent. It is low in *Xenopus* but high in *Triturus*, the model system most studied in the early days. The occurrence of unspecific induction is a straightforward consequence of the proposed activator–inhibitor scheme. By an unspecific manipulation one can only lower a substance concentration, for instance, by a leakage through an injury or by the activation of degrading enzymes. At larger distances from an existing activator maximum, the inhibitor concentration is low anyway. Any further decrease may result in the onset of autocatalysis (Fig. 17D). A second activator maximum will be triggered that becomes independent of the triggering stimulus since the activator concentration increases until it is in equilibrium with the surrounding cloud of inhibition. The resulting maximum is indistinguishable from a maximum triggered by application of the genuine autocatalytic substance.

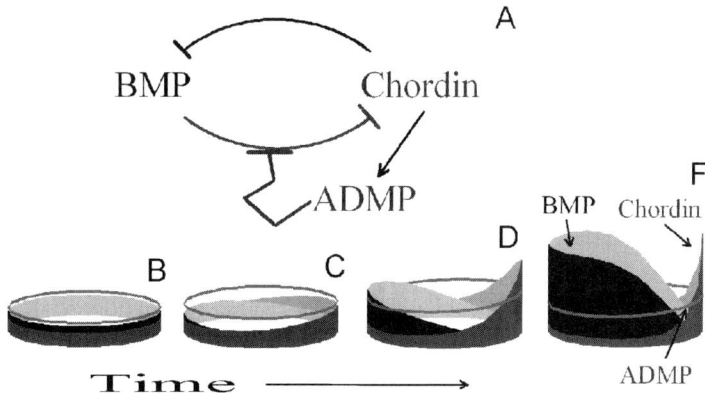

Figure 15 Self-enhancement realized by in inhibition of an inhibition—a simplified model for pattern formation in the Spemann organizer. (A) *Bmp* and *Chordin* mutually repress each other. Such an interaction is equivalent to a self-enhancement. On its own, such a system behaves as a switch (a classical example is the switching between the lytic and lysogenic stage of the Lambda phage). A balanced ratio between the regions in which either the *Bmp* or the *Chordin* level is high is achieved by a diffusible substance, *ADMP*, which is under the same control as *Chordin*. Thus, *ADMP* is expressed in the organizer although its function is to down-regulate the organizer-generating components, as observed (Moos *et al.*, 1995; Lele *et al.*, 2001). The balance between the three components is easily understandable. For instance, an elevation of *Chordin* production would lead to an elevation of *ADMP* and thus to a back-regulation. Or an elevation of *BMP* would lead to a down-regulation of *Chordin* thus to a down-regulation of *ADMP* thus to an increase of *Chordin* causing a return to the steady state. (B–F) Simulation of *BMP/Chordin/ADMP* patterning using Eqs. (5a)–(5c). The *Chordin* distribution is sharp since it also gives rise to the inhibitory component, *ADMP*. In contrast, *BMP* is complementary to *Chordin* and has, therefore, a more plateau-like distribution.

H. The AP Patterning of the Trunk: A Time-Based Sequential Posterior Transformation and a Ring-to-Rod Conversion

Trunk formation in vertebrates occurs by the so-called convergence and extension mechanism (reviewed in Keller, 2005; Solnica-Krezel, 2005). Cells near the blastoporal ring move towards the organizer and the incipient midline, causing a conversion of the huge blastoporal ring perpendicular to the AP axis into the axial structures parallel to the AP axis. In this ring-to-rod conversion cells of the ring move towards the organizer and towards the organizer-derived midline-forming cells.

As mentioned, the blastoporal ring is the source of a posteriorizing signal for the patterning of the brain. Available evidence suggests that the acquisition of even more posterior determinations during trunk formation is a time-driven process. The sequential activation of new Hox genes occurs close to the blastopore excluding the organizer region, as shown for the chick (Gaunt and Strachan, 1996) and for *Xenopus* (Wacker *et al.*, 2004a, 2004b). Involved in this sequential posteriorization is an oscillation that leads also to the formation of the periodic structures (see Fig. 24). The vertebrate organizer is not required for the time-dependent posteriorization but it

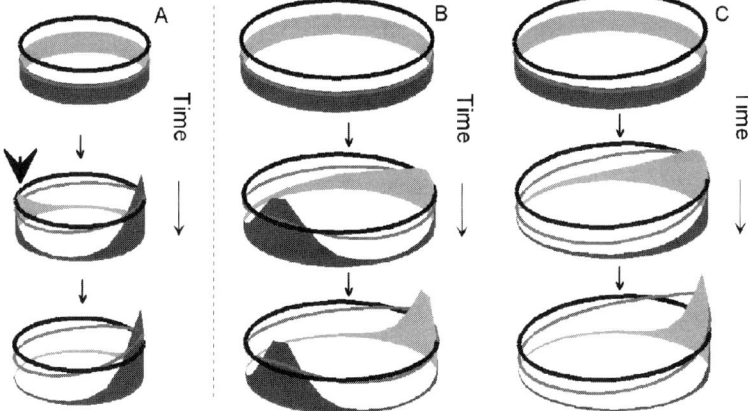

Figure 16 Maternal asymmetries as a mean to avoid supernumerary organizers. (A) In a small field only a single organizer will be formed even in the absence of maternal asymmetries (gray area: activator distribution, gray line: inhibitor distribution; black line: maternal influence, the source density ρ in terms of Eqs. (1a), (1b)); shown are the initial (top), an intermediate (middle) and the final stable state (bottom). Note that a transient activation (arrowhead) opposite to the more advanced maximum eventually disappears. (B) In a larger area, two maxima can coexist. (C) An asymmetry in the source density, e.g., due to maternally supplied determinants, makes sure that only a single maximum is formed. In even larger fields, stronger asymmetries are required to maintain a single maximum.

is crucially involved in the termination of posteriorization in the correct order. During convergence-extension movement, the cells obtain an increasing distance to the blastoporal ring and to the organizer (Fig. 12D–12F). Therewith they leave the zone in which posteriorization can take place. Cells originally closer to the organizer are the first to escape this posteriorization and form, therefore, anterior structures. This scheme accounts for the nontrivial fact that cells antipodal to the *dorsal* organizer form *posterior* structures (and not primarily *ventral* structures as it is frequently indicated in textbooks). The antipodal cells have to move the longest way to come close to the organizer and the midline. Therefore, they are exposed longest to the sequential activation of more posterior Hox genes.

After implantation of a secondary organizer, the new organizer attracts cells in the same way as does the normal organizer. Therefore, cells close to the newly implanted organizer are integrated much earlier into the supernumerary axial structure and are therewith in the same way protected from the sequential posteriorization at a much earlier stage (Meinhardt, 2001; Wacker *et al.*, 2004a, 2004b). Thus, cells that would form posterior structures in the unperturbed situation will form anterior structures instead. According to the model, this is not a posterior-to-anterior reprogramming but is based on an earlier relief from the posteriorization.

It should be kept in mind that this mode of axis formation holds only for vertebrates. In insects, for instance, formation of mesoderm occurs in a stripe that has from the

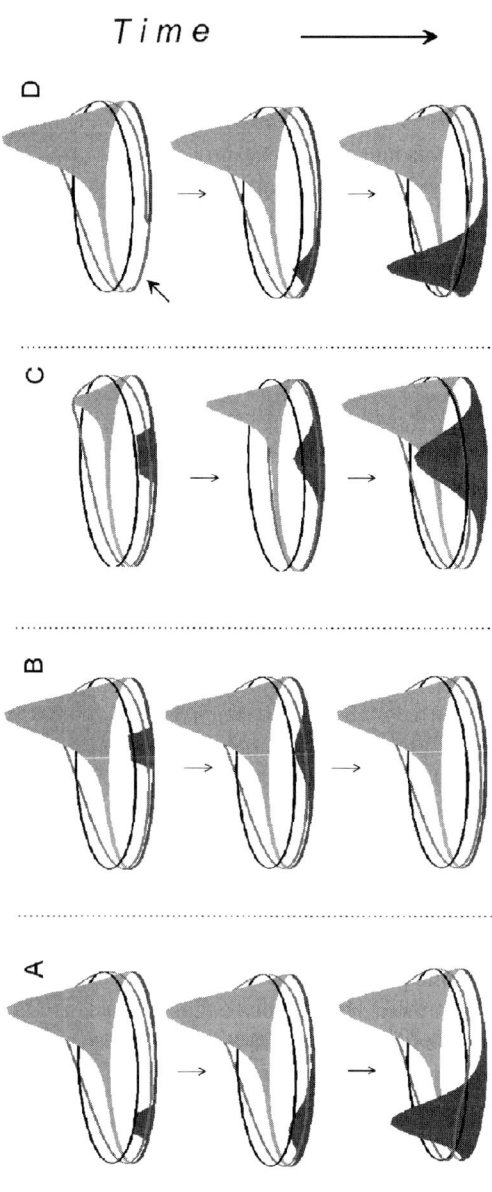

Figure 17 Simulations demonstrating regulatory properties of an organizing region. (A) Implantation of somewhat activated tissue at a position opposite to the endogenous organizer can induce a full activation. (B) When implanted closer to the organizing region, even a stronger activation will be down-regulated. (C) After partial removal of the existing organizer, the implanted activation has a better chance to survive. Whether one or two maxima survive depends on their distance and the total size of the field into which the inhibitor can escape. These simulations account for transplantation experiments in the early chick embryo (Khaner and Eyal-Giladi, 1989). (D) Unspecific induction. Any temporary reduction of the inhibitor (arrow) can lead to the trigger of a second organizing region that would have the same properties as the endogenous one. Thus, induction of secondary organizing regions is a straightforward consequence of our theory of pattern formation. That unspecific induction is frequent in some animals (as in *Triturus*) but rare in others (as in *Xenopus*) could depend on the competence of the tissue antipodal to the organizer, on the level of a baseline inhibitor production [ρ_b in Eq. (2b)] or on how rapidly the wound is closed. (Figure partially from Meinhardt, 2001.)

beginning an anteroposterior orientation, A ring-to-rod conversion as in vertebrates is not required (Meinhardt, 2004a, 2006).

I. The Left–Right Polarity: A Second Pattern Is Squeezed to the Side

Considerable progress has been made in understanding the generation of the left–right (LR) asymmetry of vertebrates (for review see Cooke, 2004; Raya and Izpisúa Belmonte, 2006). A key question is how a reliable bias is generated that determines left and right in a reproducible way (Levin and Palmer, 2007). However, even in the absence of this bias, an LR-type asymmetry emerges, although with a random orientation (Nonaka et al., 1998). Many observations can be accounted for by assuming that the midline system induces on long range and represses on short range a second patterning system that marks the left side (Fig. 18) (Meinhardt, 2001). The 'left'-system again depends on the interplay between short-ranging activation and long-ranging inhibition. It is induced by the midline system, but becomes shifted to the side. Only in the presence of a systematic bias, is the side to which the LR-activator shifted nonrandom. This model has found recently support from the demonstration that *Nodal* acts as the activator and *Lefty2* as the inhibitor (Nakamura et al., 2006). The repression of *Nodal* by the midline system, i.e., the notochord, has been shown (Lohr et al., 1998; Bisgrove et al., 1999). According to the model, if the activity of the midline activator is reduced, the shift may no longer work and the 'left'-signal can remain in the center. The model accounts for the observation that in a right fragment not only the midline signal but also the 'left'-signal regenerates (Fig. 18). In Siamese twins one of the embryos has frequently a *situs inversus* (Newman, 1928), an observation also made by Spemann with his artificially double embryos after organizer transplantation (see Sander and Faessler, 2001). According to the model, the mutual inhibition of the two 'left' signals can be stronger than the bias towards the left (Fig. 18F).

The left–right patterning is an obvious feature of vertebrates but not of insects (Cooke, 2004). In vertebrates, due to the local elongation of the midline by the node, a reliable signal can be generated next to the node (Fig. 18A). In contrast, the insect midline has from the beginning a long AP extension (Fig. 13). A squeezing out of the 'left'-signal by the midline could lead to signals on the left and on the right in an alternating sequence (Fig. 18B). Thus, the different mechanisms discussed above for midline formation provides a rationale as to why a LR-pattern is easier to realize in vertebrates.

IV. Subpatterns

The spatial complexity of higher organisms requires a finer and finer subdivision along the main body axes. There are several possibilities:

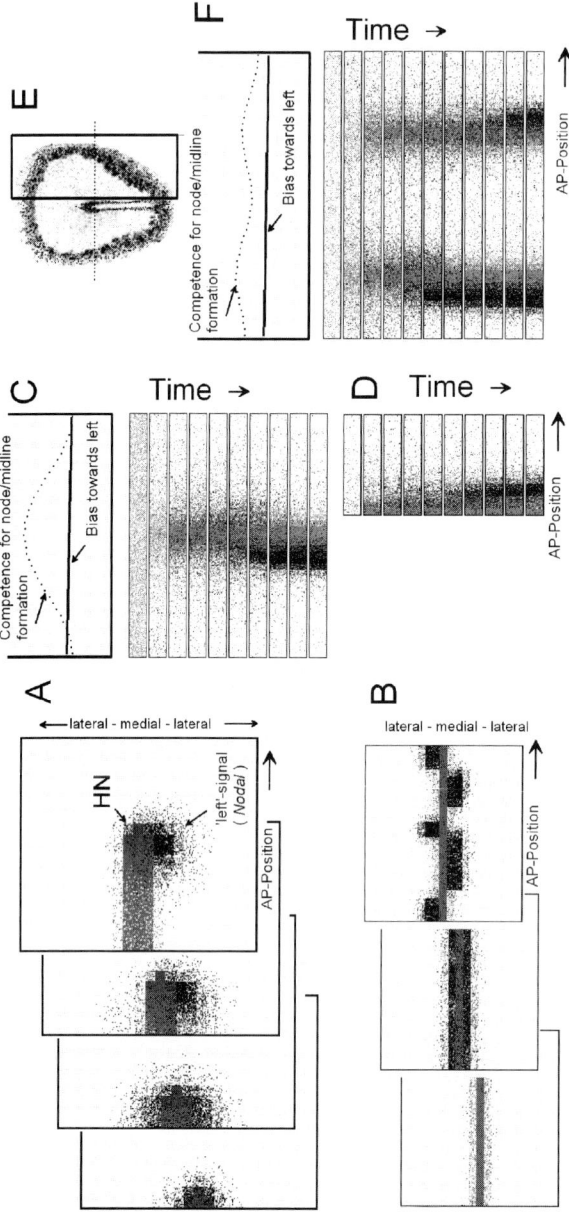

Figure 18 Model for left–right patterning. The midline/node (gray, HN) is assumed to induce on long range a second activator–inhibitor system that is responsible for the 'left'-signal (black; *Nodal/Lefty2*). (A) Originally centered on the midline the 'left' signal becomes shifted to a lateral position. (B) This mechanism would not work if the midline had the full AP extension from the beginning as it is the case in insects (see Fig. 13). The 'left' signal would appear at alternating positions on the left and the right side. (C) Simulation of the shift; shown is a cross-section (indicated by the dotted line in E). Midline formation (gray) triggers *Nodal* activation (black) and repels *Nodal* activation at high levels of the midline signal. A minor bias (oblique line) is sufficient for determining that *Nodal* displacement occurs reproducibly to the left. In the absence of this bias, the shift would occur at random to the left or to the right. (D) In a right fragment of the embryo (Levin and Mercola, 1998), first a new midline signal is reformed. This triggers again the 'left' signal (black) that normally appears on the left side. It is now shifted to the right since the left side no longer exists (Meinhardt, 2001). The bias no longer plays a role since a competition between the two sides is no longer possible. (E) Schematic drawing of the experiment (Levin and Mercola, 1998). The square indicates the right fragment. (F) Simulation of the generation of a *situs inversus* in one of a pair of Siamese twins. If two midlines are close to each other, the inhibitory effect of one 'left' signal onto the other could be stronger than the bias. The 'left' signal appears on the right side in the relatively delayed embryo.

(i) Concentration-dependent gene activation: sharply confined regions in which particular genes are active emerge under the influence of the graded distribution of a 'morphogen.' The gradients are generated as described above or by a 'cooperation of compartments' (see below).
(ii) Mutual induction of structures that locally exclude each other: cells of type A induce cells in their neighborhood to become cells of type B, which may, in turn, induce further cells to become C-cells, and so on. In such mechanisms there is no need for a global signal.

These mechanisms have different regulatory properties, and both are realized, in some cases even within the same organ. For instance, the sequence of leg segments in insects seems to be under the control of a gradient, while the internal structures *within* each leg segment is better described by the induction mechanism. The molecular ingredients required for both mechanisms and their different regulatory properties will be briefly discussed. Both mechanisms lead to sharply confined regions of gene expression. Borders between such regions can become new signaling centers that allow the generation of new coordinate systems, as required for the initiation of legs and wings. In the present paper, these mechanisms can be discussed only very briefly.

A. Switch-Like Gene Activation Requires Autocatalytic Genes

Signals generated by diffusible molecules are necessarily transient since in the enlarging tissue the communication between different parts would require more and more time. Moreover, the slope of a gradient depends on the half-life and the diffusion rate of the signaling molecule, and would not automatically adapt to natural changes in the field size. Therefore, signals generated at small scales have to be translated at a certain stage into more stable states of cell determination that can be maintained independent of the inducing signals. The obvious means is a concentration- (and thus space-) dependent activation of genes.

There is an interesting formal analogy between gene activation and pattern formation. Pattern formation requires an activation at a particular position and the inhibition in the remaining part. The selection of a particular pathway requires the activation of a particular gene and the suppression of the alternative genes. Thus, essential steps in development can be regarded as a sequence of pattern formation processes in real space coupled with a pattern formation among alternative genes. This formal analogy was the rationale for predicting autoregulatory gene activation underlying cell differentiation (Meinhardt, 1976, 1978). The long-range inhibition in pattern formation corresponds to the repression of the set of alternative genes in gene activation.

A stable switch-like activation of a single gene can result from a nonlinear autocatalytic feedback of a gene product on the activation of its own gene (Fig. 19). Equation (6) provides a possible interaction.

$$\frac{\partial g}{\partial t} = \frac{cg^2}{1 + \kappa g^2} - rg + m. \tag{6}$$

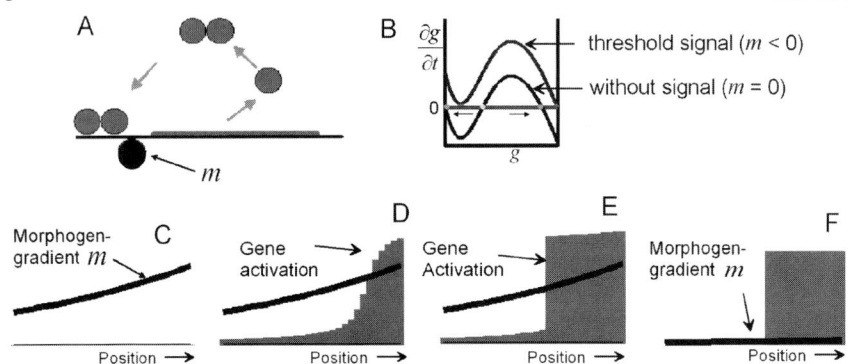

Figure 19 Model for a stable switch-like activation. (A) Assumed is a gene that codes for a gene product g that activates autocatalytically its own gene. Binding of g to the DNA requires a homo-dimerization. A signal m elicits an additional production of g. (B) In the absence the signal two stable steady states ($\partial g/\partial t = 0$; Eq. (6)) exist, one at a low and one at a high g level. With a signal m above the threshold level, cells in the low state switch to a high g-level. (C–E) Stages in gene activation. Cells exposed to a certain signal concentration m (black) make the switch; those exposed to higher m make this switch earlier. The region of high g concentration sharpens in the course of time. (F) After the signal is gone, the cells in which once the gene was activated maintain this activation due to the autoregulatory loop (Meinhardt, 1976).

The activation of this gene, i.e., the production rate of the self-activating gene product, $\partial g/\partial t$, depends on g-dimers (therefore g^2). At low g-levels the chance of finding a partner for building a dimer is low. Therefore, the linear decay term $-rg$ is dominating and the level of the gene product will remain low. The morphogen signal m is assumed to have an additional activating influence on this gene activation. Cells exposed to a certain threshold level of m obtain such a g-level that the autoactivation becomes larger than the decay rate. The gene activation switches from an OFF into an ON-state (Fig. 19). Due to the saturation term κ in the denominator the activation reaches a stable high level.

Meanwhile many genes with positive autoregulation are known. The gene *deformed* in *Drosophila* is an example (Regulski *et al.*, 1991). Due to autoregulation, a short activation of the *deformed* gene under heat shock control is sufficient for a long-lasting activation of this gene (Kuziora and McGinnis, 1988). As predicted, *deformed* acts as a dimer. Other examples for autoregulatory genes are *hunchback* (Simpson-Brose *et al.*, 1994) and *twist* (Leptin, 1991); *deficiens* and *globosa* are examples from plants (Zachgo *et al.*, 1995).

B. Activation of Several Genes by a Morphogen Gradient: A Step-Wise Promotion

According to a classical view, graded 'morphogenetic' signals lead to a space-dependent determination due to a concentration-dependent response of the cells to this 'positional information' (Wolpert, 1969). The question is then, however, how cells measure the local concentration with such precision. Imagine two adjacent cells. While the cell exposed to the lower concentration should activate, e.g., the gene 1, the neighboring cell, exposed to only a slightly higher signal concentration, should activate the gene 2. An analysis of ligation experiments with non-*Drosophila* insects suggested that cells do not measure different levels in a single step but that they compare their achieved state of determination with the strength of the external signal (Meinhardt, 1978). A sequential transition from one gene activation to the next will occur as long as the signal is still high enough. Each of the subsequent activations requires a higher concentration and each step requires a certain time (Fig. 20). According to this model the cells obtain a determination according to the highest concentration they were exposed to in their past. Therefore, a fading signaling strength due to the increasing distance between a cell and the signaling source has no effect.

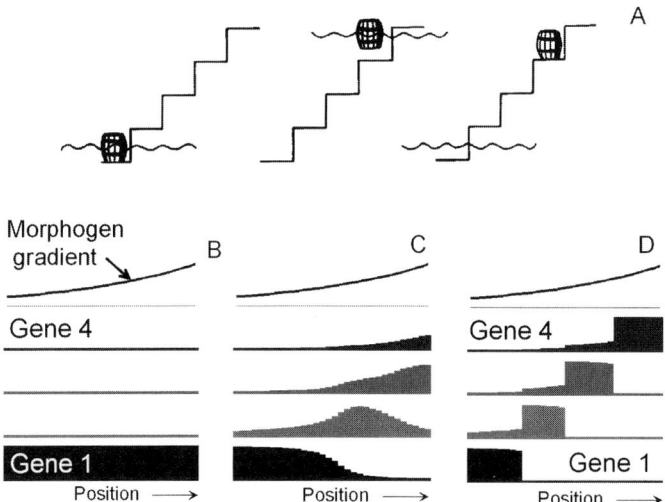

Figure 20 Model for the space-dependent activation of several genes. (A) An analogy: a barrel at the base of a staircase may be lifted up by a flood (morphogen signal). After lowering of the flood, the barrel can only remain at a few discrete levels (activation of particular genes). A later, higher flood can lift the barrel up even further; a second lower flood would have no effect. (B–D) Assumed are genes (1, ..., 4) with their gene products feeding back positively on activation of their own gene. Due to their mutual repression, only one gene of the set can be active within one cell. The morphogen is assumed to accomplish a stepwise and irreversible transition from one gene to the next. Each step requires a higher signal concentration (Meinhardt, 1978, 1982). Note that the gene that becomes activated at the highest morphogen concentration (e.g., gene 4) is the gene that is least sensitive to the signal. Modeling predicted that these less sensitive genes are more effective in autoregulation, otherwise they could not win against the more sensitive feedback loops. The first gene that becomes active in the region of high signaling level is expressed later in a region of low signaling level, in agreement, for example, with the sequence of digit determination in vertebrates.

A characteristic feature of such systems is that determination can be changed only in a unidirectional way ('distal' or 'posterior' transformation). Therefore, cells transplanted from a region of high to a region of low concentration maintain their already achieved determination. In contrast, after a low-to-high transplantation, the cells change their determination according to the new level. Strong evidence for such a unidirectional promotion exists for the hindbrain (Gould *et al.*, 1998; Grapin-Botton *et al.*, 1998) and for the response to activin signaling in the early amphibian gastrula (Gurdon *et al.*, 1995). A stepwise posterior transformation was proposed for the AP-specification in the anterior neural tube (Nieuwkoop, 1952). In this mechanism there is no direct communication between adjacent structures. The correct neighborhood depends solely on the graded signal that evokes these structures. This has the consequence that mismatches caused by transplantation at later stages might be neither detected nor repaired.

Thus, the maintenance of the determined cell state by feedback of a gene on its own activity, combined with a repression of alternative genes, seems to be a widespread mechanism. It is, however, not the only one. Another mechanism is based on changes in the chromatin packaging, e.g., by DNA methylation. This leads to a different accessibility of particular genes in the chromatin—a mechanism that will be not considered there (see Schwartz and Pirrotta, 2007; Ringrose and Paro, 2007).

C. Mutual Induction

The statement 'A induces B' seems to be straightforward. A closer look, however, shows that this is not that simple if A and B are derived from the same tissue. If A is the source of a B-inducing molecule, how it is achieved that A cells themselves do not become converted into B cells, although the A cells are exposed to the highest level of the inducing signal? A possibility is that A cells generate on long range the precondition for B, but a short-range exclusion makes sure that both structures do not merge. The long-ranging activation can be reciprocal in that A not only activates B but that, vice versa, B also activates A. Examples will be provided for both cases.

D. Hydra Tentacles as Example

Hydra tentacles appear in a ring around the primary organizer, the hypostome. This is a frequent occurrence. Leaves appear likewise at a certain distance from the organizer, the shoot apical meristem. The periodic nature of tentacle spacing indicates that their formation is under control of a separate pattern-forming system and does not occur by a simple gradient that is generated at the tip. This can be explained by the assumption that the primary organizer makes the upper part of the body column competent for tentacle activation. Locally, however, tentacle activation is excluded from the hypostome (Meinhardt, 1993).

After head removal, the tentacle signal appears first at the very tip and becomes subsequently displaced to the final subhypostomal position. In contrast, during bud formation, the tentacle signal appears directly in a ring at the correct distance from the tip. This large ring resolves later on into small rings that form the base of the tentacles (Fig. 21A and 21B) (Bode et al., 1988; Smith et al., 2000). The different dynamics in tentacle formation during head regeneration and during budding was predicted by early modeling (Fig. 21C and 21D) (Meinhardt, 1993).

These observations and their modeling allow several conclusions. (i) The primary organizer causes the tissue to become competent for the activation of a secondary organizer. (ii) The tentacle-competent zone is much larger than the zone in which tentacles are actually formed (since the formation of the tentacle signal can precede the formation of the head organizer, Fig. 21A and 21C). The fact that the remaining body column is free of tentacles results from the dominance of those tentacles that are formed in the region of relatively highest competence, not because the competence is restricted to a narrow zone (similar as the primary organizer obtains its 'apical dominance' by the graded competence, see Fig. 10). This model found support from experiments in which β-catenin was stabilized by treating animals with Alsterpaullone, which causes the entire body column to obtain the properties of the tissue that is located close to the head (Broun et al., 2005). In terms of the model, Alsterpaullone causes the whole body column to become competent for tentacle formation. Indeed, after such treatment tentacle formation occurs with narrow spacing almost everywhere (Broun et al., 2005).

E. Sequences of Structures and their Dynamic Control: Planarians as Examples

In planarians, the correct neighborhood of structures is maintained over the complete body axes in a dynamical way (for review see Saló and Baguñà, 2002; Newmark and Alvarado, 2002; Agata et al., 2003; Saló, 2006). Planarians are well-known for an almost unlimited capability for regeneration. Any missing internal part becomes intercalated. In planarians, regeneration and the replacement of old cells occur by a population of stem-cells, the neoblasts. If all neoblasts are killed by irradiation in an anterior portion, the neoblasts that survived in the posterior part can regenerate the anterior part (Saló and Baguñà, 1985). The dynamic maintenance of the correct neighborhood led Chandebois to a comparison with a human community she termed 'cell sociology' (Chandebois, 1976). Although many genes are known to be involved, the molecular basis of this dynamic regulation as a whole is still unclear.

In the model for tentacle formation given above, a mechanism for the maintenance of two structures in the correct neighborhood was discussed. In the following this mechanism will be extended to several coupled pattern-forming reactions. Each system locally excludes the others. On longer range, however, each system activates the loops required to form the adjacent structures (Meinhardt and Gierer,

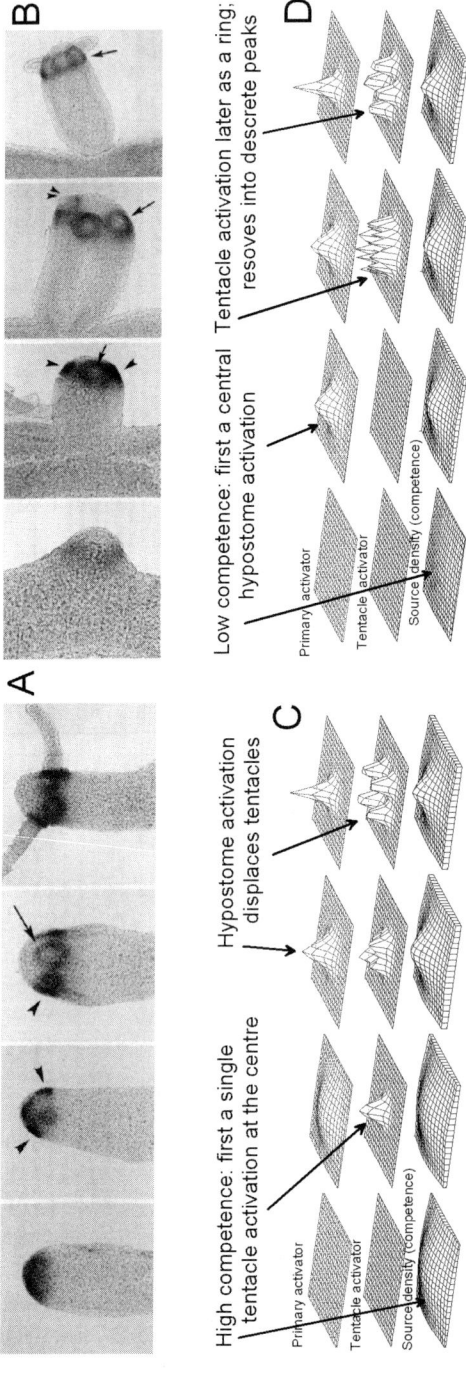

Figure 21 Dynamic regulation of a secondary organizing region: hydra tentacles as an example. The *hy-Alx* gene is a marker for tentacles (Smith *et al.*, 2000). In vertebrates, *Alx* is involved in brain patterning (for instance, Miura *et al.*, 1997). (A) After head removal, *Hy-Alx* first appears as a ring directly at the tip, to be later displaced to the final position where it disintegrates into separate signals. (B) In contrast, during bud formation *Hy-Alx* appears as a ring directly at the correct position that splits subsequently into individual rings that marks the base of tentacles. (C) Model for tentacle formation during regeneration: in a region of high source density (high competence for hypostome and tentacle activation, see Fig. 10), a single tentacle activation appears first at the position of highest source density. Only subsequently does the peak of the primary head activator form, displacing the tentacle activation. The tentacle signal widens to a ring that decays into individual spots. (D) Model for tentacle formation during bud formation: If the source density is low (as during bud formation), tentacles can be formed only after a sufficient increase of the source density under the influence of the primary head activation (Fig. 10). Since primary head (hypostome) and tentacle activator exclude each other, tentacle activation appears directly in a ring that surrounds head activation. Later this ring also resolves into isolated maxima (Meinhardt, 1993).

1980). Such a network is able to generate a dynamically stable sequence of structures in which the correct neighborhood is regulated. Mutual activation can also result, indirectly, from a self-inhibition since in competing systems the self-inhibition of one component is equivalent to promotion of the neighbors. While in the direct activator–inhibitor mechanisms discussed above there are few winners and many losers, in the mutual activation scheme in each cell one of the several alternative loops wins.

In the simulation in Fig. 22 it is assumed that the activators cross-react in such a way that each inhibits the self-enhancement of the other activators by competitive inhibition. The inhibitors also show competitive inhibition with the inhibitors that are responsible for the adjacent structures. Since an inhibition of an inhibition is in fact an activation, this leads to a long-range activation of the adjacent structures. To give an example, the type-2-inhibitor restricts the extension of the activity in which the system 2 is active, but simultaneously it has an activating influence on the structures 1 and 3 by undermining the action of the genuine 1 and 3 inhibitors. Since the inhibitors have a long range, the cross-activation is of long range too. The emerging neighborhoods are determined by these cross-reactions. Each of the feedback loops saturates at high levels. This leads to a plateau-like activation and enables size regulation such that each structure occupies an appropriate share. As shown in Fig. 22, the complete sequence is generated from any partial sequence, the polarity being maintained. If individual cells migrate into a zone in which a different subsystem is active, they adapt to their new surrounding. This respecification can go in both anterior and posterior directions; there is nothing like a unidirectional 'distal transformation.' This type of regulation also occurs in growing and budding hydra. Cells born in the middle of the animal eventually move either towards the head or to the foot, where they function according to their actual position.

Information of the underlying molecular network is sparse. Direct evidence for a reciprocal induction has been observed in the planarian head region (Bogdanova *et al.*, 1998). A hint could be the sequence of nested but overlapping expression domains of different *Wnt* genes in the sea anemone *Nematostella*, suggesting a *Wnt*-code as a possible candidate for such a system of linked pattern forming systems (Kusserow *et al.*, 2005).

F. Segmentation: A Superposition of a Periodic and a Sequential Pattern

Segmentation is a common theme in higher organisms with the trunk consisting of repetitive structures such as the alternation of anterior and posterior compartments in *Drosophila* or somites in vertebrates. Superimposed onto basically repetitive structures is a second pattern of Hox gene activation that causes the individual segments to undergo specific developments. Both patterns are precisely in register. Usually the border of Hox gene expression coincides with parasegmental organization (Zhang *et al.*, 2005; Damen, 2002) (a parasegment being defined as the overlapping unit ranging

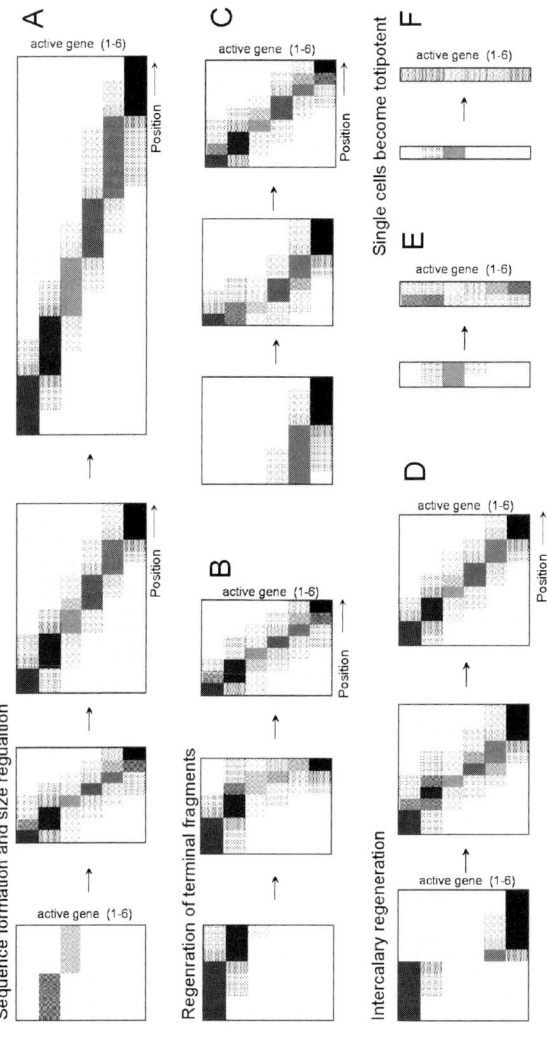

Figure 22 Generation of a dynamically stable sequence of structures by mutual activation and local exclusion of several pattern-forming systems. Arbitrarily six activator–inhibitor systems are assumed. Activity is mutually exclusive. Systems that are required for adjacent structures activate each other on a long range by a competitive interference of the inhibitors (only the activators are shown). (A) The complete sequence is generated from any partial sequence. Each region obtains a proportional share, maintained during further growth. Due to some activator diffusion, the regions have a substantial overlap. This overlap facilitates the transitions from one activation to the next and is required that each activation maintains its correct extension during growth or regeneration. (B, C) The sequence is restored after removal of terminal structures. Note that first the missing structures are formed in a compressed form, in agreement with observations (Saló, 2006). The regulation by which each part obtains its appropriate share requires a much longer period. (D) Replacement of internal structures. (E) In very small fragments positional specificity is lost, but a residual polarity is maintained (in starving planarians, the body size can shrink to a minute fraction of the normal body size, but after feeding the body pattern is reestablished with the original polarity). (F) An isolated cell activates all feedback loops, although at a low level. This is independent of the original activity. In other words, the cells are multipotent, although they obtain a position-specific activation during self-organizing patterning. In these simulations, all feedback loops have identical properties. The pattern-forming systems that generate the terminal structures may play in reality a more dominant role.

from the *P* compartment of one segment to the subsequent posterior *A* compartment, see Martinez Arias and Lawrence, 1985). In *Drosophila*, segmentation proceeds normally even in the absence of Hox genes (Lewis, 1978). This suggests that periodic patterning is the primary event and that the sequential activation of Hox genes is coupled to this.

As revealed by classical transplantation experiments, the neighborhood of structures *within* insect segments is strictly regulated (Locke, 1959; Lawrence, 1970; Bohn, 1970). Our model of mutual induction of locally exclusive cell states was in fact developed to describe this controlled neighborhood, including situations in which intercalary regeneration occurs with a reversed polarity (Meinhardt and Gierer, 1980). Molecular details that lead to the fine structure within a segment are still not completely understood. It is unclear, for instance, whether this pattern is organized by one graded quantity or by a fine-grained activation of different genes, i.e., different qualities. However, experimental evidence shows that the basic machinery causing subdivision into compartments works by mutual long-range activation and local exclusion: The gene *engrailed (en)*, the key gene for posterior compartmental specification, is autocatalytically activated. Via the diffusible molecule *hedgehog (hh)*, *en* activates in addition the gene *wingless (wg)* that is crucial for the anterior compartment (Ingham and Hidalgo, 1993). The gene *sloppy paired* is involved in the *wg*-autoregulation. The *wg* protein can reach adjacent cells via vesicle transport and is required there to stabilize *en* (Baker, 1987; van den Heuvel *et al.*, 1989). As expected from the theory, the *en* gene activity requires a functional *wg* gene in its neighborhood and *vice versa*, although both genes are transcribed in non-overlapping regions. The gene *cuD* (Eaton and Kornberg, 1990), on which *wingless*-expression depends, is involved in the local exclusion of *en* and *wg* expression. The prediction of such a complex molecular interaction by a theory could hardly be more precise (Fig. 23A).

The mechanism of long-ranging activation and short-ranging exclusion is appropriate to generate stripe-like distributions. A long common border between narrow bands of different cell states allows a most effective mutual support. As mentioned, an exchange of the activator between cells facilitates stripe formation by counteracting decay into isolated patches. In this respect it is interesting that *engrailed* and some other homeobox genes can be actively exchanged between cells (Maizel *et al.*, 2002).

Segments have an internal polarity, in contrast to a periodic alternation of two compartments . . . *APAPAP* . . . . For this reason I have proposed that in addition to the anterior and posterior compartment at least one additional element termed *S* must be present, such that a sequence . . . *SAP/SAP/SAP* . . . results. The region *S* separates one *A–P* pair from the next (Meinhardt, 1982). Now it is generally assumed that the AP pattern of each parasegment is founded by four cells (see Ingham, 1991). Four cells also establish the parasegments in crustaceans (Scholtz *et al.*, 1994). Thus, there is only one *A/P* border per segment. This is important since the *A/P* border also generates the preconditions to form legs or wings (see Figs. 26E and 26F).

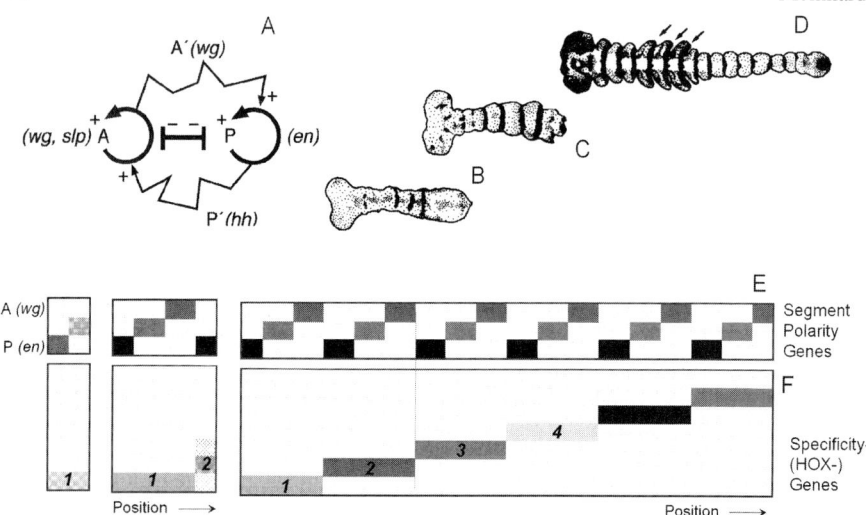

Figure 23 Segmentation—the formation of a periodic and a sequential pattern in register. (A) The basic network enabling segmentation. Proposed is that (at least two but presumably more) feedback loops activate each other on long range and exclude each other locally. This enforces a controlled neighborhood of structures (Meinhardt and Gierer, 1980; Meinhardt, 1982). (B–D) Stages in the development of a grasshopper embryo. In the course of development, more and more segments are added at the posterior pole. In the posterior part of each segment the gene *engrailed* (black) is transcribed Patel *et al.*, 1989. The segments are different from each other. Legs (arrows) are formed only in the three thoracic segments and are restricted to the *A/P* border (see Fig. 26). (E, F) Model: By proliferation of cells at the posterior pole, the respective terminal regions (*S*, *A*, or *P*) become enlarged until the long-ranging stabilization becomes insufficient and a transition into the subsequent cell state occurs. A periodic pattern with an intrinsic polarity results. Further, cells are in the *A*-stage produce a component that is required to activate the subsequent gene of the sequential pattern (1, 2, . . .). The actual transition, however, is blocked and can only occur after switch from the *A* to the *P* state. The resulting sequential (Hox-) gene activation 1, 2, 3, . . . is precisely in register with the parasegmental . . . *A/PSA/P* . . . pattern.

In short-germ insects, in contrast to *Drosophila*, new segments are added at the posterior pole. In the mutual activation scheme, the activation can be cyclic. For simplicity let us regard only two alternating components and an initial *AP* pattern. Whenever, due to growth, the *P*-region becomes too large, the support of the preceding *A*-cells will be too low (or the self-inhibition of the *P*-cells too high) and *P*-cells distant from the *AP* border will switch from *P* to *A*, elongating the *AP*- to an *APA*-pattern. A regular periodic pattern results (Fig. 23). This model predicted that the most posterior cells alternate between (at least) two states. This has been clearly demonstrated, e.g., by an ON/OFF alternation of *hairy* activation at the posterior pole of the spider embryo (Damen *et al.*, 2000). An oscillation of this type is crucial for somite formation discussed further below. The periodic pattern generated during outgrowth can be, but need not to be, accomplished on the level of the pair rule genes (Sommer and Tautz, 1993; Patel *et al.*, 1994).

G. Formation of a Precise Number of Different Segments during Terminal Outgrowth

In many cases the number of segments is precisely controlled. The polychaet *Clymenelly torquata*, for instance, has 22 segments. It regenerates removed segments such that the number of 22 segments will be restored independent of the number of segments removed (Moment, 1951). In the leech, more than the final 32 segments are initially formed. The surplus segments are later removed by programmed cell death (Fernandez and Stent, 1982; Shankland, 1991a, 1991b). These observations indicate that some sort of counting mechanism exists. As mentioned above, trunk formation requires a sequential step-wise posteriorization (Figs. 12D–12F). I have shown that the periodic alternation of cell states at the outgrowing posterior pole can be used as a counting mechanism on the gene level (Meinhardt, 1982). The mechanism is easily explained by an analogy. Imagine a grandfather's clock. The periodic movement of the pendulum drives the mechanism that allows a sequential advancement of the pointer with each full cycle by one and only one unit. In terms of gene switching, the transition from one *AP*-specifying gene to the next is prepared in one state (*A*). However, the actual transition is blocked. After switch to the other state (*P*), the prepared transition takes place but the preparation of an even further transition has to wait for a switch back to *A*. The simulation in Fig. 23 demonstrates the generation of a sequential pattern of (Hox-) gene activation (1, 2, 3, ...) that is precisely in register with the periodic reiteration of cell states, the parasegmental ... *A/P/S/A/S* ... pattern. According to present knowledge, this early model has to be modified and extended because particular Hox genes are known to be active in several segments. In *Drosophila*, for instance, *Abdominal-B* is expressed in the abdominal segments 5–9. However, there are segment-specific *cis*-regulatory units on the chromosome, which are separated by chromosomal insulators (Drewell *et al.*, 2002; Estrada *et al.*, 2002). In terms of the model, each oscillatory cycle may cause the advancement from one such insulator to the next. Thus, a certain number of insulators have to be passed until the activation of the subsequent Hox gene takes place.

H. Somite Formation: The Conversion of a Periodic Pattern in Time into a Periodic Pattern in Space

Somites are the most conspicuous segmented structures in vertebrates. In *Amphioxus* (hemichordate) somites form in a similar way to the segments in short germ insects, in a growth zone at the posterior pole (Schubert et al., 2001a, 2001b). In contrast, in higher vertebrates, somite formation occurs at a considerable distance from the posterior pole by a sequential separation from a nonsegmented presomitic mesoderm in an anterior-to-posterior sequence. Fig. 24 shows chicken embryos at

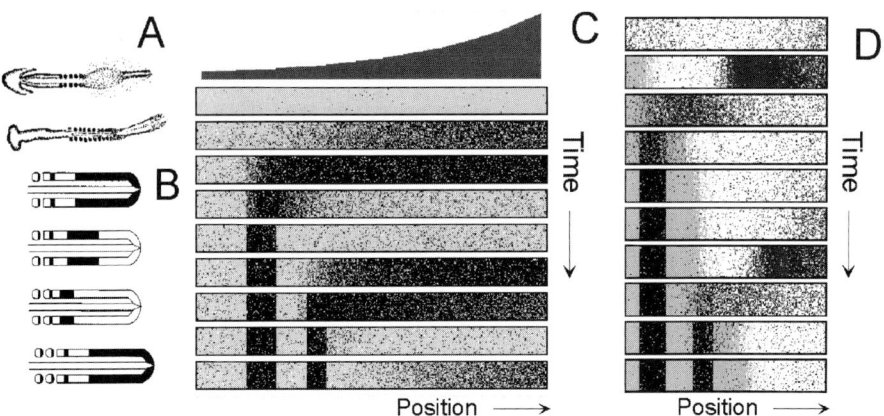

Figure 24 Somite formation. (A) Chicken embryos at 25 and 37 hour with 5 and 12 pairs of somites respectively. (B) Observed *c-hairy* pattern in the chick (Palmeirim *et al.*, 1997; Pourquie, 2004). (C) An early model (Meinhardt, 1982, 1986a). Predicted was that an oscillation between two cell-states takes place (gray and black) in the posterior presomitic part. Activity waves move towards the anterior. Each full cycle adds one pair of half-somites. Only cells that are exposed to a minimum level of a gradient (identified now as *FGF*, top) can participate in this oscillation. Cells below this level form a stable AP pattern in which *A* and *P* cells stabilize each other mutually. (D) A modified version based on *mesp2* expression (Takahashi *et al.*, 2000; Morimoto *et al.*, 2006). In each cycle, a somite-wide region of *mesp2* expression (light gray) is hit by a wave (dark gray), causing a suppression of *mesp2* in the posterior half of this region and its determination to a posterior half-somite (black). The remaining part forms the anterior half-somite. Subsequently *mesp2* is activated in a new somite-wide region. For this model, elementary networks as described in the text are employed involving gene activation with mutual exclusion (Fig. 20), and traveling waves (Fig. 9).

different stages. Somite formation has been reviewed extensively (Pourquie, 2004; Dubrulle and Pourquie, 2004a; Aulehla and Herrmann, 2004; Holley and Takeda, 2002; Takahashi *et al.*, 2000).

To be compatible with classical observations, the counter-intuitive prediction was made that segment formation occurs by an oscillation between two cell states at the posterior pole. The main reason to assume this oscillation was that, as in segmentation, the somites obtain with high precision different identities along the AP axis. As mentioned, an oscillation allows for the generation of periodic structures that are nevertheless different from each other by a counting mechanism on the gene level. According to the model for somite formation, these oscillating activities spread in a wave-like manner towards the anterior, coming to rest there and forming pairs of anterior and posterior half-somites (Fig. 24C). Each full cycle was predicted to add one pair of anterior and posterior half-somites: '*This model would obtain strong support if the postulated oscillations in the mesoderm before somite formation could be detected. One full cycle of this oscillation should take precisely the same time as that required for the formation of one somite*' (Meinhardt, 1982, 1986a). The prediction was confirmed fifteen years later by the observation of the oscillation and wave-like spread of the *hairy1* gene in the chick (Palmeirim *et al.*, 1997) (Fig. 24B). At the positions where *c-hairy1* waves come to rest, new posterior half-somites are formed. At the time the model was proposed, the subdivision of somites into anterior and posterior half somites was still a prediction that has been verified only two years later (Keynes and Stern, 1984). The predicted coupling of the (Hox-) gene activation and oscillation has been, in the meantime, also established (Zákány *et al.*, 2001; Dubrulle *et al.*, 2001; Dubrulle and Pourquie, 2002).

It was assumed that the very same molecular interaction that leads to the periodic patterns in time is also responsible for the periodic pattern in space (see also Fig. 9). While, as explained above, A- and P-cells stabilize each other in the region of a common boundary, it is a property of such an interaction that groups of cells consisting of one type only (A or P) can oscillate back and forth between the two states. For instance, if all cells are in A-state, the P-state obtains substantial help from the activated A state while the A-state itself is not supported, much as the *engrailed* activation would not be maintained without *wg*-expressing cells in the neighborhood. Therefore, after a certain time, the cells switch from A to P. Later, for the same reason, the cells switch back to A, and so on. Such a spatially homogeneous oscillating system can be converted into a periodic pattern that is stable in time if an A–P border has been formed. For a reliable separation of the region in which the oscillation can take place from the region in which stable patterning can occur, a controlling gradient was assumed. The waves were assumed to come to rest whenever the gradient level became too low. The gradient is now identified as *FGF* (Dubrulle *et al.*, 2001; Dubrulle and Pourquie, 2004b).

Meanwhile it has turned out that the *Notch* pathway is crucial for the oscillation. *Notch* was known before only as a component involved in the formation of spatial patterns. If the very same reaction is assumed to generate the pattern in space *and* in time, the model works only in a narrow parameter range. This is presumably the reason why different components of the *Notch* pathway are involved in oscillation and in the stable patterning respectively. Further it was assumed that the oscillation is not only of an ON–OFF type but that an oscillation occurs between the activation of two different cell states. In addition to the oscillating *Notch* activation, an out-of-phase oscillation of the *Wnt*-pathway has been found (Aulehla *et al.*, 2003).

More recent observations revealed that each new somite is initiated by a somite-wide activation of a gene (*Mesp2*) (Takahashi *et al.*, 2000). The *Notch*-driven wave hits the posterior part of this region, causing its transformation into a posterior half-somite and a restriction of the *Mesp2* expression to the remaining anterior half. A simulation is given in Fig. 24D. There are many other features that have to be integrated into forthcoming models, e.g., the formation of the somitic clefts and the dramatic slowing down and sharpening of the waves before they come to rest (see also the contribution of Baker *et al.* in this volume).

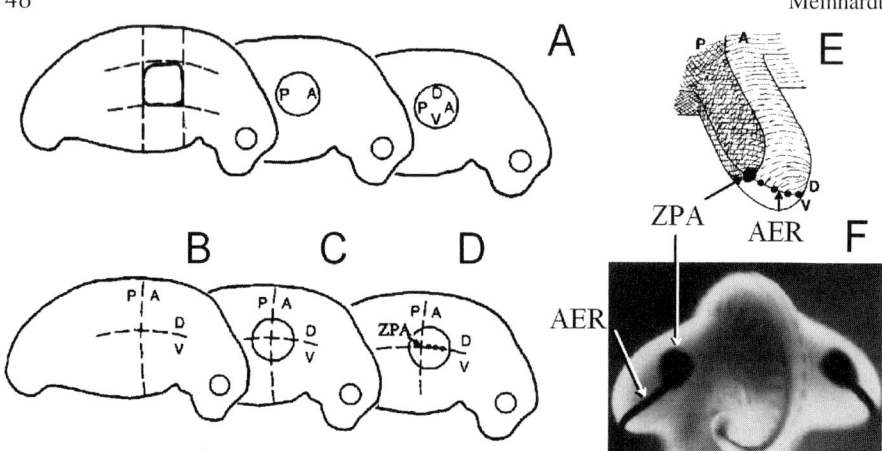

Figure 25 Models for vertebrate limb patterning. (A) The classical limb field model. First a nest of cells is determined for limb development that becomes subsequently patterned along the AP, and later along the DV axes (Harrison, 1918). (B, C) Boundary model (Meinhardt, 1982, 1983b): proposed was that the primary event is a subdivision along the axes. By cooperation of two pairs of differently determined cells (A/P and D/V) in the generation of new signals, two orthogonal borders become the organizing regions to set up the coordinate system for the substructure. Limbs are formed around previously established borders. (D, E) For vertebrate limbs only the anterior part is used. The intersection of both borders forms the organizing 'Zone of Polarizing Activity' (ZPA). The DV border forms the apical ectodermal ridge (AER), defining the plane along which the digits are formed. The outer and the inner face of the hand are different from each other since the former is formed from the dorsal and the latter from the ventral region next to the border (see Martin, 1995; Tanaka *et al.*, 1997; Martin, 2001; Niswander, 2003, for detailed experimental evidence and review). (F) View of a chick embryo from the tail onto the two wing buds (Grieshammer *et al.*, 1996). The thick black lines result from a staining of *FGF8* that is produced at the DV border (a *Wnt7-a/En-1* border) that forms the AER. The dark round spots mark high concentrations of *Sonic hedgehog* that is, according to the model, produced at an intersection of an A/P- and a D/V border. (F kindly supplied by Dr. Uta Grieshammer.)

I. Borders and Intersections of Borders Become the New Organizing Region for Secondary Structures such as Legs and Wings

Substructures such as legs or wings have their own coordinate system, with an AP, a DV and a proximodistal axis. Based on experiments with limb initiation in axolotl, it was proposed that first a limb field forms (Harrison, 1918): Cells are set aside that eventually form the future limb. At later stages an anteroposterior and subsequently a dorsoventral axis becomes determined within this initially more or less uniform cell population (Fig. 25A). It was further assumed that cells have an intrinsic polarity that is aligned with the overall body axes of the embryo (like any fragment of a magnet retains the original polarity), and that these cell polarities are used to organize the final polarity in the limb field. This model was very influential over many decades. However, Harrison himself realized problems in his views. Transplantation of a limb field to a more posterior position without any rotation led to symmetrical limbs or to limbs with AP polarity reversals, i.e., to results that were incompatible with his model (Harrison, 1921).

The observation that an organizing region is located at the posterior end of the limb field (Gasseling and Saunders, 1964) was a big step forwards. Many transplantation experiments were explicable by the assumption that this 'Zone of Polarizing Activity' (ZPA) generates a morphogen gradient that leads to the sequence of digits with the little finger close to the source (Tickle *et al.*, 1975). Although corresponding drawings were very convincing, important questions remained unanswered. How is the ZPA generated in the first place? Why are the digits formed along a line and not as a series of concentric rings, as expected if a local source generates a cone-shaped morphogen distribution? An alternative model was proposed shortly later, the 'Polar Coordinate' model (French *et al.*, 1976), postulating that positional identities are arranged in a circular fashion and that distal transformation occurs if this ring is complete. Though the model was completely formal and no explanation was given how such a ring was generated during early development, the model was surprisingly successful in predicting in detail the geometry and handedness of supernumerary limbs as they occur, e.g., after reimplantation of limbs with an angular mismatch. It was clear that any molecularly feasible model has to solve the problems inherent in the ZPA model and must be compatible with the handedness predicted by the Polar Coordinate model.

In 1980 I proposed that borders between differently determined cells can obtain organizing functions. (Meinhardt, 1980, 1983a, 1983b). Imagine that a primary pattern-forming process leads to a subdivision into several discrete regions by region-specific gene activation (see Figs. 19, 20, and 23). Among them are the adjacent regions A and P. If, for instance, in the P region a co-factor is produced that is required in the A cells to produce a new morphogen, its synthesis is restricted to a position close to the A/P border. The concentration of this morphogen provides a measure for the distance of a cell from the border and is therefore suitable for the internal organization of the A and the P region. Although the positional information is symmetric about both sides of the border, the resulting pattern can be asymmetric since A and P cells can respond differently. In vertebrate limb formation, for instance, only the A-cells are able to respond, which leads to the polar arrangement of the digits (Figs. 25D and 25E).

A border that separates two cell types along the anteroposterior axis surrounds an embryo in a belt-like fashion. To determine the DV-position of a limb, an intersection with a second border is required. Limbs are assumed to be initiated around the intersection of two borders (Fig. 25B and 25C). This explains the handedness. Since the embryo has essentially a cylindrical geometry, any subdivision of the embryo along the anteroposterior and the dorsoventral (or better mediolateral) axis leads to paired intersections, one at the right and one at the left side of the embryo. They have opposite handedness, a feature most characteristic for the formation of legs, wings, eyes etc. The four quadrants or three sectors placed around the intersection of two borders provide coarse information about the angular position. The proximodistal organization results from a cooperation of the two signals generated at each border. Therefore, the 'Complete Circle' rule of the Polar Coordinate model is to be substituted by a 'Complete set of Compartment' rule, with at least three members of the set. The model was derived from classical observations with insect and vertebrate appendages (Meinhardt, 1983a, 1983b). It took about twelve years for a direct demonstration on the molecular-genetic level (reviewed in Vincent and Lawrence, 1994; Martin, 1995; Irvine and Rauskolb, 2000). Examples for vertebrate and insect appendages are given in Figs. 25 and 26.

According to this model a homogeneous limb field never exists since the preceding subdivision is the prerequisite of limb formation, much in contrast to Harrison's view. In view of the overwhelming evidence that now exists for this mechanism, this model seems at present to be straightforward if not trivial. In those times, however, it was very difficult to publish this idea. The paper was accepted only in the fourth journal tried (Meinhardt, 1980). In retrospect it seems difficult to understand the resistance against this model since it provides a clue to why development is so reproducible: the interpretation of a first positional information leads to borders, which, in turn give rise to new positional information that leads to a finer subdivision of the new parts. Such sequential formation of organizing borders is crucial, for instance, to regionalize the brain (reviewed in Joyner, 1996; Puelles, 2001; Prakash and Wurst, 2004; Rhinn and Brand, 2001).

J. Filamentous Branching Structures: Traces behind Moving Signals

Net-like structures are a common pattern element in all higher organisms. Blood vessels, lungs, tracheae of insects, axons or veins of leaves are examples. They are used to supply the tissue with nutrition, water, oxygen, information and drain the tissue. How can such complex patterns emerge? Local signals, generated e.g., by activator–inhibitor systems, may direct the elongation of the filament. In turn, the extending filaments cause a displacement of the signals and thus further elongation (Meinhardt, 1976, 1982). The orientation of the elongating filament is controlled by a tropic factor that is removed by the filaments, and which is required for signal generation. Its concentration is a measure of how urgently the cells need the approach of new filaments, for instance, to remove an oxygen deficiency. In the simulation Fig. 27, it is assumed that local signals are generated by an activator (gray squares)–inhibitor system. Cells respond by differentiating into members of the filament system. Differentiated cells are assumed to remove the tropic substance (hatched in Fig. 27), which is assumed to be a cofactor in the generation of the elongating signal. The resulting valleys of the factor around the filaments orients further extension, normally away from the existing filament, i.e., in the forward direction. Due to the lateral inhibition in the signal generation, the elongation is sensitive to minute concentration differences. Branch formation occurs in two ways. Either new elongating signals are triggered along the filaments behind the tip, causing lateral branching. Alternatively, the signal at the tip can split, causing a dichotomous branching.

1. Models for the Organization of the Embryonic Body Axes

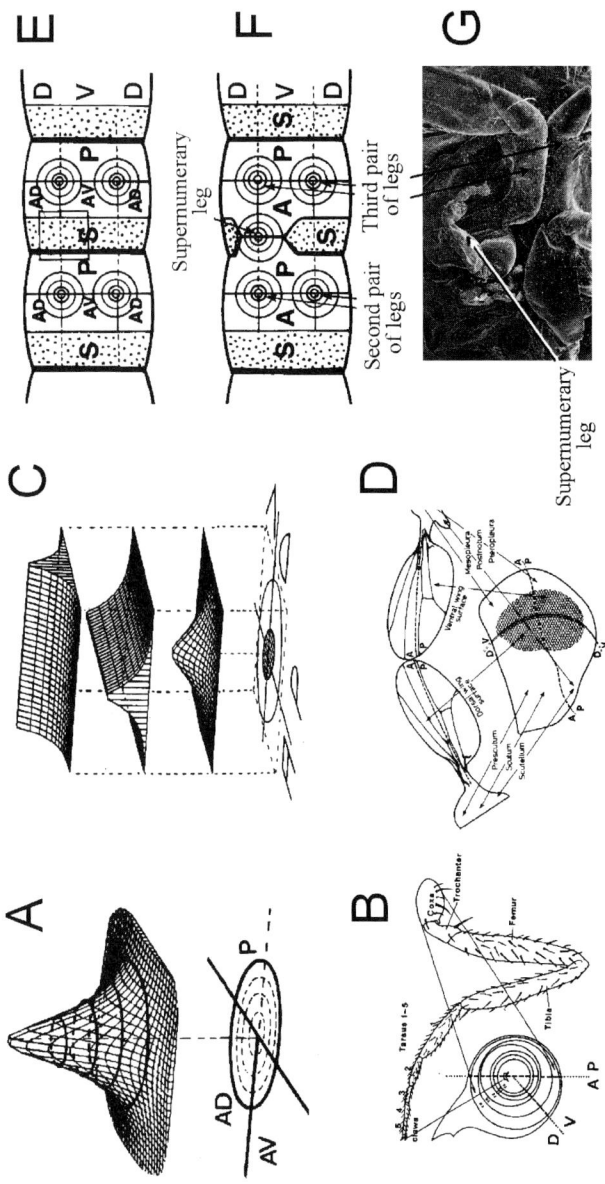

Figure 26 A boundary model for insect appendages (Meinhardt, 1982, 1983a, 1986b). The AP subdivision that occurs during primary segment formation (Fig. 23) is assumed to be the precondition for the formation of imaginal disks. (A, B) For the leg disk, a ventral compartment within the anterior compartment leads to a subdivision into three sectors. The morphogen for distal outgrowth is produced only at the region where the two borders intersect, i.e., where the three sectors are close to each other. The three sectors are characterized by *wingless* (AD), *dpp* (AV) and *en* (P). The cone-shaped distribution is appropriate for the circular fate map of the leg disk shown in B (for the molecular details see Diaz-Benjumena et al., 1994; Irvine and Rauskolb, 2000). (C–D) For the wing disk, the posterior compartment also has a DV subdivision. The DV border gives rise to the wing margin. The cone-shaped proximodistal signal leads to the separation of the wing blade. (E–F) Induction of a seventh leg by removal of nonleg-forming epidermis (Bohn, 1974). In terms of the model, after removal of tissue (square in E) that separates one AP-boundary region from the next, a new AP confrontation is generated (F) that has the opposite polarity (PA). If the operation is done on the right side, a left-handed supernumerary limb emerges since the DV polarity remains unchanged. (G) A supernumerary limb generated in this way (specimen kindly provided by Horst Bohn) (Figures partially from Meinhardt, 1982, 1986b.)

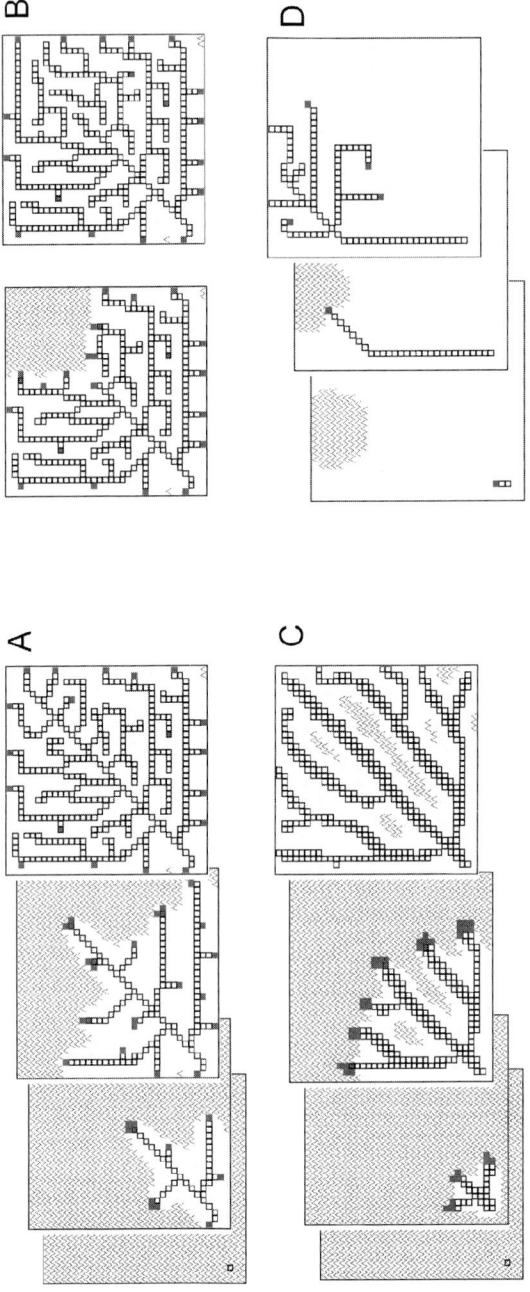

Figure 27 Formation of filamentous branched structures as a trace behind moving signals. Assumed is that local signals (filled gray squares) are generated by an activator–inhibitor mechanism. The local signal leads to cell differentiation (open squares; see Fig. 19). The differentiated cells remove a trophic substrate (hatched) that is required for signal generation. Thus, lowering of the substrate concentration by a newly differentiated cell leads to a destabilization of the signal and to its shift to that adjacent region that has the highest concentration of the trophic substrate. (A) Starting with a single cell, the filament becomes elongated towards a region of high substrate concentration, i.e., towards regions that are not sufficiently supplied by the veins. Branches can be formed by triggering new elongation signals along existing filaments. This occurs if the inhibition from active tips is low enough and the concentration of the tropic factor is sufficiently high. Branching occurs preferentially with a 90° angle. (B) Regeneration: After partial removal of a branched network, substrate accumulates again, causing new filaments to extend into the filament-deprived region. The regenerated pattern is similar but not identical to the original one. (C) If the activator autocatalysis saturates, branch formation can occur by peak splitting at the growing tip (see Fig. 8). (D) If the tropic substrate is only locally produced, filaments extend towards this region. Branching may be restricted to the region of high substrate production (Meinhardt, 1976, 1982). Addition of new cells is only one of the possible modes. Local signals also can elicit cell protrusions that lead to long filaments that consists of single cells (Fig. 11F).

The recruitment of new cells is one of the mechanisms for the elongation of filaments. Alternatively, single cells can form filaments by local extensions. A mechanism for the localization of signals on single cells under the influence of a trophic factor was given above (Fig. 11F–11J).

For the trachea of insects, most of the expected components now have been found. The following list compares the expected components and the corresponding genes so far isolated. An autoregulatory gene is expected, which allows cells to remember that they belong to the tracheal network. The corresponding gene is *trachealess* (Wilk *et al.*, 1996). A second autocatalytic loop is expected that is responsible for the generation of the local signal for filament elongation. A candidate is the transcription factor *drifter* (Anderson *et al.*, 1996). *branchless*, an *FGF*, is involved in the signal generation in front of the tracheae (Sutherland *et al.*, 1996). The corresponding receptor is *breathless* (Glazer and Shilo, 1991). In the generation of tracheae (Ghabrial and Krasnow, 2006) and of the vascular system (Hellström *et al.*, 2007; Siekmann and Lawson, 2007a, 2007b; Leslie *et al.*, 2007), the tip cells are selected by the Delta–Notch system, i.e., by a lateral inhibition mechanism as predicted. Only the selected cells send out protrusions for elongation. The trophic molecule in the formation of blood vessels is the vascular endothelial growth factor, VEGF (reviewed in Risau and Flamme, 1995; Coultas *et al.*, 2005). The secreted form of VEGF is synthesized by organs that attract blood vessels such as the brain or the kidney. Local overexpression causes hypervascularization. High amounts of this factor have been detected during embryonic and tumor angiogenesis. It orients chemotactically the elongation of the vessels.

Lung branching occurs at an epithelial–mesenchymal interface (for review see Cardoso and Lü, 2006; Chuang and McMahon, 2003). Local *FGF* maxima next to the tips attract the tips, which, in turn, cause a displacement of the *FGF* maxima. The tips are characterized by a high *Shh* level. Dichotomous and lateral branching does occur. *BMP4* seems to be one of the inhibitory components.

In leaf venation, the trophic factor seems to be auxin which is actively transported towards the roots. Veins extend towards higher auxin concentrations and local applications of auxin can attract new veins (Sachs, 1981), as shown in the simulation in Fig. 27D. The mutual competition of the cells for vein elongation seems to be based on a feedback loop that involves polarization of Auxin transport, not by a long-ranging inhibition as expected in the original model. This allows the formation of closed loops (Scarpella *et al.*, 2006). However, many details that would be crucial for more realistic modeling are still unknown, for instance, the auxin concentration within the veins. Thus, a satisfying model of the complex leaf venation pattern based on auxin transport is not yet available.

V. Conclusion

Modeling reveals the minimum requirements for essential steps in development. Self-enhancement and long-range inhibition were proposed to be the driving forces in pattern formation and in the generation of organizing regions. Self-enhancement and competition were found also to be decisive for space-dependent gene activation. Boundaries generated on this basis organize emergent substructures. Hydra was originally chosen as a model system to get insights into basic mechanisms underlying biological pattern formation. More recent modeling suggests that this radially-symmetric animal can also provide key insights into the establishment of a secondary axis and the evolution of the bilateral body plan of higher animals. There are still many important aspects that wait for future modeling such as growth control, proportioning of the body parts, pattern formation within cells and organization of cell migration in the course of development. Mathematically based modeling combined with computer simulations is expected to contribute to uncovering more of the basic principles on which development is based and to keep the complex web of interactions comprehensible.

Acknowledgments

I wish to express my sincere thanks to Prof. Alfred Gierer. Much of the basic work described in this paper emerged from a fruitful collaboration over many years. He also provided helpful comments on earlier versions of this paper.

References

Agata, K., Tanaka, T., Kobayashi, C., Kato, K., and Saitoh, Y. (2003). Intercalary regeneration in planarians. *Dev. Dyn.* **226**, 308–316.

Akiyama-Oda, Y., and Oda, H. (2006). Axis specification in the spider embryo: dpp is required for radial-to-axial symmetry transformation and sog for ventral patterning. *Development* **133**, 2347–2357.

Anderson, M. G., Certel, S. J., Certel, K., Lee, T., Montell, D. J., and Johnson, W. A. (1996). Function of the Drosophila POU domain transcription factor Drifter as an upstream regulator of Breathless receptor kinase expression in developing trachea. *Development* **122**, 4169–4178.

Aulehla, A., and Herrmann, B. G. (2004). Segmentation in vertebrates: Clock and gradient finally joined. *Genes Dev.* **18**, 2060–2067.

Aulehla, A., Wehrle, C., Brand-Saberi, B., Kemler, R., Gossler, A., Kanzler, B., and Herrmann, B. G. (2003). Wnt3a plays a major role in the segmentation clock controlling somitogenesis. *Dev. Cell* **4**, 395–406.

Baker, N. E. (1987). Molecular cloning of sequences from wingless, a segment polarity gene in Drosophila the spatial distribution of a transcript in embryos. *EMBO J.* **6**, 1765–1774.

Bisgrove, B. W., Essner, J. J., and Yost, H. J. (1999). Regulation of midline development by antagonism of lefty and nodal. *Development* **126**, 3253–3262.

Bode, H. R. (2003). Head regeneration in hydra. *Dev. Dyn.* **226**, 225–236.

Bode, P. M., Awad, T. A., Koizumi, O., Nakashima, Y., Grimmelikhuijzen, C. J. P., and Bode, H. R. (1988). Development of the two-part pattern during regeneration of the head in hydra. *Development* **102**, 223–235.

Boettger, T., Knoetgen, H., Wittler, L., and Kessel, M. (2001). The avian organizer. *Int. J. Dev. Biol.* **45**, 281–287.

Bogdanova, E., Matz, M., Tarabykin, V., Usman, N., Shagin, D., Zaraisky, A., and Lukyanov, S. (1998). Inductive interactions regulating body patterning in planarian, revealed by analysis of expression of novel gene scarf. *Dev. Biol.* **194**, 172–181.

Bohn, H. (1970). Interkalare Regeneration und segmentale Gradienten bei den Extremitäten von Leucophaea-Larven (Blattaria). I. Femur und Tibia. *Wilhelm Roux' Arch.* **165**, 303–341.

Bohn, H. (1974). Extent and properties of the regeneration field in the larval legs of cockroaches (Leucophaea maderae). I. Extirpation experiments. *J. Embryol. Exp. Morphol.* **31**, 557–572.

Broun, M., Gee, L., Reinhardt, B., and Bode, H. R. (2005). Formation of the head organizer in hydra involves the canonical wnt pathway. *Development* **132**, 2907–2916.

Broun, M., Sokol, S., and Bode, H. R. (1999). Cngsc, a homologue of goosecoid, participates in the patterning of the head, and is expressed in the organizer region of hydra. *Development* **126**, 5245–5254.

Browne, E. N. (1909). The production of new hydrants in hydra by insertion of small grafts. *J. Exp. Zool.* **7**, 1–23.

Buikema, W. J., and Haselkorn, R. (2001). Expression of the Anabaena hetR gene from a copper-regulated promoter leads to heterocyst differentiation under repressing conditions. *Proc. Natl. Acad. Sci. USA* **98**, 2729–2734.

Cardoso, W. V., and Lü, J. N. (2006). Regulation of early lung morphogenesis: Questions, facts and controversies. *Development* **133**, 1611–1624.

Carmany-Rampey, A., and Schier, A. F. (2001). Single-cell internalization during zebrafish gastrulation. *Cur. Biol.* **11**, 1261–1265.

Chandebois, R. (1976). Cell sociology: A way of reconsidering the current concepts of morphogenesis. *Acta Biotheor. Leiden* **25**, 71–102.

Chandebois, R. (1979). The dynamics of wound closure and its role in the programming of planarian regeneration. *Dev. Growth Differ.* **21**, 195–204.

Charest, P. G., and Firtel, R. A. (2006). Feedback signaling controls leading-edge formation during chemotaxis. *Curr. Opin. Gen. Dev.* **16**, 339–347.

Chen, G., Handel, K., and Roth, S. (2000). The maternal nf-kappa b/dorsal gradient of Tribolium castaneum: Dynamics of early dorsoventral patterning in a short-germ beetle. *Development* **127**, 5145–5156.

Chen, Y., and Schier, A. F. (2002). Lefty proteins are long-range inhibitors of squint-mediated nodal signaling. *Curr. Biol.* **12**, 2124–2128.

Christian, J. L., and Moon, R. T. (1993). Interactions between xwnt-8 and Spemann organizer signaling pathways generate dorsoventral pattern in the embryonic mesoderm of Xenopus. *Genes Dev.* **7**, 13–28.

Chuang, P. T., and McMahon, A. P. (2003). Branching morphogenesis of the lung: New molecular insights into an old problem. *Trends Cell Biol.* **13**, 86–91.

Clark, S. E. (2001). Cell signaling at the shoot meristem. *Nat. Rev. Mol. Cell Biol.* **2**, 276–284.

Cooke, J. (2004). The evolutionary origins and significance of vertebrate left–right organization. *BioEssays* **26**, 413–421.

Coultas, L., Chawengsaksophak, K., and Rossant, J. (2005). Endothelial cells and VEGF in vascular development. *Nature* **438**, 937–945.

Damen, W. G. M. (2002). Parasegmental organization of the spider embryo implies that the parasegment is an evolutionary conserved entity in arthropod embryogenesis. *Development* **129**, 1239–1250.

Damen, W. G. M., Weller, M., and Tautz, D. (2000). Expression patterns of hairy, even-skipped, and runt in the spider Cupiennius salei imply that these genes were segmentation genes in a basal arthropod. *Proc. Natl. Acad. Sci. USA* **97**, 4515–4519.

De Robertis, E., Larraín, J., Oelschläger, M., and Wessely, O. (2000). The establishment of Spemann's organizer and patterning of the vertebrate embryo. *Nat. Rev. Genet.* **1**, 171–181.

De Robertis, E. M., and Kuroda, H. (2004). Dorsal–ventral patterning and neural induction in Xenopus embryos. *Ann. Rev. Cell Dev. Biol.* **20**, 285–308.

de Jong, D. M., Hislop, N. R., Hayward, D. C., Reece-Hoyes, J. S., Pontynen, P. C., Ball, E. E., and Miller, D. J. (2006). Components of both major axial patterning systems of the Bilateria are differentially expressed along the primary axis of a 'radiate' animal, the anthozoan cnidarian Acropora millepora. *Dev. Biol.* **298**, 632–643.

Deblandre, G., Koyano, N., and Kintner, C. (1998). The differentiation of ciliated cells in the ectoderm of Xenopus embryos is regulated by lateral inhibition. *Dev. Biol.* **198**, 210.

Deblandre, G. A., Wettstein, D. A., Koyano-Nakagawa, N., and Kintner, C. (1999). A two-step mechanism generates the spacing pattern of the ciliated cells in the skin of Xenopus embryos. *Development* **126**, 4715–4728.

Diaz-Benjumea, F. J., Cohen, B., and Cohen, S. M. (1994). Cell-interaction between compartments establishes the proximal–distal axis of Drosophila legs. *Nature* **372**, 175–179.

Dorsky, R. I., Itoh, M., Moon, R. T., and Chitnis, A. (2003). Two tcf3 genes cooperate to pattern the zebrafish brain. *Development* **130**, 1937–1947.

Dosch, R., Gawantka, V., Delius, H., Blumenstock, C., and Niehrs, C. (1997). Bmp-4 acts as a morphogen in dorsoventral mesoderm patterning in Xenopus. *Development* **124**, 2325–2334.

Drewell, R. A., Bae, E., Burr, J., and Lewis, E. B. (2002). Transcription defines the embryonic domains of *cis*-regulatory activity at the Drosophila bithorax complex. *Proc. Natl. Acad. Sci. USA* **99**, 16853–16858.

Dubrulle, J., McGrew, M. J., and Pourquie, O. (2001). Fgf signaling controls somite boundary position and regulates segmentation clock control of spatiotemporal hox gene activation. *Cell* **106**, 219–232.

Dubrulle, J., and Pourquie, O. (2002). From head to tail: Links between the segmentation clock and anteroposterior patterning of the embryo. *Curr. Opin. Genet. Dev.* **12**, 519–523.

Dubrulle, J., and Pourquie, O. (2004a). Coupling segmentation to axis formation. *Development* **131**, 5783–5793.

Dubrulle, J., and Pourquie, O. (2004b). fgf8 mRNA decay establishes a gradient that couples axial elongation to patterning in the vertebrate embryo. *Nature* **427**, 419–422.

Eaton, S., and Kornberg, T. (1990). Repression of cubitus interruptus Dominant expression in the posterior compartment by engrailed. *Genes Dev.* **4**, 1074–1083.

Estrada, B., Casares, F., Busturia, A., and Sanchez-Herrero, E. (2002). Genetic and molecular characterization of a novel iab-8 regulatory domain in the abdominal-b gene of Drosophila melanogaster. *Development* **129**, 5195–5204.

Fagotto, F., Guger, K., and Gumbiner, B. M. (1997). Induction of the primary dorsalizing center in Xenopus by the wnt/gsk/beta-catenin signaling pathway, but not by vg1, activin or noggin. *Development* **124**, 453–460.

Fernandez, J., and Stent, G. S. (1982). Embryonic-development of the hirudinid leech Hirudo medicinalis—structure, development, and segmentation of the germinal plate. *J. Embryol. Exp. Morphol.* **72**, 71–96.

French, V., Bryant, P. J., and Bryant, S. V. (1976). Pattern regulation in epimorphic fields. *Science* **193**, 969–981.

Gasseling, M. T., and Saunders Jr., J. W. (1964). Effect of the "Posterior Necrotic Zone" on the early chick wing bud on the pattern and symmetry of limb outgrowth. *Am. Zool.* **4**, 303–304.

Gaunt, S. J., and Strachan, L. (1996). Temporal colinearity in expression of anterior hox genes in developing chick embryos. *Dev. Dyn.* **207**, 270–280.

Gerhart, J. (2002). Changing the axis changes the perspective. *Dev. Dyn.* **225**, 380–383.

Ghabrial, A. S., and Krasnow, M. A. (2006). Social interactions among epithelial cells during tracheal branching morphogenesis. *Nature* **441**, 746–749.

Gierer, A. (1977). Biological features and physical concepts of pattern formation exemplified by hydra. *Curr. Top. Dev. Biol.* **11**, 17–59.

Gierer, A., Berking, S., Bode, H., David, C. N., Flick, K., Hansmann, G., Schaller, H., and Trenkner, E. (1972). Regeneration of hydra from reaggregated cells. *Nat. New Biol.* **239**, 98–101.

Gierer, A., and Meinhardt, H. (1972). A theory of biological pattern formation. *Kybernetik* **12**, 30–39.

Glazer, L., and Shilo, B. Z. (1991). The Drosophila FGF-R homolog is expressed in the embryonic tracheal system and appears to be required for directed tracheal extensions. *Genes Dev.* **5**, 697–705.

Gould, A., Itasaki, N., and Krumlauf, R. (1998). Initiation of rhombomeric HoxB4 expression requires induction by somites and a retinoic acid pathway. *Neuron* **21**, 39–51.

Grapin-Botton, A., Bonnin, M. A., Sieweke, M., and Le Douarin, N. M. (1998). Defined concentrations of a posteriorizing signal are critical for MadB/Kreisler segmental expression in the hindbrain. *Development* **125**, 1173–1181.

Grens, A., Gee, L., Fisher, D. A., and Bode, H. R. (1996). Cnnk-2, an nk-2 homeobox gene, has a role in patterning the basal end of the axis in hydra. *Dev. Biol.* **180**, 473–488.

Grieshammer, U., Minowada, G., Pisenti, J. M., Abbott, U. K., Martin, G. R. (1996). The chick limbless mutation causes abnormalities in limb bud dorsal–ventral patterning: Implication for apical ridge formation. *Development* **122**, 3851–3861.

Gritsman, K., Talbot, W. S., and Schier, A. F. (2000). Nodal signaling patterns the organizer. *Development* **127**, 921–932.

Gurdon, J. B., Mitchell, A., and Mahony, D. (1995). Direct and continuous assessment by cells of their position in a morphogen gradient. *Nature* **376**, 520–521.

Haeckel, E. (1874). The Gastraea-theory, the phylogenetic classification of the animal kingdom and the homology of the germ lamellae. *Q. J. Microsc. Sci.* **14**, 142–165.

Harland, R., and Gerhart, J. (1997). Formation and function of Spemann's organizer. *Ann. Rev. Cell Dev. Biol.* **13**, 611–667.

Harris, M. P., Williamson, S., Fallon, J. F., Meinhardt, H., and Prum, R. O. (2005). Molecular evidence for an activator–inhibitor mechanism in development of embryonic feather branching. *Proc. Natl. Acad. Sci. USA* **102**, 11734–11739.

Harrison, R. G. (1918). Experiments on the development of the fore-limb of Amblystoma, a self-differentiating equipotential system. *J. Exp. Zool.* **25**, 413–446.

Harrison, R. G. (1921). On relations of symmetry in transplanted limbs. *J. Exp. Zool.* **32**, 1–136.

Hellström, M., Phng, L. K., Hofmann, J. J., Wallgard, E., Coultas, L., Lindblom, P., Alva, J., Nilsson, A. K., Karlsson, L., Gaiano, N., Yoon, K., Rossant, J., Iruela-Arispe, M. L., Kalén, M., Gerhardt, H., and Betsholtz, C. (2007). Dll4 signaling through Notch1 regulates formation of tip cells during angiogenesis. *Nature* **445**, 776–780.

Hild, M., Dick, A., Rauch, G. J., Meier, A., Bouwmeester, T., Haffter, P., and Hammerschmidt, M. (1999). The smad5 mutation somitabun blocks Bmp2b signaling during early dorsoventral patterning of the zebrafish embryo. *Development* **126**, 2149–2159.

Hirth, F., Kammermeier, L., Frei, E., Walldorf, U., Noll, M., and Reichert, H. (2003). An urbilaterian origin of the tripartite brain: Developmental genetic insights from Drosophila. *Development* **130**, 2365–2373.

Hobmayer, B., Rentzsch, F., Kuhn, K., Happel, C. M., Cramer von Laue, C., Snyder, P., Rothbacher, U., and Holstein, T. W. (2000). Wnt signaling molecules act in axis formation in the diploblastic metazoan hydra. *Nature* **407**, 186–189.

Holland, L. Z., Holland, N. D., and Schubert, M. (2000). Developmental expression of amphiwnt1, an amphioxus gene in the wnt1/wingless subfamily. *Dev. Genes Evol.* **210**, 522–524.

Holland, L. Z., Panfilio, K. A., Chastain, R., Schubert, M., and Holland, N. D. (2005). Nuclear beta-catenin promotes nonneural ectoderm and posterior cell fates in amphioxus embryos. *Dev. Dyn.* **233**, 1430–1443.

Holley, S. A., and Takeda, H. (2002). Catching a wave: The oscillator and wavefront that create the zebrafish somite. *Semin. Cell Dev. Biol.* **13**, 481–488.

Huang, X., Dong, Y., and Zhao, J. (2004). HetR homodimer is a DNA-binding protein required for heterocyst differentiation, and the DNA-binding activity is inhibited by PatS. *Proc. Natl. Acad. Sci. USA* **101**, 4848–4853.

Hülskamp, M. (2004). Plant trichomes: A model for cell differentiation. *Nat. Rev. Mol. Cell Biol.* **5**, 471–480.

Ingham, P., and Hidalgo, A. (1993). Regulation of wingless transcription in the Drosophila embryo. *Development* **117**, 283–291.

Ingham, P. W. (1991). Segment polarity genes and cell patterning within the Drosophila body segment. *Curr. Opin. Gen. Dev.* **1**, 261–267.

Irvine, K. D., and Rauskolb, C. (2000). Boundaries in development: Formation and function. *Ann. Rev. Cell Dev. Biol.* **17**, 189–214.

Jaffe, F. (1968). Localization in the developing Fucus egg and the general role of localizing currents. *Adv. Morphogen.* **7**, 295–328.

Joyner, A. L. (1996). Engrailed, wnt and pax genes regulate midbrain hindbrain development. *Trends Genet.* **12**, 15–20.

Kamm, K., Schierwater, B., Jakob, W., Dellaporta, S. L., and Miller, D. J. (2006). Axial patterning and diversification in the Cnidaria predate the hox system. *Curr. Biol.* **16**, 920–926.

Kato, K., Orii, H., Watanabe, K., and Agata, K. (2001). Dorsal and ventral positional cues required for the onset of planarian regeneration may reside in differentiated cells. *Dev. Biol.* **233**, 109–121.

Keller, R. (2005). Cell migration during gastrulation. *Curr. Opin. Cell Biol.* **17**, 533–541.

Keynes, R. J., and Stern, C. D. (1984). Segmentation in the vertebrate nervous system. *Nature* **310**, 786–789.

Khaner, O., and Eyal-Giladi, H. (1989). The chick's marginal zone and primitive streak formation. I. Coordinative effect of induction and inhibition. *Dev. Biol.* **134**, 206–214.

Kiecker, C., and Niehrs, C. (2001a). A morphogen gradient of wnt/beta-catenin signaling regulates anteroposterior neural patterning in Xenopus. *Development* **128**, 4189–4201.

Kiecker, C., and Niehrs, C. (2001b). The role of prechordal mesendoderm in neural patterning. *Curr. Opin. Neurobiol.* **11**, 27–33.

Kusserow, A., Pang, K., Sturm, C., Hrouda, M., Lentfer, J., Schmidt, H. A., Technau, U., Haeseler, A., von Hobmayer, B., Martindale, M. Q., and Holstein, T. W. (2005). Unexpected complexity of the Wnt gene family in a sea anemone. *Nature* **433**, 156–160.

Kuziora, M. A., and McGinnis, W. (1988). Autoregulation of a Drosophila homeotic selector gene. *Cell* **55**, 477–485.

Lawrence, P. A. (1970). Polarity and patterns in the postembryonic development of insects. *Adv. Insect Physiol.* **7**, 197–266.

Lee, H. X., Ambrosio, A. L., Reversade, B., and De Robertis, E. M. (2006). Embryonic dorsal–ventral signaling: Secreted frizzled-related proteins as inhibitors of tolloid proteinases. *Cell* **124**, 147–159.

Lele, Z., Nowak, M., and Hammerschmidt, M. (2001). Zebrafish admp is required to restrict the size of the organizer and to promote posterior and ventral development. *Dev. Dyn.* **222**, 681–687.

Lenhoff, H. (1991). Ethel Browne, Hans Spemann, and the discovery of the organizer phenomenon. *Biol. Bull.* **181**, 72–80.

Leonetti, M., Dubois-Violette, E., and Homble, F. (2004). Pattern formation of stationary transcellular ionic currents in Fucus. *Proc. Natl. Acad. Sci. USA* **101**, 10243–10248.

Leptin, M. (1991). Twist and snail as positive and negative regulators during Drosophila mesoderm development. *Genes Dev.* **5**, 1568–1576.

Leslie, J. D., Ariza-McNaughton, L., Bermange, A. L., McAdow, R., Johnson, S. L., and Lewis, J. (2007). Endothelial signaling by the Notch ligand Delta-like 4 restricts angiogenesis. *Development* **134**, 839–844.

Levin, M., and Mercola, M. (1998). Evolutionary conservation of mechanisms upstream of asymmetric Nodal expression: Reconciling chick and Xenopus. *Dev. Genet.* **23**, 185–193.

Levin, M., and Palmer, A. R. (2007). Left–right patterning from the inside out: Widespread evidence for intracellular control. *BioEssays* **29**, 271–287.

Lewis, E. B. (1978). A gene complex controlling segmentation in Drosophila. *Nature* **276**, 565–570.

Locke, M. (1959). The cuticular pattern in an insect, Rhodnius prolixus Stal. *J. Exp. Biol.* **36**, 459–477.

Lohr, J. L., Danos, M. C., Groth, T. W., and Yost, H. J. (1998). Maintenance of asymmetric nodal expression in Xenopus laevis. *Dev. Genet.* **23**, 194.

Lowe, C. J., Wu, M., Salic, A., Evans, L., Lander, E., Stange-Thomann, N., Gruber, C. E., Gerhart, J., and Kirschner, M. (2003). Anteroposterior patterning in hemichordates and the origins of the chordate nervous system. *Cell* **113**, 853–865.

Lutz, H. (1949). Sur la production experimentale de la polyembryonie et de la monstruosite double ches lez oiseaux. *Arch. d'Anat. Microsc. Morphol. Exp.* **38**, 79–144.

Maeda, Y., and Maeda, M. (1974). Heterogeneity of the cell population of the cellular slime mold Dictyostelium discoideum before aggregation and its relation to subsequent locations of the cells. *Exp. Cell Res.* **84**, 88–94.

Maizel, A., Tassetto, M., Filhol, O., Cochet, C., Prochiantz, A., and Joliot, A. (2002). Engrailed homeoprotein secretion is a regulated process. *Development* **129**, 3545–3553.

Martin, G. R. (1995). Why thumbs are up. *Nature* **374**, 410–411.
Martin, G. (2001). Making a vertebrate limb: New players enter from the wings. *BioEssays* **23**, 865–868.
Martindale, M. Q. (2005). The evolution of metazoan axial properties. *Nat. Rev. Genet.* **6**, 917–927.
Martinez Arias, A., and Lawrence, P. A. (1985). Parasegments and compartments in the Drosophila embryo. *Nature* **313**, 639–642.
Martyn, U., and Schulte-Merker, S. (2003). The ventralizing ogon mutant phenotype is caused by the zebrafisch homologue of Sizzled, a secreted Frizzled-related protein. *Dev. Biol.* **260**, 58–67.
Mayer, K. F. X., Schoof, H., Haecker, A., Lenhard, M., Jürgens, G., and Laux, T. (1998). Role of wuschel in regulating stem cell fate in the arabidopsis shoot meristem. *Cell* **95**, 805–815.
Meinhardt, H. (1976). Morphogenesis of lines and nets. *Differentiation* **6**, 117–123.
Meinhardt, H. (1978). Space-dependent cell determination under the control of a morphogen gradient. *J. Theor. Biol.* **74**, 307–321.
Meinhardt, H. (1980). Cooperation of compartments for the generation of positional information. *Z. Naturforsch.* **35c**, 1086–1091.
Meinhardt, H. (1982). Models of Biological Pattern Formation. Academic Press, London (freely available at http://www.eb.tuebingen.mpg.de/meinhardt).
Meinhardt, H. (1983a). Cell determination boundaries as organizing regions for secondary embryonic fields. *Dev. Biol.* **96**, 375–385.
Meinhardt, H. (1983b). A boundary model for pattern formation in vertebrate limbs. *J. Embryol. Exp. Morphol.* **76**, 115–137.
Meinhardt, H. (1983c). A model for the prestalk/prespore patterning in the slug of the slime mold Dictyostelium discoideum. *Differentiation* **24**, 191–202.
Meinhardt, H. (1986a). Models of segmentation. *In* "Somites in Developing Embryos." (R. Bellairs, D. A. Ede, and J. W. Lash, Eds.). *In* "Nato ASI Series A," Vol. 118. Plenum, New York, pp. 179–189.
Meinhardt, H. (1986b). The threefold subdivision of segments and the initiation of legs and wings in insects. *Trends Genet.* **2**, 36–41.
Meinhardt, H. (1989). Models for positional signaling with application to the dorsoventral patterning of insects and segregation into different cell types. *Development* (Suppl.), 169–180.
Meinhardt, H. (1993). A model for pattern-formation of hypostome, tentacles, and foot in hydra: How to form structures close to each other, how to form them at a distance. *Dev. Biol.* **157**, 321–333.
Meinhardt, H. (1995). Growth and patterning—dynamics of stripe formation. *Nature* **376**, 722–723.
Meinhardt, H. (1999). Orientation of chemotactic cells and growth cones: Models and mechanisms. *J. Cell Sci.* **112**, 2867–2874.
Meinhardt, H. (2001). Organizer and axes formation as a self-organizing process. *Int. J. Dev. Biol.* **45**, 177–188.
Meinhardt, H. (2002). The radial-symmetric hydra and the evolution of the bilateral body plan: An old body became a young brain. *BioEssays* **24**, 185–191.
Meinhardt, H. (2003). "The Algorithmic Beauty of Sea Shells," 3rd ed. Springer-Verlag, Heidelberg/New York.
Meinhardt, H. (2004a). Different strategies for midline formation in bilaterians. *Nat. Rev. Neurosci.* **5**, 502–510.
Meinhardt, H. (2004b). Out-of-phase oscillations and traveling waves with unusual properties: The use of three-component systems in biology. *Physica D* **199**, 264–277.
Meinhardt, H. (2006). Primary body axes of vertebrates: Generation of a near-Cartesian coordinate system and the role of Spemann-type organizer. *Dev. Dyn.* **235**, 2907–2919.
Meinhardt, H., and Gierer, A. (1980). Generation and regeneration of sequences of structures during morphogenesis. *J. Theor. Biol.* **85**, 429–450.
Meinhardt, H., and Gierer, A. (2000). Pattern formation by local self-activation and lateral inhibition. *BioEssays* **22**, 753–760.
Meinhardt, H., and Klingler, M. (1987). A model for pattern formation on the shells of molluscs. *J. Theor. Biol.* **126**, 63–89.

Meinhardt, H., and de Boer, P. A. J. (2001). Pattern formation in E. coli: A model for the pole-to-pole oscillations of Min proteins and the localization of the division site. *Proc. Natl. Acad. Sci. USA* **98**, 14202–14207.
Miura, H., Yanazawa, M., Kato, K., and Kitamura, K. (1997). Expression of a novel aristaless related homeobox gene arx in the vertebrate telencephalon, diencephalon and floor plate. *Mech. Dev.* **65**, 99–109.
Moment, G. B. (1951). Simultaneous anterior and posterior regeneration and other growth phenomena in maldanid polychaetes. *J. Exp. Zool.* **117**, 1–13.
Moos, M., Wang, S. W., and Krinks, M. (1995). Antidorsalizing morphogenetic protein is a novel tgf-beta homolog expressed in the Spemann organizer. *Development* **121**, 4293–4301.
Morimoto, M., Kiso, M., Sasaki, N., Saga, Y. (2006). Cooperative mesp activity is required for normal somitogenesis along the anterior–posterior axis. *Dev. Biol.* **300**, 687–698.
Moussian, B., and Roth, S. (2005). Dorsoventral axis formation in the Drosophila embryo—shaping and transducing a morphogen gradient. *Curr. Biol.* **15**, R887–R899.
Nakamura, T., Mine, N., Nakaguchi, E., Mochizuki, A., Yamamoto, M., Yashiro, K., Meno, C., and Hamada, M. (2006). Generation of robust left–right asymmetry in the mouse embryo requires a self-enhancement and lateral-inhibition system. *Dev. Cell* **11**, 495–504.
Newman, H. H. (1928). Studies of human twins. II. Asymmetry reversal, of mirror imaging in identical twins. *Biol. Bull.* **55**, 298–315.
Newmark, P. A., and Alvarado, A. S. (2002). Not your father's planarian: A classic model enters the era of functional genomics. *Nat. Rev. Genet.* **3**, 210–219.
Niehrs, C. (2004). Regionally specific induction by the Spemann–Mangold organizer. *Nat. Rev. Genet.* **5**, 425–434.
Nieuwkoop, P. D. (1952). Activation and organization of the central nervous system in amphibians. III. Synthesis of a new working hypothesis. *J. Exp. Zool.* **120**, 83–108.
Nieuwkoop, P. D. (1992). The formation of the mesoderm in urodelean amphibians. VI. The self-organizing capacity of the induced mesoendoderm. *Roux's Arch. Dev. Biol.* **201**, 18–29.
Niswander, L. (2003). Pattern formation: Old models out on a limb. *Nat. Rev. Genet.* **4**, 133–143. 54-9.
Nonaka, S., Tanaka, Y., Okada, Y., Takeda, S., Harada, A., Kanai, Y., Kido, M., and Hirokawa, N. (1998). Randomization of left–right asymmetry due to loss of nodal cilia generating leftward flow of extraembryonic fluid in mice lacking KIF3B motor protein. *Cell* **95**, 829–837.
Nordström, U., Jessell, T. M., and Edlund, T. (2002). Progressive induction of caudal neural character by graded Wnt signaling. *Nat. Neurosci.* **5**, 525–532.
Ober, E. A., and Schulte-Merker, S. (1999). Signals from the yolk cell induce mesoderm, neuroectoderm, the trunk organizer, and the notochord in zebrafish. *Dev. Biol.* **215**, 167–181.
Ogawa, K., Ishihara, S., Saito, Y., Mineta, K., Nakazawa, M., Ikeo, K., Gojobori, T., Watanabe, K., and Agata, K. (2002). Induction of a noggin-like gene by ectopic DV interaction during planarian regeneration. *Dev. Biol.* **250**, 59–70.
Orii, H., Kato, K., Agata, K., and Watanabe, K. (1998). Molecular cloning of bone morphogenetic protein (BMP) gene from the planarian Dugesia japonica. *Zool. Sci.* **15**, 871–877.
Palmeirim, I., Henrique, D., Ish-Horowicz, D., and Pourquie, O. (1997). Avian hairy gene-expression identifies a molecular clock linked to vertebrate segmentation and somitogenesis. *Cell* **91**, 639–648.
Parichy, D. M. (2006). Evolution of danio pigment pattern development. *Heredity* **97**, 200–210.
Patel, N. H., Condron, B. G., and Zinn, K. (1994). Pair-rule expression patterns of even-skipped are found in both short-germ and long-germ beetles. *Nature* **367**, 429–434.
Patel, N. H., Kornberg, T. B., and Goodman, C. S. (1989). Expression of engrailed during segmentation in grasshopper and crayfish. *Development* **107**, 201–212.
Pesch, M., and Hülskamp, M. (2004). Creating a two-dimensional pattern *de novo* during arabidopsis trichome and root hair initiation. *Curr. Opin. Genet. Dev.* **14**, 422–427.
Pourquie, O. (2004). The chick embryo: A leading model in somitogenesis studies. *Mech. Dev.* **121**, 1069–1079.
Prakash, N., and Wurst, W. (2004). Specification of midbrain territory. *Cell Tissue Res.* **318**, 5–14.

Puelles, L. (2001). Brain segmentation and forebrain development in amniotes. *Brain Res. Bull.* **55**, 695–710.

Raya, A., and Izpisúa Belmonte, J. C. (2006). Left–right asymmetry in the vertebrate embryo: From early information to higher-level integration. *Nat. Rev. Genet.* **7**, 283–293.

Regulski, M., Dessain, S., McGinnis, N., and McGinnis, W. (1991). High-affinity binding-sites for the deformed protein are required for the function of an autoregulatory enhancer of the deformed gene. *Genes Dev.* **5**, 278–286.

Reinhardt, D., Frenz, M., Mandel, T., and Kuhlemeier, C. (2003). Microsurgical and laser ablation analysis of interactions between the zones and layers of the tomato shoot apical meristem. *Development* **130**, 4073–4083.

Reversade, B., and De Robertis, E. (2005). Regulation of ADMP and BMP2/4/7 at opposite embryonic poles generates a self-regulating morphogenetic field. *Cell* **123**, 1147–1160.

Reversade, B., Kuroda, H., Lee, H., Mays, A., and De Robertis, E. M. (2005). Depletion of bmp2, bmp4, bmp7 and Spemann organizer signals induces massive brain formation in Xenopus embryos. *Development* **132**, 3381–3392.

Rhinn, M., and Brand, M. (2001). The midbrain–hindbrain boundary organizer. *Curr. Opin. Neurobiol.* **11**, 34–42.

Ringrose, L., and Paro, R. (2007). Polycomb/trithorax response elements and epigenetic memory of cell identity. *Development* **134**, 223–232.

Risau, W., and Flamme, I. (1995). Vasculogenesis. *Annu. Rev. Cell Dev. Biol.* **11**, 73–91.

Sachs, T. (1981). The control of the patterned differentiation of vascular tissues. *Adv. Bot. Res.* **9**, 151–262.

Saló, E. (2006). The power of regeneration and the stem-cell kingdom: Freshwater planarians (platyhelminthes). *BioEssays* **28**, 546–559.

Saló, E., and Baguñà, J. (1985). Proximal and distal transformation during intercalary regeneration in the planarian Dugesia. *Roux's Arch. Dev. Biol.* **194**, 364–368.

Saló, E., and Baguñà, J. (2002). Regeneration in planarians and other worms: New findings, new tools, and new perspectives. *J. Exp. Zool.* **292**, 528–539.

Sander, K., and Faessler, P. E. (2001). Introducing the Spemann–Mangold organizer: Experiments and insights that generated a key concept in developmental biology. *Int. J. Dev. Biol.* **45**, 1–11.

Saùde, L., Woolley, K., Martin, P., Driever, W., and Stemple, D. L. (2000). Axis-inducing activities and cell fates of the zebrafish organizer. *Development* **127**, 3407–3417.

Scarpella, E., Marcos, D., Friml, J., and Berleth, T. (2006). Control of leaf vascular patterning by polar auxin transport. *Genes Dev.* **20**, 1015–1027.

Schier, A. E., and Talbot, W. S. (2005). Molecular genetics of axis formation in zebrafish. *Annu. Rev. Genet.* **39**, 561–613.

Scholtz, G., Patel, N. H., and Dohle, W. (1994). Serially homologous engrailed stripes are generated via different cell lineages in the germ band of amphipod crustaceans (Malacostraca, Peracarida). *Int. J. Dev. Biol.* **38**, 471–478.

Schubert, F. R., Dietrich, S., Mootoosamy, R. C., Chapman, S. C., and Lumsden, A. (2001a). Lbx1 marks a subset of interneurons in chick hindbrain and spinal cord. *Mech. Dev.* **101**, 181–185.

Schubert, M., Holland, L. Z., Stokes, M. D., and Holland, N. D. (2001b). Three amphioxus wnt genes (amphiwnt3, amphiwnt5, and amphiwnt6) associated with the tail bud: The evolution of somitogenesis in chordates. *Dev. Biol.* **240**, 262–273.

Schwartz, Y. B., and Pirrotta, V. (2007). Polycomb silencing mechanisms and the management of genomic programmes. *Nat. Rev. Genet.* **8**, 9–22.

Shankland, M. (1991a). Leech segmentation: Cell lineage and the formation of the complex body pattern. *Dev. Biol.* **144**, 21–231.

Shankland, M. (1991b). Leech segmentation—cell lineage and the formation of complex body patterns. *Dev. Biol.* **144**, 221–231.

Shimizu, H., and Fujisawa, T. (2003). Peduncle of hydra and the heart of higher organisms share a common ancestral origin. *Genesis* **36**, 182–186.

Siekmann, A. F., and Lawson, N. D. (2007a). Notch limits angiogenic cell behavior in developing zebrafish arteries. *Nature* **455**, 781–784.

Siekmann, A. F., and Lawson, N. D. (2007b). Notch signaling limits angiogenic cell behavior in developing zebrafish arteries. *Nature* **445**, 781–784.

Simpson-Brose, M., Treisman, J., and Desplan, C. (1994). Synergy between the hunchback and bicoid morphogens is required for anterior patterning in Drosophila. *Cell* **78**, 855–865.

Smith, K. M., Gee, L., Blitz, I. L., and Bode, H. R. (1999). CnOtx, a member of the Otx gene family, has a role in cell movement in hydra. *Dev. Biol.* **212**, 392–404.

Smith, K. M., Gee, L., and Bode, H. R. (2000). Hyalx, an aristaless-related gene, is involved in tentacle formation in hydra. *Development* **127**, 4743–4752.

Smith, W. C., and Harland, R. M. (1992). Expression cloning of noggin, a new dorsalizing factor localized to the Spemann organizer in Xenopus embryos. *Cell* **70**, 829–840.

Solnica-Krezel, L. (2005). Conserved patterns of cell movements during vertebrate gastrulation. *Curr. Biol.* **15**, R213–R228.

Sommer, R. J., and Tautz, D. (1993). Involvement of an orthologue of the Drosophila pair-rule gene hairy in segment formation of the short germ-band embryo of Tribolium (Coleoptera). *Nature* **361**, 448–450.

Spemann, H., and Mangold, H. (1924). Über Induktion von Embryonalanlagen durch Implantation artfremder Organisatoren. *Wilhelm Roux' Arch. Entw. Mech. Org.* **100**, 599–638.

Spratt, N. T., and Haas, H. (1960). Integrative mechanisms in development of the early chick blastoderm. I. Regulative potentiality of separated parts. *J. Exp. Zool.* **145**, 97–137.

Sprecher, S. G., and Reichert, H. (2003). The urbilaterian brain: Developmental insights into the evolutionary origin of the brain in insects and vertebrates. *Arthropod Struct. Dev.* **32**, 141–156.

Stern, C. D. (2001). Initial patterning of the central nervous system: How many organizers? *Nat. Rev. Neurosci.* **2**, 92–98.

Steward, R. M., and Gerhart, J. C. (1990). The anterior extent of dorsal development of the Xenopus embryonic axis depends on the quantity of organizer in the late blastula. *Development* **109**, 363–373.

Sutherland, D., Samakovlis, C., and Krasnow, M. A. (1996). Branchless encodes a Drosophila FGF homolog that controls tracheal cell-migration and the pattern of branching. *Cell* **87**, 1091–1101.

Szymanski, D. B., Lloyd, A. M., and Marks, M. D. (2000). Progress in the molecular genetic analysis of trichome initiation and morphogenesis in arabidopsis. *Trends Plant Sci.* **5**, 214–219.

Takahashi, Y., Koizumi, K., Takagi, A., Kitajima, S., Inoue, T., Koseki, H., and Saga, Y. (2000). Mesp2 initiates somite segmentation through the notch signaling pathway. *Nat. Genet.* **25**, 390–396.

Tanaka, M., Tamura, K., Noji, S., Nohno, T., Ide, H. (1997). Induction of additional limb at the dorsal–ventral boundary of a chick-embryo. *Dev. Biol.* **182**, 191–203.

Technau, U., and Bode, H. R. (1999). HyBra1, a Brachyury homologue, acts during head formation in hydra. *Development* **126**, 999–1010.

Tickle, C., Summerbell, D., and Wolpert, L. (1975). Positional signaling and specification of digits in chick limb morphogenesis. *Nature* **254**, 199–202.

Tonissen, K. F., Drysdale, T. A., Lints, T. J., Harvey, R. P., and Krieg, P. A. (1994). XNkx-2.5, a Xenopus gene related to Nkx-2.5 and tinman: Evidence for a conserved role in cardiac development. *Dev. Biol.* **192**, 325–328.

Trembley, A. (1744). "Memoires pour servir a l'histoire d'un genre de polypes d'eau douce." J. u. H. Verbeek, Leiden.

Tsiantis, M., and Hay, A. (2003). The time of the leaf? *Nat. Rev. Genet.* **4**, 169–180.

Turing, A. (1952). The chemical basis of morphogenesis. *Philos. Trans. B* **237**, 37–72.

van den Heuvel, M., Nusse, R., Jonston, P., and Lawrence, P. (1989). Distribution of the wingless gene product in Drosophila embryos: A protein involved in cell–cell communication. *Cell* **59**, 739–749.

Vincent, J. P., and Lawrence, P. A. (1994). It takes three to distalize. *Nature* **372**, 132–133.

von Rosenhof, R. (1755). "Insektenbelustigung: Historie der Polypen und anderer kleiner Wasserinsecten." Vol. III. Fleishmann, Nürnberg. p. 175.

1. Models for the Organization of the Embryonic Body Axes

Wacker, S. A., Jansen, H. J., McNulty, C. L., Houtzager, E., and Durston, A. J. (2004a). Timed interactions between the Hox expressing nonorganizer mesoderm and the Spemann organizer generate positional information during vertebrate gastrulation. *Dev. Biol.* **268**, 207–221.

Wacker, S. A., McNulty, C. L., and Durston, A. J. (2004b). The initiation of Hox gene expression in Xenopus laevis is controlled by Brachyury and BMP-4. *Dev. Biol.* **266**, 123–137.

Waddington, C. H., Needham, J., and Brachet, J. (1936). Studies on the nature of the amphibian organizing center. III. The activation of the evocator. *Proc. R. Soc. London B* **120**, 173–190.

Wigglesworth, V. B. (1940). Local and general factors in the development of "pattern" in Rhodnius prolixus. *J. Exp. Biol.* **17**, 180–200.

Wilby, O. K., and Webster, G. (1970a). Experimental studies on axial polarity in hydra. *J. Embryol. Exp. Morphol.* **24**, 595–613.

Wilby, O. K., and Webster, G. (1970b). Studies on the transmission of hypostome inhibition in hydra. *J. Embryol. Exp. Morphol.* **24**, 583–593.

Wilk, R., Weizman, I., and Shilo, B. Z. (1996). Trachealess encodes a bHLH-PAS protein that is an inducer of the tracheal cell fate in Drosophila. *Genes Dev.* **10**, 93–102.

Willot, V., Mathieu, J., Lu, Y., Schmid, B., Sidi, S., Yan, Y. L., Postlethwait, J. H., Mullins, M., Rosa, F., and Peyrieras, N. (2002). Cooperative action of ADMP- and BMP-mediated pathways in regulating cell fates in the zebrafish gastrula. *Dev. Biol.* **241**, 59–78.

Wolpert, L. (1969). Positional information and the spatial pattern of cellular differentiation. *J. Theor. Biol.* **25**, 1–47.

Yamaguchi, M., Yoshimoto, E., and Kondo, S. (2007). Pattern regulation in the stripe of zebrafish suggests an underlying dynamic and autonomous mechanism. *Proc. Natl. Acad. Sci. USA* **104**, 4790–4793.

Yao, J., and Kessler, D. S. (2001). Goosecoid promotes head organizer activity by direct repression of xwnt8 in Spemann's organizer. *Development* **128**, 2975–2987.

Yoon, H. S., and Golden, J. W. (1998). Heterocyst pattern-formation controlled by a diffusible peptide. *Science* **282**, 935–938.

Yu, J. K., Satou, Y., Holland, N. D., Shin-I, T., Kohara, Y., Satoh, N., Bronner-Fraser, M., and Holland, L. Z. (2007). Axial patterning in cephalochordates and the evolution of the organizer. *Nature* **445**, 613–617.

Yuan, S. P., and Schoenwolf, G. C. (1998). *De novo* induction of the organizer and formation of the primitive streak in an experimental-model of notochord reconstitution in avian embryos. *Development* **125**, 201–213.

Zachgo, S., Silva, E. D., Motte, P., Tröbner, W., Saedler, H., and Schwarz-Sommer, Z. (1995). Functional-analysis of the Antirrhinum floral homeotic deficiens gene *in vivo* and *in vitro* by using a temperature-sensitive mutant. *Development* **121**, 2861–2875.

Zhang, H. J., Shinmyo, Y., Mito, T., Miyawaki, K., Sarashina, I., Ohuchi, H., and Noji, S. (2005). Expression patterns of the homeotic genes Scr, Antp, Ubx, and abd-A during ernbryogenesis of the cricket Gryllus bimaculatus. *Gene Expr. Patterns* **5**, 491–502.

Zhukovskaya, N. V., Fukuzawa, M., Yamada, Y., Araki, T., and Williams, J. G. (2006). The Dictyostelium bZIP transcription factor DimB regulates prestalk-specific gene expression. *Development* **133**, 439–448.

Zákány, J., Kmita, M., Alarcon, P., de la Pompa, J. L., and Duboule, D. (2001). Localized and transient transcription of hox genes suggests a link between patterning and the segmentation clock. *Cell* **106**, 207–217.

2

Robustness of Embryonic Spatial Patterning in *Drosophila melanogaster*

David Umulis,* Michael B. O'Connor,† and Hans G. Othmer‡
*Department of Chemical Engineering and Materials Science, University of Minnesota, Minneapolis, Minnesota 55455
†Department of Genetics, Cell Biology and Development and Howard Hughes Medical Institute, University of Minnesota, Minneapolis, Minnesota 55455
‡School of Mathematics and Digital Technology Center, University of Minnesota, Minneapolis, Minnesota 55455

I. Introduction
  A. *Drosophila melanogaster* as a Model System
II. Robustness in the Developmental Context
  A. Structural Stability in Dynamical Systems
  B. Sensitivity Analysis
  C. The Turing Theory of Spatial Pattern Formation and Scale Invariance
III. Scaling of AP Patterning in *Drosophila*
  A. Anterior–Posterior Patterning
  B. Intra- and Interspecies Variations in the Bicoid Distribution
  C. The Effects of Discrete Nuclei and Diffusion in the Bulk Cytoplasm
IV. Models of the Segment Polarity Network
  A. The Continuous-State Analysis of the Segment Polarity Gene Network
  B. A Boolean Model for Control of the Segment Polarity Genes
  C. The Segment Polarity Network is a Simple Switching Network
V. Dorsal–Ventral Patterning in *Drosophila*
  A. The Model for DV Patterning
  B. The Role of Positive Feedback
VI. Conclusions
  Note added in proof
  Acknowledgments
  References

I. Introduction

Much is known about the molecular components involved in signal transduction and gene expression in a number of model systems in developmental biology, and the focus is now shifting to understanding how these components are integrated into networks, and how these networks transduce the inputs they receive and produce the desired pattern of gene expression. The major question is how the correct genes are turned on at the correct point in space at the correct time in development to produce the numerous cell types present in an adult. Gene expression during embryonic development is not a cell-autonomous process, because cell fate in a multicellular embryo usually

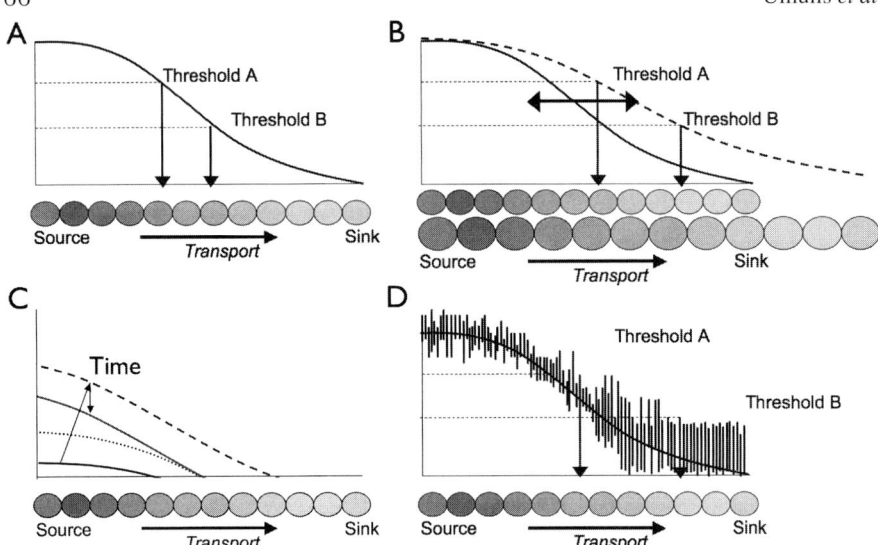

Figure 1 The classical paradigm for patterning along an axis. (A) A morphogen is produced by source cells at the left and diffuses away from the source, and diffusion and degradation of the morphogen establishes a decreasing distribution. Cells respond to the local concentration of the morphogen, which is assumed to be time-invariant, and adopt one of three distinct fates in a threshold-dependent manner. (B) Patterning mediated by morphogens can be sensitive to changes in the level of morphogen secretion or the overall size of the system, but with suitable modifications the model can produce scale-invariance. (C) The time-evolution of the morphogen distribution shown overshoots the steady state, which may not even be reached in the time available. (D) Morphogen fluctuations due to stochastic effects may lead to a noisy signal, particularly at low concentrations. (See color insert.)

depends on the cell's location in the embryo. A spatially-graded distribution of factors that influence development can be used to induce spatially-varying differentiation, and this idea played a central role throughout the early history of theoretical work in the field and was later developed into the theory of positional information by Wolpert (1969). Formally the theory posits that a cell must 'know' its position relative to other cells in order to adopt the correct developmental pathway, but of course what a cell 'knows' is determined by the information it extracts from the past and current signals received. Pattern formation in development refers to the spatially- and temporally-organized expression of genes in a multicellular array, and positional information is viewed as a necessary part of this process. Frequently pattern formation results from the response of individual cells to a spatial pattern of chemicals called *morphogens*, a term coined by the British logician Alan Turing in a fundamental paper on pattern formation (Turing, 1952) (cf. Fig. 1). Currently morphogens are defined as secreted signaling molecules that (i) are produced in a restricted portion of a tissue, (ii) are transported by diffusion, active transport, relay mechanisms, or other means to the remainder of the tissue, (iii) are detected by specific receptors or bind to specific sites on DNA, and (iv) initiate an intracellular signal transduction cascade that initiates or terminates the expression of target genes in a concentration-dependent manner. Perhaps the earliest example of a morphogen was the Bicoid protein that is involved in anterior–posterior (AP) patterning in *Drosophila* (Frohnhöfer and Nüsslein-Volhard, 1986), but many more examples are now known, including Activin, Hedgehog, Wingless, and various members of the transforming growth factor family. Theoretical studies of how different modes of transport and transduction affect patterning are reported in Kerszberg and Wolpert (1998); Kerszberg (1999); Lander *et al.* (2002); Strigini (2005); Umulis *et al.* (2006) and many others.

Thus the problem of pattern formation in a given system becomes that of discovering the mechanisms of localized production and transport that generate positional information. The classic paradigm for this is the model shown in Fig. 1A, wherein a source at one boundary of a one-dimensional domain produces the morphogen, which diffuses throughout the domain and initiates gene transcription and cell differentiation in a threshold-dependent manner. This is a static viewpoint, in that cells simply sense the local concentration and respond to it, and neither the transient dynamics of the signal nor the history of exposure to it play any role in patterning. As we show in Sections II and III, patterning by this mechanism is sensitive to changes in length unless some special mechanisms are used to compensate for such changes, and we show in Section V that the history of exposure is important in some contexts.

In many developing systems the outcome is buffered to numerous perturbations, ranging from major ones such as separation of the cells at the 2-cell stage in *Xenopus* (which can lead to one smaller, but normal adult, and an amorphous mass of tissue), to less severe ones such as changes in the ambient temperature or the loss of one copy of a gene. Indeed, many loss-of function mutations of important developmental genes in higher organisms show weak or no phenotypic effects, as will be discussed later in the context of *Drosophila*. There are many other instances in which the developmental outcome is buffered to changes in environmental variables that affect reaction rates, transport rates, and other factors that control the morphogen distribution, and the general question is how systems are buffered against variations in such factors. Said otherwise, how robust are developmental processes, and what does robustness even mean in this context? Other mathematical questions regarding morphogen patterning that arise are illustrated in Fig. 1. For instance, how do different organisms within a species preserve proportion even though they vary substantially in body size? How do cells respond to a 'noisy' morphogen signal caused by low levels of the morphogen, as in some of the patterning events discussed later? How do cells respond to a morphogen that is evolving in time during the course of development? Mathematical models and analysis can shed light on these issues and suggest mechanisms for mitigating the deleterious effects of some of these factors.

In the following section we discuss robustness in the developmental context in a general framework, and thereafter we discuss specific examples in the context of anterior–posterior patterning (Section III), parasegment patterning (Section IV), and of dorsal–ventral patterning (Section V), all in *Drosophila*.

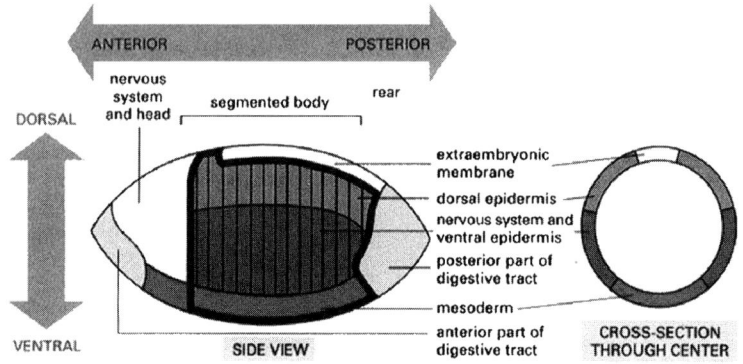

Figure 2 The definitions of the anterior–posterior (AP) and dorsal–ventral (DV) axes and the tissue types in *Drosophila*. (From Alberts *et al.*, 1994, with permission.) See color insert.

A. *Drosophila melanogaster* as a Model System

Drosophila melanogaster, which is the common fruit fly, has served as a model system to study many aspects of development for the past 100 years. *Drosophila* has a short life cycle, it is easily grown, the genome is sequenced, and many of the components of the signal transduction and gene control networks involved in patterning are known. However, less is known about how these networks produce the desired spatiotemporal pattern of gene expression. Development is a sequential process in which later stages build on earlier stages, but within stages there are often multiple feedback loops in signaling and gene control networks that may serve to buffer against perturbations caused by fluctuations in morphogen concentration and other components. Understanding the structural features in a network that ensure reliable patterning in the face of various perturbations is a major unresolved problem.

The *Drosophila* oocyte, or egg, is an approximately prolate ellipsoid that forms from a germline cell in the ovary (cf. Fig. 2). The egg is surrounded by a thin fluid-filled shell, the perivitelline (PV) space, that is bounded on the outside by the vitelline membrane. A coordinate system in the egg is first established by gradients of maternally-inherited cytoplasmic factors in the AP direction and by gradients of factors in the PV space in the dorsal–ventral (DV) direction. About fifty maternal genes set up the AP and DV axes, which provide the early positional information in the zygote. Zygotic gene expression is initiated by transcription factors produced from maternal RNA. Spatial patterning proceeds in stages: first the AP and DV axes are set up, then the whole domains are divided into broad regions, and finally smaller domains are established in which a unique set of zygotic genes is transcribed. The successive stages of patterning are initiated by a strictly-controlled hierarchy of gene expression in both the AP and DV directions. The former is shown in Fig. 3, which shows some of the interactions within and between levels. Both AP and DV patterning will be discussed in detail later, but first we discuss robustness in general terms.

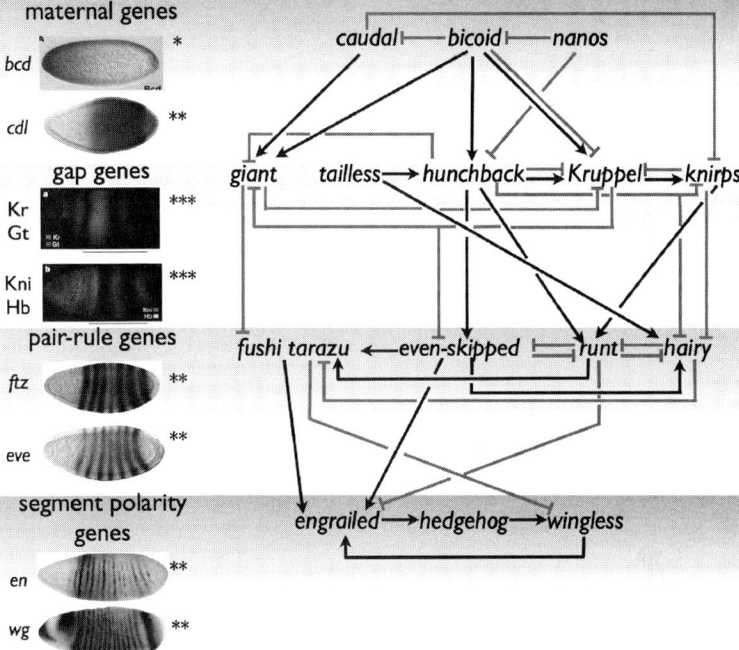

Figure 3 The expression patterns for some of the genes in the different classes (left column) and a representation of some of the feedback loops and cross interactions (right column). [Images on the left are taken from Crauk and Dostatni (2005) (*), from Fly-Base SDB-Online (**), and from Jaeger *et al.* (2004) (***).] See color insert.

II. Robustness in the Developmental Context

We define robustness as follows: a component, system or process is robust with respect to a given class of perturbations if its output or response is unchanged to within some tolerance by these perturbations, i.e., if the system is unlikely to 'fail' in the face of these perturbations. The perturbations can be in the inputs to the component, system or process, or they can be perturbations or alterations in the internal structure of the component or system. Explicit in this definition is that robustness can be defined at many different levels, that it can only be defined with respect to a specified class of perturbations of the inputs or internal structure, and that a system may be robust even if it contains nonrobust components. Conversely, a system comprised of robust components need not be robust itself. In the developmental context the system might be an entire developing embryo, or a portion thereof, the output could be the expression of a certain gene, the spatial pattern of tissue types, etc., and the perturbations might be knock-outs or overexpression of an upstream gene, the removal of a portion of the system, a mutation that interferes with a signal transduction pathway, variations in the amount of cellular constituents received at division time, and so on.

A. Structural Stability in Dynamical Systems

Since robustness is defined as insensitivity of the output, which is usually a prescribed function of the state of a system, to a defined class of perturbations, it should be possible to test for robustness if one has a model for the system. Thus consider a system whose state u evolves according to the finite system of ordinary differential equations of the form

$$\frac{du}{dt} = F\big(u, p, \Phi(t)\big), \quad u(0) = u_0. \tag{1}$$

Here u represents all the concentrations and other variables that define the state, p is a vector of constant parameters that appear in the model equations, and Φ represents time-dependent inputs. There are at least three classes of 'perturbations' we can consider, and these lead to three types of insensitivity or robustness.

1. Insensitivity with respect to inputs, by which we mean that *in the long run* the system ignores a certain class of inputs. This is a well-established notion in control theory and it is known that the system will ignore a given class of inputs if and only if it contains a subsystem that can generate the given class of inputs (Sontag, 2003). This is an example of a characteristic common to many sensory or signal transduction systems, and in that context the system is said to adapt to the specified inputs. In the developmental context this would imply that the system only responds to the transient changes in a signal and subtracts out the background signal.
2. Insensitivity with respect to changes in the model itself, which are reflected in changes to the function F. This is captured in the notion of coarseness or structural stability developed for dynamical systems: the system (1) is said to be structurally stable if its associated flow is orbit equivalent to the flow generated by any system in a sufficiently small neighborhood of it, using a suitable measure of closeness. In more prosaic terms this means that the family of solution curves in phase space 'look alike.' A global criterion for this kind of robustness, i.e., one valid for all values of the state u, is known for linear systems, but only local results are known in general. Said otherwise, one can test whether the output or state of the system is insensitive to small perturbations near a known solution, but in general not globally. These might include deletion of paths in a signal transduction network if the flux of signal or material flow along that pathway is sufficiently small.
3. Insensitivity with respect to changes in parameters. Strictly speaking this is a subset of 2, but this is the concept of robustness most frequently applied to signal transduction and gene control systems.

B. Sensitivity Analysis

There are well-established techniques for examining robustness or sensitivity of a mathematical model to changes in the parameters of the model. Suppose that the equations for the local dynamics, which could arise from a system of chemical reactions, are written as the system

$$\frac{du}{dt} = F(u, \Phi). \tag{2}$$

Here we suppose that there are no time-dependent inputs and simply lump p and Φ in (1) together. At a steady state $F(\mathbf{u}, \Phi) = 0$, and we shall assume that $\det(F_u) \neq 0$, where $\det(\cdot)$ denotes the determinant and the subscript denotes the partial derivative. This implies that locally there is a unique function $u^s : R^m \longrightarrow R^n$ such that $F(\mathbf{u}^s(\Phi), \Phi) = 0$. Differentiating the steady-state equation

$$F(u, \Phi) = 0 \tag{3}$$

with respect to Φ leads to

$$F_u u^s_\Phi + F_\Phi = 0,$$

where u^s_Φ is an $n \times m$ matrix when Φ has m components. By the above assumption F_u is invertible, so the steady-state sensitivity to Φ is

$$u^s_\Phi = -F_u^{-1} F_\Phi. \tag{4}$$

Obviously one can determine from this the sensitivity of any particular component.

Now consider a distributed problem described by a system of reaction–diffusion equations, and for simplicity consider only the steady-state problem. Let $\Omega \subset R^n$ be a given region of space with boundary $\partial \Omega$, and write the governing equations as

$$D \Delta u + F(u, \Phi(x)) = 0, \quad \text{in } \Omega,$$
$$-D \frac{\partial u}{\partial n} = B(u, \Phi_B), \quad \text{on } \partial \Omega, \tag{5}$$

where B incorporates the fluxes at the boundary. If we assume this has a unique solution u^s and differentiate the equation with respect to Φ as before, we obtain the following system.

$$D \Delta u^s_\Phi + F_u u^s_\Phi + F_\Phi = 0, \quad \text{in } \Omega,$$
$$-D \frac{\partial u^s}{\partial n} - B_u(u^s, \Phi_B) u^s_\Phi = B_\Phi(u^s, \Phi_B), \quad \text{on } \partial \Omega. \tag{6}$$

One could easily compute the solution of the variational problem (6) in parallel with the solution of (5). From this one could see how different components vary at different points in space as the input fluxes or kinetic parameters are varied. As a result, one could, for example, study the sensitivity of the spatial location of a particular concentration level to changes in the kinetic parameters.

C. The Turing Theory of Spatial Pattern Formation and Scale Invariance

An alternative to the localized-source, distributed-response paradigm built into the French flag model of patterning, which requires preexisting differentiation of those cells that serve as sources, is to suppose that pattern formation is a self-organizing process. The classical model based on this idea is due to Turing (1952). A Turing model involves two or more chemical species, which he called morphogens, that react together and diffuse throughout the system. In Turing's original analysis no cells were distinguished *a priori*; all could serve as sources or sinks of the morphogen. Moreover, Turing only considered periodic systems or closed surfaces, in which case no boundary conditions are needed. More generally, we call any system of reaction–diffusion equations for which the boundary conditions are of the same type for all species a Turing system.[1] Turing showed for a two-component system that under suitable conditions on the kinetic interactions and diffusion coefficients a spatially-homogeneous stationary state can, as a result of slow variation in parameters such as kinetic coefficients, become unstable with respect to small nonuniform disturbances. Such instabilities, which Turing called symmetry-breaking because the homogeneous locally-isotropic stationary state becomes unstable and therefore dynamically inaccessible, can lead to either a spatially nonuniform stationary state or to more complicated dynamical behavior. Such transitions from uniform stationary states to spatially- and/or temporally-ordered states might in turn lead, via an unspecified 'interpretation' mechanism, to spatially-ordered differentiation. For mathematical simplicity most analyses of Turing models deal with instabilities of uniform stationary states, since numerical analysis is generally required for more general reference states. However, Turing himself recognized the biological unreality of this in stating that 'most of an organism, most of the time is developing from one pattern to another, rather than from homogeneity into a pattern' (Turing, 1952). In a classic example of theory preceding experiment, it took until 1990 to verify Turing's theory in a system comprising chlorite, iodide and malonic acid (Castets *et al.*, 1990), and there is still no conclusive evidence of biological examples that can be explained by his theory. Theoretical models for patterning of hair follicles related to Turing's mechanism are well known (Claxton, 1964; Nagorcka and Mooney, 1982), but only recently have specific morphogens been suggested. Sick *et al.* (2006) suggest that Wnt and Dkk may function as morphogens in determining the hair follicle spacing in mice, and while this is potentially an important advance, much remains to be explained. For example, the underlying network governing expression of these species is certainly much more complex than the simple two-species model employed in the computations. Extension of Turing's analysis leads to the structural conditions that guarantee Turing instabilities for any number of components (Othmer, 1980), and it may turn out after a more detailed analysis that this is the first biological system in which patterning is self-organized.

---

[1] Of course if there are sources or sinks at the boundary, as in (6), then there is also preexisting differentiation of some cells.

While Turing patterns show a degree of insensitivity to parameter variations, particularly if the condition that all species satisfy the same boundary conditions is relaxed (Dillon et al., 1994), it is quite sensitive to changes in the size of the system, particularly for more complex patterns that correspond to higher modes (Othmer and Pate, 1980). Thus it is usually difficult to guarantee that a given spatial pattern will persist under significant changes in diffusion coefficients, the kinetic parameters, or the length scale of the system, because the underlying instability that leads to a spatial pattern carries with it an intrinsic length scale defined by the kinetic parameters and the diffusion coefficients. This can be understood by realizing that the rate of diffusion depends on the gradient of concentration, hence on the spatial scale, whereas the kinetic interactions in a homogeneous system do not.

In a number of contexts to be discussed later, patterning in developing embryos occurs reliably even when the size changes by a factor of 2–3, and neither the Turing model nor the French flag model will produce a correctly-proportioned pattern under such changes. In the following section we propose a model that explains recent experimental results on scale-invariance of the Bicoid distribution in *Drosophila*, and as background for that we describe a general method for producing scale-invariant distributions of morphogen distributions set up by reaction and diffusion. Consider again the region Ω and now suppose that reaction and diffusion can occur both in the interior of the region and on the boundary. Let u (resp., v) denote the concentration of species in the interior (resp., on the boundary), the latter measured in units of $M/L^2 T$, and write the governing equations as follows.

$$\frac{\partial u}{\partial t} = D\Delta u + \kappa R(u, x) \quad \text{in } \Omega, \tag{7}$$

$$-D\frac{\partial u}{\partial n} = B_0(u, x) - B_1(u, v, x) \quad \text{on } \partial\Omega, \tag{8}$$

$$\frac{\partial v}{\partial t} = D_b\Delta v + R_b(v, x) + B_1(u, v, x) \quad \text{on } \partial\Omega. \tag{9}$$

Here $R(u)$ denotes the reaction in the interior of the region, κ^{-1} is a characteristic time scale for the reactions, $B_0(u, x)$ denotes a specified input flux, $B_1(u, v, x)$ denotes reactions that occur on the boundary and involve u, $R_b(v)$ denotes other reactions on the boundary, and n denotes the inward normal. All reaction rates are allowed to depend on x, since these may vary throughout the system. For simplicity we have omitted parameters in the rate functions.

By scale invariance of the morphogen distribution we mean that the solution of (7)–(9) (or at least one component) is unchanged, both in amplitude and spatial variation, when the system is dilated, either positively or negatively, within a given range. This implies that when the equations are transformed to relative coordinates (i.e., a scaled version of the original coordinates), the solution is unchanged in amplitude and spatial variation under specified changes in the size measure. To understand what must occur to produce scale-invariant morphogen distributions, we reduce (7)–(9) to a form that corresponds to the simplest model for positional information. Consider the one-dimensional, steady-state problem on the interval [0, L], with a specified flux at $x = 0$, linear decay in the interval [0, L], and zero flux at $x = L$. Thus

$$D\frac{d^2u}{dx^2} = \kappa u, \quad x \in (0, L),$$
$$-D\frac{du}{dx} = j, \quad x = 0, \qquad (10)$$
$$\frac{du}{dx} = 0, \quad x = L.$$

In scaled coordinates this becomes

$$\frac{d^2u}{d\xi^2} = \lambda^2 u, \quad \xi \in (0, 1),$$
$$-\frac{du}{d\xi} = J, \quad \xi = 0, \qquad (11)$$
$$\frac{du}{d\xi} = 0, \quad \xi = 1,$$

where $\lambda^2 \equiv \kappa L^2/D$ and $J = jL/D$. The solution is

$$u(\xi) = \frac{J}{\lambda}\left[\frac{e^{\lambda(2-\xi)} + e^{\lambda\xi}}{e^{2\lambda} - 1}\right] \equiv \frac{J}{\lambda}\phi(\xi) = \frac{j}{\sqrt{kD}}\phi(\xi). \qquad (12)$$

If the flux j is fixed, the factor J/λ is independent of L, since both J and λ are proportional to L. However, the shape function $\phi(\xi)$ depends on L, and as a result, so does the amplitude. Thus neither the shape nor the amplitude of the morphogen distribution is scale invariant; the distribution decays proportionally more rapidly the larger L is.

The shape $\phi(\xi)$ of the distribution is invariant under changes in L if λ is independent of L, and this can be achieved if either $D \propto L^2$ or $\kappa \propto L^{-2}$, but then the amplitude is not invariant. One can understand the origin of the problem from (7)–(9), for there one sees that in converting these equations to scaled coordinates the diffusion terms scale as L^{-2}, whereas the boundary terms involving the normal derivative scale as L^{-1}. Thus it is difficult to achieve invariance, even if the diffusion or kinetic coefficients scale properly, without scaling the input flux j. If the system is closed this problem disappears, and a mechanism for producing the required dependence of D or κ on L in the context of a generalized Turing problem is described elsewhere (Othmer and Pate, 1980).

Suppose that the morphogens are produced by all cells and that the boundary is closed, and there are no reactions on the boundary. Suppose also that all cells produce a control species at a constant rate R_0, that the concentration of this species is zero at the boundary, and that this control species modulates the diffusion and/or kinetic

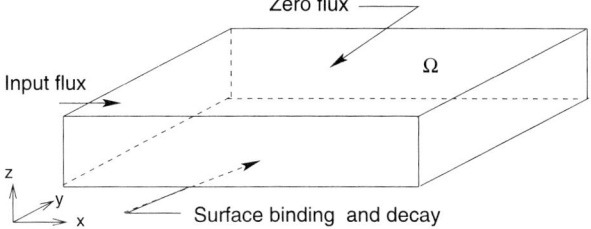

Figure 4 The geometry of a thin fluid layer over receptors embedded in a surface. See color insert.

coefficients. This leads to the system of equations

$$\frac{\partial c}{\partial t} = \nabla \cdot (D(C)\nabla c) + \kappa(C)R(c) \quad \text{in } \Omega,$$

$$\frac{\partial C}{\partial t} = D_c \nabla^2 C + R_0 \quad \text{in } \Omega,$$

$$D(C)\frac{\partial c}{\partial n} = 0 \quad \text{on } \partial\Omega,$$

$$C = 0 \quad \text{on } \partial\Omega,$$

wherein $c = (c_1, c_2, \ldots, c_n)$ is the vector of concentrations for the pattern-inducing species, C is the concentration of the control species that modulates either the diffusion coefficients or the kinetic coefficients, and $D(C)$ and $\kappa(C)$ reflect the dependence on C. It is easy to see that the steady state concentration of C scales as L^2, and thus perfect scale-invariance is possible if $D_i(\cdot) \propto C$ or $\kappa(\cdot) \propto C^{-1}$. In effect the control species can change the underlying metric by changing the diffusion coefficient, it can change the time scale by modulating the reaction rates, or it can do both. By relaxing the assumption that the control species is zero on the boundary, one can incorporate an influx of the control species and still achieve very weak dependence of the morphogen distributions on L. One can thereby achieve a controllable level of robustness with respect to changes in the size of the system (Othmer and Pate, 1980).

As a second example that incorporates receptor dynamics on one surface, suppose we specialize the general problem described by (7)–(9) to the geometry shown in Fig. 4. Assume that the surface reactions involve only binding to a receptor and decay of the receptor–ligand complex, and to simplify the analysis, suppose that whenever a receptor–ligand complex is internalized it is replaced by a bare receptor (Umulis et al., 2006). The surface $z = 0$ can be thought of as the receptor-filled outer mebrane that divides intra- and extracellular environments, as in *Drosophila* dorsal surface patterning, or as the cytoplasmic side of the inner membrane in which the nuclei are embedded in the *Drosophila* blastoderm. Let L_x, L_y, and L_z be the lengths in the x, y, and z directions, respectively, and let C be the concentration of a morphogen in the fluid and R the concentration of receptor on the surface $z = 0$. Suppose there is a fixed influx of C on the boundary $x = 0$, and zero flux on the remaining faces except $z = 0$;

then the governing equations can be written as follows.

$$\frac{\partial C}{\partial t} = D\Delta C \quad \text{in } \Omega, \tag{13}$$

$$\frac{\partial R}{\partial t} = -k_{\text{on}} RC + (k_{\text{off}} + k_e)\overline{RC} \quad \text{on } z = 0, \tag{14}$$

$$\frac{\partial \overline{RC}}{\partial t} = k_{\text{on}} RC - (k_{\text{off}} + k_e)\overline{RC} \quad \text{on } z = 0, \tag{15}$$

$$-D\frac{\partial C}{\partial z} = -k_{\text{on}} RC + k_{\text{off}}\overline{RC} \quad \text{on } z = 0, \tag{16}$$

$$-D\frac{\partial C}{\partial x} = j \quad \text{on } x = 0. \tag{17}$$

Equation (14) reflects the assumption that the steps by which an occupied receptor is internalized and a free receptor is recycled to the surface reach a steady state rapidly compared with other processes. As a result, the total receptor density is constant at every point on the surface $z = 0$, i.e., $R + \overline{RC} = R_T$ where R_T is a constant.

To simplify this system we define the dimensionless variables $u = C/C_0$, $v = R/R_T$, $w = \overline{RC}/R_T$, the scaled coordinates $\xi = x/L_x$, $\eta = y/L_y$ and $\zeta = z/L_z$, and the dimensionless time $\tau = t/T$. The system then becomes

$$\frac{\partial u}{\partial \tau} = \frac{DT}{L_x^2}\left(\frac{\partial^2 u}{\partial \xi^2} + \frac{L_x^2}{L_y^2}\frac{\partial^2 u}{\partial \eta^2} + \frac{L_x^2}{L_z^2}\frac{\partial^2 u}{\partial \zeta^2}\right) \quad \text{in } \Omega, \tag{18}$$

$$\frac{\partial v}{\partial \tau} = -Tk_{\text{on}} C_0 R_T uv + T(k_{\text{off}} + k_e)w \quad \text{at } \zeta = 0, \tag{19}$$

$$\frac{\partial w}{\partial \tau} = Tk_{\text{on}} R_T C_0 uv - T(k_{\text{off}} + k_e)w \quad \text{at } \zeta = 0, \tag{20}$$

$$-\left(\frac{DC_0}{L_z}\right)\frac{\partial u}{\partial \zeta} = -k_{\text{on}} R_T C_0 uv + k_{\text{off}} w \quad \text{at } \zeta = 0, \tag{21}$$

$$-\left(\frac{DC_0}{L_x}\right)\frac{\partial u}{\partial \xi} = j \quad \text{at } \xi = 0. \tag{22}$$

If we assume that $L_z \ll L_x, L_y$ the first equation can be averaged over ζ. In view of the boundary conditions, the solution must be constant in the η direction at steady state and the equations reduce to

$$\frac{d^2 u}{d\xi^2} = \gamma^2 \frac{u}{K + u} \quad \text{in } \Omega, \tag{23}$$

$$-\frac{du}{d\xi} = J \quad \text{at } \xi = 0, \tag{24}$$

where u now stands for the average over ζ, and

$$K = \frac{k_{\text{off}} + k_e}{k_{\text{on}} C_0}, \qquad \gamma^2 = \frac{k_e R_T L_x^2}{DC_0 L_z}, \qquad J = \frac{jL_x}{DC_0}.$$

If $u \ll K$ then this reduces to

$$\frac{d^2 u}{d\xi^2} = \lambda^2 u \quad \text{in } \Omega, \tag{25}$$

$$-\frac{du}{d\xi} = J \quad \text{at } \xi = 0, \tag{26}$$

where

$$\lambda^2 = \frac{k_e k_{\text{on}}}{k_{\text{off}} + k_e} \frac{R_T L_x^2}{DL_z} \equiv k_s \frac{R_T L_x^2}{DL_z}.$$

This system is identical to (11), given the assumed zero-flux boundary condition at $\xi = 1$, and therefore the dimensionless solution u is given by (12). Again we can ask under what conditions the shape and amplitude are scale-invariant, and now the answer may be different. Suppose that the total number of receptors is fixed, even though the lengths L_x and L_y may change. It is clear that the total number N_T is

$$N_T = L_x L_y R_T$$

under the assumption that the density is uniform, and therefore

$$\lambda^2 = \frac{k_s N_T L_x}{DL_y L_z}. \tag{27}$$

Since k_s and D are constants, it is clear that λ will be independent of the length scale if (i) N_T is constant, (ii) L_x and L_y scale by the same amount, and (iii) L_z is constant. Thus a uniform dilation in the x and y directions, no dilation in the z direction, and a constant total number of receptors produces a scale-invariant shape function. This is conceptually equivalent to embedding the receptors in a rubber sheet and stretching or shrinking the sheet uniformly in x and y without changing the thickness of the fluid layer. Given that λ is constant, it follows that the amplitude will be constant under size changes if J is constant, which holds if the total input flux over the surface $x = 0$ is constant under uniform dilations in x and y, since in that case j scales as L_y^{-1}.

III. Scaling of AP Patterning in *Drosophila*

A. Anterior–Posterior Patterning

Here and in later sections we describe only the primary genes and the effects of their products; a more complete description of the patterning process can be found in Gilbert (2006). When referring to genes and their products we italicize the former and capitalize the latter.

Three groups of maternal gene products, anterior, posterior and terminal, are involved in the first stage of AP patterning, in which anterior is distinguished from posterior. Four messenger RNAs (*bicoid*, *hunchback*, *nanos*, and *caudal*) are critical in this stage. These encode transcriptional and translational regulatory proteins that activate or repress the expression of certain zygotic genes. The maternally-inherited *bicoid* mRNA is localized at the anterior end of the unfertilized egg and is translated after fertilization. The Bicoid protein diffuses throughout the syncytial blastoderm and establishes a concentration gradient along the AP axis.

The posterior pattern is primarily controlled by the levels of Nanos and Caudal. *nanos* mRNA is localized at the posterior pole of the egg and Nanos suppresses translation of *hunchback* in the posterior. *hunchback* mRNA is both maternally-inherited and transcribed in the zygote, the former distributed uniformly throughout the egg. Zygotic transcription of *hunchback* is activated by high Bicoid levels and inhibited by Nanos and Pumilio. *caudal* mRNA is distributed uniformly, but a posterior-to-anterior gradient of Caudal is established by inhibition of Caudal synthesis by Bicoid.

Thus in the initial stage of AP patterning, Bicoid and maternal Hunchback activate *hunchback* expression throughout the anterior half of the embryo. By midcycle 14, this primary expression pattern of *hunchback* is replaced by a secondary pattern consisting of a variable anterior domain, a stripe at the position of parasegment 4 (described later), and a posterior cap (Margolis *et al.*, 1995). In the next stage, which is initiated by Bicoid and Hunchback, the AP axis is divided into broad regions by the expression of the gap genes *giant*, *kruppel*, and *knirps*. *kruppel* is activated by a combination of Bicoid and low levels of Hunchback, but is repressed by high levels of Hunchback. This localizes *kruppel* expression to the center of the embryo, in parasegments 4–6. *knirps* is repressed by high levels of Hunchback, and therefore is expressed in parasegments 7–12. The details of how the boundaries of expression domains are determined are complex, but by an elaboration of further localization of expression the initial gradients of morphogens lead to the establishment of regions within the syncytial blastoderm that foreshadow the onset of segmentation, which follows formation of parasegments.

Expression of pair-rule genes defines the boundaries of the parasegments. These genes are expressed in 7 transverse stripes corresponding to every second parasegment: *even-skipped* in odd parasegments and *fushi-tarazu* in even parasegments. Their expression begins just before cellularization (midcycle 14) and after cellularization, each pair-rule gene is restricted to a few cells in seven stripes. Again, the details, which can be found in Gilbert (2006), are more complex, since each stripe is controlled by combinations of transcription factors. Segments are made from the posterior part of one parasegment and the anterior of the next. The last step for our purposes is expression of the segment polarity genes, which are discussed later. These pattern the 14 parasegments and stabilize parasegment and segment boundaries, thus defining the segmented structure of the abdomen.

B. Intra- and Interspecies Variations in the Bicoid Distribution

As we saw in Section II.C, a preset morphogen threshold in the French flag model described by (10) does not scale properly under changes in length, and unless there are downstream mechanisms for adjusting the response, there will be variation in the proportioning of an embryo into the different cell types. Since Bicoid is a primary determinant of AP patterning, it is important to determine how much variation there is in the spatial location of a threshold in Bicoid for activation of *hunchback*. It has been shown experimentally that there is significant variation in this threshold from embryo-to-embryo at early cycle 14, both in *Drosophila* and other closely-related dipteran species (Gregor *et al.*, 2005). However, the shape and amplitude of the Bicoid distribution is essentially independent of the embryo size over a four-fold change in length when comparing different species (*ibid*). It was found that Hunchback expression is less sensitive to embryo size in *Drosophila*, which suggested that there may be other mechanisms that cooperate or compete with Bicoid to control target gene expression (Aegerter-Wilmsen *et al.*, 2005; Howard and ten Wolde, 2005; Houchmandzadeh *et al.*, 2005). Other data suggests that at midcycle 14 the variation in the Bicoid threshold location in *Drosophila* is no greater than that in the boundary of *hunchback* expression, and thus Bicoid can control target gene expression precisely without additional factors or control schemes (Crauk and Dostatni, 2005). The high embryo-to-embryo variability of the Bicoid distribution observed in early cycle 14 may originate from the antibody staining methods (Houchmandzadeh *et al.*, 2002; Crauk and Dostatni, 2005), or the distribution may sharpen in midcycle 14 by some unknown mechanism.

The analysis of Section II.C shows that the interspecies scaling of the Bicoid distribution could be achieved by scaling of the diffusion coefficients, scaling of the reaction rates, or a combination of the two. The experimental results in Gregor *et al.* (2005) show that the characteristic decay length $\sqrt{D/\kappa}$ in the notation of (11) scales as the length of the AP axis, which implies that $\lambda = \sqrt{\kappa l^2/D}$ in (10) is constant across species. These authors also show that the diffusion of similarly-sized molecules between different embryos does not change appreciably, and therefore the scale-invariance of the Bicoid distribution must be achieved by scaling the pseudo-first-order decay κ of Bicoid (Gregor *et al.*, 2005). As we showed in Section II.C, this can be achieved if (i) the total number of binding sites is constant, (ii) the integrated input flux remains constant under scaling, and (iii) the morphogen is confined to a thin layer and the lateral dimensions of this layer each scale as L. The first of these conditions has been verified experimentally: the number of nuclei at cell cycle 14 is given by $\log_2 N_{\text{nuc}} = 12.8 \pm 0.2$ nuclei for the species studied (Gregor *et al.*, 2005). Condition (ii) will be met if the net production of Bicoid in an egg is constant across species, which can be achieved if either the amount of maternal RNA deposited in an egg is constant across species or the transcription rate is adjusted appropriately. The last condition has not been tested experimentally, but support for it stems from the fact that in embryos stained for Bicoid, the protein accumulates near the nuclei at the periphery and there appears to be less in the bulk cytoplasm (Gregor *et al.*, 2005; Houchmandzadeh *et al.*, 2002).

The proposed mechanism based on the simplified configuration in Fig. 4, in which the receptors are uniformly distributed, can account for perfect invariance. In the following section we show that in applying this to *Drosophila*, the discreteness of the nuclei does not affect this conclusion, and we show that some of the embryo-to-embryo variation in the location of a threshold may be accounted for by diffusion in the bulk cytoplasm.

C. The Effects of Discrete Nuclei and Diffusion in the Bulk Cytoplasm

To investigate the applicability of the analytical solution to nonuniform uptake and release of transcription factors, we developed a computational model in which nuclei are arranged in a regular two-dimensional hexagonal array and the transcription factors are free to diffuse unhindered over the field of nuclei. Nuclei are round and their size remains constant under changes in the overall system lengths. The x- and y-directions can be scaled independently of each other and the nuclei spacing is determined by the centers of the dilated hexagonal packing. Fig. 5 shows a small section of the nuclei distribution for total lengths that vary 3-fold from the base case L_x to $3XL_x$. As in the analytical system, the steady-state distribution of morphogens in the nonuniform field exhibit perfect scale invariance as long as the ratio L_y/L_x remains constant. In the example here, this is demonstrated with a system that has regular hexagonal packing, but the details of the grid are not relevant: the solution remains scale invariant as long as L_y/L_x is constant. As expected, perfect scale invariance is not achieved when L_y/L_x is not constant. For instance, if only L_x is scaled while L_y remains fixed, the amplitude scales as $\sqrt{L_x}$ while the dimensionless decay length scales as $1/\sqrt{L_x}$.

Fig. 6 shows the effect of incorporating diffusion in the bulk cytoplasm as well as in the thin layer at the periphery in a full 3D model of the embryo. One sees there that the profiles no longer scale perfectly, as they would without diffusion in the bulk, which suggests that one may be able to explain the embryo-to-embryo variability within species by effects such as this. A more detailed analysis of this and other aspects of scaling will be reported elsewhere (Umulis *et al.*, 2007).

IV. Models of the Segment Polarity Network

As described earlier, the genes involved in spatial patterning are expressed in a temporal sequence that leads to successively more refined spatial patterns of expression. The gap and pair-rule genes are only expressed transiently, but the segment polarity genes, which pattern the parasegments, are expressed throughout the life of the fly. A parasegment comprises four cells, and the gene expression pattern is repeated in each of the fourteen parasegments. The network of gene and protein interactions at this stage are shown in Fig. 7. The stable expression pattern of *wingless* and *engrailed*, defines and maintains the borders between different parasegments (Wolpert *et al.*, 2002). The segment polarity genes are first expressed after cellularization, and intercellular communication is an essential component at this stage.

2. Robustness of Embryonic Spatial Patterning in *Drosophila melanogaster*

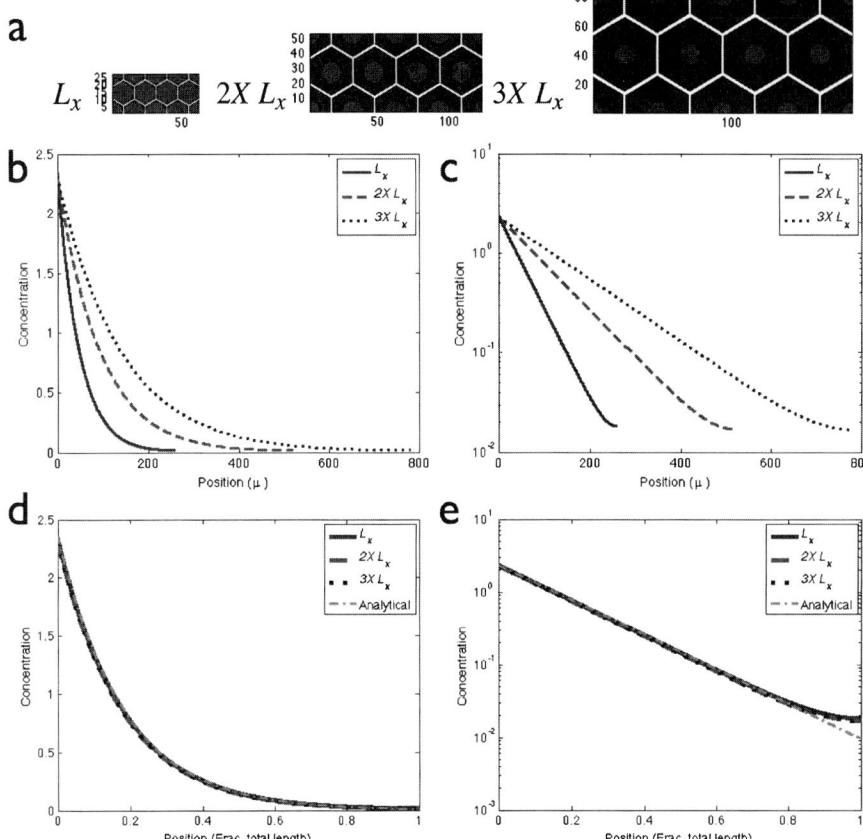

Figure 5 (a) Section of nonuniform distributions of nuclei shows spacing for base case (left) L_x, L_y; and rescaled cases (middle) $2 \times L_x$, $2 \times L_y$; and (right) $3 \times L_x$, $3 \times L_y$. (b, c) shows the scaling of the decay length and amplitude for L_x (solid), $2XL_x$ (dashed), and $3XL_x$ (dotted). (d, e) same as (b, c) mapped to the interval [0, 1] and plotted logarithmically. The approximate analytical solution gotten by ignoring the growing exponential term in (12) is also shown (red dot–dash). The superimposed lines for L_x, $2 \times L_x$, and $3 \times L_x$ in (d) and (e) demonstrate scale invariance of the nonuniform system. The nuclei were taken to be 2 μm in diameter, the diffusion coefficient was 20 μ²/s and $k_{on} R_{tot} = 5.6 \times 10^{-3}$. See color insert.

The major components involved are the transcription factor engrailed (EN), the cytosolic protein Cubitus Interruptus (CI), the secreted proteins Wingless (WG) and Hedgehog (HH), and the transmembrane proteins Patched (PTC) and Smoothened (SMO), the latter two of which are the receptor and an auxiliary protein involved in transduction of the HH signal. The pair-rule gene *sloppy paired* (*slp*) is activated earlier and expressed constitutively thereafter (Grossniklaus *et al.*, 1992), and therefore serves as a constant input. A detailed description of the network and the biological

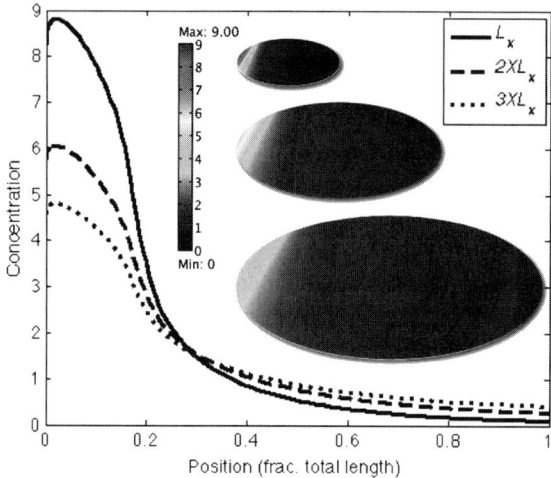

Figure 6 The effect of diffusion throughout the cytoplasm on the scaling of Bicoid distributions. See color insert.

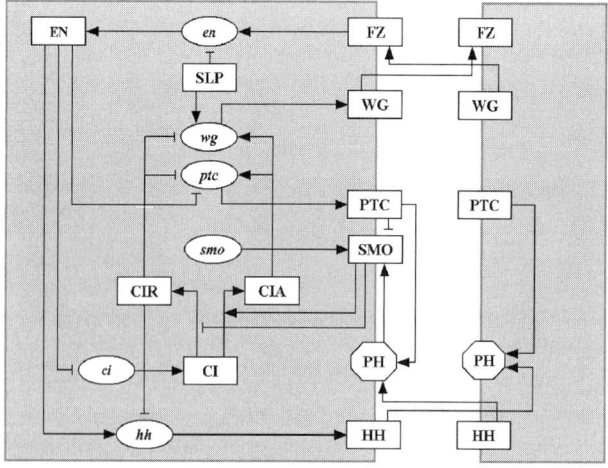

Figure 7 The network of interactions between the segment polarity genes. The shape of the nodes indicates whether the corresponding substances are mRNAs (ellipses), proteins (rectangles) or protein complexes (octagons). The edges of the network signify either biochemical reactions (e.g., translation) or regulatory interactions (e.g., transcriptional activation). The edges are distinguished by their signatures, i.e., whether they are activating or inhibiting. Terminating arrows (→) indicate translation, post-translational modifications (in the case of CI), transcriptional activation or the promotion of a post-translational modification reaction (e.g., SMO determining the activation of CI). Terminating segments (⊣) indicate transcriptional inhibition or in the case of SMO, the inhibition of the post-translational modification reaction CI → CIR. From Albert and Othmer (2003) with permission. See color insert.

2. Robustness of Embryonic Spatial Patterning in *Drosophila melanogaster*

justification for the activation and inhibition steps depicted in Fig. 7 is given in Albert and Othmer (2003).

The essential function of this network can be understood as follows. Label the full cell shown in Fig. 7 as i and the cell (shown partially) to the right of it as $i + 1$. If $i + 1$ expresses wingless protein WG, this upregulates EN production in i via the *frizzled* receptor and the intermediate protein Disheveled (not shown), and this in turn upregulates HH production in i and downregulates CI and ptc in i. Conversely, if i expresses HH, this upregulates CIA and WG production in $i + 1$, and thus there is a self-reinforcing feedback loop that encompasses both cells. The qualitative description suggests (a) that one should not see simultaneous expression of wg and either en or hh in the same cell, and that, (b) since all cells have the same network, the initial conditions will determine which cells express wg and which express en, hh. The results shown later will confirm these expectations.

The first detailed model for the segment polarity gene network was analyzed by von Dassow et al. (2000) and von Dassow and Odell (2002), who developed a continuous-state model of the core network of five genes (en, wg, ptc, ci, and hh) and their proteins. This is described in the following section. In this model and the Boolean model described later no account is taken of concentration variations within cells, but cell–cell interactions are included.

A. The Continuous-State Analysis of the Segment Polarity Gene Network

The model described in von Dassow *et al.* (2000) leads to differential equations of the following general form for mRNAs and proteins

$$\frac{d[hh]_i}{dt} = T_{\max} \rho_{hh} \left[\frac{[EN]^\alpha}{K_{EHhh}^\alpha + [EN]^\alpha} \right] - \frac{[hh]_i}{H_{hh}},$$

$$\frac{d[HH]_{i,j}}{dt} = \frac{P_{\max}}{6} \sigma([hh]_i) - \frac{[HH]_{i,j}}{H_{HH}} - k_{PTCHH}[HH]_{i,j}[PTC]_{n,j}.$$

Here $[hh]_i$ denotes the hedgehog mRNA in the ith cell and $[HH]_{i,j}$ denotes the corresponding protein between the ith and jth cells. When a single transcription factor controls gene expression, the transcription rate is given by a Hill function, but when multiple factors are involved the transcription rate is either a rational function of such functions, or a composition of them (cf. the supplemental material to von Dassow *et al.*, 2000). These assumptions embody enormous simplifications of the underlying biochemical events, and a more realistic description would involve the probabilities of the various configurations of the promoter in the presence of activators and inhibitors. The transcription rate at any time could be set to a linear combination of these configurations, each weighted according to the rate of transcription in that configuration.

Suitable modifications of the initial network topology led to correct patterns of wild-type gene expression under wide variations in the kinetic constants in the rate laws, provided that the cooperativity in the Hill functions was sufficiently large. This insensitivity to parameters suggests that the essential features involved are the topology of the segment polarity network and the signatures of the interactions in the network (i.e., whether they are activating or inhibiting). In the next section we describe a Boolean model developed in Albert and Othmer (2003) that substantiates this prediction.

B. A Boolean Model for Control of the Segment Polarity Genes

The Boolean description of the segment polarity network is a discrete-time, discrete state model in which the state of each mRNA and protein is either 1 (ON) or 0 (OFF), and time is discretized into steps approximately equal to the duration of a transcription or translation event. Each mRNA or protein is represented by a node of a network, and the interactions between them are encoded as directed edges (see Fig. 7). The state of each node is 1 or 0, according as the corresponding substance is expressed or not, and the next state of node i is determined by a Boolean (logical) function $\mathcal{F}_i$ of its state and the state of those nodes that have edges incident on it. The regulatory influences in the network are encoded in the connections shown in Fig. 7 and their signatures, by which we mean whether an edge corresponds to activation or inhibition of the terminal node. The Boolean interaction functions are then constructed from these interactions: the rules governing the transcription of an mRNA, for example, are determined by a Boolean function of the states of its transcriptional activators and inhibitors.

The Boolean rules used in the model are given in Table I; a detailed justification for them is given in Albert and Othmer (2003). The key assumptions underlying the updating rules are (i) the effect of transcriptional activators and inhibitors is never additive, but rather, inhibitors are dominant; (ii) the dependence of transcription and translation is an ON/OFF function of the state; (iii) mRNAs decay in one step if not transcribed; and (iv) transcription factors and proteins undergoing post-translational modification decay in one step if their transcript is not present. The rules give the expression of a node at time $t + 1$ as a function of the expression of its effector nodes at time t, except that we assume the expression of SLP does not change, and that the activation of SMO and the binding of PTC to HH are instantaneous.

If x_{ij}^t is the state of the ith node ($i = 1, 15$) in the jth cell at time t, and $x = (x_{11}, \ldots, x_{nN})$, then the next state of the network is $x^{t+1} = \mathcal{F}(x^t)$, which defines a discrete dynamical system whose iteration determines the evolution of the state of all nodes. This discrete dynamical system is much easier to analyze than the differential equations; one simply prescribes an initial state for each node, and the state at the next time step is determined by the Boolean function $\mathcal{F}_i$ for that node. A fixed point of $\mathcal{F}$ is a time-invariant state of the system, whereas a fixed point of $\mathcal{F}^p$ for some $p > 1$ represents a state that repeats periodically with least period p. The test for whether or not the model correctly predicts the observed spatial patterns is whether, starting from wild-type initial conditions, the state evolves to a fixed point that corresponds to this pattern. Since the objective is to describe the effect of the segment polarity genes in maintaining the parasegment border, the patterns of segment polarity genes formed

2. Robustness of Embryonic Spatial Patterning in *Drosophila melanogaster*

Table I The Boolean rules used in the model

Variable	Logical Functions Used for Updating
SLP_i	$SLP_i^{t+1} = SLP_i^t = \begin{cases} 0 & \text{if } i \bmod 4 = 1 \text{ or } i \bmod 4 = 2 \\ 1 & \text{if } i \bmod 4 = 3 \text{ or } i \bmod 4 = 0 \end{cases}$
wg_i	$wg_i^{t+1} = (CIA_i^t \text{ and } SLP_i^t \text{ and not } CIR_i^t)$ or $[wg_i^t \text{ and } (CIA_i^t \text{ or } SLP_i^t) \text{ and not } CIR_i^t]$
WG_i	$WG_i^{t+1} = wg_i^t$
en_i	$en_i^{t+1} = (WG_{i-1}^t \text{ or } WG_{i+1}^t) \text{ and not } SLP_i^t$
EN_i	$EN_i^{t+1} = en_i^t$
hh_i	$hh_i^{t+1} = EN_i^t \text{ and not } CIR_i^t$
HH_i	$HH_i^{t+1} = hh_i^t$
ptc_i	$ptc_i^t = CIA_i^t \text{ and not } EN_i^t \text{ and not } CIR_i^t$
PTC_i	$PTC_i^{t+1} = ptc_i^t \text{ or } (PTC_i^t \text{ and not } HH_{i-1}^t \text{ and not } HH_{i+1}^t)$
PH_i	$PH_i^t = PTC_i^t \text{ and } (HH_{i-1}^t \text{ or } HH_{i+1}^t)$
SMO_i	$SMO_i^t = \text{not } PTC_i^t \text{ or } HH_{i-1}^t \text{ or } HH_{i+1}^t$
ci_i	$ci_i^{t+1} = \text{not } EN_i^t$
CI_i	$CI_i^{t+1} = ci_i^t$
CIA_i	$CIA_i^{t+1} = CI_i^t \text{ and } SMO_i^t \text{ or } hh_{i-1}^t \text{ or } hh_{i+1}^t$
CIR_i	$CIR_i^{t+1} = CI_i^t \text{ and not } SMO_i^t \text{ and not } hh_{i\pm1}^t$

before stage 8 are considered as initial states for the network, and the final stable state should reflect the wild type patterns observed in stage 11.

The initial state, based on the experimental observations of stage 8 embryos, includes a two-cell-wide SLP stripe in the posterior half of the parasegment, a single-cell-wide *wg* stripe in the most posterior part of the parasegment, single-cell-wide *en* and *hh* stripes in the most anterior part of the parasegment, and *ci* and *ptc* expressed in the posterior three-fourths of the parasegment [see Albert and Othmer (2003) and references therein]. The one-dimensional representation of the mRNA and protein patterns for this initial condition is shown in Fig. 8a. Beginning with this initial distribution, one finds that after 6 time steps the system reaches the stable fixed point represented by the pattern shown in Fig. 8b). This pattern agrees with the experimental observations of the wild type expression of the segment polarity genes after stage 11. Note that *wg* and WG are expressed in the most posterior cell of each parasegment, while *en*, EN, *hh* and HH are expressed in the most anterior cell of each parasegment, as is observed experimentally (Tabata *et al.*, 1992; Ingham *et al.*, 1991).

One can also analyze the patterns that result from initial conditions other than wild-type, and for various mutants. Analysis of the steady states and their domains of attraction shows that the minimal prepatterning that leads to wild type stable expression is as follows:

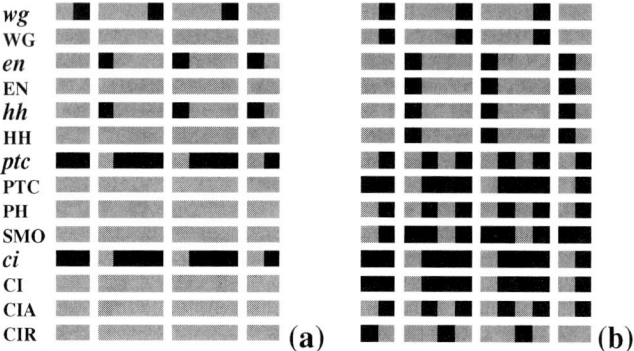

Figure 8 Wild-type expression patterns of the segment polarity genes. (a) The initial state before stage 8 based on experimental observations. (b) The wild type stable state of the model if initialized with the pattern in (a). From Albert and Othmer (2003) with permission. See color insert.

- wg is wild type,
- en and hh are not expressed,
- ptc is expressed in the third cell of the parasegment,
- ci and the proteins are not expressed.

Broader initial expression of any node except PH, SMO, ci and CIR leads to a pattern with broad stripes in about 90% of the total number of initial states.

It is shown in Albert and Othmer (2003) that the family of possible stable steady states includes three well-known states corresponding to the wild type, broadened and nonsegmented patterns, and three states with very limited basins of attraction. Nevertheless, the existence of the latter steady states suggests the adaptability of the network to produce patterns other than those required in *Drosophila* embryogenesis. Additionally, if we relax the assumption of identical parasegments, the steady state patterns corresponding to individual parasegments can be combined to form diverse patternings of the whole ectoderm. For example, the wild type steady state and its double wg variant can be seamlessly integrated.

C. The Segment Polarity Network is a Simple Switching Network

The robustness observed in the continuous-state description suggested that the essential feature of the segment polarity network is the topology and signature of the interactions, not the details of the rate laws for the kinetic steps, and the Boolean model confirms this in most details. It reproduces the wild type expression pattern of these genes, as well as the ectopic expression patterns observed in various mutants. Furthermore, the Boolean representation enables one to do a complete analysis of the possible steady states, and allows for a more precise identification of the basin of attraction of each steady state. While both models reproduce the observed steady states of gene expression, the temporal evolution may not reflect the *in vivo* evolution of expression patterning for either model, since there is no data to compare the predictions with: both models can attain the steady state in the time available for patterning.

Various extensions of each model are possible. Several that can be tested in the continuous-state description are (i) the suitability of the simplified description of gene expression, and (ii) the effect of time delays in various steps. Recent analysis of the original ODE system shows that a number of components must function in an ON–OFF manner to achieve robustness (Ingolia, 2004), which provides indirect validation of the Boolean model. In the Boolean model it is assumed that the expression of proteins decays in one time step if their mRNAs are switched off, and this assumption induces on–off flickering and rearrangements in the expression pattern that slowly stabilize. An extension of the model that assumes that protein expression decays in two timesteps after the disappearance of the mRNA has been analyzed, and this two-step model leads to the same steady states as the original model, and all the conclusions regarding the initial state and gene mutations are preserved. The only change is in the transient dynamics, and generally the number of steps leading to a striped steady state decreases in the two-step model (cf. Albert and Othmer, 2003). One can also introduce a stochastic aspect into the Boolean functions by assigning probabilities that given Boolean functions are realized. For example, if one relaxes the rigid 'inhibition dominates activation' rule, and chooses this output probabilistically, then the network functions correctly if the output realizes the deterministic rule at least 75–80% of the time.

The success of the Boolean model of the segment polarity network suggests that the network has evolved to function as a simple switching network wherein genes (nodes) are either on or off; no graded response is needed and thus the details of the kinetic steps are not so important for understanding the steady state behavior. This may be true of many other gene control networks, such as the delta-notch signaling network. In any case, a Boolean description of a network facilitates the integration of qualitative observations on gene interactions into a coherent picture, and provides an easy verification of the sufficiency of these interactions (i.e., whether some steps are missing or whether some postulated interactions are incorrect).

V. Dorsal–Ventral Patterning in *Drosophila*

Establishment of a coordinate frame for DV patterning in the oocyte begins when the nucleus migrates from a central posterior position to a dorsal, anterior position. The nucleus expresses *gurken*, which establishes a cortical region of high *gurken* mRNA and protein levels. Gurken is a secreted growth factor that binds to receptors on the follicle cells surrounding the egg and results in a DV gradient of EGF receptor (Egfr) activation and up- or downregulation of target genes in the EGF pathway (Reeves *et al.*, 2006). At least three genes expressed in follicle cells are involved: *pipe*, *nudel*, and *windbeutel*. Expression of *pipe* is repressed by Gurken, with the result that it is only

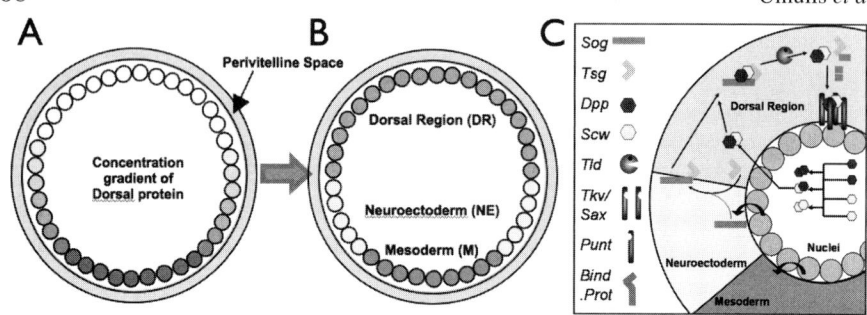

Figure 9 A schematic of a cross-section of an embryo, showing the distribution of Dorsal, which sets the DV polarity (A), the major tissue types in a DV section (B), and the profiles of the major determinants of dorsal surface patterning and the interactions between these components. See color insert.

expressed in the ventral 40% of the follicle cells. *pipe* encodes a protein similar to heparan sulfate 2-*O*-sulfotransferase that may modify ECM components to produce sites for the initiation of reactions that produce a ventralizing signal (Morisato and Anderson, 1994; Sen *et al.*, 1998, 2000). Four maternal effect genes whose products are secreted into the PV space (*gastrulation defective*, *snake*, *easter*, and *spatzle*) are involved from the germline. The first three are serine proteases that act in a cascade whose end product is the activated form of the diffusible ligand *Spätzle*, which is secreted in an inactive form and cleaved in the ventral region. The active C-terminal fragment, denoted Spa*, binds to the receptor Toll, which is uniformly expressed at the surface of the embryo (Hashimoto *et al.*, 1991). Toll in turn activates an intracellular cascade that stimulates the nuclear uptake of the transcription factor Dorsal (Dl). The graded distribution in the PV space of the diffusible ligand *Spätzle* results in graded activation of its receptor and graded nuclear uptake of the transcription factor Dorsal, which, combined with threshold responses for gene activation, subdivides the DV cross-section into three major regions: the ventral-most prospective mesoderm, the intermediate lateral ectoderm and neuroectoderm, and the dorsal ectoderm, where the amnioserosa is determined in the dorsal-most subregion (Morisato and Anderson, 1994). Absence of Dorsal allows expression of *tolloid* (*tld*), and *decapentaplegic* (*dpp*) genes in the lateral dorsal region, while *zerknüllt* (*zen*) and its protein are expressed in the dorsal-most region that defines the amnioserosa (Wolpert *et al.*, 2002; Levine and Davidson, 2005) (cf. Fig. 9).

Maternal factors determine the position of lateral ectoderm relative to dorsal ectoderm, but zygotic genes *sog* and *dpp*, which are expressed uniformly in the former and latter domains, *respectively*, are required to maintain that initial subdivision and to define the amnioserosa. The bone morphogenetic proteins (BMPs), Sog and Dpp, are free to diffuse in the PV space (for a more detailed review of the biology of dorsal surface patterning see O'Connor *et al.*, 2006). Dpp maintains expression of dorsally-acting genes, including itself, and Sog prevents Dpp from activating Dpp production in the lateral ectoderm (Biehs *et al.*, 1996). At least five secreted zygotic gene products are required to specify dorsal ectoderm and the amnioserosa. These include Dpp and Screw (Scw) (which is another signaling BMP), two BMP inhibitors Sog and Twisted gastrulation (Tsg), and the protease Tolloid (Tld), which cleaves complexes of BMPs with Sog and/or Tsg, thereby freeing BMPs to bind to receptors (Biehs et al., 1996; Shimmi and O'Connor, 2003). The Type I receptors for Dpp (Thickvein: Tkv) and Scw (Saxophone: Sax) and the common Type II receptor Punt are all uniformly expressed in the embryo (Dorfman and Shilo, 2001). BMP signals are transduced by a heteromeric complex of Type I and II receptors, which phosphorylate members of the R-Smad family of transcription factors (Massague and Chen, 2000). Once phosphorylated, the R-Smads form a complex with a common Smad (co-Smad) and translocate to the nucleus where they regulate target gene activity. Signaling is assayed by the level of phosphorylated Mad (p-Mad), which initially appears in a dorsal strip 18–20 cells wide (20% of the circumference), but within ~40 minutes is refined to a strip 8–10 cells wide (Ross et al., 2001). Despite the fact that Scw is uniformly expressed, p-Mad expression only appears in the dorsal domain where Dpp is expressed, which shows that Mad phosphorylation requires simultaneous activation by Scw and Dpp (Nguyen et al., 1998). In fact, the early p-Mad pattern is not seen when Sax, Tkv, or Dpp are absent. However, several studies show that patterning of the dorsal region is robust to changes in the concentrations of some of the components in the signaling network. For example, heterozygous scw, sog, tld, or tsg mutant embryos are viable and do not show any apparent macroscopic phenotype (Mason et al., 1997; Eldar et al., 2002), despite broadened p-Mad expression for sogmutants (Shimmi et al., 2005; Wang and Ferguson, 2005).

A major question concerning patterning in the dorsal region is how the BMP activity gradient is established. It is known that Sog and Tsg can form a trimeric complex with Dpp, thereby sequestering it, and that Sog in this complex can be cleaved by Tld, thereby releasing the ligand (Marques et al., 1997; Shimmi and O'Connor, 2003). The suggestion that facilitated transport of BMP may underly pattern formation within the dorsal domain of the embryo (Holley et al., 1995) could account for the fact that Sog and Tsg can have both positive and negative effects on the patterning process (Ross et al., 2001; Decotto and Ferguson, 2001). It has been widely believed that Sog, Tsg, and Tld can act to produce localization of BMPs at the dorsal midline, but how this occurs has remained unresolved until recently. A mathematical model by Eldar et al. (2002) took into account the formation of BMP–Sog dimers, incorporated the diffusion of Sog, BMP, and BMP–Sog, and allowed for the cleavage of Sog and BMP–Sog by Tld. Computational results showed that the model could produce a steady state distribution of BMPs that corresponds to the observed p-Mad distribution for certain choices of parameters, but analysis of the steady-state BMP distributions for wide ranges of the parameters showed that in most cases a prescribed threshold position of BMP shifted significantly with changes in the parameters. Halving the expression of tld, scw, and sog changed the threshold by less than 10% in only ~0.3% of the networks, and these were termed robust networks. Analysis of the robust networks showed insensitivity to most parameters, provided the following conditions were met

(i) processing of Sog by Tld is BMP-dependent; (ii) unbound BMP does not diffuse; (iii) Dpp and Scw homodimer patterning is decoupled by the formation of the inhibitor complex Sog/Tsg; (iv) receptor binding is irreversible; and (v) Sog removes BMP from receptors (Eldar et al., 2002).

The first condition has been validated in vitro which suggests that analyzing a network for robustness properties can yield significant biological insight (Holley et al., 1996; Marques et al., 1997; Shimmi and O'Connor, 2003). Conditions (iv) and (v) have not been tested yet, but affinity assays may provide some insight into the validity of these requirements. Recently, it was demonstrated that the primary signaling molecule is a Dpp/Scw heterodimer, and that the loss of Scw precludes the formation of an extracellular Dpp gradient (Shimmi et al., 2005). Condition (ii) is more controversial: early experimental evidence suggested that BMP is indeed localized and this suggestion has been bolstered by extracellular Dpp-GFP staining using perivitelline injection (Eldar et al., 2002; Wang and Ferguson, 2005). However, in another study it was suggested that Dpp has an effective range of 15–20 cell diameters (Mizutani et al., 2005), while Scw can act over even larger distances (Wang and Ferguson, 2005).

To investigate establishment of the spatial pattern when the primary BMP can diffuse, a simple model was developed that could reproduce many of the observed phenotypes (Mizutani et al., 2005). Furthermore, it was demonstrated that embryos are not as robust with respect to changes in the level of Sog as previously reported, and that the mathematical model could not account for the large changes that are actually observed (Mizutani et al., 2005). Instead of limiting the diffusion of BMPs by reducing the diffusion coefficient, the extracellular BMP rapidly binds to immobile receptors and is degraded. This effectively limits the diffusion length of BMP to between 1 and 9 cell diameters and thus BMP is highly localized in the presence of diffusion.

However one can ask whether freely diffusible Dpp can be redistributed by extracellular regulators to form the narrow, high-level signal at the dorsal midline of the embryo, even in the absence of receptors, or whether other transport mechanisms have to be invoked? If so, do the extracellular regulators Sog, Tsg, and Tld confer robustness with respect to changes in gene copy number, as is observed in heterozygous genetic mutants? The objectives in this section are three-fold. Firstly, the contributions of reaction and diffusion of Dpp with Sog, Tsg, and Tld are investigated to determine the balance of processes that lead to the transient evolution and steady-state spatial distribution of BMPs in the PV space. Since patterning occurs on a relatively short time scale, it is important to study both the transient evolution of the morphogen profile and the steady-state profile. Secondly, two models of extracellular transport are studied to determine what kinetic steps lower the sensitivity of the BMP distribution to reductions in gene copy number, as occurs in heterozygous mutants of *sog*, *tsg*, and *tld*. Also, since this is a fairly well-studied system, large-scale parameter screens of two patterning models are used to determine how sensitive the conclusions on robustness are to the specific choice of measurement method used, and the results are compared to those previously published for similar models (Eldar et al., 2002). Here it is shown that small changes in the kinetic structure of the model can enhance the

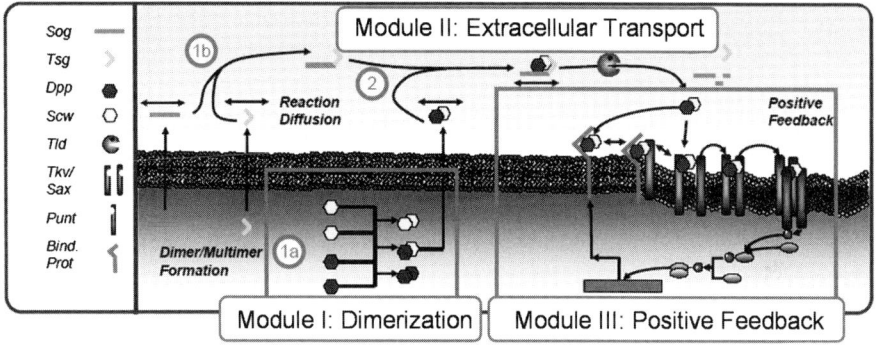

Figure 10 The modular description underlying the DV patterning model. 1a, 1b, and 2 show a two-stage dimerization cascade. (From Umulis et al., 2006, with permission.) See color insert.

robustness of extracellular patterning to changes in input. The cell-autonomous formation of Dpp/Scw before secretion into the PV-space along with dimerization of Sog and Tsg in the extracellular space buffer out changes in input independent of the specific method used to analyze the robustness. Lastly, we show that certain key aspects of the temporal evolution can be explained by incorporation of positive feedback in the signal transduction network. In order to investigate these aspects systematically we decompose the complete patterning system into three modules, as shown in Fig. 10.

A. The Model for DV Patterning

1. Dimerization of BMP Ligands and Extracellular Inhibitors

Mechanisms that confer robustness to changes in parameters, gene copy number, and other environmental conditions can be embedded in any stage in a hierarchical signal transduction system. For dorsal surface patterning, robustness can be gained or lost before secretion, during extracellular ligand patterning, or during receptor-signaling and feedback. Robustness of the full patterning model can be determined by analysis of each stage or module independently, followed by analysis of the linked modules. The first step at which compensation for reductions in the levels of ligand or inhibitor arises is during formation of the respective active ligand or inhibitor complexes. Dpp/Scw formation occurs intracellularly prior to secretion, whereas Sog/Tsg formation occurs in the PV-space. The local dynamics for heterodimer formation have been analyzed previously (Shimmi et al., 2005) and we review the main conclusions here.

Since the intracellular dimerization steps for Dpp/Scw formation occur in small volumes with high concentrations relative to the extracellular space, we assume that those processes equilibrate rapidly. This leads to the algebraic system

$$0 = \phi_D(x) - K_1 D \cdot W - K_3 D^2, \tag{28}$$

$$0 = \phi_W(x) - K_1 D \cdot W - K_2 W^2, \tag{29}$$

$$0 = K_1 D \cdot W - \gamma_1 \overline{DW}_{\text{in}}, \tag{30}$$

$$0 = \frac{1}{2} K_2 W^2 - \gamma_2 \overline{W_2}_{\text{in}}, \tag{31}$$

$$0 = \frac{1}{2} K_3 D^2 - \gamma_3 \overline{D_2}_{\text{in}}. \tag{32}$$

By defining the dimensionless variables $u = \sqrt{K_2/\phi_W^{wt}}\, W$, $v = \sqrt{K_3/\phi_W^{wt}}\, D$, $\Omega \equiv K_1/2\sqrt{K_2 K_3}$, $\beta \equiv \phi_D/\phi_W^{wt}$, and $\lambda \equiv \phi_W^{\text{perturbed}}/\phi_W^{wt}$, the Eqs. (28)–(32) are simplified to:

$$u^2 + 2\Omega uv = \lambda, \tag{33}$$

$$v^2 + 2\Omega uv = \beta, \tag{34}$$

$$u^2 = \lambda \frac{-b(\lambda, \beta, \Omega) - \sqrt{b(\lambda, \beta, \Omega)^2 - 4a(\Omega)}}{2a(\Omega)}, \tag{35}$$

$$v = -\Omega u + \sqrt{\Omega^2 u^2 + \beta}, \tag{36}$$

wherein $a(\Omega) = 1 - 4\Omega^2$, and $b(\lambda, \beta, \Omega) = 4\Omega^2(1 - \beta/\lambda) - 2$. When $\Omega = 1/2$, the relative change in production of the heterodimer to perturbations of the input is given by

$$\gamma_1 \overline{DW} = \frac{\phi_D \phi_W}{\phi_D + \phi_W}, \tag{37}$$

$$\gamma_2 W_2 = \frac{1}{2} \frac{\phi_W^2}{\phi_W + \phi_D}, \tag{38}$$

$$\gamma_3 D_2 = \frac{1}{2} \frac{\phi_D^2}{\phi_W + \phi_D}, \tag{39}$$

$$\frac{\overline{DW}^{\text{perturbed}}}{\overline{DW}^{wt}} = \lambda \frac{1+\beta}{\lambda+\beta}, \tag{40}$$

$$\frac{W_2^{\text{perturbed}}}{W_2^{wt}} = \lambda^2 \frac{1+\beta}{\lambda+\beta}, \tag{41}$$

$$\frac{D_2^{\text{perturbed}}}{D_2^{wt}} = \frac{1+\beta}{\lambda+\beta}. \tag{42}$$

When heterodimer formation occurs according to this scheme, the level of compensation conferred on the heterodimer increases as the ratio of input production of monomer (β) decreases. As β approaches zero the ratio approaches 1, which means that the output production of the heterorodimer is unaffected. Formation of the heterodimer Dpp and Scw would significantly enhance the robustness of Dpp/Scw levels for changes in the production rate of Scw, but not Dpp, if Scw is present in slight excess. Example results showing the level of compensation for changes in the production

2. Robustness of Embryonic Spatial Patterning in *Drosophila melanogaster* 93

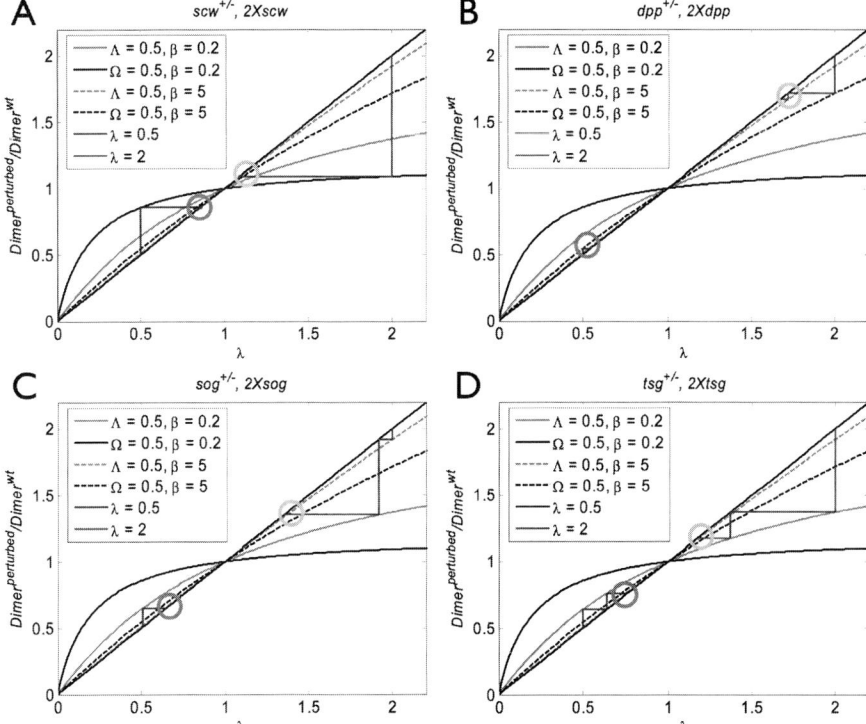

Figure 11 Heterodimer formation can reduce the effect of changes in the input, and a sequence of dimerization steps increases the robustness additively in the number of steps. Each curve in (A–D) shows the level of output heterodimer for changes in input monomer levels for different values of β with $\Omega = 0.5$. The stairs indicate the level of compensation achieved for a sequence of dimerization reactions and the circles indicate the final level of heterodimer for reductions (red lower circle) and increases (green upper circle) for increases in the monomer input. Results are shown for reductions and increases in *scw* (A), *dpp* (B), *sog* (C), and *tsg* (D). See color insert.

of Dpp and Scw are shown in Figs. 11A and 11B. The final compensation achieved by sequential dimerization of Dpp/Scw and Sog/Tsg is shown for 50% reductions in monomer production (lower red circle) and 100% increases in monomer production (upper green circle). The closer the output of the dimerization cascade is to 1, the greater the robustness of heterodimer output is to perturbations of the input.

The local dynamics for the production of the Sog/Tsg heterodimer is similar to the dimerization of Dpp and Scw with two key differences: Sog and Tsg binding is highly reversible since the molecules do not form a covalent bond, and secondly, Sog and Tsg do not form Sog–Sog and Tsg–Tsg homodimers once secreted (note: the Tsg molecule is actually a homodimer of two Tsg subunits which form before secretion and is irreversible). The local dynamics for Sog/Tsg formation can be represented by three ordinary differential equations:

$$\frac{dS}{dt} = \phi_S - k_2 S \cdot T + k_{-2} I - \delta_S S, \tag{43}$$

$$\frac{dT}{dt} = \phi_T - k_2 S \cdot T + k_{-2} I - \delta_T T, \tag{44}$$

$$\frac{dI}{dt} = k_2 S \cdot T - k_{-2} I - \gamma_I I. \tag{45}$$

We can solve the steady-state equations for Sog/Tsg and deduce the changes in output relative to changes in input. To simplify the results we define the dimensionless variables $s = \delta_S/\phi_T^{wt} \cdot S$, $g = \delta_T/\phi_T^{wt} \cdot T$ and the dimensionless parameters $\lambda \equiv \phi_{Tsg}^{perturbed}/\phi_{Tsg}^{wt}$, $\beta \equiv \phi_{Sog}^{wt}/\phi_{Tsg}^{wt}$, $\Lambda = K_{ST}\phi_T^{wt}/(2\delta_S\delta_T)$, where $K_{ST} = k_2\gamma_I/(k_{-2} + \gamma_I)$. This leads to the following dimensionless equations

$$g + 2\Lambda gs = \lambda, \tag{46}$$

$$s + 2\Lambda gs = \beta, \tag{47}$$

and the solution of these is

$$g = \frac{-b(\lambda, \beta, \Lambda) + \sqrt{b(\lambda, \beta, \Lambda)^2 + 8\lambda\Lambda}}{4\Lambda}, \tag{48}$$

$$s = \frac{\beta}{1 + 2\Lambda g}, \tag{49}$$

wherein $b = 1 + 2\Lambda(\beta - \lambda)$. Now the output ratio is

$$\frac{ST^{mut}}{ST^{wt}} = \frac{\lambda - u(\lambda, \beta, \Lambda)}{1 - u(1, \beta, \Lambda)}$$

and the output level of heterodimer can be computed as before. The output for a two-level cascade of Sog/Tsg formation and binding to Dpp/Scw is shown in Figs. 11C and 11D. It is clear that dimerization reduces the effect of changes in the levels of the input, but the magnitude of the effect depends on the input that is changed. For example, if Scw is in slight excess to Dpp, the output is less sensitive to changes in the level of Scw than Dpp and vice versa, and the compensation for changes in the level of Scw provides a filter in an early stage of patterning. The local dynamics for Sog and Tsg also suggest that dimerization could increase the level of robustness, but this has to be tested when transport is also included, and this is done in the following section.

2. Extracellular Patterning

We assume that the embryo is symmetric across a plane through the DV axis and orthogonal to the cross-section, and develop equations for the evolution of the species involved (Sog, Tsg, etc.) in an axial cross-section midway between the anterior and

2. Robustness of Embryonic Spatial Patterning in *Drosophila melanogaster*

Table II Dependent variables for the heterodimer-based model described by Eqs. (50)–(57) and the model due to Eldar *et al.* (2002) (to be defined later and referred to as Model I hereafter)

Name	Description
Heterodimer model	
D	Dpp monomer
W	Scw monomer
DW_{in}	Dpp/Scw (internal)
B	Dpp/Scw (PV space)
S	Sog
T	Tsg
I	Sog/Tsg
IB	Sog/Tsg–Dpp/Scw
Model I	
W	Scw monomer
$W2_{in}$	Scw homodimer (internal)
$W2$	Scw homodimer (PV space)
S	Sog
$SW2$	Sog–Scw/Scw

posterior poles of the embryo. In addition, since the PV space is narrow (∼0.5 microns) in the direction normal to the inner membrane, we neglect variations in this direction and reduce the problem to one space dimension. The governing equations are as follows and descriptions of the variables are given in Table II.

$$0 = \phi_D(x) - K_1 D \cdot W - K_2 D^2, \tag{50}$$

$$0 = \phi_W(x) - K_1 D \cdot W - K_3 W^2, \tag{51}$$

$$0 = K_1 D \cdot W - \gamma_1 \overline{DW_{in}}, \tag{52}$$

$$\frac{\partial B}{\partial t} = D_B \frac{\partial^2 B}{\partial x^2} + \gamma_1 \frac{V_{in}}{V_{PV}} \overline{DW_{in}} - k_3 I \cdot B + k_{-3} IB + \lambda Tol \cdot IB - \delta_B B, \tag{53}$$

$$\frac{\partial S}{\partial t} = D_S \frac{\partial^2 S}{\partial x^2} + \phi_S - k_2 S \cdot T + k_{-2} I - \delta_S S, \tag{54}$$

$$\frac{\partial T}{\partial t} = D_T \frac{\partial^2 T}{\partial x^2} + \phi_T - k_2 S \cdot T + k_{-2} I + \lambda Tol \cdot IB - \delta_T T + \lambda_2 Tol \cdot I, \tag{55}$$

$$\frac{\partial I}{\partial t} = D_I \frac{\partial^2 I}{\partial x^2} + k_2 S \cdot T - k_{-2} I - k_3 I \cdot B + k_{-3} IB - \lambda_2 Tol \cdot I, \tag{56}$$

$$\frac{\partial IB}{\partial t} = D_{IB} \frac{\partial^2 IB}{\partial x^2} + k_3 I \cdot B - k_{-3} IB - \lambda Tol \cdot IB. \tag{57}$$

Numerical simulations of these equations were done with initial conditions set to zero for the levels of all species but Tld, which was uniformly distributed in the dorsal region. The results for the transient evolution of Dpp/Scw, assuming first-order degradation of Dpp/Scw, are shown at 15, 30, 45, and 60 minutes in Fig. 12A. One sees that the early profile (15 minutes) is broad and weak with a shallow peak near the dorsal midline. The shape at this time reflects the fact that Sog/Tsg is not yet present in sufficient quantities to sequester Dpp/Scw near the NE boundary. At 30 minutes the level of Sog/Tsg is high enough to foster localization of Dpp/Scw near the DM, and by 60 minutes, when the profile has essentially reached steady state, there are two distinct regions. Dpp/Scw levels are high near the DM, and with a suitable threshold in the downstream signal transduction pathway, the super-threshold region could specify the presumptive amnioserosa. Directly adjacent to the DM the profile is shallow and flat, which would produce a low level signal to specify the dorsal ectoderm. The evolution of the profile generally corresponds with the transient evolution of p-Mad in the embryo, but, as in previous models with time-independent BMP production, it does not capture the sharpening and contraction of the gradient (cf. Fig. 12A) (cf. Eldar *et al.*, 2002; Mizutani *et al.*, 2005, and the review in O'Connor *et al.*, 2006).

3. Morphogen Gradient Shape Arises by Transport and Inhibitor/Protease Competition

We indicated earlier that the net rate of change of a species is the sum of contributions from various processes, and a plot of the spatial profiles of these individual contributions at 60 minutes is shown in Fig. 12C. The largest kinetic contributions to the local rate of change of Dpp/Scw in the dorsal region are from binding of Sog/Tsg and resupply from Dpp/Scw/Sog/Tsg after Sog is cleaved by Tld (Fig. 12C). The net contribution of these two processes (Fig. 12F) removes Dpp/Scw from the majority of the dorsal region (negative values) except near the dorsal midline where the net contribution is positive. Balancing the release of Dpp/Scw near the midline is diffusion and degradation (Fig. 12C).

The steady-state diffusion profiles of Dpp/Scw, and Dpp/Scw/Sog/Tsg are shown in Fig. 12D. One sees that the net Dpp/Scw flux is negative near the DM (i.e., Dpp/Scw diffuses away from the midline), positive near the AS boundary, and negative again near the NE boundary. In contrast, the Dpp/Scw/Sog/Tsg net flux is large and positive near the DM and the NE region, and negative in between. Thus there is a complicated pattern of diffusion of Dpp/Scw and its complexes. Since the complexes provide a source of Dpp/Scw after Tld processing, the Dpp/Scw profile reflects the balances between the diffusion of Dpp/Scw away from the midline and the facilitated diffusion of Dpp/Scw complexed with Sog/Tsg towards the midline. The spatial distributions of the various forms of Dpp/Scw are shown in Fig. 12E, and one sees there that the unbound form of Dpp/Scw is present at the lowest level of the two forms.

To understand the transient evolution of the Dpp/Scw profiles in Fig. 12A, we also examined the transient evolution of the kinetic terms for Sog/Tsg binding and Tld processing, as well as the evolution of the diffusion fluxes of all forms of Dpp/Scw (Fig. 12F). At 15 minutes, the net kinetic balance is negative (red, dashed) because

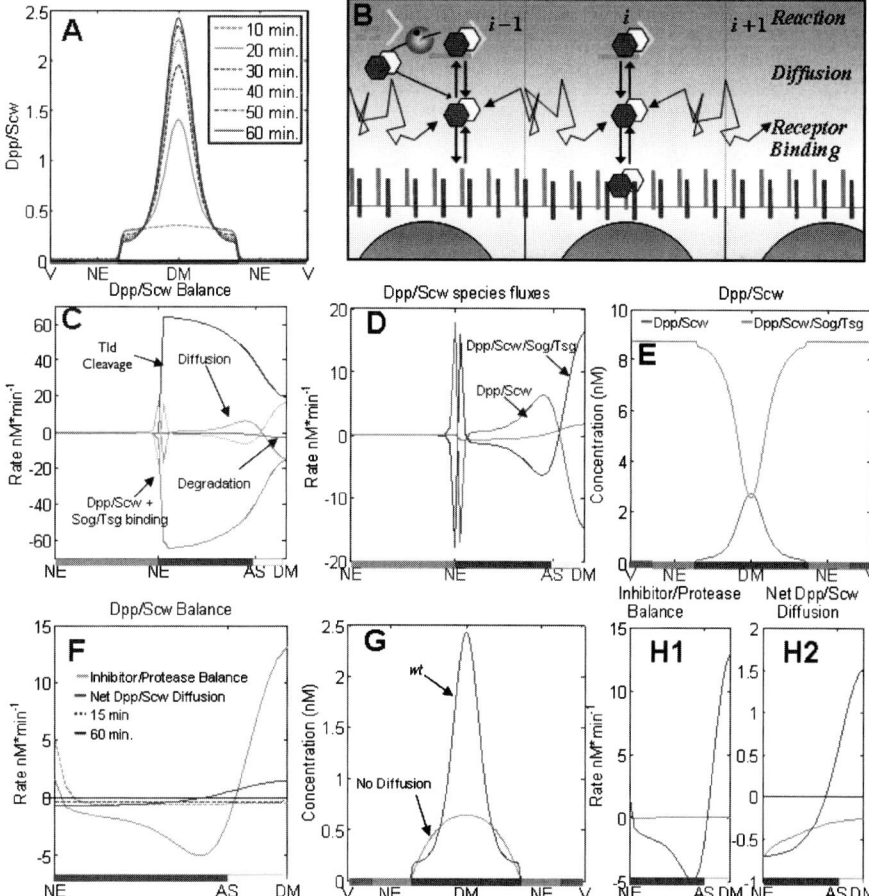

Figure 12 The transient Dpp/Scw activity gradient is established by interacting processes that limit receptor binding laterally and redistribute Dpp/Scw dorsally. (A) The temporal evolution of the spatial profile of Dpp/Scw/Tkv/Sax at the times indicated. (B) Schematic of reactions controlling localization of Dpp/Scw activity (for key see Fig. 10). (C–H) The spatial profile of Dpp/Scw, which determines the amount of bound receptor, results from a balance of Sog/Tsg binding, Tsg binding, and diffusion. (C) Spatial distribution of reaction and diffusion contributions to the shape of the Dpp/Scw activity distribution. Kinetic contributions from Sog/Tsg binding (red, rescaled 0.25×), Tsg (blue, rescaled 0.25×) are shown. Diffusion (blue/green) and degradation (purple) of Dpp/Scw reduces levels near the midline. (D) Dpp/Scw flux distribution in the dorsal region shows that diffusion of Dpp/Scw and Dpp/Scw/Tsg moves Dpp/Scw from the DM region toward the lateral regions while Dpp/Scw/Sog/Tsg redistributes Dpp/Scw towards the midline. (E) Spatial distribution of Dpp/Scw in various bound forms. (F) Net of Sog/Tsg binding and resupply from Tsg release after Tld cleavage (red) and net flux of Dpp/Scw in all forms (blue). Data is shown for 15 (dashed) and 60 (solid) minutes. (G) wt (blue) results are compared with solutions where the diffusion of Dpp/Scw/Sog/Tsg is set to zero (red). (H1) Inhibitor/Protease balance contributions to Dpp/Scw for cases described in (G). (H2) Contributions of combined Dpp/Scw redistribution for cases described in (G). The parameters used in the computations for these figures are given in the supplement to Umulis et al. (2006). See color insert.

Dpp/Scw binds rapidly to Sog/Tsg, but later the balance between binding and release is reversed. In addition, the net flux of Dpp/Scw near the midline is initially negative, which reflects the fact that all forms of Dpp/Scw diffuse away from the DM due to the localized binding to Sog at the NE boundary, which creates a sink. However this also reverses at later times (blue, solid). Later, the shoulders of the Dpp/Scw profile diminish as the peak near the dorsal midline increases, which is consistent with the transient evolution of the inhibitor/protease balance and net Dpp/Scw diffusion (Fig. 12F). To confirm our interpretation of the evolution, we set the diffusion coefficient of the Dpp/Scw/Sog/Tsg complex to zero. In Fig. 12G we show the profiles of Dpp/Scw after 60 minutes, for the *wt* and the nondiffusible Dpp/Scw/Sog/Tsg. Removing the diffusion of Dpp/Scw/Sog/Tsg reduces the maximum amplitude at the DM and abolishes the shoulder near the NE boundary. The changes in the balance between Tld processing and Sog/Tsg binding in the two cases, as well as the change in the net diffusion fluxes are shown in Figs. 12H1 and 12H2, respectively.

4. Simulations of Mutants Match Observed Phenotypes at Steady-State

Next we examine the effect of mutations on patterning. We model the homozygous mutants $sog^{-/-}$, $tsg^{-/-}$, $scw^{-/-}$, and $tld^{-/-}$ by computing Dpp/Scw profiles with the production rates for the appropriate component set to zero (e.g., Sog in a $sog^{-/-}$ mutant). The profiles for homozygous mutants are shown in Fig. 13A at 60 minutes, and the corresponding experimentally-observed phenotypes are listed in Fig. 13E. Sog mutants cannot redistribute the Dpp/Scw ligand, and this leads to broad and low morphogen distribution between the levels necessary to specify the dorsal ectoderm and presumptive amnioserosa. In *sog* and *tsg* mutants, the high level target *race* is lost, but, medium level *rho*, and low level *pnr* are expanded (Fig. 13E) (Jazwinska *et al.*, 1999; Nguyen *et al.*, 1998; Shimmi and O'Connor, 2003; Shimmi *et al.*, 2005). This suggests that in mutant embryos the dorsal ectoderm enlarges at the expense of the presumptive amnioserosa, which does not form. Consistent with this finding is the observation that in a *sog* mutant, enough BMP diffuses into the neurogenic ectoderm to turn on dorsal genes (Biehs *et al.*, 1996). In this case the only mechanism for redistribution is diffusion of Dpp/Scw (Fig. 13B). In the *tld* mutant, p-Mad expression is low and embryos are unable to fully specify dorsal ectoderm or amnioserosa. In the absence of Tld, the gradient of Dpp/Scw/Sog/Tsg that leads to transport toward the DM is lost and the net Dpp/Scw flux is slightly negative (Fig. 13B). Here Sog/Tsg sequesters Dpp/Scw throughout the PV space and facilitates the transport of Dpp/Scw out of the dorsal region. This effectively removes Dpp/Scw from the vicinity of the dorsal region, resulting in a strongly ventralized embryo.

It was suggested previously that the p-Mad profile in *Drosophila* embryos was unchanged in heterozygous mutant embryos, but in that study the variability was not addressed (Eldar *et al.*, 2002). We measured the width in cell numbers of the p-Mad stripe in various heterozygous mutant backgrounds *wt* (12.2), $sog^{+/-}$ (14.1), $tsg^{+/-}$ (12.0), and $scw^{+/-}$ (10.8) in 13, 10, 13, and 9 embryos, respectively, (cf. Figs. 13F–

13K). The embryos are largely insensitive to reductions in copy number for these genes, but we note that $sog^{+/-}$ mutant embryos show an expansion in the width of the p-Mad stripe by about 2–3 cells (15–25%) on average. Others have also noted that the system is somewhat sensitive to *sog* gene copy number (Mizutani *et al.*, 2005; Wang and Ferguson, 2005). To compare these data with the model predictions, we simulated heterozygous mutant profiles by reducing the input or level of the heterozygous factor by 50%. For certain choices of parameters, the Dpp/Scw morphogen distribution is insensitive to the heterozygous mutant genotypes for all reductions except Dpp (cf. Fig. 13C). Analysis of Model I showed that this insensitivity could only be achieved if (i) Dpp and Scw homodimer patterning is decoupled by formation of Sog/Tsg, and (ii) free BMP does not diffuse (Eldar *et al.*, 2002), but these conditions are not necessary in the present model. Here the diffusion constant for Dpp/Scw was estimated from standard correlations to be 73 μ^2 s^{-1} (Young *et al.*, 1980), and all binding steps are reversible, which implies that free BMP is less tightly localized than in earlier models (Eldar *et al.*, 2002).

5. Selective Use of Heterodimers Enhances Robustness: Extracellular Patterning

The spatial distributions of BMPs predicted by the heterodimer patterning model can also be tested for their sensitivity to changes in other kinetic components. To determine whether changes in the kinetic interactions of Dpp and Scw in the steps prior to secretion, or changes in the steps for the formation of Sog/Tsg heterodimers, enhance or reduce robustness of the spatial distributions of BMPs, a wide range of parameters was tested, and the results were compared to those predicted by Model I, given by Eqs. (58)–(62) below. The heterodimer-based model contains three heteromer formation steps, namely Sog + Tsg, Dpp + Scw, and Sog/Tsg + Dpp/Scw, whereas Model I contains only one, formation of BMP/Sog. We did not explicitly compare the model developed by Mizutani *et al.* (2005), but we expect the degree of robustness with respect to variations in Sog, Tsg, and Tld to be the same as in the heterodimer model, since both rely on the selective use of the Sog/Tsg heterodimer that binds to BMP and is cleaved by Tld.

$$\frac{\partial W}{\partial t} = \phi_W(x) - K_3 W^2, \tag{58}$$

$$\frac{\partial W_{2\text{in}}}{\partial t} = \frac{1}{2} K_3 W^2 - \gamma_3 W_{2\text{in}}, \tag{59}$$

$$\frac{\partial W_2}{\partial t} = D_{W_2} \nabla^2 W_2 + \gamma_3 \frac{V_{\text{in}}}{V_{PV}} W_2 - k_1 S \cdot W_2 + k_{-1} \overline{SW_2}$$
$$+ \lambda Tol \cdot \overline{SW_2} - \delta_{W_2} W_2, \tag{60}$$

$$\frac{\partial S}{\partial t} = D_S \nabla^2 S + \phi_S - k_1 S \cdot W_2 + k_{-1} \overline{SW_2} - \delta_S S, \tag{61}$$

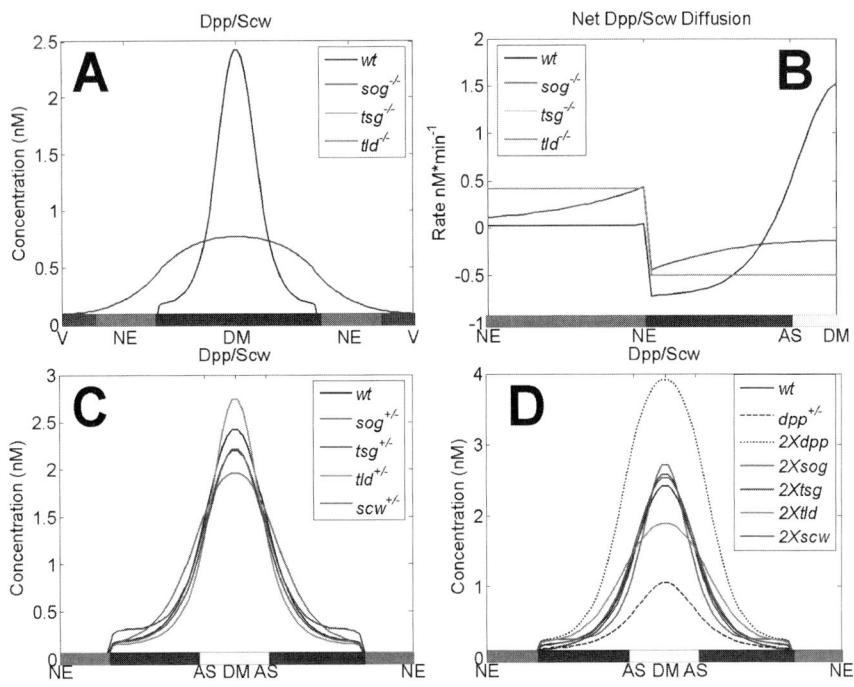

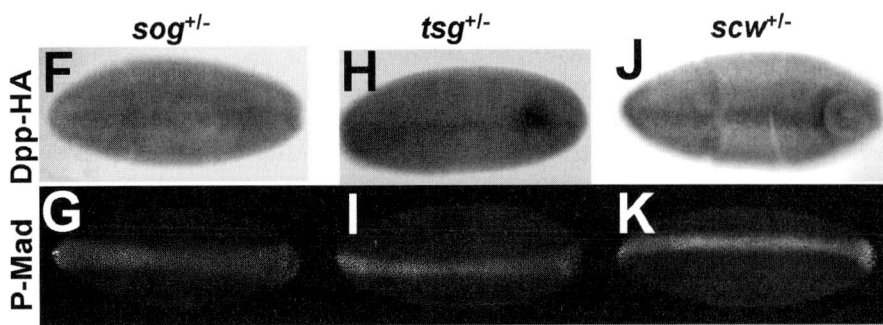

2. Robustness of Embryonic Spatial Patterning in *Drosophila melanogaster*

Figure 13 Homozygous and heterozygous computation results correspond well with observed phenotypes. (A) Dpp/Scw/Tkv/Sax profile is shown for *wt* (blue), and homozygous mutant sog- (red), tsg- (green), scw- (purple), and tld- (blue/green) after 60 minute of patterning time. (B) Net diffusion of Dpp/Scw in homozygous mutants. Dpp/Scw/Tkv/Sax profile is shown for (C) *wt* and heterozygous mutant $sog^{+/-}$, $tsg^{+/-}$, $tld^{+/-}$, and $scw^{+/-}$. (D) *wt*, $dpp^{+/-}$, 2X *dpp*, and 2X *scw*. (E) Top: Summary of homozygous mutant target gene expression profiles (1, 11, 23, 26) (* denotes data not shown). Bottom: Width of p-Mad activity for heterozygous mutants. (F–K) Dpp-HA accumulates in a narrow stripe that corresponds well with p-Mad signaling in $sog^{+/-}$, $tsg^{+/-}$, and $scw^{+/-}$ heterozygotes. Staining techniques are as described in Shimmi *et al.* (2005). See color insert.

$$\frac{\partial \overline{SW_2}}{\partial t} = D_{\overline{SW_2}} \nabla^2 \overline{SW_2} + k_1 S \cdot \overline{W_2} - k_{-1} \overline{SW_2} - \lambda Tol \cdot \overline{SW_2}. \tag{62}$$

In this case the balance on the homodimer is trivial, and leads to an input of BMP to the extracellular patterning of $V_{in}\phi_W/2V_{PV}$. Here $V_{in}/V_{pv} = 1 \times 10^{-3}$ and $\phi_W = 10\ \mu M\ min^{-1}$.

The protocol for the computations and comparisons is as follows. Parameters in both sets of equations were chosen randomly from uniform distributions, ranging over 5 orders of magnitude for the forward and reverse binding reactions (10^{-3}–$10^1\ min^{-1}\ nM^{-1}$ or min^{-1}), 4 orders of magnitude for the degradation/internalization rates (10^{-3}–$10^0\ min^{-1}$), 3 orders of magnitude (0.01–1 $nM\ s^{-1}\ cell^{-1}$) for the production rates, and 2 orders of magnitude (1 $\mu^2\ s^{-1}$–100 $\mu^2\ s^{-1}$) for the diffusion coefficients. Once a set of random parameter values is chosen, a base case or *wt* profile for each component is calculated, and the spatial distribution of each species is normalized by the maximum concentration of that component. It is observed experimentally that p-Mad staining occurs in the dorsal-most 10–15% cells, corresponding to the localization of Dpp/Scw (Figs. 3F–3K), and therefore we consider computational solutions that show at least a 2-fold change in the Dpp/Scw concentration within the dorsal-most 20% of the embryo suitable for specification of the amnioserosa. All other components are ignored in this criterion.

A total of 99,186 steady-state solutions were calculated for the heterodimer system and 99,399 solutions for Model I, and of these, 3385 (3.4%) were admissible in the heterodimer model, whereas 2452 (2.5%) for Model I met the criterion. Then for each admissible set of parameters, the spatial distributions for the heterozygous genotypes for $sog^{+/-}$ and $tsg^{+/-}$ were computed after reducing the production of each protein to 50% of the *wt* level. Each mutant profile was normalized by the *wt* maximum and then compared with the *wt* to determine the sensitivity of the BMP distribution to that perturbation of the input. Usually robustness is measured by the magnitude of the spatial shift of the 'perturbed' case versus the base case at a specified threshold value of the morphogen. However morphogens usually specify multiple targets in a concentration-dependent manner, and thus basing robustness on a single level is inappropriate: the entire shape of profile should be considered in a criterion for robustness. To illustrate

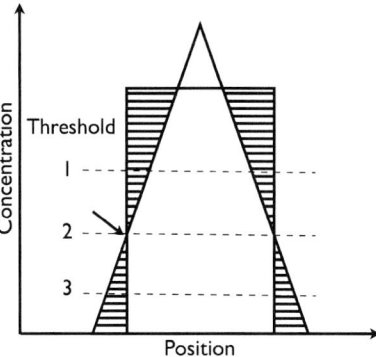

Figure 14 An illustration of how fixing a single morhpogen threshold as the criterion for robustness can be misleading. See color insert.

the inadequacy of tests based on a single threshold, Fig. 14 shows an example 'base case' distribution (the triangle) and a 'perturbed' or mutant profile (the rectangle). If the threshold is set at the morphogen level 2 one would conclude that the profiles are robust, but they clearly are not, since threshold levels at position 1 (resp., 3) would indicate that the mutant profile is wider (resp., narrower) than the base case. There are many ways to measure the difference between two profiles and we use the following one for comparing *wt* and mutant profiles. Regard the morphogen level as the independent variable, invert the position–concentration relation for each profile to obtain position as a function of concentration, integrate the absolute value of the difference in the positions for the profiles, and divide by the concentration range. A mutant profile was judged admissible if the region of high intensity p-Mad staining, determined by computing the integral average of the absolute value of the differences between the *wt* and heterozygous cases over the central 60% (from 20–80%) of the *wt* profile, changed by less than 10 µ in the half-width of the embryo. This corresponds to a change of less than 4 cells with high intensity p-Mad staining in the entire dorsal region.

6. Sog/Tsg Formation Compensates for Reductions in Both Sog and Tsg

In the heterodimer-based model, 71% of the $sog^{+/-}$ solutions shifted by less than 10 µm, whereas in Model I only 44% met the same criterion. More (92%) passed in the heterodimer-based model for reductions in the level of Tsg and 65% passed when compensation for 50% reductions in both Sog *and* Tsg were considered independently. This is still about 50% more than the monomer-based model. These data suggest that transport by the heteromeric complex Sog/Tsg enhances compensation for reductions in the levels of both Sog and Tsg (Figs. 15A–15C).

Our previous analysis of heterodimer production suggested that Sog/Tsg formation can compensate for reductions of Tsg, and to a lesser extent Sog (Shimmi *et al.*, 2005).

2. Robustness of Embryonic Spatial Patterning in *Drosophila melanogaster*

Summary of Computations	Heterodimer Model		Model I	
A Total Simulations	99,186		99,399	
Solutions that vary 2-fold in dorsal 20%	3,385	(3.4% of total)	2,452	(2.5% of total)
Sog-Tsg Compensation for $sog^{+/-}$, $tsg^{+/-}$				
Number (%) passed sog+/- for 0.2-0.8	2,403	(71%)	1081	(44%)
Number (%) passed tsg+/- for 0.2-0.8	3,122	(92%)	-	
Number (%) passed sog+/- and tsg+/-	2,195	(65%)	1081	(44%)
$tld^{+/-}$, $scw^{+/-}$ Compensation				
Number (%) passed tld+/- for 0.2-0.8	2,823	(83%)	2102	(86%)
Number (%) passed scw+/- for 0.2-0.8	600	(18%)	11	(0.5%)
Threshold Dependence				
Threshold at 0.2 (% of 2-fold change)	1,925	(57%)	252	(10%)
Threshold at 0.3	2,138	(63%)	175	(7%)
Threshold at 0.4	1,874	(55%)	107	(4%)
Threshold at 0.5	1,399	(41%)	60	(2%)
Threshold at 0.6	937	(28%)	22	(0.9%)
Threshold at 0.7	575	(17%)	7	(0.3%)
Threshold at 0.8	94	(3%)	2	(0.1%)
Average 0.4-0.6	1,282	(38%)	32	(1.3%)
Average 0.3-0.7	938	(28%)	7	(0.3%)
Average 0.2-0.8	227	(7%)	3	(0.1%)

Figure 15 Robustness of the heterodimer-based model. (A) Summary of computation results comparing the simplified heterodimer-based model (50)–(57) with a previous model based on monomers and homodimers. (B–C) Histograms show that the heterodimer-based model is less sensitive to most types of heterozygous mutant perturbations (B), and produces more robust hits at every threshold position chosen between 20 and 80% of *wt* (C). (D) The *wt* distribution of Sog, Tsg, and Sog/Tsg. (E–G) *wt* and $sog^{+/-}$ and $tsg^{+/-}$ distributions of Sog, Tsg, and Sog/Tsg. (H–I) Solutions robust with respect to $scw^{+/-}$ (red) vs nonrobust (blue) are shown for different values of Ω and β (H) and also for Dpp/Scw^{perturbed}/Dpp/Scw^{wt} vs β (I). See color insert.

The results here reflect this imbalance, in that there are many more solutions that are robust to decreases in Tsg and fewer for Sog. However, when compared to Model I, which incorporates only Sog, Sog/Tsg formation enhances robustness overall, and leads to more robust solutions when both heterozygotes $sog^{+/-}$ and $tsg^{+/-}$ are considered. Since Sog and Tsg diffuse in the PV space and form a Sog/Tsg complex that has a high affinity for BMP, the shift in their profiles under changes in their inputs provides insight into the origin of robustness. Fig. 15D shows the *wt* profiles of Sog, Tsg, and Sog/Tsg, and the mutant distributions for Sog, Tsg, and Sog/Tsg are shown in Figs. 15E–15G, respectively. Sog and Tsg compensate for the partial loss of the other component by shifting in the same direction, thereby reducing the effect of the perturbation on the formation of Sog/Tsg, and thus on the Sog/Tsg distribution. As expected, the buffering is not symmetric and the Sog/Tsg profile is more sensitive to reductions of Sog than Tsg, which agrees with the experimentally observed shift in p-Mad signaling. However, it is noteworthy that the Sog distribution itself is less sensitive to perturbations in the presence of Tsg, which suggests that perturbations are distributed to Tsg by the Sog/Tsg complex.

It was previously reported that patterning by BMPs is also insensitive to knockdowns in *scw*, i.e.,, there is no phenotypic change for $scw^{+/-}$ mutant genotypes (Eldar et al., 2002). We found that $scw^{+/-}$ embryos actually narrowed by about 1 nucleus on average (Fig. 13E). Computationally we found that in the heterodimer-based model 18% of the solutions were robust under reductions in Scw. This stands in sharp contrast to the results for Model I, for which only 0.5% of solutions passed the same test (Figs. 15A–15B).

When considering reductions in Tld, 83% of the perturbed met the criterion for insensitivity vs Model I's 86%. Since Tld interactions in the two models are nearly equivalent, we would expect roughly the same number of robust solutions.

7. Robustness Depends on Threshold Position and Measurement

The number of solutions that passed the test for Model I (here 0.1%) is fewer than reported previously (0.3%), and the difference stems from the difference in the specific measure of robustness used. Here robustness is measured using an integral average over a majority of the *wt* profile, rather than judging robustness by the shift in location of a chosen threshold value. Indeed, if a narrower range or a specific threshold value is used, the number of robust solutions increases for both models. For a narrower range of the integral average and for threshold choices in 10% increments from 20% of the *wt* profile up to 80% (Figs. 15B–15C) the number of robust profiles increases from 7% (0.1%) to a maximum of 63% (10%) for the heterodimer model and Model I, respectively. The heterodimer-based model has significantly more robust solutions at all thresholds chosen between 0.2 and 0.8 of the *wt* maximum, ranging from 5.7–93 times, depending on the threshold position or range used (Fig. 15A).

B. The Role of Positive Feedback

Module I (dimerization) and II (extracellular transport) enhance the robustness by limiting the range of ligand diffusion and buffering changes in monomer levels by forming active heterodimers. However, the current extracellular patterning model with constant ligand production cannot reproduce the experimentally-observed transient evolution of p-Mad staining observed in *wt* embryos. One way to overcome this is to have a pulse or transient input of the BMP ligand (Eldar *et al.*, 2002), but this is not likely to occur during development since *dpp* expression persists throughout cell cycle 14. Another possibility suggested by Wang and Ferguson (2005) is that BMP signaling initiates expression of a protein that enhances binding of Dpp to the receptors by localizing Dpp at the surface. When positive feedback of a surface bound BMP-binding protein is included in the BMP patterning model, the level of BMP-bound receptors contracts in time to produce a sharp distribution that corresponds well with embryonic p-Mad staining (Umulis *et al.*, 2006). Positive feedback and receptor interactions make up the third module for dorsal surface patterning, which can be modeled as an autonomous unit since the receptors and other complexes do not diffuse. However, analysis of the third module's contribution to overall patterning is more complex than for modules 1 and 2, since the extracellular distribution of Dpp and receptor processes are highly coupled (Umulis *et al.*, 2006).

Interestingly, inclusion of positive feedback does not reduce the upstream contributions to robustness of patterning, and actually enhances the robustness of the mechanism to changes in the level of receptors (Umulis *et al.*, 2006). Inclusion of positive feedback leads to a distribution of BMP bound receptors that contracts in time and ultimately produces a step-like profile due to a bi-stability in the local dynamics (Umulis *et al.*, 2006) (see Fig. 16). Another interesting consequence of positive feedback is that the level of signaling is determined by the temporal course of exposure to extracellular BMP, rather than by the level at a fixed time. Thus the history of the ligand exposure rather than only the instantaneous level is important for the transient response in this patterning mechanism, as in other systems (Dillon and Othmer, 1999; Harfe *et al.*, 2004). Dynamic interpretation of the morphogen has also been suggested in the context of anterior–posterior patterning (Bergmann *et al.*, 2007).

VI. Conclusions

Mathematical models of embryonic development can provide insights into the complex interactions between spatial variations in morphogens, signal transduction, and intracellular response in patterning processes. Mechanistic models of specific processes based on our current understanding of them provide an additional tool to help in understanding the complex regulation of development. In addition to the fundamental biological questions concerning the structure of components and pathways, mathematical models introduce a host of new questions. For example, how do signal

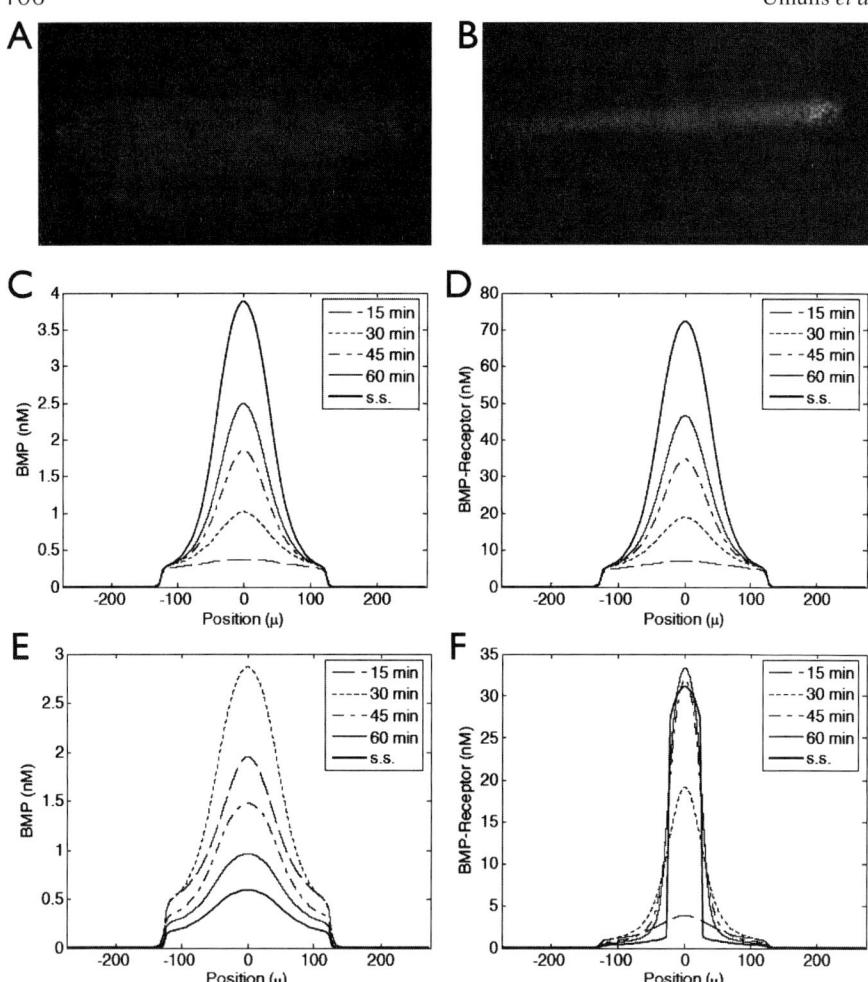

Figure 16 Positive feedback of a surface-bound BMP-binding protein leads to contraction of BMP-receptors towards the dorsal-midline in time. (A) early, and (B) late p-Mad distributions, showing the contraction in later stages. (C–F) Simulation results for the evolution of extracellular BMP and BMP-bound receptor levels in time for two versions of module III: (C–D) receptor mediated endocytosis, and (E–F) positive feedback of a BMP-binding protein. The level of BMP-bound receptors widens for any chosen threshold in (D). (F) The distribution of BMP-bound receptors contracts in time for low-levels, while growing in amplitude near the dorsal midline. See color insert.

transduction networks respond to perturbations in the levels of molecules, and how do changes in the length of the system affect patterning? How can the often nonintuitive behavior observed *in vivo* be reconciled with *in vitro* data? What is the potential function of a novel gene? Incorporating mechanistic models during experimental design and data analysis will enrich our understanding of individual biological processes and of the system as an integrated unit rather than just the sum of its parts.

We have discussed a number of different aspects of robustness in *Drosophila* embryonic patterning, and have shown how the models lead to new insights concerning scale-invariance in AP patterning, the role of network topology and signature in the switching network used for control of the segment polarity genes, and the role of signaling via heterodimers in DV patterning. The modular decomposition of DV patterning described herein is based on the current understanding of the kinetic interactions between morphogens and inhibitors in the PV space, and incorporates the various dimeric forms of inhibitors and signaling BMPs.

Scale-invariance is a basic feature of a number of developmental processes. The conservation of morphogen binding sites and production between different size organisms may be a natural way to ensure scale-invariance without requiring additional levels of regulation such as feedback, opposing gradients or rescaling of the diffusion coefficients. A side effect of achieving scale-invariance by modulating the time scale of the kinetics (done here by invariance of the number of binding sites for different-sized embryos) is that the time to reach a steady-state distribution increases in proportion to the length of the system squared.

A rather surprising result is that the robustness of DV spatial patterning can essentially be predicted from an analysis of the local dynamics. Transport certainly plays a role in the establishment of the spatial distribution of all components, but the key property of the kinetic interactions that produces robustness is the use of heterodimeric species for both the primary signaling molecule (Dpp/Scw) and the primary inhibitor (Sog/Tsg). Previously we showed that the use of heterodimers, rather than monomers or homodimers, can compensate particularly well for changes in Scw, and that a cascade of stages in which heterodimers are dominant can provide a high degree of compensation (Shimmi *et al.*, 2005). Here the analysis of the spatially-distributed system corroborates the earlier results and provides further evidence for the selective advantage of using heterodimers for signal transduction and morphogenesis. We found that robustness with respect to reductions in Scw requires that $\beta = \phi_{Dpp}/\phi_{Scw}$ be less than 0.71 (Fig. 15H), which means that Scw production must be slightly larger than Dpp production. Robustness is also favored by large values of $\Omega = K_1/2(K_2 K_3)^{1/2}$, which means that heterodimer formation must be favored over the homodimer counterparts Dpp/Dpp and Scw/Scw. These results correspond well with our earlier analysis of a simplified system (Shimmi *et al.*, 2005), as can be seen by a comparison of Fig. 15I herein and Fig. 6C in (Shimmi *et al.*, 2005).

In the absence of positive feedback, tight localization of nanomolar levels of Dpp/Scw requires a binding rate of Dpp/Scw to the Sog/Tsg complex that approaches the theoretical maximum of $\sim 10^9$ M^{-1} s^{-1} (Keizer, 1987). The kinetics of Sog binding, Tld cleavage and receptor binding operate in a similar dynamic range with on-rates approximately 1000 times lower than the level necessary to localize BMPs in the absence of receptors, or at lower rates of Dpp (Eldar *et al.*, 2002; Shimmi and O'Connor, 2003; Mizutani *et al.*, 2005). Thus, while the tight localization of BMPs by Sog and Tsg is *theoretically* possible, other mechanisms are required to either limit the diffusion of Dpp/Scw or reinterpret the shallow gradient and amplify small differences in concentration. One possible mechanism that can produce the sharp p-Mad distribution is positive feedback on a factor that enhances Dpp binding to its receptor by localizing Dpp. Addition of positive feedback does not offset the robustness gained in upstream mechanisms, but rather enhances it with respect to changes in the level of receptors, when compared to simple equilibrium receptor binding and other similar mechanisms (Umulis *et al.*, 2006). Positive feedback can lead to the contraction of p-Mad signaling in time, which is observed during dorsal surface patterning and also in a related signaling mechanism during wing vein patterning. Thus, the spatial refinement of components may be a general property of BMP pathways, and additional experiments are necessary to determine the nature of the positive feedback.

Note added in proof

Since acceptance of this paper, two papers have appeared that deal with the issues discussed herein (Gregor *et al.*, 2007a; Gregor *et al.*, 2007b). Gregor et al. developed a Bicoid–GFP fusion protein to study the dynamics, reproducibility, and other aspects of morphogen patterning in the *Drosophila syncytium*, and they discuss theoretical approaches to the reliability of threshold determination.

Acknowledgments

This research was funded in part by NIH Grant GM29123 and NSF Grants DMS-0317372 and DMS-0517884 to HGO, and by a Biotechnology Training Grant (DMU). MBO is an Investigator with the Howard Hughes Medical Institute.

References

Aegerter-Wilmsen, T., Aegerter, C. M., and Bisseling, T. (2005). Model for the robust establishment of precise proportions in the early Drosophila embryo. *J. Theor. Biol.* **234**, 13–19.
Albert, R., and Othmer, H. G. (2003). The topology of the regulatory interactions predicts the expression pattern of the segment polarity genes in Drosophila melanogaster. *J. Theor. Biol.* **223**, 1–18.
Alberts, B., Bray, D., Lewis, J., Raff, M., Roberts, K., and Watson, J. D. (1994). "Molecular Biology of The Cell." Garland Publishing Inc., New York.
Bergmann, S., Sandler, O., Sberro, H., Shnider, S., Schejter, E., Shilo, B. -Z., and Barkai, N. (2007). Presteady-state decoding of the Bicoid morphogen gradient. *PLoS Biol.* **5**, e46.
Biehs, B., Francois, V., and Bier, E. (1996). The Drosophila short gastrulation gene prevents Dpp from autoactivating and suppressing neurogenesis in the neuroectoderm. *Genes Dev.* **10**, 2922–2934.
Castets, V., Dulos, E., and Kepper, P. De. (1990). Experimental evidence of a sustained standing Turing-type nonequilibrium chemical pattern. *Phys. Rev. Lett.* **64**, 2953–2956.

Claxton, J. H. (1964). The determination of patterns with special reference to that of the central primary skin follicles in sheep. *J. Theor. Biol.* **7**, 302–317.

Crauk, O., and Dostatni, N. (2005). Bicoid determines sharp and precise target gene expression in the Drosophila embryo. *Curr. Biol.* **15**, 1888–1898. Comparative study.

Decotto, E., and Ferguson, E. L. (2001). A positive role for Short gastrulation in modulating BMP signaling during dorsoventral patterning in the Drosophila embryo. *Development* **128**, 3831–3841.

Dillon, R., and Othmer, H. G. (1999). A mathematical model for outgrowth and spatial patterning of the vertebrate limb bud. *J. Theor. Biol.* **197**, 295–330.

Dillon, R., Maini, P. K., and Othmer, H. G. (1994). Pattern formation in generalized Turing systems. I. Steady-state patterns in systems with mixed boundary conditions. *J. Math. Biol.* **32**, 345–393.

Dorfman, R., and Shilo, B. Z. (2001). Biphasic activation of the BMP pathway patterns the Drosophila embryonic dorsal region. *Development* **128**, 965–972.

Eldar, A., Dorfman, R., Weiss, D., Ashe, H., Shilo, B. Z., and Barkai, N. (2002). Robustness of the BMP morphogen gradient in Drosophila embryonic patterning. *Nature* **419**, 304–308.

Frohnhöfer, H. G., and Nüsslein-Volhard, C. (1986). Manipulating the anteroposterior pattern of the Drosophila embryo. *J. Embryol. Exp. Morphol. (October)* **97**, 169–179.

Gilbert, S. F. (2006). "Developmental Biology," 8th ed. Sinauer Associates, Inc., Sunderland, MA.

Gregor, T., Bialek, W., de Ruyter van Steveninck, R. R., Tank, D. W., and Wieschaus, E. F. (2005). Diffusion and scaling during early embryonic pattern formation. *Proc. Natl. Acad. Sci. USA* **102**, 18403–18407.

Gregor, T., Wieschaus, E. F., McGregor, A. P., Bialek, W., Tank, D. W. (2007a). Stability and nuclear dynamics of the bicoid morphogen gradient. *Cell* **130**, 141–152.

Gregor, T., Tank, D. W., Wieschaus, E. F., Bialek, W. (2007b). Probing the limits to positional information. *Cell* **130**, 153–164.

Grossniklaus, U., Pearson, R. K., and Gehring, W. J. (1992). The *Drosophila sloppy paired* locus encodes two proteins involved in segmentation that show homology with mammalian transcription factors. *Genes Dev.* **6**, 1030–1051.

Harfe, B. D., Scherz, P. J., Nissim, S., Tian, H., McMahon, A. P., and Tabin, C. J. (2004). Evidence for an expansion-based temporal Shh gradient in specifying vertebrate digit identities. *Cell* **118**, 517–528.

Hashimoto, C., Gerttula, S., and Anderson, K. V. (1991). Plasma membrane localization of the Toll protein in the syncytial *Drosophila* embryo: Importance of transmembrane signaling for dorsal–ventral pattern formation. *Development* **111**, 1021–1028.

Holley, S. A., Jackson, P. D., Sasai, Y., Lu, B., De Robertis, E. M., Hoffmann, F. M., and Ferguson, E. L. (1995). A conserved system for dorsal–ventral patterning in insects and vertebrates involving sog and chordin. *Nature* **376**, 249–253.

Holley, S. A., Neul, J. L., Attisano, L., Wrana, J. L., Sasai, Y., O'Connor, M. B., De Robertis, E. M., and Ferguson, E. L. (1996). The Xenopus dorsalizing factor noggin ventralizes Drosophila embryos by preventing DPP from activating its receptor. *Cell* **86**, 607–617.

Houchmandzadeh, B., Wieschaus, E., and Leibler, S. (2002). Establishment of developmental precision and proportions in the early Drosophila embryo. *Nature* **415**, 798–802.

Houchmandzadeh, B., Wieschaus, E., and Leibler, S. (2005). Precise domain specification in the developing Drosophila embryo. *Phys. Rev. E Stat. Nonlin. Soft Matter. Phys.* **72**, 061920.

Howard, M., and ten Wolde, P. R. (2005). Finding the center reliably: Robust patterns of developmental gene expression. *Phys. Rev. Lett.* **95**, 208103.

Ingham, P. W., Taylor, A. M., and Nakano, Y. (1991). Role of the *Drosophila patched* gene in positional signaling. *Nature* **353**, 184–187.

Ingolia, N. T. (2004). Topology and robustness in the *Drosophila* segment polarity network. *PLoS Biol.* **2**, E123.

Jaeger, J., Surkova, S., Blagov, M., Janssens, H., Kosman, D., Kozlov, K. N., Myasnikova, E., Vanario-Alonso, C. E., Samsonova, M., Sharp, D. H., and Reinitz, J. (2004). Dynamic control of positional information in the early *Drosophila* embryo. *Nature* **430**, 368–371.

Jazwinska, A., Rushlow, C., and Roth, S. (1999). The role of brinker in mediating the graded response to Dpp in early Drosophila embryos. *Development* **126**, 3323–3334.

Keizer, J. (1987). Diffusion effects on rapid bimolecular chemical reactions. *Chem. Rev.* **87**, 167–180.
Kerszberg, M. (1999). Morphogen propagation and action: Towards molecular models. *Semin. Cell Dev. Biol.* **10**, 297–302. Review.
Kerszberg, M., and Wolpert, L. (1998). Mechanisms for positional signaling by morphogen transport: A theoretical study. *J. Theor. Biol.* **191**, 103–114.
Lander, A., Nie, Q., and Wan, F. Y. M. (2002). Do morphogen gradients arise by diffusion? *Dev. Cell* **2**, 785–796.
Levine, M., and Davidson, E. H. (2005). Gene regulatory networks for development. *Proc. Natl. Acad. Sci. USA* **102**, 4936–4942.
Margolis, J. S., Borowsky, M. L., Steingrimsson, E., Shim, C. W., Lengyel, J. A., and Posakony, J. W. (1995). Posterior stripe expression of hunchback is driven from two promoters by a common enhancer element. *Development* **121**, 3067–3077.
Marques, G., Musacchio, M., Shimell, M. J., Wunnenberg-Stapleton, K., Cho, K. W., and O'Connor, M. B. (1997). Production of a DPP activity gradient in the early *Drosophila* embryo through the opposing actions of the SOG and TLD proteins. *Cell* **91**, 417–426.
Mason, E., Williams, S., Grotendorst, G., and Marsha, J. (1997). Combinatorial signaling by twisted gastrulation and decapentaplegic. *Mech. Dev.* **64**, 61–75.
Massague, J., and Chen, Y. G. (2000). Controlling TGF-beta signaling. *Genes Dev.* **14**, 627–644.
Mizutani, C. M., Nie, Q., Wan, F. Y., Zhang, Y. T., Vilmos, P., Sousa-Neves, R., Bier, E., Marsh, J. L., and Lander, A. D. (2005). Formation of the BMP activity gradient in the Drosophila embryo. *Dev. Cell* **8**, 915–924.
Morisato, D., and Anderson, K. V. (1994). The spatzle gene encodes a component of the extracellular signaling pathway establishing the dorsal–ventral pattern of the Drosophila embryo. *Cell* **76**, 677–688.
Nagorcka, B. N., and Mooney, J. R. (1982). The role of a reaction–diffusion system in the formation of hair fibres. *J. Theor. Biol.* **98**, 575–607.
Nguyen, M., Park, S., Marques, G., and Arora, K. (1998). Interpretation of a BMP activity gradient in Drosophila embryos depends on synergistic signaling by two type I receptors, SAX and TKV. *Cell* **95**, 495–506.
O'Connor, M. B., Umulis, D. M., Othmer, H. G., and Blair, S. S. (2006). Shaping BMP morphogen gradients in the Drosophila embryo and pupal wing. *Development* **133**, 183–193.
Othmer, H. G. (1980). Synchronized and differentiated modes of cellular dynamics. *In* "Dynamics of Synergetic Systems" (H. Haken, Ed.), Springer-Verlag, Berlin/London.
Othmer, H. G., and Pate, E. (1980). Scale-invariance in reaction–diffusion models of spatial pattern formation. *Proc. Natl. Acad. Sci. USA* **77**, 4180–4184.
Reeves, G. T., Muratov, C. B., Schupbach, T., and Shvartsman, S. Y. (2006). Quantitative models of developmental pattern formation. *Dev. Cell* **11**, 289–300.
Ross, J. J., Shimmi, O., Vilmos, P., Petryk, A., Kim, H., Gaudenz, K., Hermanson, S., Ekker, S. C., O'Connor, M. B., and Marsh, J. L. (2001). Twisted gastrulation is a conserved extracellular BMP antagonist. *Nature* **410**, 479–483.
Sen, J., Goltz, J. S., Stevens, L., and Stein, D. (1998). Spatially restricted expression of pipe in the Drosophila egg chamber defines embryonic dorsal–ventral polarity. *Cell* **95**, 471–481.
Sen, J., Goltz, J. S., Konsolaki, M., Schupbach, T., and Stein, D. (2000). Windbeutel is required for function and correct subcellular localization of the Drosophila patterning protein Pipe. *Development* **127**, 5541–5550.
Shimmi, O., and O'Connor, M. B. (2003). Physical properties of Tld, Sog, Tsg, and Dpp protein interactions are predicted to help create a sharp boundary in Bmp signals during dorsoventral patterning of the Drosophila embryo. *Development* **130**, 4673–4682.
Shimmi, O., Umulis, D., Othmer, H. G., and O'Connor, M. B. (2005). Facilitated transport of a Dpp/Scw heterodimer by Sog/Tsg leads to robust patterning of the Drosophila blastoderm embryo. *Cell* **120**, 873–886.
Sick, S., Reinker, S., Timmer, J., and Schlake, T. (2006). WNT and DKK determine hair follicle spacing through a reaction–diffusion mechanism. *Science* **314**, 1447–1450.

Sontag, E. D. (2003). Adaptation and regulation with signal detection implies internal model. *Syst. Control Lett.* **50**, 119–126.
Strigini, M. (2005). Mechanisms of morphogen movement. *J. Neurobiol.* **64**, 324–333.
Tabata, T., Eaton, S., and Kornberg, T. B. (1992). The *Drosophila hedgehog* gene is expressed specifically in posterior compartment cells and is a target of engrailed regulation. *Genes Dev.* **6**, 2635–2645.
Turing, A. M. (1952). The chemical basis of morphogenesis. *Philos. Trans. R. Soc. London Ser. B Biol. Sci.* **237**, 37–72.
Umulis, D. M., Serpe, M., O'Connor, M. B., and Othmer, H. G. (2006). Robust, bistable patterning of the dorsal surface of the Drosophila embryo. *Proc. Natl. Acad. Sci. USA* **103**, 11613–11618.
Umulis, D.M., O'Connor, M.B., and Othmer, H.G. (2007). Scale-invariance, Embryon. Dev. (in preparation).
von Dassow, G., Meir, E., Munro, E. M., and Odell, G. M. (2000). The segment polarity network is a robust developmental module. *Nature* **406**, 188–192.
von Dassow, G., and Odell, G. M. (2002). Design and constraints of the *Drosophila* segment polarity module: Robust spatial patterning emerges from intertwined cell state switches. *J. Exp. Zool.* **294**, 179–215.
Wang, Y. C., and Ferguson, E. L. (2005). Spatial bistability of Dpp-receptor interactions during Drosophila dorsal–ventral patterning. *Nature* **434**, 229–234.
Wolpert, L. (1969). Positional information and the spatial pattern of cellular differentiation. *J. Theor. Biol.* **25**, 1–47.
Wolpert, L., Beddington, R., Jessel, T., Lawrence, P., Meyerowitz, E., and Smith, J. (2002). "Principles of Development." Oxford Univ. Press, New York, NY.
Young, M. E., Carroad, P. A., and Bell, R. L. (1980). Estimation of diffusion coefficients of proteins. *Biotechnol. Bioeng.* **22**, 947–955.

3

Integrating Morphogenesis with Underlying Mechanics and Cell Biology

Lance A. Davidson
Department of Bioengineering, University of Pittsburgh, Pittsburgh, Pensylvania 15260

I. Introduction
II. *Xenopus laevis* as a Model System
III. Distinct and Separable Tissue-Scale Processes
IV. Complex Trajectories through Dynamic Microenvironments
    A. Mid-Blastula Transition/Cleavage Cycles (FN Fibril Assembly)
    B. Epiboly (Radial Intercalation)
    C. Reach Base of Cleft (Avidity Change–Tissue-Separation)
    D. Deep Cell Involution (Restorative Radial Intercalation)
    E. Convergent Extension (Mediolateral Cell Intercalation)
V. At the Cell-Scale: Act Locally Move Globally
VI. Molecular-Scale: Mechanics, Adhesion, and Traction
VII. Modeling Morphogenesis: A Grand Challenge
    Acknowledgments
    References

Morphogenesis integrates a wide range of cellular processes into a self-organizing, self-deforming tissue. No single molecular "magic bullet" controls morphogenesis. Wide ranging cellular processes, often without parallels in conventional cell culture systems, work together to generate the architecture and modulate forces that produce and guide shape changes in the embryo. In this review we summarize the early development of the frog *Xenopus laevis* from a biomechanical perspective. We describe processes operating in the embryo from whole embryo scale, the tissue-scale, to the cellular and extracellular matrix scale. We focus on describing cells, their behaviors and the unique microenvironments they traverse during gastrulation and discuss the role of tissue mechanics in these processes. © 2008, Elsevier Inc.

I. Introduction

Consider the challenges met by any organism that seeks a multicellular rather than a solo lifestyle. Many essential processes of a single cell take place as material crosses the cell's surface. Nutrients must be taken in and signals must be sensed at the surface. Waste must be eliminated through secretion or gas exchange. These processes scale unfavorably when multiple cells join together; however, advantages to the organism such as defense against the environment, adaptability, and specialization can offset the disadvantages. A cell within a multicellular organism still retains the ability to alter its environment, change its shape, or move. Multicellular organisms have evolved several strategies in their development to offset the unfavorable surface area deficiencies of larger organisms by increasing the surface area available to their cells.

Gastrulation is one of the earliest and most crucial strategies to increase surface area and involves a major topological transformation, turning a ball of cells into hollow thick-walled cylinder (Stern, 2004). Gastrulation serves to double the effective surface area, places endoderm in the inner lining of the cylinder, ectoderm on the outer surface, and mesoderm in between. As gastrulation proceeds the short thick-walled cylinder is transformed into an elongate cylinder through a process known as *axis elongation*. This long hollow cylinder is then elaborated during organogenesis as large surface areas vital to functions such as nutrition (stomach and intestines), gas exchange (heart, gills, and vasculature), and waste removal (liver and kidney) are created. Morphogenesis and organogenesis brings "external" surfaces to otherwise deeply buried cells to carry out these surface-dependent functions as if they were single cells. One of the greatest challenges in modern biology is unraveling the complex details of morphogenesis along with the cell, molecular, and biophysical mechanisms responsible for shaping the embryo.

Most research on the mechanisms driving morphogenesis is carried out on model organisms that have been selected based on their unique experimental qualities or their particular modes of development (Bolker, 1995; Gilbert, 2006). Several invertebrate organisms such as *Dictyostelium discoides* (slime mold), *Drosophila melanogaster* (fruit fly), *Caenorhabditis elegans* (nematode worm), and the sea urchin are well characterized due to their small size, rapid development, stereotyped cell divisions, or abundance. Fly and worm have proven extremely adaptable genetic model systems. Vertebrate model systems such as *Xenopus laevis* (frog), *Gallus gallus* (chicken), and *Danio rerio* (zebrafish) are studied due to their ease of use, large size, and accessibility to classic embryological techniques such as microsurgery. Zebrafish is also a well developed genetic model. *Mus musculus* (house mouse) is a mammalian model system for development and is as genetically malleable as invertebrate genetic models. Each model system has advantages and disadvantages but modern molecular techniques have unified these systems into a single test bed for studying the role of many genes that are conserved between humans and invertebrates (Alberts *et al.*, 2002).

II. *Xenopus laevis* as a Model System

The *Xenopus laevis* amphibian model has been useful for studying the molecular mechanisms for patterning cell identity and cell movements in a vertebrate embryo due to its large size and tolerance of both molecular and microsurgical manipulation. Since eggs are fertilized *in vitro*, the one-cell stage embryo can be injected with protein, mRNA, DNA, lineage tracers or other prepared molecules (Sive *et al.*, 2000). Exogenous proteins can be overexpressed from injected synthetic mRNA and

3. Integrating Morphogenesis with Mechanics

protein translation in the embryo can be selectively knocked-down after injection of morpholino oligonucleotides. Since the large fertilized *Xenopus* egg undergoes predictable cell divisions individual blastomeres can be injected with mRNA or morpholinos to create embryos with scattered individual or whole patches of injected cells targeted to specific territories (Blitz *et al.*, 2006). Techniques for microsurgery developed in other large amphibian species over the last century have proven to be completely transferable to *Xenopus laevis* (Keller, 1991; Malacinski *et al.*, 2000). These classical microsurgery techniques allow either removal of tissue explants or grafts of injected cells or tissues at controlled times in development into specified locations in prepared host embryos. Microsurgery uniquely allows recombination of tissues that would be impossible to achieve through genetic means (Ariizumi *et al.*, 2000). Tissue explants can be cultured and incubated in growth factors, placed into contact with other tissues, or visualized with high resolution confocal microscopy. Combining microsurgery, molecular manipulation, and confocal microscopy provides unique opportunities to test cell and molecular mechanisms of morphogenesis and organogenesis. Studies using *Xenopus laevis* have been used to expose the roles and functions of many genes that specify the three germ layers and cell polarity within those layers (De Robertis and Kuroda, 2004; Heasman, 2006; Wallingford *et al.*, 2002; Wardle and Smith, 2006).

Molecular control of morphogenesis has three primary functions: (i) establishment of the cell identity, (ii) establishment of cell polarity, and (iii) programming cell behavior. Specification, or establishment of cell identity, lays out mesoderm, endoderm, and ectoderm territories in the early embryo. These territories are generally indicated by domains of gene expression (e.g., chordin indicates the notochord field); however, gene expression patterns are malleable and can change through both cell movement and refinement of expression pattern. A cell can have a specific identity but still not recognize its position and orientation. Positional cues create subfields within germ layers while polarity provides information on its orientation within a field. Long range factors and short range microenvironmental cues can polarize a cell within a field. In recent years there has been significant progress in understanding the establishment of cell identity and polarity; in contrast we still have a poor understanding of the molecular control of both the microenvironment and cell behaviors during morphogenesis. We will focus on the vertebrate model *Xenopus laevis* which has proven useful for cell biological studies in addition to its classical use in experimental embryology.

III. Distinct and Separable Tissue-Scale Processes

Descriptive cell and molecular studies complemented by classical amphibian embryology allow the definition of early landmarks of the embryonic axis and establishment of a timeline of tissue movements (see Table I; Nieuwkoop and Faber, 1967). The *animal pole*, marking the center of the pigmented half of the embryo, and the *vegetal pole*

Table I Timeline of morphogenetic processes active during gastrulation in *Xenopus laevis*

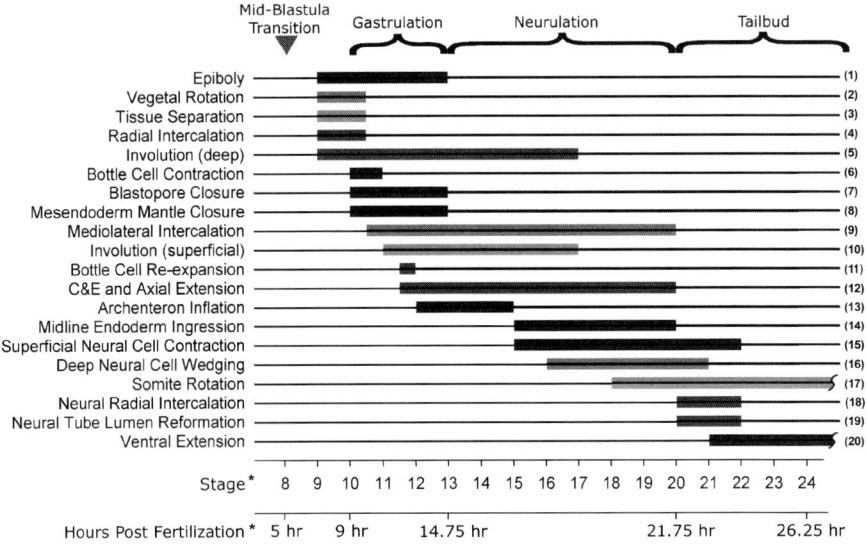

1. Epiboly (Keller, 1980; Marsden and DeSimone, 2001). 2. Vegetal rotation (Winklbauer and Schurfeld, 1999). 3. Tissue separation (Wacker *et al.*, 2000; Winklbauer *et al.*, 2001). 4. Radial intercalation (Keller, 1980; Longo *et al.*, 2004; Marsden and DeSimone, 2001). 5. Involution (deep) (Papan *et al.*, 2007). 6. Bottle cell contraction (Hardin and Keller, 1988). 7. Blastopore closure (Nieuwkoop and Faber, 1967). 8. Mesendoderm mantle closure (Davidson *et al.*, 2002; Ibrahim and Winklbauer, 2001). 9. Mediolateral intercalation (Keller *et al.*, 2000). 10. Involution (superficial) (Papan *et al.*, 2007). 11. Bottle cell reexpansion (Ewald *et al.*, 2004; Nieuwkoop and Faber, 1967). 12. Convergence and extension and axial extension (Keller 2002, 2006). 13. Archenteron inflation (Ewald *et al.*, 2004). 14. Midline endoderm ingression (Minsuk and Keller, 1997; Shook *et al.*, 2002, 2004). 15. Superficial neural cell contraction (Davidson and Keller, 1999; Lee *et al.*, 2007; Schroeder, 1970). 16. Deep neural cell wedging (Davidson and Keller, 1999; Schroeder, 1970). 17. Somite rotation (Afonin *et al.*, 2006; Keller, 2000). 18. Neural radial intercalation (Davidson and Keller, 1999). 19. Neural tube lumen reformation (Davidson and Keller, 1999). 20. Ventral extension (Larkin and Danilchik, 1999). * Stages and timing based on Nieuwkoop and Faber (1967).

mark, marking the center of the unpigmented vegetal half, define the animal–vegetal axis (Fig. 1A). These poles are present in the unfertilized egg and remain useful landmarks until the last exposed vegetal cells of the yolk plug are enclosed toward the end of gastrulation (Fig. 1A). Shortly into the rapid cell divisions that follow fertilization a fluid filled space, the *blastocoel*, forms directly under the animal pole (Fig. 2A). The blastocoel remains a distinct landmark throughout gastrulation (Figs. 2B and 2C). The *marginal zone* consists of an "equatorial belt" of tissue midway between the animal and vegetal poles that contributes to the dorsal axial tissues in the tadpole (Fig. 1A). The vegetal portion of the marginal zone that carries mesoderm and endoderm into the embryo is the involuting marginal zone (IMZ) and the animal portion of the marginal zone that contributes tissue to the neural tube and dorsal epidermis is the noninvoluting marginal zone (NIMZ) (Keller *et al.*, 2003).

3. Integrating Morphogenesis with Mechanics

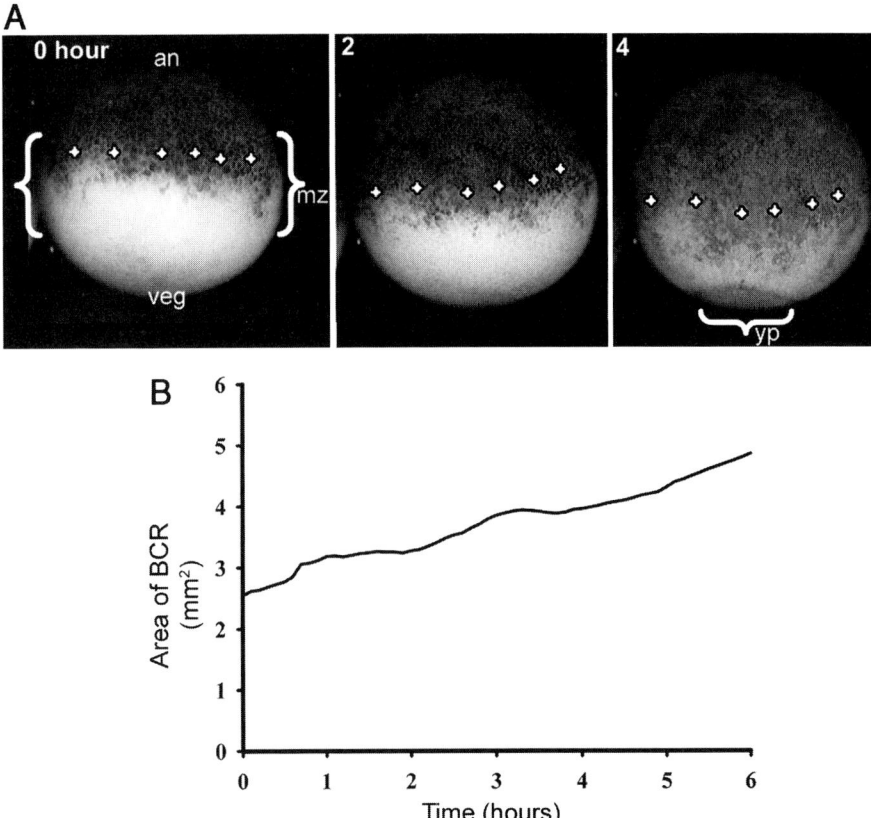

Figure 1 Epiboly spreads the animal cap ectoderm over the embryo. (A) Frames from a low magnification time-lapse sequence track the movements of animal cap ectoderm during epiboly. Based on histological data (not shown), the lower limits of the blastocoel are identified with "+'s" in the first frame (0 hours). Movement of these fiduciary marks is noted with "+'s" in subsequent frames. Over the course of gastrulation these fiduciary marks are tracked on frames from a time-lapse sequence collected with a stereoscope modified to lie on its side. (B) Calculation of the three-dimensional spherical shape of the ectoderm shows the surface area increases 15% per hour over 6 hours. Brackets indicate the location of the marginal zone (mz) at 0 hours (late blastula stage) and the yolk plug (yp) 4 hours later (stage 11.5).

Gastrulation begins with the contraction of a patch of epithelial cells into *bottle cells* at a site that marks the future anterior end of the endoderm (Hardin and Keller, 1988; Nieuwkoop and Faber, 1967). The edge of the exposed tissue immediately adjacent to the vegetal cells is the *blastopore lip*. This site where bottle cells first contract is referred to as the "dorsal" blastopore. [Note: The term "dorsal" is a term in common usage, along with the use of the term "ventral" to describe the opposite point, however, the resulting "dorsal–ventral" axis of the three germ layers in the early gastrula embryo do not always coincide with the true dorsal–ventral axis of the germ layers

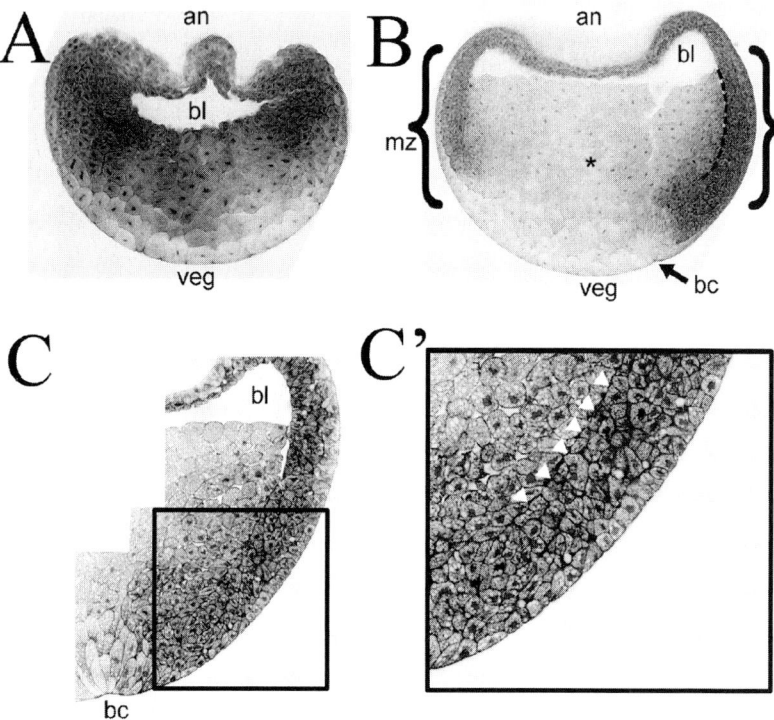

Figure 2 Microenvironments during gastrulation. Mid-sagittal sections collected using a confocal laser scanning microscope of rhodamine dextran labeled embryos (Davidson and Wallingford, 2005) show the location and organization of microenvironments from late blastula (stage 8) to mid-gastrula (stage 11.5). (A) The blastula stage embryo has a well defined gradient of cell sizes from large cells in the vegetal endoderm (veg) to the small cells of the animal cap ectoderm (an). Across the marginal zone the cells are evenly sized. (B) By gastrula stage cell size becomes more heterogeneous; small cells are found underlying the marginal zone (brackets; mz) with large cells in the core of the embryo (asterisk). (C and C') The bottle cells (bc) form at the "dorsal" midline (arrow) and a cleft forms under the dorsal marginal zone (dotted line). (D and D') Involution is well underway by mid-gastrulation (stage 11.5). A yolk plug is still prominent but is in the process of being engulfed by the blastopore lip (bpl). Cells reside transiently in the blastopore lip as they transition from pre- to postinvolution (pre- and post-). Superficial epithelial cells (black arrow) move around the outer surface of the blastopore lip whereas deep cells transiently enter a mixing region as they move around the base of the cleft (white arrowheads). By mid-gastrula stage first mesendodermal tissues (mese) to move into the embryo have reached the anterior end of the neural ectoderm (ne).

in the postgastrula embryo (Lane and Sheets, 2006).] Gastrulation proceeds as more bottle cells form laterally toward the ventral blastopore with epithelial cells at a particular "latitudes" along the animal–vegetal axis progressively contracting their apical surface. The large vegetal yolk cells enclosed by the ring of bottle cells form the *yolk plug* (Fig. 1A). The area of the yolk plug exposed on the embryo's surface decreases during the course of gastrulation until it is completely engulfed by the embryo. These

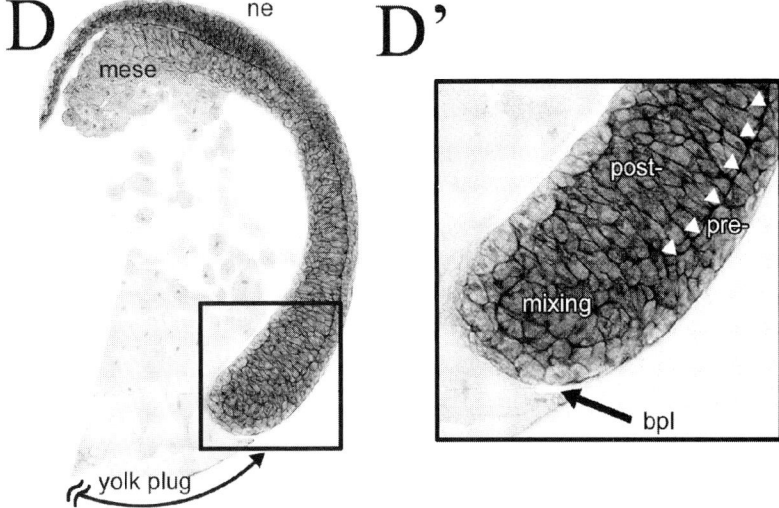

Figure 2 (*continued*)

movements can be seen in the whole embryo using a time-lapse image collection with a stereomicroscope, however, since the frog embryo is optically opaque internal movements of cells and tissues are hidden from view.

A tissue *cleft* appears coincident with the appearance of bottle cells. Morphologically, this cleft appears as a fracture between two groups of cells that had previously appeared as a coherent mass. The cleft forms initially beneath the dorsal marginal zone but like bottle cell formation, and with approximately the same progressive rate, spreads laterally, eventually separating the entire superficial marginal zone from the deeper vegetal endoderm.

Within the embryo the distinct *large scale tissue movements* of bottle cell contraction, cleft formation and tissue separation, engulfment of the yolk plug, and epiboly, and convergent extension occur simultaneously during gastrulation (Table I). *Epiboly* is defined as the process of ectodermal spreading over the entire surface of the embryo. Starting before the formation of bottle cells, epiboly continues until the end of gastrulation (Figs. 1A and 1B). The original surface of the animal cap ectoderm spreads to surround the entire embryo providing prospective epidermis (Fig. 1B). *Tissue separation* and *vegetal rotation* occur as vegetal endoderm tissues "flow" as if they were advecting from the vegetal to the animal pole. Vegetal endoderm cells reach the floor of the blastocoel and spread outward toward the marginal zone. This movement combined with the onset of adhesion changes bring about the separation of tissues across the newly forming cleft. *Mesendoderm mantle closure* occurs as tissues originally located beneath the surface of the marginal zone move as a "mantle" toward the animal pole along the surface of the animal ectoderm. The mesendoderm mantle effectively encloses the blastocoelic space and separates it from the animal ectoderm. Epiboly, tissue separation, and mesendoderm mantle closure can be observed at all "longitudes" around the embryo whereas convergent extension is confined to the gastrula's dorsal quadrant. More lateral tissues are draw-in from other quadrants and packed into the posterior dorsal axis. *Convergent extension* is the narrowing and lengthening of definitive dorsal tissues from the start of gastrulation through formation of primitive muscle from undifferentiated dorsal mesoderm. The term "convergent extension" describes the anterior–posterior lengthening and mediolateral narrowing of dorsal axial tissues from early gastrulation through neurulation. Lengthening and narrowing is accomplished as both mesodermal and neural cells rearrange within the plane in which convergent extension is confined. Movements of cells out-of-the-plane into a multilayered cell mass typically reduces the extension-generating efficiency of convergence movements and is termed "convergent thickening" (Davidson *et al.*, 2006; Keller and Shook, 2004). In *Xenopus* both mesodermal and neural tissues undergo convergent extension and together drive the lengthening of the dorsal axis (Elul and Keller, 2000). Convergent extension drives both the early movement of mesoderm into the embryo as well as the dorsal elongation of the embryonic axis in a process known as axis elongation. Epiboly, tissue separation and vegetal rotation, mesendoderm mantle closure, and convergent extension create the tissue architecture of the vertebrate body plan, establishing distinct yet undifferentiated tissues of the ectoderm, mesoderm, and endoderm.

How are these movements coordinated and coupled? The tissue movements described above are the result of multiple cellular processes interacting at a distance. Traction and cell shape change produce forces that must be transmitted to adjacent cells and beyond. How those forces are transmitted, sensed, and transduced into action dictates the course of morphogenesis. One approach is to consider the mechanical constraints on these tissue scale processes.

Just as the biochemistry of the embryo functions within physiological limits such as temperature and osmolarity, dynamic mechanical changes within the embryo are regulated by the capacity of the tissue to generate force and maintain tissue stiffness (Koehl, 1990). Furthermore, the physical mechanism of tissue shape change (i.e., the location and orientation of forces) must be compatible with the physical architecture of the embryo. This is not to say that morphogenetic processes are particularly fragile, but that we must consider the ability of the embryo to successfully carry out tissue movements within a range of physical mechanical properties. For instance, consider several possible ways that gastrulation can fail: (1) cells may not generate enough force to change the tissue shape, (2) the tissue is either too stiff to be deformed, too deformable to support cell-generated forces, or too viscous to maintain shape changes, or (3) the architecture of the embryo is malformed from the start. Malformed architecture may result from too few cells generating force or a misshapen distribution of cells leading to misdirected forces.

Consider the mechanical constraints on the process of dorsal axis extension. First, the dorsal mesodermal tissues must generate autonomous forces sufficient to deform itself. Second, those forces must be sufficient to deform surrounding tissues such as the overlying neural ectoderm, the anterior prechordal mesoderm and lateral plate mesoderm. Next, the dorsal tissues must be sufficiently stiff to apply these forces to surrounding tissues without buckling or kinking. Finally, the entire tissue must deform reproducibly in order to place tissue in the proper location for subsequent phases of induction and organogenesis. These constraints are likely to be under both physical and molecular control and mechano-chemically coupled to each other. For instance, the ability of a tissue to generate force might depend on the same molecular systems as the maintenance of tissue stiffness.

Tissue architecture can guide morphogenetic movements by coupling distinct tissues into a composite so that two tissues either deform in register or by physically uncoupling adjacent tissues to allow independent movements via shear. Tissues in the embryo are constructed from: (i) epithelial cell sheets connected apically with tight and adherens junctions, (ii) mesenchymal sheets of cells, and (iii) extracellular matrix. The stereotyped organization and development of these tissue components define embryonic development yet their maintenance and mechanical roles are poorly understood. For instance, are tissue boundaries sites of mechanical weakness or strength? Does extracellular matrix provide a stiff scaffold, a glue holding tissues together, or provide an interface for tissues to move over one another? In order to answer questions concerning the function of tissue architecture during gastrulation we will focus on the dynamic interaction of cells and the extracellular matrix during gastrulation.

IV. Complex Trajectories through Dynamic Microenvironments

The extracellular matrix (ECM) serves many functions in both the embryo and adult organism (Alberts *et al.*, 2002). A major function of the ECM is to provide a stiff mechanical scaffold for cell adhesion. Cell–matrix adhesion enables cells to apply forces to their neighbors and more distant tissues through cell shape change (Harris *et al.*, 1980; Pelham and Wang, 1997; Sawhney and Howard, 2002; Stopak and Harris, 1982). Cell–matrix adhesion can guide migratory cells and provide a tractive-mechanical substrate for cell movement (Lo *et al.*, 2000). Assembled ECM can increase tissue stiffness directly or through enhanced cell–matrix adhesions (Gildner *et al.*, 2004). Cell–matrix adhesion also plays a role in tissue homeostasis by keeping cells in an environment where a cell's proliferation or differentiation status can be maintained (Harris, 1987). In addition to the adhesive role of ECM many growth factors and molecular guidance cues bind the ECM which serves to localize these cues. These cues can repress or initiate cell differentiation or motility programs, provide positional information as well as orient the cell (Alberts *et al.*, 2002). Once cells are polarized, or motile and directed, the ECM provides a tractive substrate for cell contraction or translocation.

During gastrulation mesoderm and ectodermal tissues assemble and remodel the extracellular matrix (Davidson *et al.*, 2004; Nakatsuji *et al.*, 1985; Winklbauer, 1998). It is common to think about the ECM as a static structure, however, the ECM undergoes remodeling throughout development, organogenesis, and into adulthood (Hynes,

1990). Remodeling can take the form of polymerization of fibrils at new locations, addition or removal of specific classes of glycoproteins from existing matrices, the physical movement of fibrils, or the modification of matrix topology (Sawhney and Howard, 2002; Sivakumar *et al.*, 2006). Remodeling can be accomplished on the molecular level by localized secretion of new matrix proteins or by selective degradation by proteases. Remodeling can also occur on the macroscopic level through the physical action of cells (Sawhney and Howard, 2002; personal observations). In the case of fibronectin, an early protein forming fibrils within the ECM, polymerization may require molecular-scale strains driven by cell traction (Hocking *et al.*, 1996; Zhong *et al.*, 1998). *In vitro* studies have shown that fibronectin fibrils can even depolymerize after cell traction is inhibited (Baneyx *et al.*, 2002). The need for cellular forces to remodel ECM may provide a reason for the continuing tractive or motile activity of adult somatic cells even though they have nowhere to go. Complex cell and molecular pathways regulating ECM assembly and usage form cycles rather than a linear pathway. Cell motility is required to assemble a fibrillar matrix that then modulates cell behaviors. ECM assembly via cell-generated traction in turn creates a microenvironment required by cells migrating singly or as groups through established tissues. At the largest scale the microenvironment regulates how the entire tissue directs and channels forces into morphogenetic movement. In order to understand these processes we return to examples from the *Xenopus* model system and consider in chronological order the microenvironments crossed by mesoderm cells that will eventually contribute to the *Xenopus* embryo's dorsal axis.

A. Mid-Blastula Transition/Cleavage Cycles (FN Fibril Assembly)

Cell interactions with the ECM are subject to distinct spatial and temporal control. Fibronectin begins to be synthesized in the embryo following the mid-blastula transition but integrin $\alpha 5\beta 1$-dependent assembly of fibronectin fibrils by deep-layer prospective ectoderm cells facing the blastocoel (animal cap ectoderm) (Lee *et al.*, 1984) does not begin until just prior to gastrulation. Assembly is restricted to the cell surface facing the blastocoel even though integrin $\alpha 5\beta 1$ is expressed ubiquitously throughout the embryo and localizes to all basolateral cell surfaces from cleavage stages onward (Gawantka *et al.*, 1992; Joos *et al.*, 1995).

B. Epiboly (Radial Intercalation)

Once fibrils are deposited on the blastocoel-facing surface of the animal cap ectoderm, additional fibronectin- and integrin-dependent cell behaviors are initiated. The earliest of these behaviors is radial intercalation within the multicell layered blastocoel roof where prospective ectoderm cells receive polarity cues from FN fibrils and intercalate contributing to the forces driving epiboly (Longo *et al.*, 2004;

Marsden and DeSimone, 2001). Fibril density increases throughout the process of radial intercalation (Longo et al., 2004). As epiboly proceeds fibronectin fibrils form in the cleft under the marginal zone (Fig. 2B). Fibrils in the cleft between pre- and postinvolution marginal zone tissues may be freshly assembled or physically carried into position. Mesendoderm cells that will contribute to the heart and head of the embryo first begin to migrate toward the animal pole from the dorsal marginal zone (Figs. 2C and 2C'). Mesendoderm cells use integrin $\alpha5\beta1$ to migrate on fibronectin fibrils on the undersurface of the animal cap ectoderm (Davidson et al., 2006; Hoffstrom, 2002; Marsden and DeSimone, 2001; Ramos and DeSimone, 1996).

C. Reach Base of Cleft (Avidity Change–Tissue-Separation)

Preinvolution mesoderm cells translocate vegetally along the fibronectin fibril-lined cleft. Formation of the cleft may direct some deep-cell derived notochord cells to the inner surface of the fibril-lined cleft without the need for involution (Ibrahim and Winklbauer, 2001; Papan et al., 2007). The first signs of cell elongation in the vegetal alignment zone are seen in more superficial notochord cells approaching the blastopore (Lane and Keller, 1997). Before reaching the base of the cleft mesoderm cells are thought to exhibit a low affinity for fibronectin (Ramos et al., 1996) and a high cell–cell affinity mediated by C-cadherin (Zhong et al., 1999).

D. Deep Cell Involution (Restorative Radial Intercalation)

Immediately after cells move around the cleft both their cell–cell and cell–matrix affinity appears to change. C-cadherin mediated cell–cell adhesion is thought to drop (Zhong et al., 1999) just as cell–fibronectin adhesion mediated by integrin $\alpha5\beta1$ increases (Ramos et al., 1996). Cells approaching the base of the cleft, five to ten cells animal-ward of the blastopore lip (Figs. 2D and 2D'), have a complex relationship with the ECM lining the cleft. Cells appear to leave the surface of the cleft and mix within a matrix-free region of the blastopore lip. As these cells transition from a pre- to postinvolution array the ECM-mediated polarity cues continue to drive radial intercalation (Marsden and DeSimone, 2001). The timing and location of affinity changes in pre- and postinvolution cells have been operationally defined by the elegant tissue-separation assay developed by Winklbauer and colleagues (Wacker et al., 2000; Winklbauer et al., 2001).

E. Convergent Extension (Mediolateral Cell Intercalation)

After prospective notochord cells have involuted, the cleft marks the interface between the neural ectoderm and mesoderm and a new phase of fibronectin fibril assembly begins between the endoderm, somitic mesoderm, and the notochord (Davidson *et al.*, 2004). New classes of extracellular matrix appear around the notochord such as fibrillin (Skoglund *et al.*, 2006). Arrays of fibrillar fibronectin, sandwiching the mesoderm, continue to be modified during axial extension. In whole embryos, the relationship between polarized protrusive activity, mediolateral cell intercalation and the ECM is difficult to resolve directly but can be observed in explants of these tissues (Davidson *et al.*, 2006; Skoglund and Keller, 2007). A fibrillar ECM may regulate mediolateral cell intercalation through several mechanisms. First, fibrils could act to repress protrusions directed along the fibrillar matrix and enhance traction generated by cell–cell adhesions and prevent cells from crawling under their neighbors, an event that produces tissue-thickening rather than tissue extension. Successful assembly of a fibrillar matrix would serve to maintain the tissue architecture that guides cell rearrangement into axial extension. Fibrils could encourage "restorative" radial intercalation, i.e., confining or channeling cell intercalation along exclusively planar paths rather than cell-layer "jumping" (e.g., radial "deintercalation") that would otherwise lead to convergent thickening. A fibrillar ECM might serve a structural role by directly stiffening axial tissues.

Cells encounter diverse microenvironments as they move within the developing embryo, from the assembly of fibronectin fibrils beginning at the mid-blastula transition through the mesoderm sandwiching fibrillar arrays found during convergent extension. The concurrently operating processes of epiboly and involution appear to produce shear between different layers of mesoderm resulting in a three-dimensionally complex reorganization of cells within the blastopore lip. Observed from outside the embryo, epiboly and involution appear to produce cohesive tissue movements with little mixing (Fig. 1A). However, cell mixing and tissue shear are apparent after tissues are grafted into the anterior marginal zone prior to involution (Fig. 3A) and tracked after involution is underway (Fig. 3B). Tissue shear has been observed in whole embryos using noninvasive imaging such MRI (Papan *et al.*, 2007) but grafting experiments reveal the complex cell mixing that occurs as cells involute. Paradoxically, cell-by-cell analysis of chordin gene expression does not reflect the cell arrangements that take place within the blastopore lip as cell transition from pre- (Figs. 3C and 3C') to postinvolution (Figs. 3D and 3D'). It is likely that the domain of Chordin mRNA expression completely contains this dynamic region.

V. At the Cell-Scale: Act Locally Move Globally

In *Xenopus*, localized distinctive cellular behaviors are thought to drive large scale tissue movements. These local cell behaviors have been operationally defined by their isolation within explanted tissues. Tissue explants have correlated specific cell behaviors with epiboly, mesendoderm mantle closure, and convergent extension. *Radial cell intercalation* of multiple layers of cells into a single layer drives epiboly (Keller, 1980; Marsden and DeSimone, 2001). *Monopolar directed cell migration* drives cohesive mesendodermal tissues to spread along the ectodermal face of the blastocoel so that

3. Integrating Morphogenesis with Mechanics   125

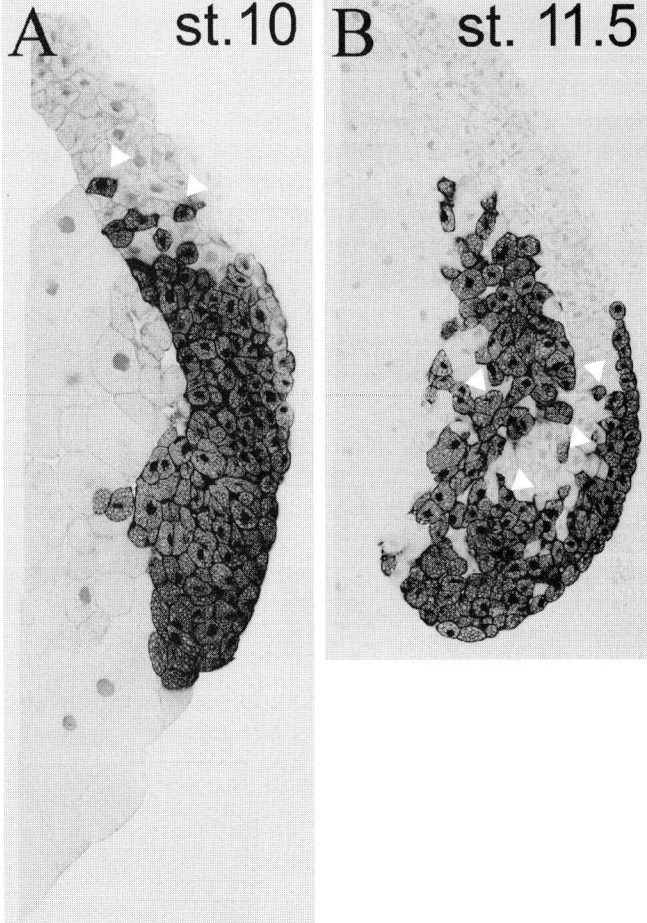

Figure 3 Shearing movements produce more diverse microenvironments. (A) A plug of lineage-labeled cells (rhodamine dextran; dark black) grafted into a host embryo just above the bottle cells at stage 10. The host is lineage-labeled with contrasting fluorophore (fluorescein dextran; light gray). A representative host with the graft is fixed 1 hour later at stage 10.25, bisected, cleared in Murray's clear, and confocal sectioned (Davidson and Wallingford, 2005). (B) By stage 11.5 the cells within the grafted plug have undergone shearing and mixed extensively along the plug's anterior and posterior boundaries (arrowheads). The resulting radial mixing leaves labeled surface cells well posterior of the scattered labeled deep cells (arrowheads). Many deep cells have already undergone deep involution whereas no superficial cells have undergone involution. (C) Confocal sagittal section of the whole embryo and a close-up (C') showing chordin gene expression in the late blastula (stage 9). At the stages shown, chordin marks prospective notochord, midline endoderm, and neural floorplate cells (Sasai *et al.*, 1994). (D) Deep marginal zone cells expressing chordin undergo deep involution around the base of the cleft (white dashed line). In contrast to the shear seen in lineage labeled cells (B) the cells expressing chordin present a sharp line (arrowheads) at their posterior margin (D'). Chordin expressing cells are visualized in from mid sagittal confocal sections collected from embryos processed for fluorescent RNA *in situ* hybridization (Davidson and Keller, 1999).

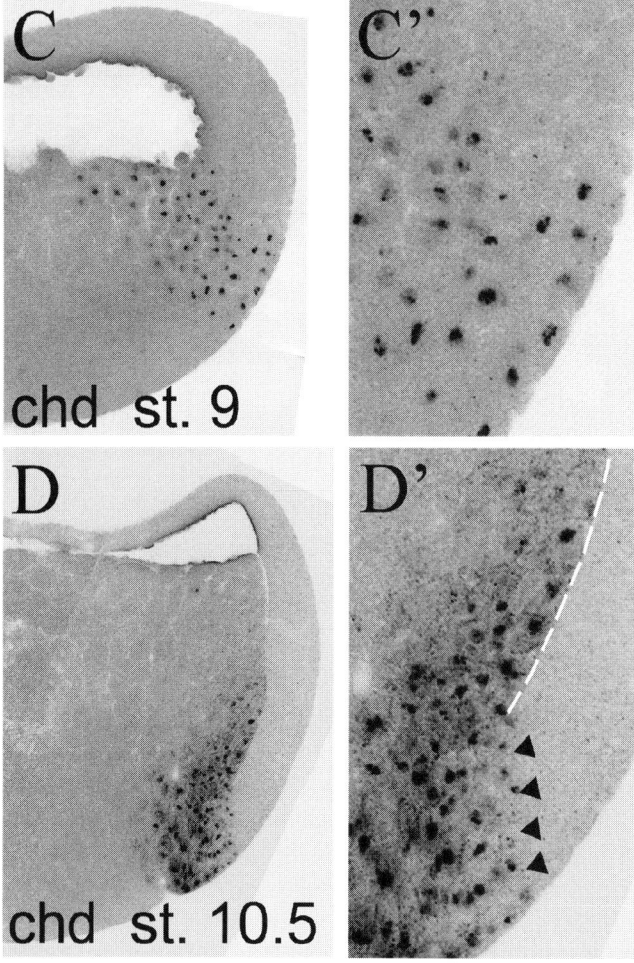

Figure 3 (*continued*)

the mantle converges to the center of the animal pole in mesendodermal mantle closure (Davidson *et al.*, 2002; Ren *et al.*, 2006; Winklbauer *et al.*, 1991). *Mediolateral cell intercalation* involves the directed rearrangement of prospective mesoderm cells that results in cells forcing their medial and lateral neighbors apart along the anterior–posterior axis driving convergent extension independently in both the prospective mesoderm and neural tissues (Keller *et al.*, 2000). Multiple cell behaviors can operate within the same cells to produce greater movements than when operating individually. For instance, radial cell intercalation occurs during mesendoderm mantle closure, bringing more cells into contact with the blastocoelar surface

3. Integrating Morphogenesis with Mechanics

of the animal ectoderm (Davidson *et al.*, 2002), continues during convergent extension to prevent multilayering of cells in converging mesoderm (Davidson *et al.*, 2006; Marsden and DeSimone, 2003), and plays a major role in establishing the single-cell layer of the neural tube during relumeration (Davidson and Keller, 1999).

Each of these cell behaviors can be thought of as a specialized instance of cell motility operating within a densely packed cellular space. The duty cycle of these motile cells (Horwitz and Parsons, 1999), e.g., protrusion, attachment, contraction, translocation, and retraction, must be oriented either uni- or bidirectionally. Radial intercalation involves the directed movement of cells either away from the outer epithelial cell layer or toward the extracellular matrix. Radially intercalating cells migrate singly and unidirectionally to insert themselves onto a fibrillar ECM leading to the spreading and thinning of the ectoderm during epiboly (Longo *et al.*, 2004; Marsden and DeSimone, 2001). Radially intercalating cells appear not to require ECM directly for this movement but instead receive cues from cells already in contact with the ECM. Mesendoderm cells also migrate unidirectionally, but do so with all of their immediate neighbors as a cohesive mass of cells (Davidson *et al.*, 2002; Winklbauer *et al.*, 1996). Mesendoderm motility requires an underlying ECM but cells take their directional cues from the geometry of their initial position. Following the only direction available they behave like the "free-edge" of a wounded sheet of cells (Rand, 1915; Trinkaus, 1984). The cells immediately behind the leading edge cells are then entrained by this free-edge-cue so that the entire mesendoderm migrates directionally as a coherent mass.

Both mesoderm and neural precursor cells undergo mediolateral cell intercalation. Within mediolaterally intercalating groups of cells there is not a coordinated directional cue but instead each cell appears to act independently. Unlike mesendoderm cell movements, movements of mediolaterally intercalating cells are episodic. For instance, immediately neighboring mesodermal cells may intercalate in opposite directions or one cell may intercalate while the adjacent cell remains transiently quiescent (Shih and Keller, 1992a, 1992b). Intercalating mesoderm cells can change direction, moving toward the midline as often as away from it. In contrast, neural cells only intercalate toward the midline (Elul *et al.*, 1998, 1997; Ezin *et al.*, 2003, 2006). Several molecular pathways appear to provide cues that regulate cell orientation and protrusion (Keller, 2002; Wallingford *et al.*, 2002). How these cues are maintained during dynamic cell rearrangement and how duty cycles of cell intercalation are coordinated between neighboring cells to bring about productive cell rearrangement are still poorly understood.

VI. Molecular-Scale: Mechanics, Adhesion, and Traction

The molecular genetic approach has produced a large list of "parts" required by the developing embryo during morphogenesis. This characterization of development has had the greatest success elaborating signal-transduction and gene-regulatory networks.

However, the process has been relatively unsatisfying in uncovering the mechanical mechanisms of morphogenesis. Genes that have been discovered have often been obvious candidates from the perspective of cell biologists studying cell motility—akin to the discovery that actin is required for development. Many genes involved in morphogenetic movements will have their effectors and the numerous upstream and downstream transduction networks exposed but the embryologist still must ask how do these proteins work together to produce cell rearrangement, cell sheet spreading, cell streams, and cell motility within the embryo.

Several classes of proteins are localized within morphogenetically active cells, yet how they direct cell behaviors, matrix assembly, and directed motility is not understood. Many pathways involved in cell motility have also been identified as downstream effectors of vertebrate tissue patterning pathways. For instance, Rho, Rac, and CDC42 of the Rho-family of GTPases identified by their effects on cell protrusions and ECM interaction (Nobes and Hall, 1995), are regulated by Disheveled (Habas *et al.*, 2003), a central component of the planar cell polarity (PCP) pathway that regulates mediolateral cell intercalation (Park *et al.*, 2005). Other elements of the PCP pathway such as frizzled, prickle, and strabismus are localized in *Xenopus* cells undergoing intercalation (Hyodo-Miura *et al.*, 2006; Jenny *et al.*, 2003). Just how localized polarity factors guide cell behaviors such as directed rearrangement is unclear.

VII. Modeling Morphogenesis: A Grand Challenge

The grand challenge to developmental biologists in the coming years will be to understand how molecular-genetic pathways such as the planar cell polarity pathway regulate the mechanics of morphogenesis. Our intuition fails us when we consider the complexity of the problem. Molecular pathways must create subcellular structures responsible not just for cell motility but for cell adhesion and the extracellular matrix. Quantitative systems modeling approaches are bringing a clearer understanding of how cell biology is integrated on this scale (Mogilner *et al.*, 2006). The structures within a cell or within its microenvironment must be coupled to other cells. Several physically-based hypotheses seek to explain experiments carried out by classical embryologists such as Johannes Holtfreter on cell sorting, mixing, and tissue engulfment (Harris, 1976; Steinberg and Gilbert, 2004). Extensive computer simulation efforts have identified numerous plausible mechanisms for converting localized factors into morphogenesis. For instance, models have shown that directed protrusive activity (Weliky *et al.*, 1991), differential adhesion (Zajac *et al.*, 2003), or differential contraction (Brodland, 2006) can all drive cell intercalation. Simulations such as these extend human intuition to the physical regime of the embryo and demonstrate proof-of-principle.

The challenge for future modeling efforts at this scale will be to allow multiple physical processes to interact, allow sensitivity analyses, and must enable the user to challenge hypotheses in the laboratory (Platt, 1964). Connecting the molecular mechanisms mediating cell behaviors to physical changes in cell shape and movement require novel biophysical approaches to measure critical parameters that can distinguish between mathematical models and resolve how molecular lesions effect those same processes. Often the simple process of formulating a model exposes gaps in the biological description. These gaps will include limited information on the values on quantitative parameters required for the model. In order to accommodate these gaps requires the systematic exploration of physiologically plausible parameters, requires running simulations under many conditions, and testing multiple "toy-hypotheses" within the same framework. Lastly, simulation efforts must feed the experimentalist with concrete predictions on cell or tissue patterning, cell behaviors and tissue movements, or physical–mechanical properties in standard biological and engineering terms (Koehl, 1990). Models must be sufficiently complex and robust that model refinement and empirical experiment can be swiftly checked in the real embryo.

Acknowledgments

This work was supported by the US Public Health Service (Grant HD44750 to the author).

References

Afonin, B., Ho, M., Gustin, J. K., Meloty-Kapella, C., and Domingo, C. R. (2006). Cell behaviors associated with somite segmentation and rotation in *Xenopus laevis. Dev. Dyn.* **235**, 3268–3279.

Alberts, B., Johnson, A., Lewis, J., Raff, M., Roberts, K., and Walter, P. (2002). "Molecular Biology of the Cell." Garland Publishing, New York.

Ariizumi, T., Takano, K., Asashima, M., and Malacinski, G. M. (2000). Bioassays of inductive interactions in amphibian development. *Methods Mol. Biol.* **135**, 89–112.

Baneyx, G., Baugh, L., and Vogel, V. (2002). Fibronectin extension and unfolding within cell matrix fibrils controlled by cytoskeletal tension. *Proc. Natl. Acad. Sci. USA* **99**, 5139–5143.

Blitz, I. L., Andelfinger, G., and Horb, M. E. (2006). Germ layers to organs: Using *Xenopus* to study "later" development. *Semin. Cell Dev. Biol.* **17**, 133–145.

Bolker, J. A. (1995). Model systems in developmental biology. *BioEssays* **17**, 451–455.

Brodland, G. W. (2006). Do lamellipodia have the mechanical capacity to drive convergent extension? *Int. J. Dev. Biol.* **50**, 151–155.

Davidson, L. A., and Keller, R. E. (1999). Neural tube closure in *Xenopus laevis* involves medial migration, directed protrusive activity, cell intercalation and convergent extension. *Development* **126**, 4547–4556.

Davidson, L. A., and Wallingford, J. B. (2005). Visualizing morphogenesis in the frog embryo. *In* "Imaging in Neuroscience and Development: A Laboratory Manual." (R. Yuste and A. Konnerth, Eds.). Cold Spring Harbor Laboratory Press, Cold Spring Harbor, NY.

Davidson, L. A., Hoffstrom, B. G., Keller, R., and DeSimone, D. W. (2002). Mesendoderm extension and mantle closure in *Xenopus laevis* gastrulation: Combined roles for integrin alpha5beta1, fibronectin, and tissue geometry. *Dev. Biol.* **242**, 109–129.

Davidson, L. A., Keller, R., and Desimone, D. W. (2004). Assembly and remodeling of the fibrillar fibronectin extracellular matrix during gastrulation and neurulation in *Xenopus laevis. Dev. Dyn.* **231**, 888–895.

Davidson, L. A., Marsden, M., Keller, R., and Desimone, D. W. (2006). Integrin alpha5beta1 and fibronectin regulate polarized cell protrusions required for *Xenopus* convergence and extension. *Curr. Biol.* **16**, 833–844.

De Robertis, E. M., and Kuroda, H. (2004). Dorsal–ventral patterning and neural induction in *Xenopus* embryos. *Annu. Rev. Cell Dev. Biol.* **20**, 285–308.

Elul, T., and Keller, R. (2000). Monopolar protrusive activity: A new morphogenic cell behavior in the neural plate dependent on vertical interactions with the mesoderm in *Xenopus* [in process citation]. *Dev. Biol.* **224**, 3–19.

Elul, T. M., Koehl, M. A. R., and Keller, R. E. (1997). Cellular mechanism underlying neural convergence and extension in *Xenopus laevis* embryos. *Dev. Biol.* **191**, 243–258.

Elul, T., Koehl, M. A., and Keller, R. E. (1998). Patterning of morphogenetic cell behaviors in neural ectoderm of *Xenopus laevis*. *Ann. NY Acad. Sci.* **857**, 248–251.

Ewald, A. J., Peyrot, S. M., Tyszka, J. M., Fraser, S. E., and Wallingford, J. B. (2004). Regional requirements for Dishevelled signaling during *Xenopus* gastrulation: Separable effects on blastopore closure, mesendoderm internalization and archenteron formation. *Development* **131**, 6195–6209.

Ezin, A. M., Skoglund, P., and Keller, R. (2003). The midline (notochord and notoplate) patterns the cell motility underlying convergence and extension of the *Xenopus* neural plate. *Dev. Biol.* **256**, 100–114.

Ezin, A. M., Skoglund, P., and Keller, R. (2006). The presumptive floor plate (notoplate) induces behaviors associated with convergent extension in medial but not lateral neural plate cells of *Xenopus*. *Dev. Biol.* **300**, 670–686.

Gawantka, V., Ellinger-Ziegelbauer, H., and Hausen, P. (1992). Beta 1-integrin is a maternal protein that is inserted into all newly formed plasma membranes during early *Xenopus* embryogenesis. *Development* **115**, 595–605.

Gilbert, S. F. (2006). "Developmental Biology." Sinauer Associates Inc., Sunderland, MA.

Gildner, C. D., Lerner, A. L., and Hocking, D. C. (2004). Fibronectin matrix polymerization increases tensile strength of model tissue. *Am. J. Physiol. Heart Circ. Physiol.* **287**, H46–H53.

Habas, R., Dawid, I. B., and He, X. (2003). Coactivation of Rac and Rho by Wnt/Frizzled signaling is required for vertebrate gastrulation. *Genes Dev.* **17**, 295–309.

Hardin, J., and Keller, R. (1988). The behavior and function of bottle cells during gastrulation of *Xenopus laevis*. *Development* **103**, 211–230.

Harris, A. K. (1976). Is cell sorting caused by differences in the work of intercellular adhesion? A critique of the Steinberg hypothesis. *J. Theor. Biol.* **61**, 267–285.

Harris, A. K. (1987). Cell motility and the problem of anatomical homeostasis. *J. Cell Sci.* **8**, 121–140.

Harris, A. K., Wild, P., and Stopak, D. (1980). Silicone rubber substrata: A new wrinkle in the study of cell locomotion. *Science* **208**, 177–179.

Heasman, J. (2006). Maternal determinants of embryonic cell fate. *Semin. Cell Dev. Biol.* **17**, 93–98.

Hocking, D. C., Smith, R. K., and McKeown-Longo, P. J. (1996). A novel role for the integrin-binding III-10 module in fibronectin matrix assembly. *J. Cell Biol.* **133**, 431–444.

Hoffstrom, B. G. (2002). Integrin function during *Xenopus laevis* gastrulation. *In* "Depart of Cell Biology." Univ. of Virginia, Charlottesville, VA, p. 137.

Horwitz, A. F., and Parsons, J. T. (1999). Cell biology: Cell migration—movin' on. *Science* **286**, 1102–1103.

Hynes, R. O. (1990). "Fibronectins." Springer-Verlag, New York.

Hyodo-Miura, J., Yamamoto, T. S., Hyodo, A. C., Iemura, S., Kusakabe, M., Nishida, E., Natsume, T., and Ueno, N. (2006). XGAP, an ArfGAP, is required for polarized localization of PAR proteins and cell polarity in *Xenopus* gastrulation. *Dev. Cell* **11**, 69–79.

Ibrahim, H., and Winklbauer, R. (2001). Mechanisms of mesendoderm internalization in the *Xenopus* gastrula: Lessons from the ventral side. *Dev. Biol.* **240**, 108–122.

Jenny, A., Darken, R. S., Wilson, P. A., and Mlodzik, M. (2003). Prickle and Strabismus form a functional complex to generate a correct axis during planar cell polarity signaling. *EMBO J.* **22**, 4409–4420.

Joos, T. O., Whittaker, C. A., Meng, F., DeSimone, D. W., Gnau, V., and Hausen, P. (1995). Integrin alpha 5 during early development of *Xenopus laevis*. *Mech. Dev.* **50**, 187–199.

Keller, R. E. (1980). The cellular basis of epiboly: An SEM study of deep-cell rearrangement during gastrulation in *Xenopus laevis*. *J. Embryol. Exp. Morphol.* **60**, 201–234.
Keller, R. (1991). Early embryonic development of *Xenopus laevis*. *Methods Cell Biol.* **36**, 61–113.
Keller, R. (2000). The origin and morphogenesis of amphibian somites. *Curr. Top. Dev. Biol.* **47**, 183–246.
Keller, R. (2002). Shaping the vertebrate body plan by polarized embryonic cell movements. *Science* **298**, 1950–1954.
Keller, R. (2006). Mechanisms of elongation in embryogenesis. *Development* **133**, 2291–2302.
Keller, R., and Shook, D. R. (2004). Gastrulation in Amphibians. *In* "Gastrulation: From Cells to Embryos." (Stern C., Ed.). Cold Spring Harbor Laboratory Press, Cold Spring Harbor, NY, pp. 171–203.
Keller, R., Davidson, L., Edlund, A., Elul, T., Ezin, M., Shook, D., and Skoglund, P. (2000). Mechanisms of convergence and extension by cell intercalation. *Philos. Trans. R. Soc. London B* **355**, 897–922.
Keller, R., Davidson, L. A., and Shook, D. R. (2003). How we are shaped: The biomechanics of gastrulation. *Differentiation* **71**, 171–205.
Koehl, M. A. R. (1990). Biomechanical approaches to morphogenesis. *Semin. Dev. Biol.* **1**, 367–378.
Lane, M. C., and Keller, R. (1997). Microtubule disruption reveals that Spemann's organizer is subdivided into two domains by the vegetal alignment zone. *Development* **124**, 895–906.
Lane, M. C., and Sheets, M. D. (2006). Heading in a new direction: Implications of the revised fate map for understanding *Xenopus laevis* development. *Dev. Biol.* **296**, 12–28.
Larkin, K., and Danilchik, M. V. (1999). Ventral cell rearrangements contribute to anterior–posterior axis lengthening between neurula and tailbud stages in *Xenopus laevis*. *Dev. Biol.* **216**, 550–560.
Lee, C., Scherr, H. M., and Wallingford, J. B. (2007). Shroom family proteins regulate {gamma}-tubulin distribution and microtubule architecture during epithelial cell shape change. *Development* **134**, 1431–1441.
Lee, G., Hynes, R., and Kirschner, M. (1984). Temporal and spatial regulation of fibronectin in early *Xenopus* development. *Cell* **36**, 729–740.
Lo, C. M., Wang, H. B., Dembo, M., and Wang, Y. L. (2000). Cell movement is guided by the rigidity of the substrate. *Biophys. J.* **79**, 144–152.
Longo, D., Peirce, S. M., Skalak, T. C., Davidson, L., Marsden, M., Dzamba, B., and DeSimone, D. W. (2004). Multicellular computer simulation of morphogenesis: Blastocoel roof thinning and matrix assembly in *Xenopus laevis*. *Dev. Biol.* **271**, 210–222.
Malacinski, G. M., Ariizumi, T., and Asashima, M. (2000). Work in progress: The renaissance in amphibian embryology. *Comp. Biochem. Physiol. B Biochem. Mol. Biol.* **126**, 179–187.
Marsden, M., and DeSimone, D. W. (2001). Regulation of cell polarity, radial intercalation and epiboly in *Xenopus*: Novel roles for integrin and fibronectin. *Development* **128**, 3635–3647.
Marsden, M., and DeSimone, D. W. (2003). Integrin-ECM interactions regulate cadherin-dependent cell adhesion and are required for convergent extension in *Xenopus*. *Curr. Biol.* **13**, 1182–1191.
Minsuk, S. B., and Keller, R. E. (1997). Surface mesoderm in *Xenopus*: A revision of the stage 10 fate map. *Dev. Genes Evol.* **207**, 389–401.
Mogilner, A., Wollman, R., and Marshall, W. F. (2006). Quantitative modeling in cell biology: What is it good for? *Dev. Cell* **11**, 279–287.
Nakatsuji, N., Smolira, M. A., and Wylie, C. C. (1985). Fibronectin visualized by scanning electron microscopy immunocytochemistry on the substratum for cell migration in *Xenopus laevis* gastrulae. *Dev. Biol.* **107**, 264–268.
Nieuwkoop, P. D., and Faber, J. (1967). "Normal Tables of *Xenopus laevis* (Daudin)." Elsevier/North-Holland Biomedical Press, Amsterdam.
Nobes, C. D., and Hall, A. (1995). Rho, rac, and cdc42 GTPases regulate the assembly of multimolecular focal complexes associated with actin stress fibers, lamellipodia, and filopodia. *Cell* **81**, 53–62.
Papan, C., Boulat, B., Velan, S. S., Fraser, S. E., and Jacobs, R. E. (2007). Formation of the dorsal marginal zone in *Xenopus laevis* analyzed by time-lapse microscopic magnetic resonance imaging. *Dev. Biol.*
Park, T. J., Gray, R. S., Sato, A., Habas, R., and Wallingford, J. B. (2005). Subcellular localization and signaling properties of dishevelled in developing vertebrate embryos. *Curr. Biol.* **15**, 1039–1044.

Pelham Jr., R. J., and Wang, Y. (1997). Cell locomotion and focal adhesions are regulated by substrate flexibility. *Proc. Natl. Acad. Sci. USA* **94**, 13661–13665.

Platt, J. R. (1964). Strong inference. *Science* **146**, 347–353.

Ramos, J. W., and DeSimone, D. W. (1996). *Xenopus* embryonic cell adhesion to fibronectin: Position-specific activation of RGD/synergy site-dependent migratory behavior at gastrulation. *J. Cell Biol.* **134**, 1–14.

Ramos, J. W., Whittaker, C. A., and DeSimone, D. W. (1996). Integrin-dependent adhesive activity is spatially controlled by inductive signals at gastrulation. *Development* **122**, 2873–2883.

Rand, H. W. (1915). Wound closure in actinian tentacles with reference to the problem of organization. *Dev. Genes Evol.* **41**, 159–214.

Ren, R., Nagel, M., Tahinci, E., Winklbauer, R., and Symes, K. (2006). Migrating anterior mesoderm cells and intercalating trunk mesoderm cells have distinct responses to Rho and Rac during *Xenopus* gastrulation. *Dev. Dyn.* **235**, 1090–1099.

Sasai, Y., Lu, B., Steinbeisser, H., Geissert, D., Gont, L. K., and Robertis, E. M. D. (1994). *Xenopus* chordin: A novel dorsalizing factor activated by organizer-specific homeobox genes. *Cell* **79**, 779–790.

Sawhney, R. K., and Howard, J. (2002). Slow local movements of collagen fibers by fibroblasts drive the rapid global self-organization of collagen gels. *J. Cell Biol.* **157**, 1083–1091.

Schroeder, T. E. (1970). Neurulation in *Xenopus laevis*. An analysis and model based upon light and electron microscopy. *J. Embryol. Exp. Morphol.* **23**, 427–462.

Shih, J., and Keller, R. (1992a). Cell motility driving mediolateral intercalation in explants of *Xenopus laevis*. *Development* **116**, 901–914.

Shih, J., and Keller, R. (1992b). Patterns of cell motility in the organizer and dorsal mesoderm of *Xenopus laevis*. *Development* **116**, 915–930.

Shook, D. R., Majer, C., and Keller, R. (2002). Urodeles remove mesoderm from the superficial layer by subduction through a bilateral primitive streak. *Dev. Biol.* **248**, 220–239.

Shook, D. R., Majer, C., and Keller, R. (2004). Pattern and morphogenesis of presumptive superficial mesoderm in two closely related species, *Xenopus laevis* and *Xenopus tropicalis*. *Dev. Biol.* **270**, 163–185.

Sivakumar, P., Czirok, A., Rongish, B. J., Divakara, V. P., Wang, Y. P., and Dallas, S. L. (2006). New insights into extracellular matrix assembly and reorganization from dynamic imaging of extracellular matrix proteins in living osteoblasts. *J. Cell Sci.* **119**, 1350–1360.

Sive, H. L., Grainger, R. M., and Harland, R. M. (2000). "Early Development of *Xenopus laevis*: A Laboratory Manual." Cold Spring Harbor Laboratory Press, Cold Spring Harbor, NY. p. 338.

Skoglund, P., and Keller, R. (2007). *Xenopus* fibrillin regulates directed convergence and extension. *Dev. Biol.* **301**, 404–416.

Skoglund, P., Dzamba, B., Coffman, C. R., Harris, W. A., and Keller, R. (2006). *Xenopus* fibrillin is expressed in the organizer and is the earliest component of matrix at the developing notochord-somite boundary. *Dev. Dyn.* **235**, 1974–1983.

Steinberg, M. S., and Gilbert, S. F. (2004). Townes and Holtfreter (1955): Directed movements and selective adhesion of embryonic amphibian cells. *J. Exp. Zoolog. A Comp. Exp. Biol.* **301**, 701–706.

Stern, C. D. (2004). "Gastrulation: From Cells to Embryo." Cold Spring Harbor Laboratory Press, Cold Spring Harbor, NY.

Stopak, D., and Harris, A. K. (1982). Connective tissue morphogenesis by fibroblast traction. I. Tissue culture observations. *Dev. Biol.* **90**, 383–398.

Trinkaus, J. P. (1984). "Cells into Organs: The Forces that Shape the Embryo." Prentice–Hall Inc., Englewood Cliffs.

Wacker, S., Grimm, K., Joos, T., and Winklbauer, R. (2000). Development and control of tissue separation at gastrulation in *Xenopus*. *Dev. Biol.* **224**, 428–439.

Wallingford, J. B., Fraser, S. E., and Harland, R. M. (2002). Convergent extension: The molecular control of polarized cell movement during embryonic development. *Dev. Cell* **2**, 695–706.

Wardle, F. C., and Smith, J. C. (2006). Transcriptional regulation of mesendoderm formation in *Xenopus*. *Semin. Cell Dev. Biol.* **17**, 99–109.

Weliky, M., Minsuk, S., Keller, R., and Oster, G. (1991). Notochord morphogenesis in *Xenopus laevis*: Simulation of cell behavior underlying tissue convergence and extension. *Development* **113**, 1231–1244.

Winklbauer, R. (1998). Conditions for fibronectin fibril formation in the early *Xenopus* embryo. *Dev. Dyn.* **212**, 335–345.

Winklbauer, R., and Schurfeld, M. (1999). Vegetal rotation, a new gastrulation movement involved in the internalization of the mesoderm and endoderm in *Xenopus*. *Development* **126**, 3703–3713.

Winklbauer, R., Selchow, A., Nagel, M., Stoltz, C., and Angres, B. (1991). Mesoderm cell migration in the *Xenopus* gastrula. *In* "Gastrulation: Movements, Patterns, and Molecules." (R. Keller, W. Clark, and F. Griffin, Eds.). Plenum Press, New York/London, pp. 147–168.

Winklbauer, R., Nagel, M., Selchow, A., and Wacker, S. (1996). Mesoderm migration in the *Xenopus* gastrula. *Int. J. Dev. Biol.* **40**, 305–311.

Winklbauer, R., Medina, A., Swain, R. K., and Steinbeisser, H. (2001). Frizzled-7 signaling controls tissue separation during *Xenopus* gastrulation. *Nature* **413**, 856–860.

Zajac, M., Jones, G. L., and Glazier, J. A. (2003). Simulating convergent extension by way of anisotropic differential adhesion. *J. Theor. Biol.* **222**, 247–259.

Zhong, C., Chrzanowska-Wodnicka, M., Brown, J., Shaub, A., Belkin, A. M., and Burridge, K. (1998). Rho-mediated contractility exposes a cryptic site in fibronectin and induces fibronectin matrix assembly. *J. Cell Biol.* **141**, 539–551.

Zhong, Y., Brieher, W. M., and Gumbiner, B. M. (1999). Analysis of C-cadherin regulation during tissue morphogenesis with an activating antibody. *J. Cell Biol.* **144**, 351–359.

4

The Mechanisms Underlying Primitive Streak Formation in the Chick Embryo

Manli Chuai and Cornelis J. Weijer
Division of Cell and Developmental Biology, Wellcome Trust Biocentre,
College of Life Sciences, University of Dundee, Dundee DD1 5EH, United Kingdom

I. Introduction
II. Structure of the Early Embryo
III. Experimental Observations of Streak Formation
IV. Mesoderm Induction
V. Cellular Mechanisms of Streak Formation
    A. Locally Restricted Cell Division
    B. Spatially Organized Cell–Cell Intercalation
    C. Chemotaxis
    D. Adhesion Gradients
    E. Contact Guidance
VI. Mechanisms of Movement
VII. Challenges for Modeling and Computational Approaches
VIII. Outlook
    References

Formation of the primitive streak is one of the key events in the early development of amniote embryos. The streak is the site where during gastrulation the mesendoderm cells ingress to take up their correct topographical positions in the embryo. The process of streak formation can be conveniently observed in the chick embryo, where the streak forms as an accumulation of cells in the epiblast in the posterior pole of the embryo and extends subsequently in anterior direction until it covers 80% of the epiblast. A prerequisite for streak formation is the differentiation of mesoderm, which is induced in the epiblast at the interface between the posterior Area Opaca and Area Pellucida in a sickle shaped domain overlying Koller's sickle. Current views on the molecular mechanisms of mesoderm induction by inducing signals from the Area Opaca and inhibitory signals from the hypoblast are briefly discussed. During streak formation the sickle of mesoderm cells transforms into an elongated structure in the central midline of the embryo. We discuss possible cellular mechanisms underlying this process, such as oriented cell division, cell–cell intercalation, chemotactic cell movement in response to attractive and repulsive signals and a combination of chemotaxis and contact following. We review current experimental evidence in favor and against these different hypotheses and outline some the outstanding questions. Since many of the interactions between cells signaling and moving are dynamic and nonlinear in nature they will require detailed modeling and computer simulations to be understood in detail. © 2008, Elsevier Inc.

I. Introduction

The early development of amniotes follows a defined series of events, after fertilization of the egg a series of rapid cell divisions are initiated which result in a mass of cells forming embryonic and extra-embryonic tissues. In amniotes all of the embryonic material derives from an epithelial layer of cells, the epiblast. However its differentiation into the three germlayers, the ectoderm, mesoderm and endoderm is dependent on interactions with the extra-embryonic tissues. During gastrulation the cells of the ectoderm, mesoderm and endoderm take up their correct topological positions in the embryo, a process that involves large scale highly coordinated cell movements. During this process the cells that are going to form the mesoderm and endoderm move inside the embryo through a structure known as the primitive streak (Stern, 2004). The genetic basis of axis development, germlayer and streak formation has been studied extensively through mutational analysis in the mouse (Beddington and Robertson, 1999; Tam et al., 2006). However the *in vivo* analysis of cell movement patterns associated with gastrulation are difficult to study in this organism so far, due to its small size and technical difficulties in culturing early mouse embryos outside the mother animal. Gastrulation movements can be conveniently studied in the chick embryo, since these embryos can be easily cultured *in vitro* and the embryo at the early stages of gastrulation consists of a flat essentially bilayered structure, which lends itself to itself to direct observation and experimental manipulation (Bortier *et al.*, 1996; Mikawa *et al.*, 2004; Stern, 2004). In this paper we will describe our current understanding of the mechanisms underlying the initial phases of gastrulation in the chick embryo and concentrate on the description of possible mechanisms underlying the formation of the primitive streak.

II. Structure of the Early Embryo

In the chick the early stages of development of the embryo takes place in the mother animal. At the time of egg laying the embryo consists of a roughly circular mass of ~10.000 cells located on the top of the yolk and covered by the vitelline membrane, which separates the yolk and the embryo from the egg white. The chick embryo has a circular flat geometry and consists of two concentric rings of cells, the inner translucent Area Pellucida and the outer thicker Area Opaca. The cells in the Area Pellucida form a rough epithelial sheet that is continuous with epithelial cells overlaying the Area Opaca. The interface between the Area Pellucida and the Area Opaca is known as the marginal zone. Beneath the epithelial cells, the Area Opaca consists of large, yolk rich cells, the deeper cells of which are in direct contact with the yolk. The outer ring of epithelial cells of the Area Opaca are specialized and migrate outward over the vitelline membrane, which is stretched over the yolk (Andries *et al.*, 1985). These cells keep the embryo under tension during its growth and development and help to expand it (Downie, 1976). Cells in the outer area proliferate more rapidly than cells elsewhere in the embryo, enabling the smooth expansion of the embryo diameter. The more central cells of the epiblast are not in contact with the vitelline membrane but stick together through cell contacts mediated in large part by E-cadherin-containing adherens junctions and are in contact with a complex basal lamina. In some embryos there is a little ridge of deep mesenchymal cells just in front of the posterior marginal zone and this thickening can be seen as a dark sickle-shaped region, a structure that is known as Koller's Sickle.

During the early stages of development individual cells start to drop out of the epiblast and form small islands of cells underneath the epiblast (Harrisson *et al.*, 1991). These cells give rise to the primary hypoblast. At stage XI–XIII cells from the deeper layer of the posterior marginal zone, which have been shown to express a specific carbohydrate epitope that is recognized by the monoclonal antibody HNK1, start to move in the anterior direction (Canning and Stern, 1988). These cells form the secondary hypoblast that will fuse with the isolated cells of the primary hypoblast. The formation of the hypoblast is finished by Stage XIV and all cells in the secondary hypoblast are HNK1 positive (Stern and Canning, 1990).

After 6–8 hours of development epiblast cells start to move and to concentrate in an area at the posterior pole of the embryo where the streak will form. The transition of cells from a single layered epithelial sheet into a structure several cell layers thick results in the macroscopically visible structure known as the streak (Fig. 1). The cells in the streak originate mainly from the epiblast. When the streak has extended about halfway over the epiblast it can be seen that the mesoderm cells start to migrate out. Streak formation is a bi-directional process; already from the very early stages of its formation the streak extends both in the anterior and posterior direction, although initially the extension in the anterior direction is much faster (Fig. 1). The first cells that start to migrate in the posterior direction are located in the deeper layers of the embryo (Cui and Weijer unpublished observations). The formation of the streak involves extensive large scale cell movement patterns that will be described in detail below.

III. Experimental Observations of Streak Formation

Early time lapse cinematography of these movements by Graeper and later by Vakaet, using carbon and carmine labeling of small groups of cells, showed that streak formation involves movement of cells overlaying Koller's Sickle along the posterior end of the embryo towards the central midline, where they meet, change direction, and move towards the center of the epiblast, a process called "Polonaise movements" (Graeper, 1929; Vakaet, 1970). More recent observations based on DiI injections as well as transfection with GFPs have shown that the cells in the epiblast start to move well before the streak becomes visible and that these movements are large scale involving all cells in the epiblast (Chuai *et al.*, 2006; Cui *et al.*, 2005). The remarkable observation is that cells appear to start to move at the same time co-ordinately and that cells move over very long distances relative to the size of the embryo, but

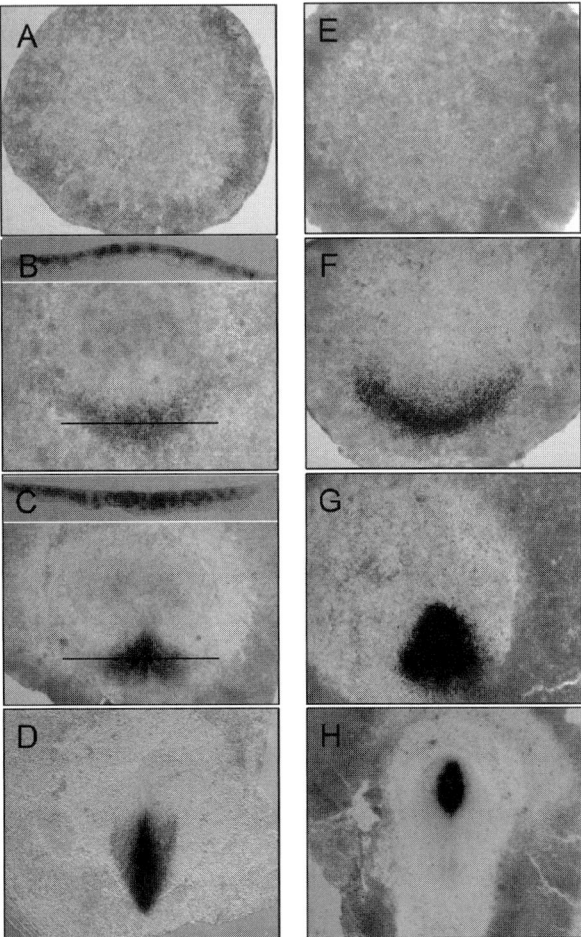

Figure 1 Expression of mesodermal markers genes and cell movement patterns during streak formation. (A–D) RNA expression pattern of Wnt8C. (A) Weak Wnt8c expression in the Area Opaca. (B) Expression of Wnt8c in Koller's Sickle, Section at the level of the black line shows that the epiblast above Koller's Sickle only one cell layer thick and that Wnt8C is only expressed in the epiblast. (C) Expression of Wnt8c in the forming streak. Section at the level of the black line shows that the streak is now several cell layers thick and that Wnt8C is expressed only in the epiblast. (D) Expression of Wnt8C in a HH stage 4 embryo. Wnt8C is only expressed in the posterior streak and in the mesodermal cells that migrate out of the streak. (E–H) Expression of the anterior mesoderm marker Chordin at similar developmental stages as shown in A–D. Note the strong expression in the epiblast overlying Koller's Sickle (F) and in the tip of the streak and forming head process (H). (I) Brightfield image of HH2 embryo in which cell movement were recorded during the early phases of streak formation. The white line outlines the shape of the forming streak. (J) Cell traces of randomly scattered cells in the epiblast that have been transfected with GFP. The trajectories show the distance that the cells migrate over a 5 hour period of development. The trajectories are color coded the distance migrated over the last hour of the experiment is shown in green and the color coding thus indicates the direction of migration. The outline of the streak is shown in red. See color insert.

4. The Mechanisms Underlying Primitive Streak Formation in the Chick Embryo

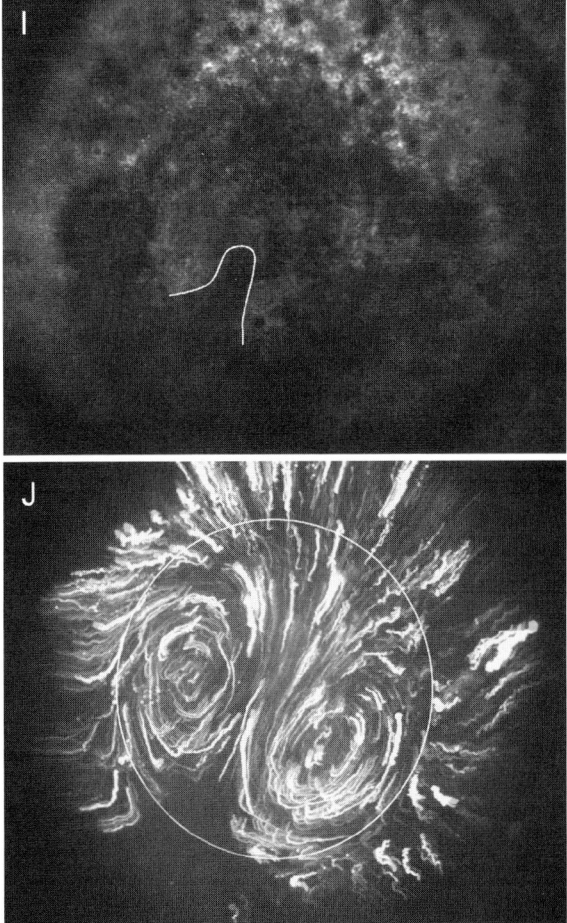

Figure 1 (*continued*)

that there seems to be very little relative cell movement. It can be clearly observed that daughter cells stay together in compact clusters for long periods of time. Cells overlying Koller's Sickle will move into the central midline, while these cells are being replaced by cells from more lateral positions in the epiblast, in agreement with fate maps for the epiblast (Hatada and Stern, 1994). Cells originally located anterior to Koller's Sickle move in the anterior direction to an area that will form the headfold and extra embryonic ectoderm (Fig. 1J). The GFP transfection experiments predominantly label cells in the epiblast, while the DiI labeling experiments label both cells in the epiblast and the deeper layers of the marginal zone (Chuai *et al.*, 2006; Cui *et al.*, 2005). Cells in the deeper layers scatter much more rapidly than cells in the epiblast. In posterior positions the deeper cells already start to move in a posterior direction at a time in which the epiblast cells move anterior, giving the first indication of the posterior extension of the embryo Fig. 1 (Cui *et al.*, 2005).

The streak becomes visible because the epiblast undergoes a transition from a single epithelial layer into a multilayered structure at the site of streak formation. It does not appear that cells in the streak undergo more divisions than outside the streak and therefore the accumulation of cells in the streak presumably reflects the fact that the "Polonaise movements" bring more cells towards the streak than cells move away in front of it, resulting in a local accumulation of cells. The transition of the simple epithelial sheet into a multilayered structure coincides with the expression of Snail2, which is a well established inhibitor of E-cadherin transcription (Barrallo-Gimeno and Nieto, 2005; Lawson *et al.*, 2001). At the same time N-cadherin expression is upregulated in the cells that are forming the streak (F. Morrison and C. J. Weijer, unpublished observations). It appears likely that the change in the expression of the adhesion molecules plays a role in this transition, which is currently under investigation.

The role of the hypoblast in streak formation has been long discussed. The hypoblast extends from the posterior pole of the embryo in an anterior direction (Spratt and Haas, 1960b) and derives apart from individual scattered cells ingressing from the epiblast, from cells that derive from the deep layer of the posterior marginal zone (Stern, 1990). It has been shown that removal of the hypoblast can result in the formation of multiple streaks, suggesting that the hypoblast secretes an inhibitor of the streak formation (Bertocchini and Stern, 2002). It has also been shown that rotation of the hypoblast can result in bending of the streak away from the forming hypoblast (Foley *et al.*, 2000). It is not known what causes the bending away of the streak but this suggests that the hypoblast secretes a factor that influences the movement behavior of cells in the epiblast, possibly a repellent.

IV. Mesoderm Induction

So far much work related to the mechanisms underlying streak formation has cent red on the investigation of the signaling mechanisms that result in the formation of a streak. Extensive grafting and manipulation experiments have shown that the marginal zone is instrumental to the induction of the streak. There is a strong anterior to posterior gradient in the ability of the marginal zone to induce streaks, the posterior marginal zone having the highest streak inducing ability (Eyal-Giladi, 1997; Khaner and Eyal-Giladi, 1989). It has also been shown that development is highly regulative: embryos can be cut in anterior and posterior halves and each piece will regenerate a complete embryo with a high degree of success (Bachvarova *et al.*, 1998; Bertocchini *et al.*, 2004; Lutz, 1949; Spratt and Haas, 1960a). Even when cut in quarters, all four pieces can regenerate into complete embryos, under the best conditions. Cells in the epithelial layer of the marginal zone normally contribute to the streak, while the deeper layers of the marginal zone contribute to the secondary hypoblast

4. The Mechanisms Underlying Primitive Streak Formation in the Chick Embryo

and endoblast. It was shown, however, that in the absence of both upper and deeper layer marginal zone cells a streak of abnormal morphology can still form in which mesoderm cells differentiate (Stern, 1990).

Mesoderm differentiation seems to be a prerequisite for streak formation, since manipulations that inhibit mesoderm formation inhibit streak formation, suggesting that mesoderm cells play an active role in the formation of the streak. For instance, inhibition of FGF signaling, through expression of a dominant negative FGF receptor, application of small molecule FGF receptor inhibitors, or through depletion of FGF receptor ligands, all result in an inhibition of mesoderm formation and are associated with the absence of a primitive streak (Chuai *et al.*, 2006). On the other hand, a number of treatments are known to be able to induce of extra streaks in the embryo. These involve localized application of the TGFβ signaling molecules cVg1, inhibitors of BMP signaling, chordin and noggin, as well as Wnts and FGF's (Bertocchini *et al.*, 2004; Shah *et al.*, 1997; Skromne and Stern, 2001; Streit *et al.*, 1998).

Vg1 a TGFβ family member is expressed in the posterior marginal zone and ectopic expression Vg1 can induce an ectopic streak (Shah *et al.*, 1997; Skromne and Stern, 2002). It has been shown that Vg1 and Wnt expression have to work in concert and the actions of these molecules induce a range of mesodermal marker genes possibly through the control of nodal expression, a well established mesodermal inducer (Bertocchini *et al.*, 2004). Wnt8C has been shown to be expressed in an anterior posterior gradient in the marginal zone and Area Opaca, with highest expression at the posterior pole of the embryo. FGF may potentiate mesoderm differentiation through a parallel pathway (Bertocchini *et al.*, 2004). It has been shown that the hypoblast expresses and most likely secretes Cerberus, an inhibitor of nodal signaling and that this inhibits formation of ectopic streaks. The displacement of the secondary hypoblast by the forming endoblast, which starts to form posteriorly and replaces the hypoblast in an anterior direction, allows streak formation to occur at the posterior pole of the embryo, while the hypoblast still present in the anterior part of the embryo prevents streak formation there (Bertocchini and Stern, 2002). In agreement with this, it has been shown that ectopic local application of Cerberus results in an inhibition of streak formation.

Mutational analysis in mice has shown that BMP and Wnt signaling are critical for the formation of a primitive streak. In the mouse embryo mutants in BMP2 and BMP4 and the BMP receptors BMPR1a, BMPR2 and the downstream signaling factors SMAD 2 and SMAD 4 are all defective in streak formation, as are mutants in which Wnt3 and the Wnt co-receptors LRP5 and LRP6 have been deleted and mutants in FGF8 (Tam *et al.*, 2006). In the chick embryo there is very early expression of BMP4 and BMP 7 in the Area Opaca of the epiblast, while there is only limited expression in the Area Pellucida. Chordin starts to be expressed in the epiblast overlying Koller's Sickle and it thus appears that inhibition of BMP signaling may be required for streak formation to occur. Misexpression of BMP4 in the posterior Area Opaca inhibits streak formation and misexpression of the BMP inhibitor Chordin in the anterior Area Opaca can induce the formation of ectopic streaks (Streit *et al.*, 1998).

In agreement with these expression patterns it has been shown that during the early stages of development high levels of SMAD activation (phosphorylation) are detected throughout the epiblast, indicating high levels of BMP signaling, but that the site of streak formation is characterized by lower levels of SMAD phosphorylation (Faure *et al.*, 2002). It has been shown recently in frogs that Vg1 is required for the expression of Chordin and Cerberus (Birsoy *et al.*, 2006) and it would appear likely that this mechanism operates in the chick embryo. So far it is unknown what regulates the expression of Vg1. It appears likely that Wnt3 signaling will also be important in streak formation in the chick embryo.

Wnt3 and Wnt3a are expressed in the forming streak (Chapman *et al.*, 2004), beta-catenin and members of the TCF/LEF family of transcription factors, essential elements of the Wnt canonical signaling pathway, are also expressed in the streak. Furthermore activation of the canonical Wnt signaling pathway with LiCl results in heavily dorsalized embryos, where many cells in the epiblast show nuclear translocation of beta-catenin and also show strong expression of the dorsal mesodermal marker Goosecoid (Roeser *et al.*, 1999; Schmidt *et al.*, 2000). In conclusion, at the molecular level a signaling cascade responsible for the formation of mesoderm is beginning to be uncovered. However it as yet is unknown what regulates the expression of BMPs and Wnt8c in the Area Opaca, processes that appear to initiate during the process of egg laying and are therefore less amenable to experimental investigation. It will be especially interesting to know what determines the gradient in Wnt8c expression and what triggers the highly localized expression of Vg1, one of the earliest markers of mesoderm formation and streak initiation.

Another interesting open question is whether a significant degree of cell sorting of mesoderm cells occurs during streak formation. It has been reported that early in development scattered cells in the epiblast are positive for the carbohydrate epitope HNK1 and that these cells are later found preferentially in mesodermal structures (Canning and Stern, 1988; Stern and Canning, 1990). This suggests that they may preferentially aggregate into the streak during its formation. It can also be readily observed that in Koller's Sickle mesodermal marker genes such as Wnt8c and Chordin are expressed in large overlapping expression domains. However, later in the streak the expression domains of these genes are widely separated. Chordin expression is found in the tip of the streak, while Wnt8 expression is found in the back of the streak and the mesoderm cells that migrate out of the posterior portion of the streak (Lawson *et al.*, 2001; Mikawa *et al.*, 2004), compare Figs. 1B, 1D and Figs. 1D, 1H. Investigation of expression of these genes at the individual cell level shows that there is a large degree of variation in expression levels between even neighboring cells. It would thus appear possible that there is initially a salt and pepper distribution of anterior and posterior mesoderm cells in Koller's Sickle, which during streak formation undergo a cell sorting process whereby the Chordin expressing cells that end up in the tip of the streak move forward relative to the Wnt8C expressing cells that will form the posterior end of the streak. It could be that inhibition of BMP signaling in the cells that are going to form the tip of the streak would allow more efficient movement of the cells that

4. The Mechanisms Underlying Primitive Streak Formation in the Chick Embryo

do not express BMPs possibly through modulation of E-cadherin activity (see below for further discussion). Only detailed analysis and correlation of gene-expression and movement of individual cells in the streak of early embryos will be able to resolve these questions.

V. Cellular Mechanisms of Streak Formation

Streak formation is a process that requires the coordination of movement behavior of many thousands of cells in the epiblast and possibly in the hypoblast. So far it has not been conclusively demonstrated by which mechanisms this is achieved. Streak formation most likely involves the active movement of most cells in the embryo. Local inhibition of the actin cytoskeleton at the tip of the streak results in an inhibition of extension of the streak, suggesting that the tip of the streak has to move actively. At the same time more lateral cells keep on flowing into the streak which as a result may extend in the posterior direction as a result of more cells piling up at the back of the streak, showing that these cells continue to move in the absence of movement of cells in the tip. Inhibition of cells at the base of the streak also results in the inhibition of streak extension formation showing that many cells that are going to form the streak have to pass through the base of the streak in accordance with the observed cell movement patterns (Chuai *et al.*, 2006; Cui *et al.*, 2005). These observations show that the tip of the streak does not act like an engine that pulls all the other cells in the epiblast behind it, nor does it push the cells in front of it. The fact that the cells more lateral to the streak move actively in response to cues that are present even when the tip cells do not move appreciably point to the existence of other than just mechanical guidance information. These could be gradients of chemical information that guide the overall movement patterns of the cells. It does not, however, rule out an intercalation or zippering mechanism possibly localized at the base of the streak. Several cell behaviors have been assumed to be involved in this process and in the following section we will describe and discuss these in more detail.

A. Locally Restricted Cell Division

During the process of streak formation the embryo is growing rapidly. It has been described that there are slightly elevated rates of proliferation in the posterior marginal zone in a stage XIII embryo (Zahavi *et al.*, 1998) and it could be imagined that localized cell division would result in a local increases in cell mass such as observed in the streak. However at the later stages of development when the streak becomes visible there is no clear indication of a significant increase in mitotic activity or cycling behavior of cells in the streak (Cui *et al.*, 2005). In cases where individual cells were transfected with viruses expressing a reporter gene, cells derived from one clone line up in the primitive streak, suggesting that division might be preferential in the direction of streak formation (Wei and Mikawa, 2000). Furthermore it was observed that the cell division planes were aligned perpendicular to the direction of streak elongation. This led to the suggestion that oriented cell division could result in streak formation. However our recent estimates of cell cycle times of cells entering the streak, which is based on recording the interval between two successive divisions of GFP labeled cells, suggest that the cell cycle times are ∼7 hours, which is not sufficient to account for the 6–8 divisions that would be necessary if the streak would form through oriented cell division alone. Recent experiments with cell cycle blockers have shown that division is necessary for proper development, but does not appear to be the driving force for development. We have suggested that the alignment of mitotic cells is a result rather than a cause of movement (Chuai *et al.*, 2006; Cui *et al.*, 2005). Since there is a local accumulation of cells in the streak, while there appears to be no significant increase in local cell proliferation it is evident that the epiblast has to expand elsewhere in order to accommodate accumulation of cells in the streak. Indeed if cell cycle progression is blocked by application of inhibitors of cell cycle progression, the epiblast tends to tear itself apart. This shows that formation of the streak is an active process that can occur in the absence of cell division, but that increase in surface area of the epiblast and thus in cell number is required for the process to work properly.

B. Spatially Organized Cell–Cell Intercalation

A well documented mechanism underlying large-scale tissue shape changes during development is cell–cell intercalation. This is an attractive mechanism since limited local cell movements can result in large scale tissue shape changes (Keller *et al.*, 2000; Keller *et al.*, 2003). Intercalation, as proposed to occur in amphibians, requires a polarization of cells where they become either bipolar as in the case of lateral mesoderm or monopolar as in the case of the overlying neural plate followed by movement of the cells in between each other, resulting in a narrowing of the tissue in the direction of intercalation and an extension in the direction perpendicular to that (Keller, 2005). The mechanisms that has been proposed to underlie intercalation is that the cells in the mesoderm develop protrusive behavior at their ends and that these protrusions are the machines that the cells use to pull themselves forward using neighboring cells and the extracellular matrix as a scaffold. It has been suggested that intercalation behavior could underlie the transition of the cells in Koller's Sickle into the forming streak (Lawson and Schoenwolf, 2001). However so far a detailed high resolution *in vivo* analysis at the level of individual cell behaviors as has been performed in fish and frogs has not yet been performed in the chick embryo.

Work especially in lower vertebrates like fish and frogs, has suggested that intercalation is controlled through the Wnt planar polarity signaling pathway. In this

4. The Mechanisms Underlying Primitive Streak Formation in the Chick Embryo   145

pathways Wnt5 and Wnt11 signal through particular frizzled receptors (Frizzled 7), which then through disheveled activate small G proteins of the Rho family (Rac, Rho, and CDC42) which signal to downstream regulatory components of the actin–myosin cytoskeleton The latter may involve the activation of Rho kinase and atypical PkC (Keller, 2002; Seifert and Mlodzik, 2007). However experiments directed to test the involvement of this pathway in streak formation in the chick embryo, by expression of dominant negative Wnt11 constructs, mutant forms of Disheveled that specifically block signaling through the planar polarity signaling pathway and application of the Rho kinase inhibitor Y27632, did not inhibit streak formation, although they all affected later development after the regression stages where convergent extension has been shown to be an important mechanism (Chuai *et al.*, 2006; Wei *et al.*, 2001).

An alternative to the case where individual cells polarize and intercalate by active movement has been suggested to occur during germ band extension in *Drosophila*. In the early stages of development most cells are organized in a tight packed epithelial sheet, where the cells are packed in an optimum space filling hexagonal pattern. However during development the cells in this epithelial sheet start to develop a polarized distribution of specific polarity proteins, cadherins and elements of the actin–myosin cytoskeleton. Intercalation results from two neighboring cells shortening a common boundary, which requires myosin II motor activity. This will bring four cells together in a T junction which is subsequently resolved in a direction perpendicular to the direction of contraction. This results in a local reshuffling of cells; cells that were initially neighbors are now one cell apart, while cells that were one cell apart are now direct neighbors, thereby resulting in intercalation (Bertet *et al.*, 2004). More recent observations have revealed that this behavior appears not only to occur at the level of single cells moving in between each other, but is often coordinated between small groups of cells that simultaneously contract some their vortices aligned along the axis of convergence to form a typical rosette like structure, where many cells border each other (Fig. 2A). Then in a second step this rosette gets resolved in a direction perpendicular to the direction of initial contraction, in the direction of extension (Blankenship *et al.*, 2006). This mechanism results in a local rearrangement of cells and when the behavior of many of these rosettes is coordinated it can result in the large scale tissue shape changes. Although there are many open questions as to the mechanisms responsible for the coordination of this behavior, it appears clear that in *Drosophila* patterning genes control the frequency and directionality of rosette formation and or resolution, since the frequency and dynamics of these structures are largely altered in patterning mutants. Interestingly cells in the epiblast of early chick embryos are also arranged in an essentially hexagonal pattern (Figs. 2B and 2C), although not quite as regular as observed in *Drosophila*. However during development rosettes in which substantially more than 3 cells share one junction are frequently observed in the epiblast of the chick embryo, especially in the forming streak (Figs. 2D and 2E). The increase in occurrence of rosette structures coincides with the time that the polonaise movements start and they are later especially prevalent in the region of the

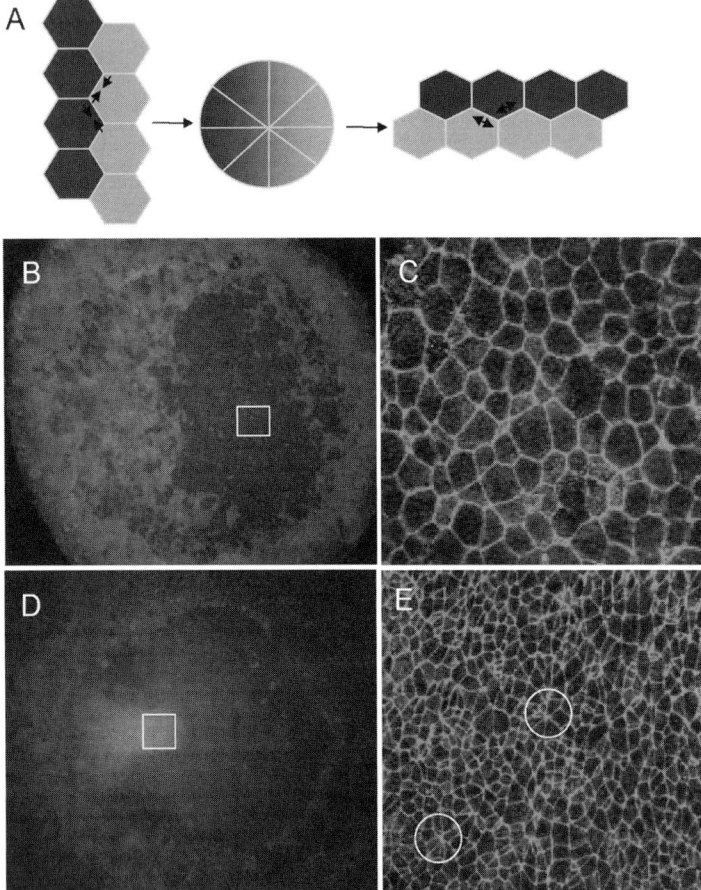

Figure 2 Intercalation of cells through rosette formation. (A) Schematic diagram of rosette formation. Cells in a two-dimensional sheet have a roughly hexagonal shape. Upon selective contraction of vertical vertices (inward pointing black arrows) between in this case 8 neighboring cells a rosette is formed where all eight cells contact each other in one point. Oriented resolution of the rosette by selective elongation of the common vertices in a horizontal direction (outward pointing black arrows) results in a reordering of cells and a concomitant tissue shape change. (B) Unincubated chick embryo stained with phalloidin that labels actin mostly localized at cell boundaries at the level of adherens junctions. (C) High magnification confocal image of the epiblast of the area indicated by a white box of the embryo shown in B. Note the predominantly hexagonal shape of the cells. (D) HH2 stage embryo stained with phalloidin, the tip of the streak is outlined by a white square. (E) High magnification confocal image of the epiblast of the area indicated by a white box of the embryo shown in (D). Note the irregular shapes of the individual cells and the complex structures that they form. There is clear evidence for many rosette-like structures of different degrees of complexity, two are outlined by white circles. Also note the smaller size of the cells in the tip of the streak compared to the much larger size of the cells in the epiblast of a stage XIV embryo (C). Rosette formation could result in contraction of the cells in the epiblast along the length of Koller's Sickle followed by resolution along the length of the extending streak. See color insert.

forming streak (Fig. 2 and Chuai and Weijer, unpublished observations). It remains to be shown how these structures are formed in the living embryo, whether they are required for streak formation to occur and how they are regulated. It is interesting to note that the contraction of epidermal cells in the *Drosophila* embryo required the action of Myosin II. Inhibition of Myosin motor activity through small molecule inhibitor blebbistatin does not result in an inhibition of streak formation in the chick embryo, while it does severely impede development after the streak regression stages.

C. Chemotaxis

A completely different way to coordinate the behavior of many cells is through chemotaxis. Chemotaxis is a well established mechanism to control the migration of individual cells, via gradients of attractants and or repellents. A paradigm for chemotactic signaling is the formation of aggregation streams and migrating slugs in *Dictyostelium*. Here propagating waves of cAMP initiated in the tip signal to the other cells to follow them (Dormann and Weijer, 2006a). However chemotaxis, has also been shown to be important in early development of higher organisms, for instance, tracheal development in *Drosophila*, the guidance of mesodermal cells during gastrulation and the guidance of primordial germ cells (Affolter and Weijer, 2005; Dormann and Weijer, 2006b). So far there is less direct evidence for this mechanism in the control of migration of cells in organized in sheets as is the case in the epiblast of the chick embryo. Observation of the cell migration pattern in the chick embryo shows that the overall cell flow pattern resembles that of closed flow lines rotating around quiescent zones (Fig. 1J). The streams converge at the site of streak formation and diverge in more anterior positions. The convergence of cells could result from chemoattraction, while the divergence of cells could result from a chemorepulsive mechanism. A combination of chemoattraction and chemorepulsion could result in large scale cell flows that are compatible with those observed to occur during the formation of the primitive streak (Fig. 3B). A possible scenario would be the existence of a chemorepellent produced in the tip of the streak, followed by a group of cells behind the tip that secrete an attractant. This would create the equivalent of a chemical dipole that could direct the observed circular cell movement patterns and direct the forward movement of the tip of the streak. Model calculations have shown that the creation of such a chemotactic dipole could account for the observed cell movement patterns, particularly if this involved mechanical interactions between the cells (Newman and Weijer, in preparation).

The attractant would be responsible for the cells coming together, and this could explain the contraction of cells in Koller's sickle. A well studied example of this type of cellular behavior is the aggregation of *Dictyostelium* cells into multicellular aggregation centers and their subsequent movement into aggregation streams mounds and slugs (Dormann and Weijer, 2006a). Here the cells secrete an attractant, cAMP,

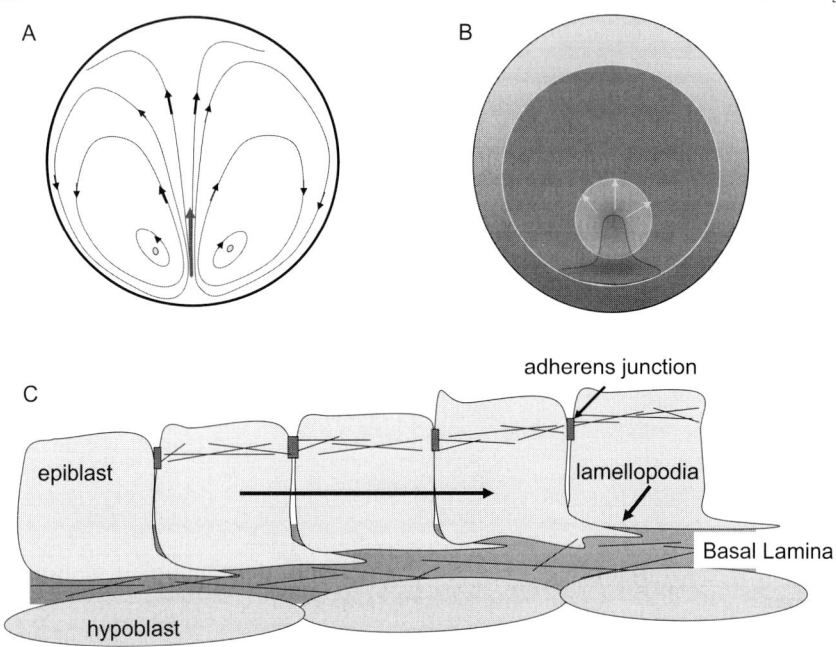

Figure 3 Mechanisms of streak formation and cell movement. (A) Schematic diagram indicating fluid flow patterns in an incompressible liquid in a circular dish after application of a force directed towards the center (red arrow). Note the existence of two counter rotating vortexes directed towards and away from the site of force application (Stokes flow). (B) Proposed mechanism to produce a chemotactic dipole. Cells in the tip of the forming streak (orange) produce a repellent (yellow) that diffuses out in the epiblast. Surrounding cells in the epiblast express receptors for this repellent and if this repellent rises above the threshold for detection they move away allowing the streak to advance in the direction of the center of the embryo. Cells in the back of the streak produce an attractant (blue) and also express a receptor for this attractant. This results in accumulation of the cells in the streak. The combination of chemoattraction and chemorepulsion results in extension of the streak and counter-rotational movements of the cells in the epiblast following a pattern as indicated in Figs. 1J and 3A. (C) Mechanism of migration of cells in the epiblast and hypoblast. Epiblast cells produce a basal lamina, which is anchored in the surrounding Area Opaca. The epiblast cells move by walking on this membrane using lamellipodia that they extend in the direction of migration. They may also exert traction on each other possibly involving the formation of filopodia. The cells stick together at their apical side through well-developed but dynamic adherens junctions. These lamellipodia may not only be important for cell movement but could possibly also detect and interpret gradients of signaling molecules such as growth factors and Wnts that are often intimately associated with the extracellular matrix. See color insert.

which they detect with cell surface receptors and to which they respond by chemotactic movement up gradients of this molecule. This coupling between cAMP by cells and their movement of gradients of cAMP creates an autocatalytic feedback loop resulting in the aggregation of cells. Aggregation of a few cells results in a local increase in cAMP secretion, which in turn will attract more cells over a larger

4. The Mechanisms Underlying Primitive Streak Formation in the Chick Embryo   149

distance, etc. The early stages of streak formation could possibly involve this mechanism resulting of the aggregation of cells in Koller's sickle into an aggregate at the posterior streak. Repulsion would be necessary to drive the extension of the streak. For this scenario to work very specific conditions have to be imposed on the distribution of the repellents and attractants and their receptors. Attraction requires the cells to express both the ligand and its receptor, while the repellent should be expressed in the tip of the streak. Moreover, cells outside the streak would have to respond to it by moving away, creating space for the cells of the tip to move forward. This requires the receptor for the repellent to be expressed in the epiblast cells outside the forming streak. Possible candidates for attractants and repellents are members of the FGF family. There are 23 FGF genes expressed in higher vertebrates and many FGFs are expressed in the streak (Bottcher and Niehrs, 2005; Karabagli *et al.*, 2002). These FGFs are detected via specific FGF receptors coded by at least four genes and for each gene several splice variants with different binding properties exist. The FGFR1 and FGFR4 receptors are expressed throughout the early epiblast and are candidates to receive attractive as well as repulsive signals (Karabagli *et al.*, 2002; Lunn *et al.*, 2007; Shamim and Mason, 1999; Walshe and Mason, 2000). Other possible molecules involved in these early processes may be Wnts, many Wnts are expressed in the streak, while their receptors the Frizzleds show more widespread distribution patterns in the epiblast (Chapman *et al.*, 2004). Until many of these candidate molecules have been tested in detail it is difficult to know their precise role in the process of streak formation.

D. Adhesion Gradients

It is conceivable that cells in the epiblast move in response to local changes in adhesiveness, achieved through differences in the expression of adhesion molecules or the quantitative modulation of adhesive cell–cell and or cell–matrix interactions. It is well established that individual fibroblast-like cells can detect gradients of stiffness in the substrate on which they move and they tend to move in the direction of stiffer substrates and move up these gradients (Choquet *et al.*, 1997; Dobereiner *et al.*, 2005). It also appears possible that cells could detect gradients in adhesiveness and move towards regions of higher adhesion. Recently it has recently been suggested that the dorsal migration of lateral mesoderm cells in the fish embryo is the result of cell migrating towards regions of higher adhesiveness, since higher adhesion would enable them to strengthen contacts with neighboring cells and allow them to apply more force needed to pull them forward (von der Hardt *et al.*, 2007; Witzel *et al.*, 2006). It was suggested that BMP signaling would affect cadherin activity, and therefore indirectly repel the cells away from ventral sides of low cadherin activity to dorsal sides of higher cadherin activity. The molecular mechanism underlying the action of BMP signaling on adhesion has not been resolved in detail. It does not act through direct transcriptional control of adhesion molecules but involves regulation of factors controlling the dynamics of adhesive interactions. BMP signaling itself is regulated through BMP antagonists such as Chordin. Expression and secretion of Chordin by the organizer inhibits BMP signaling in the organizer and makes it a center of high adhesion and attraction. It is interesting to note that in the chick embryo BMP4 and BMP7 are expressed in the Area Opaca of the early embryo. The site of streak formation is characterized by low BMP signaling which may result from high levels of Chordin expression in the epiblast overlying Koller's sickle. Extrapolating from the zebrafish results this would mean that the cells would be able to move towards the region of low BMP signaling, i.e., regions of high adhesion in the center of Koller's Sickle. This idea can be tested by experiments that locally modulate cell–cell adhesion, for instance, via local perturbation of extracellular calcium, which is required for cadherin-meditated cell adhesion or through local application of function blocking antibodies.

E. Contact Guidance

A final mechanism could be that some cells move in response to a chemical signals and that other cells will follow these cells. Since the cells are connected through adhesive cell–cell contacts movement of one cell will result in pulling by this cell on a following cell. If the mechanical pulling would in some way result in a signal for this cell to follow its predecessor, then this cell would pull on its neighbor, thereby effectively propagating a directional movement signal. A mechanism through which this could work would be if local pulling of one cell would result in locally opening of stretch-sensitive calcium channels, resulting in a local calcium influx, triggering an extension and movement on that side of the cell. This phenomenon has been proposed to underlie slow calcium and mechanical contraction waves (Jaffe, 2007). In this case the signal is propagated by local mechanical force and transduced via intracellular calcium changes. In this case only some cells interpret the primary guidance information, but other cells follow. A loose analogy to the latter are the two counter-rotational flow patterns in a liquid confined in a circular beaker, after local application of a force directed from the periphery towards the center of the beaker (Fig. 3A). The flow pattern throughout the liquid results from the adhesive interactions between the water molecules, the incompressibility of the liquid and the boundary conditions (Stokes flow). Model calculations using a combination of chemical signal detection and contact following have shown that this is a mechanism that could produce large scale cell flows under the right boundary conditions (Vasiev, Chaplain, and Weijer, submitted).

VI. Mechanisms of Movement

The structure of the epiblast is that of a typical epithelial sheet. Each cell is in contact at its apical end with other cells through E-cadherin containing adherens junctions as

4. The Mechanisms Underlying Primitive Streak Formation in the Chick Embryo

well as tight-junctions (Andries *et al.*, 1985). The basal side of each cell is in contact with a complex fibronectin- and collagen-containing basal lamina. There is little information on how cells in the epiblast move. It would appear that several mechanisms could be involved: either the cells move by pushing and pulling on each other as described above. Provided that the outer cells are anchored to the vitelline membrane this could result in a reshuffling of the cells in the more central parts of the epiblast. This mechanism would require extensive modulation of cell–cell adhesion. If this mechanism was active in a graded manner it could result in intercalation, when guided by a proper signaling system.

As an alternative, the cells could be moving on the basal lamina substrate (Fig. 3C). The cells would be in contact with the substrate on the basal side and extend lamellipodia and filopodia and use these to migrate in a manner similar to other cells moving on a substrate. Interestingly, this mode of movement has been shown to occur in MDCK cells in tissue culture during wound healing. It appears that the cells make many cryptic lamellipodia and the interface with the substrate in the direction of movement (Farooqui and Fenteany, 2005). For the cells to move on the basal lamina it is required that this forms a relatively stiff structure that is anchored somewhere in the embryo. By staining matrix components in the embryo it appears that the basement membrane is formed underneath the epiblast cells but in the Area Opaca may become more continuous and intertwine between the cells (Morrison and Weijer, unpublished observations). This basement membrane could form a matrix on which cells could move. Furthermore this matrix could also act as a substrate that would bind many signaling molecules especially of the FGF, PFGF, VEGF, and Wnt families. Experiments in which specific components of the basement membrane such as chondroitin sulfate were digested by enzymes resulted in failure of development and abnormal streak formation, underscoring the important role of this membrane in early development (Canning *et al.*, 2000). Finally it is possible that the cells use both mechanisms at the same time, i.e., push and pull on each other, a mechanism mediated through junctional contacts involving cell–cell adhesion molecules and active movement on the basal lamina which is mediated by cell matrix receptors of the integrin family (Hynes, 2002). It remains to be seen how important these various types of interactions are.

VII. Challenges for Modeling and Computational Approaches

Gastrulation as described above is a very complex process. It involves the coordinated behavior of many tens of thousands of cells, which differentiate in response to signals that they themselves generate. As a result of differentiation, cells can perform different behaviors, i.e., divide, die, change shape and or move. Cell movement in particular results in the creation of new cell–cell interactions and signaling environments, which then in turn trigger the activation of novel gene expression programs. These interactions between signaling and cell behaviors will finally result in gastrulation. Since many of these interactions are highly nonlinear it will be necessary to use more quantitative methods to understand and explore the full range of interactions and behaviors. This approach will have to involve extensive mathematical modeling. So most models that have been developed to explain aspects of gastrulation are continuous models based on partial differential equation that describe reaction–diffusion mechanisms of morphogens to explain gene expression patterns (Page *et al.*, 2001) or patterns of chemoattractants to explain movement (Painter *et al.*, 2000) mostly in one or two dimensions. Although these models can give some useful insights in some aspects of the process it is clear that what is really need are models that can deal with tens of thousands of cells, that may all have individual internal signaling and gene-expression programs as well as have the ability to physically interact with neighboring cells through chemical and mechanical signals. Furthermore these cells secrete extra cellular matrixes which they actively modulate and with which they interact chemically and mechanically. Therefore to model gastrulation it will be necessary to develop cell based models that allow the calculation of chemical as well as biomechanical interactions between cells and their noncellular environment. Many of these interactions are local while others may be long range and acting on different time scales, which poses special challenges to modeling. Attempts to develop cell based models for aspects of gastrulation have started (Bodenstein and Stern, 2005). However cell based models such as the Cellular Potts, or Subcellular Element Model that are currently being developed, where cells consists of multiple elements and a result show intracellular heterogeneity in signaling and behavior, look particularly promising (Chaturvedi *et al.*, 2005; Cickovski *et al.*, 2005; Newman, 2005). However, so far none of these models have been rigorously applied to gastrulation as a full three-dimensional problem.

VIII. Outlook

Initially mechanisms of streak formation were investigated by grafting and transplantation experiments, which gave valuable insights in the cellular origin of the streak and in the regulative potential of the early embryo. This was followed by an era of investigation of the molecular basis of the signaling systems involved in mesoderm induction, a prerequisite for streak formation. This line of research is vibrant and ongoing and, as outlined, many molecular details remain to be resolved. The big challenge for the future will be to understand how the differentiation of mesoderm cells results in their acquisition of properties that allow them to form complex structures such as the streak. It will be necessary to understand how cells detect, polarize and move in response to signals and how these movements feed back on the signaling systems to result in coordinated development. Interesting questions include to what extent differential cell movement is a part of the normal development of the early chick embryo, whether cells differentiate in a mosaic pattern, do they then sort out and if this is the case what is the cellular basis for this sorting behavior?

To gain more insight in the mechanics of the gastrulation it will be necessary to develop physical methods to measure tension and traction forces produced by individual cells in the context of the organism and their consequences for force production and distribution in tissues. A possible way forward is have embryos or parts of embryos develop on elastic substrates or on "beds of nails," arrays of elastic pillars that allow local measurements of forces with high spatial and temporal resolution (du Roure *et al.*, 2005; Ganz *et al.*, 2006; Steinberg *et al.*, 2007). As indicated above these experimental approaches will have to be complemented by advances in modeling and computations and important and exciting developments are expected to take place in this area of research.

References

Affolter, M., and Weijer, C. J. (2005). Signaling to cytoskeletal dynamics during chemotaxis. *Dev. Cell* **9**, 19–34.
Andries, L., Harrisson, F., Hertsens, R., and Vakaet, L. (1985). Cell junctions and locomotion of the blastoderm edge in gastrulating chick and quail embryos. *J. Cell Sci.* **78**, 191–204.
Bachvarova, R. F., Skromne, I., and Stern, C. D. (1998). Induction of primitive streak and Hensen's node by the posterior marginal zone in the early chick embryo. *Development* **125**, 3521–3534.
Barrallo-Gimeno, A., and Nieto, M. A. (2005). The Snail genes as inducers of cell movement and survival: Implications in development and cancer. *Development* **132**, 3151–3161.
Beddington, R. S., and Robertson, E. J. (1999). Axis development and early asymmetry in mammals. *Cell* **96**, 195–209.
Bertet, C., Sulak, L., and Lecuit, T. (2004). Myosin-dependent junction remodeling controls planar cell intercalation and axis elongation. *Nature* **429**, 667–671.
Bertocchini, F., and Stern, C. D. (2002). The hypoblast of the chick embryo positions the primitive streak by antagonizing nodal signaling. *Dev. Cell* **3**, 735–744.
Bertocchini, F., Skromne, I., Wolpert, L., and Stern, C. D. (2004). Determination of embryonic polarity in a regulative system: Evidence for endogenous inhibitors acting sequentially during primitive streak formation in the chick embryo. *Development* **131**, 3381–3390.
Birsoy, B., Kofron, M., Schaible, K., Wylie, C., and Heasman, J. (2006). Vg1 is an essential signaling molecule in Xenopus development. *Development* **133**, 15–20.
Blankenship, J. T., Backovic, S. T., Sanny, J. S., Weitz, O., and Zallen, J. A. (2006). Multicellular rosette formation links planar cell polarity to tissue morphogenesis. *Dev. Cell* **11**, 459–470.
Bodenstein, L., and Stern, C. D. (2005). Formation of the chick primitive streak as studied in computer simulations. *J. Theor. Biol.* **233**, 253–269.
Bortier, H., Callebaut, M., and Vakaet, L. C. (1996). Time-lapse cinephotomicrography, videography, and videomicrography of the avian blastoderm. *Methods Cell Biol.* **51**, 331–354.
Bottcher, R. T., and Niehrs, C. (2005). Fibroblast growth factor signaling during early vertebrate development. *Endocr. Rev.* **26**, 63–77.
Canning, D. R., and Stern, C. D. (1988). Changes in the expression of the carbohydrate epitope HNK-1 associated with mesoderm induction in the chick embryo. *Development* **104**, 643–655.
Canning, D. R., Amin, T., and Richard, E. (2000). Regulation of epiblast cell movements by chondroitin sulfate during gastrulation in the chick. *Dev. Dyn.* **219**, 545–559.
Chapman, S. C., Brown, R., Lees, L., Schoenwolf, G. C., and Lumsden, A. (2004). Expression analysis of chick Wnt and frizzled genes and selected inhibitors in early chick patterning. *Dev. Dyn.* **229**, 668–676.
Chaturvedi, R., Huang, C., Kazmierczak, B., Schneider, T., Izaguirre, J. A., Glimm, T., Hentschel, H. G., Glazier, J. A., Newman, S. A., and Alber, M. S. (2005). On multiscale approaches to three-dimensional modeling of morphogenesis. *J. R. Soc. Interface* **2**, 237–253.

Choquet, D., Felsenfeld, D. P., and Sheetz, M. P. (1997). Extracellular matrix rigidity causes strengthening of integrin–cytoskeleton linkages. *Cell* **88**, 39–48.

Chuai, M., Zeng, W., Yang, X., Boychenko, V., Glazier, J. A., and Weijer, C. J. (2006). Cell movement during chick primitive streak formation. *Dev. Biol.* **296**, 137–149.

Cickovski, T. M., Huang, C., Chaturvedi, R., Glimm, T., Hentschel, H. G., Alber, M. S., Glazier, J. A., Newman, S. A., and Izaguirre, J. A. (2005). A framework for three-dimensional simulation of morphogenesis. *IEEE/ACM Trans. Comput. Biol. Bioinform.* **2**, 273–288.

Cui, C., Yang, X., Chuai, M., Glazier, J. A., and Weijer, C. J. (2005). Analysis of tissue flow patterns during primitive streak formation in the chick embryo. *Dev. Biol.* **284**, 37–47.

Dobereiner, H. -G., Dubin-Thaler, B. J., Giannone, G., and Sheetz, M. P. (2005). Force sensing and generation in cell phases: Analyses of complex functions 1152. *J. Appl. Physiol.* **98**, 1542–1546.

Dormann, D., and Weijer, C. J. (2006a). Chemotactic cell movement during Dictyostelium development and gastrulation. *Curr. Opin. Gen. Dev.* **16**, 367–373.

Dormann, D., and Weijer, C. J. (2006b). Imaging of cell migration. *EMBO J.* **25**, 3480–3493.

Downie, J. R. (1976). The mechanism of chick blastoderm expansion. *J. Embryol. Exp. Morphol.* **35**, 559–575.

du Roure, O., Saez, A., Buguin, A., Austin, R. H., Chavrier, P., Silberzan, P., and Ladoux, B. (2005). Force mapping in epithelial cell migration. *Proc. Natl. Acad. Sci. USA* **102**, 2390–2395.

Eyal-Giladi, H. (1997). Establishment of the axis in chordates: Facts and speculations. *Development* **124**, 2285–2296.

Farooqui, R., and Fenteany, G. (2005). Multiple rows of cells behind an epithelial wound edge extend cryptic lamellipodia to collectively drive cell-sheet movement. *J. Cell Sci.* **118**, 51–63.

Faure, S., de Santa Barbara, P., Roberts, D. J., and Whitman, M. (2002). Endogenous patterns of BMP signaling during early chick development. *Dev. Biol.* **244**, 44–65.

Foley, A. C., Skromne, I., and Stern, C. D. (2000). Reconciling different models of forebrain induction and patterning: A dual role for the hypoblast. *Development* **127**, 3839–3854.

Ganz, A., Lambert, M., Saez, A., Silberzan, P., Buguin, A., Mege, R. M., and Ladoux, B. (2006). Traction forces exerted through N-cadherin contacts. *Biol. Cell.* **98**, 721–730.

Graeper, L. (1929). Die Primitiventwicklung des Hühnchens nach stereokinematographischen Untersuchungen kontrolliert durch vitale Farbmarkierung und verglichen mit der Entwicklung anderer Wirbeltiere. *Wilhelm Roux' Arch. Entwicklungsmechan. Org.* **116**, 382–429.

Harrisson, F., Callebaut, M., and Vakaet, L. (1991). Features of polyingression and primitive streak ingression through the basal lamina in the chicken blastoderm. *Anat. Rec.* **229**, 369–383.

Hatada, Y., and Stern, C. D. (1994). A fate map of the epiblast of the early chick embryo. *Development* **120**, 2879–2889.

Hynes, R. O. (2002). Integrins: Bidirectional, allosteric signaling machines. *Cell* **110**, 673–687.

Jaffe, L. F. (2007). Stretch-activated calcium channels relay fast calcium waves propagated by calcium-induced calcium influx. *Biol. Cell.* **99**, 175–184.

Karabagli, H., Karabagli, P., Ladher, R. K., and Schoenwolf, G. C. (2002). Comparison of the expression patterns of several fibroblast growth factors during chick gastrulation and neurulation. *Anat. Embryol. (Berlin)* **205**, 365–370.

Keller, R. (2002). Shaping the vertebrate body plan by polarized embryonic cell movements. *Science* **298**, 1950–1954.

Keller, R. (2005). Cell migration during gastrulation. *Curr. Opin. Cell Biol.* **17**, 533–541.

Keller, R., Davidson, L., Edlund, A., Elul, T., Ezin, M., Shook, D., and Skoglund, P. (2000). Mechanisms of convergence and extension by cell intercalation. *Philos. Trans. R. Soc. London B Biol. Sci.* **355**, 897–922.

Keller, R., Davidson, L. A., and Shook, D. R. (2003). How we are shaped: The biomechanics of gastrulation. *Differentiation* **71**, 171–205.

Khaner, O., and Eyal-Giladi, H. (1989). The chick's marginal zone and primitive streak formation. I. Coordinative effect of induction and inhibition. *Dev. Biol.* **134**, 206–214.

4. The Mechanisms Underlying Primitive Streak Formation in the Chick Embryo 155

Lawson, A., Colas, J. F., and Schoenwolf, G. C. (2001). Classification scheme for genes expressed during formation and progression of the avian primitive streak. *Anat. Rec.* **262**, 221–226.

Lawson, A., and Schoenwolf, G. C. (2001). Cell populations and morphogenetic movements underlying formation of the avian primitive streak and organizer. *Genesis* **29**, 188–195.

Lunn, J. S., Fishwick, K. J., Halley, P. A., and Storey, K. G. (2007). A spatial and temporal map of FGF/Erk1/2 activity and response repertoires in the early chick embryo. *Dev. Biol.* **302**, 536–552.

Lutz, H. (1949). Sur le production experimentale de la polyembryonie et de la monstruosite double chez les oiseaux. *Arch. Anat. Microsc. Morph. Exp.* **38**, 79–144.

Mikawa, T., Poh, A. M., Kelly, K. A., Ishii, Y., and Reese, D. E. (2004). Induction and patterning of the primitive streak, an organizing center of gastrulation in the amniote. *Dev. Dyn.* **229**, 422–432.

Newman, T. J. (2005). Modeling multicellular systems using subcellular elements. *Math. Biosci. Eng.* **2**, 611–622.

Page, K. M., Maini, P. K., Monk, N. A., and Stern, C. D. (2001). A model of primitive streak initiation in the chick embryo. *J. Theor. Biol.* **208**, 419–438.

Painter, K. J., Maini, P. K., and Othmer, H. G. (2000). A chemotactic model for the advance and retreat of the primitive streak in avian development. *Bull. Math. Biol.* **62**, 501–525.

Roeser, T., Stein, S., and Kessel, M. (1999). Nuclear beta-catenin and the development of bilateral symmetry in normal and LiCl-exposed chick embryos. *Development* **126**, 2955–2965.

Schmidt, M., Tanaka, M., and Munsterberg, A. (2000). Expression of beta-catenin in the developing chick myotome is regulated by myogenic signals. *Development* **127**, 4105–4113.

Seifert, J. R., and Mlodzik, M. (2007). Frizzled/PCP signaling: A conserved mechanism regulating cell polarity and directed motility. *Nat. Rev. Genet.* **8**, 126–138.

Shah, S. B., Skromne, I., Hume, C. R., Kessler, D. S., Lee, K. J., Stern, C. D., and Dodd, J. (1997). Misexpression of chick Vg1 in the marginal zone induces primitive streak formation. *Development* **124**, 5127–5138.

Shamim, H., and Mason, I. (1999). Expression of Fgf4 during early development of the chick embryo. *Mech. Dev.* **85**, 189–192.

Skromne, I., and Stern, C. D. (2001). Interactions between Wnt and Vg1 signaling pathways initiate primitive streak formation in the chick embryo. *Development* **128**, 2915–2927.

Skromne, I., and Stern, C. D. (2002). A hierarchy of gene expression accompanying induction of the primitive streak by Vg1 in the chick embryo. *Mech. Dev.* **114**, 115–118.

Spratt, N. T., and Haas, H. (1960a). Integrative mechanisms in the development of the early chick blastoderm. I. Regulative potentiality of separated parts. *J. Exp. Zool.* **145**, 97–137.

Spratt, N. T., and Haas, H. (1960b). Morphogenetic movements in the lower surface of the unincubated and early chick blastoderm. *J. Exp. Zool.* **144**, 139–158.

Steinberg, T., Schulz, S., Spatz, J. P., Grabe, N., Mussig, E., Kohl, A., Komposch, G., and Tomakidi, P. (2007). Early keratinocyte differentiation on micropillar interfaces. *Nano Lett.* **7**, 287–294.

Stern, C. D. (1990). The marginal zone and its contribution to the hypoblast and primitive streak of the chick embryo. *Development* **109**, 667–682.

Stern, C. D. (2004). "Gastrulation, Form Cells to Embryo's." Cold Spring Harbor Laboratory Press, New York.

Stern, C. D., and Canning, D. R. (1990). Origin of cells giving rise to mesoderm and endoderm in chick embryo. *Nature* **343**, 273–275.

Streit, A., Lee, K. J., Woo, I., Roberts, C., Jessell, T. M., and Stern, C. D. (1998). Chordin regulates primitive streak development and the stability of induced neural cells, but is not sufficient for neural induction in the chick embryo. *Development* **125**, 507–519.

Tam, P. P., Loebel, D. A., and Tanaka, S. S. (2006). Building the mouse gastrula: Signals, asymmetry and lineages. *Curr. Opin. Genet. Dev.* **16**, 419–425.

Vakaet, L. (1970). Cinephotomicrographic investigations of gastrulation in the chick blastoderm. *Arch. Biol. (Liege)* **81**, 387–426.

von der Hardt, S., Bakkers, J., Inbal, A., Carvalho, L., Solnica-Krezel, L., Heisenberg, C. P., and Hammerschmidt, M. (2007). The Bmp gradient of the zebrafish gastrula guides migrating lateral cells by regulating cell–cell adhesion. *Curr. Biol.* **17**, 475–487.

Walshe, J., and Mason, I. (2000). Expression of FGFR1, FGFR2 and FGFR3 during early neural development in the chick embryo. *Mech. Dev.* **90**, 103–110.

Wei, L., Roberts, W., Wang, L., Yamada, M., Zhang, S., Zhao, Z., Rivkees, S. A., Schwartz, R. J., and Imanaka-Yoshida, K. (2001). Rho kinases play an obligatory role in vertebrate embryonic organogenesis. *Development* **128**, 2953–2962.

Wei, Y., and Mikawa, T. (2000). Formation of the avian primitive streak from spatially restricted blastoderm: Evidence for polarized cell division in the elongating streak. *Development* **127**, 87–96.

Witzel, S., Zimyanin, V., Carreira-Barbosa, F., Tada, M., and Heisenberg, C. P. (2006). Wnt11 controls cell contact persistence by local accumulation of Frizzled 7 at the plasma membrane. *J. Cell Biol.* **175**, 791–802.

Zahavi, N., Reich, V., and Khaner, O. (1998). High proliferation rate characterizes the site of axis formation in the avian blastula-stage embryo. *Int. J. Dev. Biol.* **42**, 95–98.

5

Grid-Free Models of Multicellular Systems, with an Application to Large-Scale Vortices Accompanying Primitive Streak Formation

Timothy J. Newman
Department of Physics and School of Life Sciences, Arizona State University, Tempe, Arizona 85287

I. Introduction
II. Grid-Free Models of Multicellular Systems
III. Recent Experimental Results Concerning Primitive Streak Formation
IV. Components of the Planar Cell Polarity Mechanism
V. Phenomenological Cell-Based Model of the Chick Epiblast
VI. Computer Simulations: Formation and Maintenance of Vortices during Streak Formation
VII. Discussion and Conclusions
    Acknowledgments
    Appendix A. Details of the Model
    Appendix B. The XY Model of Ferromagnetism
    References

This paper is comprised of two parts. In the first we provide a brief overview of grid-free methods for modeling multicellular systems. We focus on an approach based on Langevin equations, in which inertia is ignored, and stochastic effects on cell motion are included. The discussion starts with simpler models, in which cells are modeled as adhesive spheres. We then turn to more sophisticated approaches in which nontrivial cell shape is accommodated, including the recently introduced Subcellular Element Model, in which each cell is described as a cluster of adhesively coupled over-damped subcellular elements, representing patches of cytoskeleton. In the second part of the paper we illustrate the use of a standard grid-free cell-based model to computationally probe interesting new features associated with primitive streak formation in the chick embryo. Streak formation is a key developmental step in amniotes (i.e., birds, reptiles, and mammals), and can be observed in detail in the chick embryo, where the streak extends across a tightly-packed two-dimensional sheet (the epiblast) comprised of about 50,000 cells. The Weijer group [Cui, Yang, Chuai, Glazier, and Weijer, Dev. Biol. 284 (2005) 37–47] recently observed that streak formation is accompanied by coordinated cell movement lateral to the streak, resulting in two large counter-rotating vortices. We study a mechanism based on cell polarity (in the plane of the epiblast) that provides an explanation for these vortices, and test it successfully using computer simulations. This mechanism is robust, since the emergent vortex formation depends only on the gross features of the initial spatial distribution of planar polarity in the epiblast. © 2008, Elsevier Inc.

I. Introduction

The complexity of coordinated cell motion in developmental processes is such that computational approaches can be very helpful in elucidating the emergent dynamics of thousands of cells given simpler mechanistic hypotheses. There are several diverse approaches to constructing computational models, ranging from more traditional continuum models based on differential equations (Murray, 2004), to individual based models, which have been instantiated using cellular automata (Deutsch and Dormann, 2004), the cellular Potts model (Graner and Glazier, 1993), and grid-free models. Given that the cellular automata and Potts model methods will be described elsewhere in this book, we will provide here a brief overview of some grid-free models.

As an illustration of the power and flexibility of grid-free models, we will also present an application of a simple grid-free model to the recent observation of large-scale vortices accompanying primitive streak extension in the chick embryo.

II. Grid-Free Models of Multicellular Systems

It is helpful to start the discussion with a biological example, namely, aggregation of the unicellular amoeba *Dictyostelium discoideum* (*Dd*). This well-known biological phenomenon, in which *Dd* cells emit waves of chemoattractant (cAMP) and thereby spatially cluster, was first modeled in a continuum approximation using two coupled differential equations (Keller and Segel, 1970). One equation describes the dynamics of the cell population density, denoted by ρ; the second the dynamics of the cAMP concentration, denoted by φ. The Keller–Segel (KS) equations have become influential as a prime example of a more general set of phenomena termed reaction–diffusion processes. The KS equations have the explicit form

$$\frac{\partial \rho}{\partial t} = D_1 \nabla^2 \rho - \alpha \nabla (\rho \nabla \varphi),$$
$$\frac{\partial \varphi}{\partial t} = D_2 \nabla^2 \varphi - \lambda \varphi + \beta \rho, \tag{1}$$

where D_1 and D_2 are the effective diffusion constants for the cells and cAMP, respectively, α is a measure of the strength of chemotaxis, λ is the rate of cAMP degradation, and β is a measure of the rate of production of cAMP by the cells. These equations contain only the most basic biology of the system, and yet have given useful insights into the aggregation process, such as the existence of a cAMP concentration threshold for large-scale aggregation.

In 1994 two physicists considered a similar problem of self-aggregating agents, but retained the discrete nature of the agents, modeling their dynamics using Langevin

5. Grid-Free Models and Primitive Streak Formation

equations (Schweitzer and Schimansky-Geier, 1994). Their equations take the form

$$\dot{x}_i = \sqrt{D_1}\xi_i(t) + \alpha \nabla_i \varphi,$$
$$\frac{\partial \varphi}{\partial t} = D_2 \nabla^2 \varphi - \lambda \varphi + \beta \sum_i \delta(x - x_i(t)). \qquad (2)$$

The second equation for the cAMP concentration is still a continuum equation (i.e., one is not modeling individual cAMP molecule dynamics, but just tracking the concentration), and the rationale for this is that on the length and time scales of cells, only the "mesoscopic" dynamics of the cAMP is relevant. The key difference is in the source term for the concentration—there are discrete sources located at the positions $x_i(t)$ of the cells. These positions are updated by the first equation (which is really a set of equations for the N cells, labeled here by $i = 1, 2, \ldots, N$). The first term on the right-hand-side is called "noise" and represents the random motion of individual cells in the absence of external stimuli. Equations of motion which include a noise term are, loosely speaking, referred to as Langevin equations (van Kampen, 2007). The second term represents chemotaxis; i.e., a linear response to the spatial gradient of the cAMP concentration.

It was later shown that for large populations of identical cells, and moderately weak cell–cell interactions, such discrete models have the same long-time behavior as the KS equations (Stevens, 2000). It turns out that for smaller cell populations, the discrete nature of the cells is important, and novel dynamical effects can occur (Newman and Grima, 2004; Grima, 2005a).

The Langevin framework for updating the cell positions is very flexible and various other types of interactions can be added linearly to the right-hand-side of the equation for the cell velocity. The most important is a term representing the finite size and adhesive properties of the cells—this can be encapsulated by using a "potential function" V whose spatial gradient is a force. The equation of motion for cells interacting through biomechanics only (disregarding chemotaxis for simplicity) takes the form

$$\dot{x}_i = \sqrt{D_1}\xi_i(t) - \nabla_i \sum_j V(|x_i - x_j|). \qquad (3)$$

The potential is usually taken to be short-ranged (i.e., exponentially decreasing beyond a separation of approximately one cell diameter), as cells separated by more than a cell diameter have no biomechanical interaction. A typical form for the potential is illustrated in Fig. 1. When two cells approach within a certain range r_{eq}, the gradient of the potential is negative and the cells are repelled—this models the excluded volume of cells (i.e., cells do not interpenetrate). When two cells have a separation a little beyond r_{eq} the gradient of the potential is positive and the cells are weakly attracted—this models cell adhesion. Note, this form of biomechanical interaction, based on a single isotropic potential, implicitly assumes that cells are spherical in shape.

Despite the fact that continuum models often capture the essence of discrete cell-based models, a real strength of the latter comes into play when one wishes to

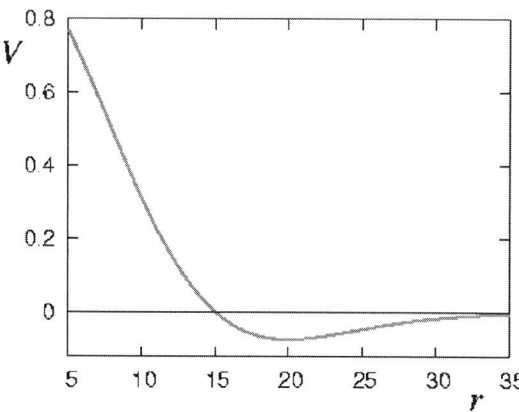

Figure 1 The generic form of the short-ranged potential used to model local biomechanical interactions between cells. The functional form used for the simulations in this paper is given in Eq. (A.2). In this plot, the equilibrium separation is $r_{\text{eq}} = 20.0$. See color insert.

describe heterogeneous cell populations. Continuing with the example of *Dd*, after the main aggregation phase the cells form a mound phase, and then a slug phase. In these later phases there is cell differentiation into pre-stalk and pre-spore type cells. If these two cell types have strong spatial heterogeneity it becomes problematic to model the system using continuum cell densities. Cells modeled in a discrete framework can be uniquely labeled and given a range of properties reflecting a degree of the true biological complexity of the system. Grid-free Langevin equations (coupled to continuum equations for various chemotactic signals) have been successfully used to describe the mound and slug phases of *Dictyostelium* (Bretschneider *et al.*, 1997, 1999). This work did rely on grids to numerically update the diffusion equation for the dynamics of the cAMP concentration. It was later shown that Green function methods can be very effective in updating the concentration field of signaling agents (or other chemical fields, such as nutrients) without the use of grids, thereby avoiding possible lattice artifacts (Newman and Grima, 2004; Grima, 2005b).

The Langevin equations for cells can be efficiently implemented on a computer, using grid-free algorithms based on neighbor tables or sectors. These types of algorithms are akin to molecular dynamics codes for simulation of various "many-body" systems, such as liquids, and have been developed in those contexts for decades (Allen and Tildesley, 1989). Given that for cells, the mechanical interactions are short-ranged, these algorithms are straightforward to implement and are very computationally efficient. The CPU time to simulate the system scales linearly with the number of cells in the system, regardless of the spatial dimension.

Drasdo and coworkers have used similar grid-free techniques to model a range of multicellular systems, including tissues (Drasdo *et al.*, 1995) and avascular tumors

(Drasdo and Höhme, 2003, 2005). Instead of using explicit Langevin equations, they have utilized Monte Carlo methods, in which the increment of a given cell's displacement is drawn for a probability distribution, in which the various mechanical interactions of the cell's neighborhood are weighted appropriately.

In recent years several methods have been introduced which attempt to go beyond the "spherical cell" approximation. This is not a simple problem, as shape is a difficult property to characterize. Othmer and co-workers used ellipsoidal cells to model the mound and slug phases of *Dictyostelium* (Palsson and Othmer, 2000; Dallon and Othmer, 2004). In a technical *tour de force*, Meyer-Hermann and co-workers have constructed computationally efficient algorithms to uniquely partition the three-dimensional space occupied by points representing the centers of cells (Schaller and Meyer-Hermann, 2005; Beyer *et al.*, 2005). Each compartmentalized region of space represents the cell body, and each facet of a given compartment represents an interface between two neighboring cells. The dynamics of these interfaces is governed by biomechanical equations including effects such as cell adhesion and friction terms. A third approach introduced recently by the author is the Subcellular Element Model (SEM) (Newman, 2005, 2007).

The SEM is a simple generalization of the original grid-free cell-based methods. Instead of regarding each point as a cell, in the SEM local clusters of points, or more appropriately, subcellular elements, are labeled as belonging to a given cell. The interactions between the elements belonging to the same cell are short-ranged and quite strongly adhesive, thereby giving the cell mechanical integrity. Neighboring elements belonging to different cells can be given interactions appropriate to the type of cell–cell interactions one is interested in. For example, for cell adhesion, neighboring surface elements of two cells will have weak short-ranged attraction. Labeling a given cell by i and its elements by α_i, the Langevin equations for the elements have the following form, which is a simple generalization of Eq. (3):

$$\dot{x}_{\alpha_i} = \sqrt{D}\xi_{\alpha_i}(t) - \nabla_{\alpha_i} \sum_{\beta_i \neq \alpha_i} V_{\text{intra}}\left(|x_{\beta_i} - x_{\alpha_i}|\right)$$
$$- \nabla_{\alpha_i} \sum_{j \neq i} \sum_{\beta_j} V_{\text{inter}}\left(|x_{\beta_j} - x_{\alpha_i}|\right). \qquad (4)$$

The first term on the right-hand-side again represents noise acting on the given element α_i (e.g., due to random polymerization events within the cell). The second term represents biomechanical interactions between that element and other elements within the same cell. The third term represents biomechanical interactions between that element and elements in neighboring cells. This term is only relevant if the element in question lies on the surface of the cell, such that it is potentially in close proximity to elements belonging to other cells. Note, when the number of elements per cell is reduced to one, the intracellular potential terms drop out, and Eq. (4) reduces to the well-studied spherical cell model described by Eq. (3).

The subcellular elements can be interpreted as small three-dimensional regions of cytoskeleton. On applying a force to a cell comprised of elements, the elements will respond and redistribute themselves spatially, resulting in the overall cell shape deforming in response to the force. Computer simulations on single cells composed of hundreds or thousands of elements indicate that visco-elastic cell biomechanics is reproduced rather well by this model (Sandersius and Newman, in preparation). The SEM has several other attractive features for modeling multicellular systems:

- cell-shape is adaptive to local forces.
- "seamless multiscalability," as the number of elements per cell can be varied according to how much intracellular detail is required.
- flexible accommodation of cell biology, as elements can be "functionalized," e.g., to represent actin cortex, or bulk cytoskeleton.
- computationally efficient—the SEM shares the computational properties of the original grid-free models, scaling linearly with the number of elements in three dimensions.

Regarding the last point, the high efficiency of the algorithm is not sensitive to the rate at which neighbors are changed. Determining the neighbor relationships is a very efficient process, and is performed each time step.

Cell biological processes such as cell growth and cytokinesis are readily implemented using the SEM (Newman, 2007). In brief, elements in the core of the cell have a small probability to replicate, if there is sufficient space nearby. A new element within the cell will cause a redistribution of other elements within the cell, thus increasing the volume of the cell. Element replication can be regulated by concentration of nutrients to model resource-limited cell growth. Once a cell has reached an appropriate size, cell division can be triggered. The simplest plausible mechanism for this is to discern the "long axis" of the cell and define the plane which is perpendicular to this line, and which passes through the center of mass of the cell. Elements on either side of this plane are relabeled as belonging to two new daughter cells, and division is completed by subsequent mechanical equilibration. The representation of a cell as a cluster of elements is shown in Fig. 2, along with the iso-surface of the cell. Growth and division are illustrated in Fig. 3. Repeated growth and division will lead to the formation of either epithelial sheets or spherical tumor-like cell clusters, depending on the boundary conditions imposed on the cells. An example of a large cluster (comprising ~1000 cells) is shown in Fig. 4. Each cell is constructed from several hundred elements, and so the entire simulation must keep track of hundreds of thousands of grid-free elements in three-dimensional space. A cross-sectional view (Fig. 5) illustrates the adaptive cell shape which emerges from the SEM algorithm. The codes used to implement the SEM are written in Fortran 90 modules. As a gauge of efficiency, the large cell cluster shown in Fig. 4 was created from a single cell in about ten hours on a single processor workstation.

Having given a sense of the various approaches to grid-free cell-based modeling, we proceed now to illustrate the method in its simplest form, and investigate the collec-

5. Grid-Free Models and Primitive Streak Formation

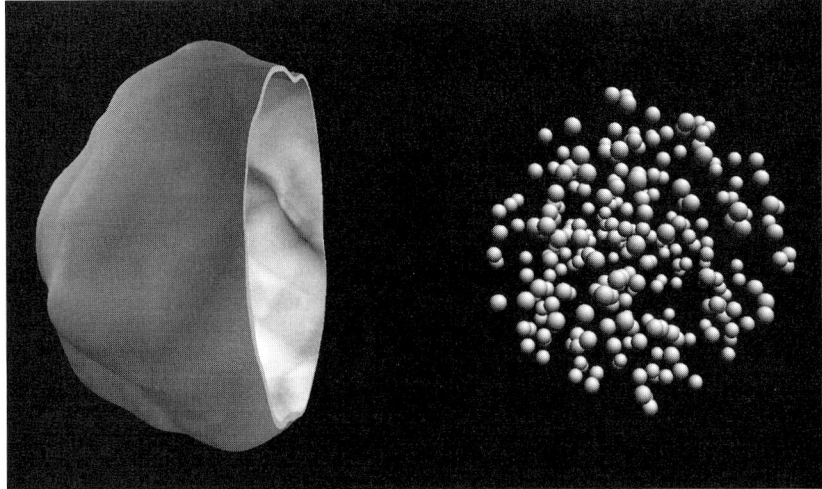

Figure 2 Two views of a cell created in the SEM. On the right is shown the cluster of 256 subcellular elements comprising the cell. It is difficult to see from this static image, but on rotating the cluster one observes that the elements are approximately equidistant from one another. The left panel shows an iso-surface constructed from this cluster of elements. (Image reproduced from "Single Cell-Based Models in Biology and Medicine," A. Anderson, M. Chaplain, and K. Rejniak, Eds., 2007, Birkhaüser.)

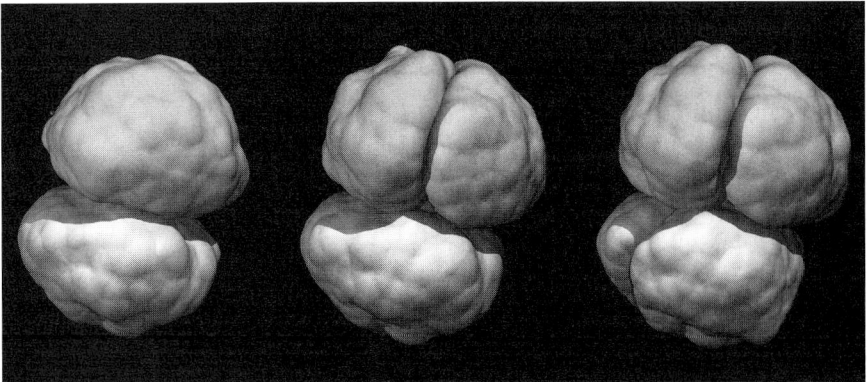

Figure 3 Cell growth is modeled using element replication in the cell core. Once a cell reaches a critical size it is triggered to undergo cytokinesis. Shown here are snapshots of two successive cell division processes. (Image reproduced from "Single Cell-Based Models in Biology and Medicine," A. Anderson, M. Chaplain, and K. Rejniak, Eds., 2007, Birkhaüser.)

tive vortex motion of cells that has recently been seen to accompany primitive streak formation in the chick embryo (Cui *et al.*, 2005). We study here a putative mechanism based on a coupling between cell motion and planar cell polarity.

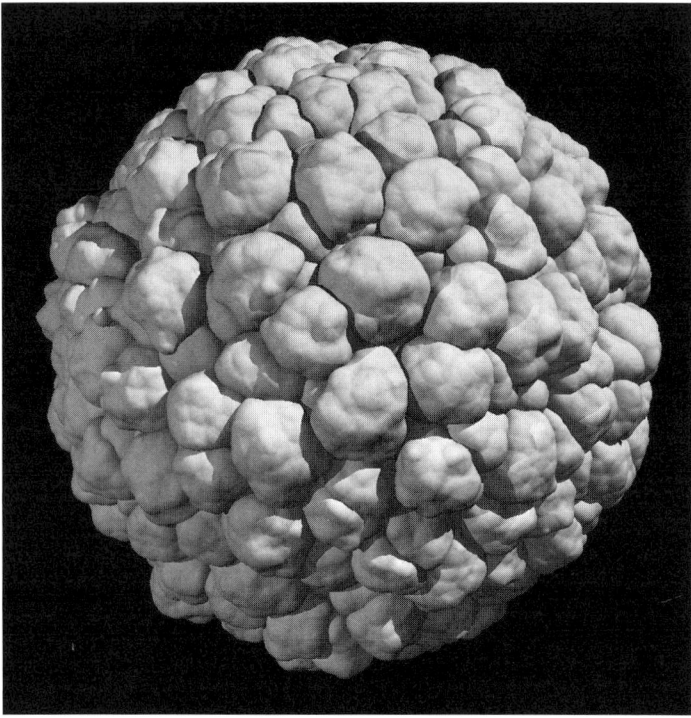

Figure 4 Repeated cell growth and division leads to a large contiguous cluster of cells. The cluster shown here has approximately 1000 cells, each with several hundred elements. (Image reproduced from "Single Cell-Based Models in Biology and Medicine," A. Anderson, M. Chaplain, and K. Rejniak, Eds., 2007, Birkhaüser.)

III. Recent Experimental Results Concerning Primitive Streak Formation

The chick embryo is a classic model of development, and has been studied using various forms of microscopy for over a century (Wolpert *et al.*, 2002). The chick is not just a model system for development in birds. The amniotes, a group of vertebrate organisms encompassing birds, reptiles, and mammals, have certain commonalities in the early stages of development. Thus, for example, processes seen in chick gastrulation (occurring after approximately 24 hours) have direct relevance to gastrulation in humans (occurring after approximately 12 days) (Gilbert, 2003). The most striking dynamical process of early development in the amniotes is the extension of the primitive streak, which in the chick proceeds for a period of roughly ten hours (Schoenwolf, 2001). The streak is composed of several layers of rather loosely attached cells, and is embedded in the epiblast, a two-dimensional sheet of tightly connected epithelial-like cells. The streak has two major developmental functions: (i) as a portal for the in-

5. Grid-Free Models and Primitive Streak Formation 165

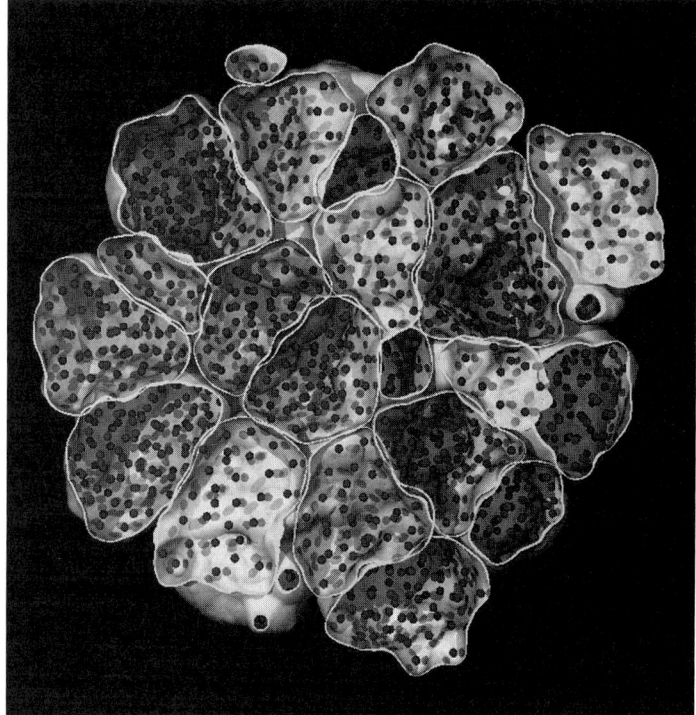

Figure 5 A cross-sectional view of a cell cluster (smaller than that shown in Fig. 4 for clarity). This image gives a particularly clear view of the range of cellular shapes that emerge from the SEM. Also shown are individual elements making up the cells lying in the cross-sectional plane. (Image reproduced from "Single Cell-Based Models in Biology and Medicine," A. Anderson, M. Chaplain, and K. Rejniak, Eds., 2007, Birkhaüser.) See color insert.

gression of pre-endodermal and pre-mesodermal cells, recruited from the epiblast and moving into the subgerminal cavity, and (ii) as a geometric delineation of the future vertebrate axis.

Recent developments in the Weijer laboratory in live imaging and fluorescent labeling of cells have allowed exciting new visualizations and insights into the coordinated cell movement which accompanies extension of the primitive streak. In particular, it has been observed that as the streak extends across the epiblast, there is long-range coordinated cell movement lateral to the streak in the form of two large counter-rotating vortices which span the entire epiblast (Cui *et al.*, 2005). Fluorescent cell tracks from these vortices are shown in Fig. 6. At first glance, these "flow patterns" are reminiscent of fluid flow.

To aid the subsequent discussion, Fig. 7 (left panel) shows a schematic of the chick epiblast midway through extension of the primitive streak, with labeling of important features referred to in the text.

Figure 6 A time-lapse image of cell-tracks in the epiblast during primitive streak formation. For orientation, Koller's sickle is at the lower left corner of the picture. Note the two large vortices lateral to the streak. Image courtesy of C. J. Weijer. See color insert.

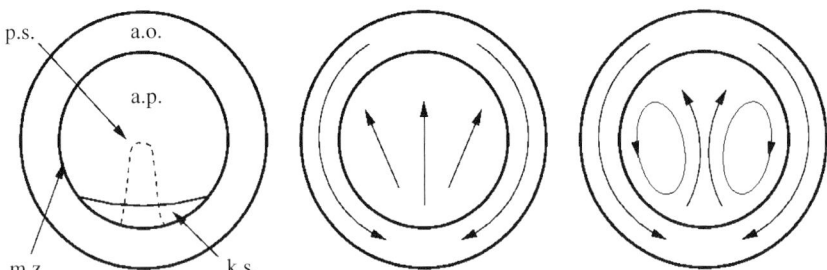

Figure 7 (Left panel) A schematic figure of the chick embryo epiblast, with labels of important structures referred to in the text: marginal zone (m.z.), area opaca (a.o.), area pellucida (a.p.), Koller's sickle (k.s.), and the primitive streak (p.s.). (Middle panel) A schematic figure indicating the necessary initial distribution of planar polarity in the area opaca and area pellucida. These distributions are assumed to be established by prior morphogen gradients emanating from Koller's sickle and the area opaca. (Right panel) A schematic of the "equilibrated" planar polarity distribution, achieved through maximizing local alignment of polarity between neighboring groups of cells.

There has been previous work on modeling streak extension, especially using continuum models (Schnell *et al.*, 2000). In this paper we will focus on the dynamics of the vortices, which we assume to be cooperative with, or even causative of, the streak extension, rather than simply an effect of streak extension.

5. Grid-Free Models and Primitive Streak Formation

A biologically plausible explanation of the vortices is complicated by the fact that at this early stage of chick development (meaning, just prior to extension of the streak) expression patterns for genes known to be important for gastrulation are localized to two regions: Koller's sickle (KS), and the caudal boundary region of the area opaca (Chapman *et al.*, 2002). For example, *Nodal* is expressed only in Koller's sickle, *Fgf8* is expressed on the caudal midline of Koller's sickle, *Bmp7* is expressed in the caudal region of the area opaca, and *Chordin* is expressed just rostral to Koller's sickle. Thus, from a modeling viewpoint there is no evidence to support a mechanism which requires active cell movement or function in localized areas of the area pellucida, except in the caudal region near Koller's sickle. Some influence, be it biomechanical, biochemical, or cell biological, is exerted on the area pellucida cells from cells in Koller's sickle, and perhaps cells in the area opaca near to the marginal zone, resulting in collective vortex-like cell motion.

We have recently investigated a mechanism for this vortex pattern based on the most intuitively appealing analog, namely, fluid flow. We used computer simulations of a grid-free model, in which cell–cell interactions (namely, contact following of cells) were designed to closely mimic viscous interactions in a fluid. Although such interactions are capable of producing vortices, the region of active forcing of the "fluid" needs to be in the area pellucida, rostral to Koller's sickle. Thus, we are less confident that the fluid analogy is in fact sound. An alternative mechanism invoking chemotaxis, with cells at the node of the streak acting as signaling centers, has also been pursued and is more promising (Newman *et al.*, in preparation).

In this paper we explore a third mechanism by which counter-rotating vortices may be formed using an influence on area pellucida cells that originates entirely from the marginal zone and Koller's sickle. This mechanism relies on an interaction between two cellular distributions of vectors (i.e., "vector fields"): one unmistakable, namely, the velocity field of the cells, and the second putative, namely, a vector field formed from an assumed planar polarity of the epiblast cells. (By "planar polarity" we mean a directionality of a particular cell phenotype, with that directionality lying in the plane of the epiblast.) The assumption that such a planar polarity vector field exists, along with three further assumptions of a simple nature which will be discussed in the next section, are sufficient to robustly reproduce the observed cell movement patterns.

IV. Components of the Planar Cell Polarity Mechanism

The mechanism we propose does not depend on the particular manifestation of the planar polarity, which may be in terms of cell shape, distribution of membrane receptors or adhesion, internal organization of the cell, or some other phenotype of the cell. For our purposes, we require only that the planar polarity exists for each cell. We note that the epiblast cells are epithelial in nature and therefore certainly have apical polarity.

That these cells can also have planar polarity is not without precedent, as evidenced by the cell movement patterns observed in the epithelial cell layer of the *Drosophila* embryo, which have been shown recently to be strongly coupled to the cells' planar polarity (Blankenship *et al.*, 2006). Planar cell polarity has a well-established role in other developmental systems, most notably the hexagonal packing of hairs on the *Drosophila* wing (Eaton, 1997).

Given the epiblast is composed of discrete cells, it is more accurate to describe the vector fields in terms of the planar polarity and velocity vectors of each cell. A given cell, labeled by i, is situated at position x_i, has planar polarity p_i, and velocity v_i. Each of these three quantities typically depends explicitly on time. During streak formation, the velocity distribution (or "field") can be measured experimentally (Fig. 6), and, as has already been described, is dominated by two large counter-rotating vortices lateral to the streak. The putative planar polarity distribution has not, to the knowledge of the author, been systematically studied in this stage of the chick embryo; however, recent work appears to have ruled out the role of planar polarity through a cell–cell intercalation mechanism (well-known in *Xenopus* and *Drosophila*) in streak formation (Chuai *et al.*, 2006).

The key to our hypothesized mechanism is an intimate coupling between the velocity and polarity distributions, which we now describe. On viewing movies of streak formation in the chick embryo (such movies courtesy of the Weijer group) one is struck by the instantaneous nature in which the entire area pellucida initiates the vortex motion. There is no discernible time lag between vortex motion near to the node of the streak and vortex motion at the outer edge of the area pellucida, some hundreds of cell diameters distant. Either a spatially complex chemotactic signal for such motion is rapidly communicated among the cells, or else, the cells have previously created a vector field for the subsequent velocity distribution, such that a "start" signal communicated to the cells allows them to follow the pre-defined pattern without instantaneous self-organization. Here we focus on the latter mechanism, and hypothesize that the pre-pattern for the velocity field is contained within the planar polarity field. The onset of large-scale cell motion could be triggered by a rapidly diffusible signal, originating from Koller's sickle, that instructs each cell to move with a local velocity parallel to the vector defined by the local planar polarity. In mathematical notation, the signal instructs $v_i \parallel p_i$. We note that recent modeling of lymphocyte motility has invoked a similar coupling between cell velocity and polarity (Meyer-Hermann and Maini, 2005).

It is convenient, though actually not necessary, to deconstruct this mechanism into two nontrivial stages: (i) the establishment of the planar polarity pre-pattern mapping out the large-scale vortices, and (ii) the maintenance of these vortex structures in the planar polarity field once large-scale cell motion is initiated. The establishment stage requires two assumptions. The first is that all cells in the epiblast have planar polarity and, after an appropriate signal, the direction of a given area pellucida cell's polarity will be aligned with the average polarity of the cells in its neighborhood. This will have the effect of producing a distribution of planar polarity which

5. Grid-Free Models and Primitive Streak Formation

is smooth over the scale of several cell diameters. The polarities of the cells in the area opaca are taken as fixed. The second assumption addresses the initial cell polarity distribution—namely, a distribution that can be plausibly established with simple cues (e.g., chemical gradients from Koller's sickle or from regions of the marginal zone), and upon which the alignment interaction can work to establish the more complex polarity distribution which contains large-scale vortices. We assume that the polarities of all cells in the area pellucida are initially directed away from Koller's sickle. We assume that the polarity of a given cell in the area opaca is initially directed caudally, but parallel to the marginal zone. A schematic figure illustrating this initial polarity distribution is shown in Fig. 7. The maintenance stage requires no additional assumptions—from computer simulations, it is found that the vortices in the planar-polarity field are maintained throughout cell motion without the need for additional constraints. This emergent behavior can be seen as a strength of the hypothesis.

Starting from this initial polarity distribution, we see that local alignment of polarity is fulfilled except, by construction, at the marginal zone. On instructing the area pellucida cells to align their polarity with the average polarity of their cellular neighborhood, the cells near the marginal zone will attempt to rotate their polarity to point caudally, and this will have a knock-on effect to cells further into the interior. It is not simple to intuit the future state of this alignment dynamics, but the system is driven to two large static vortex distributions of polarity. This will be discussed explicitly in the results section. This emergence of the polarization vortices from the alignment process is very robust, and is essentially preset by the fixed circumferential polarity of the area opaca cells.

We mentioned above that it is not strictly necessary to deconstruct this mechanism into two separate stages of establishment of vortices in planar polarity and subsequently vortices of cell velocities. The relaxation of the planar polarity field due to local alignment presumably occurs on a much faster timescale than large-scale movement of cells, simply because the movement of cells is inhibited by spatial crowding. It is therefore possible to recover similar results from this model by simultaneously switching on alignment of cell polarity and active movement of cells parallel to polarity. Though simultaneous, the dynamics of the alignment will quickly establish planar polarity vortices while the cells have moved only short distances, after which the cell motion will eventually map out the familiar large scale vortex flow. We have presented the mechanism in terms of two successive stages primarily for ease of conceptualization.

There exists an interesting analog from condensed matter physics: the alignment process of planar polarity is analogous to the dynamics of magnetic moments in planar ferromagnetism, and the final state in this case is similar and corresponds to a state of lowest internal energy. We give more details of this analog for interested readers in Appendix B.

V. Phenomenological Cell-Based Model of the Chick Epiblast

Having stated in detail the underlying assumptions of the mechanism, we proceed to instantiate them using computer simulations. We describe the cell-based computer model in this section, and present simulation results in the next section. A slightly more detailed mathematical discussion is given in Appendix A.

We use a simple grid-free cell-based model, akin to that described in Eq. (3). Given we are trying to model an epithelial-like sheet of the epiblast, each cell is envisioned as an adhesive cylinder of radius r_c, and labeled by its center of mass in the two-dimensional plane of the epiblast; i.e., the position of cell i is represented by the two-dimensional position vector x_i. We assume no spatial structure in the dimension perpendicular to the sheet, and so only keep track of the two-dimensional "sheet" coordinates of each cell. Intrinsic random motion of cells is not utilized in this work. Consistent with Eq. (3), inertia of cells is ignored, and so cell motion is over-damped. Clearly, modeling epithelial cells as cylinders is a strong approximation, and will lead to certain artifacts (such as local regions of hexagonal close packing). This can be improved by incorporating cell shape dynamics, e.g., through use of the SEM; however, in this work we are aiming for a qualitative description of a large-scale mechanism, which should be relatively insensitive to the finer details of how we model individual cells. If the mechanism is sensitive to details at the cellular level, this could indicate a lack of robustness, and a requirement for detailed cell biological regulation. The mechanism described here is robust, and, indeed, is strongly so, given that the creation of vortices in the planar polarity field is essentially determined by the macroscopic boundary conditions at the marginal zone.

The planar polarity is modeled by associating with each cell a polarity vector p_i. The polarity vector is defined to have unit magnitude and to lie in the plane of the epiblast; thus the planar polarity for cell i can be described by a single angle θ_i—the angle subtended by the polarity vector and the posterior–anterior axis of the epiblast. The dynamics of the planar polarity vectors is also taken to be over-damped, with interactions between cells that tend to align the polarity vectors of cells in the same neighborhood. As mentioned above, a mathematical description of both the cell motion and planar polarity dynamics is provided in Appendix A.

VI. Computer Simulations: Formation and Maintenance of Vortices during Streak Formation

The equations of motion for the cell positions (A.3) and polarities (A.4) are implemented in a grid-free algorithm written in Fortran 90 modules. Typical parameter values used may be found in Table I. Location of nearby cells is achieved using the standard method of sectors (Allen and Tildesley, 1989), allowing linear scaling of the efficiency of the algorithm with cell number. In the largest simulations presented in this paper, we track ~3500 cells for hundreds of thousands of iterations, which takes

5. Grid-Free Models and Primitive Streak Formation

Table I Typical values for model parameters used in the simulations presented in this paper. Length scales have dimensions of microns. All other parameters have dimensionless units

Model Parameter (symbol)	Description	Typical Value
V_0	Repulsive potential prefactor	11.0
U_0	Attractive potential prefactor	10.0
ξ_1	Range of repulsive core	12.9
ξ_2	Range of attractive tail	13.3
r_c	Cell radius	10.0
γ	Coupling of polarity to velocity	0.002
λ	Coupling of polarity alignment	0.01
dt	Time step for integration	0.1
q	Rate of cell replication attempt	0.02

several hours on a single processor workstation. We have preformed larger simulations using ~30,000 cells and the results are very similar. We present images of the intermediate sized system here, for clarity of presentation.

The epiblast is prepared by growing a layer of cells within a disk of predetermined radius, with a repulsive potential at the boundary. Once this disk is prepared, cells are labeled according to their positions as being area opaca cells, area pellucida cells (excluding Koller's sickle), and Koller's sickle cells. The adhesive interactions between the area opaca cells and the area pellucida cells are taken to be significantly weaker than the "intraspecific" adhesions within these two groups of cells. Once the epiblast is prepared, the repulsive potential at the disk boundary is removed. This allows the epiblast to adjust its global shape during cell movement. We do not try to model the effects of pressure on the periphery of the epiblast, although this would be important for a future study in order to correctly capture the global shape changes of the embryo.

First, we consider the establishment of the vortices in the cells' polarities. It is found that by initializing the polarity of the cells in the area opaca as shown in Fig. 7, and ensuring that the polarities of the area pellucida cells are initially aligned, two large vortices in the cell polarity distribution are robustly and spontaneously formed. If the polarities of the area pellucida cells were initially nonaligned (e.g., randomized), the polarity distribution would smoothen to some extent, but be composed of multiple small vortices. (This phenomenon is well known in the physical analog of the planar ferromagnet, and is responsible for the phase transition in that system.) Interestingly, we find that the location of the vortices in the polarity distribution is sensitive to whether or not the initial distribution of area pellucida cell polarities are splayed or not. In Fig. 8 we show two examples: in the first, the area pellucida cells have polarities which are splayed, as would be the case if they are initially aligned along a morphogen gradient emanating from Koller's sickle. One finds that the emergent vortices have centers which lie in the posterior half of the epiblast. In the second example, the initial distribution of polarities is initially directed along the posterior–anterior axis of the embryo. In this case, the emergent vortices lie on a lateral line

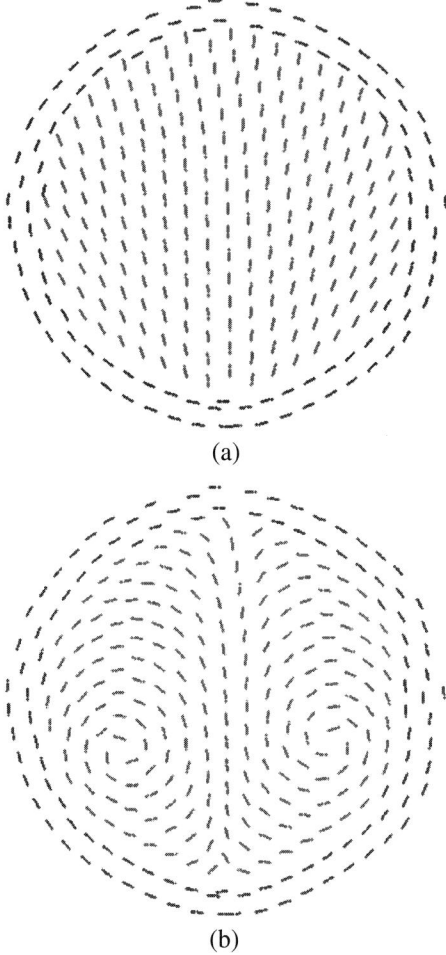

(a)

(b)

Figure 8 Sensitivity of the positions of the vortices to the initial distribution of cell polarity, here shown in a small system of a few hundred cells. In (a) the initial polarity of a given cell in the area pellucida is directed along a line joining that cell to the Koller's sickle region—as if the polarities are aligned along the gradient of a morphogen originating from that region. The resulting vortices after 10,000 iterations in (b) are positioned caudal to the lateral midline of the epiblast, and are slightly tilted with respect to the posterior–anterior axis. In (c) the initial polarity of cells in the area pellucida are parallel and pointing along the posterior–anterior direction. The resulting vortices after 10,000 iterations in (d) are positioned along the lateral midline of the epiblast. See color insert.

midway along this axis. Assuming that the later cell flow vortices mirror the position of the polarity vortices (which does indeed turn out to be the case), then the first vortex placement is more consistent with the experimental findings (Fig. 6).

5. Grid-Free Models and Primitive Streak Formation

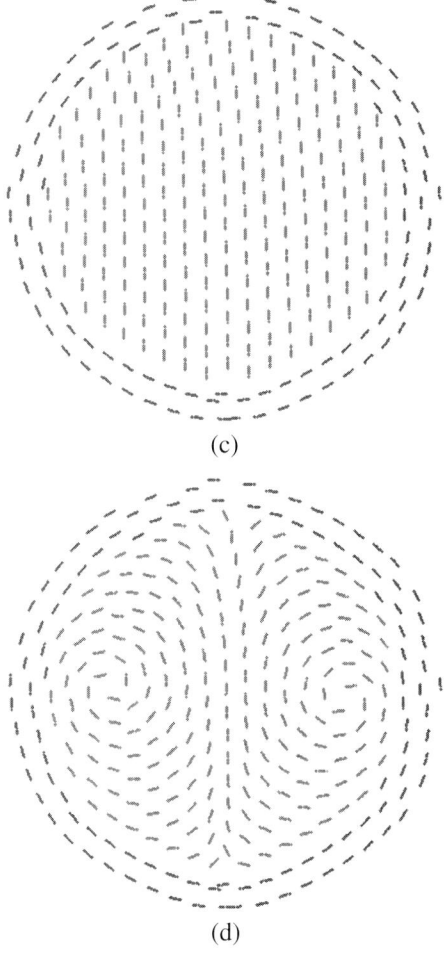

(c)

(d)

Figure 8 (*continued*)

The vortices of the polarity field are, in fact, metastable, although their lifetime increases rapidly with system size. On simulating rather small systems (of a few hundred cells) one finds that the vortices, being composed of only tens of cells, are sensitive to the precise cell arrangement and after formation begin to drift in the epiblast, eventually annihilating themselves at the anterior and posterior poles. On increasing the number of cells to several thousand, this vortex drifting is barely noticeable, and the vortices stay in their metastable equilibrium positions for periods longer than is required to reproduce the experimental data on cell flow.

In the second stage of the simulation, cells are instructed to actively move with a local velocity parallel to their planar polarity. This in principle could destabilize the vortices, as their component parts are now motile. However, we find that vortices are extremely stable to cell movement—much like the stability of fluid vortex, despite the constant flux of fluid through the vortex. There is a transient period during which the positions of the vortices shift slightly due to the coupling to the cell velocities, but thereafter the vortex positions are steady, and can be seen as the epicenters of the vortices observed from tracking cell flow. We illustrate the modest movement of vortices during cell movement over hundreds of thousands of iterations (Fig. 9). The integrity of the epiblast remains in tact throughout the simulation (Fig. 10). In the simulation, we allow a small background rate of cell division, which occurs in regions of relatively lower density. After 600,000 iterations, the number of cells has increased by about 15%. Without this division process, tearing of the epiblast would occur—rostral to the midpoint between the vortices, where cell tracks are divergent. Similar tearing has been observed in the chick embryo during streak formation when cell division is inhibited (Weijer, private communication). We have highlighted cells initially in the region of Koller's sickle. Over the course of the simulation, these cells are "extruded" through the vortex motion into the epiblast, forming a structure reminiscent of the primitive streak. In Fig. 11 we show cell tracks formed from cell movement along the anterior–posterior axis, and circulation about the vortices. This numerical result is in good qualitative agreement with the experimental data (Fig. 6).

VII. Discussion and Conclusions

In this paper we have given a brief overview of a range of grid-free cell-based computational models, and have proceeded to apply a simple version of such models to a contemporary problem in chick embryogenesis.

As stated at the beginning of this paper, there are a number of general approaches to the modeling of developmental systems, each with its own advantages and disadvantages. Advantages of the grid-free cell-based approaches are (i) the ability to define multiple cell types, (ii) the absence of a grid, the grid being biologically unrealistic and a potential source of artifacts in the dynamics (Grima, 2005b), and (iii) the natural computational efficiency of grid-free algorithms, that allows simulation of these systems to scale linearly with the number of cells in three dimensions. The major weakness of the cell-based approaches in general is their analytic intractability. It is very difficult to evaluate the behavior of the system without a full-scale numerical simulation, although there are some techniques to either reduce the cell equations to a cell-density approximation (Stevens, 2000) and even to calculate corrections to this using many-body techniques (Newman and Grima, 2004; Grima, 2005a).

The simpler grid-free models essentially treat cells as slightly compressible adhesive spheres. We discussed recent developments in which more sophisticated models had been proposed to incorporate some degree of cell shape, the key methods being

5. Grid-Free Models and Primitive Streak Formation 175

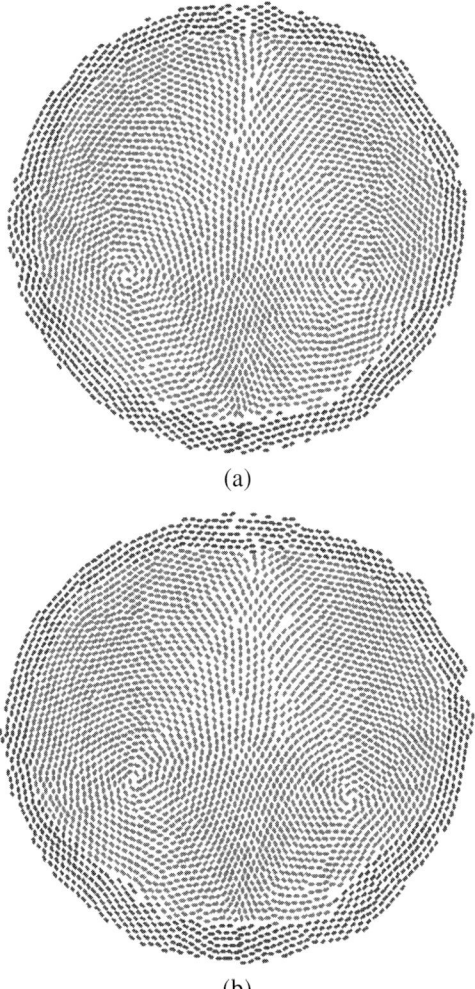

(a)

(b)

Figure 9 This figure and Figs. 10 and 11 were created from a larger simulation of ∼3500 cells. Successive views of the polarity distribution. Panels (a–d) correspond to the system after 1.5, 3.0, 4.5, and 6.0 × 10^5 iterations, respectively. The vortices in the planar polarity remain relatively undisturbed by the cell motion. See color insert.

the use of ellipsoidal cells (Palsson and Othmer, 2000), the Voronoi–Delaunay triangulation of space in three dimensions (Schaller and Meyer-Hermann, 2005), and the Subcellular Element Model, which invokes a representation of cells as clusters of adhesive elements (Newman, 2005, 2007). The use of these more sophisticated models will probably be focused in the intermediate future on modeling smaller systems (10^2–

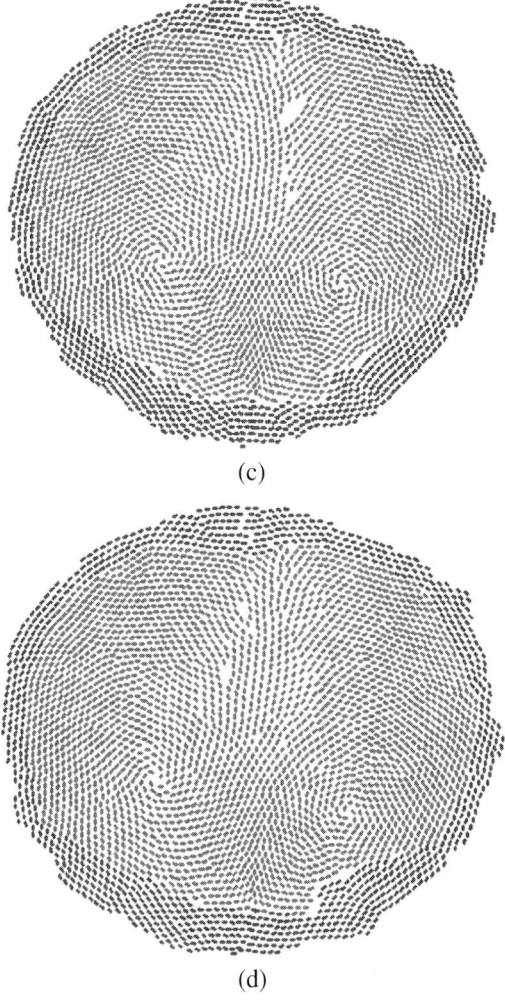

Figure 9 (*continued*)

10^4 cells) in more detail. Systems of 10^5–10^6 cells are at the limit of computational power for the simpler grid-free models.

Following our overview of grid-free models, we applied a simple grid-free model to the recent conundrum of large-scale vortex formation in the epiblast of the chick, concurrent with primitive streak formation (Cui *et al.*, 2005). We use the word conundrum, since, as explained in more detail in the main text, expression of key genes linked to gastrulation indicate that the cells in the area pellucida (excluding Koller's sickle) are homogeneous in their gene expression, and thus one cannot meaningfully

5. Grid-Free Models and Primitive Streak Formation

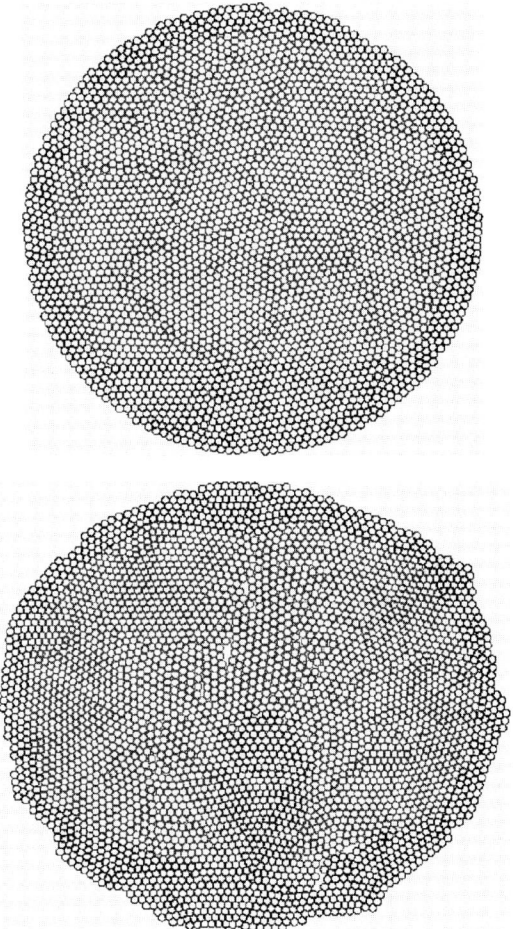

Figure 10 The cell distributions (with circumferences drawn at a radius of r_c) of the initial system, and after 6.0×10^5 iterations of the algorithm, showing that the epiblast remains in tact throughout the vortex motion. Red, blue, and purple cells corresponds to cells in the area pellucida, area opaca, and Koller's sickle, respectively. Cells initially in the Koller's sickle region (upper panel) have been transported by the cell flow into a structure resembling the primitive streak (lower panel). See color insert.

define a localized set of cells in the area pellucida as a source of biomechanical or biochemical influence responsible for the vortices. It is not so simple to imagine mechanisms for the vortices that could be controlled by cells in either Koller's sickle and/or the marginal zone. A mechanism that is consistent with this condition was presented in this paper, and relies upon a coupling of the cell velocities and a presumed planar cell polarity. The existence of planar cell polarity in the epithelial-like sheet of the epiblast has some precedent from recent work on the epithelial sheet surrounding

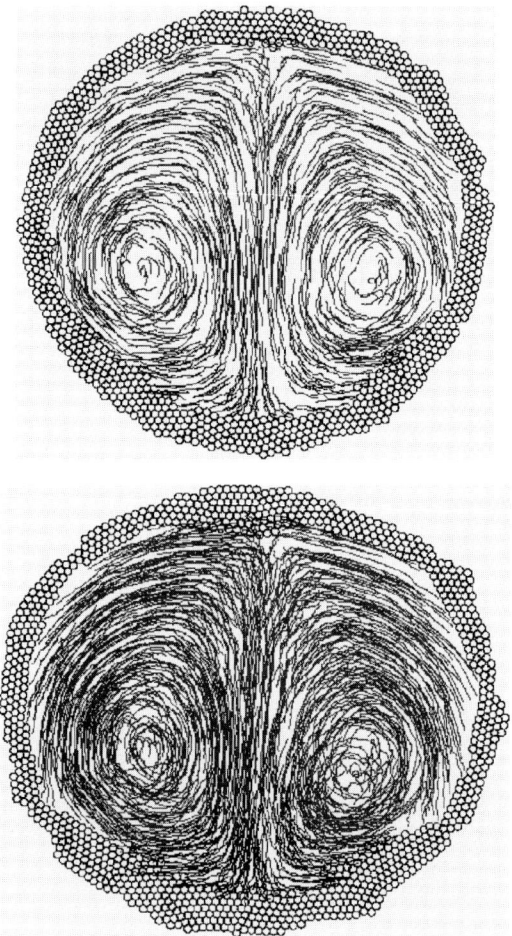

Figure 11 Cell tracks produced over the course of the simulation. Upper and lower panels correspond to 1.5 and 6.0×10^5 iterations, respectively. The distribution of cell tracks qualitatively mirrors the experimental results of Fig. 6. See color insert.

the Drosophila embryo (Blankenship *et al.*, 2006). This mechanism was instantiated using a stripped down grid-free biomechanical cell model coupled to relaxational dynamics of the cells' planar polarities. Simulations revealed that (i) the establishment of large-scale vortices in the planar polarity distribution is robust given certain boundary conditions in the marginal zone, and (ii) once cells are instructed to move, with a velocity of a given cell parallel to its planar polarity, the large-scale vortex patterns are reproduced by the cell flow, thus making contact with the cell movement patterns observed in experiment (Cui *et al.*, 2005).

A key question concerning the primitive streak and the accompanying vortices concerns cause and effect: are the vortices a result of the streak motion, or is the streak a result of, or at least guided by, the vortices? The analog of the epiblast as a viscous fluid sits firmly in the first camp, while the mechanism described here is firmly in the second. The planar polarity distribution is responsible for pre-patterning the vortices, and cells trace out this pattern once instructed to move parallel to their local polarity. The primitive streak is then extruded somewhat like a "rack" between the "pinion-wheels" of the vortices. The vortices continue to "turn," as cells follow their polarity, and the cells in the Koller's sickle region continue to be guided through the gap separating the vortices.

Acknowledgments

It is a pleasure to thank Cornelius Weijer for critical discussions and generous access to experimental data of vortex formation in the chick. I also thank members of my group: Ralph DeSimone, Sebastian Sandersius, Erick Smith, and Jens Weismüller, for helpful discussions of this mechanism. This work was partially supported by NSF Grant IOB-0450680.

Appendix A. Details of the Model

We use here a "stripped down" grid-free cell-based model, as we are investigating a qualitative phenomenon involving collective motion of tens of thousands of cells, which is thus, to a first approximation, likely to be insensitive to the precise modeling details at the level of an individual cell. To model a sheet of cells, such as the epiblast, a given cell i is represented by an adhesive circular region, of mean radius r_c, centered at position vector x_i. The velocity of the cell has a fundamental contribution arising from the mechanical interactions between this cell and its neighbors. By "neighbors" we mean those cells whose centers are within approximately two cell radii of cell i. The reason for using the words "mean" and "approximately" in describing length scales of cells is that a given cell is not modeled as having an exact, rigid radius. Rather, a cell is modeled as a compressible/extensible object. Two neighboring cells interact mechanically via a potential function V, such that the preferred separation between two cells is given by $r_{eq} = 2r_c$, indicating that r_c is a measure of the cell radius. Accounting only for this primary biomechanical interaction between cells leads to the equations of motion:

$$v_i = \dot{x}_i = -\nabla_i \sum_j V(|x_i - x_j|). \tag{A.1}$$

The precise form of the potential used here is:

$$V(r) = V_0 e^{-r^2/\xi_1^2} - U_0 e^{-r^2/\xi_2^2}. \tag{A.2}$$

This potential is short-ranged, and has a form similar to the generic potential illustrated in Fig. 1. The parameters in the potential are chosen such that the potential minimum is located at the equilibrium distance r_{eq}. The forces generated in the simulation are weak enough such that the distance between cell centers does not decrease below about $0.95 r_{\text{eq}}$.

There are a number of terms that can be added to the right-hand-side of Eq. (A.1) representing various biological influences on the cell. Examples are the effects on the cell velocity from random cell motion (i.e., noise), chemotaxis (Schweitzer and Schimansky-Geier, 1994; Newman and Grima, 2004), and cell polarity. We consider here only the last of this list. The planar polarity of cell i is denoted by $\boldsymbol{p}_i$. If cells are instructed to attempt to move along an axis defined by their planar polarity, then Eq. (A.1) is modified to

$$\dot{\boldsymbol{x}}_i = -\nabla_i \sum_j V(|\boldsymbol{x}_i - \boldsymbol{x}_j|) + \gamma \boldsymbol{p}_i, \tag{A.3}$$

where γ is the strength of the coupling between cell velocity and planar polarity. Equation (A.3) is the equation of motion used for the simulations contained in this paper. We also need to define the dynamics of the planar polarity. This is most conveniently expressed in terms of the polar angle θ_i, defined via $\boldsymbol{p}_i = (\cos\theta_i, \sin\theta_i)$. To model relaxational dynamics of planar polarity alignment, we use over-damped dynamics to a steady-state characterized by locally aligned cell polarities:

$$\dot{\theta}_i = \lambda \sum_j \sin(\theta_j - \theta_i), \tag{A.4}$$

where λ sets the rate of relaxation. In the limit that neighboring planar polarities are closely aligned, Eq. (A.4) has the form of linear diffusion. The nonlinear form allows stable relaxation to alignment even if neighboring cells are poorly aligned initially, as is the case for cells either side of the marginal zone.

A given cell may produce a new cell in its immediate neighborhood (a crude approximation of cytokinesis) if there is sufficient space available. This replication is attempted at a rate q, yet typically fails due to the close-packed nature of the modeled epiblast. Cell replication tends to occur in lower density regions that appear as the vortex dynamics proceeds—e.g., in rostral regions of the epiblast, and in the caudal edge of Koller's sickle.

Table I provides typical values of the model parameters used in the simulations. Length scales have dimensions of microns. All other parameters have dimensionless units. In principle, these parameters can be calibrated by comparing mechanical properties of cell–cell interactions with experimentally measured values. A good example of such a model calibration, for the epidermis, has recently appeared (Schaller and Meyer-Hermann, 2007).

The equations of motion for cell movement (A.3) and planar-polarity relaxation (A.4) are implemented in Fortran 90 using a grid-free algorithm. Equations are temporally discretized in the simplest Euler fashion, with a time-step $\delta t = 0.1$. Location of neighboring cells, which, for a population of N cells, naively takes order N^2 operations, is performed using the sector method (Allen and Tildesley, 1989) which scales as order N. This allows large systems to be simulated.

Appendix B. The XY Model of Ferromagnetism

It is worth mentioning a physical analog of the planar polarity field—namely, the magnetic "spin" field of the planar ferromagnet, often termed the XY model (Kadanoff, 2000). The XY model consists of a square lattice, at each node of which is placed a "spin." This spin is essentially a unit vector that can point in any direction in the plane of the lattice. Just as with the planar polarity of cells described in this paper, a given spin's direction can be described a single angle θ_i. Spins interact with their four nearest neighbors on the grid, and energetically prefer to be aligned. The total internal energy of the system can be written as $H = -J \sum_{i,j} \cos(\theta_i - \theta_j)$, where the sum is over nearest neighbors. The equilibrium configurations of this system are known to be dominated by the presence of vortices (Kosterlitz and Thouless, 1973), similar to those described in this paper. The relaxational dynamics of the system have also been studied, and the equation of motion for spins has the form of Eq. (A.3), which follows from demanding that the total energy of an isolated system does not increase over time (Newman, Bray, and Moore, 1990). The XY model is a very important model in the theory of phase transitions and has deep connections to many other problems in statistical physics, e.g., the Coulomb gas and roughening of atomic surfaces (Kadanoff, 2000).

References

Allen, M. P., and Tildesley, D. J. (1989). "Molecular Simulation of Liquids." Oxford Univ. Press, Oxford.
Beyer, T., Schaller, G., Deutsch, A., and Meyer-Hermann, M. (2005). Parallel dynamic and kinetic regular triangulation in three dimensions. *Comput. Phys. Commun.* **172**, 86–108.
Blankenship, J. T., Backovic, S. T., Sanny, J. S., Weitz, O., and Zallen, J. A. (2006). Multicellular rosette formation links planar cell polarity to tissue morphogenesis. *Dev. Cell* **11**, 459–470.
Bretschneider, T., Vasiev, B., and Weijer, C. J. (1997). Cell movement during *Dictyostelium* mound formation. *J. Theor. Biol.* **189**, 41–51.
Bretschneider, T., Vasiev, B., and Weijer, C. J. (1999). A model for *Dictyostelium* slug movement. *J. Theor. Biol.* **199**, 125–136.
Chapman, S. C., Schubert, F. R., Schoenwolf, G. C., and Lumsden, A. (2002). Analysis of spatial and temporal gene expression patterns in blastula and gastrula stage chick embryos. *Dev. Biol.* **245**, 187–199.
Chuai, M., Zeng, W., Yang, X., Boychenko, V., Glazier, J. A., and Weijer, C. J. (2006). Cell movement during chick primitive streak formation. *Dev. Biol.* **296**, 137–149.
Cui, C., Yang, X., Chuai, M., Glazier, J., and Weijer, C. J. (2005). Analysis of tissue flow patterns during primitive streak formation in the chick embryo. *Dev. Biol.* **284**, 37–47.
Dallon, J. C., and Othmer, H. G. (2004). How cellular movement determines the collective force generated by the Dictyostelium discoideum slug. *J. Theor. Biol.* **231**, 299–306.

Deutsch, A., and Dormann, S. (2004). "Cellular Automata Modeling of Biological Pattern Formation." Birkhäuser, Boston.
Drasdo, D., and Höhme, S. (2003). Individual based approaches to birth and death in avascular tumors. *Math. Comput. Model.* **37**, 1163–1175.
Drasdo, D., and Höhme, S. (2005). A single-cell based model of tumor growth in vitro: Monolayers and spheroids. *Phys. Biol.* **2**, 133–147.
Drasdo, D., Kree, R., and McCaskill, J. S. (1995). Monte Carlo approach to tissue-cell populations. *Phys. Rev. E* **52**, 6635–6657.
Eaton, S. (1997). Planar polarization in *Drosophila* and vertebrate epithelia. *Curr. Opin. Cell Biol.* **9**, 860–866.
Gilbert, S. F. (2003). "Developmental Biology," 7th ed. Sinauer Associates, Sunderland, MA.
Graner, F., and Glazier, J. A. (1993). Simulation of biological cell-sorting using a two-dimensional extended Potts model. *Phys. Rev. Lett.* **69**, 2013–2016.
Grima, R. (2005a). Strong coupling dynamics of a multicellular chemotactic system. *Phys. Rev. Lett.* **95**. Art. No. 128103.
Grima, R. (2005b). Multiscale modeling of chemotactic interactions, Ph.D. thesis, Arizona State University.
Kadanoff, L. P. (2000). "Statistical Physics: Statics, Dynamics, and Renormalization." World Scientific, Singapore.
Keller, E. F., and Segel, L. A. (1970). Initiation of slime mold aggregation viewed as an instability. *J. Theor. Biol.* **26**, 399–415.
Kosterlitz, J. M., and Thouless, D. J. (1973). Ordering, metastability, and phase transitions in a two-dimensional system. *J. Phys. C* **6**, 1181–1203.
Meyer-Hermann, M. E., and Maini, P. K. (2005). Interpreting two-photon imaging data of lymphocyte motility. *Phys. Rev. E* **71**. Art. No. 061912.
Murray, J. D. (2004). "Mathematical Biology. II. Spatial Models and Biomedical Applications." Springer-Verlag.
Newman, T. J. (2005). Modeling multicellular systems using subcellular elements. *Math. Biosci. Eng.* **2**, 611–622.
Newman, T. J. (2007). Modeling multicellular structures using the subcellular element model. *In* "Single Cell-Based Models in Biology and Medicine." (A. Anderson, M. Chaplain, and K. Rejniak, Eds.), Birkhäuser.
Newman, T. J., and Grima, R. (2004). Many-body theory of chemotactic cell–cell interactions. *Phys. Rev. E* **70**. art. no. 051916.
Newman, T. J., Bray, A. J., and Moore, M. A. (1990). Growth of order in vector spin systems and self-organized criticality. *Phys. Rev. B* **42**, 4514–4523.
Palsson, E., and Othmer, H. O. (2000). A model for individual and collective cell movement in *Dictyostelium discoideum*. *Proc. Natl. Acad. Sci. USA* **97**, 10446–10453.
Schaller, G., and Meyer-Hermann, M. (2005). Multicellular tumor spheroid in an off-lattice Voronoi–Delaunay cell model. *Phys. Rev. E* **71**. Art. No. 051910.
Schaller, G., and Meyer-Hermann, M. (2007). Epidermal homeostasis control in an off-lattice agent-based model. *J. Theor. Biol.* **244**, 656–669.
Schnell, S., Painter, K. J., Maini, P. K., and Othmer, H. G. (2000). Spatiotemporal pattern formation in early development: A review of primitive streak formation and somitogenesis. *In* "Mathematical Models for Biological Pattern Formation." (P. K. Maini and H. G. Othmer, Eds.), Springer-Verlag, New York.
Schoenwolf, G. C. (2001). "Laboratory Studies of Vertebrate and Invertebrate Embryos," 8th ed. Benjamin Gummings, San Francisco.
Schweitzer, F., and Schimansky-Geier, L. (1994). Clustering of active walkers in a two-component system. *Phys. A* **206**, 359–379.
Stevens, A. (2000). The derivation of chemotaxis equations as limiting dynamics of moderately interacting stochastic many-particle systems. *SIAM J. Appl. Math.* **61**, 183–212.
Van Kampen, N. G. (2007). "Stochastic Processes in Physics and Chemistry." Elsevier, Amsterdam.
Wolpert, L., Beddington, R., Jessell, T., Lawrence, P., Meyerowitz, E., and Smith, J. (2002). "Principles of Development," 2nd ed. Oxford Univ. Press, Oxford.

6

Mathematical Models for Somite Formation

Ruth E. Baker,* Santiago Schnell,† and Philip K. Maini*,‡
*Centre for Mathematical Biology, Mathematical Institute, University of Oxford,
24-29 St. Giles,' Oxford OX1 3LB, United Kingdom
†Indiana University School of Informatics and Biocomplexity Institute, 1900 East 10th Street, Eigenmann Hall 906, Bloomington, Indiana 47406
‡Oxford Centre for Integrative Systems Biology, Department of Biochemistry, University of Oxford, South Parks Road, Oxford OX1 3QU, United Kingdom

- I. Introduction
- II. Models for Somite Formation
    - A. The Clock and Wavefront Model
    - B. Transcriptional Regulation Models for the Segmentation Clock
    - C. Models for the FGF8 Gradient
    - D. Models Including Cell Adhesion
- III. Discussion
- IV. Perspective
    - Acknowledgments
    - References

Somitogenesis is the process of division of the anterior–posterior vertebrate embryonic axis into similar morphological units known as somites. These segments generate the prepattern which guides formation of the vertebrae, ribs and other associated features of the body trunk. In this work, we review and discuss a series of mathematical models which account for different stages of somite formation. We begin by presenting current experimental information and mechanisms explaining somite formation, highlighting features which will be included in the models. For each model we outline the mathematical basis, show results of numerical simulations, discuss their successes and shortcomings and avenues for future exploration. We conclude with a brief discussion of the state of modeling in the field and current challenges which need to be overcome in order to further our understanding in this area. © 2008, Elsevier Inc.

I. Introduction

The dramatic advances made in genetic and molecular biology in recent years have allowed us to produce detailed descriptions of a number of complex processes that arise on many different spatial and temporal scales during development. This unparalleled flood of experimental data may well enable us to understand how genes and proteins work collectively in a cell, and from this how multicellular organisms develop. But, therein exists one of the great challenges of modern science; all too often our knowledge remains in isolated pockets, lacking a conceptual framework tying the fragmented data together and allowing ideas and hypotheses to be explored. This is where mathematical modeling can play a fundamental role, comparable with any laboratory research tool: it allows us to combine the effects of multiple nonlinear processes into a coherent structure which can be used for hypothesis testing and making experimentally testable predictions.

Somitogenesis, the segmentation of the vertebrate anteroposterior (AP) axis, is a perfect paradigm for studying such issues (Gossler and Hrabě de Angelis, 1998); it has a long history of experimental investigation at the cellular level, and we have observed recently a huge explosion in the rate of identification of the underlying molecular components. Somitogenesis serves as a model process for studying pattern formation during embryogenesis, encompassing a wide range of mechanisms and processes, such as the role of biological clocks (Dale *et al.*, 2003, 2006; Dale and Pourquié, 2000; Maroto and Pourquié, 2001; Pourquié 1998, 2001b), positional information gradients (Diez del Corral *et al.*, 2002, 2003; Diez del Corral and Storey, 2004; Dubrulle *et al.*, 2001; Dubrulle and Pourquié, 2002, 2004b), cell migration and adhesion (Drake and Little, 1991; Duband *et al.*, 1987; Duguay *et al.*, 2003; Kulesa and Fraser, 2002), all of which are tightly coupled, both spatially and temporally.

Somites are tightly bound blocks of cells that lie along the AP axis of vertebrate embryos (Gossler and Hrabě de Angelis, 1998; McGrew and Pourquié, 1998; Pourquié, 2001b). They are transient structures, and further differentiation of the somites gives rise to the vertebrae, ribs and other associated features of the trunk. Somites segment from the presomitic mesoderm (PSM); thick bands of tissue that lie on either side of the notochord, along the AP axis. At regular time intervals, groups of cells at the anterior ends of the PSM undergo changes in their adhesive and migratory behavior and condense together to form an epithelial block of cells known as a somite (Stockdale *et al.*, 2000). As a result, somites form in a very strict AP sequence. As the body axis extends, the budding of cells anteriorly is compensated by the addition of cells at the posterior ends of the PSM (Pourquié, 2003). Each band of the PSM stays approximately constant in length throughout the process of somite formation and a wave of patterning appears to sweep along the AP axis leaving the somites in its wake (Collier *et al.*, 2000).

From the molecular point of view, formation of the somites begins with the establishment of a prepattern of gene expression. In mouse and zebrafish, members of the Mesp family are periodically activated in regions corresponding to the future somites (Haraguichi *et al.*, 2001; Morimoto *et al.*, 2006; Saga *et al.*, 2001; Takahashi *et al.*, 2000; Takashi *et al.*, 2005). Overt signs of segmentation occur during subsequent cell cycles when the prepattern is converted to morphological somites. Cells undergo changes in their adhesive and migratory properties and a series of extensive rearrangements and shape-changes, mediated by cell adhesion molecules (CAMs), take place (Drake and Little, 1991; Duband *et al.*, 1987; Duguay *et al.*, 2003; Foty and Steinberg, 2004; Kalcheim and Ben-Yair, 2005).

6. Models for Somite Formation

Although the physical somite boundaries form at the anterior-most end of the PSM, early scanning microscope images showed that the PSM is not a homogeneous tissue (Meier, 1979; Tam and Meier, 1982). Prior to somite formation early signs of segmentation were seen: the posterior PSM was shown to display metameric arrangements of cells known as somitomeres, which seem to be precursors of the somites (Gossler and Hrabĕ de Angelis, 1998; Jacobson and Meier, 1986). The existence of this prepattern was confirmed by microsurgical experiments in which isolated parts of the PSM formed somites in strict isolation (Chernoff and Hilfer, 1982; Packard, 1976). However, this somitomeric prepattern was shown to be undetermined and susceptible to modification by external agents (Packard *et al.*, 1993). More recently, existence of the prepattern has been further contested by microsurgical experiments conducted by Dubrulle and co-workers: AP inversions of somite-length regions of the posterior PSM resulted in normal segmentation whilst inversions of the anterior PSM resulted in somites with reversed polarity (Dubrulle *et al.*, 2001). These results suggest that the anterior-most portion of the PSM is determined with regard to its segmentation program, whilst the posterior-most portion is labile in this respect. The different regions of the PSM were found to correspond to regions of varying FGF signaling; we discuss this more in following paragraphs.

There are several genes expressed dynamically in the PSM. Some are known as cyclic genes, because they oscillate with the same frequency it takes to form one somite (Dale *et al.*, 2006; Ishikawa *et al.*, 2005; McGrew *et al.*, 1998; Palmeirim *et al.*, 1997; Pourquié, 2001a, 2003). In the chick, gene expression bands of *c-hairy1* and *l-fng* sweep along the AP axis: beginning as wide bands in the tail end of the embryo, they narrow as they travel in an anterior direction, until they come to rest in the newly forming somite. Expression is considered to arise as a result of an underlying segmentation clock in the PSM cells: several cell-autonomous components of the clock have been identified (Aulehla and Johnson, 1999), together with other molecular pathways which couple the oscillations in neighboring PSM cells (del Barco Barrantes *et al.*, 1999; Jiang *et al.*, 1998, 2000; Jiang and Lewis, 2001; Pourquié, 1999; Rida *et al.*, 2004).

An additional gene with dynamic expression in the PSM is fgf8. A gradient of FGF8 exists along the AP axis, peaking at the posterior end of the embryo (Dubrulle *et al.*, 2001 Dubrulle and Pourquié, 2002, 2004a, 2004b). As the body axis elongates, the gradient of FGF8 moves in a posterior direction so that it stays at a constant axial level, relative to the PSM. Cells are initially part of a region where FGF8 signaling prevails, but they gradually exit this region as they move up through the PSM. Dubrulle and co-workers have shown that downregulation of FGF signaling is necessary for cells to become part of a somite, and they term the threshold level of FGF signaling at which this ability is gained the *determination front* (Dubrulle *et al.*, 2001).

Following the experimental discoveries described above, Pourquié and co-workers have postulated that it is the interaction between the segmentation clock and the FGF8 gradient that gates the cells into somites: the segmentation clock defines *when* somite boundaries form, whilst the FGF8 gradient defines *where* they form (Dubrulle *et al.*,

2001; Tabin and Johnson, 2001). This is now the generally accepted "model" for somite formation.

II. Models for Somite Formation

In this section we will present a number of mathematical models which describe different stages of the formation of somites. We begin by presenting a mathematical formulation of the Clock and Wavefront model, which integrates information from the segmentation clock and the FGF8 wavefront to produce a somitic prepattern. Next we consider individual models for the segmentation clock and FGF8 gradient, which may be integrated into the Clock and Wavefront model. Finally we present a model which includes the role of cell adhesion molecules and the cell rearrangements which take place during the generation of epithelial somites. In each case we introduce the model and its mathematical formulation. We show the results of numerical simulations, discuss the successes and shortcomings of the model and avenues for extension and refinement.

A. The Clock and Wavefront Model

With the recent experimental findings concerning the existence of a segmentation clock and a wavefront of FGF8 along the AP of vertebrate embryos, the Clock and Wavefront model has received a large amount of support. First proposed by Cooke and Zeeman (Cooke, 1975, 1998; Cooke and Zeeman, 1976; Zeeman, 1974), the model assumed the existence of a longitudinal positional information gradient along the AP axis which interacts with a smooth cellular oscillator (the clock) to set the times at which cells undergo a *catastrophe*. In this context, Cooke and Zeeman were referring to the changes in adhesive and migratory behavior of cells as they form somites.

Pourquié and co-workers recently proposed a revised version of the model which involves the interaction of the segmentation clock and FGF8 wavefront in gating the cells into somites (Dubrulle *et al.*, 2001; Tabin and Johnson, 2001). Specifically, they assume that the clock sets the times at which new somite boundaries form whilst the position of the determination front sets where they form (Dubrulle *et al.*, 2001). We developed a mathematical formulation of the Clock and Wavefront model based on these assumptions (Baker *et al.*, 2006a, 2006b). We also made the following further assumptions: (i) upon reaching the determination front, a cell acquires the ability to segment by becoming competent to produce a somitic factor—which could be, for example, Mesp2 (Morimoto *et al.*, 2006; Saga *et al.*, 2001; Takashi *et al.*, 2000, 2005)—in response to the presence of a signaling molecule; (ii) one clock oscillation after reaching the determination front, cells become able to produce the signaling molecule; (iii) after responding to the signaling molecule, a cell is specified as somitic and becomes refractory to FGF8 signaling; (iv) a cell forms part of a somite with nearby cells which produce somitic factor at a similar time.

6. Models for Somite Formation

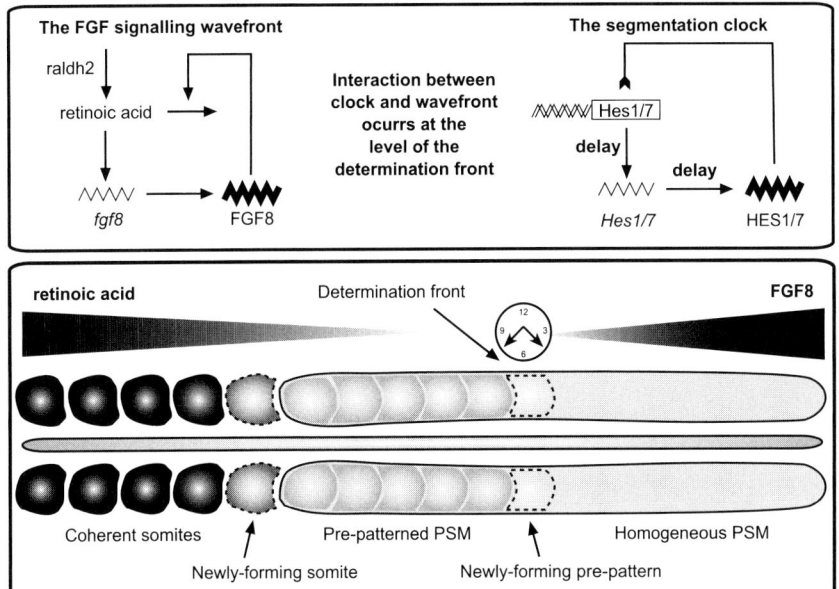

Figure 1 The mechanisms underlying somite formation. The lower box illustrates the AP axis and the various stages of somite formation. The posterior PSM (gray) is homogeneous and cells are undetermined with respect to their developmental pathway. At the level of the determination front, the interaction between the clock and wavefront specifies the chemical prepattern. Cells of the PSM which lie anterior to the determination front will follow a specific developmental pathway which cannot be altered by subsequent perturbation of the clock or wavefront. At the anterior end of the PSM cells undergo changes in their morphological properties and condense to form coherent somites (black segments). The upper box shows schematics of the networks underlying the clock and the gradient. On the left, FGF acts in a negative feedback loop with retinoic acid to control gradient progression whilst on the right a negatively-regulating transcriptional delay network consisting of Hes1/7 controls clock oscillations.

Our mathematical model is based on the signaling model for somite formation, originally proposed by Primmett and co-workers (Primmett *et al.*, 1988, 1989), and then developed by Maini and co-workers (Baker *et al.*, 2003; Collier *et al.*, 2000; McInerney *et al.*, 2004; Schnell *et al.*, 2002). In this model (see Fig. 1), a small group of cells at the anterior end of the PSM will reach the determination front and undergo a whole cycle of the segmentation clock. These *pioneer* cells produce and emit a diffusible signaling molecule which spreads along the PSM. Any other cell which has reached the determination front, but not yet been specified as somitic, will respond to the signal by producing somitic factor. At this point the cell becomes allocated to a somite along with other cells which are behaving in a similar manner. The process begins once again when the new cohort of cells at the anterior end of the PSM becomes able to produce signaling molecule.

The mathematical model consists of a system of three nonlinear partial differential equations (PDEs). The variables are somitic factor concentration (only cells with a high level of somitic factor can go on to be specified as somitic), signaling molecule levels and FGF8 concentration. We choose to model the gradient by assuming that FGF8 is produced only in the tail of the embryo and that it undergoes linear decay (see Baker et al., 2006a, 2006b, for more details). As the embryo extends the gradient regresses posteriorly, conferring the ability on cells to be able to produce somitic factor. A time t_s later cells gain the ability to produce signaling molecule; t_s is the period of the segmentation clock. Somitic factor production is activated in response to a pulse in signaling molecule, which is emitted from cells at the anterior end of the PSM. However, rapid inhibition of signal production means that the peaks in signal concentration are transient and produced at regular intervals: this ensures that a series of coherent somites is generated (McInerney et al., 2004).

The system of nonlinear PDEs can be written as follows:

$$\frac{\partial u}{\partial t} = \frac{(u + \mu v)^2}{\gamma + u^2} \chi_u - u,$$
$$\frac{\partial v}{\partial t} = \kappa \left(\frac{\chi_v}{\varepsilon + u} - v \right) + D_v \frac{\partial^2 v}{\partial x^2}, \qquad (1)$$
$$\frac{\partial w}{\partial t} = \chi_w - \eta w + D_w \frac{\partial^2 w}{\partial x^2},$$

where u represents the concentration of somitic factor, v the concentration of signaling molecule and w the concentration of FGF8. μ, γ, κ, ε, η, D_v, and D_w are positive constants and production of u, v, and w are controlled by the respective Heaviside functions[1]

$$\chi_u = H(w - w_*),$$
$$\chi_v = H(t - t_*(w_*, t)), \qquad (2)$$
$$\chi_w = H(x - x_n - c_n t),$$

where w_* is the level of FGF8 at the determination front, $t_*(w_*, t)$ is the time at which a cell at x reaches the determination front (i.e., $w(x, t_*) = w_*$), t_s is the period of the segmentation clock, x_n is the initial position of the tail and c_n represents the rate at which the AP axis is extending. In short, the system of PDEs describes the following behavior: (i) somitic factor production is activated by the signaling molecule and is self-regulating; (ii) signaling molecule is produced rapidly in areas where somitic factor concentration is low and it is able to diffuse along the AP axis; (iii) FGF8 is produced in the tail, which is constantly regressing along the x axis.

We solve the mathematical model on the interval $(-\infty, \infty)$, where $x \to \infty$ corresponds to the tail of the embryo and $x \to -\infty$ the head of the embryo. We assume

---

[1] The Heaviside function $H(x)$ is equal to unity if $x > 0$ and zero otherwise: it acts as a switch.

6. Models for Somite Formation

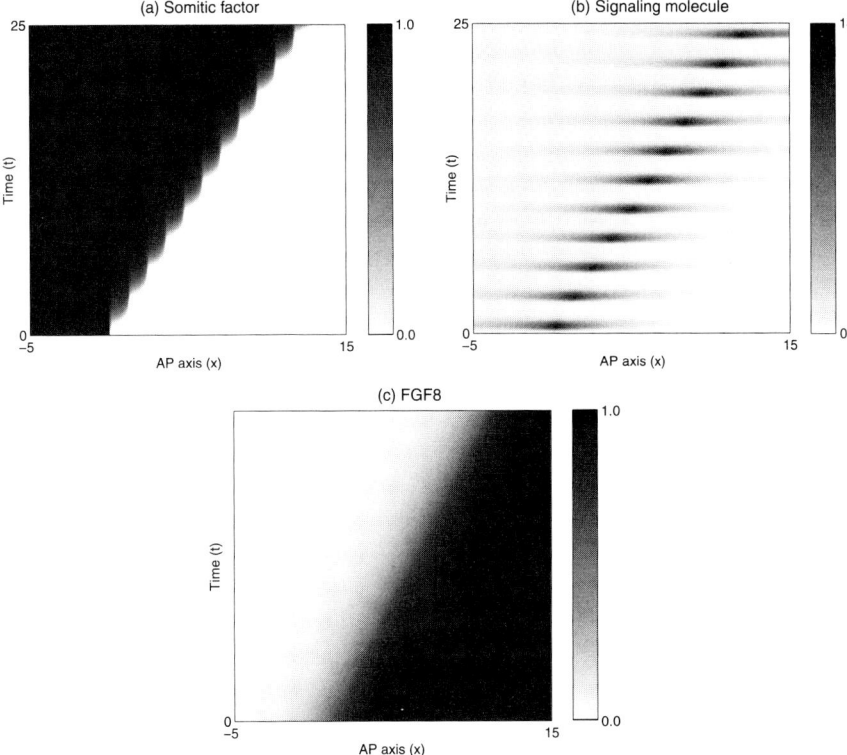

Figure 2 Numerical solution of the clock and wavefront model in one spatial dimension given by Eqs. (1) and (2). Continuous regression of the wavefront (c), is accompanied by a series of pulses in the signaling molecule (b), and coherent rises in the level of somitic factor (a). Parameters are as follows: $\mu = 10^{-4}$, $\gamma = 10^{-3}$, $\kappa = 10$, $\varepsilon = 10^{-3}$, $\eta = 1.0$, $\phi = 0.0$, $D_v = 50$, $D_w = 20$, $x_n = 0.0$, $c_n = 0.5$, $x_b = 7.5$ and $\xi = 0.0$.

that: (i) the levels of somitic factor, signaling molecule and FGF8 are bounded as $x \to \pm\infty$; (ii) initially the region $x < 0$ has already been specified as somitic; and (iii) the system is between pulses in the signaling molecule.

The mathematical model can be solved numerically using the NAG library routine D03PCF and the results plotted using the MATLAB function `imagesc`. The results are shown in Fig. 2. As time progresses, the region of high FGF8 signaling regresses along the AP axis with constant speed. In its wake a series of pulses in signaling molecule result in a series of coherent rises in somitic factor, enabling the formation of discrete somites.

The model has been tested against experimental data. Dubrulle and co-workers used experimental techniques to make local perturbations to the FGF8 wavefront (Dubrulle et al., 2001). By implanting beads soaked in FGF8 alongside or into the PSM they generated a series of somite anomalies. Typical anomaly patterns consisted of series of small somites ahead of the bead, followed by a large somite behind the bead and normal segmentation thereafter. The mathematical model can be adapted to take this into account by adding a term into the equation for FGF8, which represents a stationary source of FGF8 (Baker et al., 2006a, 2006b):

$$\frac{\partial w}{\partial t} = \chi_w + \phi \chi_b - \eta w + D_w \frac{\partial^2 w}{\partial x^2}, \qquad (3)$$

where the term

$$\chi_b = H(\xi - x_b + x) H(\xi + x_b - x), \qquad (4)$$

represents a source of FGF8 from a bead implant. ϕ is a measure of the strength of the bead source, x_b is the position of the bead and ξ is a measure of the width of the bead. The model can be solved as before, and the results are shown in Fig. 3. As time progresses, the region of high FGF8 signaling regresses along the AP axis as in the control case, but the speed of wavefront progression is disrupted around the bead implant. This results in a series of small somites ahead of the bead and a large somite behind the bead, as observed experimentally.

The model can also be adapted to include other possible experimental perturbations, such as downregulation of FGF8 signaling and perturbation of the segmentation clock (see Baker et al., 2006a, 2006b, for more details). Testing of these kinds of hypotheses is vital for furthering our understanding of the mechanisms underlying somite formation.

At present the drawback to this kind of modeling approach is that the clock and gradient are modeled at a very high (phenomenological) level. We incorporate the segmentation clock by using a single parameter to represent the period of the clock. Likewise, it is simply assumed that a generic FGF8 molecule makes up the gradient controlling the position of the determination front. Whilst this approach has many benefits—we can make simplifications and predictions regarding the effects of many experimental perturbations (Baker et al., 2003, 2006a, 2006b)—we are unable to infer exactly how each biological perturbation affects somite formation. For example, vitamin A deficient embryos have somites which are reduced in size (Diez del Corral et al., 2003; Molotkova et al., 2005; Moreno and Kinter, 2004). In our model we could explain this observation by assuming that the determination front progresses more slowly along the AP axis. This requires us to decrease the parameter c_n, which controls the rate at which the Heaviside function controlling FGF8 production moves [see Eq. (2)]. However from our model, we cannot deduce why this change in c_n occurs. Explicit modeling of the segmentation clock and FGF8 wavefront will allow us to address these kinds of questions.

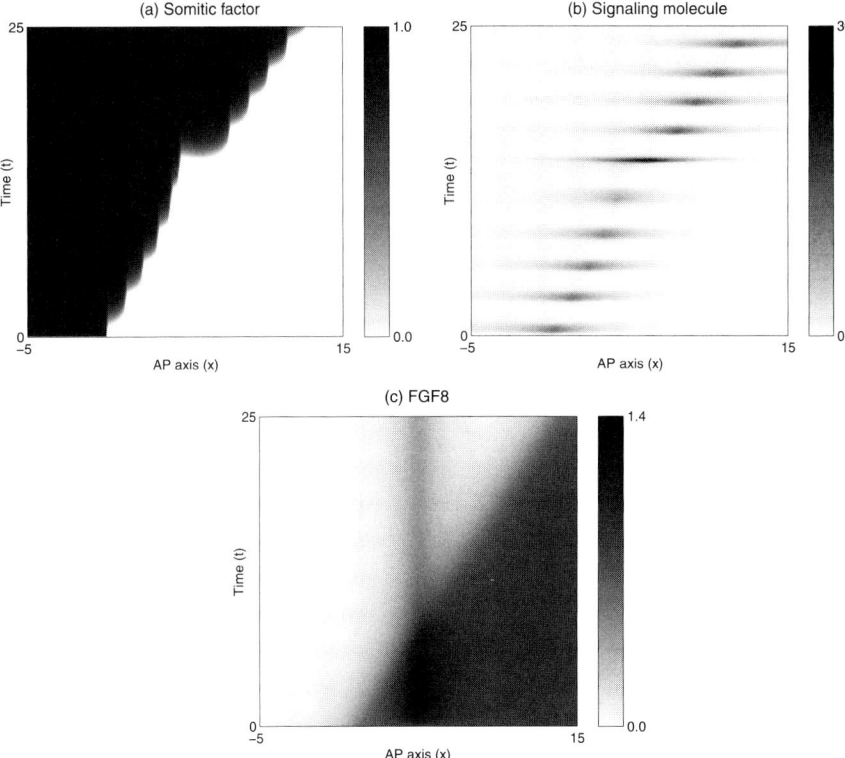

Figure 3 Numerical solution of the clock and wavefront model in one spatial dimension given by Eqs. (1)–(4). Continuous regression of the wavefront (c), is accompanied by a series of pulses in the signaling molecule (b), and coherent rises in the level of somitic factor (a). With a source of FGF8 implanted in the PSM the somite anomalies are obvious. Parameters are as follows: $\mu = 10^{-4}$, $\gamma = 10^{-3}$, $\kappa = 10$, $\varepsilon = 10^{-3}$, $\eta = 1.0$, $\phi = 1.5$, $D_v = 50$, $D_w = 20$, $x_n = 0.0$, $c_n = 0.5$, $x_b = 7.5$ and $\xi = 0.5$.

B. Transcriptional Regulation Models for the Segmentation Clock

The segmentation clock has been modeled using networks of transcriptional regulatory modules (Giudicelli and Lewis, 2004; Hirata et al., 2004; Horikawa et al., 2006; Lewis, 2003; Monk, 2003; Pourquié and Goldbeter, 2003). In each model, the simplest possible feedback loop is considered: binding of a transcription factor to the regulatory region of its own gene inhibits expression (see Fig. 1). Such a model can be written in the form

$$\frac{dp}{dt} = am(t - \tau_p) - bp(t),$$
$$\frac{dm}{dt} = f\big(p(t - \tau_m)\big) - cm(t),$$

(5)

where p represents protein (transcription factor) concentration and m represents mRNA concentration. The models incorporate delays of length τ_p and τ_m to reflect the time taken for synthesis of mature protein and mRNA, respectively. a is the rate of protein production, b and c are the decay rates of protein and mRNA, respectively, and the function $f(\cdot)$ is the rate of production of new mRNA molecules. To satisfy the requirement that protein binding inhibits transcription, $f(\cdot)$ should be a decreasing function of p, which we assume to be

$$f(p) = \frac{k}{(1+\frac{p^n}{p_0^n})}, \qquad (6)$$

where k, n, and p_0 are positive constants. In general we assume $n = 2$; this represents the inhibitory action of a protein which acts as a dimer (Lewis, 2003).

To consider a more realistic model for the behavior of a single cell, Lewis also includes a characteristic property of gene regulation: noise (Lewis, 2003). Binding and dissociation of the protein to/from the regulatory DNA site is considered to behave in a stochastic manner (Turner et al., 2004). The regulatory site makes transitions between bound and unbound states with a certain probability; these probabilities can be calculated by considering the rates of binding and unbinding and the numbers of protein molecules in the cell.

Fig. 4 shows the results of simulation of the system. In Fig. 4a, we plot the results of numerical simulation of the deterministic system using the MATLAB solver dde23. After initial large-amplitude oscillations in protein and mRNA levels, the system settles down to oscillate in a regular fashion with a period of approximately 50 minutes. The remaining plots, Figs. 4b–4d, show the results of numerical simulation of the noisy system. Simulations were carried out in the following fashion: at each time step, a random number drawn from a uniform distribution was used to determine whether binding or unbinding events occurred and then numbers of protein and mRNA molecules were updated accordingly. Fig. 4b is the stochastic equivalent of Fig. 4a—the system oscillates in much the same manner as the deterministic case, with a similar period but a small amount of variability in peak levels of mRNA and protein. As the protein synthesis rate is decreased the stochastic nature of gene regulation becomes more prominent. The period of oscillation remains robust even at a tenth of its original value [see Fig. 4c] but the peaks become increasingly noisy. At 100th of its original value, the oscillations are hard to discern and protein numbers fluctuate randomly and persistently (see Fig. 4d) (Lewis, 2003).

These models neatly capture the dynamics of the segmentation clock and reproduce oscillations with a period relatively close to that observed experimentally. The period of oscillation is robust in the sense that it remains practically unchanged even when the rate of protein synthesis is drastically reduced (Pourquié and Goldbeter, 2003). However, this generally requires the Hill coefficient, n, to be higher than that reported experimentally (Monk, 2003). In other words, one must assume that rather than a dimer, regulation of mRNA transcription requires binding of a multimer, such as a tetramer or larger, to the regulatory DNA site.

6. Models for Somite Formation

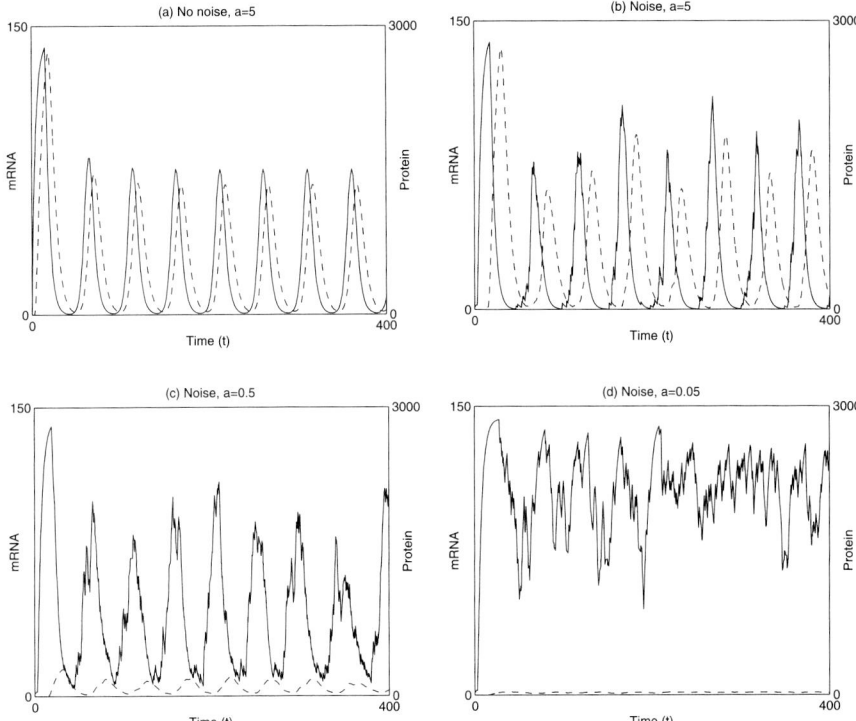

Figure 4 Numerical solution of the segmentation clock model given by Eqs. (5) and (6). (a) The case with no noise. (b–d) The case in which binding of the protein to and from its site on the DNA are stochastic processes. As the rate of protein synthesis is gradually decreased the oscillations become more irregular, becoming almost undetectable as synthesis rates reach 1/100th of their original value. In each case protein concentration/number is indicated by the solid line and mRNA concentration/number by the dashed line. Parameters are as follows: $dt = 0.1$ min, $\tau_m = 12.0$ min, $\tau_p = 2.8$ min, $p_0 = 40$, $k = 33$ min^{-1}, $k_{\text{off}} = 1$ min^{-1}, $b = 0.23$ and $c = 0.23$.

The models also assume that (i) translation is nonsaturating, (ii) the movement of protein molecules between the cytoplasm and cell nucleus is instantaneous and (iii) the delays in transcription and translation take discrete values. Monk (2003) suggests that a more appropriate assumption would be that the delay is uniformly distributed on some finite interval; this results in a revised equation for mRNA dynamics of the form

$$\frac{dm}{dt} = a \int_{\tau_1}^{\tau_2} k(\sigma) f\big(p(t-\tau)\big) \, d\sigma - bm(t), \tag{7}$$

where

$$\int_{\tau_1}^{\tau_2} k(\sigma)\, d\sigma = 1. \tag{8}$$

One prediction of the model is that oscillations are crucially dependent on instability of the protein. Hirata and co-workers (Hirata *et al.*, 2004) generated mice expressing mutant Hes7 with a half-life of approximately 30 minutes (compared to approximately 22 minutes in the wild-type mouse). Somite segmentation and oscillatory gene expression became severely disorganized after a few cycles. This observation is consistent with the results observed in their mathematical models and provides support for this mechanism in the segmentation clock. In a similar manner, Lewis uses his stochastic models to postulate that the effect of reduced protein synthesis may be to cause the progressively irregular oscillations seen in mutant embryos (Jiang *et al.*, 2000; Lewis, 2003). It would be interesting to see if this could be tested experimentally.

The models discussed in this section consider oscillations in protein and mRNA numbers in a single cell, where the environment has been assumed to be spatially homogeneous. The reaction kinetics of gene transcription can vary substantially in spatially heterogeneous environments (Schnell and Turner, 2004; Turner *et al.*, 2004). Although Lewis and Horikawa and co-workers briefly discuss models which incorporate the role of the Notch–Delta pathway in coupling adjacent PSM cells (Horikawa *et al.*, 2006; Lewis, 2003), future work in this area will need to consider the possibilities of coupling large numbers of cells, as is the case with the segmentation clock.

C. Models for the FGF8 Gradient

Less well studied have been the mechanisms underlying maintenance of the FGF8 gradient, which has been shown to control the position of the determination front (Dubrulle *et al.*, 2001). Our mathematical formulation of the Clock and Wavefront model (Baker *et al.*, 2006a, 2006b) assumed simply that FGF8 is produced in the tail of the embryo, which is extending at a constant rate. In our model, diffusion and linear decay are combined to set up a monotonic gradient, which regresses along the AP axis as time proceeds and hence is a constantly moving determination front. In this case, the rate of progression of the determination front is solely controlled by a parameter chosen in the simulations.

We consider a revised mechanism for FGF8 gradient formation (Baker and Maini, 2007) based on the finding that FGF8 acts in a negative feedback loop with RA (Diez del Corral *et al.*, 2002, 2003; Diez del Corral and Storey, 2004). We show a schematic representation of this feedback in Fig. 1. In the negative feedback loop, it is possible that FGF8 might downregulate RA production or upregulate RA decay, or both. The same occurs for RA. In order to account for both possibilities, we investigated the

6. Models for Somite Formation

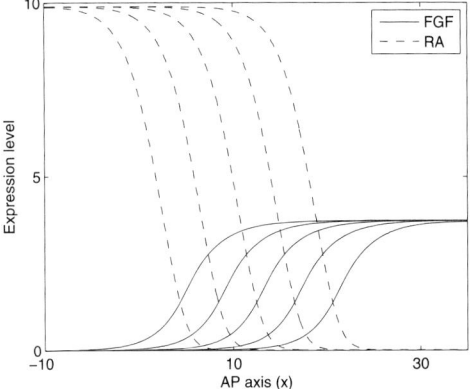

Figure 5 Numerical solution of the FGF8/RA model given by Eq. (9) over the series of successive time steps $t = 10, 20, 30, 40, 50$. The fronts move from left to right as time proceeds. Parameters are as follows: $r_f = 2.0$, $r_a = 5.0$, $\lambda_f = 1.0$, $\lambda_a = 1.0$, $\eta_f = 0.5$, $\eta_a = 0.5$, $s_f = 0.5$, $s_a = 0.5$, $\beta_a = 1.0$, $\beta_f = 1.0$, and $D = 5.0$.

phenomenon using the following system of nonlinear PDEs:

$$\frac{\partial F}{\partial t} = r_f \frac{F^2}{1+F^2} \frac{\lambda_f^2}{\lambda_f^2 + A^2} - \left(\eta_f + s_f \frac{A^2}{A^2 + \beta_f^2}\right) F + D \frac{\partial^2 F}{\partial x^2},$$
$$\frac{\partial A}{\partial t} = r_a \frac{A^2}{1+A^2} \frac{\lambda_a^2}{\lambda_a^2 + F^2} - \left(\eta_a + s_a \frac{F^2}{F^2 + \beta_a^2}\right) A + \frac{\partial^2 A}{\partial x^2}, \tag{9}$$

where F denotes the concentration of FGF8, A the concentration of RA and $r_f, r_a, \lambda_f, \lambda_a, \eta_f, \eta_a, s_f, s_a$, and D are positive constants. In briefly, the first equation assumes: (i) the rate of production of FGF8 is decreased by RA; (ii) FGF8 production saturates for high FGF8 concentration; (iii) FGF8 undergoes linear decay; (iv) RA augments FGF8 decay; and (v) FGF8 diffuses at rate D compared to RA. Similar statements can be made about the dynamics of RA.

The system of equations describing RA and FGF8 dynamics is mathematically intractable and so we carry out numerical simulation of the system using the MATLAB function pdepe, in order to analyze the behavior. Fig. 5 shows the results of numerical simulation for a specific set of parameter values. The initial conditions are chosen so that FGF8 obtains its maximal value on the positive x axis and is zero elsewhere and *vice versa* for RA. In this case, traveling waves (constant shape) form, with the region of high FGF8 signaling moving in a positive x direction at a constant rate. Variation of the model parameters produces changes in both wave speed and shape (Baker and Maini, 2007).

Despite the fact that analytical solution of the system is impossible, it can be shown that the presence of RA, acting in negative feedback loop with FGF8, makes the model more robust. Robustness is meant in the sense that it increases the parameter range for which the wave moves in a positive x direction (since we wish the determination front to regress along the AP axis over time).

Goldbeter and co-workers have also recently proposed a model for the FGF8 gradient which considers the negative interactions between RA and FGF8 (Goldbeter et al., 2007). In this model, negative feedback results in windows of bistability: regions along the PSM in which both high and low levels of FGF8 signaling are possible. Goldbeter and co-workers postulate that the action of the clock within this window could result in a shift from high to low levels of FGF signaling and the specification of a presumptive somite.

At present the model of Goldbeter and co-workers does not consider extension of the AP axis or movement of the bistability window along the AP axis. Also it does not explicitly consider interaction of the clock with this window in gating cells into somites. On the other hand, our model does not explicitly couple gradient formation to the rate of AP axis extension. Moreover it needs to be coupled to our previous formulation of the Clock and Wavefront model in order to predict somite patterns.

D. Models Including Cell Adhesion

All the models discussed thus far are concerned purely with the formation of a prepattern in gene expression, which is later converted to physical somites by a series of changes in cell morphology. However, it is possible to write down a model capturing the essential interactions between cells and cell adhesion molecules (CAMs), which leads to formation of a series of peaks in cell density—the somites. To date, the only mathematical model attempting to describe the bulk movement of PSM cells to form a somite is by Schnell and co-workers (Schnell et al., 2002). This is an extension of the signaling model for somite formation by Maini and co-workers, which we have discussed previously. Here we present a simpler model, which can explain the bulk movements of cells to form a somite.

Letting n denote cell density and a denote the concentration of a morphogen that stimulates expression of CAMs, the following system of nonlinear PDEs can be used to describe somite formation:

$$\frac{\partial n}{\partial t} = \frac{\partial}{\partial x}\left[\frac{\partial n}{\partial x} - f(n)\frac{\partial a}{\partial x}\right],$$
$$\frac{\partial a}{\partial t} = D\frac{\partial^2 a}{\partial x^2} + n - a. \tag{10}$$

The first term on the right-hand side of the equation for cell density represents cell diffusion and the second term the tendency of cells to move up gradients in CAM activity (resulting in cell aggregations). $f(n)$ is a decreasing function of n which for illustrative purposes we take to be $\chi n(N - n)$. The constant N is chosen to be large enough to ensure that $f(n)$ does not become negative. The equation for morphogen concentration also contains a diffusion term, with diffusion of CAM-triggering morphogen

6. Models for Somite Formation

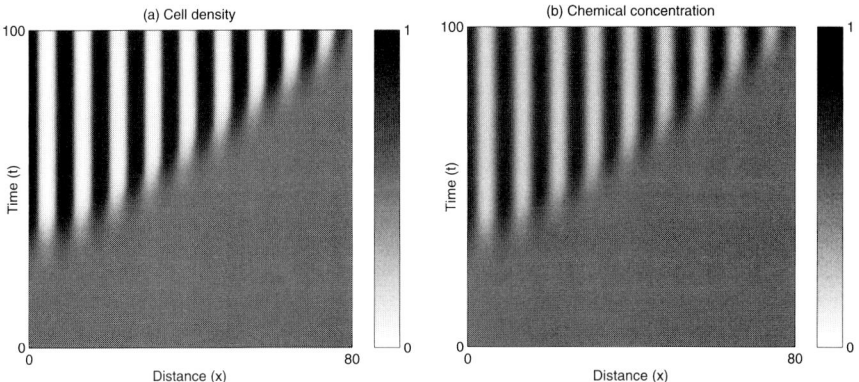

Figure 6 Numerical solution of the cell-chemotaxis model for somite formation in one spatial dimension given by Eq. (10). Initially the field is supposed to be homogeneous throughout, with a small perturbation made to the cell density at $x = 0$. The pattern of somites propagates across the field over time, from left to right. Parameters are as follows: $D = 1.0$, $\chi = 10.0$, $N = 1.0$ and $n_0 = 0.5$.

occurring at rate D relative to PSM cells. Production of CAM-triggering morphogen occurs at a rate proportional to the number of cells present, whilst we assume that no cell birth or death occurs.

In this model we have assumed the presence of a diffusive morphogen that triggers CAM response. In this simplified model we have combined the effects of both morphogen and CAMs into one term, a, in order to facilitate mathematical analysis.

Once again, it is not possible to solve the model directly: however, it is possible to undertake a linear stability analysis of the system to find parameter regions in which patterns of cell aggregations may arise. The condition for spatial patterns in cell density is

$$\chi n_0 (N - n_0) > 1, \tag{11}$$

where n_0 is the initial (uniform) cell density.

The system can be solved numerically using the MATLAB solver `pdepe`. We assume that initially both PSM cells and CAM-triggering morphogen are distributed uniformly along the AP axis and that a small perturbation is made to the cell density at the anterior end of the PSM (in our simulations this corresponds to $x = 0$). Fig. 6 shows the results of numerical simulation for a set of parameter values which satisfy Eq. (10). From an initially homogeneous field we see the sequential development of a series of peaks and troughs in cell density with equivalent patterns in CAM-triggering morphogen concentration. The patterns arise on a timescale and with a wavelength that is dependent on the choice of parameters. In other words, each new somite forms at a specific time interval after the last somite and with a defined length, which depends upon the model parameters.

An extension of this model needs to address a number of issues. Firstly, how can the model be made consistent with the observations of genetic prepatterns along the AP axis and how can we incorporate the role of the segmentation clock? Secondly, mathematical analysis shows that the initial (unpatterned) field of cells is unstable to perturbation: should a disturbance arise in a region ahead of the somite pattern, a second patterning envelope would initiate from the point of disturbance. Due to the inherently stochastic nature of gene transcription and translation it is likely this will occur—extension of the model to take account of this is the subject of current work. It is also important to point out that the major drawback of this model is that it does not take into account the intercellular mechanical forces involved in the process of somite formation. As a consequence, it cannot account for the morphological changes observed during somitogenesis.

III. Discussion

We began this article by outlining the processes involved in somite formation, from creation of a genetic prepattern to the mechanisms involved in generating morphological somites. We described the patterns of oscillating gene expression which are readouts of the segmentation clock and the traveling patterns of gene expression which control segmental determination. In the middle section we described a series of mathematical models which aim to capture various aspects of somite formation. The mathematical formulation of the Cooke–Zeeman–Pourquié Clock and Wavefront model integrates simple readouts from the clock and gradient to describe the formation of a genetic prepattern which seems to guide development of the physical somites. Models for the segmentation clock assume an underlying regulatory transcription network, characterized by delays in transcription and translation, drives oscillations in mRNA and protein levels. Models for the FGF8 gradient assume similar types of negative feedback (but without delay) to describe the formation of traveling gradients of RA and FGF8. Finally we postulated a model which, rather than characterizing the formation of a genetic prepattern, describes the formation of cellular aggregations—the physical somites.

The current challenge is to build on these models in order to drive forward knowledge in this area. Development of more sophisticated models for both clock and gradient will allow us to elucidate the crucial mechanisms underlying each system. These models can be integrated back into a generalized "Clock and Wavefront" framework and used to make predictions and test hypotheses. Below we outline the key challenges in each area.

Current models for the segmentation clock generally consist of simple transcriptional oscillators with delay terms allowing the intermediate synthesis steps, such as transport, elongation and splicing, to be encompassed by a single parameter. Such models are easily solved numerically and allow predictions on the period of the clock to be made based on model parameters. However, there is now a wealth of information available regarding the different molecular players involved in regulation of the segmentation clock (Dequéant *et al.*, 2006): how they interact and what happens when embryos mutant for crucial components are generated. Larger scale models are required in order to integrate this new data into a concrete framework. This will entail developing a "Systems Biology" approach with the deployment of, for example, Boolean and network models, and multiscale simulation techniques (Schnell *et al.*, 2007). Another important aspect of the segmentation clock that has not yet been fully explored is the coupling between adjacent PSM cells that arises from signaling via the Notch–Delta pathway. A mean-field approach generally assumes equal coupling between all cells of a network, which is clearly not the case in the somitogenesis oscillator. Development of a theoretical approach for dealing with large systems of coupled oscillators is clearly important here, and also in a wider biological context.

Modeling of the gradient which underlies the determination front has only recently begun to acquire attention and models for this are still at an early stage. It seems highly probable that the formation and maintenance of such gradients are tightly coupled to the events surrounding axial elongation. It has recently been shown, in mouse, that the FGF8 signaling gradient is mirrored by a gradient in Wnt3a, and that in fact, Wnt3a acts upstream of FGF8 (Aulehla and Herrmann, 2004; Aulehla *et al.*, 2003). Wnt3a has also been shown to play a major role in controlling cell proliferation, axis elongation and oscillations of Axin2 in the PSM (Aulehla *et al.*, 2003; Galli *et al.*, 2004). This yields the exciting possibility that Wnt3a may provide an underlying link between the segmentation clock, specification of positional information and body axis elongation (Aulehla and Herrmann, 2004; Dubrulle and Pourquié, 2004a).

The final piece of the story is the transition from genetic prepattern to epithelial somites. Most current models for somitogenesis are concerned with the formation of a genetic prepattern along the AP axis and few have attempted to model the series of cell rearrangements and shape changes that accompany the transition from prepattern to coherent somite. Probably this is linked to the enormity of the task: our understanding of the biological processes involved is far from complete—numerous complex, interacting processes are taking place—and theoretical tools have not yet been developed to answer some of the questions which need to be addressed. The cell–CAM model presented here is a simple attempt to address this process. It neglects many of the salient features of the epithelialization process and cannot yet replicate many experimental observations. However, it is a first step on the road to constructing a framework for this complex process and development of the model is the subject of current work. Glazier et al. (in this volume) developed a computational model of the spatiotemporal pattern of adhesive and repulsive forces at work during somite formation. Sherratt and co-workers are also interested in applying their theoretical studies of differential cell adhesion (Armstrong *et al.*, 2006) to somite formation, using the genetic prepattern to drive the process (Armstrong *et al.*, 2007), whilst Schnell and Grima have proposed a model in which chemotaxis, differential cell adhesion and minimization of tissue surface tension drive the cell sculpting process (Grima and Schnell, 2007).

IV. Perspective

The goal of multiscale modeling is to investigate events happening on the microscopic scale and integrate results into models on a macroscopic level (Schnell *et al.*, 2007). Although most models for somite formation are still firmly based on events taking place at the macroscopic level, the wealth of new experimental data becoming available and the generation of new experimental techniques is beginning to allow the generation of microscale models, thereby paving the way for truly multiscale models in the future.

The future of interdisciplinary research in developmental biology lies in the attitudes of both theoretical and experimental communities. Commitment to communication across the boundaries of increasingly specialist research areas, alongside the development of tools and methods to facilitate the achievement of common goals, is crucial. Mathematical modeling should be used as a tool to drive and expand knowledge. It has the power to integrate multiple hypotheses and ideas into a rigorous framework in which they can be submitted to analysis and *in silico* experiments, used to elucidate pertinent questions and devise future experiments. In turn, results from these experiments can be applied to adjust and refine models. This feedback loop of modeling, testing and fine-tuning is essential for the development of biologically-accurate models and the acceleration of our understanding in the biomedical sciences.

Acknowledgments

R.E.B. thanks Research Councils UK for an RCUK Fellowship in Mathematical Biology, Lloyds Tercentenary Foundation for a Lloyds Tercentenary Foundation Fellowship, Microsoft for a European Postdoctoral Research Fellowship, St. Hugh's College, Oxford for a Junior Research Fellowship and the Australian Government, Department for Science, Education and Training for an Endeavor Postdoctoral Research Fellowship to visit the Department of Mathematics and Statistics, University of Melbourne. S.S. acknowledges support from NIH Grant R01GM76692. P.K.M. was partially supported by a Royal Society Wolfson Merit Award.

References

Armstrong, N. J., *et al.* (2006). A continuum approach to modeling cell–cell adhesion. *J. Theor. Biol.* **243**, 98–113.
Armstrong, N. J., *et al.* (2007). Modeling the role of cell adhesion in somite formation (work in progress).
Aulehla, A., and Herrmann, B. G. (2004). Segmentation in vertebrates: Clock and gradient finally joined. *Genes Dev.* **18**, 2060–2067.
Aulehla, A., and Johnson, R. L. (1999). Dynamic expression of lunatic fringe suggests a link between notch signaling and an autonomous cellular oscillator driving somite segmentation. *Dev. Biol.* **207**, 49–61.
Aulehla, A., *et al.* (2003). Wnt3a plays a major role in the segmentation clock controlling somitogenesis. *Dev. Cell* **4**, 395–406.

6. Models for Somite Formation

Baker, R.E. and Maini, P.K. (2007). Traveling gradients in interacting morphogen systems. *Math. Biosci.*, doi:10.1016/j.mbs.1007.01.006 (to appear).

Baker, R. E., *et al.* (2003). Formation of vertebral precursors: Past models and future predictions. *J. Theor. Med.* **5**, 23–35.

Baker, R. E., *et al.* (2006a). A clock and wavefront mechanism for somite formation. *Dev. Biol.* **293**, 116–126.

Baker, R. E., *et al.* (2006b). A mathematical investigation of a clock and wavefront model for somitogenesis. *J. Math. Biol.* **52**, 458–482.

Chernoff, E. A. G., and Hilfer, S. R. (1982). Calcium dependence and contraction in somite formation. *Tiss. Cell* **14**, 435–449.

Collier, J. R., *et al.* (2000). A cell cycle model for somitogenesis: Mathematical formulation and numerical solution. *J. Theor. Biol.* **207**, 305–316.

Cooke, J. (1975). Control of somite number during morphogenesis of a vertebrate, Xenopus laevis. *Nature* **254**, 196–199.

Cooke, J. (1998). A gene that resuscitates a theory—somitogenesis and a molecular oscillator. *Trends Genet.* **14**, 85–88.

Cooke, J., and Zeeman, E. C. (1976). A clock and wavefront model for control of the number of repeated structures during animal morphogenesis. *J. Theor. Biol.* **58**, 455–476.

Dale, K. J., and Pourquié, O. (2000). A clock-work somite. *BioEssays* **22**, 72–83.

Dale, J. K., *et al.* (2003). Periodic notch inhibition by lunatic fringe underlies the chick segmentation clock. *Nature* **421**, 275–278.

Dale, J. K., *et al.* (2006). Oscillations of the Snail genes in the presomitic mesoderm coordinate segmental patterning and morphogenesis in vertebrate somitogenesis. *Dev. Cell* **10**, 355–366.

del Barco Barrantes, I., *et al.* (1999). Interaction between Notch signaling and Lunatic Fringe during somite boundary formation in the mouse. *Curr. Biol.* **9**, 470–480.

Dequéant, M. -L., *et al.* (2006). A complex oscillating network of signaling genes underlies the mouse segmentation clock. *Science* **314**, 1595–1598.

Diez del Corral, R., *et al.* (2002). Onset of neural differentiation is regulated by paraxial mesoderm and requires attenuation of FGF8 signaling. *Development* **129**, 1681–1691.

Diez del Corral, R., *et al.* (2003). Opposing FGF and retinoid pathways control ventral neural pattern, neuronal differentiation and segmentation during body axis extension. *Neuron* **40**, 65–79.

Diez del Corral, R., and Storey, K. (2004). Opposing FGG and retinoid pathways: A signaling switch that controls differentiation and patterning onset in the extending vertebrate body axis. *BioEssays* **26**, 957–969.

Drake, C. J., and Little, C. D. (1991). Integrins play an essential role in somite adhesion to the embryonic axis. *Dev. Biol.* **143**, 418–421.

Duband, J. L., *et al.* (1987). Adhesion molecules during somitogenesis in the avian embryo. *J. Cell Biol.* **104**, 1361–1374.

Dubrulle, J., and Pourquié, O. (2002). From head to tail: Links between the segmentation clock and anteroposterior patterning of the embryo. *Curr. Opin. Genet. Dev.* **12**, 519–523.

Dubrulle, J., and Pourquié, O. (2004a). Coupling segmentation to axis formation. *Development* **131**, 5783–5793.

Dubrulle, J., and Pourquié, O. (2004b). fgf8 mRNA decay establishes a gradient that couples axial elongation to patterning in the vertebrate embryo. *Nature* **427**, 419–422.

Dubrulle, J., *et al.* (2001). FGF signaling controls somite boundary position and regulates segmentation clock control of spatiotemporal Hox gene activation. *Cell* **106**, 219–232.

Duguay, D., *et al.* (2003). Cadherin-mediated cell adhesion and tissue segregation: Qualitative and quantitative determinants. *Dev. Biol.* **253**, 309–323.

Foty, R. A., and Steinberg, M. S. (2004). Cadherin-mediated cell–cell adhesion and tissue segregation in relation to malignancy. *Int. J. Dev. Biol.* **48**, 397–409.

Galli, L. M., *et al.* (2004). A proliferative role for Wnt-3a in chick somites. *Dev. Biol.* **269**, 489–504.

Giudicelli, F., and Lewis, J. (2004). The vertebrate segmentation clock. *Curr. Opin. Genet. Dev.* **14**, 407–414.

Goldbeter, A., Gonze, D., and Pourquie, O. (2007). Sharp developmental thresholds defined through bistability by antagonistic gradients of retinoic acid and FGF signaling. *Dev. Dyn.* **236**, 1495–1508.

Gossler, A., and Hrabĕ de Angelis, M. (1998). Somitogenesis. *Curr. Top. Dev. Biol.* **38**, 225–287.

Grima, , R., and Schnell, , S. (2007). Can tissue surface tension drive somite formation?. *Dev. Biol.* **307**, 248–257.

Haraguichi, S., *et al.* (2001). Transcriptional regulation of Mesp1 and Mesp1 genes: Differential usage of enhancers during development. *Mech. Dev.* **108**, 59–69.

Hirata, H., *et al.* (2004). Instability of Hes7 protein is crucial for the somite segmentation clock. *Nat. Genet.* **36**, 750–754.

Horikawa, K., *et al.* (2006). Noise-resistant and synchronized oscillation of the segmentation clock. *Nature* **441**, 719–723.

Ishikawa, A., *et al.* (2005). Mouse Nkd1, a Wnt antagonist, exhibits oscillatory gene expression in the PSM under the control of Notch signaling. *Mech. Dev.* **121**, 1443–1453.

Jacobson, A., and Meier, S. (1986). Somitomeres: The primordial body segments. *In* "Somites in Developing Embryos" (R. Bellairs, *et al.*, Ed.), Plenum, New York, pp. 1–16.

Jiang, Y. -J., *et al.* (2000). Notch signaling and the synchronization of the segmentation clock. *Nature* **408**, 475–479.

Jiang, Y.-J., and Lewis, L. (2001). Notch as a synchronizer in somite segmentation. *In* "Proceedings of the NATO Advanced Research Workshop on the Origin and Fate of Somites" (C. P. Ordahl Ed.). Vol. 329. IOS Press, Amsterdam, pp. 71–79.

Jiang, Y. J., *et al.* (1998). Vertebrate segmentation: The clock is linked to Notch signaling. *Curr. Biol.* **8**, R868–R871.

Kalcheim, C., and Ben-Yair, R. (2005). Cell rearrangements during development of the somite and its derivatives. *Curr. Opin. Genet. Dev.* **15**, 371–380.

Kulesa, P. M., and Fraser, S. E. (2002). Cell dynamics during somite boundary formation revealed by time-lapse analysis. *Science* **298**, 991–995.

Lewis, J. (2003). Autoinhibition with transcriptional delay: A simple mechanism for the zebrafish somitogenesis oscillator. *Curr. Biol.* **13**, 1398–1408.

Maroto, M., and Pourquié, O. (2001). A molecular clock involved in somite segmentation. *Curr. Top. Dev. Biol.* **51**, 221–248.

McGrew, M. J., and Pourquié, O. (1998). Somitogenesis: Segmenting a vertebrate. *Curr. Op. Genet. Dev.* **8**, 487–493.

McGrew, M. J., *et al.* (1998). The Lunatic Fringe gene is a target of the molecular clock linked to somite segmentation in avian embryos. *Curr. Biol.* **8**, 979–982.

McInerney, D., *et al.* (2004). A mathematical formulation for the cell cycle model in somitogenesis: Parameter constraints and numerical solutions. *IMA J. Math. Appl. Med. & Biol.* **21**, 85–113.

Meier, S. (1979). Development of the chick embryo mesoblast: Formation of the embryonic axis and establishment of the metameric pattern. *Dev. Biol.* **73**, 24–45.

Molotkova, N., *et al.* (2005). Requirement of mesodermal retinoic acid generated by Raldh2 for posterior neural transformation. *Mech. Dev.* **122**, 145–155.

Monk, N. A. M. (2003). Oscillatory expression of Hes1, p53, and NF-κB driven by transcriptional time delays. *Curr. Biol.* **13**, 1409–1413.

Moreno, T. A., and Kinter, C. (2004). Regulation of segmental patterning by retinoic acid signaling during Xenopus somitogenesis. *Dev. Cell* **6**, 205–218.

Morimoto, M., *et al.* (2006). Cooperative Mesp activity is required for normal somitogenesis along the anterior–posterior axis. *Dev. Biol.* **300**, 687–698.

Packard, D. S. (1976). The influence of axial structures on chick formation. *Dev. Biol.* **53**, 36–48.

Packard, D. S., *et al.* (1993). Somite pattern regulation in the avian segmental plate mesoderm. *Development* **117**, 779–791.

6. Models for Somite Formation

Palmeirim, I., *et al.* (1997). Avian hairy gene expression identifies a molecular clock linked to vertebrate segmentation and somitogenesis. *Cell* **91**, 639–648.
Pourquié, O. (1998). Clocks regulating developmental processes. *Curr. Opin. Neurobiol.* **8**, 665–670.
Pourquié, O. (1999). Notch around the clock. *Curr. Opin. Gen. Dev.* **9**, 559–565.
Pourquié, O. (2001a). The vertebrate segmentation clock. *J. Anat.* **199**, 169–175.
Pourquié, O. (2001b). Vertebrate somitogenesis. *Ann. Rev. Cell Dev. Biol.* **17**, 311–350.
Pourquié, O. (2003). The segmentation clock: Converting embryonic time into spatial pattern. *Science* **301**, 328–330.
Pourquié, O., and Goldbeter, A. (2003). Segmentation clock: Insights from computational models. *Curr. Biol.* **13**, R632–R634.
Primmett, D. R. N., *et al.* (1988). Heat-shock causes repeated segmental anomalies in the chick-embryo. *Development* **104**, 331–339.
Primmett, D. R. N., *et al.* (1989). Periodic segmental anomalies induced by heat-shock in the chick-embryo are associated with the cell-cycle. *Development* **105**, 119–130.
Rida, P. C., *et al.* (2004). A Notch feeling of somite segmentation and beyond. *Dev. Biol.* **265**, 2–22.
Saga, Y., *et al.* (2001). Mesp2: A novel mouse gene expressed in the presegmented mesoderm and essential for segmentation initiation. *Genes Dev.* **2**, 835–845.
Schnell, S., *et al.* (2007). Multiscale modeling in biology. *Am. Sci.* **95**, 134–142.
Schnell, S., and Turner, T. E. (2004). Reaction kinetics in intracellular environments with macromolecular crowding: Simulations and rate laws. *Prog. Biophys. Mol. Biol.* **85**, 235–260.
Schnell, S., *et al.* (2002). Models for pattern formation in somitogenesis: A marriage of cellular and molecular biology. *C. R. Biol.* **325**, 179–189.
Stockdale, F. E., *et al.* (2000). Molecular and cellular biology of avian somite development. *Dev. Dyn.* **219**, 304–321.
Tabin, C. J., and Johnson, R. L. (2001). Developmental biology: Clocks and Hox. *Nature* **412**, 780–781.
Takahashi, Y., *et al.* (2000). Mesp2 initiates somite segmentation through the Notch signaling pathway. *Nat. Gen.* **25**, 390–396.
Takashi, Y., *et al.* (2005). Differential contributions of Mesp1 and Mesp2 to the epithelializations and rostro-caudal patterning of somites. *Development* **132**, 787–796.
Tam, P. P. L., and Meier, S. (1982). The establishment of a somitomeric pattern in the mesoderm of the gastrulating mouse embryo. *Am. J. Anat.* **164**, 209–225.
Turner, T. E., *et al.* (2004). Stochastic approaches for modeling in vivo reactions. *Comput. Biol. Chem.* **28**, 165–178.
Zeeman E. C. (1974). Primary and secondary waves in developmental biology: Some mathematical questions in biology, VI. *In* "Proc. Eighth Symp., Mathematical Biology." San Francisco, CA, 1974. *In* "Lectures on Mathematics in the Life Sciences," Vol. 7, Am. Math. Soc., Providence, RI, pp. 69–161.

7

Coordinated Action of N-CAM, N-cadherin, EphA4, and ephrinB2 Translates Genetic Prepatterns into Structure during Somitogenesis in Chick

James A. Glazier,*,† Ying Zhang,* Maciej Swat,* Benjamin Zaitlen,* and Santiago Schnell†,*

*Biocomplexity Institute and Department of Physics, 727 East Third Street, Indiana University, Bloomington, Indiana 47405
†Indiana University School of Informatics, 1900 East Tenth Street, Indiana University, Bloomington, Indiana 47406

    I. Introduction
       A. Nomenclature
   II. Patterns of Gene Expression and Protein Distribution during Somitogenesis
       A. Cell Adhesion Molecules
       B. Eph/ephrin-Induced Cell "Repulsion"
       C. Interaction Between Adhesion and Repulsion during Somitogenesis
  III. From Genetic Oscillators to Adhesion/Repulsion-Protein Patterns
  IV. From Adhesion-Protein Patterns to Segmentation
       A. Segmentation Model
   V. Computer Simulation of Segmentation
       A. The Glazier–Graner–Hogeweg Model
       B. GGH Somitogenesis Simulation
       C. Simulation Implementation
       D. Parameter Values
       E. Global Parameters
       F. Cell-Adhesion Energies
       G. Adhesion Hierarchies
       H. Target Areas and Volumes
       I. Time
  VI. Results and Discussion
       A. Parameter Choices
       B. Segmentation Requires EphA4/ephrinB2 Repulsion
       C. Segmentation Requires Multiple Levels of EphA4/ephrinB2 Expression
       D. Dynamic Morphological Changes and Error Correction during Segmentation
 VII. Conclusion
      Acknowledgments
      Introduction to Appendices
      Appendix A. Python Code to Execute Somitogenesis Simulations (somite.py)
      Appendix B. CC3D ML Code to Execute Somitogenesis Simulations (somite.xml)
      Appendix C. Python Steppables for Somitogenesis Simulations (somiteSteppables.py)
      References

During gastrulation in vertebrates, mesenchymal cells at the anterior end of the *presomitic mesoderm* (PSM) periodically compact, transiently epithelialize and detach from the posterior PSM to form somites. In the prevailing *clock-and-wavefront* model of somitogenesis, periodic gene expression, particularly of Notch and Wnt, interacts with an FGF8-based thresholding mechanism to determine cell fates. However, this model does not explain how cell determination and subsequent differentiation translates into somite morphology. In this paper, we use computer simulations of chick somitogenesis to show that experimentally-observed temporal and spatial patterns of adhesive N-CAM and N-cadherin and repulsive EphA4–ephrinB2 pairs suffice to reproduce the complex dynamic morphological changes of somitogenesis in wild-type and N-cadherin $(-/-)$ chick, including intersomitic separation, boundary-shape evolution and sorting of misdifferentiated cells across compartment boundaries. Since different models of determination yield the same, experimentally-observed, distribution of adhesion and repulsion molecules, the patterning is independent of the details of this mechanism. © 2008, Elsevier Inc.

I. Introduction

Somitogenesis, during which initially-continuous anterior–posterior (AP) bands of loosely-bound cells on either side of the medial primitive streak, the *presomitic mesoderms* (PSM), break apart sequentially and periodically (at intervals of ∼90 minutes in chick) in AP sequence into a spatially-regular series of separated, tightly-bound *somites* is the classical example of *segmentation* during vertebrate embryogenesis. In vertebrates, somites are the precursors of vertebrae, muscle and skin derivatives, and provide a scaffold for assembly of the peripheral vasculature and nervous system (Gossler and Hrabe de Angelis, 1998).

Somite formation requires: (i) Physical separation of somitic tissue from the initially-continuous PSM, (ii) coalescence of cells in the forming somite, and (iii) the establishment of a stable border between the somite and the PSM.

Experimentally, animal species differ in the cell rearrangements which create the somite–PSM border and how aggressively new somites pull apart from the PSM (Kulesa *et al.*, 2007). In *Xenopus* embryos, two short, discrete fissures start from both the medial and lateral edges of the PSM and expand gradually towards the middle of the PSM to form a stable somite–PSM border (Afonin *et al.*, 2006). In zebrafish embryos, cell rearrangements within forming somites are minimal and the forming-somite–PSM border develops when cells in the forming somite gently detach from their PSM neighbors, forming a medial notch which spreads laterally (Wood and Thorogood, 1994; Henry *et al.*, 2000; Jiang *et al.*, 2000). In both animals, the cells then retract towards the center of the forming somite or towards the PSM depending on the side of the somite–PSM boundary.

In chick embryos, cell rearrangements during somite–PSM border formation are more dramatic. Somitogenesis occurs in a complex spatiotemporal pattern, not via simple cleavage of the PSM (Kulesa and Fraser, 2002). Time-lapse analyses reveal that the forming somite–PSM boundary develops a dynamic *ball-and-socket* shape and that some cells cross the presumptive somite–PSM boundary (Kulesa and Fraser, 2002). Tissue transplantation studies have shown that cells in the region near the posterior border of the forming somite possess border-forming signals mediated by Notch and reinforced by Lunatic Fringe (Sato *et al.*, 2002). Transplantation of the ventral-most cells in the posterior of the forming somite induces formation of ectopic borders and somite subdivisions in more dorsal PSM tissues (Sato and Takahashi, 2005). Thus, in chick, the initial separation of a forming somite from the PSM appears to occur in a ventral-to-dorsal (VD) direction. However, in line with most experimental and computational studies, this paper will treat the PSM and somites as essentially two-dimensional (2D), neglecting VD and medial-to-lateral (ML) variation.

To explain the complex cell rearrangements in chick somitogenesis we employ computer simulations, which allow us to study how previously-determined spatiotemporal variations in gene expression (*prepatterns*) lead to variations in cell adhesion and local microenvironment, which induce cell rearrangement into coherent somites. The rearrangements, in turn, can further affect gene expression (*feedback*).

Essentially all models of somitogenesis to date have neglected the properties and movements of individual cells, and concentrated on mechanisms to generate periodic patterns of gene expression in the PSM, i.e., *somite specification*. This focus on early gene expression seemed reasonable, because determination of cell fate occurs early, about three somitic-clock cycles (about $4\frac{1}{2}$ hours) before physical segmentation (Dubrulle and Pourquié, 2004). The existence of a prepattern is consistent with the finding that reversal of the AP axis of the PSM leads to reversed somites (Keynes and Stern, 1988). The numerous theoretical models of gene-expression patterning include (Schnell and Maini, 2000; Baker *et al.*, 2003, 2006), the *clock-and-wavefront* model (Cooke and Zeeman, 1976; Dubrulle *et al.*, 2001), *reaction–diffusion* models (Meinhardt, 1996), *cell-cycle* models (Primmett *et al.*, 1988, 1989; Stern *et al.*, 1988; Collier *et al.*, 2000), and the *clock-and-induction* model (Schnell and Maini, 2000). Each of these models includes certain key features of the underlying biology and predicts that the PSM develops spatially-periodic patterns of gene expression in tissue blocks, but fails to explain some experimental observations. Pourquié and co-workers' version of the clock-and-wavefront model (Durbrulle and Pourquié, 2002; Pourquié, 2004) is, perhaps, the most successful (Baker *et al.*, 2006). For a detailed review of mathematical models of somitic prepatterning, see the paper by Baker *et al.* in this volume.

The actual process of somite formation—how a somite pulls apart from the PSM and the ensuing morphological changes—is not well understood. The only mathematical model attempting to describe the bulk movement of somitic cells to form a somite (Schnell *et al.*, 2002) does not account for the intercellular mechanical forces involved in somite formation. Grima and Schnell (2007) have investigated the possibility that minimization of tissue surface tension drives the subsequent morphological changes leading to rounded somites. This paper addresses another question—given a prepattern of gene expression, can known biological mechanisms give rise to the patterns of cell movement and morphological changes observed in both wild-type and gene-knockout experiments? Our model is completely agnostic about the origin of this pattern of gene expression (it works equally well with any of the somite-specification models in the preceding paragraph).

A. Nomenclature

Because somitogenesis proceeds in a temporally-periodic and spatially-progressive fashion, the identity of groups of cells changes in time, making nomenclature somewhat confusing (see Fig. 2). We use the nomenclature most common in the experimental community. At the beginning of a 90-minute cycle of somite formation, the anterior-most portion of the PSM becomes the site of a *newly-forming somite*, which we refer to as S_0; we will then refer to the portion of PSM immediately posterior to the region which contains the forming somite as the *anterior* PSM or S_{-1}. Initially, these two regions are contiguous; the forming somite S_0 then gradually separates from the remaining PSM along the *presumptive somite–PSM boundary*. When somite separation is complete, the forming somite S_0 becomes somite S_1, the anterior portion of the remaining PSM, S_{-1}, becomes S_0 and the cycle repeats (Ordahl, 1993; Pourquié and Tam, 2001). We call the regions of the PSM successive somite lengths behind S_{-1}, S_{-2}, S_{-3}, ... Each somite also has an *anterior* and *posterior compartment*, which we will denote S_{-1A} and S_{-1P}, respectively, and a *central* (or *core*) and *peripheral* region.

II. Patterns of Gene Expression and Protein Distribution during Somitogenesis

To understand how the gene-expression prepattern translates into changes in tissue morphology, we first review the functions of key molecules and their expression patterns.

A. Cell Adhesion Molecules

N-CAM and N-cadherin are *homophilic* membrane-bound proteins which contribute to contact adhesion between cells. The strength of adhesion increases with protein density on each cell, though the form of the dependence is not completely clear (Foty and Steinberg, 2005). While both molecules are homotypically adhesive as isolated monomers, they normally associate into groups which then aggregate in

7. Coordinated Action of N-CAM, N-cadherin, EphA4, and ephrinB2

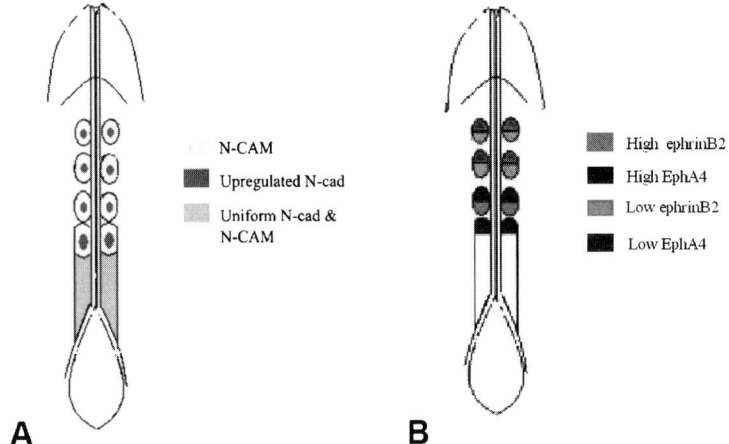

Figure 1 Schematic diagram of the distribution of adhesion and repulsion molecules during chick somitogenesis. (A) N-cadherin and N-CAM protein distributions based on immunocytochemistry experiments (Linask et al., 1998). The PSM has a uniform low background level of N-cadherin and N-CAM. N-CAM levels do not change significantly during segmentation, while N-cadherin levels increase at the core of the forming somite. (B) EphA4 and ephrinB2 levels based on in-situ-hybridization experiments (Baker et al., 2003; Aulehla and Pourquié, 2006). During segmentation EphA4 expresses in the anterior compartment of the forming somite and the anterior of the PSM; ephrinB2 expresses in the posterior compartment of the forming somite. Both EphA4 and ephrinB2 mRNA levels decrease after the somite has separated from the PSM. See color insert.

the cell membrane to form clusters (e.g., clusters of trimers of dimers) which increases the effective binding strength per molecule. Both molecules bind to the actin cytoskeleton via β-catenin. This binding affects their effective adhesivity directly, possibly through changes in conformation of their extracellular domains, and indirectly, because a functional actin cytoskeleton is necessary for their clustering. In general, for a given receptor density, N-CAM results in weaker adhesion than N-cadherin.

In avian and mouse embryos, somite formation follows *compaction* (i.e., a reduction in the intercellular space between cells) and the heightened expression of N-CAM and N-cadherin (Duband et al., 1987; Linask et al., 1998), which are expressed at lower levels in the rest of the PSM (Fig. 1A). In mice (Kimura et al., 1995), cadherin-11 strictly correlates with S_0; it is not expressed in other parts of the PSM. Fig. 1A schematically illustrates the dynamic changes of cell-adhesion-molecule types and concentrations during somite segmentation in chick embryos. Before segmentation, mesenchymal cells in the PSM weakly express N-CAM and N-cadherin. During segmentation, the condensing cells in S_0 significantly increase their N-CAM and N-cadherin expression.

Before and during somite separation, the cells at the periphery of S_0 *epithelialize* (Dubrulle and Pourquié, 2004). During this epithelialization, N-CAM remains uniformly distributed over the entire surfaces of epithelial cells, whereas N-cadherins concentrate predominantly on the apical surfaces (which somewhat counter-intuitively face towards the center of the somite). Cells located in the core of the somite remain mesenchymal and continue to have essentially uniform surface distributions of both N-CAM and N-cadherin (Duband *et al.*, 1987; Linask *et al.*, 1998).

B. Eph/ephrin-Induced Cell "Repulsion"

Ephs and ephrins are families of heterotypically-active cell-surface receptors that can lead to effective "repulsion" between a cell expressing an Eph and a cell expressing the corresponding ephrin. How contact leads to effective repulsion is still under active investigation. A plausible mechanism is that pairing of complementary Ephs and ephrins on apposing cells triggers bidirectional signaling, which results in the local collapse of the actin cytoskeleton in both cells near their point of contact (Harbott and Nobes, 2005). This collapse then locally disrupts the condensation and pairing of cell adhesion molecules like N-CAM and N-cadherin, which reduces, but need not completely eliminate, the strength of their homotypic intercellular binding (as suggested in Kasemeier-Kulesa *et al.*, 2006; Cooke *et al.*, 2005). Thus EphA4–ephrinB2 "repulsion" in somites may actually result from a reduction in effective cell–cell adhesion. Since Eph–Eph and ephrin–ephrin apposition has no effect on the cytoskeleton, cell adhesion molecules remain fully functional within the interior of an Eph-expressing or ephrin-expressing domain. Thus boundaries between domains expressing an Eph and its complementary ephrin are structurally weak, while the domains themselves can be strong, allowing tissue to pull apart along Eph–ephrin contact lines. Collapse of the actin cytoskeleton also destroys the *pseudopods* or *leading edges* which cells use to move in a particular direction (Mellitzer *et al.*, 1999; Xu *et al.*, 1999; Poliakov *et al.*, 2004), preventing a cell expressing Eph from moving into a cluster of cells expressing the corresponding ephrin and *vice versa* (*contact inhibition*). The net effect is to establish compartmental boundaries between clusters of cells expressing Eph and clusters of cells expressing the complementary ephrin. Such boundaries are clearly visible using ordinary microscopy.

Eph/ephrin signaling is responsible for boundary formation in the developing hindbrain (Mellitzer *et al.*, 1999; Xu *et al.*, 2000) and is necessary for the formation of intersomitic boundaries and subsequent epithelialization (Durbin *et al.*, 1998). During somite segmentation, EphA4 expresses in the anterior half of somites (S_{0A}) and in the anterior tip of the PSM (S_{-1A}), while EphrinB2 expresses in the posterior half of somites (S_{0P}) (Fig. 1B) (Nieto *et al.*, 1992; Bergemann *et al.*, 1995; Durbin *et al.*, 1998; Baker and Antin, 2003; Baker *et al.*, 2003). EphA4/ephrinB2 signaling also regulates the mesenchymal-to-epithelial transition of the PSM during somitogenesis (Barrios *et al.*, 2003).

C. Interaction Between Adhesion and Repulsion during Somitogenesis

That the adhesive interactions of N-CAM and N-cadherin are homophilic while EphA4 and ephrinB2 produce an effective *heterorepulsion* (repulsion between cells of two complementary types) is crucial to the observed mechanics of somite formation. Cells expressing EphA4 and cells expressing ephrinB2 meet both at the presumptive somitic boundary (S_{0P} to S_{-1A}) and inside each somite (at the center of S_0, i.e., S_{0A} to S_{0P}). If contact between cells expressing EphA4 and cells expressing ephrinB2 causes the separation of the posterior end of S_0 (S_{0P}) from the anterior end of the PSM (S_{-1A}), why does it not cause a similar boundary to form inside S_0 (between S_{0A} and S_{0P}), subdividing it into two noncontacting smaller somites? N-cadherin in the center of the somite seems to be essential, since, in mouse, knocking out N-cadherin results in segmentation of normal somites into two separated sub-somites (Kimura *et al.*, 1995; Radice *et al.*, 1997; Horikawa *et al.*, 1999). One question we will investigate through simulation is what interactions of adhesion and repulsion lead to segmentation without fragmentation of individual somites.

III. From Genetic Oscillators to Adhesion/Repulsion-Protein Patterns

Molecular signaling during segmentation-prepattern specification is still a subject of intensive research. The following highly-simplified and speculative description provides at least a working hypothesis for these mechanisms. The embryo elongates primarily through division of cells in the extreme posterior of the PSM and in the tail bud. Cells leave the tail bud, then cease to transcribe fgf8 DNA into mRNA. This mRNA slowly degrades, but continues to be translated into FGF8 protein (Dubrulle and Pourquié, 2004); thus the average level of FGF8 in more anterior PSM cells is lower than in more posterior cells. In addition, the precursor to retinoic acid (RA), RALDH2, diffuses into the PSM from the anterior of the embryo. FGF8 and RA couple antagonistically within cells to select either high-RA or high-FGF8 states in a bistable manner (Diez del Corral *et al.*, 2003; Goldbeter *et al.*, 2007). Above a threshold level of FGF8, individual cells in the PSM exhibit spontaneous oscillations in expression levels of many genes, especially those belonging to the Notch, Wnt and FGF cascades (Palmeirim *et al.*, 1997; Dequeant *et al.*, 2006; Kulesa *et al.*, 2007), with a period which depends on cytoplasmic FGF8 levels. Neighboring cells' oscillators synchronize via Delta–Notch signaling (Horikawa *et al.*, 2006). The oscillation of FGF8 activity is superimposed on the background decrease of FGF8, a somewhat unusual circuit that results in a whole block of cells (the size of one somite, or about 200 microns in chick) switching to the high-RA state nearly simultaneously. The combination of this switch with the local phase of the oscillation determines the cell's later differentiation (Aulehla *et al.*, 2003; Aulehla and Herrmann, 2004). Oscillations cease some time after this switch, but their

Somitic Clock
Anterior
Body Axis
RA Level
Notch
Wnt
Somite Region
Translational Region
Presumptive Somite Boundary
Prepattern Region
Expression Strength
Region of Wnt Oscillation
90 Minutes ~100 μm
Region of Cell Division
Hensen's Node
FGF Level
Repulsion Strength
Adhesion Strength
Posterior
EphA4
N-cadherin
ephrinB2
Threshold for Wnt Oscillation
Threshold for Cell Division
RA
Wnt Oscillation
FGF
Wnt
beta-Catenin
N-cadherin
Somite Center
A/P Compartmentalization
Nkd1
axin2
Mesp2
EphA4
Notch
Anterior Somite Periphery
Dll1
ephrinB2
Posterior Somite Periphery

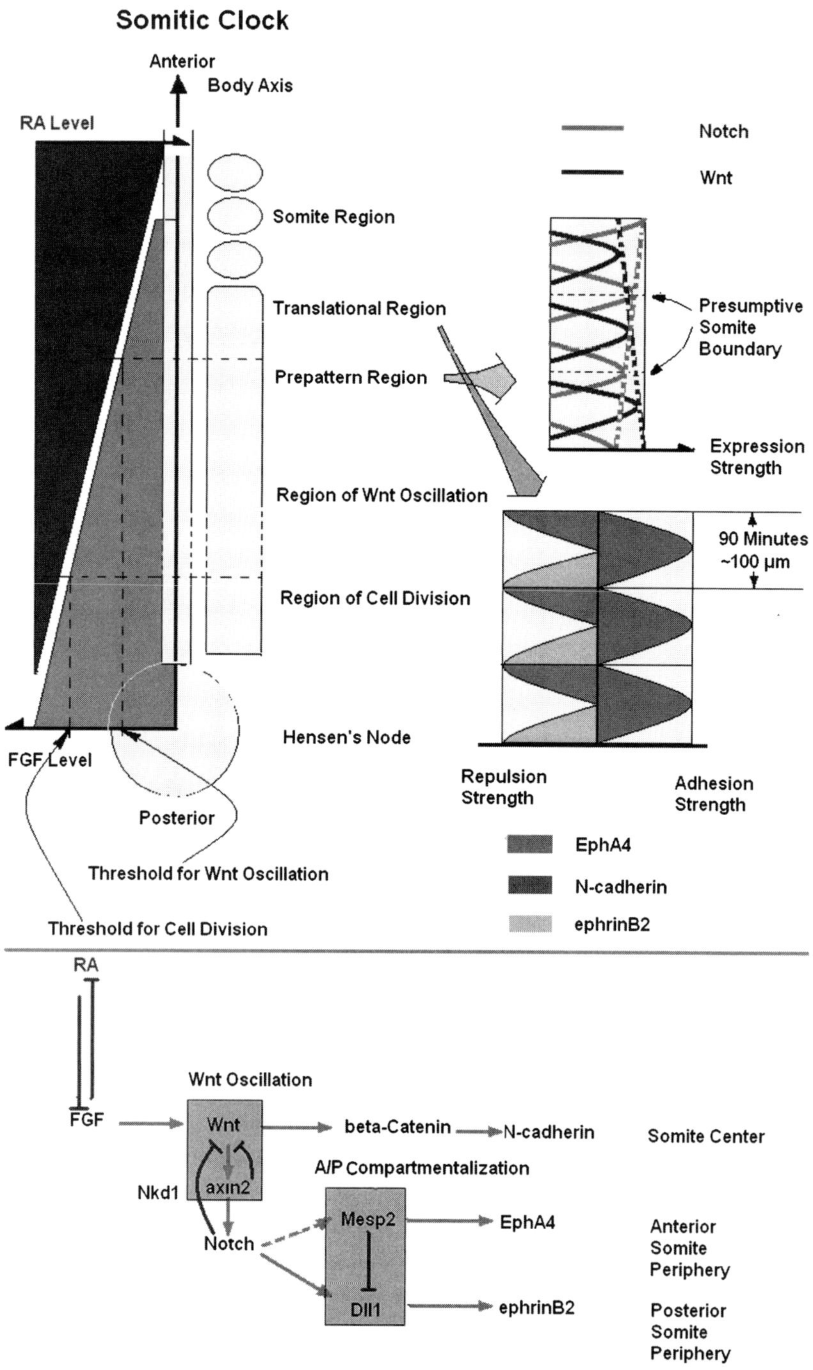

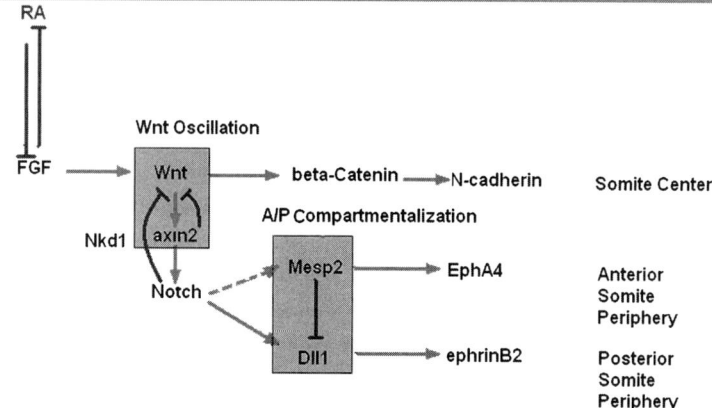

Figure 2 Schematic diagram of the translation of FGF8, Wnt and Notch cyclic expression into spatially-periodic patterns of N-cadherin, EphA4, and ephrinB2. (Top left) Within the PSM, fgf8-mRNA decay produces a PA decrease in FGF8 protein levels. Above a threshold FGF8 concentration, the intracellular Wnt concentration oscillates with a period of ~90 minutes. Intercellular Delta-Notch signaling (not shown) synchronizes neighboring cells, producing coherent oscillations in Wnt concentration. (Top right) When the FGF8 concentration falls below a threshold, Wnt oscillations cease, creating a spatial oscillation in Wnt levels. The inhibitory interaction between Wnt and Notch leads to out-of-phase expression of Wnt and Notch. (Middle right) N-cadherin, EphA4, and ephrinB2 levels. N-cadherin expression is maximal at the cores of somites. EphA4 is expressed in anterior somite compartments and ephrinB2 in posterior compartments. Both are maximal at somite boundaries. (Bottom) Plausible regulatory links from Wnt and Notch to N-cadherin, EphA4, and ephrinB2. The mutual inhibition of Wnt and Notch by axin2 and Nkd1 causes the Wnt and Notch oscillations. In the anterior somite, Mesp2 suppresses Notch signaling and also induces EphA4 expression. In the posterior somite, DLL1 maintains ephrinB2. See color insert.

persistence has no known effect on cell differentiation (Pourquié, 2007, personal communication).

Phase read-out during cell determination seems to depend on Wnt signaling (Aulehla and Pourquié, 2006), though the full regulatory cascades are not known. Fig. 2 summarizes regulatory interactions (bottom) and AP expression patterns (top). The fundamental prepatterning mechanism is the inhibitory coupling from Notch to Wnt, via axin2, Nkd1, and other pathways, which leads to elevated levels of Wnt in the centers of presumptive somites and of Notch at presumptive somite boundaries (Pourquié, 2007, personal communication). The expression domains are fairly broad, with substantial overlap in the regions midway between a presumptive somite's center and its boundaries. Wnt stabilizes cytoplasmic β-catenin, which acts as a transcription factor increasing expression of N-cadherin, which accumulates in a broad area around the presumptive somite's core (Nelson and Nusse, 2004). Notch similarly transiently upregulates Eph-ephrin expression and activity at the presumptive somite boundaries. The selective expression of the transcription factor Mesp2, activated by RA (Moreno and Kintner, 2004), in the anterior half of each presumptive somite, defines anterior and posterior compartments by inhibiting DLL expression in the anterior compartments (Takahashi *et al.*, 2003). Mesp2 further leads to transient expression of EphA4 in the anterior compartments (Nakajima *et al.*, 2006), which peaks near the presumptive anterior boundary. In the posterior compartments, DLL maintains ephrinB2 expression (De Bellard *et al.*, 2002), peaking near the presumptive boundary. Other protein level changes result from unknown mechanisms. The time lag from determination to segmentation seems primarily to result from transcriptional and translational delays and the time required for EphA4, ephrinB2, N-CAM and N-cadherin to reach functional levels (Kulesa, 2007, private communication). EphrinB2 levels seems to increase more slowly than do the levels of the other molecules. We have included this difference in the timing of molecular level changes in our simulations; however, the exact sequence is not crucial to our results.

IV. From Adhesion-Protein Patterns to Segmentation

Based on experiments that show that cells in the PSM condense into somites by changing their adhesive and migratory properties (Gossler and Hrabe de Angelis, 1998), our key hypothesis is that the primary ways that the cells' internal differentiation states translate into mechanical activity are through:

i) Differential expression and binding of the cell adhesion molecules N-CAM and N-cadherin.
ii) Differential expression and bidirectional activation of EphA4 receptors and ephrinB2 proteins.

Segmentation then results from the spatiotemporal coordination of N-CAM, N-cadherin and Eph-ephrin expression.

A. Segmentation Model

We use a 2D approximation, neglecting DV variations. We also neglect mediolateral (ML) variations in cell properties (which may be quite important in many cases), assume that the intrinsic level of cytoskeletal cell motility is constant in all cells and that cells do not divide or die.

Based on the biological observations we described above, our model assumes that somitogenesis depends predominantly on four molecular species: N-CAM, N-cadherin, EphA4 and ephrinB2. We make a number of simplifying assumptions concerning the behavior and spatiotemporal expression patterns of these molecules. We assume that the distribution of these adhesive and repulsive species is uniform over cell membranes and that the primary effect of peripheral epithelialization is to change the relative cell sizes and adhesions in the somite core and periphery rather than redistributing adhesion molecules over cell surfaces. We assume that binding of EphA4 and ephrinB2 on apposing cells reduces the effective adhesion due to their cell adhesion molecules. However, we do include the variation in levels of molecular expression between the core and periphery of each somite and between the anterior and posterior halves of each somite compartment.

We assume the following temporal sequence of molecular distributions (see Fig. 3). All molecular levels turn on abruptly at the beginning of a segmentation cycle (times $t_0, t_1, t_2, t_3, \ldots$) and remain the same until the end of the simulation. In reality, these levels would change further as the somites matured, but the chief focus of this paper is the initial formation of the somites rather than their later maturation. We do not attempt to model the origin of the somite-cycle timing. Within the posterior PSM ($S_{-2}, S_{-3}, \ldots$), all cells express a uniform background level of N-CAM and N-cadherin, keeping the cells loosely connected. At the beginning of each segmentation cycle, N-cadherin levels in the cells at the core of S_{-1} (the somite posterior to the forming somite) increase substantially, N-cadherin levels in cells at the periphery of S_{-1} decrease, high levels of EphA4 appear in the anterior half of the anterior compartment (S_{-1A})

7. Coordinated Action of N-CAM, N-cadherin, EphA4, and ephrinB2

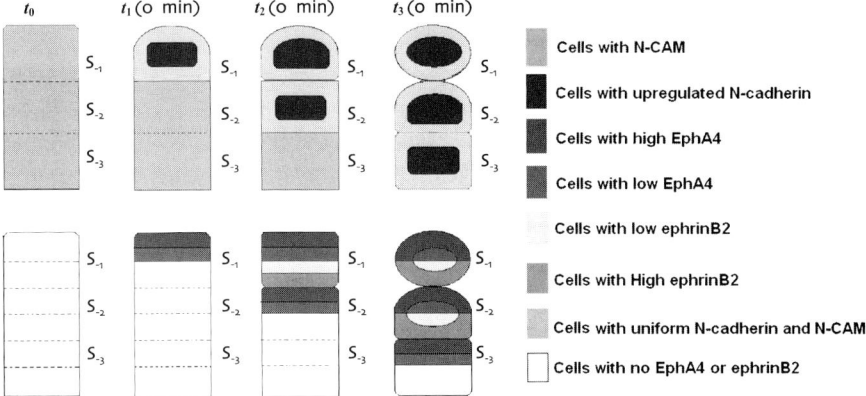

Figure 3 Schematic diagram of the spatiotemporal activation of N-cadherin, N-CAM, EphA4, and ephrinB2 during somite segmentation. (Upper panels) At t_0, all cells have uniform N-cadherin and N-CAM levels. At t_1, N-cadherin levels increase in the core of the anterior domain of the PSM (the new somite S_{-1}). At t_2, when the somitic boundary between S_0 and S_{-1} starts to form, N-cadherin levels increase in the core of somite S_{-1}. (Lower panels) At t_0, no cells have EphA4 or ephrinB2. At t_1 the cells in the anterior compartment of S_{-1} increase their EphA4 level, with a high level in the anterior half of the compartment and a low level in the posterior half. At t_2 the cells in the anterior compartment of the new S_{-1} increase their EphA4 level, with a high level in the anterior half of the compartment and a low level in the posterior half, while the cells in the posterior compartment of S_0 increase their levels of ephrinB2, with a high level in the posterior half of the compartment and a low level in the anterior half. (Both panels) At t_3, an intersomitic boundary forms between S_0 and S_{-1}, and the expression of N-cadherin, EphA4, and ephrinB2 reiterates caudally. See color insert.

and low levels in the posterior half of the anterior compartment (S_{-1A}). To represent the delayed appearance of ephrinB2, we also turn on a low level of ephrinB2 in the anterior half of S_{0P}, and a higher level in the anterior half of S_{0P}.

We assume that cells in the PSM initially have the same size. We represent the epithelialization and compaction of the peripheral cells into *epithelial cells*, by increasing the size of peripheral cells and decreasing the size of core cells slightly when a given compartment begins to express either EphA4 or ephrinB2.

At the following cycle ($t_2, t_3, \ldots$), this entire pattern repeats, shifted posteriorly by one somite length.

V. Computer Simulation of Segmentation

A. The Glazier–Graner–Hogeweg Model

To simulate our somite-segmentation model, we implement it using the *Glazier–Graner–Hogeweg* model (GGH) (also known as the *Extended* or *Cellular Potts Model*) (Glazier and Graner, 1992, 1993). Extensive comparisons between experiments and

GGH simulations have validated GGH methods (Mareé and Hogeweg, 2001; Zeng et al., 2004; Chaturvedi et al., 2005; Poplawski et al., 2007; Merks et al., 2006; Merks and Glazier, 2006) for multicell morphogenesis modeling.

Cells in the GGH are extended domains of pixels on a lattice which share the same *cell index* σ. They can represent either biological cells or *generalized cells* like subregions of extracellular matrix (ECM). We define an *Effective Energy*, which includes the key biological features we wish to simulate (in this case, cell adhesion and cells' volumes and membrane areas):

$$H = \sum_{\substack{\vec{i},\vec{i}' \\ \text{neighbors}}} J\left(\tau(\sigma(\vec{i})), \tau(\sigma(\vec{i}'))\right)\left(1 - \delta(\sigma(\vec{i}), \sigma(\vec{i}'))\right)$$

$$+ \sum_{\sigma} \lambda\left(V(\sigma(\vec{i})) - V_t\left(\tau(\sigma(\vec{i}))\right)\right)$$

$$+ \sum_{\sigma} \beta\left(S(\sigma(\vec{i})) - S_t\left(\tau(\sigma(\vec{i}))\right)\right), \qquad (1)$$

where J is the total adhesion energy per unit surface area between cells of type $\tau(\sigma)$ and $\tau(\sigma')$ (negative for adhesion), V_t the target volume of cells of type τ, V the actual volume of each cell, λ the strength of the volume constraint (the bigger λ the smaller the cells' volume fluctuations), S_t the target membrane area for cells of type τ, S the actual membrane area of each cell, and β the membrane elasticity.

We also define a dynamics which represents cell motility. We pick a lattice site, $\vec{i}$, at random and have one of the cells occupying a neighboring lattice site, $\vec{i}'$, attempt to displace the cell currently occupying that site. If the displacement lowers H, we accept it. Otherwise, we accept the displacement with a probability decaying exponentially in the Effective Energy cost, rescaled by the *Effective Cell Motility* $T_m > 0$:

$$P\left(\sigma(\vec{i}) \to \sigma(\vec{i}')\right) = \left\{\exp(-\Delta H/T_m: \Delta H > 0; \; 1: \Delta H \leq 0)\right\}. \qquad (2)$$

If our lattice has N pixels, we define one *Monte-Carlo Step* (MCS) to be N displacement attempts. We discuss the conversion between MCSs and experimental time below.

B. GGH Somitogenesis Simulation

To translate our model into a GGH simulation, we must define the GGH parameters and their temporal changes for each cell.

Fig. 4 shows the initial condition of our 2D rectangular lattice (170 × 450 pixels). We begin with a segment of PSM, with ECM surrounding the PSM in the anterior and lateral directions. The segment of PSM, which extends to the bottom of the lattice, is long enough to allow more than three somites to form and the ECM layer is thick enough that the cells do not interact with the lattice edges in the region of somite

7. Coordinated Action of N-CAM, N-cadherin, EphA4, and ephrinB2

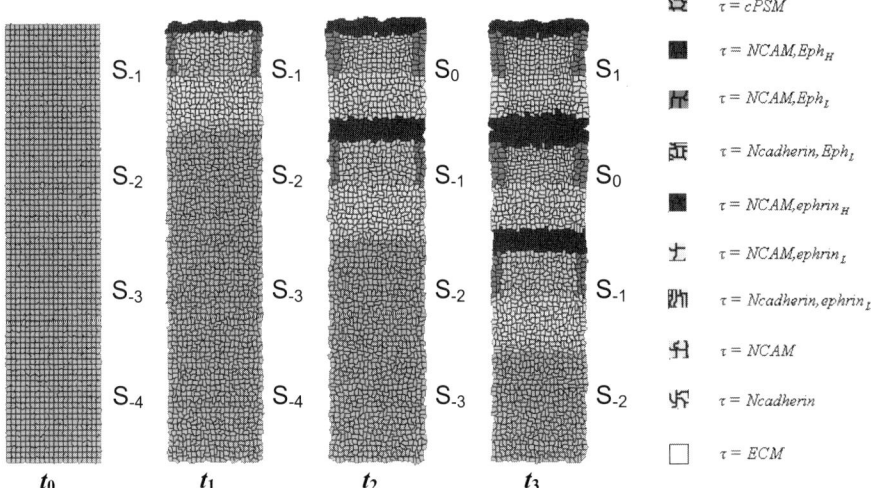

Figure 4 Simulated spatiotemporal levels of N-cadherin, N-CAM, EphA4, and ephrinB2 during somitogenesis. At $t_0 = 0$ MCS, the simulation begins with a regular array of PSM cells expressing background levels of N-CAM and N-cadherin, and surrounded by ECM. At $t_1 = 5000$ MCS, the N-cadherin level increases in the core of S_{-1} and the N-CAM level increases at the periphery. EphA4 levels increase to High in the anterior half of S_{-1A} and to Low in the posterior half, while the peripheral cells in S_{-1A} grow slightly in volume and the core cells shrink slightly in volume. At $t_2 = 7000$ MCS this process repeats in the new S_{-1} and the ephrinB2 level increases to High in the posterior half of S_{-0P} and to Low in the anterior half, while the peripheral cells in S_{-0P} grow slightly in volume and the core cells shrink slightly in volume. At $t_3 = 9000$ MCS, the process repeats for the new S_0 and S_{-1} somites. See color insert.

formation. The long axis of the rectangle corresponds to the AP axis in the embryo and the short axis to the ML axis. Each cell initially occupies 5×5 pixels and each somite contains 20×20 cells, which corresponds to the approximately 400 cells in chick somites (Kulesa and Fraser, 2002).

Fig. 4 shows our 10 cell types:

1. ECM substrate (a single large generalized cell with unconstrained volume), $\tau = ECM$.
2. Mesenchymal cells in the posterior PSM with uniform volumes and low background levels of N-cadherin and N-CAM, $\tau = cPSM$.
3. Epithelial cells in the periphery of the anterior half of the anterior compartments of somites, with increased volumes, high levels of EphA4, background levels of N-CAM and low levels of N-cadherin, $\tau = NCAM, Eph_H$.
4. Epithelial cells in the periphery of the posterior half of the anterior compartments of somites, with increased volumes, low levels of EphA4, background levels of N-CAM and low levels of N-cadherin, $\tau = NCAM, Eph_L$.

5. Mesenchymal cells in the core of the posterior half of the anterior compartments of somites, with decreased volumes, low levels of EphA4, background levels of N-CAM and high levels of N-cadherin, $\tau = Ncadherin, Eph_L$.
6. Epithelial cells in the periphery of the posterior half of the posterior compartments of somites, with increased volumes, high levels of ephrinB2, background levels of N-CAM and low levels of N-cadherin, $\tau = NCAM, ephrin_H$.
7. Epithelial cells in the periphery of the anterior half of the posterior compartments of somites, with increased volumes, low levels of ephrinB2, background levels of N-CAM and low levels of N-cadherin, $\tau = NCAM, ephrin_L$.
8. Mesenchymal cells in the core of the anterior half of the posterior compartments of somites, with decreased volumes, low levels of ephrinB2, background levels of N-CAM and high levels of N-cadherin, $\tau = Ncadherin, ephrin_L$.
9. Mesenchymal cells in the anterior region of the PSM corresponding to the core of somite S_{-1}, with their original volumes, background levels of N-CAM and high levels of N-cadherin, $\tau = Ncadherin$.
10. Mesenchymal cells in the anterior region of the PSM which will correspond to the periphery of somite S_{-1}, with their original volumes, background levels of N-CAM and low levels of N-cadherin, $\tau = NCAM$.

A clock controls the temporal distribution of cell types. All cells except the ECM begin as type *PSM*. At the beginning of the first somite cycle in the simulation, at t_1, cells in the anterior region of the PSM, which we define as beginning as S_{-1A}, change their types from type *cPSM* to type *NCAM, Eph_H, NCAM, Eph_L* or *Ncadherin, Eph_L*, and those in S_{-1P}, change their types from type *cPSM* to type *Ncadherin* or *NCAM*. At the beginning of each succeeding somite cycle, the cells in the new S_{0P} change their types from types *Ncadherin* or *NCAM* to types *NCAM, ephrinH, NCAM, ephrinL* or *Ncadherin, ephrinL*, cells in S_{-1A} change from type *cPSM* to type *NCAM, Eph_H, NCAM, Eph_L* or *Ncadherin, Eph_L*, and cells in S_{-1P}, change from type *cPSM* to type *Ncadherin* or *NCAM*. Fig. 4 shows the spatiotemporal activation of NCAM, N-cadherin, Eph, and ephrin during the simulation, which schematically reproduces the biological pattern in Fig. 3.

C. Simulation Implementation

We implemented our simulations using the open-source software package CompuCell3D, downloadable from (https://simtk.org/home/compucell3d and http://compucell3d.org/), which allows rapid translation of complex biological models into simulations using a combination of CC3D ML and Python scripting. The great advantage of this framework is that it allows compact description of models and hence their publication and validation. We provide our simulation code in Appendices A, B, and C.

7. Coordinated Action of N-CAM, N-cadherin, EphA4, and ephrinB2

Table I Initial and Target Values for Surface Areas and Volumes of Specific Cell Types

τ	λ	V	Biological Value	V_t	Biological Value	β	S	S_t
ECM	0	NA	NA	NA	NA	0	NA	NA
NCAM	20	25	100 µm²	25	100 µm²	20	20	20
Ncadherin	20	25	100 µm²	25	100 µm²	20	20	20
NCAM, Eph_H	20	25	100 µm²	36	144 µm²	20	20	24
NCAM, Eph_L	20	25	100 µm²	36	144 µm²	20	20	24
Ncadherin, Eph_L	20	25	100 µm²	16	64 µm²	20	20	16
NCAM, $ephrin_H$	20	25	100 µm²	36	144 µm²	20	20	24
Ncadherin, $ephrin_L$	20	25	100 µm²	16	64 µm²	20	20	16
NCAM, $ephrin_L$	20	25	100 µm²	36	144 µm²	20	20	24
cPSM	20	25	100 µm²	25	100 µm²	20	20	20

Table II Global Simulation Parameters

	Simulation Value	Biological Value
T_m	100	
t_1	5000 MCS	0 minutes
t_2	7000 MCS	90 minutes
t_3	9000 MCS	180 minutes
t_{total}	15,000 MCS	270 minutes
δ	7.0	14 µm

D. Parameter Values

Our GGH simulation has a substantial number of parameters, most of which are not known quantitatively from experiments. Fortunately, the patterning depends primarily on the hierarchy of the adhesion energies. Since our primary goals in this paper are qualitative reproduction of experiments, we retain a good deal of flexibility in choosing parameter values.

E. Global Parameters

The parameter T_m determines the intrinsic cell motility and rescales all of the other parameters in the Effective Energy. We are therefore free to fix it and vary the scale of our other parameters or *vice versa*. In this case we fix $T_m = 100$ unless we specify otherwise. We then choose the remaining parameters so that $\Delta H / T_m$ in Eq. (2) is neither too large nor too small (typically between about 0.5 and 5). If $\Delta H / T_m$ is very large, the cells will not move at all. If $\Delta H / T_m$ is large, cells will interact with the lattice, producing unnatural shapes, while, if $\Delta H / T_m$ is too small, cells may fall apart. In

Table III Initial Adhesion Energies (J) for Mesenchymal and Epithelial Cell Sorting during Somitogenesis (m indicates a mesenchymal cell type, e an epithelial cell type)

	τ	ECM	NCAM	Ncadherin	NCAM, Eph_H	NCAM, Eph_L	Ncadherin, Eph_L	NCAM, $ephrin_H$	Ncadherin, $ephrin_L$	NCAM, $ephrin_L$	cPSM
		m	m	e	e	e	e	e	e	e	m
ECM	0										
NCAM	15	15	−20.25	−24							
Ncadherin	15	−38.44	−24	−20.25							
NCAM, Eph_H	15	−20.25	−24	−38.44	15						
NCAM, Eph_L	15	−20.25	−20.25	−24	−24	15					
Ncadherin, Eph_L	15	−38.44	−38.44	−20.25	−20.25	−24	15				
NCAM, $ephrin_H$	15	−20.25	−24	−24	−20.25	−20.25	−38.44	15			
Ncadherin, $ephrin_L$	15	−38.44	−20.25	−38.44	−20.25	−20.25	−24	−38.44	15		
NCAM, $ephrin_L$	15	−20.25	−24	−24	−20.25	−20.25	−24	−20.25	−24	15	
cPSM	15	−20.25	−20.25	−38.44	−20.25	−20.25	−38.44	−20.25	−38.44	−20.25	−20.25

principle, we can calculate T_m from experiments measuring the diffusion constants of cells in the tissue, but these measurements are not yet available for somites. We chose the values of $\beta = 20$ and $\lambda = 20$ for the cell-volume and surface-area constraints to keep surface and volume fluctuations relatively small, without causing cells to stop moving. We present the parameters in Tables I, II, III.

F. Cell-Adhesion Energies

We have two types of adhesion in the simulation: cell–ECM adhesion and cell–cell adhesion.

G. Adhesion Hierarchies

The biology of the somite largely determines our adhesion hierarchy (where we have no information at all, we will set parameters equal to a default value).

In general, we will assume that adhesion between cells with higher levels of N-CAM or N-cadherin is stronger and that both N-CAM and N-cadherin are homotypically adhesive and that N-cadherin is more cohesive than N-CAM:

For our two classes of *epithelial* (*e*) and *mesenchymal* (*m*) cells surrounded by ECM, we need an initially-random mixture of the two cell types to sort stably, with the epithelial cells at the surface and mesenchymal cells condensing in the core, which requires the following relations among the adhesion energies (Glazier and Graner, 1993):

$$J(m,m) < \bigl(J(m,m) + J(e,e)\bigr)/2 < J(m,e) < J(e,e). \tag{3}$$

To prevent cells disassociating into the ECM, rather than sticking together, requires that (Glazier and Graner, 1993):

$$J(e,e) < J(e, ECM) \leqslant J(m, ECM). \tag{4}$$

To prevent cells from dispersing into the ECM, we set $J(cell, ECM) = 15$ initially.

The interaction energy per unit contact length between cells includes both adhesion and *effective repulsion*. Binding between EphA4 and ephrinB2 on apposing cells reduces the effective adhesion (increases J) by an *effective repulsion energy*, J_r, compared to the adhesion energy, J_a, which we would predict based on the number and type of cell-adhesion molecules. The decrease in adhesion need not scale linearly with the number of bound Eph/ephrin pairs. The simplest approximation to such an effect is to write the *net adhesion*, J, as a sum of the adhesion and repulsion alone plus a bilinear perturbation of strength c:

$$J(\tau_1, \tau_2) = J_a(\tau_1, \tau_2) + J_r(\tau_1, \tau_2) + c J_a(\tau_1, \tau_2) J_r(\tau_1, \tau_2). \tag{5}$$

When c is positive, the interaction weakens adhesion compared to the additive case. When c is negative, the interaction strengthens adhesion compared to the additive case.

Additionally, cells expressing EphA4 should not mix with cells expressing ephrinB2. In this case, the energy for cell–cell interaction with repulsion must be larger than that for cell–cell interaction with adhesion:

$$J(Eph, cadherin, ephrin, cadherin)$$
$$> \bigl(J(Eph, cadherin, ECM) + J(ephrin, cadherin, ECM)\bigr)/2$$
$$= \bigl(J(cadherin, ECM) + J(cadherin, ECM)\bigr)/2. \tag{6}$$

Since domains of EphA4 and ephrinB2 cells separate but remain compact as the ECM furrow forms between them, the ECM must engulf both domains:

$$J(Eph, cadherin, Eph, cadherin) < J(Eph, caderin, ECM),$$
$$J(ephrin, cadherin, ephrin, cadherin) < J(ephrin, cadherin, ECM). \tag{7}$$

H. Target Areas and Volumes

In 2D, the ratio of a cell's target membrane area squared to its target volume determines how "floppy" the cell is. If the ratio is small, cells will be round and stiff; in the opposite limit, cells will be floppy and extended, like an uninflated beach ball.

I. Time

The conversion between MCS and experimental time will depend on the average value of $\Delta H / T_m$, and hence on the choices of parameters. We make this assignment empirically, by observing the time in MCS that somite reorganization takes after we switch on the pattern of N-CAM, N-cadherin, EphA4, and ephrinB2 and set it to the corresponding experimental time. For the values of parameters we have chosen, 2000 MCS = 90 minutes (or 1 MCS = 2.7 seconds).

Because our initial configuration uses nonbiological, rectangular cells, we set $t_0 = 5000$ MCS to allow the PSM to relax before we turn on the initial pattern of N-CAM and N-cadherin. Since the somitic clock interval is 2000 MCS, $t_1 = 7000$ MCS, $t_2 = 9000$ MCS, and so on.

VI. Results and Discussion

A. Parameter Choices

All viable developmental mechanisms must be relatively insensitive to fluctuations in key parameters. Biochemical redundancy is one approach to such robustness. Our

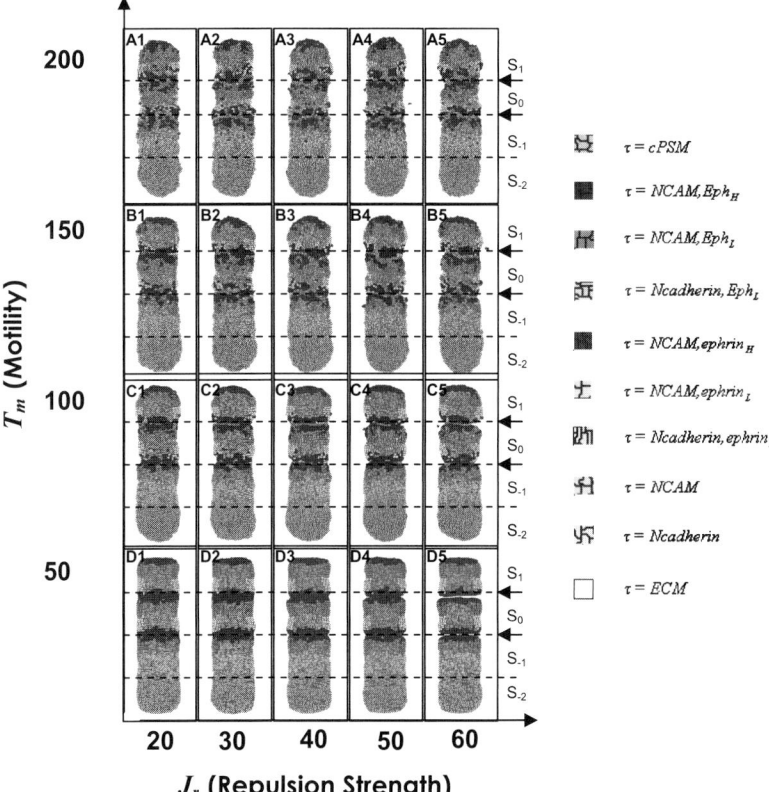

Figure 5 Morphology of somites for different cell motilities and EphA4/ephrinB2 repulsion strengths. Arrows (→) mark intersomitic boundaries. (A1–A5), (B1–B5). For very high motilities, coherent somites do not form. (C1–C3), (D1–D2). Weak repulsion does not produce sharp intersomitic boundaries. (C5, D5). Segmentation with somite separation requires strong repulsion and limited cell motility. All simulations are shown after 15,000 MCS. See color insert.

model lacks such redundancy. However, even reasonable agreement with experiment would be unsatisfactory if it required very tight parameter tuning. Instead, we expect that a viable model will have a relatively broad range over which parameters have relatively little effect on segmentation. We therefore conducted parameter sweeps to explore parameter dependencies and optimize our choices.

Fig. 5 shows the long-time morphologies ($t = 15,000$ MCS) for cell motility T_m between 50 to 200 and repulsion energy J_r between 20 to 60. For very high cell motility $T_m \geqslant 150$ (Fig. 5A1–A5), coherent somites fail to form. Similarly, if the effect of EphA4/ephrinB2 binding is too weak ($J_r < 50$), the simulation does not form a clear intersomitic boundary. Only for strong repulsion ($J_r = 60$) and low motility ($T_m \leqslant 100$) does the somitic furrow form (Figs. 5C5 and 5D5). Thus, the sensitiv-

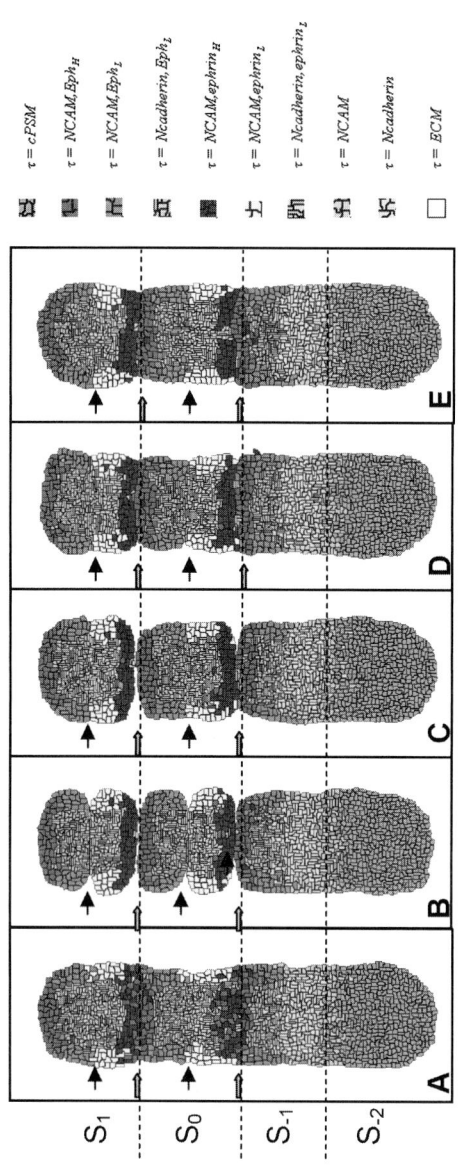

Figure 6 Effects of adhesion/repulsion coordination on somite morphology. Heavy arrows (⇒) mark intersomitic boundaries. Light arrows (→) mark intrasomitic compartment boundaries. (A) Without EphA4 and ephrinB2, cells mix between the posterior of a somite and the anterior of the following somite. (B) Linear interaction of adhesion and repulsion with uniform repulsion results in both intersomitic and intrasomitic separation. (C) Linear interaction of adhesion and repulsion with graded repulsion results in intersomitic separation without intrasomitic separation. (D) Weak nonlinear interaction of adhesion and repulsion with uniform repulsion results in both intersomitic and intrasomitic segmentation. (E) Nonlinear interaction of adhesion and repulsion with graded repulsion results in intersomitic segmentation without intrasomitic segmentation. All simulations are shown after 15,000 MCS. See color insert.

ity to motility is fairly low (we can vary T_m by a factor of 2), but the sensitivity to variations in repulsion strength requires further study.

B. Segmentation Requires EphA4/ephrinB2 Repulsion

Fig. 6A shows the long-time configuration ($t = 15{,}000$ MCS) of a simulation in the absence of EphA4 and ephrinB2. The epithelial cells from different somites and mesenchymal cells from the anterior and posterior domains of each somite mix. Segmentation fails, as is observed in experiments in zebrafish which lack EphA4 (Barrios et al., 2003).

C. Segmentation Requires Multiple Levels of EphA4/ephrinB2 Expression

We now explore the effect of EphA4/ephrinB2 repulsion on segmentation.

Our model includes two nonzero levels of EphA4 and ephrinB2: *High* and *Low*. We define J_{rHH} to be the repulsion energy between a cell with a High level of EphA4 and a cell with a High level of ephrinB2, J_{rHL} to be the repulsion energy between a cell with a High level of EphA4 and a cell with a Low level of ephrinB2 or between a cell with a Low level of EphA4 and a cell with a High level of ephrinB2, and J_{rLL} to be the repulsion energy between a cell with a Low level of EphA4 and a cell with a Low level of ephrinB2.

First, we test whether correct segmentation requires two levels of EphA4 and ephrinB2 expression within each somite compartment. We assume an additive relation between adhesion and repulsion ($c = 0$ in Eq. (5)) and that levels of EphA4 and ephrinB2 are uniform, $J_{rHH} = J_{rLL} = J_{rHL} = 60$. Fig. 6B shows that an intersomitic furrow (⇒) forms correctly. However an obvious intrasomitic notch (→) develops between the anterior and posterior somite compartments, which does not occur in normal somite segmentation. Therefore, correct segmentation appears to require multiple levels of EphA4 and ephrinB2.

Next, we assume that the effect of the higher level of expression of EphA4 and ephrinB2 in the somite and PSM periphery results in effective repulsion energies of $J_{rHH} = J_{rHL} = 3 J_{rLL} = 60$, still assuming an additive relation between adhesion and repulsion ($c = 0$ in Eq. (5)). Fig. 6C shows that an intersomitic furrow (⇒) forms correctly. A very small intrasomitic notch (→) develops between the anterior and posterior somite compartments, as is observed experimentally. The resulting morphology is very close to normal somite segmentation.

Now we examine the effect of the bilinear term on the final pattern morphology. We assume a cooperative relation between adhesion and repulsion ($c = 1/50$ in Eq. (5)) and assume that High and Low levels of EphA4 and ephrinB2 are equal: $J_{rHH} = J_{rLL} = J_{rHL} = 60$. Even this small perturbation greatly changes the morphology from Fig. 6B. Fig. 6D shows that an intersomitic notch forms (⇒) but the somites do not separate. Somewhat surprisingly, the size of the intrasomitic notch (→) is much smaller. Again, correct somite formation seems to require multiple levels of EphA4 and ephrinB2.

Finally, we assume a cooperative relation between adhesion and repulsion ($c = 1/50$ in Eq. (5)) and that the higher level of expression of EphA4 and ephrinB2 in the somite and PSM periphery results in effective repulsion energies of $J_{rHH} = J_{rHL} = 3J_{rLL} = 60$, as in Fig. 6C. Fig. 6E shows only a slight intersomitic notch ($\Rightarrow$) and no intrasomitic notch ($\rightarrow$). As in Figs. 6A and 6D, the somites do not separate.

Thus we will use additive repulsion ($c = 0$) at two levels, with $J_{rHH} = J_{rHL} = 3J_{rLL} = 60$ for our remaining simulations if we do not specify otherwise.

The Takeichi group (Horikawa et al., 1999) has observed the separation of the anterior and posterior somite compartments during somite formation in N-cadherin ($-/-$) mouse embryos. We simulated this experiment by setting the adhesion energy for cell types expressing N-cadherin equal to the adhesion energy for the corresponding cell types expressing N-CAM, effectively removing N-cadherin from our simulation. We kept the repulsion at the same levels as the values used for Fig. 6B. We observed both intersomitic ($\Rightarrow$) and intrasomitic ($\rightarrow$) furrows as seen in Fig. 7D. Thus N-cadherins seem essential to keeping the two compartments of a somite fused during segmentation.

D. Dynamic Morphological Changes and Error Correction during Segmentation

Time-lapse movies of shape changes during segmentation show intriguing effects (Kulesa and Fraser, 2002) (Figs. 8A–8F). In particular, the boundary between S_0 and S_{-1} forms a characteristic *ball-and-socket* or *W* shape (Fig. 8C), with the groove between S_0 and S_{-1} first opening up at the dips of the W, followed by the arms of the W retracting and sometimes folding inwards. Initially, the PSM envelops the forming somite, forming a *sleeve*. During segmentation, the sleeve cells fold back into the PSM along the S_0–S_{-1} boundary and the somite eventually rounds up.

One way to reproduce this dynamics would be for the PSM to cohere more to ECM than other cell types, i.e., $J(cPSM, ECM) < J(\ldots, ECM)$, in which case we expect that PSM cells will partially envelop the S_0 somite. A possible mechanism for such behavior would be a reduction in membrane levels of integrins in the posterior compartments of S_{-1} and later (S_0, S_1, ...) somites.

In Figs. 8G–8L, black arrows ($\rightarrow$) mark intersomitic boundaries (corresponding to the white arrows in Figs. 8A–8B). During the clock cycle before segmentation proper, and thus before the initiation of ephrinB2 expression in S_{-1} (Figs. 8G–8H), as in the experiment, we observe that anterior cells at the sides of S_{-2} move anteriorly and wrap around the posterior end of S_{-1}. During the following clock cycle (Figs. 8I–8J), the appearance of ephrinB2 in S_{0P} (formerly S_{-1P}) causes the formation of a W-shaped intersomitic furrow between S_0 and S_{-1}, as in Figs. 8C–8D. During the following cycle, the wings of the W retract and the somite rounds up (Figs. 8K–8L and 8E–8F).

7. Coordinated Action of N-CAM, N-cadherin, EphA4, and ephrinB2 227

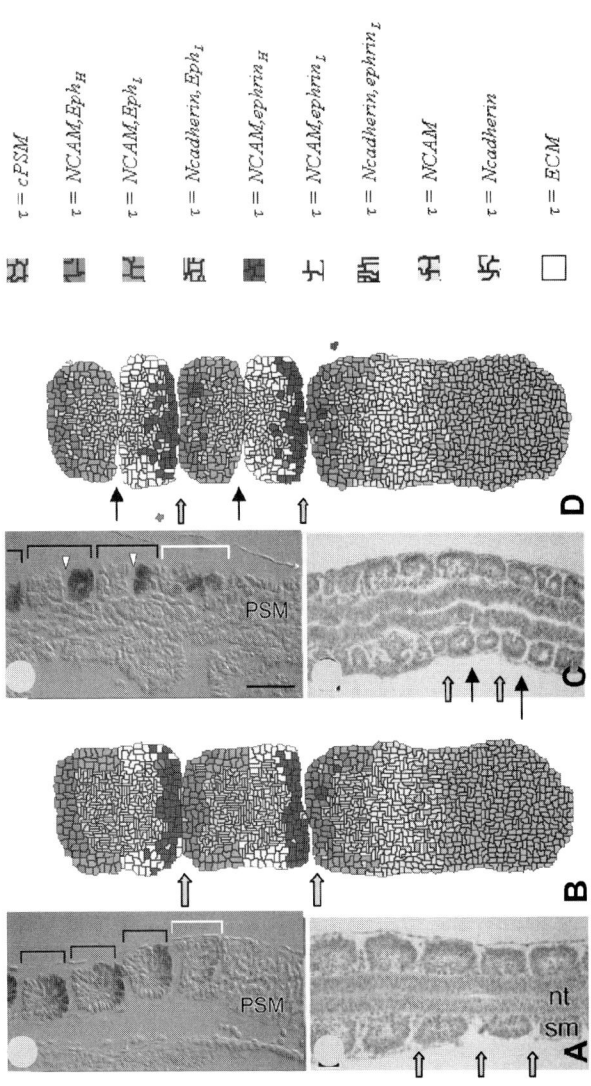

Figure 7 Comparison of somite structures in wild-type and N-cadherin-knockout experiments and simulations. Heavy arrows (→) mark intersomitic boundaries. Light arrows (→) mark intrasomitic compartment boundaries. (A) Experimental wild phenotype (reprinted from Horikawa *et al.*, 1999, with permission from Elsevier). (B) Simulated wild phenotype after 15,000 MCS. (C) Experimental N-cadherin-double-knockout phenotype (reprinted from Horikawa *et al.*, 1999, with permission from Elsevier). The somites separate into *Uncx4.1*-positive and *Uncx4.1*-negative regions. *Uncx4.1* is a specific marker for the posterior compartments of somites. (D) Simulated N-cadherin-double-knockout phenotype at 10,000 MCS. Both somites and somite compartments separate. See color insert.

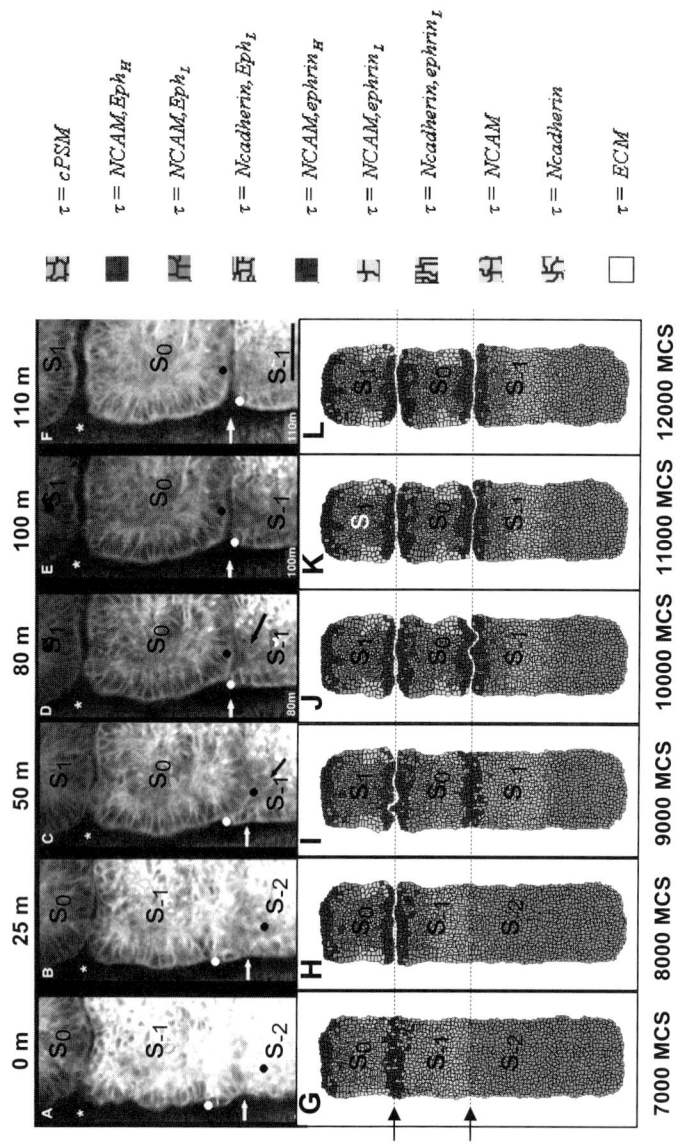

Figure 8 Simulated and experimental somite-segmentation dynamics. (A–F) Confocal time-lapse images of vitally-stained tissue (from Kulesa and Fraser, 2002; reprinted with permission from AAAS). White arrows indicate intersomitic boundaries. The white and black dots label cells which cross the presumptive boundary. (G–L) Simulated somite segmentation with preferential adhesion between PSM and ECM and AP indeterminacy in somite differentiation. Black arrows (→) mark intersomitic boundaries. Scale bar in experimental image F is 50 μm and $J_{rHH} = J_{rHL} = 3 J_{rLL} = 80$. See color insert.

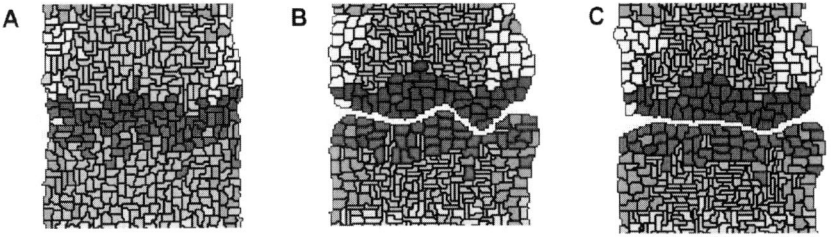

Figure 9 Detail of boundary crossing of misdifferentiated cells from Fig. 8. (A) 9000 MCS. (B) 10,000 MCS. (C) 11,000 MCS. See color insert.

The details of the retraction differ somewhat from some experiments, with the wings of the W retracting but not folding inwards, as is seen in other experiments.

Because biological signaling is noisy, some cells will slightly misread the somitic clock during determination. As a result they will differentiate inappropriately for their location (*misdifferentiate*). Such misdifferentiation occurs primarily near compartment boundaries, where cells are near a biological threshold between differentiation states and a small amount of noise can throw the switch. Initial fuzzy compartment boundaries followed by boundary refinement are common during development. Kulesa and Fraser may have observed such effects in their movies, where cells initially on one side of the presumptive intersomitic boundary cross the boundary during segmentation (Kulesa and Fraser, 2002). The white and black dots in Figs. 8A–8F denote cells that are not simply carried along as the W extends and retracts, but actually cross from S_{-2} to S_{-1} and from S_{-1} to S_{-2}, respectively.

Our simulations show that adhesion-based cell sorting provides a viable mechanism for the correction of minor errors in differentiation. In our simulation in Figs. 8G–8L, we have allowed for misdifferentiation by having each cell read its AP position with a small error chosen from a Gaussian distribution with a width δ of 14 microns when it selects its type, making the initial compartment boundaries blurry. Thus, some cells in S_{0P} assume types (red) that should be in S_{-1A}, etc., ... If we follow these red cells during their subsequent time evolution (Figs. 9A–9C), we find that adhesion-driven cell sorting causes most (though not all) of them to cross the compartment boundary into the correct compartment, in a manner identical to that in the experiments. The greater the cell motility compared to the rate of furrow formation, the more complete the correction will be.

VII. Conclusion

In this paper, we have shown that multilevel homotypic adhesion and heterotypic repulsion can reproduce many of the phenomena of normal somite segmentation, including the ball-and-socket dynamics and compartment crossing by misdifferentiated cells. Simulated N-cadherin knockouts produce an intrasomitic furrow as observed in experiments.

Acknowledgments

We acknowledge support from Grants NIGMS 1R01 GM076692-01, NSF IBN-0083653 and NASA NAG 2-1619, and from the Biocomplexity Institute, the Office of the Vice President for Research and the College of Arts and Sciences at Indiana University Bloomington. We thank Dr. A. W. Neff of the Indiana University School of Medicine for his critical reading of the manuscript, and Dr. O. Pourquié and his group members and Dr. P. Kulesa, all at the Stowers Institute for Medical Research, for helpful discussions. Y. Z. thanks the Department of Physics of Indiana University Bloomington for financial support.

Introduction to Appendices

To facilitate the validation, extension and reuse of complex multicell models, the interagency IMAG Consortium, which includes the NIGMS which primarily funded this research, has requested that all publications include source listings and related materials sufficient to allow a reader to duplicate and verify the primary computational claims made. In support of this goal, we provide in the following appendices, documented code which should suffice to replicate our results.

Running a GGH simulation using CompuCell3D requires initial installation of the CompuCell3D package (from www.compucell3d.org), followed by loading the appropriate CC3D ML and/or Python scripts into the CompuCell3D Player. The somitogenesis simulations we present in this paper employ both CC3D ML and Python scripts. The CC3D ML file contains the basic description of GGH parameters, such as lattice size, simulation duration in Monte Carlo Steps, Intrinsic Cell Motility (*Temperature*), definitions of adhesion energies between different types of cells, instructions to load Volume and Surface energy terms (the parameters of which Python script will initialize) and modules tracking cells' centers of mass and a Cell Lattice initialization routine (UniformInitializer).

The two Python scripts include one, somite.py that sets up the simulation (instantiating C++ and Python objects and implementing the main GGH loop) and a script that contains implementations of *Steppables* (somiteSteppables.py). Steppables define and manage transitions between cell types (SomiteMaskSteppable) and changes in Volume and Surface parameters (SomiteVolumeSurfaceSteppable).

The listings include brief comments explaining the significance of the code modules and indicating how to change the parameters to implement the different simulations in this paper. For convenience, complete files for implementing the somitogenesis simulations are also available for download from http://compucell3d.org/Models/somiteSimulationFiles.tgz.

Appendix A. Python Code to Execute Somitogenesis Simulations (somite.py)

```python
def mainfcn():
    ### Initialization section
    import sys
    from os import environ
    import string
    sys.path.append(environ["PYTHON_MODULE_PATH"])

    import SystemUtils
    SystemUtils.setSwigPaths()
    SystemUtils.initializeSystemResources()

    import CompuCell
    import PlayerPython
    # This function wraps up the plugin initialization code.
    CompuCell.initializePlugins()

    # Create a Simulator. This section returns a Python object
    # that wraps Simulator.
    sim = CompuCell.Simulator()

    simthread=PlayerPython.getSimthreadBasePtr();
    simthread.setSimulator(sim)
    simulationFileName=simthread.getSimulationFileName()

    # Add the Python-specific extensions.
    reg = sim.getClassRegistry()

    CompuCell.parseConfig(simulationFileName, sim)
    ### End of Initialization section.

    ###############################
    # Registering objects that will allow adding
    # cell attributes during simulation runtime.
    # For more information see CompuCell3D Python Scripting
    # tutorials at www.compucell3d.org
    from CompuCell import PyAttributeAdder
    from PyListAdder import ListAdder
    from sys import getrefcount
    adder=PyAttributeAdder()
    adder.registerRefChecker(getrefcount)
    listAdder=ListAdder()
    adder.registerAdder(listAdder)
    potts=sim.getPotts()
    potts.registerAttributeAdder(adder.getPyAttributeAdderPtr())
```

```
### Further initialization of Player and CompuCell3D C++
# code.
dim=sim.getPotts().getCellFieldG().getDim()
# After all CC3D ML steppables and plugins have been loaded
# we call extraInit to complete the initialization.
sim.extraInit()
simthread.preStartInit()
sim.start()
simthread.postStartInit()
screenUpdateFrequency=simthread.getScreenUpdateFrequency()

### Instantiating and registering Python-based plugins.
# Notice that all Python steppables are registered
# with steppableRegistry,
# e.g., steppableRegistry.registerSteppable
# (somiteMaskSteppable) from PySteppables import
# SteppableRegistry

steppableRegistry=SteppableRegistry()

from somiteSteppables import SomiteMaskSteppable
somiteMaskSteppable=SomiteMaskSteppable
(_simulator=sim,_frequency=10)
# Sweep as [T0, T0+T, T0+2T],
# T0=2000, 3000, 400, 5000,6000;
#  T=1000,2000, 3000, 4000, 5000;
# Here you specify the MC steps at which transitions
# take place.
transitionStepsList=[5000,7000,9000]
somiteMaskSteppable.setTransitionStepsList
(transitionStepsList)
somiteMaskSteppable.blur(0.0)
steppableRegistry.registerSteppable(somiteMaskSteppable)

from somiteSteppables import SomiteVolumeSurfaceSteppable
somiteVolumeSurfaceSteppable=SomiteVolumeSurfaceSteppable(\
_simulator=sim,_frequency=100)
# Here you specify the max MCS at which changes take place -
# this way you do not execute code in this steppable after
# all transitions have taken place.
somiteVolumeSurfaceSteppable.setMaxTransitionStep(\
somiteMaskSteppable.getMaxTransitionStep())
steppableRegistry.registerSteppable
(somiteVolumeSurfaceSteppable)

steppableRegistry.init(sim)
```

7. Coordinated Action of N-CAM, N-cadherin, EphA4, and ephrinB2

```
    steppableRegistry.start()

    ### Main GGH algorithm loop.
    for i in range(sim.getNumSteps()):
        sim.step(i)
        steppableRegistry.step(i)
        if not i % screenUpdateFrequency:
            simthread.loopWork(i)
            simthread.loopWorkPostEvent(i)
    sim.finish()
    steppableRegistry.finish()
mainfcn()
```

Appendix B. CC3D ML Code to Execute Somitogenesis Simulations (somite.xml)

```xml
--<CompuCell3D>
 <!-- This section defines the basic parameters of the
   GGH model.-->
 <Potts>
    <Dimensions x="170" y="1" z="450"/>
    <Anneal>1</Anneal>
    <Steps>100</Steps>
    <Temperature>100</Temperature>
    <Flip2DimRatio>1</Flip2DimRatio>
 </Potts>

<!-- Each CompuCell3D CC3D ML file must include a list
of all cell types used in the simulation.-->
<Plugin Name="CellType">
    <CellType TypeName="Medium" TypeId="0"/>
    <CellType TypeName="ncam" TypeId="1"/>
    <CellType TypeName="ncad" TypeId="2" />
    <CellType TypeName="eph-rec" TypeId="3"/>
    <CellType TypeName="eph-rec-ncam" TypeId="4" />
    <CellType TypeName="eph-rec-ncad" TypeId="5" />
    <CellType TypeName="eph-lig-ncam" TypeId="6" />
    <CellType TypeName="eph-lig-ncad" TypeId="7" />
    <CellType TypeName="eph-lig" TypeId="8" />
    <CellType TypeName="psm" TypeId="9" />
</Plugin>

<!-- This plugin tells the player which lattice
projection should be displayed at the start.-->
<Plugin Name="PlayerSettings">
    <Project2D XZProj="1"/>
```

```xml
    <InitialProjection Projection="xz"/>
</Plugin>

<!-- Additional initialization of volume and surface target
values and lambdas is required. This is done in the
Python scripts.-->
<Plugin Name="VolumeLocalFlex"/>
<Plugin Name="SurfaceLocalFlex"/>

<!-- List of contact energies between different cell types.-->
<Plugin Name="Contact">
    <Energy Type1="Medium" Type2="Medium">0</Energy>
    <Energy Type1="Medium" Type2="ncam">15</Energy>
    <Energy Type1="Medium" Type2="ncad">15</Energy>
    <Energy Type1="Medium" Type2="eph-rec">15</Energy>
    <Energy Type1="Medium" Type2="eph-rec-ncam">15</Energy>
    <Energy Type1="Medium" Type2="eph-rec-ncad">15</Energy>
    <Energy Type1="Medium" Type2="eph-lig-ncam">15</Energy>
    <Energy Type1="Medium" Type2="eph-lig-ncad">15</Energy>
    <Energy Type1="Medium" Type2="eph-lig">15</Energy>
    <Energy Type1="Medium" Type2="psm">15</Energy>
    <Energy Type1="ncam" Type2="ncam">-20.25</Energy>
    <Energy Type1="ncam" Type2="ncad">-24</Energy>
    <Energy Type1="ncam" Type2="eph-rec">-20.25</Energy>
    <Energy Type1="ncam" Type2="eph-rec-ncam">-20.25</Energy>
    <Energy Type1="ncam" Type2="eph-rec-ncad">-24</Energy>
    <Energy Type1="ncam" Type2="eph-lig-ncam">-20.25
    </Energy>
    <Energy Type1="ncam" Type2="eph-lig-ncad">-24</Energy>
    <Energy Type1="ncam" Type2="eph-lig">-20.25</Energy>
    <Energy Type1="ncam" Type2="psm">-20.25</Energy>
    <Energy Type1="ncad" Type2="ncad">-38.44</Energy>
    <Energy Type1="ncad" Type2="eph-rec">-24</Energy>
    <Energy Type1="ncad" Type2="eph-rec-ncam">-24</Energy>
    <Energy Type1="ncad" Type2="eph-rec-ncad">-38.44</Energy>
    <Energy Type1="ncad" Type2="eph-lig-ncam">-24</Energy>
    <Energy Type1="ncad" Type2="eph-lig-ncad">-38.44</Energy>
    <Energy Type1="ncad" Type2="eph-lig">-24</Energy>
    <Energy Type1="ncad" Type2="psm">-38.44</Energy>
    <Energy Type1="eph-rec" Type2="eph-rec">-20.25</Energy>
    <Energy Type1="eph-rec" Type2="eph-rec-ncam">-20.25
    </Energy>
    <Energy Type1="eph-rec" Type2="eph-rec-ncad">-24</Energy>
    <Energy Type1="eph-rec" Type2="eph-lig-ncam">-20.25</Energy>
    <Energy Type1="eph-rec" Type2="eph-lig-ncad">-24.0</Energy>
    <Energy Type1="eph-rec" Type2="eph-lig">-20.25</Energy>
    <Energy Type1="eph-rec" Type2="psm">-20.25</Energy>
```

7. Coordinated Action of N-CAM, N-cadherin, EphA4, and ephrinB2

```xml
        <Energy Type1="eph-rec-ncam" Type2="eph-rec-ncam">-20.25
        </Energy>
        <Energy Type1="eph-rec-ncam" Type2="eph-rec-ncad">-24
        </Energy>
        <Energy Type1="eph-rec-ncam" Type2="eph-lig-ncam">-20.25
        </Energy>
        <Energy Type1="eph-rec-ncam" Type2="eph-lig-ncad">-24.0
        </Energy>
        <Energy Type1="eph-rec-ncam" Type2="eph-lig">-20.25
        </Energy>
        <Energy Type1="eph-rec-ncam" Type2="psm">-20.25</Energy>
        <Energy Type1="eph-rec-ncad" Type2="eph-rec-ncad">-38.44
        </Energy>
        <Energy Type1="eph-rec-ncad" Type2="eph-lig-ncam">-24
        </Energy>
        <Energy Type1="eph-rec-ncad" Type2="eph-lig-ncad">-38.44
        </Energy>
        <Energy Type1="eph-rec-ncad" Type2="eph-lig">-24.0</Energy>
        <Energy Type1="eph-rec-ncad" Type2="psm">-38.44</Energy>
        <Energy Type1="eph-lig-ncam" Type2="eph-lig-ncam">-20.25
        </Energy>
        <Energy Type1="eph-lig-ncam" Type2="eph-lig-ncad">-24
        </Energy>
        <Energy Type1="eph-lig-ncam" Type2="eph-lig">-20.25
        </Energy>
        <Energy Type1="eph-lig-ncam" Type2="psm">-20.25</Energy>
        <Energy Type1="eph-lig-ncad" Type2="eph-lig-ncad">-38.44
        </Energy>
        <Energy Type1="eph-lig-ncad" Type2="eph-lig">-24</Energy>
        <Energy Type1="eph-lig-ncad" Type2="psm">-38.44</Energy>
        <Energy Type1="eph-lig" Type2="eph-lig">-20.25</Energy>
        <Energy Type1="eph-lig" Type2="psm">-20.25</Energy>
        <Energy Type1="psm" Type2="psm">-20.25</Energy>
        <Depth>2.3</Depth>
</Plugin>

<!-- This plugin tracks the center of mass of each cell
and is necessary for laying out the prepattern of cadherin
expression.-->
<Plugin Name="CenterOfMass"/>

<!-- UniformInitializer lays out the initial pattern of cells.
In our case, a rectangular block corresponding to the
presomitic mesoderm (psm).-->
<Steppable Type="UniformInitializer">
    <Region>
        <BoxMin x="35" y="0" z="30"/>
```

```
            <BoxMax x="135" y="1" z="430"/>
            <Gap>0</Gap>
            <Width>5</Width>
        <Types>psm</Types>
    </Region>
</Steppable>
</CompuCell3D>
```

Appendix C. Python Steppables for Somitogenesis Simulations (somiteSteppables.py)

```
from PySteppables import *
import CompuCell
import sys
import CompuCell
import random

# The function forEachCellInInventory takes as arguments
# the inventory of cells and a function that will operate on a
# single cell. It runs singleCellOperation on each cell from
# the cell inventory.
def forEachCellInInventory(inventory,singleCellOperation):
    invItr=CompuCell.STLPyIteratorCINV()
    invItr.initialize(inventory.getContainer())
    invItr.setToBegin()
    cell=invItr.getCurrentRef()
    while (not invItr.isEnd()):
        cell=invItr.getCurrentRef()
        singleCellOperation(cell)
        invItr.next()

# TypeTransition is a class that describes a
# cell-type transition (self.typeIdSource, self.typeIdTarget)
# and at what time (in MCS) the transition should take place.
class TypeTransition:
    def __init__(self,_typeIdSource,_typeIdTarget,_step):
        self.typeIdSource=_typeIdSource
        self.typeIdTarget=_typeIdTarget
        self.step=_step

# The SomiteMaskSteppable defines a set of masks that describe
# cell types. Overlaying a mask is equivalent to defining new
# type transitions for a given cell. Masks are overlaid before
# simulation begins, which is why we can use fixed mask
# coordinates (e.g., self.x\_rect\_low=45). As discussed
# in the paper, the development of the prepattern of cadherin
```

7. Coordinated Action of N-CAM, N-cadherin, EphA4, and ephrinB2   237

```python
# expression can be thought of as a series of cell-type
# transitions.
class SomiteMaskSteppable(SteppablePy):
    def __init__(self,_simulator,_frequency=1):
        SteppablePy.__init__(self,_frequency)
        self.simulator=_simulator
        self.cellFieldG=self.simulator.getPotts().getCellFieldG()
        self.dim=self.cellFieldG.getDim()
        self.inventory=self.simulator.getPotts().
        getCellInventory()
        self.typeNameTable={"Medium":0,"AL":1,"RH":2,"3":3,"4":4,
        "5":5,\
        "6":6,"7":7,"8":8,"PSM"':9}
        
        #base mask parameters
        self.z_base_low=30
        self.z_base_size=100
        #mask rectangle parameters
        self.x_rect_low=45
        self.z_rect_low=40
        self.x_rect_size=80
        self.z_rect_size=80
        #mask0 parameters
        self.z_low_mask0=30
        self.z_size_mask0=10
        #mask1 parameters
        self.z_low_mask1=40
        self.z_size_mask1=40
        #mask2 parameters
        self.z_low_mask2=80
        self.z_size_mask2=40
        #mask3 parameters
        self.z_low_mask3=120
        self.z_size_mask3=10
        # The above determine the shifts in the masks'
          z_low positions -
        # They should be applied to all masks in the correct
        # order.
        self.shift=100
        self.maskMargin=10
        self.stepForTransition=0
        self.transitionStepsList=[]
        self.maxTransitionStep=0
        self.sigmaBlur=0.0
    
    def getMaxTransitionStep(self):
        return self.maxTransitionStep
```

```python
def setTransitionStepsList(self,_transitionStepsList):
    self.transitionStepsList=_transitionStepsList
    self.maxTransitionStep=max(self.transitionStepsList)

# Whenever we overlay a mask, we attach to each cell a
# transition from its current type to a target type.
# Later, when we make the transition, we simply read from
# the attached list of transitions and perform those for
# which the current time (in MCS) matches the times defined
# in the transition-description list.

# The functions below implement a series of masks that are
# used to construct the prepattern.
def maskBase(self,cell,xCM,yCM,zCM,attrib):
    if (zCM >= self.z_base_low and zCM<self.z_base_low
        +self.z_base_size):
      sourceId=cell.type
      cell.type=self.typeNameTable["AL"]
      # To record a type transition.
      attrib.append(TypeTransition(sourceId,cell.type,
                    self.stepForTransition))

def maskRectangle(self,cell,xCM,yCM,zCM,attrib):
    if(xCM>=self.x_rect_low and xCM <self.x_rect_low
        +self.x_rect_size\
    and zCM >= self.z_rect_low and zCM<self.z_rect_low
        +self.z_rect_size and\
    cell.type==self.typeNameTable["AL"]):
      sourceId=cell.type
      cell.type=self.typeNameTable["RH"]
      attrib.append(TypeTransition(sourceId,cell.type,
      self.stepForTransition))

def mask0(self,cell,xCM,yCM,zCM,attrib):
    if(zCM>=self.z_low_mask0 and zCM <self.z_low_mask0+
        self.z_size_mask0):
      if(cell.type==self.typeNameTable["AL"]):
        sourceId=cell.type
        cell.type=self.typeNameTable["3"]
        attrib.append(TypeTransition(sourceId,cell.type,\
        self.stepForTransition))

def mask1(self,cell,xCM,yCM,zCM,attrib):
    if(zCM>=self.z_low_mask1 and zCM <self.z_low_mask1+
        self.z_size_mask1):
      if(cell.type==self.typeNameTable["AL"]):
        sourceId=cell.type
```

7. Coordinated Action of N-CAM, N-cadherin, EphA4, and ephrinB2

```
            cell.type=self.typeNameTable["4"]
            attrib.append(TypeTransition(sourceId,cell.type,\
            self.stepForTransition))
        elif(cell.type==self.typeNameTable["RH"]):
            sourceId=cell.type
            cell.type=self.typeNameTable["5"]
            attrib.append(TypeTransition(sourceId,cell.type,\
            self.stepForTransition))

    def mask2(self,cell,xCM,yCM,zCM,attrib):
        if(zCM>=self.z_low_mask2-self.maskMargin and zCM \
        <self.z_low_mask2+self.z_size_mask2):
            if(cell.type==self.typeNameTable["AL"]):
                sourceId=cell.type
                cell.type=self.typeNameTable["6"]
                attrib.append(TypeTransition(sourceId,cell.type,\
                self.stepForTransition))
            elif(cell.type==self.typeNameTable["RH"]):
                sourceId=cell.type
                cell.type=self.typeNameTable["7"]
                attrib.append(TypeTransition(sourceId,cell.type,\
                self.stepForTransition))

    def mask3(self,cell,xCM,yCM,zCM,attrib):
        if(zCM>=self.z_low_mask3 and zCM <self.z_low_mask3+
            self.z_size_mask3):
            sourceId=cell.type
            cell.type=self.typeNameTable["8"]
            attrib.append(TypeTransition(sourceId,cell.type,
            self.stepForTransition))

    def maskPSM(self,cell,xCM,yCM,zCM,attrib):
        cell.type=self.typeNameTable["PSM"]

    def blur(self,_sigmaBlur):
    self.sigmaBlur=_sigmaBlur

    # The imposeMask function is a Closure that takes as
    # its first argument mask and returns a function
    # that operates on a single cell, making use
    # of the mask object. For more information on Closures
    # in Python please consult, e.g., the "Python Cookbook,"
    # or search using a web search-engine, using "python Closure"
    # as keywords. Notice that as a result (return
    # imposeMaskForSingleCell) we obtain a function that operates
    # on a single cell exactly as required by the
    # forEachCellInInventory algorithm.
```

```python
def imposeMask(self,mask):
    def imposeMaskForSingleCell(cell):
        xCM=cell.xCM/float(cell.volume)
        yCM=cell.yCM/float(cell.volume)
        zCM=cell.zCM/float(cell.volume)
        pyAttrib=CompuCell.getPyAttrib(cell)
        mask(cell,xCM,yCM,zCM,pyAttrib)
    return imposeMaskForSingleCell

# Overlaying masks can be coded very elegantly in just
# two lines making use of the imposeMask Closure and the
# forEachCellInInventory algorithm.
def overlayMasks(self,mask):
    imposeMaskFunction=self.imposeMask(mask)
    forEachCellInInventory(inventory=self.inventory,\
    singleCellOperation=imposeMaskFunction)

# The singleCellTransition Closure returns a function
# that operates on a single cell and implements a type
# transition for a single cell. We had to use a Closure
# because we are using an extra argument _mcs in addition
# to the cell variable.
def singleCellTransition(self,_mcs):
    def transition(cell):
        pyAttrib=CompuCell.getPyAttrib(cell)
        attribSize=len(pyAttrib)
        if(attribSize>1):
            for i in xrange(1,attribSize):
                # The transition will take place if and only if
                # the current time (in MCS) (_mcs)
                # matches the step variable defined in the
                # transition object.
                if(pyAttrib[i].step==_mcs):
                    cell.type=pyAttrib[i].typeIdTarget
    return transition

# The doTransitions function iterates over each cell
# and makes type transitions as necessary.
def doTransitions(self,_mcs):
    singleCellTransitionFunction=
    self.singleCellTransition(_mcs)
    forEachCellInInventory(inventory=self.inventory,\
    singleCellOperation=singleCellTransitionFunction)

# The following implements the misdifferentiation of cells
# due to inaccurate positional information.
# The doBlurSingleCell Closure returns a function operating
```

7. Coordinated Action of N-CAM, N-cadherin, EphA4, and ephrinB2   241

```python
    # on a single cell object that calculates the center of mass
    # cCM of a given cell, adds a random vector V to it
    # (the coordinates of which are chosen from a Gaussian
    # distribution) and changes the type of the current cell
    # to the type of a cell located at cCM+V (or leaves cell
    # type untouched if the cell at cCM+V happens to be medium).

    def doBlurSingleCell(self,_mcs):
       def blurFunction(cell):
          xCM=cell.xCM/float(cell.volume)
          yCM=cell.yCM/float(cell.volume)
          zCM=cell.zCM/float(cell.volume)
        if self.sigmaBlur!=0.0:
           # print "Will do the bluring ",
              random.gauss(0.0,self.sigmaBlur)
           xCM+=random.gauss(0.0,self.sigmaBlur)
           yCM+=random.gauss(0.0,self.sigmaBlur)
           zCM+=random.gauss(0.0,self.sigmaBlur)
          pt=CompuCell.Point3D()
       pt.x=int(xCM)
       pt.y=int(yCM)
       pt.z=int(zCM)
       neighborCell=self.cellFieldG.get(pt)
       if neighborCell:
            cell.type=neighborCell.type
       return blurFunction

    # Iterate over each cell and apply the blur
    # (misdifferentiation).
    def doBlur(self,_mcs):
       doBlurSingleCellFunction=self.doBlurSingleCell(_mcs)
       forEachCellInInventory(inventory=self.inventory,\
       singleCellOperation=doBlurSingleCellFunction)

    # This function is run before the simulation begins.
    # It lays out the cadherin prepattern, i.e., it initializes
    # a set of potential type transitions for each cell.
    def start(self):

       if(len(self.transitionStepsList)<3):
          print "You need to provide list with MC steps \
          at which spin reassignment takes place"
          sys.exit()
       # Initialize the anterior compartment (low z coordinates)
       # initial prepattern.
       self.stepForTransition=self.transitionStepsList[0]
       self.overlayMasks(self.maskBase)
```

```
      self.overlayMasks(self.maskRectangle)
      self.overlayMasks(self.mask0)
      self.overlayMasks(self.mask1)

      # Initialize transitions for anterior (low z) middle
      # (medium z) and posterior (high z) somite cells.
      # Notice that we go from low values of z to higher
      # values of z by manipulating class variables
      # such as self.z_base_low and self.z_low_mask0 and
      # shifting them for each of the transition steps.
      for i in xrange(1,3):
         self.stepForTransition=self.transitionStepsList[i]
         self.overlayMasks(self.mask2)
         self.overlayMasks(self.mask3)
         # z_low for mask 2 and 3 is shifted after the call
         # to overlay masks.
         self.z_low_mask2+=self.shift
         self.z_low_mask3+=self.shift
         self.z_base_low+=self.shift
         self.z_rect_low+=self.shift
         self.overlayMasks(self.maskBase)
         self.overlayMasks(self.maskRectangle)
         self.z_low_mask0+=self.shift
         self.z_low_mask1+=self.shift
         self.overlayMasks(self.mask0)
         self.overlayMasks(self.mask1)
      # Because the overlayMasks function has the side effect
      # of changing cell types, we need to reset the cell
      # types after we impose the masks. At the beginning
      # of the simulation all the cells are of type PSM
      # (presomitic mesoderm).
      self.overlayMasks(self.maskPSM)
   # This function is run every 10 MCS -
   # see somiteMaskSteppable=
   # SomiteMaskSteppable(_simulator=sim,_frequency=10)in
   # somite.py.
   def step(self,mcs):
      if(mcs<=self.maxTransitionStep):
         if mcs in self.transitionStepsList:
            self.doTransitions(mcs)
            self.doBlur(mcs)
            return
      return

   # SomiteVolumeSurfaceSteppable is responsible for
   # periodically assigning new volume and surface energy
   # parameters. Those parameters are local to each cell,
```

7. Coordinated Action of N-CAM, N-cadherin, EphA4, and ephrinB2

```
    # so when cells change types these parameters need to be
    # updated as well. We have hard-coded the parameters
    # (targerVolume, lambda) (targetSurface, lambda).
    class SomiteVolumeSurfaceSteppable(SteppablePy):
        def __init__(self,_simulator,_frequency=1):
            SteppablePy.__init__(self,_frequency)
            self.simulator=_simulator
            self.cellFieldG=self.simulator.getPotts().
            getCellFieldG()
            self.dim=self.cellFieldG.getDim()
            self.inventory=self.simulator.getPotts().
            getCellInventory()
            self.maxTransitionStep=0
            #format
            #type:[targetVolume,lambda]
            #example 2:[25.0,20.0]
            self.typeVolumeParamMap={1:[25.0,20.0], 2:[25.0,20.0],
            3:[36.0,20.0],\
            4:[36.0,20.0],5:[16.0,20.0],6:[36.0,20.0],
            7:[16.0,20.0],
            8:[36.0,20.0],9:[25.0,20.0]}
            #format
            #type:[targetSurface,lambda]
            #example 2:[20.0,20.0]

self.typeSurfaceParamMap={1:[20.0,20.0],2:[20.0,20.0],
3:[24.0,20.0],4:[24.0,20.0],\
5:[16.0,20.0],6:[24.0,20.0],7:[16.0,20.0],
8:[24.0,20.0],9:[20.0,20.0]}


      def setMaxTransitionStep(self,_maxTransitionStep):
          self.maxTransitionStep=_maxTransitionStep

      # At the beginning of the simulation all cells have
      # these volume parameters.
      def volumeInitSet(self,cell):
          cell.targetVolume=25.0
          cell.lambdaVolume=20.0
      # At the beginning of the simulation all cells have
      # these surface parameters.
      def surfaceInitSet(self,cell):
          cell.targetSurface=20.0
          cell.lambdaSurface=20.0

    # At later stages we will change the volume and
    # surface parameters using
```

```
# self.typeVolumeParamMap and self.typeSurfaceParamMap.
def volumeParamSet(self,cell):
   par=self.typeVolumeParamMap[cell.type]
   cell.targetVolume=par[0]
   cell.lambdaVolume=par[1]
def surfaceParamSet(self,cell):
   par=self.typeSurfaceParamMap[cell.type]
   cell.targetSurface=par[0]
   cell.lambdaSurface=par[1]

def setParameters(self,paramSetFcn):
   forEachCellInInventory(inventory=self.inventory,\
   singleCellOperation=paramSetFcn)

def start(self):
   self.setParameters(self.volumeInitSet)
   self.setParameters(self.surfaceInitSet)

# This function is run every 100 MCS-
# see somiteVolumeSurfaceSteppable=
# SomiteVolumeSurfaceSteppable(_simulator=sim,_frequency
# =100) in somte.py. Notice that we run this function
# whether a transition took place or not, which is redundant,
# but makes the code easier to read.
def step(self,mcs):
   if(mcs<=self.maxTransitionStep):
      self.setParameters(self.volumeParamSet)
      self.setParameters(self.surfaceParamSet)
   return
```

References

Afonin, B., Ho, M., Gustin, J. K., Meloty-Kapella, C., and Domingo, C. R. (2006). Cell behaviors associated with somite segmentation and rotation in Xenopus laevis. *Dev. Dyn.* **235**, 3268–3279.

Aulehla, A., and Herrmann, B. G. (2004). Segmentation in vertebrates: Clock and gradient finally joined. *Genes Dev.* **18**, 2060–2067.

Aulehla, A., and Pourquié, O. (2006). On periodicity and directionality of somitogenesis. *Anat. Embryol.* **211**, S3–S8.

Aulehla, A., Wehrle, C., Brand-Saberi, B., Kemler, R., Gossler, A., Kanzler, B., and Herrmann, B. G. (2003). Wnt3a plays a major role in the segmentation clock controlling somitogenesis. *Dev. Cell* **4**, 395–406.

Baker, R. K., and Antin, P. B. (2003). Ephs and ephrins during early stages of chick embryogenesis. *Dev. Dynam.* **228**, 128–142.

Baker, R. E., Schnell, S., and Maini, P. K. (2003). Formation of vertebral precursors: Past models and future predictions. *J. Theor. Med.* **5**, 23–35.

Baker, R. E., Schnell, S., and Maini, P. K. (2006). A clock and wavefront mechanism for somite formation. *Dev. Biol.* **293**, 116–126.

Barrios, A., Poole, R. J., Durbin, L., Brennan, C., Holder, N., and Wilson, S. W. (2003). Eph/Ephrin signaling regulates the mesenchymal-to-epithelial transition of the paraxial mesoderm during somite morphogenesis. *Curr. Biol.* **13**, 1571–1582.

Bergemann, A. D., Cheng, H. J., Brambilla, R., Klein, R., and Flanagan, J. G. (1995). ELF-2, a new member of the Eph ligand family, is segmentally expressed in mouse embryos in the region of the hindbrain and newly forming somites. *Mol. Cell Biol.* **15**, 4921–4929.

Chaturvedi, R., Huang, C., Kazmierczak, B., Schneider, T., Izaguirre, J. A., Glimm, T., Hentschel, H. G. E., Glazier, J. A., Newman, S. A., and Alber, M. S. (2005). On multiscale approaches to three-dimensional modeling of morphogenesis. *J. R. Soc. Interface* **2**, 237–253.

Collier, J. R., McInerney, D., Schnell, S., Maini, P. K., Gavaghan, D. J., Houston, P., and Stern, C. D. (2000). A cell cycle model for somitogenesis: Mathematical formulation and numerical simulation. *J. Theor. Biol.* **207**, 305–316.

Cooke, J., and Zeeman, E. C. (1976). A clock and wavefront model for control of the number of repeated structures during animal morphogenesis. *J. Theor. Biol.* **58**, 455–476.

Cooke, J. E., Kemp, H. A., and Moens, C. B. (2005). EDhA4 is required for cell adhesion and rhombomere-boundary formation in the zebrafish. *Curr. Biol.* **15**, 536–542.

De Bellard, M. E., Ching, W., Gossler, A., and Bronner-Fraser, M. (2002). Disruption of segmental neural crest migration and ephrin expression in delta-1 null mice. *Dev. Biol.* **249**, 121–130.

Dequeant, M. L., Glynn, E., Gaudenz, K., Wahl, M., Chen, J., Mushegian, A., and Pourquié, O. (2006). A complex oscillating network of signaling genes underlies the mouse segmentation clock. *Science* **314**, 1595–1598.

Diez del Corral, R., Olivera-Martinez, I., Goriely, A., Gale, E., Maden, M., and Storey, K. (2003). Opposing FGF and retinoid pathways control ventral neural pattern, neuronal differentiation, and segmentation during body axis extension. *Neuron* **40**, 65–79.

Duband, J. L., Dufour, S., Hatta, K., Takeichi, M., Edelman, G. M., and Thiery, J. P. (1987). Adhesion molecules during somitogenesis in the avian embryo. *J. Cell Biol.* **104**, 1361–1374.

Dubrulle, J., and Pourquié, O. (2002). From head to tail: Links between the segmentation clock and antero-posterior patterning of the embryo. *Curr. Opin. Genet. Dev.* **12**, 519–523.

Dubrulle, J., and Pourquié, O. (2004). Coupling segmentation to axis formation. *Development* **131**, 5783–5793.

Dubrulle, J., McGrew, M. J., and Pourquié, O. (2001). FGF signaling controls somite boundary position and regulates segmentation clock control of spatiotemporal Hox gene activation. *Cell* **106**, 219–232.

Durbin, L., Brennan, C., Shiomi, K., Cooke, J., Barrios, A., Shanmugalingam, S., Guthrie, B., Lindberg, R., and Holder, N. (1998). Eph signaling is required for segmentation and differentiation of the somites. *Genes Dev.* **12**, 3096–3109.

Foty, R. A., and Steinberg, M. S. (2005). The differential adhesion hypothesis: A direct evaluation. *Dev. Biol.* **278**, 255–263.

Glazier, J. A., and Graner, F. (1993). Simulation of the differential adhesion driven rearrangement of biological cells. *Phys. Rev. E* **47**, 2128–2154.

Goldbeter, A., Gonze, D., and Pourquié, O. (2007). Sharp developmental thresholds defined through bistability by antagonistic gradients of retinoic acid and FGF signaling. *Dev. Dyn.* **236**, 1495–1508.

Gossler, A., and Hrabe de Angelis, M. (1998). Somitogenesis. *Curr. Top. Dev. Biol.* **38**, 225–287.

Graner, F., and Glazier, J. A. (1992). Simulation of biological cell sorting using a two-dimensional extended Potts model. *Phys. Rev. Lett.* **69**, 2013–2016.

Grima, R., and Schnell, S. (2007). Can tissue surface tension drive somite formation? *Dev. Biol.* **307**, 248–257.

Harbott, L. K., and Nobes, C. D. (2005). A key role for Abl family kinases in EphA receptor-mediated growth cone. *Mol. Cell. Neurosci.* **30**, 1–11.

Henry, C. A., Hall, L. A., Burr Hille, M., Solnica-Krezel, L., and Cooper, M. S. (2000). Somites in zebrafish doubly mutant for knypek and trilobite form without internal mesenchymal cells or compaction. *Curr. Biol.* **10**, 1063–1066.

Horikawa, K., Radice, G., Takeichi, M., and Chisaka, O. (1999). Adhesive subdivisions intrinsic to the epithelial somites. *Dev. Biol.* **215**, 182–189.

Horikawa, K., Ishimatsu, K., Yoshimoto, E., Kondo, S., and Takeda, H. (2006). Noise-resistant and synchronized oscillation of the segmentation clock. *Nature* **441**, 719–723.

Jiang, Y. J., Aerne, B. L., Smithers, L., Haddon, C., Ish-Horowicz, D., and Lewis, J. (2000). Notch signaling and the synchronization of the somite segmentation clock. *Nature* **408**, 475–479.

Kasemeier-Kulesa, J. C., Bradley, R., Pasquale, E. B., Lefcort, F., and Kulesa, P. M. (2006). Eph/ephrins and N-cadherin coordinate to control the pattern of sympathetic ganglia. *Development* **133**, 4839–4847.

Keynes, R. J., and Stern, C. D. (1988). Mechanisms of vertebrate segmentation. *Development* **103**, 413–429.

Kimura, Y., Matsunami, H., Inoue, T., Shimamura, K., Uchida, N., Ueno, T., Miyazaki, T., and Takeichi, M. (1995). Cadherin-11 expressed in association with mesenchymal morphogenesis in the head, somite, and limb bud of early mouse embryos. *Dev. Biol.* **169**, 347–358.

Kulesa, P. M., and Fraser, S. E. (2002). Cell dynamics during somite boundary formation revealed by time-lapse analysis. *Science* **298**, 991–995.

Kulesa, P. M., Schnell, S., Rudloff, S., Baker, R. E., and Maini, P. K. (2007). From segment to somite: Segmentation to epithelialization analyzed within quantitative frameworks. *Dev. Dyn.* **236**, 1392–1402.

Linask, K. K., Ludwig, C., Han, M. D., Liu, X., Radice, G. L., and Knudsen, K. A. (1998). N-cadherin/catenin-mediated morphoregulation of somite formation. *Dev. Biol.* **202**, 85–102.

Marée, A. F., and Hogeweg, P. (2001). How amoeboids self-organize into a fruiting body: Multicellular coordination in Dictyostelium discoideum. *Proc. Natl. Acad. Sci. USA* **98**, 3879–3883.

Meinhardt, H. (1996). Models of biological pattern formation: Common mechanism in plant and animal development. *Int. J. Dev. Biol.* **40**, 123–134.

Mellitzer, G., Xu, Q., and Wilkinson, D. G. (1999). Eph receptors and ephrins restrict cell intermingling and communication. *Nature* **400**, 77–81.

Merks, R. M., and Glazier, J. A. (2006). Dynamic mechanisms of blood vessel growth. *Nonlinearity* **19**, C1–C10.

Merks, R. M., Brodsky, S. V., Goligorksy, M. S., Newman, S. A., and Glazier, J. A. (2006). Cell elongation is key to in silico replication of in vitro vasculogenesis and subsequent remodeling. *Dev. Biol.* **289**, 44–54.

Moreno, T. A., and Kintner, C. (2004). Regulation of segmental patterning by retinoic acid signaling during Xenopus somitogenesis. *Dev. Cell* **6**, 205–218.

Nakajima, Y., Morimoto, M., Takahashi, Y., Koseki, H., and Saga, Y. (2006). Identification of Epha4 enhancer required for segmental expression and the regulation by Mesp2. *Development* **133**, 2517–2525.

Nelson, W. J., and Nusse, R. (2004). Convergence of Wnt, beta-catenin, and cadherin pathways. *Science* **303**, 1483–1487.

Nieto, M. A., Gilardi-Hebenstreit, P., Charnay, P., and Wilkinson, D. G. (1992). A receptor protein tyrosine kinase implicated in the segmental patterning of the hindbrain and mesoderm. *Development* **116**, 1137–1150.

Ordahl, C. P. (1993). Myogenic lineages within the developing somite. *In* "Molecular Basis of Morphogenesis." (M. Bernfield, Ed.). Wiley, New York, pp. 165–176.

Palmeirim, I., Henrique, D., Ish-Horowicz, D., and Pourquié, O. (1997). Avian hairy gene expression identifies a molecular clock linked to vertebrate segmentation and somitogenesis. *Cell* **91**, 639–648.

Poliakov, A., Cotrina, M., and Wilkinson, D. G. (2004). Diverse roles of eph receptors and ephrins in the regulation of cell migration and tissue assembly. *Dev. Cell* **7**, 465–480.

Poplawski, N. J., Swat, M., Gens, J. S., and Glazier, J. A. (2007). Adhesion between cells, diffusion of growth factors, and elasticity of the AER produce the paddle shape of the chick limb. *Physica A* **373C**, 521–532.

Pourquié, O. (2004). The chick embryo: A leading model in somitogenesis studies. *Mech. Dev.* **121**, 1069–1079.

Pourquié, O., and Tam, P. P. (2001). A nomenclature for prospective somites and phases of cyclic gene expression in the presomitic mesoderm. *Dev. Cell* **1**, 619–620.

Primmett, D. R., Stern, C. D., and Keynes, R. J. (1988). Heat shock causes repeated segmental anomalies in the chick embryo. *Development* **104**, 331–339.

Primmett, D. R., Norris, W. E., Carlson, G. J., Keynes, R. J., and Stern, C. D. (1989). Periodic segmental anomalies induced by heat shock in the chick embryo are associated with the cell cycle. *Development* **105**, 119–130.

Radice, G. L., Rayburn, H., Matsunami, H., Knudsen, K. A., Takeichi, M., and Hynes, R. O. (1997). Developmental defects in mouse embryos lacking N-cadherin. *Dev. Biol.* **181**, 64–78.

Sato, Y., and Takahashi, Y. (2005). A novel signal induces a segmentation fissure by acting in a ventral-to-dorsal direction in the presomitic mesoderm. *Dev. Biol.* **282**, 183–191.

Sato, Y., Yasuda, K., and Takahashi, Y. (2002). Morphological boundary forms by a novel inductive event mediated by Lunatic fringe and Notch during somitic segmentation. *Development* **129**, 3633–3644.

Schnell, S., and Maini, P. K. (2000). Clock and induction model for somitogenesis. *Dev. Dynam.* **217**, 415–420.

Schnell, S., Maini, P. K., McInerney, D., Gavaghan, D. J., and Houston, P. (2002). Models for pattern formation in somitogenesis: A marriage of cellular and molecular biology. *C. R. Biol.* **325**, 179–189.

Stern, C. D., Fraser, S. E., Keynes, R. J., and Primmett, D. R. (1988). A cell lineage analysis of segmentation in the chick embryo. *Development* **104S**, 231–244.

Takahashi, Y., Inoue, T., Gossler, A., and Saga, Y. (2003). Feedback loops comprising Dll1, Dll3 and Mesp2, and differential involvement of Psen1 are essential for rostrocaudal patterning of somites. *Development* **130**, 4259–4268.

Wood, A., and Thorogood, P. (1994). Patterns of cell behavior underlying somitogenesis and notochord formation in intact vertebrate embryos. *Dev. Dynam.* **201**, 151–167.

Xu, Q. L., Mellitzer, G., Robinson, V., and Wilkinson, D. G. (1999). In vivo cell sorting in complementary segmental domains mediated by Eph receptors and ephrins. *Nature* **399**, 267–271.

Xu, Q. L., Mellitzer, G., and Wilkinson, D. G. (2000). Roles of Eph receptors and ephrins in segmental patterning. *Philos. Trans. Roy. Soc. London B Biol. Sci.* **355**, 993–1002.

Zeng, W., Thomas, G. L., and Glazier, J. A. (2004). NonTuring stripes and spots: A novel mechanism for biological cell clustering. *Physica A* **341**, 482–494.

8

Branched Organs: Mechanics of Morphogenesis by Multiple Mechanisms

Sharon R. Lubkin[*,†]

[*]Department of Mathematics, North Carolina State University, Raleigh, North Carolina 27695-8205
[†]Department of Biomedical Engineering, North Carolina State University, Raleigh, North Carolina 27695-8205

I. Introduction
II. Background
    A. Tissue Types
    B. Morphogenesis Overview
III. Candidate Physical Mechanisms
    A. Growth and Apoptosis
    B. Cell Size Changes
    C. Cell Shape Changes
    D. Cell Rearrangement and Migration
    E. Cell–ECM interactions
    F. Basal Lamina
    G. Traction
    H. Adhesion
IV. Models of Branching
    A. Importance of Viscosity
    B. Fingering Instability
    C. Mesenchymal Force
V. Discussion
    Acknowledgments
    References

Branching morphogenesis is ubiquitous and important in creating bulk transport systems. Branched ducts can be generated by several different mechanisms including growth, cell rearrangements, contractility, adhesion changes, and other mechanisms. We have developed several models of the mechanics of cleft formation, which we review. We discuss the implications of several candidate mechanisms and review what has been found in models and in experiments. © 2008, Elsevier Inc.

I. Introduction

Every organism has transport mechanisms which allow it to function. The smallest scale mechanisms, useful at the single-cell level, are based on diffusion, osmotic

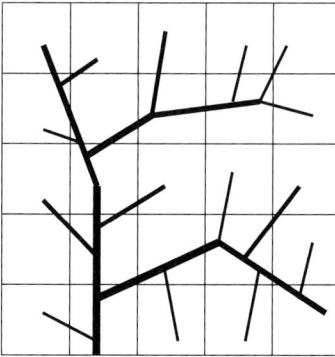

Figure 1 A branched system of ducts can transport to and from every neighborhood of a tissue using very little material. In this schematic illustration each square neighborhood has some vascular connection to the main vessel.

forces, and motor proteins. In general these small-scale transport mechanisms have a useful range of less than a millimeter. Beyond approximately 1 mm, diffusion and osmotic pumps are too slow to be major transport mechanisms, with the exception of some spectacular reproductive organs of plants and fungi which rely on slow osmotic pumps to load ballistic dispersal devices.

Plants get most of their bulk transport needs met through capillary forces, but the fast-paced world of metazoans requires a system of macroscopic ducts. Since the cell that has to benefit from bulk transport can only receive transported substances by small-scale mechanisms, the metazoan duct systems must direct transport to a continuum of very small locations. A branched system can, with very little material, transport to and from each location simultaneously (Fig. 1). This is the reason for the ubiquity of branched systems of ducts.

Branched ducts, carrying flow driven by muscles, cilia, or other mechanisms, are found in vascular systems, lungs, kidneys, glands, and other organs. Their importance cannot be overemphasized. If branched ducts are ubiquitous, diverse, and important, then the morphogenesis of branched ducts is also important. However, understanding morphogenesis in such a diverse set of organs and organisms complicates the analysis and makes it extremely difficult to determine what the generic mechanisms of branching are.

By "generic mechanisms," we do not necessarily mean a common genetic pathway. Because the evolutionary advantage of branched ducts is so great, it is likely that branched ducts, like eyes and sexual reproduction, have evolved multiple times from multiple precursors.

When asked the intriguing and important question "What makes an airplane fly?" most people focus on the relationship between thrust and lift, perhaps with some concern for circulation, separation, and stalling. Another possible conceptual framework for an answer focuses on the switches that the pilot manipulates, to turn on the engines, to regulate the fuel–air mix, to calibrate the altimeter, etc. In the case of the airplane, the controls are quite different in different models (e.g., in propeller vs jet planes), but the general principles of lift, thrust, and circulation are universal. It is not very different in branching morphogenesis. The diversity of branched systems in several phyla makes it difficult to find unifying principles among the switches—the gene networks. However there are, we believe, underlying physical principles which unify most examples of branching morphogenesis. *In this paper, we focus on the physical forces responsible for the morphogenesis of branched ducts.*

II. Background

While the details differ in the different branched organs, there are commonalities in the morphogenesis of the various systems, which we will review.

A. Tissue Types

The generation of a branched organ *in vivo* involves interactions between epithelium and mesenchyme (Fig. 2). Developmentally, these tissue types behave quite differently from each other, and their behaviors must be considered separately (Bard, 1990; Trinkaus, 1984). In both tissue types, mitosis and apoptosis and cell migration take place, but the largest differences between the tissue types are in the interactions of

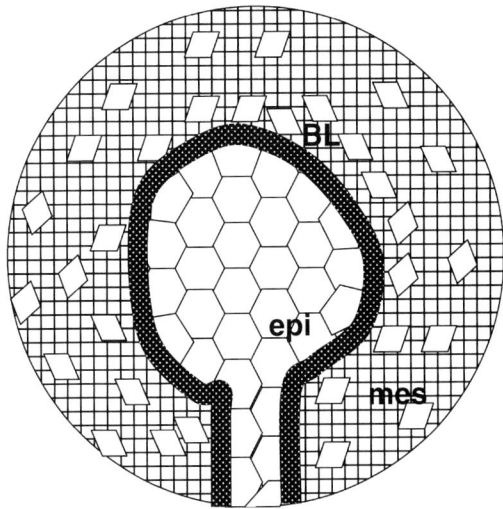

Figure 2 Schematic of structure of branching rudiment. A finger of epithelium is surrounded by a layer of mesenchyme. A basal lamina separates the two tissues. Hexagons and rhombi represent cells; mesh represents ECM.

cells with each other and with the extracellular matrix. In epithelium, there is virtually no space between cells, so the movement of cells relative to each other is due to direct mechanical interaction of cell on cell. By contrast, in mesenchyme, there is significant extracellular matrix (ECM) between cells so that most motion does not involve cell–cell contacts; cell motion is highly dependent on the properties of the ECM. In epithelium, there is significant interaction between the basal cells and the basal lamina, which provides some stability and an adhesive surface, and which also interacts with the surrounding mesenchyme.

B. Morphogenesis Overview

Branching morphogenesis generally proceeds the same way in salivary gland (or submandibular gland, SMG), lung, kidney, mammary gland, etc. and is illustrated in Fig. 3. A finger of epithelium, initially with no lumen, surrounded by mesenchyme, expands. One or more clefts pinch the epithelium and inhibit its expansion in the cleft region. This clefting forms two or more lobules, which continue to grow and subsequently branch—in the case of the human lung, approximately 23 times. Branching morphogenesis is usually dichotomous, or bifid, but has been shown in the kidney to be trifid in 18% of branching locations (Watanabe and Costantini, 2004). These trifid branchings are transient initialization events; subsequent modification of the resulting branches usually resolves the branches into dichotomous pairs (Watanabe and Costantini, 2004). The mesenchyme condenses in the stalk and cleft regions as the organ develops. Collagen is denser in stalks and clefts, and less dense around the lobules, which expand (Grobstein and Cohen, 1965).

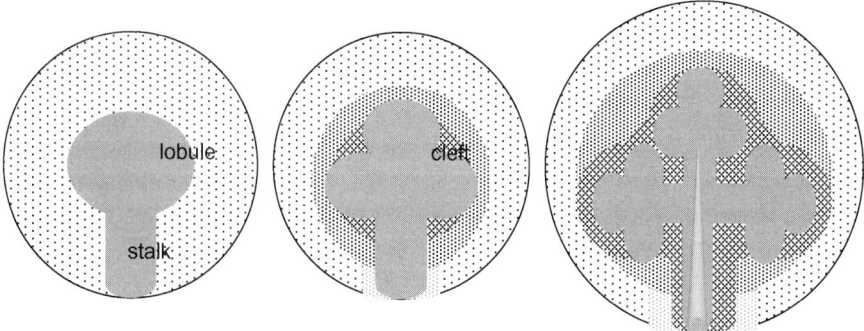

Figure 3 Schematic of procedure of branching morphogenesis. Unbranched epithelium grows and clefts form, dividing lobule into multiple lobules. Resulting lobules grow and cleft. Previous generations of branches become stalks. Mesenchyme condenses around epithelium, stabilizing stalks and clefts. In SMG, lumen forms after branching, as stalks mature. Density of ECM components roughly indicated by density of dots.

8. Mechanics of Branching Morphogenesis 253

III. Candidate Physical Mechanisms

It is far easier to speculate on physical mechanisms of morphogenesis than to establish whether or not they play a role, and if so, how big a role, when the various mechanisms are active, and how they interact with other mechanisms. We will begin by listing and discussing morphogenetic mechanisms which may or may not play a substantial role in branching morphogenesis.

A. Growth and Apoptosis

One obvious morphogenetic mechanism is growth. Uniform growth does not lead to shape change; shape change requires differential growth (Fig. 4). Growth in branching morphogenesis is regulated by reciprocal tissue interactions. In lung, the mesenchyme regulates epithelial growth with excitatory (EGF, FGF, and others) and inhibitory (TGF-β) factors. Conversely, the epithelium produces factors which regulate the mesenchyme (Shh) (Nakanishi *et al.*, 2001). The counterpart of growth is apoptosis, which does not appear to play a role in cleft formation, though it may in lumen formation (Fig. 3). It is self-evident that growth is required in order to generate a whole branched organ, but branching and cell proliferation are separable processes. Branching morphogenesis proceeds normally in SMG when growth is slowed or halted by irradiation, aphidicolin, or tunicamycin (Nakanishi *et al.*, 1987; Spooner *et al.*, 1989). Bud formation also occurs independent of mitosis in the lung (Nogawa *et al.*, 1998).

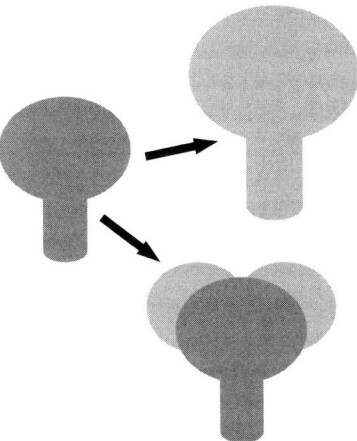

Figure 4 Growth is necessary in branching morphogenesis simply to gain more tissue, but it does not play a role in cleft formation. Uniform growth (top) does not lead to shape changes; growth must be nonuniform (bottom) to effect shape change.

B. Cell Size Changes

Swelling and/or shrinkage of cells and/or ECM could conceivably play a role in morphogenesis, but these mechanisms are not known to play a role in branching morphogenesis.

C. Cell Shape Changes

If cuboidal cells in an epithelium change their aspect ratio in a coordinated fashion, a squamous or columnar epithelium will result. Coordinated contraction of actin microfilaments, known as "purse stringing," can heal epidermal wounds (Martin and Lewis, 1992) and may play a role in morphogenesis. Mesenchymal cells around 16-day mouse lung are cuboidal in the stalk (bronchi) region and flattened around the lobes (Hilfer, 1996). Gastrulation and neurulation have been modeled by coordinated cell shape changes in the blastula (Odell *et al.*, 1980). It is conceivable that coordinated cell-shape changes in an epithelium could lead to branching morphogenesis (Fig. 5), but this mechanism has not been documented in this context.

D. Cell Rearrangement and Migration

Cell rearrangements may be based primarily in the epithelium or mesenchyme, and may be driven by adhesive or chemotactic gradients, or both. There are models of epithelial morphogenesis by cell rearrangements driven by differential adhesion (Mittenthal and Mazo, 1983), but none specifically modeling branching morphogenesis. Epithelial cells of embryonic SMG engage in vigorous migration and rearrangement (Fig. 6), of the order of microns per hour, which can be clearly seen in 3D confocal microscopy (Larsen *et al.*, 2006). It is not clear, however, what the net effect or morphogenetic function of the movements is. Despite sophisticated microscopic

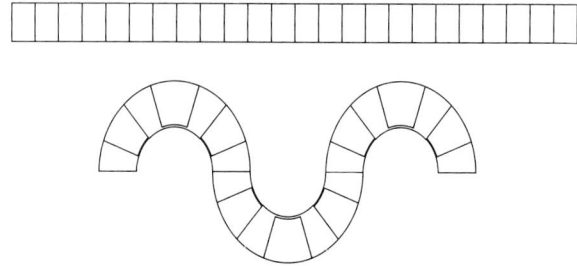

Figure 5 Coordinated cell shape changes can change the overall shape of an epithelium by changing the local curvature.

8. Mechanics of Branching Morphogenesis

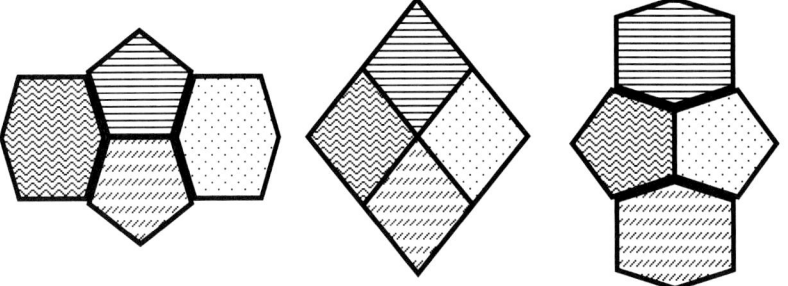

Figure 6 Cell rearrangements are driven by cytoskeletal forces and by changes in adhesion at the cell surface and have been shown to be a potent morphogenetic mechanism in many contexts.

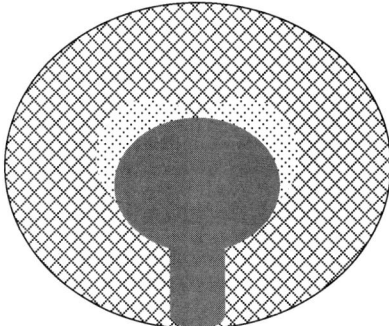

Figure 7 Secretion of collagenases and MMPs by the epithelium can in theory digest the stromal ECM sufficiently for an expanding epithelium to "melt" its way through. ECM represented by solid and broken grids (original and partially digested ECM, respectively).

and analytical techniques, Larsen *et al.* (2006) did not find any distinguishing characteristics of the migratory pattern of epithelial cells in the cleft region of the SMG versus other regions, nor did they find any relationship of mitosis to migratory pattern. SMG epithelium has a remarkable ability to self-organize when disaggregated and can reaggregate and form branches (Wei *et al.*, 2007).

E. Cell–ECM interactions

There is a great amount of cell movement in both the epithelium and mesenchyme of branching morphogenesis. Cell–ECM interactions may consist primarily of cells migrating in the ECM or rearranging the ECM through digestion by enzymes (Fig. 7), secretion of ECM components, or mechanical alteration by traction forces (Fig. 8). Cells and ECM fibers in embryonic salivary mesenchyme are oriented parallel to the epithelial surface (Flint, 1902, 1903, 1906; Moral, 1913, 1916). Mesenchymal cells

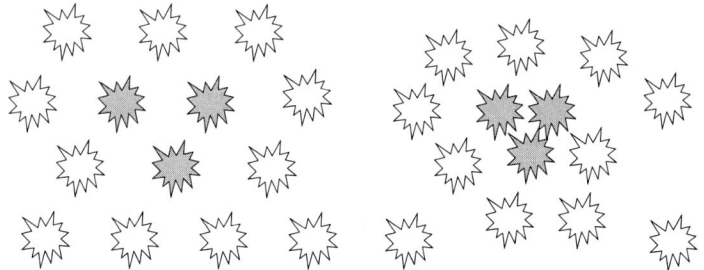

Figure 8 Contractile fibroblasts advect towards each other, locally increasing the cell density of contractile cells (gray) and also of neighbor cells which are not contractile (white), but which are mechanically linked to the contractile cells by the ECM.

in SMG rapidly move along the epithelial surface and in the cleft, where ECM fibers are aligned (Hieda and Nakanishi, 1997). Early researchers on glandular branched morphogenesis believed that the orientation of stromal components was due to "mechanical factors such as traction and pressure" [Flint (1902, 1906); Moral (1913, 1916) cited in Borghese (1950a, 1950b)]. Stiffness of the ECM has been shown to affect cell morphology and motility (Pelham and Wang, 1997), tissue formation (Guo and Wang, 2004), and differentiation of stem cells (Engler *et al.*, 2006). Larsen *et al.* (2006) propose that the morphogenetic branching force in SMG is from fibronectin (FN) production on the lobular surface pushing a "wedge" of FN into the cleft. However, no model of the physics of this conceptual model has been published.

F. Basal Lamina

Removal of the basal lamina from branched salivary epithelium leads to loss of epithelial surface features, which can be recovered if the lamina is allowed to regenerate. If the epithelium is recombined with its mesenchyme before the regeneration of the basal lamina, the branched morphology is lost (Banerjee *et al.*, 1977). This suggests that the basal lamina's mechanical role is to stabilize epithelial morphology, which can be altered by forces produced by the mesenchyme. Previous conceptual models of branching morphogenesis emphasized the role of basal lamina turnover and epithelial cell wedging (Spooner *et al.*, 1986b; Spooner and Wessells, 1972), though the mechanical implications of these conceptual models have never been examined.

G. Traction

Traction forces are generated by cells which may be distributed spatially (Fig. 8). A uniform spatial distribution of contractile forces, in many cases, leads to a uniform

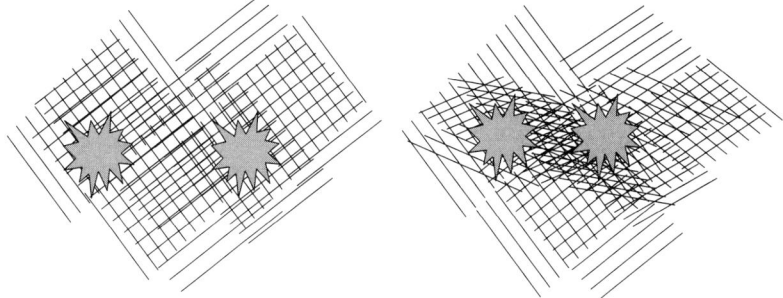

Figure 9 Contractile cells pull each other closer and also dramatically change the alignment of ECM fibers. The alignment can extend many cell diameters away.

deformation (Barocas *et al.*, 1995) and in others leads to pattern formation by a mechanical instability (Manoussaki *et al.*, 1996). Neither of these cases is appropriate to the case of branching morphogenesis, where the clefting is at a single specific location (or small number of locations, as in the cases of salivary gland and kidney morphogenesis). Fibroblasts are well known to exert strong traction forces on their surrounding ECM (Harris *et al.*, 1980). What may appear to be small contractile motions by fibroblasts (∼10 microns) can in fact generate substantial long-distance realignment of ECM (∼1 mm away) (Sawhney and Howard, 2002) (Fig. 9).

Branching morphogenesis in lung and kidney is inhibited by several agents which interfere with cell contractility through the Rho-ROCK (Rho-associated kinase) pathway; conversely, branching morphogenesis is enhanced by Rho activation. The inhibition of branching can be quantified (by time-dependent morphometry), and is found to be dose-dependent in the concentration of Rho inhibitors (Michael *et al.*, 2005; Moore *et al.*, 2005). Myosin is required for branching in kidney (Michael *et al.*, 2005). In kidney branching, F-actin is seen to be abundant in the core of the lobular epithelium and not abundant in the surface layer of cells (Michael *et al.*, 2005). There are a large number of radially-oriented actin filaments extending from the epithelial surface; it is not clear from light microscopy if these filaments are epithelial or fibroblastic (Michael *et al.*, 2005).

It has been suggested (Discher *et al.*, 2005) that cells adapt their traction forces to the stiffness of the local ECM, so that the local ECM strains due to cell traction are of the order of a few percent. Contractile stress in some contexts regulates cell proliferation (Nelson *et al.*, 2005).

Nogawa and Nakanishi (1987) measured the contractility of several mesenchymal cell types (2×10^5 cells/ml) in floating disks of collagen gel (1.5 mg/ml). They report submandibular mesenchyme contracting gel diameters to 88, 69, and 59% of initial diameter by days 1, 2, and 3. This corresponds to 16% diameter contraction per day or 29% per day area contraction. Lung mesenchyme is found to have quantitatively comparable contractility. Although mesenchymal contractile strength is positively correlated with branching activity in recombination experiments (Nogawa and Nakanishi, 1987), there is not a known, quantified pattern of mesenchymal cell movement in branching morphogenesis.

H. Adhesion

Regional differences in adhesion have been shown to be a robust morphogenetic mechanism (reviewed in Forgacs and Newman, 2005). There are substantial changes in the expression of adhesion molecules during branching morphogenesis. Fibronectin (FN) is necessary for branching morphogenesis in salivary gland, lung, and kidney. Branching responds to FN in a dose-dependent fashion (buds per gland as a function of FN concentration). FN is abundant in mesenchyme before branching, accumulates in clefts during branching, and gradually decreases in concentration in clefts as clefts widen. In mesenchyme-free culture, SMG clefting proceeds and FN is localized in the *bottom* of clefts, spreading somewhat as a cleft matures (Sakai *et al.*, 2003). We hypothesize that this may set up a haptotactic gradient inducing fibroblast migration into the cleft by biased random motion. In lung, FN is also localized in the bottom of the cleft, on the epitheliomesenchymal surface [Sakai *et al.* (2003), supp. Fig. 5]. In kidney, FN is at the bottom of the cleft and also on the stalk, but not on the lobe. In many cases, the epithelial cells adjacent to high-FN areas are columnar (Sakai *et al.*, 2003).

IV. Models of Branching

The general form of a mechanical model for tissue dynamics is some combination of force balance equations and conservation equations for each "species." Branching SMG epithelia do not have tight junctions (Hieda *et al.*, 1996). This suggests that embryonic branching epithelium behaves mechanically, as many other embryonic tissues do, like a viscoelastic fluid. The viscoelastic coefficients and surface tension of many different embryonic tissue types have been measured (Forgacs *et al.*, 1998; Foty *et al.*, 1994), however no mechanical measurements of the specific tissues involved in branching morphogenesis have been published. On the time scale of branching morphogenesis, the shorter-term elastic component in the tissues can be neglected, and the tissues may be modeled as fluids.

A. Importance of Viscosity

We have explored the implications of the fluid nature of the embryonic tissues involved in branching morphogenesis, in order to clarify the significance of epithelial

8. Mechanics of Branching Morphogenesis

branching with and without mesenchyme. Epithelia of branching organs can be separated from their mesenchymes and, in an appropriate cell-free ECM with appropriate growth factors, will form branches (Costantini, 2006; Nogawa and Takahashi, 1991; Takahashi and Nogawa, 1991). However, the branches are not morphologically characteristic of the tissue of epithelial origin. In numerous recombination experiments [reviewed in Doljanski (2004); Nelson and Bissell (2006)], the form of a combination structure is governed by the mesenchyme of origin, and the function is determined by the epithelium. Mammary epithelium recombined with submandibular mesenchyme forms a branched structure with lobes characteristic of SMG and stalks characteristic of mammary gland (Kratochwil, 1969), which can produce milk (Parmar and Cunha, 2004; Sakakura et al., 1976). Lung epithelium cultured with salivary mesenchyme forms salivary-like branches (Iwai et al., 1998). When SMG mesenchyme from developmental days 13 and 14 is recombined with SMG epithelium from days 13 or 14, the stage of the mesenchyme determines the form of the resulting branching (Nogawa, 1983).

We hypothesize that these morphological differences may be due to mechanical differences. Mesenchymal tissue properties are primarily due to the composition of the ECM, and altering the quantity of ECM components and the degree of crosslinking affect the tissue's resistance to deformation and can dramatically affect cell behavior (Brekken et al., 2003; Discher et al., 2005; Engler et al., 2006). We hypothesize that the morphological differences among branched organs may be due to mechanical effects, specifically the effect of the viscosity of the material surrounding the epithelium.

In order to clarify the mechanical implications of mesenchymal viscosity and the implications of the mesenchyme-free branching systems, we constructed a model of the mechanical response of embryonic tissues to a branch-forming force (Lubkin and Li, 2002). We modeled epithelium and mesenchyme as immiscible Stokes fluids. The same model can be applied to a mesenchyme-free situation, where mesenchyme is replaced by some culture medium, which is also reasonably modeled as a Stokes fluid. In each region,

$$\nabla \cdot \left(\mu\left(\nabla \mathbf{v} + (\nabla \mathbf{v})^T\right)\right) - \nabla p = 0, \tag{1}$$

$$\nabla \cdot \mathbf{v} = 0, \tag{2}$$

where p is a pressure, $\mathbf{v}$ the tissue velocity, and $\mu = \mu_e$ or μ_m the viscosity in epithelium and mesenchyme (or medium), respectively. We assume that viscosity is constant within each tissue. On the interface between the two tissues (the basal lamina), there are jump boundary conditions

$$[\mathbf{v}] = 0, \tag{3}$$

$$[p] = 2\left[\frac{\partial(\mu \mathbf{v})}{\partial \mathbf{n}} \cdot \mathbf{n}\right] - 2\gamma\kappa, \tag{4}$$

$$\left[\frac{\partial p}{\partial \mathbf{n}}\right] = \left[\nabla^2 \mu \mathbf{v} \cdot \mathbf{n}\right], \tag{5}$$

and the additional condition (no Marangoni stresses)

$$\frac{\partial \gamma}{\partial \mathbf{t}_i} = 0, \quad i = 1, 2, \tag{6}$$

where $\mathbf{n}$, $\mathbf{t}_1$, and $\mathbf{t}_2$ are the local unit outward normal and tangents at the interface, γ is the surface tension, and κ is the mean curvature. In the interests of parsimony and clarity, we focused solely on the tissues' mechanical response to forces, and ignored many of the known features of branching systems, such as growth. Since branching morphogenesis is robust, it is fair to assume that the effect of forces far from the epithelio-mesenchymal interface is small; we used periodic boundary conditions on the outside of the mesenchyme/culture medium. Finally, the clefting force is input to the model as a point force at a specified value in specific locations and held constant in magnitude and direction for the duration of clefting (Fig. 10).

This model demonstrated several mechanical features:

1. The clefting force strongly affects the time course of branching morphogenesis. Decreasing the clefting force by half an order of magnitude increases the clefting time by 2 orders of magnitude.
2. Mesenchymal viscosity strongly affects the time course of branching morphogenesis. Increasing the viscosity of the material the epithelium is embedded in by an order of magnitude increases the clefting time by half an order of magnitude.
3. The ratio of viscosities of epithelium and mesenchyme strongly affects morphology of the clefts, with more viscous mesenchyme yielding wider clefts.

These observations suggest new interpretation of the recombination experiments and the mesenchyme-free experiments. The form of a combination branched organ is determined by the mesenchyme (see review above); we suggest that the form is determined specifically by the viscosity of that mesenchyme. If the material in which the

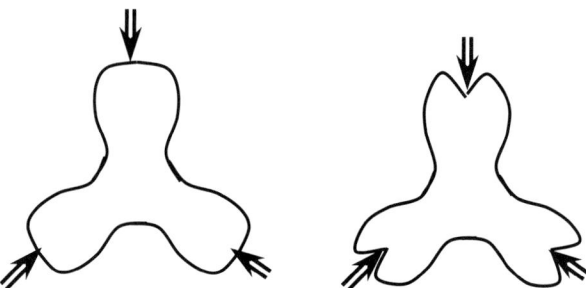

Figure 10 Branching modeled by Stokes fluids deformed by a fixed force. Epithelium and mesenchyme each have constant viscosity. Growth is omitted. Deformations depend on viscosity ratio, surface tension, and clefting force (arrows). Results suggest that forces from epithelium alone are insufficient to generate observed shape changes when epithelium is embedded in mesenchyme. Figure schematic after simulations from (Lubkin and Li, 2002).

(mesenchyme-free) epithelium is embedded is DMEM or some gel such as Matrigel, the viscosity of that material is many orders of magnitude smaller than that of mesenchyme, so the mechanical situation is very different from the mechanical situation when epithelium branches inside mesenchyme, either its own or that of some other organ. Replacing an epithelium's native mesenchyme with a low-viscosity material should:

1. Dramatically speed up clefting.
2. Dramatically narrow the resulting clefts.

If these are not observed, then removal of the mesenchyme does not solely remove a resistance to deformation. Removing the mesenchyme must also be removing a clefting force. Indeed, in some beautifully controlled experiments it can be seen that while mesenchyme-free SMG epithelium will branch *in vitro*, the addition of mesenchyme substantially increases branching morphogenesis in the same time period (Wei et al., 2007).

To put it another way, epithelium is capable of generating a certain amount of clefting force. How much deformation results from that force depends on the material that is being deformed. A pedestrian can make excellent progress in the fresh air, slow progress in a swimming pool, and no progress in a tar pit. For epithelium to deform mesenchyme just as well as it deforms Matrigel, it must be getting help from the mesenchyme.

We conclude from the viscosity model (Lubkin and Li, 2002) that there are multiple forces contributing to branching morphogenesis, some from the epithelium, some from the mesenchyme, and perhaps some from the interaction of the tissues.

B. Fingering Instability

A passive fluid, left to itself, will respond to its surface tension by minimizing its surface area. This property of embryonic tissues is counterproductive to generating a structure of ducts, the primary role of which is to *increase* effective surface area. What is the driving force causing an increase in curvature?

Motivated by observations in the late fetus that transpleural pressure has a strong effect on the development of the lung (Alcorn et al., 1977; Fewell et al., 1983; Moessinger et al., 1990), and by other observations that epithelia tend to modify their configuration and regulate their growth in order to maintain a uniform tangential stress (Kolega, 1981; Takeuchi, 1979), we modeled the developing lung as a viscous fluid (mesenchyme) surrounding a less viscous fluid (lumen contents) with a surface tension due to the epithelium (Lubkin and Murray, 1995). A combination of lumenal pressure and pressure resulting from growth, interacting with the epithelial surface tension, gave a pattern-forming instability which yielded a branched hollow structure of the same morphology and on the same developmental time scale as an embryonic lung.

Since (1) the length and time scales, and the resulting morphology, are the same in the viscous fingering model as in the embryonic lung, and (2) the proposed mechanism is plausible, clearly the proposed mechanism is a sufficient mechanism for generating branching morphogenesis. However, that does not imply that it is the primary mechanism in any real tissue system (certainly not in the SMG, which during branching has no lumen). This mechanism cannot be required for branching morphogenesis, because (a) branching morphogenesis can proceed without mesenchyme, (b) branching morphogenesis can proceed without growth (hence without growth pressure), and (c) the robustness of branching morphogenesis both developmentally and evolutionarily suggests that the branching process is more likely to be a highly controlled transient phenomenon rather than a near-equilibrium instability.

C. Mesenchymal Force

Our previous model (Lubkin and Li, 2002) revealed the importance of the passive mechanical role of mesenchyme in branching morphogenesis, and indicated that it must play an active role for normal morphogenesis to take place. What specific active morphogenetic force does mesenchyme contribute? We know several relevant experimental observations.

Disruption of collagen or collagen synthesis interferes with lung and salivary gland morphogenesis (Cutler, 1990; Spooner and Faubion, 1980). Collagenase disrupts lung (Fukuda et al., 1988; Spooner and Faubion, 1980; Wessells and Cohen, 1968) and salivary gland branching (Fukuda et al., 1988; Grobstein and Cohen, 1965; Spooner et al., 1986a), and collagenase inhibitors enhance salivary branching (Nakanishi et al., 1986b). The effect is mechanical: in recombination experiments where SMG epithelia were physically separated from SMG mesenchyme by a filter, collagenases did not affect branching morphogenesis, although they did in systems where the epithelium and mesenchyme were in physical contact (Mori et al., 1994).

The bottom of clefts in 12–13 day mouse submandibular glands contains an array of roughly parallel fibrils (Nakanishi et al., 1986a). These cleft-bottom bundles are thicker (0.3–1.0 micron diameter) if the SMG rudiment is treated with a collagenase inhibitor (Fukuda et al., 1988). Mesenchymal cells in SMG rapidly move along the epithelial surface and in the cleft, where ECM fibers are aligned (Hieda and Nakanishi, 1997). Collagenase interferes with this movement (Nakanishi and Ishii, 1989).

Supported by the above observations, Nakanishi and colleagues developed a conceptual model of clefting driven by traction force from mesenchymal cells in the presumptive cleft region (Nakanishi et al., 1986a; Nogawa and Nakanishi, 1987). We generalized the two-viscous fluid model (Lubkin and Li, 2002) to include a realistic description of the hypothesized mesenchymal contractility (Wan et al., 2007).

We modeled both epithelium and mesenchyme as Stokes fluids of constant viscosities μ_e and μ_m, respectively. The force balance and volume conservation equations for the epithelium are as described in (1)–(2), but the force balance in the mesenchyme is

8. Mechanics of Branching Morphogenesis

modified to include a term for traction force

$$\nabla \cdot \left(\mu_m\left(\nabla \mathbf{v} + (\nabla \mathbf{v})^T\right)\right) - \nabla p + \nabla(\tau c) = 0, \tag{7}$$

where τ is the averaged contractile stress per cell, and where c is the concentration of contractile fibroblasts in the mesenchyme.

It remains to model the transport of contractile fibroblasts. Manoussaki et al. (Manoussaki *et al.*, 1996) modeled the mechanics of vasculogenesis by contractility, and included a term for biased random movement of cells in the direction of ECM alignment. They found that the pattern formation was independent of any random cell motion. Furthermore, when the model is completely isotropic, the vasculogenesis proceeds as before. Therefore, in the interests of simplicity, we model branching morphogenesis with isotropy in all components. This means that the active contractile stresses take the form of a pressure in (7), and the contractile cells can be tracked with a simple advection–diffusion equation

$$\frac{\partial c}{\partial t} + \nabla \cdot (c\mathbf{v}) = D\nabla^2 c,$$

where $\mathbf{v}$ is the local tissue velocity and D is the coefficient of random motility. The initial condition for the location of contractile fibroblasts is unknown. We hypothesize an arc of contractility around an initially spherical lobe. Simulations using physically justified parameter values in 2D (Fig. 11) and 3D (Wan *et al.*, 2007) show cleft formation on realistic time and length scales, with morphology dependent on viscosity and surface tension, as previously observed (Lubkin and Li, 2002). Our model clearly shows that a hypothetical distribution of contractile fibroblasts can generate a cleft on the correct time scale and of the correct morphology.

V. Discussion

Branching morphogenesis, a ubiquitous phenomenon, displays different features in different organs and organisms. The gene regulation in various branching systems has been richly explored. Unfortunately knowing all the genes and their signaling networks will not explain what the physical forces are which generate the form of a branched organ. Development is not outside the realm of physics; it is just that the physics of development is (at the present moment) new and challenging. It's not "rocket science"—it's harder.

It is possible to imagine at least a dozen different mechanisms which are capable of producing a branched duct structure. It is also possible to imagine completely plausible branching mechanisms which are not actually physically possible. A conceptual model, usually expressed in words or in a diagram containing arrows, is a beginning. Translation of these words and diagrams into careful definitions and justifiable equations is the realm of mathematical modeling.

(a)

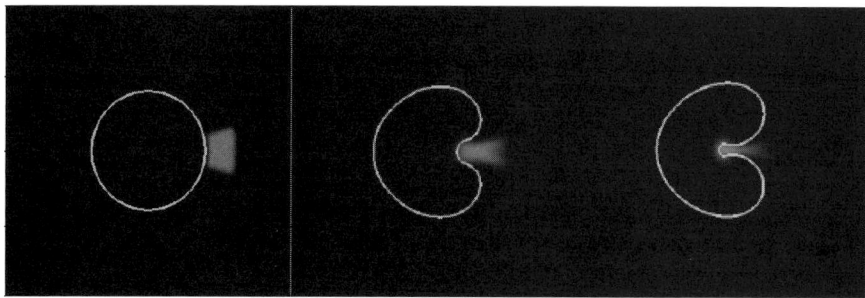

(b)

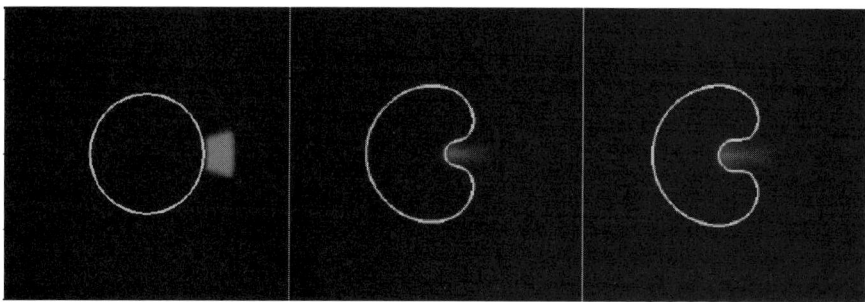

Figure 11 Morphogenesis of 2D 2-fluid model simulation with explicitly modeled mesenchymal contractility. Interface between epithelium and mesenchyme (basal lamina) in green. Lobe modeled as initially spherical. Contractile region of mesenchyme in red. Box 400 μm on a side. Contractility 500 Pa. Surface tension 10^{-6} Pa-m = 1 Pa-μm. Epithelium assumed to be initially round. Viscosity of epithelium and mesenchyme (a) equal, $\mu_{epi} = \mu_{mes} = 5 \times 10^5$ poise ~ 5 Pa-day; (b) $\mu_{epi} = 5$ Pa-day, $\mu_{mes} = 50$ Pa-day. Times (a) 0, 6, 12 hours; (b) 0, 24, 48 hours. Larger mesenchymal viscosity (b) slows clefting and yields a wider cleft. Images courtesy of Xiaohai Wan. See color insert.

We have systematically examined the physical consequences of several hypothesized morphogenetic mechanisms (Lubkin and Jackson, 2002; Lubkin and Li, 2002; Lubkin and Murray, 1995; Manoussaki *et al.*, 1996; Wan *et al.*, 2007). We have built a collection of relatively simple, physically justifiable models with measurable parameters. Analysis of these models has clarified the magnitude of deformations in response to specific forces.

It has become clear that branching morphogenesis *in situ* is generated by forces from both epithelium and mesenchyme (Lubkin and Li, 2002). The epithelial mechanisms can be observed most clearly in mesenchyme-free preparations *in vitro*. The mesenchymal mechanisms are harder to separate from the epithelial mechanisms, but abundant experimental observations and our modeling analyses indicate that the role of the mesenchyme is substantial. Clarifying the contributions and regulation of the mesenchyme could perhaps be fruitfully explored in epithelium-free preparations, where the epithelium is replaced by a cell-free material of similar passive physical properties.

Branching morphogenesis is likely the result of multiple mechanisms operating in tandem. It is one of the great strengths of mathematical modeling that it can, if carefully done, quantify the relative contributions of different mechanisms, and the sensitivity of a system to variations in the physical parameters of the system. Complete understanding of a system requires a quantifiable framework; this is no less true in developmental biology. We aim to build that quantifiable framework, one component at a time.

Acknowledgments

This work has been supported by National Science Foundation Grant DMS-0201094. Images in Fig. 11 courtesy of Xiaohai Wan.

References

Alcorn, D., Adamson, T. M., Lambert, T. F., Maloney, J. E., Ritchie, B. C., and Robinson, P. M. (1977). Morphological effects of chronic tracheal ligation and drainage in the fetal lamb lung. *J. Anat.* **123**, 649–660.

Banerjee, S. D., Cohn, R. H., and Bernfield, M. R. (1977). Basal lamina of embryonic salivary epithelia—production by epithelium and role in maintaining lobular morphology. *J. Cell Biol.* **73**, 445–463.

Bard, J. B. L. (1990). "Morphogenesis: The Cellular and Molecular Processes of Developmental Anatomy." Cambridge Univ. Press, Cambridge.

Barocas, V., Moon, A., and Tranquillo, R. (1995). The fibroblast-populated collagen microsphere assay of cell traction force. Part 2. Measurement of the cell traction parameter. *J. Biomech. Eng.* **117**, 161–170.

Borghese, E. (1950a). Explantation experiments on the influence of the connective tissue capsule on the development of the epithelial part of the submandibular gland of Mus musculus. *J. Anat.* **84**, 303–318.

Borghese, E. (1950b). The development in vitro of the submandibular and sublingual glands of Mus musculus. *J. Anat.* **84**, 287–300.

Brekken, R. A., Puolakkainen, P., Graves, D. C., Workman, G., Lubkin, S. R., and Sage, E. H. (2003). Enhanced growth of tumors in sparc null mice is associated with changes in the ecm. *J. Clin. Invest.* **111**, 487–495.

Costantini, F. (2006). Renal branching morphogenesis: Concepts, questions, and recent advances. *Differentiation* **74**, 402–421.

Cutler, L. S. (1990). The role of extracellular matrix in the morphogenesis and differentiation of salivary glands. *Adv. Dent. Res.* **4**, 27–33.

Discher, D. E., Janmey, P., and Wang, Y. L. (2005). Tissue cells feel and respond to the stiffness of their substrate. *Science* **310**, 1139–1143.

Doljanski, F. (2004). The sculpturing role of fibroblast-like cells in morphogenesis. *Perspect. Biol. Med.* **47**, 339–356.

Engler, A. J., Sen, S., Sweeney, H. L., and Discher, D. E. (2006). Matrix elasticity directs stem cell lineage specification. *Cell* **126**, 677–689.

Fewell, J. E., Hislop, A. A., Kitterman, J. A., and Johnson, P. (1983). Effect of tracheotomy on lung development in fetal lambs. *J. Appl. Physiol.* **55**, 1103–1108.

Flint, J. M. (1902). The development of the reticulated basement membrane in the submaxillary gland. *Am. J. Anat.* **2**, 1–11.

Flint, J. M. (1903). The angiology, angiogenesis, and organogenesis of the submaxillary gland. *Am. J. Anat.* **2**, 417–444.

Flint, J. M. (1906). The development of the lungs. *Am. J. Anat.* **6**, 1–136.

Forgacs, G., Foty, R. A., Shafrir, Y., and Steinberg, M. S. (1998). Viscoelastic properties of living embryonic tissues: A quantitative study. *Biophys. J.* **74**, 2227–2234.

Forgacs, G., and Newman, S. A. (2005). "Biological Physics of the Developing Embryo." Cambridge Univ. Press, Cambridge.

Foty, R. A., Forgacs, G., Pfleger, C. M., and Steinberg, M. S. (1994). Liquid properties of embryonic tissues: Measurement of interfacial tensions. *Phys. Rev. Lett.* **72**, 2298–2301.

Fukuda, Y., Masuda, Y., Kishi, J. I., Hashimoto, Y., Hayakawa, T., Nogawa, H., and Nakanishi, Y. (1988). The role of interstitial collagens in cleft formation of mouse embryonic submandibular-gland during initial branching. *Development* **103**, 259–267.

Grobstein, C., and Cohen, J. (1965). Collagenase—effect on morphogenesis of embryonic salivary epithelium in vitro. *Science* **150**, 626–628.

Guo, W., and Wang, Y. (2004). The substrate rigidity regulates the formation and maintenance of tissues. *Mol. Biol. Cell* **15**, 3A.

Harris, A. K., Wild, P., and Stopak, D. (1980). Silicone-rubber substrata—new wrinkle in the study of cell locomotion. *Science* **208**, 177–179.

Hieda, Y., and Nakanishi, Y. (1997). Epithelial morphogenesis in mouse embryonic submandibular gland: Its relationships to the tissue organization of epithelium and mesenchyme. *Dev. Growth Differ.* **39**, 1–8.

Hieda, Y., Iwai, K., Morita, T., and Nakanishi, Y. (1996). Mouse embryonic submandibular gland epithelium loses its tissue integrity during early branching morphogenesis. *Dev. Dynam.* **207**, 395–403.

Hilfer, S. R. (1996). Morphogenesis of the lung: Control of embryonic and fetal branching. *Annu. Rev. Physiol.* **58**, 93–113.

Iwai, K., Hieda, Y., and Nakanishi, Y. (1998). Effects of mesenchyme on epithelial tissue architecture revealed by tissue recombination experiments between the submandibular gland and lung of embryonic mice. *Dev. Growth Differ.* **40**, 327–334.

Kolega, J. (1981). The movement of cell clusters in vitro: Morphology and directionality. *J. Cell Sci.* **49**, 15–32.

Kratochwil, K. (1969). Organ specificity in mesenchymal induction demonstrated in embryonic development of mammary gland of mouse. *Dev. Biol.* **20**, 46–71.

Larsen, M., Wei, C., and Yamada, K. M. (2006). Cell and fibronectin dynamics during branching morphogenesis. *J. Cell Sci.* **119**, 3376–3384.

Lubkin, S. R., and Jackson, T. (2002). Multiphase mechanics of capsule formation in tumors. *J. Biomech. Eng.* **124**, 237–243.

Lubkin, S. R., and Li, Z. (2002). Force and deformation on branching rudiments: Cleaving between hypotheses. *Biomech. Model. Mechanobiol.* **1**, 5–16.

Lubkin, S. R., and Murray, J. D. (1995). A mechanism for early branching in lung morphogenesis. *J. Math. Biol.* **34**, 77–94.

Manoussaki, D., Lubkin, S. R., Vernon, R. B., and Murray, J. D. (1996). A mechanical model for the formation of vascular networks in vitro. *Acta Biotheor.* **44**, 271–282.

Martin, P., and Lewis, J. (1992). Actin cables and epidermal movement in embryonic wound healing. *Nature* **360**, 179–183.

Michael, L., Sweeney, D. E., and Davies, J. A. (2005). A role for microfilament-based contraction in branching morphogenesis of the ureteric bud. *Kidney Int.* **68**, 2010–2018.

Mittenthal, J. E., and Mazo, R. M. (1983). A model for shape generation by strain and cell–cell adhesion in the epithelium of an arthropod leg segment. *J. Theor. Biol.* **100**, 443–483.

Moessinger, A. C., Harding, R., Adamson, T. M., Singh, M., and Kiu, G. T. (1990). Role of lung fluid volume in growth and maturation of the fetal sheep lung. *J. Clin. Invest.* **86**, 1270–1277.

8. Mechanics of Branching Morphogenesis

Moore, K. A., Polte, T., Huang, S., Shi, B., Alsberg, E., Sunday, M. E., and Ingber, D. E. (2005). Control of basement membrane remodeling and epithelial branching morphogenesis in embryonic lung by rho and cytoskeletal tension. *Dev. Dynam.* **232**, 268–281.

Moral, H. (1913). Über die ersten Entwickelungsstadien der Glandula submaxillaris. *Anat. Embryol.* **47**, 277–382.

Moral, H. (1916). Zur Kenntnis von der Speicheldrüsenentwicklung der Maus. *Anat. Embryol.* **53**, 351–679.

Mori, Y., Yoshida, K., Morita, T., and Nakanishi, Y. (1994). Branching morphogenesis of mouse embryonic submandibular epithelia cultured under three different conditions. *Dev. Growth Differ.* **36**, 529–539.

Nakanishi, Y., Hieda, Y., Cardoso, W. V., Lubkin, S. R., and Daniel, C. W. (2001). Epithelial Branching. In "Encyclopedia of Life Sciences." Wiley.

Nakanishi, Y., and Ishii, T. (1989). Epithelial shape change in mouse embryonic submandibular gland: Modulation by extracellular matrix components. *BioEssays* **11**, 163–167.

Nakanishi, Y., Morita, T., and Nogawa, H. (1987). Cell proliferation is not required for the initiation of early cleft formation in mouse embryonic submandibular epithelium in vitro. *Development* **99**, 429–437.

Nakanishi, Y., Sugiura, F., Kishi, J. -I., and Hayakawa, T. (1986a). Scanning electron microscopic observation of mouse embryonic submandibular glands during initial branching: Preferential localization of fibrillar structures at the mesenchymal ridges participating in cleft formation. *J. Embryol. Exp. Morph.* **96**, 65–77.

Nakanishi, Y., Sugiura, F., Kishi, J. I., and Hayakawa, T. (1986b). Collagenase inhibitor stimulates cleft formation during early morphogenesis of mouse salivary-gland. *Dev. Biol.* **113**, 201–206.

Nelson, C. M., and Bissell, M. J. (2006). Of extracellular matrix, scaffolds, and signaling: Tissue architecture regulates development, homeostasis, and cancer. *Annu. Rev. Cell Dev. Biol.* **22**, 287–309.

Nelson, C. M., Jean, R. P., Tan, J. L., Liu, W. F., Sniadecki, N. J., Spector, A. A., and Chen, C. S. (2005). Emergent patterns of growth controlled by multicellular form and mechanics. *Proc. Natl. Acad. Sci. USA* **102**, 11594–11599.

Nogawa, H. (1983). Determination of the curvature of epithelial-cell mass by mesenchyme in branching morphogenesis of mouse salivary-gland. *J. Embryol. Exp. Morph.* **73**, 221–232.

Nogawa, H., and Nakanishi, Y. (1987). Mechanical aspects of the mesenchymal influence on epithelial branching morphogenesis of mouse salivary-gland. *Development* **101**, 491–500.

Nogawa, H., and Takahashi, Y. (1991). Substitution for mesenchyme by basement-membrane-like substratum and epidermal growth-factor in inducing branching morphogenesis of mouse salivary epithelium. *Development* **112**, 855–861.

Nogawa, H., Morita, K., and Cardoso, W. V. (1998). Bud formation precedes the appearance of differential cell proliferation during branching morphogenesis of mouse lung epithelium in vitro. *Dev. Dynam.* **213**, 228–235.

Odell, G., Oster, G., Burnside, B., and Alberch, P. (1980). A mechanical model for epithelial morphogenesis. *J. Math. Biol.* **9**, 291–295.

Parmar, H., and Cunha, G. R. (2004). Epithelial-stromal interactions in the mouse and human mammary gland in vivo. *Endocr. Relat. Cancer* **11**, 437–458.

Pelham, R. J., and Wang, Y. (1997). Cell locomotion and focal adhesions are regulated by substrate flexibility. *Proc. Natl. Acad. Sci.* **94**, 13661–13665.

Sakai, T., Larsen, M., and Yamada, K. M. (2003). Fibronectin requirement in branching morphogenesis. *Nature* **423**, 876–881.

Sakakura, T., Nishizuka, Y., and Dawe, C. J. (1976). Mesenchyme-dependent morphogenesis and epithelium-specific cytodifferentiation in mouse mammary gland. *Science* **194**, 1439–1441.

Sawhney, R. K., and Howard, J. (2002). Slow local movements of collagen fibers by fibroblasts drive the rapid global self-organization of collagen gels. *J. Cell Biol.* **157**, 1083–1091.

Spooner, B. S., and Faubion, J. M. (1980). Collagen involvement in branching morphogenesis of embryonic lung and salivary gland. *Dev. Biol.* **77**, 84–102.

Spooner, B. S., and Wessells, N. K. (1972). Analysis of salivary gland morphogenesis: Role of cytoplasmic microfilaments and microtubules. *Dev. Biol.* **27**, 38–54.

Spooner, B. S., Thompson, H. A., Stokes, B., and Bassett, K. (1986a). Extracellular matrix macromolecules and epithelial branching morphogenesis. *Am. Zool.* **26**, 545–547.

Spooner, B. S., Thompson-Pletscher, H. A., Stokes, B., and Bassett, K. E. (1986b). Extracellular matrix involvement in epithelial branching morphogenesis. *Dev. Biol. (NY 1985)* **3**, 225–260.

Spooner, B. S., Bassett, K. E., and Spooner Jr., B. S. (1989). Embryonic salivary gland epithelial branching activity is experimentally independent of epithelial expansion activity. *Dev. Biol.* **133**, 569–575.

Takahashi, Y., and Nogawa, H. (1991). Branching morphogenesis of mouse salivary epithelium in basement membrane-like substratum separated from mesenchyme by the membrane-filter. *Development* **111**, 327–335.

Takeuchi, S. (1979). Wound healing in the cornea of the chick embryo. 4. Promotion of the migratory activity of isolated corneal epithelium in culture by the application of tension. *Dev. Biol.* **70**, 232–240.

Trinkaus, J. P. (1984). "Cells into Organs: The Forces that Shape the Embryo." Prentice–Hall, Englewood Cliffs.

Wan, X., Li, Z., and Lubkin, S.R. (2007). Mechanics of mesenchymal contribution to clefting force in branching morphogenesis (in review).

Watanabe, T., and Costantini, F. (2004). Real-time analysis of ureteric bud branching morphogenesis in vitro. *Dev. Biol.* **271**, 98–108.

Wei, C., Melinda Larsen, P. D., Hoffman, M. P., Bds, P. D., and Yamada, K. M. (2007). Self-organization and branching morphogenesis of primary salivary epithelial cells. *Tissue Eng.* **13**.

Wessells, N. K., and Cohen, J. H. (1968). Effects of collagenase on developing epithelia in vitro: Lung, ureteric bud, and pancreas. *Dev. Biol.* **18**, 294–309.

ism # 9

Multicellular Sprouting during Vasculogenesis

Andras Czirok,*,† Evan A. Zamir,* Andras Szabo,†
and Charles D. Little*

*Department of Anatomy and Cell Biology, University of Kansas Medical Center, Kansas
City, Kansas 66160
†Department of Biological Physics, Eotvos University, Budapest 1117, Hungary

I. Introduction
   A. Vasculogenesis
   B. Self-Organization vs Genetic Hard-Wiring
   C. Theories of Self-Assembly
II. Empirical Data, *in vivo*
   A. Vascular Drift
   B. Vasculogenic Sprouts
   C. Cell–Cell Adhesion
III. Elongated Structures, *in vitro*
IV. Mathematical Model of Sprout Formation
   A. Equation of Motion
   B. Interaction with the Environment
   C. Simulation Results
V. Conclusions
   Acknowledgments
   References

Living organisms, from bacteria to vertebrates, are well known to generate sophisticated multicellular patterns. Numerous recent interdisciplinary studies have focused on the formation and regulation of these structures. Advances in automated microscopy allow the time-resolved tracking of embryonic development at cellular resolution over an extended area covering most of the embryo. The resulting images yield simultaneous information on the motion of multiple tissue components—both cells and extracellular matrix (ECM) fibers. Recent studies on ECM displacements in bird embryos resulted in a method to distinguish tissue deformation and cell-autonomous motion. Patterning of the primary vascular plexus results from a collective action of primordial endothelial cells. The emerging "polygonal" vascular structure is shown to be formed by cell–cell and cell–ECM interactions: adhesion and protrusive activity (sprouting). Utilizing avb3 integrins, multicellular sprouts invade rapidly into avascular areas. Sprout elongation, in turn, depends on a continuous supply of endothelial cells. Endothelial cells migrate along the sprout, towards its tip, in a vascular endothelial (VE) cadherin-dependent process. The observed abundance of multicellular sprout formation in various *in vitro* and *in vivo* systems can be explained by a general mechanism based on preferential attraction to elongated structures. Our interacting particle model exhibits robust sprouting dynamics and results in patterns with morphometry similar to native primordial vascular plexuses—without ancillary assumptions involving chemotaxis or mechanochemical signaling. © 2008, Elsevier Inc.

I. Introduction

A. Vasculogenesis

It is widely assumed that adhesion-based activities such as exertion of traction and compressional forces, shape-change and motility are the physical means by which tissues and organs are formed (Trinkaus, 1984; Forgacs and Newman, 2005). However, our knowledge is limited about how collective cell behavior creates a specific physical tissue or organ (Keller et al., 2003).

Vasculogenesis of warm blooded vertebrates, the formation of a primary vascular pattern from mesodermal-derived precursors (angioblasts), is an excellent system in which to address tissue pattern emergence. In bird embryos, well before the onset of circulation, hundreds of essentially identical vascular endothelial cells create a polygonal network within a relatively uncomplicated, sheet-like anatomical environment (Risau and Flamme, 1995). Each link in the polygonal network is a cord consisting of 3–10 endothelial cells (Drake et al., 1997).

It has been known for nearly a century that endothelial cells differentiate from solitary angioblasts within the avian embryo (Area Pellucida) (Reagan, 1915; Sabin, 1920). Committed angioblasts display a random spatial distribution within the mesoderm, without an observable *preexisting* pattern at length scales comparable to that of the future primary vascular polygons (Drake et al., 1997).

Vasculogenic processes may also occur in certain pathophysiologies. Recent data provided evidence that vascular endothelial cell progenitors exist in the adult and may become bloodborne, enter extravascular tissues, and promote *de novo* vessel formation (Zammaretti and Zisch, 2005; Rumpold et al., 2004). For that reason, endothelial progenitors, mobilized in situ or transplanted, are a major target of therapeutic vascularization approaches to prevent ischemic disease and control endothelial injury. Moreover, endothelial progenitors represent a potential target for strategies to block tumor growth, and a requisite mechanism for tissue engineering (Wu et al., 2004). The capacity of less than fully differentiated cells to assemble into vascular-like tubes is also manifested in various tumors. Best characterized are highly malignant melanomas in which tumor cells are assembling into tubes to secure blood supply (Hendrix et al., 2003). Thus, arguably, the least understood and most important question facing vascular developmental biologists and tissue engineers today is—what are the general principles guiding morphogenesis of an endothelial tube network?

B. Self-Organization vs Genetic Hard-Wiring

Conventional models of vasculogenesis often assume that endothelial cells migrate to *pre- and well-defined* positions following extracellular guidance cues or chemoattractants (Ambler *et al.*, 2001; Cleaver and Krieg, 1998; Poole and Coffin, 1989). However, the capacity of endothelial cells to form a polygonal pattern is preserved in various *in vitro* systems, where the presence of a genetic prepattern is unlikely. The mouse allantois, when explanted, forms a vascular network very similar to the primary vascular pattern (LaRue *et al.*, 2003)—instead of a pair of umbilical vessels. Similarly, a polygonal vascular network emerges when endothelial cells are placed in three-dimensional collagen gels (Montesano and Orci, 1985; Davis *et al.*, 2000). We argue that the ubiquitous polygonal networks are *self-organized* in the sense that we do not expect a direct correspondence between gene expression patterns and the position of individual segments within the primary vascular network.

This concept of self-organization is markedly different from that of genetic hard-wiring. During insect segmentation the position and identity of body segments is directly correlated with a gene expression pattern. Similarly, genetic prepatterning is clearly involved in the determination of major vessels in the developing vascular network. In fish, where the major vessels assemble directly, without forming an intermediate vascular plexus, specific vascular malformations are correlated with genetic defects (Weinstein, 1999). Thus, endothelial progenitors presumably respond to various environmental cues, including the presence of other endothelial cells as well as genetically preprogrammed "zip codes."

C. Theories of Self-Assembly

Various hypotheses are proposed to explain the self-organized aspect of vasculogenesis. The *mechanochemical* mechanism assumes cells to exert mechanical stress on the underlying substrate, and the resulting stress to guide cell motility (Murray *et al.*, 1998; Murray, 2003). A variant of the mechanochemical model proposes that angioblasts first segregate into compact clusters and engage the surrounding ECM fibers. As a result of traction forces, ECM bundles develop, which in turn later route the motile primordial endothelial cells between clusters (Drake *et al.*, 1997; Manoussaki *et al.*, 1996; Vernon *et al.*, 1995).

A recent body of research focused on pattern emergence based on *autocrine* chemotactic signaling (Gamba *et al.*, 2003; Serini *et al.*, 2003). The suggested chemoattractant, $VEGF_{165}$, is unlikely to fit the model assumptions in embryonic vasculogenesis as $VEGF_{165}$ is expressed throughout the embryo except in endothelial cells (Flamme *et al.*, 1995; Poole *et al.*, 2001). However, patterning guided by an unspecified chemoattractant continues to serve as the basis of biologically plausible models resolving individual cells (Merks *et al.*, 2006) as well as freely diffusive or matrix-bound chemoattractants (Bauer *et al.*, 2007).

Recently we proposed that mechanical effects may also modulate cell migration guided by adjacent cells, a view we will further explore in Chapter III.

II. Empirical Data, *in vivo*

Most of the experimental biology underpinning the mathematical models has relied on *in vitro* studies. However, vascular morphogenesis occurs at times and places in avian embryos that are readily observed and manipulated *in vivo* (Little and Drake, 2000). Due to recent improvements in digital microscopy, it is now possible to address directly how new blood vessels form *de novo* (Czirok *et al.*, 2002; Rupp *et al.*, 2003a). Scanning time-lapse microscopy allows both the global (tissue-scale) and local (cell-autonomous) characterization of primordial endothelial cell behavior during the initial stages of avian vasculogenesis.

Endothelial cells are tracked by use of microinjected and fluorochrome-conjugated QH1 antibodies, specific for quail vascular endothelium (Pardanaud *et al.*, 1987). At each time point images can be taken in multiple optical modes: one for QH1 fluorescence, another visualizing the extracellular matrix (ECM) environment by using an immunolabeled ECM component. A third image, bright field or differential interference contrast (DIC), is also acquired to provide an anatomical frame of reference. The resulting images are aligned such that certain anatomical reference points, e.g., the intersomitic clefts remain stationary. Furthermore, due to the availability of images from multiple focal planes, no object is lost or rendered out of focus.

A. Vascular Drift

Early vasculogenesis is a time of vigorous rearrangements in the embryo, with parallel processes occurring at various length scales from the migration of individual cells to global morphogenic events, such as gastrulation or neurulation. These tissue movements, which can be mapped using ECM fibrils as passive tracers (Zamir *et al.*, 2006), profoundly influence vascular pattern formation. As Rupp *et al.* (2004) and Fig. 1 demonstrates, there is a substantial medial movement of the forming vascular plexus, a phenomenon predicted by Coffin and Poole (1988). This process, *vascular drift*, was shown to occur within almost the entire nascent vasculature. A comparison of changes in JB3 antifibrillin 2 (Rongish *et al.*, 1998) and endothelium-specific QH1 immunofluorescence reveals that the motion of vascular segments is largely coincident with the global tissue flow accompanying gastrulation and elongation of the notochord (Fig. 1). Therefore, vascular structures appear as embedded in a mechanical continuum.

Morphogenetic flows in the embryo have a profound impact on tissue structure. One experimentally accessible measure of tissue structure is its two-dimensional (2D) ECM or cell density. This measure gives the number of objects, projected onto a frontal plane, per unit area. As we demonstrated (Czirok *et al.*, 2006), morphogenetic

9. Vasculogenic Sprouts

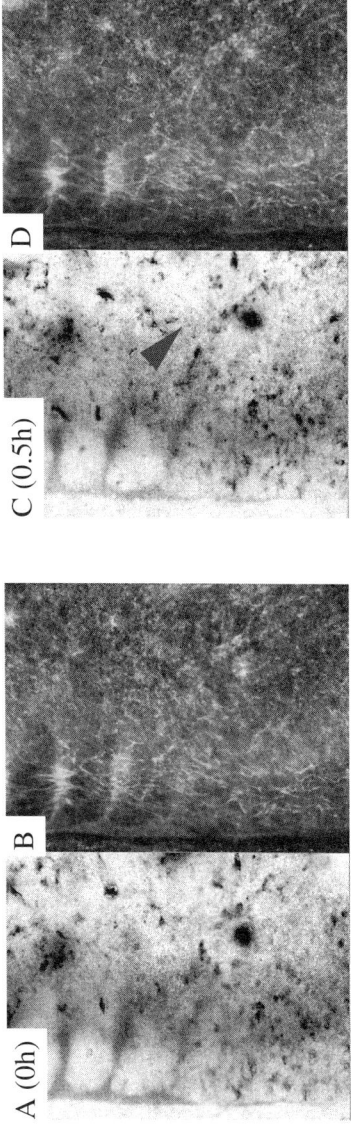

Figure 1 Medial drift of the vasculature and the surrounding extracellular matrix. Endothelial cells are shown in a HH stage 6 quail embryo (Hamburger and Hamilton, 1951), labeled by microinjected, fluorophore-conjugated QH1 antibodies. The fibrillin-2 extracellular matrix (ECM) component was labeled by JB3 antibodies. Images capturing the fluorescence of both kind of dyes allow the characterization of the endothelial cell movements relative to the ECM surrounding. Panels A–J depict the early vascular plexus forming from disconnected endothelial cell clusters within a time course of 3 hours. The left panels (A, C, E, G, and I) depict the vasculature (QH1 fluorescence, inverted for better visibility). The endothelium- (red) and fibrillin-2-specific (green) fluorescence patterns are shown superimposed on the right panels (B, D, F, H, and J). The blue arrowheads point to a vasculogenic sprout, which extends in a medial-caudal direction. Panel K depicts the fluorescence intensities along the dotted line in panel J, collected from each time-lapse image within a 6 hour-long time interval. The topmost line of panel K is the pixel-array—constituting the marked line in panel J—from the first image of the sequence. Pixel-arrays from the consecutive image sequence are then placed under each other. Tilted lines thus represent the gradual medial motion of ECM (green) and vascular (red) tissue components. The two components usually generate lines with similar slopes, indicating co-movement. However, vasculogenic sprouts (marked by the blue arrowheads) move medially much faster then the surrounding ECM. See color insert.

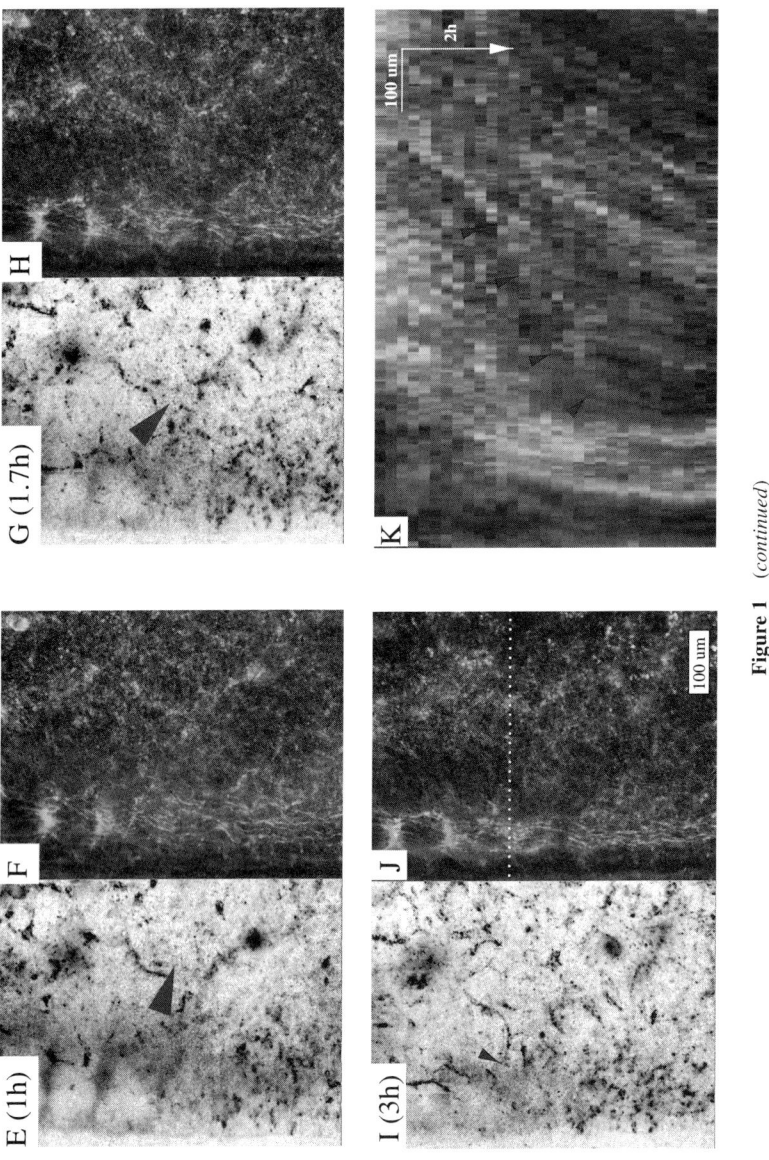

Figure 1 (continued)

movements accumulate tissue components around the somites. This increase in 2D density can reflect both an accumulation of cells and ECM, and a general expansion in tissue thickness. As Fig. 2 demonstrates, large-scale tissue movements indeed accumulate endothelial cells. As there is a strong correlation between 2D cell density and network morphology (LaRue et al., 2003), changes in endothelial cell density are associated with structural rearrangements of the vascular network. The most frequently observed manifestation of a structural rearrangement is *vascular fusion* (Drake and Little, 1999), whereby distinct polygons fuse to form a common lumen, as in the case of large vessels such as the sinus venosus, or the dorsal aortae (see Fig. 1 of Rupp et al., 2004). Tissue displacements also alter network morphology by imposing substantial strain along the embryonic axis aligning the segments of the nascent plexus (Fig. 2).

B. Vasculogenic Sprouts

Endothelial cells are also capable of active motion, relative to their ECM surroundings. The early vascular plexus is characterized by disconnected endothelial clusters. To establish the network, endothelial cells send out extensions across the avascular ECM. As demonstrated in Fig. 1, the protrusions (*vasculogenic sprouts*) typically move medially, and much faster than the drift of the ECM. This type of protrusive behavior is reminiscent of angiogenic sprouting, a process thought to be characteristic of later vascular development.

As analyzed in detail by Rupp *et al.* (2004), vasculogenic sprouts can contact neighboring extensions and thus establish endothelial cords. A stabilized protrusion may later be reinforced by subsequent addition of cells—making thereby new vertices and segments in the primary vascular pattern. As sprouts can extend hundreds of micrometers (Fig. 3), they are presumably multicellular structures and thus sprout extension involves the coordinated activity of several (3–10) endothelial cells. Conversely, existing connections are also observed to retract, albeit with much less frequency.

The empirical motility data thus reveal that protrusive activity or sprouting is the mechanism used to generate new vascular cords resulting in the polygonal vascular pattern. Only a small fraction of endothelial cells exhibit this type of protrusive behavior, and within a well-defined developmental stage (Rupp *et al.*, 2004). Furthermore, simultaneous monitoring of endothelial sprouts and changes in the surrounding ECM configuration revealed no evidence for sprout guidance by local ECM deformations (Fig. 1)—in contrast with the predictions of the mechanochemical models.

C. Cell–Cell Adhesion

By subtracting the overall medial drift of the vascular pattern, trajectories of individual endothelial cells, relative to the surrounding endothelial structures, can be established

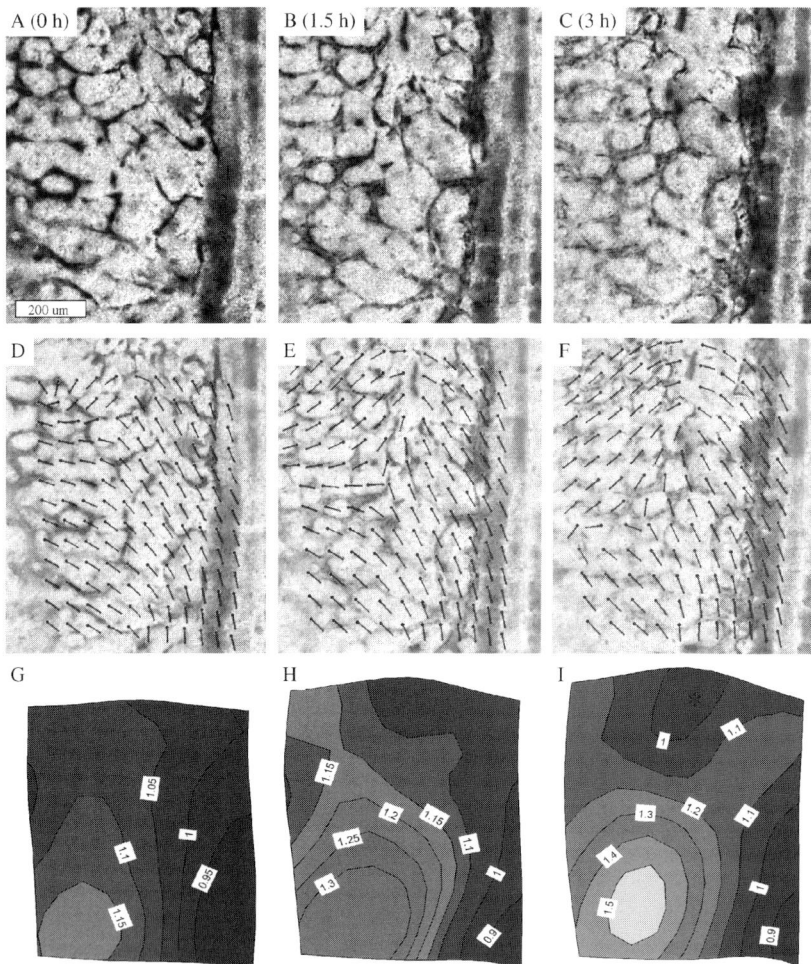

Figure 2 Large-scale tissue movements contribute to the rearrangement of the nascent vascular plexus. Panels A–C depict changes in the QH1-labeled endothelium during a time course of 3 hours within a HH stage 7 quail embryo. The somites (marked from 1 to 4) appear as a nonspecific background. The chosen time interval includes the formation of the dorsal aortae, major vessels assembling at both sides of the embryonic axis. Large vessels form by incorporating cells from adjacent smaller endothelial structures and by vascular fusion. The vasculature, together with the surrounding tissue, undergoes substantial deformations, calculated by particle image velocimetry (PIV) as described by Zamir *et al.* (2005). Panels D–F show the direction of the largest principal strain component, calculated from the tissue displacement PIV data. The area of the dorsal aortae is strained along the anterior–posterior axis. Tissue displacements also change the density (number of cells projected onto a unit area of the frontal plane) of the endothelial cells. Panels G–I show the relative change in the area of the traced tissue as a contour line plot (delineated by the markers in the corresponding panels D–F). Numbers below 1 indicate density increase. The condensation at the area marked by the asterisk in panel I is the place of the forming sinus venosus—also characterized by extensive vascular fusion. See color insert.

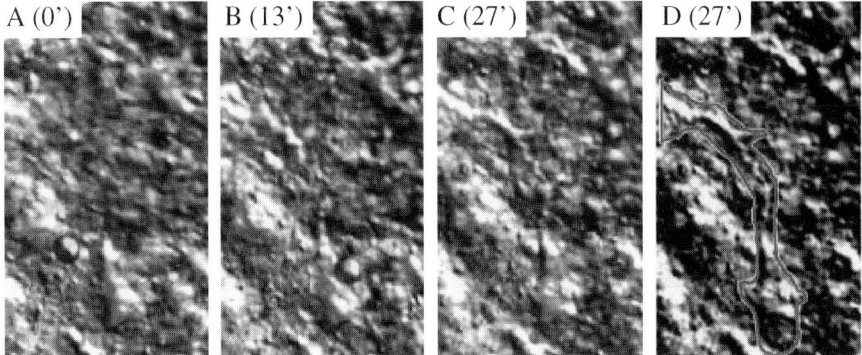

Figure 3 High resolution view of vasculogenic sprouting. Panels A–C depict changes in endothelial cell configuration during a time interval of 27 minutes. DIC images are shown superimposed with QH1 immunofluorescence (red). The extending cell protrusions eventually form a stable connection between two endothelial clusters, delineated with red in panel D. After Rupp *et al.* (2004). See color insert.

(Fig. 4). Each labeled cell can be traced back either to a cluster of endothelial cells or to an avascular area, where they appear *de novo*. These newly recruited cells move quickly until incorporation into an existing vascular structure occurs. Once part of a vascular cord, their speed is usually reduced, and when moving, they remain in close vicinity of other endothelial cells. The course of individual motile cells was found to be highly persistent along the vascular segments. There is a significant variation of motile activity within the endothelial cell population, as the substantially differing trajectories demonstrate.

Most endothelial cells move in a medial direction (Rupp *et al.*, 2003b, 2004), which is also preferred by vasculogenic sprouts. This combination of medial migration along vascular segments and medial sprouting increases the apparent speed of vascular drift above that of tissue movements (Rupp *et al.*, 2004). This increase occurs despite the fact that the vascular structure is embedded into, and moves with, the surrounding ECM as Fig. 1 demonstrates. The two types of endothelial cell motility (sprout vs individual) differ mainly with respect to their substrates: while the protrusive activity is integrin dependent and requires an active engagement of the ECM (Rupp *et al.*, 2004), endothelial cell motility along existing vascular structures appears to rely more upon vascular endothelial (VE)-cadherin mediated cell–cell interactions (Perryn *et al.*, in preparation; Ph.D. dissertation, 2006).

III. Elongated Structures, *in vitro*

The formation of linear segments via sprout-like activity, however, is not restricted to vascular endothelial cells (Szabo *et al.*, 2007). As Fig. 5 demonstrates, nonvascular C6 cells, along with muscle- or fibroblast-related cells (data not shown) also exhibit

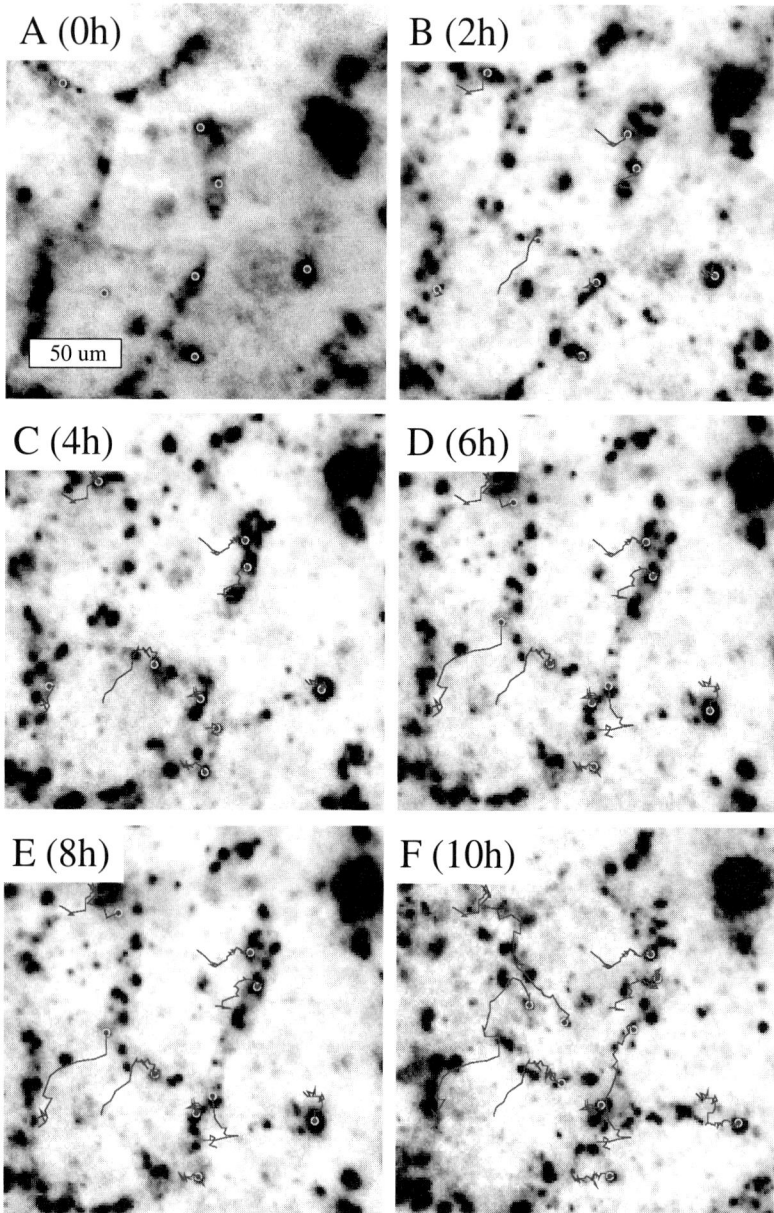

Figure 4 Trajectories of individual primordial endothelial cells, relative to the surrounding vascular structures. The panels encompass a 10 hour long period of normal development. Yellow circles indicate the current position of representative QH1 positive cells; lines show their trajectory up to the corresponding time point. The embryonic midline is towards the right. After Rupp et al. (2004). See color insert.

9. Vasculogenic Sprouts

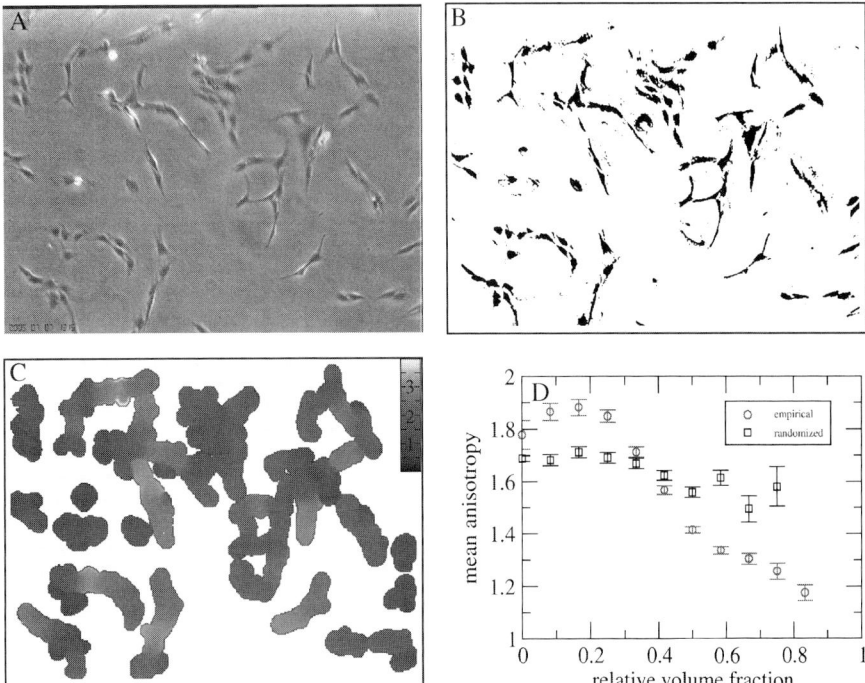

Figure 5 Various cells, grown under standard culture conditions, can also exhibit "sprouts", i.e., long arrays of interconnected cells. Panel A depicts a phase contrast image of a C6 (rat glioma) cell line. To assay the nonrandom cell positions, the phase contrast image was segmented in a two-step procedure according to Wu *et al.* (1995). As a result, cell bodies appear as black clusters in panel B. The anisotropy of the local cell configuration (panel C) is characterized by a method employing a diffusion process where the diffusion coefficient is distributed according to a long-pass filtered version of panel B. The obtained local anisotropy is dependent on the local cell density (panel D). Cell density is characterized by the relative volume fraction, the local volume fraction normalized by the maximal observed value (0.6 for C6 cells). Thus, the unit of relative volume fraction corresponds to confluent cells. As C6 cells are themselves elongated, we compared the obtained data to simulated configurations where each cell body is randomly placed on the plane. While the cell bodies of C6 cells are anisotropic themselves, the configuration of adjacent cells significantly increase the anisotropy if the local cell density is sufficiently sparse (below 30% confluency). See color insert.

linear structures when grown on a rigid plastic tissue culture substrate in the presence of a continuously shaken culture medium. Depending on the cell density, the linear segments merge and form a network. Compared to the primary vascular pattern, the resulting network appears to be more irregular due to the wider distribution of cell-free areas or segment lengths and widths. Nevertheless, the distribution of cells is far from random, and favors linear configurations.

The nonrandom placement of cell bodies can be characterized by the following morphometric procedure for *local anisotropy*. A diffusive process is started from various points of the segmented, and subsequently smoothened (long pass-filtered), image (see

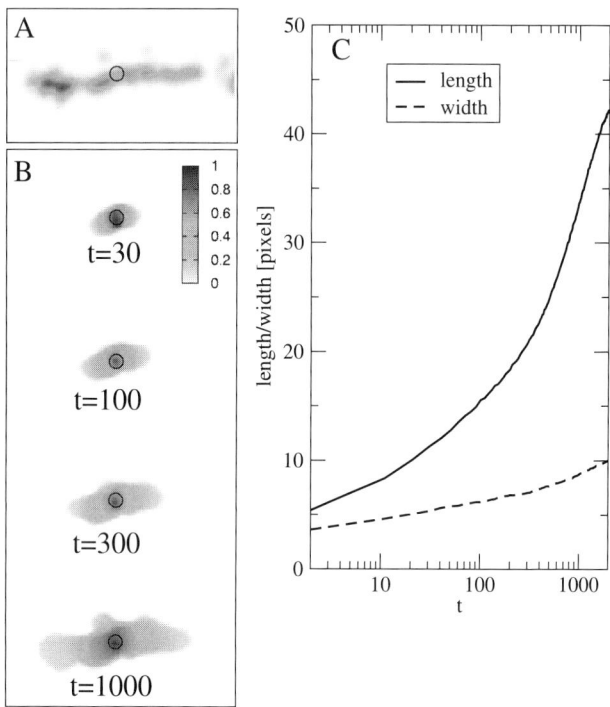

Figure 6 Calculation of anisotropy index. A diffusive process is started from a selected point of the image (circle). The diffusion coefficient changes between 0 and 1 according to the pixel values in the segmented and smoothened image (panel A). The selected point acts as a point-source of the diffusive field, where the concentration is kept steady. During the simulated diffusion process a gradually increasing area emerges around the point source (panel B). The growing area is characterized by its length and width (panel C), calculated from the principal moments of the concentration field depicted in panel B. The local anisotropy is characterized as a ratio between the area length and width at the time point when the width reaches the typical size of a single cell. See color insert.

the diagram in Fig. 6). The selected point acts as steady point-source of the diffusive field. The diffusion coefficient is zero below a concentration threshold, thus a well-defined front emerges which encloses a gradually increasing area around the point source. The growing area is characterized by the principal moments of its inertia, λ_1 and λ_2, where $\lambda_1 > \lambda_2$. If the area is circular or isotropic, these principal moments are of a similar magnitude. In contrast, highly elongated anisotropic structures result in dissimilar principal moments. The area is grown until its width, $\lambda_2^{3/8}\lambda_1^{-1/8}$, reaches a predefined value, the typical width of a single cell. The local anisotropy of the location at the point-source is then characterized as $\sqrt{\lambda_1/\lambda_2}$. Repeating the procedure for various positions results in an anisotropy map, as shown in Fig. 5c.

To correlate the average anisotropy index with local cell density, anisotropy maps were obtained from 15 independent microscopic fields of C6 cells. Cell density was characterized by the *local volume fraction* (area covered by cells on the segmented image, divided by the size of the local area). For convenience, the volume fraction values are rescaled such that a confluent cell layer has a relative volume fraction value of 1. As Fig. 5d reveals, areas with high cell densities have an isotropic structure. Very low cell densities prohibit the formation of multicellular clusters. Between these two extremes, at 20% confluency (relative volume fraction 0.2), cell arrangements have a significantly higher anisotropy index ($p < 0.01$) than in the corresponding randomized sample where each cell body is randomly placed, without overlaps, within the same area.

IV. Mathematical Model of Sprout Formation

As shown in Section II and by Szabo *et al.* (2007), linear multicellular structures form via sprouting both during *in vivo* vascular patterning, and in simple *in vitro* cell cultures. Patterning through sprouting is markedly different from the gradual coarsening of an initially uniform density field, and its possible arrest, characteristic for colloid gels (see, e.g., Foffi *et al.*, 2005) or for several models proposed to describe vasculogenesis. In particular, a "frozen" pattern was reported to gradually emerge, with an increasing spread in cell density, in the mechanochemical model (see Fig. 6, of Manoussaki *et al.* (1996), or Fig. 7 of Namy *et al.*, 2004). Gradually increasing avascular area sizes were reported in the autocrine chemoattractant model (Fig. 1 of Serini *et al.*, 2003). Except for a recent model with autocrine chemotaxis and contact inhibition (Merks *et al.*, 2007), none of these models were reported to produce sprouts—most likely because they aimed to reproduce a specific *in vitro* model system where endothelial cells are placed on the surface of a malleable gel (Vernon *et al.*, 1995).

While both the mechanochemical and chemoattractant mechanisms may be biologically relevant under certain circumstances, neither mechanism is expected to operate within the simple *in vitro* experimental setup of Section III. The rigid substrate excludes the mechanochemical mechanism. A specific chemotactic response is unlikely to be shared by such a variety of cell types. Finally, convection currents in the culture medium, generated by temperature inhomogeneities within the incubator and the vibrations of microscope stage motion, are expected to hamper the maintenance of concentration gradients, or impose a strong directional bias upon the chemotaxis-related cell movements (Szabo *et al.*, 2007).

Motivated by the empirical observations on sprout expansion guided by adjacent projections of other cells or elongated multicellular structures, recently we proposed that multicellular sprouting may employ a general mechanism, the *preferential attraction to elongated structures*. While the molecular basis of such a behavior is unknown, one may conjecture that cells in elongated structures are under mechanical tension, and strained cells can have a stiffer cytoskeleton (Xu *et al.*, 2000). Cells are able to respond to variations in extracellular matrix stiffness (Gray *et al.*, 2003), and an analogous mechanotaxis utilizing cell–cell contacts is also feasible. For example, VE-cadherin, a major cell–cell adhesion receptor of vascular endothelial cells, was recently shown to be incorporated in cell surface mechanosensing complexes (Tzima *et al.*, 2005).

A. Equation of Motion

To assess the collective behavior of cells exhibiting the proposed preferential attraction property, a simple model was studied in which individual cells were represented as particles. Cell motility is often approximated as a persistent, Ornstein–Uhlenbeck diffusion process (Stokes *et al.*, 1991; Selmeczi *et al.*, 2005), where the velocity $\vec{v}_k$ of cell k is described by the Langevin equation

$$\frac{d\vec{v}_k}{dt} = -\vec{v}_k/\tau + \sqrt{D}\xi_k + \vec{M}_k, \qquad (1)$$

where τ and D are parameters specifying the persistence time and the randomness of motion, respectively. The variable ξ represents an uncorrelated white noise: $\langle \xi \rangle = 0$ and $\langle \xi_k(t) \xi_l(t') \rangle = \delta_{kl}\delta(t - t')$. Term $\vec{M}$ is a deterministic bias, representing interaction with the environment, that is, with adjacent particles.

While Eq. (1) describes the motion of a Brownian particle at finite temperatures, animal cell motility is driven by complicated molecular machinery, and it is *not* thermal fluctuation driven. Thus, parameters τ and D depend substantially on cell type and molecular state. Measurements performed with noninteracting endothelial cells and fibroblasts resulted τ and D values in the 0.1–5 h and 100–2000 $\mu m^2/h^3$ range, respectively (Dunn, 1983; Stokes *et al.*, 1991).

B. Interaction with the Environment

Interactions among mobile agents are usually modeled as a sum of pair interactions (Helbing, 2001). In this spirit, $\vec{M}$ is factored into

$$\vec{M}_k = \sum_{\{j\}} \frac{\vec{x}_j - \vec{x}_k}{d_{kj}} \left[f_1(d_{kj}) + w_j f_2(d_{kj}) \right], \qquad (2)$$

where the sum is taken over the Voronoi neighbors of particle k, and $d_{kj} = |\vec{x}_j - \vec{x}_k|$ (see Fig. 7). The repulsion term f_1 ensures that model cells are impenetrable. The range of repulsion is the size R_1 of the organelle-packed region around the cell nucleus. The preferential attraction response is incorporated in the f_2 term. Cells are expected to explore their surroundings with protrusions as proposed by Flamme *et al.* (1993), and respond when protrusions contact elongated structures. Filopodia typically extend from R_2, the cell surface ($R_1 \leqslant R_2$), to a maximal distance of R. Thus, $f_1(d) = 0$ for $d > R_1$ and $f_2(d) = 0$ for $d < R_2$ and $d > R$. We estimate $R_1 = 10$ μm, $R_2 = 30$ μm, and $R = 40$ μm. These values, however, can vary by

9. Vasculogenic Sprouts

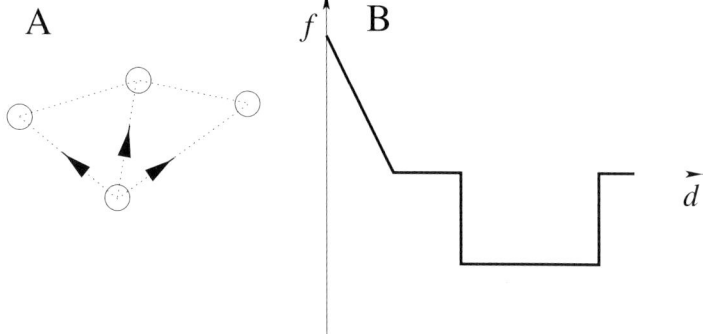

Figure 7 Schematic representation of the model. Panel A depicts particles (circles) interacting with their neighbors (dashed lines). The interaction results in deterministic drift motion (arrows) along the line connecting the particles. The direction and magnitude f of the drift is determined by the distance d between the particles. If particles are close, the repel each other ($f > 0$). If the distance is within a predetermined range, they tend to move closer ($f < 0$). The strength of the attraction is modulated by the anisotropy of the environment (not depicted).

at least a factor of two, depending on the cell types, shapes and experimental conditions.

There is little empirical guidance on the choice of functions f_1 and f_2. Among other functions, Szabo et al. (2007) assumed a linear repulsion $f_1(d) = -A(R_1 - d)$ and a zonal, distance-independent attraction $f_2(d) = B$ for $R_2 < d < R$, where A and B are parameters. The representative simulation results of Figs. 8 and 9 were obtained with parameters $A = 160 \, \text{h}^{-2}$ and $B = 130 \, \mu\text{m/h}^2$. These values represent a strong response to external cues: the ratio of the directed and random velocity components is $B\tau/\sqrt{D\tau} \approx 3$. As a comparison, a similar measure for chemotactic response of endothelial cells was found slightly larger than one (Stokes et al., 1991).

In the case of generic cell adhesion ($w_k = 1/2$) the model was reported to behave as a two-dimensional, colloid-like system (Szabo et al., 2007). Most colloid systems with large enough attraction range exhibit transient gels; the pattern coarsens and eventually collapses into globular clusters (Butler et al., 1995). This behavior is indeed exhibited by some cell types, as cell sorting experiments demonstrate (Beysens et al., 2000).

For the sake of simplicity, cell shape is not explicitly resolved in the model. Therefore, elongation or local anisotropy is inferred from the configuration of particles. To represent an attraction to cells within anisotropic structures, the weights w_k are constructed as

$$w_k = \frac{1}{n_k} \left| \sum_{\{j:d_{jk}<R\}} e^{2i\varphi_{jk}} \right|^2, \tag{3}$$

where the sum is taken over all n_k particles that are within a circle of radius R around particle k. The angle between $\vec{x}_k - \vec{x}_j$ and an arbitrary reference direction is denoted

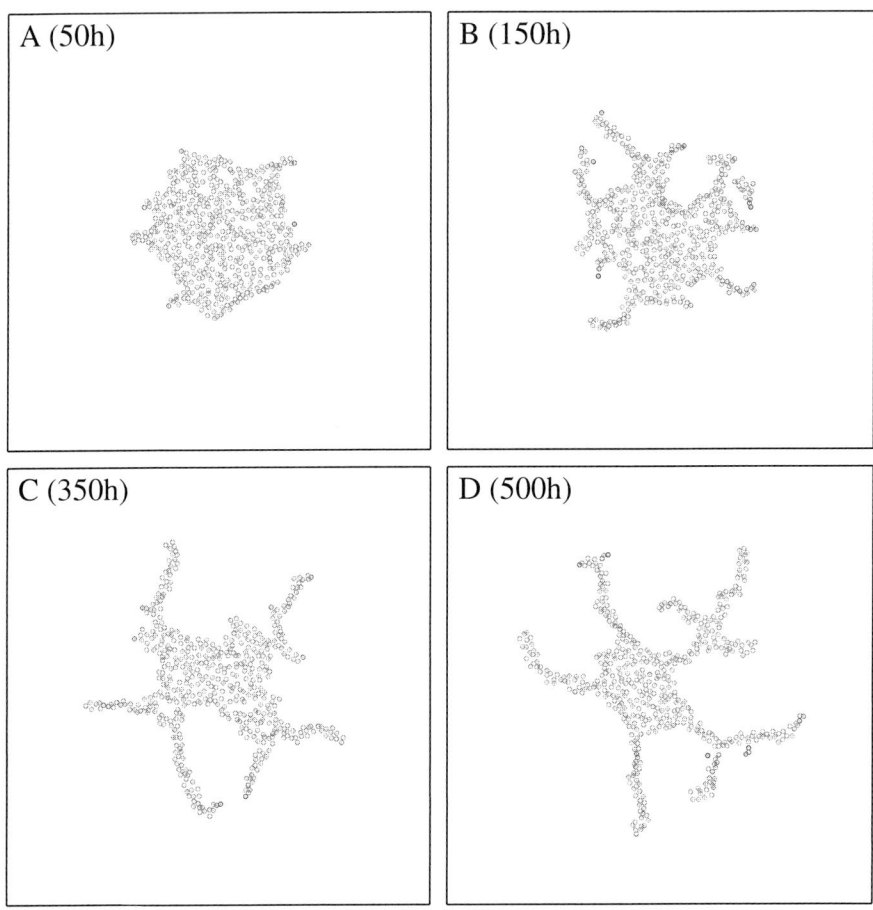

Figure 8 Time development of the model consisting of particles preferentially attracted to anisotropic structures. When $N = 500$ particle are started from a dense cluster (A), branches form (B) which slowly elongate (C and D). The gray level of the particles indicate the anisotropy of their local environment. The linear size of the area shown is $L = 40R \approx 1.6$ mm.

by φ_{jk}. Thus, $w = 0$ for particles in an isotropic environment and $w = 1$ for particles in a highly elongated, linear configuration. Nonuniform weights result in asymmetric pair-interactions, which is feasible as $\vec{M}$ represents a bias in cell activity instead of physical forces.

C. Simulation Results

When the initial condition is a single, dense cluster, stable branches form and elongate, reminiscent of the sprout formation observed *in vivo* or *in vitro* (Fig. 8). If the

9. Vasculogenic Sprouts

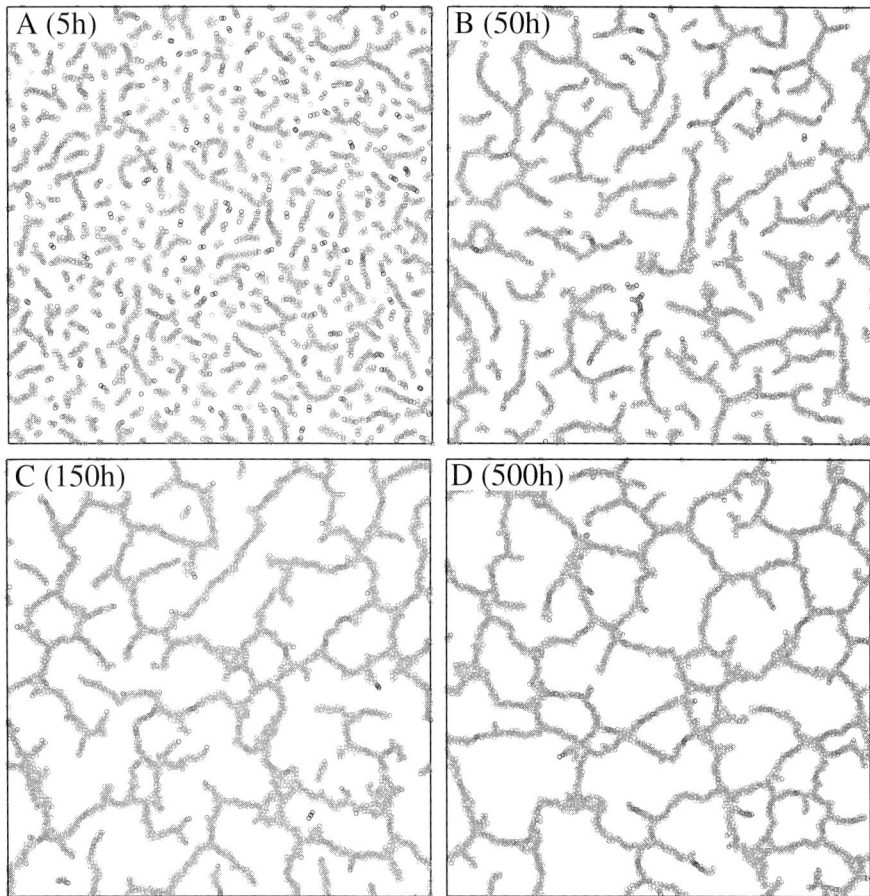

Figure 9 Time development of the model consisting of particles preferentially attracted to anisotropic structures, started from a random initial configuration. $N = 5000$ particles assemble in growing linear segments (panels A and B). An interconnected network develops if the number of particles is sufficiently large (C). After an initial transient period a quasistationary state emerges where the formation of new branches offset the occasional damage of random particle motion, and the characteristic size of avascular areas do not change (compare panels C and D). The linear size of the area shown is $L = 80R \approx 3.2$ mm.

particles are randomly positioned at $t = 0$, then long, linear arrays form after an initial coarsening process (Fig. 9). At sufficiently high particle densities the linear segments interconnect and form a polygonal network (Fig. 10). After the initial transient the pattern becomes quasistationary: the generation of new branches balances the surface-tension driven coarsening and elimination of holes. The characteristic pattern size was found to be insensitive to the cell density (Szabo et al., 2007), in good agreement

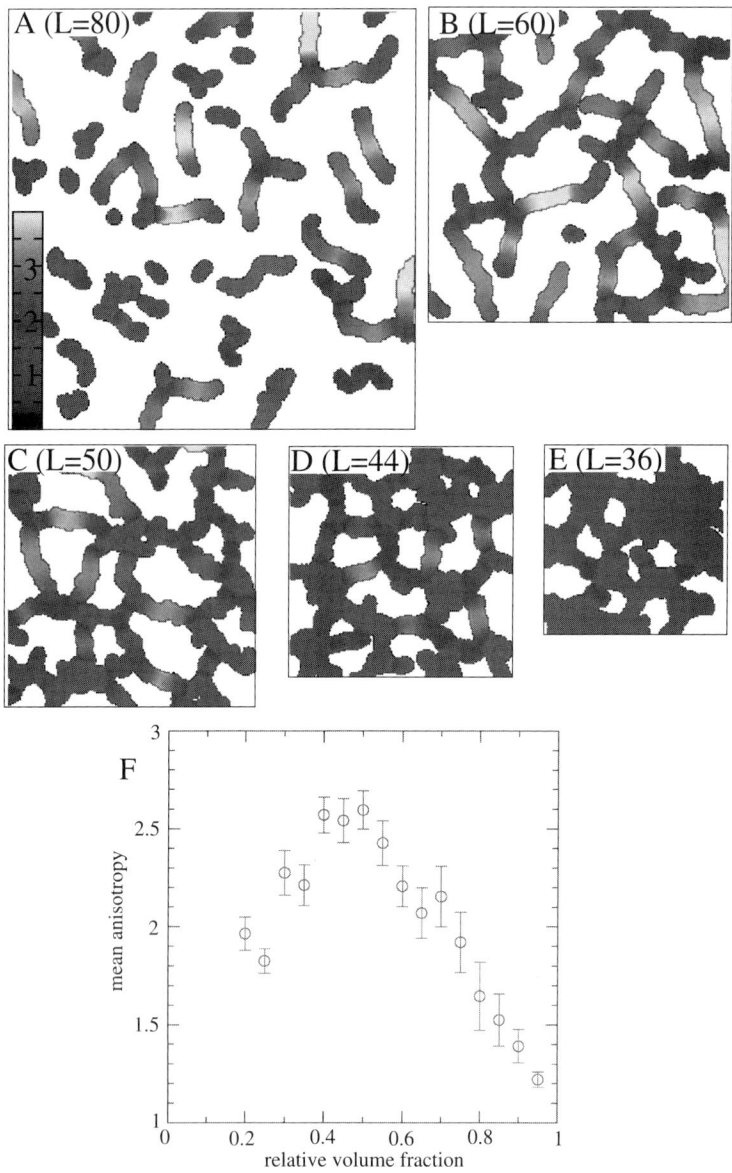

Figure 10 Depending on particle density, the model yields long elongated clusters or an interconnected network. $N = 2000$ particles are placed in a square area of size $L = 80R$ (A), $60R$ (B), $50R$ (C), $44R$ (D), and $36R$ (E), with periodic boundary conditions. The calculated local anisotropy map is superimposed on the configurations obtained at the quasistationary state of the simulations. Such simulations allow the determination of the average anisotropy of the configurations as a function of local particle density (F), to be compared with panel D of Fig. 5. See color insert.

with the somewhat limited morphometric data available for the vasculature of quail embryos (LaRue et al., 2003).

V. Conclusions

This chapter demonstrates that the primary vascular plexus of warm blooded vertebrates is formed through processes operating on various length scales: including sprouting and tissue movements. The formation and rapid expansion of multicellular sprouts is the key mechanism by which disconnected endothelial cell clusters join to form an interconnected network. As the simulation results demonstrate, the proposed hypothesis of cellular attraction to elongated structures can explain a number of empirically observed features of multicellular sprout formation. Thus, it is likely to complement other mechanisms such as chemotaxis and interaction with the surrounding ECM in the determination of endothelial cell behavior.

Acknowledgments

We are grateful to Tracey Cheuvront, Mike Filla, and Alan Petersen for their technical expertise. This work was supported by the G. Harold and Leila Y. Mathers Charitable Foundation (to C.D.L.), the NIH (R01 HL68855 to C.D.L., R01 HL87136 to A.C.), the American Heart Association (Scientist Development Grant 0535245N to A.C., Heartland Affiliate postdoctoral fellowship to E.A.Z.) and the Hungarian Research Fund (OTKA T047055, to A.C.).

References

Ambler, C. A., Nowicki, J. L., Burke, A. C., and Bautch, V. L. (2001). Assembly of trunk and limb blood vessels involves extensive migration and vasculogenesis of somite-derived angioblasts. *Dev. Biol.* **234**, 352–364.
Bauer, A. L., Jackson, T. L., and Jiang, Y. (2007). A cell-based model exhibiting branching and anastomosis during tumor-induced angiogenesis. *Biophys. J.* **92**, 3105–3121.
Beysens, D. A., Forgacs, G., and Glazier, J. A. (2000). Cell sorting is analogous to phase ordering in fluids. *Proc. Natl. Acad. Sci.* **97**, 9467–9471.
Butler, B., Hanley, H., Hansen, D., and Evans, D. (1995). Dynamic scaling in an aggregating 2D Lennard–Jones system. *Phys. Rev. Lett.* **74**, 4468–4471.
Cleaver, O., and Krieg, P. A. (1998). Vegf mediates angioblast migration during development of the dorsal aorta in Xenopus. *Development* **125**, 3905–3914.
Coffin, D., and Poole, T. (1988). Embryonic vascular development: Immunohistochemical identification of the origin and subsequent morphogenesis of the major vessel primordia in quail embryos. *Development* **102**, 735–748.
Czirok, A., Rupp, P., Rongish, B., and Little, C. (2002). Multifield 3D scanning light microscopy of early embryogenesis. *J. Microsc.* **206**, 209–217.
Czirok, A., Zamir, E. A., Filla, M. B., Little, C. D., and Rongish, B. J. (2006). Extracellular matrix macroassembly dynamics in early vertebrate embryos. *Curr. Top. Dev. Biol.* **73** (in press).

Davis, G. E., Black, S. M., and Bayless, K. J. (2000). Capillary morphogenesis during human endothelial cell invasion of three-dimensional collagen matrices. *In Vitro Cell Dev. Biol. Anim.* **36**, 513–519.
Drake, C. J., and Little, C. D. (1999). Vegf and vascular fusion: Implications for normal and pathological vessels. *J. Histochem. Cytochem.* **47**, 1351–1356.
Drake, C. J., Brandt, S. J., Trusk, T. C., and Little, C. D. (1997). Tal1/scl is expressed in endothelial progenitor cells/angioblasts and defines a dorsal-to-ventral gradient of vasculogenesis. *Dev. Biol.* **192**, 17–30.
Dunn, G. A. (1983). Characterizing a kinesis response: Time averaged measures of cell speed and directional persistence. *Agents Actions* **12**(Suppl.), 14–33.
Flamme, I., Baranowski, A., and Risau, W. (1993). A new model of vasculogenesis and angiogenesis in vitro as compared with vascular growth in the avian area vasculosa. *Anat. Rec.* **237**, 49–57.
Flamme, I., Breier, G., and Risau, W. (1995). Vascular endothelial growth factor (vegf) and vegf receptor 2 (flk-1) are expressed during vasculogenesis and vascular differentiation in the quail embryo. *Dev. Biol.* **169**, 699–712.
Foffi, G., De Michele, C., Sciortino, F., and Tartaglia, P. (2005). Arrested phase separation in a short-ranged attractive colloidal system: A numerical study. *J. Chem. Phys.* **122**, 224903.
Forgacs, G., and Newman, S. A. (2005). "Biological Physics of the Developing Embryo." Cambridge Univ. Press, Cambridge.
Gamba, A., Ambrosi, D., Coniglio, A., de Candia, A., Di Talia, S., Giraudo, E., Serini, G., Preziosi, L., and Bussolino, F. (2003). Percolation, morphogenesis, and burgers dynamics in blood vessels formation. *Phys. Rev. Lett.* **90**. 118101.
Gray, D., Tien, J., and Chen, C. (2003). Repositioning of cells by mechanotaxis on surfaces with micropatterned Young's modulus. *J. Biomed. Mater. Res. A* **66**, 605–614.
Hamburger, V., and Hamilton, H. (1951). A series of normal stages in the development of the chick embryo. *J. Morphol.* **88**, 49–92.
Helbing, D. (2001). Traffic and related self-driven many-particle systems. *Rev. Mod. Phys.* **73**, 1067–1141.
Hendrix, M. J. C., Seftor, E. A., Hess, A. R., and Seftor, R. E. B. (2003). Vasculogenic mimicry and tumour-cell plasticity: Lessons from melanoma. *Nat. Rev. Cancer* **3**, 411–421.
Keller, R., Davidson, L. A., and Shook, D. R. (2003). How we are shaped: The biomechanics of gastrulation. *Differentiation* **71**, 171–205.
LaRue, A. C., Mironov, V. A., Argraves, W. S., Czirok, A., Fleming, P. A., and Drake, C. J. (2003). Patterning of embryonic blood vessels. *Dev. Dyn.* **228**, 21–29.
Little, C., and Drake, C. (2000). Whole-mount immunolabeling of embryos by microinjection. *Methods Mol. Biol.* **135**, 183–189.
Manoussaki, D., Lubkin, S., Vernon, R., and Murray, J. (1996). A mechanical model for the formation of vascular networks in vitro. *Acta Biotheor.* **44**(3–4), 271–282.
Merks, R., Perryn, E., and Glazier, J. (2007). Contact-inhibited chemotactic motility: Role in de novo and sprouting blood vessel growth. arXiv, 0505033.
Merks, R. M., Brodsky, S. V., Goligorksy, M. S., Newman, S. A., and Glazier, J. A. (2006). Cell elongation is key to in silico replication of in vitro vasculogenesis and subsequent remodeling. *Dev. Biol.* **289**, 44–54.
Montesano, R., and Orci, L. (1985). Tumor-promoting phorbol esters induce angiogenesis in vitro. *Cell* **42**, 469–477.
Murray, J. D. (2003). "Mathematical Biology," 2 ed. Springer-Verlag, Berlin.
Murray, J. D., Manoussaki, D., Lubkin, S. R., and Vernon, R. (1998). A mechanical theory of in vitro vascular network formation. *In* "Vascular Morphogenesis: In vivo, in vitro, in mente." (C. D. Little, V. Mironov, and E. H. Mironov, Eds.). Birkhäuser, Boston, pp. 223–239.
Namy, P., Ohayon, J., and Tracqui, P. (2004). Critical conditions for pattern formation and in vitro tubulogenesis driven by cellular traction fields. *J. Theor. Biol.* **227**, 103–120.
Pardanaud, L., Altmann, C., Kitos, P., Dieterlen-Lievre, F., and Buck, C. (1987). Vasculogenesis in the early quail blastodisc as studied with a monoclonal antibody recognizing endothelial cells. *Development* **100**, 339–349.

Poole, T., and Coffin, J. (1989). Vasculogenesis and angiogenesis: Two distinct morphogenetic mechanisms establish embryonic vascular pattern. *J. Exp. Zool.* **251**, 224–231.

Poole, T. J., Finkelstein, E. B., and Cox, C. M. (2001). The role of fgf and vegf in angioblast induction and migration during vascular development. *Dev. Dyn.* **220**, 1–17.

Reagan, F. (1915). Vascularization phenomena in fragments of embryodic bodies completely isolated from yolk sac blastoderm. *Anat. Rec.* **9**, 329–341.

Risau, W., and Flamme, I. (1995). Vasculogenesis. *Annu. Rev. Cell Dev. Biol.* **11**, 73–91.

Rongish, B., Drake, C., Argraves, W., and Little, C. (1998). Identification of the developmental marker, JB3-antigen, as fibrillin-2 and its de novo organization into embryonic microfibrous arrays. *Dev. Dyn.* **212**, 461–471.

Rumpold, H., Wolf, D., Koeck, R., and Gunsilius, E. (2004). Endothelial progenitor cells: A source for therapeutic vasculogenesis?. *J. Cell Mol. Med.*, 509–518.

Rupp, P., Rongish, B., Czirok, A., and Little, C. (2003a). Culturing of avian embryos for time-lapse imaging. *Biotechniques* **34**, 274–278.

Rupp, P. A., Czirok, A., and Little, C. D. (2003b). Novel approaches for the study of vascular assembly and morphogenesis in avian embryos. *Trends Cardiovasc. Med.* **13**, 283–288.

Rupp, P. A., Czirok, A., and Little, C. D. (2004). alphavbeta3 integrin-dependent endothelial cell dynamics in vivo. *Development* **131**, 2887–2897.

Sabin, F. (1920). Studies on the origin of the blood vessels and of red blood corpuscles as seen in the living blastoderm of chick during the second day of incubation. *Carnegie Contrib. Embryol.* **9**, 215–262.

Selmeczi, D., Mosler, S., Hagedorn, P. H., Larsen, N. B., and Flyvbjerg, H. (2005). Cell motility as persistent random motion: Theories from experiments. *Biophys. J.* **89**, 912–931.

Serini, G., Ambrosi, D., Giraudo, E., Gamba, A., Preziosi, L., and Bussolino, F. (2003). Modeling the early stages of vascular network assembly. *EMBO J.* **22**, 1771–1779.

Stokes, C. L., Lauffenburger, D. A., and Williams, S. K. (1991). Migration of individual microvessel endothelial cells: Stochastic model and parameter measurement. *J. Cell Sci.* **99**, 419–430.

Szabo, A., Perryn, E. D., and Czirok, A. (2007). Network formation of tissue cells via preferential attraction to elongated structures. *Phys. Rev. Lett.* **98**. 038102.

Trinkaus, J. (1984). "Cells into Organs: The Forces that Shape the Embryo," 2nd ed. Prentice–Hall, Englewood Cliffs, NJ.

Tzima, E., Irani-Tehrani, M., Kiosses, W. B., Dejana, E., Schultz, D. A., Engelhardt, B., Cao, G., DeLisser, H., and Schwartz, M. A. (2005). A mechanosensory complex that mediates the endothelial cell response to fluid shear stress. *Nature* **437**, 426–431.

Vernon, R., Lara, S., Drake, C., Iruela-Arispe, M., Angello, J., Little, C., Wight, T., and Sage, E. (1995). Organized type I collagen influences endothelial patterns during "spontaneous angiogenesis in vitro": Planar cultures as models of vascular development. *In Vitro Cell Dev. Biol. Anim.* **31**(3), 120–131.

Weinstein, B. (1999). What guides early embryonic blood vessel formation? *Dev. Dynam.* **215**, 2–11.

Wu, K., Gauthier, D., and Levine, M. D. (1995). Live cell image segmentation. *IEEE T. Biomed. Eng.* **42**, 1–11.

Wu, X., Rabkin-Aikawa, E., Guleserian, K. J., Perry, T. E., Masuda, Y., Sutherland, F. W. H., Schoen, F. J., Mayer, J. E. J., and Bischoff, J. (2004). Tissue-engineered microvessels on three-dimensional biodegradable scaffolds using human endothelial progenitor cells. *Am. J. Physiol. Heart Circ. Physiol.* **287**, H480–H487.

Xu, J., Tseng, Y., and Wirtz, D. (2000). Strain hardening of actin filament networks. Regulation by the dynamic cross-linking protein alpha-actinin. *J. Biol. Chem.* **275**, 35886–35892.

Zamir, E. A., Czirók, A., Cui, C., Little, C. D., and Rongish, B. J. (2006). Mesodermal cell displacements during avian gastrulation are due to both individual cell-autonomous and convective tissue movements. *Proc. Natl. Acad. Sci. USA* **103**, 19806–19811.

Zamir, E. A., Czirok, A., Rongish, B. J., and Little, C. D. (2005). A digital image-based method for computational tissue fate mapping during early avian morphogenesis. *Ann. Biomed. Eng.* **33**, 854–865.

Zammaretti, P., and Zisch, A. H. (2005). Adult 'endothelial progenitor cells' renewing vasculature. *Int. J. Biochem. Cell Biol.* **37**, 493–503.

10

Modeling Lung Branching Morphogenesis

Takashi Miura*,†

*Department of Anatomy and Developmental Biology, Kyoto University Graduate School of Medicine. Yoshida Konoe-chou, Sakyo-Ku 606-8501, Japan
†JST CREST

I. Introduction
   A. Lung Branching Morphogenesis
   B. Branching Morphogenesis in Nature
   C. Modeling Methods for Branching Morphogenesis
II. Modeling *in vitro* Lung Branching Morphogenesis
   A. *In vitro* Lung Branching Morphogenesis
   B. Modeling *in vitro* Lung Branching Morphogenesis
   C. Experimental Verification of the Model
III. Functional Modeling—Structure and Air Flow
IV. Future Directions
V. Numerical Simulations of Branching Morphogenesis Models
   A. Programming Simulations with *Mathematica*
   B. Diffusion-Limited Aggregation
   C. L-System
   D. Reaction–Diffusion Model
   References

Vertebrate lung has tree-like structure which facilitates gas exchange. After discovery of the involvement of several key toolkit genes—FGF10, BMP4, and Shh, huge amount of molecular information on lung development is now available. However, how their interactions result in a branched structure has not been elucidated. Recently, some studies have utilized mathematical models to understand the mechanism of branching morphogenesis, and we now have some models which are reliable enough to make experimental predictions in the *in vitro* system. In addition, a different type of modeling, which generates tree-like branching pattern by repeatedly applying a set of simple rules iteratively, is also utilized to model lung function. In this review, I focus on how these models can contribute to understand pattern formation phenomena from experimental biologist's point of view. © 2008, Elsevier Inc.

I. Introduction

A. Lung Branching Morphogenesis

There are many branched epithelial structures in the animal body. Lacrimal gland (Makarenkova *et al.*, 2000), salivary gland (Kashimata and Gresik, 1996), lung

(Gilbert, 2003), pancreas (Kim and MacDonald, 2002), prostate gland (Davies, 2006), and kidney (Shah *et al.*, 2004) all undergo branching morphogenesis during development (for review, see Davies, 2006). Although epithelial–mesenchymal interaction plays an important role in these systems, whether there is a common molecular mechanism among these organs remain to be elucidated (Davies, 2002).

Among them, lung branching morphogenesis is one of the most extensively studied systems (Cardoso and Lue, 2006; Kumar *et al.*, 2005; Takaki, 2005; Shannon and Hyatt, 2004; Chuang and McMahon, 2003; Warburton *et al.*, 2000; Warburton *et al.*, 1999; Hogan, 1999). In human and mouse, the lung first appears as a protrusion from ventral part of gastrointestinal tract, which is called the lung bud. The lung bud consists of epithelial tube surrounded by mesenchymal tissue. Then it undergoes branching morphogenesis, which eventually generates the bronchial tree (Sadler, 2004).

Earlier studies concentrated on effect of extracellular matrix and cell behavior using organ culture system (Shannon and Hyatt, 2004). At the very beginning of branching morphogenesis, dense deposits of collagen fiber are observed at the cleft (Heine *et al.*, 1990). When the extracellular matrix component was digested by Type I collagenase, branching morphogenesis was impaired and resulted in cyst-like morphology (Ganser *et al.*, 1991). These observations lead to a model in which surrounding mesenchyme cells exert traction forces on collagen fibers to generate clefts in the epithelium in the very beginning of the branching morphogenesis (Nakanishi *et al.*, 1986).

Since the discovery of "toolkit" molecules in development (e.g., Sonic Hedgehog (Shh), Fibroblast Growth Factor (FGF), Bone Morphogenetic Protein (BMP), etc.), recent studies have concentrated on these extracellular signaling molecules (Hogan, 1999; Chuang and McMahon, 2003; Fig. 1). Among them, the most important example is FGF10. FGF10 is expressed at mesenchyme tissue surrounding distal tip of the growing epithelium (Bellusci *et al.*, 1997b), and when FGF10 is locally applied by beads, lung epithelium moves toward the source of FGF10 (Bellusci *et al.*, 1997b; Park *et al.*, 1998). FGF10 knockout mice show lung agenesis (Ohuchi *et al.*, 2000; Sekine *et al.*, 1999; Min *et al.*, 1998), indicating FGF10 is the key regulator of branching morphogenesis. The receptor for FGF10 is FGFR2 (Ohuchi *et al.*, 2000), which is also critical for lung branching morphogenesis (Arman *et al.*, 1999). Heparan sulfate proteoglycan, which mediates binding of FGF and FGF receptor, is expressed at the tip of the branching lung and promotes branching Izvolsky *et al.* (2003a, 2003b). Sprouty, a downstream component of FGF signaling pathway, also influences lung branching (Mailleux *et al.*, 2001; Perl *et al.*, 2003; Tefft *et al.*, 2002). Large-scale screening of FGF10 target molecules showed various biological activities to be modulated by FGF10 (Lue *et al.*, 2005).

BMP4 is expressed at the growing tip of the epithelium (Bellusci *et al.*, 1996) and inhibits epithelial proliferation *in vitro* (Weaver *et al.*, 2000; Hyatt *et al.*, 2002) and *in vivo* (Bellusci *et al.*, 1996). Therefore, it is thought to be involved in lateral inhibition during branching morphogenesis (Hogan, 1999). However, several other studies show that BMP4 promotes cell proliferation under some *in vitro* situations (Eblaghie

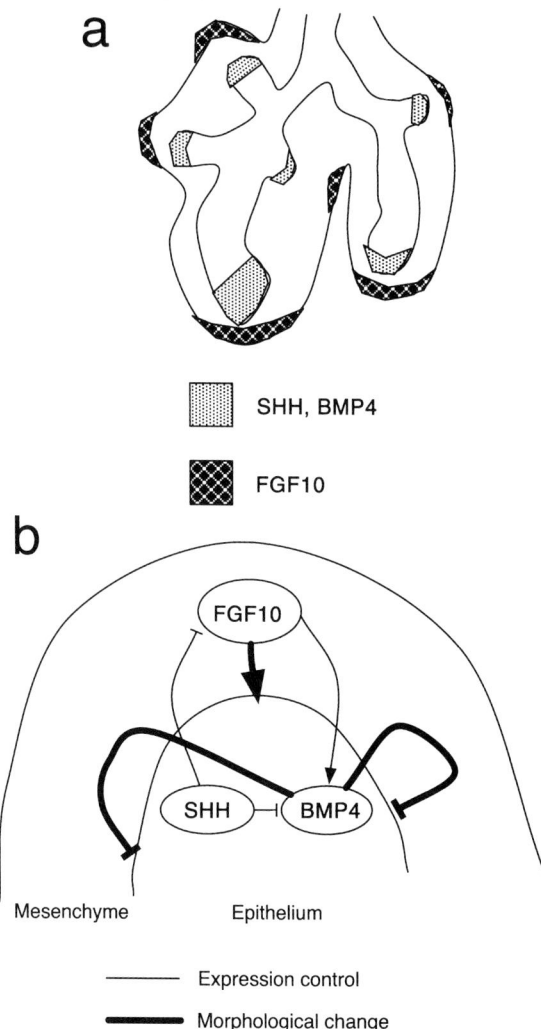

Figure 1 (a) Expression of "toolkit" molecules during development. SHH is expressed throughout the epithelium, most strongly at the tip of the growing epithelium. FGF10 is expressed at mesenchyme surrounding growing tip. BMP4 is expressed at tip of the epithelium. (b) Genetic interaction of toolkit molecules during limb development. The thin line represents expression control at mRNA level, and the thick line represents influence on epithelial morphology.

et al., 2006; Chen et al., 2005; Bragg et al., 2001). BMP4 expression is upregulated by FGF10 (Hyatt et al., 2004; Lebeche et al., 1999).

Sonic hedgehog (Shh) is expressed in the lung epithelium, most strongly at the tip (Urase et al., 1996; Bellusci et al., 1997a) and shown to repress FGF10 and BMP4 expression (Pepicelli *et al.*, 1998). Its receptor Ptc is expressed in the mesenchyme (Bellusci *et al.*, 1997a), indicating that the signal influences lung mesenchyme. Gli3, a component of the intracellular Shh signaling pathway, also influences branching morphogenesis (Li *et al.*, 2004).

Other extracellular signaling molecules are also involved in modulating effects of these toolkit genes (reviewed by Cardoso and Lue, 2006). FGF9 is expressed in the mesothelium surrounding mesenchymal tissue and induces mesenchymal cell proliferation, and targeted disruption of FGF9 results in reduced FGF10 expression (Colvin *et al.*, 2001; del Moral *et al.*, 2006; White *et al.*, 2006). The Wnt pathway is also involved in modulating balance between FGF and BMP signal (Dean *et al.*, 2005; Li *et al.*, 2005; Pongracz and Stockley, 2006; Shu *et al.*, 2005). Retinoic acid induces FGF10 expression via retinoic acid receptor beta (RARβ) (Desai *et al.*, 2006; Desai *et al.*, 2004; Malpel *et al.*, 2000).

However, this classic molecular approach will not ultimately lead to an understanding of *how* the branch pattern is generated from a seemingly characterless initial form. At the distal part of the lung, it is highly probable that the branch pattern is stochastic rather than genetically determined. There is already a huge amount of molecular information (Cardoso and Lue, 2006), and adding one or two molecules to this large collection would not seem to improve our understanding. We need to utilize a different approach for this problem.

B. Branching Morphogenesis in Nature

Significantly, formation of branched structure has been extensively studied in physics and chemistry (for review, see Ball, 1999). The example includes crystal formation (Ball, 1999), snowflakes (Bentley and Humphreys, 1962), viscous fingering of fluids (Cross and Hohenberg, 1993), and bacterial colonies (Hartmann, 2004; Ben-Jacob and Levine, 2000; Matsushita *et al.*, 1998; Kawasaki *et al.*, 1997).

There is a very simple experiment to reproduce branching morphogenesis in a nonbiological system (Prof. Sharon Lubkin of North Carolina State University, personal communication). Take a Petri dish and its lid, and put a small pool of glycerine in the lid. Place the smaller half of the dish inside, so the flat sides of the dishes are pressed against each other with the glycerine between them. Allow the glycerine to spread to the edges. Observe that as you press the dishes together the interface between the air and the glycerine stays smooth and circular. Then pull the dishes apart, and observe that the interface forms fingers as you pull. This is one example of physical phenomena called *viscous fingering*. When less viscous liquid (air) is pressed into more viscous fluid (glycerine), it does not spread evenly and various branches are formed. Although we cannot say from this experiment that lung branching is a viscous fingering phenomenon in a physical sense, we can see that complex molecular interactions are not a necessary condition for the formation of branched structure.

At an abstract level, the mechanism of branching morphogenesis in these systems can be explained by "protrusion grows faster" tendency. For example, in the case of viscous fingering, a protruding region of air bubble can invade a viscous fluid more easily than a concave region. In the case of crystal formation, heat is released more efficiently from the tip of the protrusion, which further promotes formation of crystal structure at the tip. In the case of bacterial colonies, a protruding tip of the colony edge will be exposed to higher concentration of nutrient, which results in bacterial cell proliferation and further protrusion at that point. This positive feedback loop is the origin of the interface instability in all cases.

As the generated pattern in these systems is quite similar to that in lung branching morphogenesis, we can assume that the pattern formation mechanism has something in common, though at an abstract level. The morphological similarity does not imply that the actual mechanism of branching morphogenesis is exactly the same—for example, applying viscous fingering phenomenon directly to lung branching morphogenesis does not sound very convincing because of our knowledge of the many key molecular players as described in the previous section. Understanding the formal properties of branching morphogenesis is necessary in order to formulate appropriate models, so in the next section we will describe various computational models which can generate branched patterns *in silico*.

C. Modeling Methods for Branching Morphogenesis

There are several ways to numerically implement branching morphogenesis. One method is diffusion-limited aggregation, which was originally used in fractal geometry [airway tree structures have noninteger fractal dimensions (Nelson *et al.*, 1990)]. In this model, an initial "seed" point is defined at the center of the field, and small particles, which move randomly, are released from the periphery of the field. When a particle reaches one of the neighboring grids of the seed, it becomes part of the seed, and new particles are released again from the periphery. In this system, a slightly protruded region has a higher probability of meeting a moving particle, which results in interfacial instability (Section V). Although this method is simple and suitable for generating a fractal structure, implementation of the surface tension effect is cumbersome and it is difficult to reproduce the branch to cyst transition which is observed experimentally in lung system.

The other method is the reaction–diffusion based model (Turing, 1952, Section V). An introduction to numerical simulation of this system is described in Miura and Maini (2004). In this system we define two variables, for example, concentration of some diffusing chemical and cell density. If we assume that a cell grows by consuming the diffusing chemical, a protruded area is exposed to higher chemical concentration, resulting in more protrusion at that region. There are several advantages in this model—in the first place, this system is well studied and we have many mathematical tools to analyze this system. With these tools, we can make very strong predictions.

For example, we can determine in what parameter range pattern formation happens, and how the pattern size will change if we increase or decrease one of the model parameters.

Another completely different way of modeling is L-system-like implementation. This method was originally described by Aristid Lindenmayer (Lindenmayer, 1975) and frequently utilized to model the branching form of trees (Honda, 1971; Honda and Fisher, 1979; Honda et al., 1997; Prusinkiewicz, 2004). This method describes formal rules which decide the way a tip branches—for example, number and length of sister branches, angle and diameter change of these branches, etc. Then one defines the location of initial branch and applies the rule repeatedly, which results in tree-like structure arising from a very simple set of rules (Section V). This method is rather descriptive compared to the other two and contains no information on the mechanism of tip bifurcation. However, this level of model has its own advantage—since the resulting structure contains information of connections of the branches, it is easy to analyze functional aspects of the resulting structure.

II. Modeling *in vitro* Lung Branching Morphogenesis

A. *In vitro* Lung Branching Morphogenesis

As we have seen in previous sections, there is an abundance of factors known to be involved in *in vivo* lung branching. One obvious strategy is to use a simplified experimental system which reproduces key features of branching morphogenesis. Such an experimental system was developed by Nogawa and Ito (1995). They isolated epithelial part of the developing lung, and embedded it in Matrigel, a gel which mimics the extracellular matrix component of basement membrane. They added FGF1 to the culture medium and reproduced branching morphogenesis even without mesenchymal tissue.

Local application of signaling molecules also showed that FGF plays an important role in branching morphogenesis. Application of FGF10 by beads induced chemotactic movement of epithelial tissue toward the FGF source (Park et al., 1998; Weaver et al., 2000). We could not observe any filopodia or lamellipodia during this process (Miura and Shiota, 2000), so the mechanism of how a cystic epithelial structure moves toward a source of FGF10 remains unknown. Local application of FGF1 induced similar effects, but enlargement of lung bud epithelium also occurred, indicating FGF1 induces both cell proliferation and chemotaxis while FGF10 mainly induces chemotaxis.

Other signaling molecules also influence the pattern in this culture system. For example, the subtype of FGF affected the final pattern. FGF10 showed similar effects as FGF1, but a higher concentration was needed to induce branch when added in culture medium. When FGF7 was added to the culture medium instead of FGF1, the result was cyst-like morphology instead of branches (Cardoso et al., 1997). BMP4 was shown to inhibit epithelial cell proliferation in this system (Weaver *et al.*, 2000), but we still have not observed a corresponding effect—application of BMP4 by beads does not show local inhibition but overall decrease of cell proliferation (Miura and Shiota, 2002).

The effect of extracellular matrix seems to be retained in this system. It has been shown that developing lung assumes cystic structure as a result of type I collagenase treatment *in vitro* (Ganser *et al.*, 1991). Moreover, a similar effect can be obtained in mesenchyme-free system (Miura and Shiota, 2002), indicating the cyst–branch selection mechanism is similar both with and without mesenchyme.

B. Modeling *in vitro* Lung Branching Morphogenesis

The mesenchyme-free system provides a simple model for the behavior of the epithelium. There are two main factors that are involved in morphogenesis—epithelial cells and FGF. FGF10, not FGF1, plays a major role *in vivo*, but in this case we use FGF1 in order to utilize previous *in vitro* data. From Section I.B, we know that "protrusion grows faster" tendency can generate a branching pattern, so we formulate such interactions which are biologically plausible. We made following assumptions:

- We consider two factors, epithelial cells $c(x, y, t)$ and FGF molecules $n(x, y, t)$ in this model.
- Epithelial cells grow by consuming FGF molecules.
- Epithelial cells move to keep certain cell density.
- FGF molecules move in Matrigel by diffusion.

The scheme of this interaction and governing equations are shown in Figs. 2a–2b. We can reproduce the early phase of branching morphogenesis using this model (Fig. 2c). An intuitive explanation of this phenomenon is as follows: the initial form of a lung explant cannot be completely spherical, so there will be slightly convex or concave regions. Then, a slightly protruded area of explant should be exposed to higher concentration of FGF because it is away from FGF-consuming regions. As a result, the slightly protruded region should grow and protrude further. This positive feedback loop results in interfacial instability and generation of a branched structure (Fig. 2d).

Movement of epithelial cells is mathematically difficult to formulate, so our first version model is discrete, for which very few mathematical analysis methods are available. Therefore, recently we use a different model to utilize mathematical analysis (Hartmann and Miura, 2006a). In this model we define a region of epithelial explant, and define the growth speed of the interface according to cell growth induced by FGF consumption. We can derive a dispersion relation in this model under specific conditions—which means we can roughly predict the size of each bud, and bud. The behavior of this model was analyzed in detail in Hartmann and Miura (2006a) and Hartmann (2007).

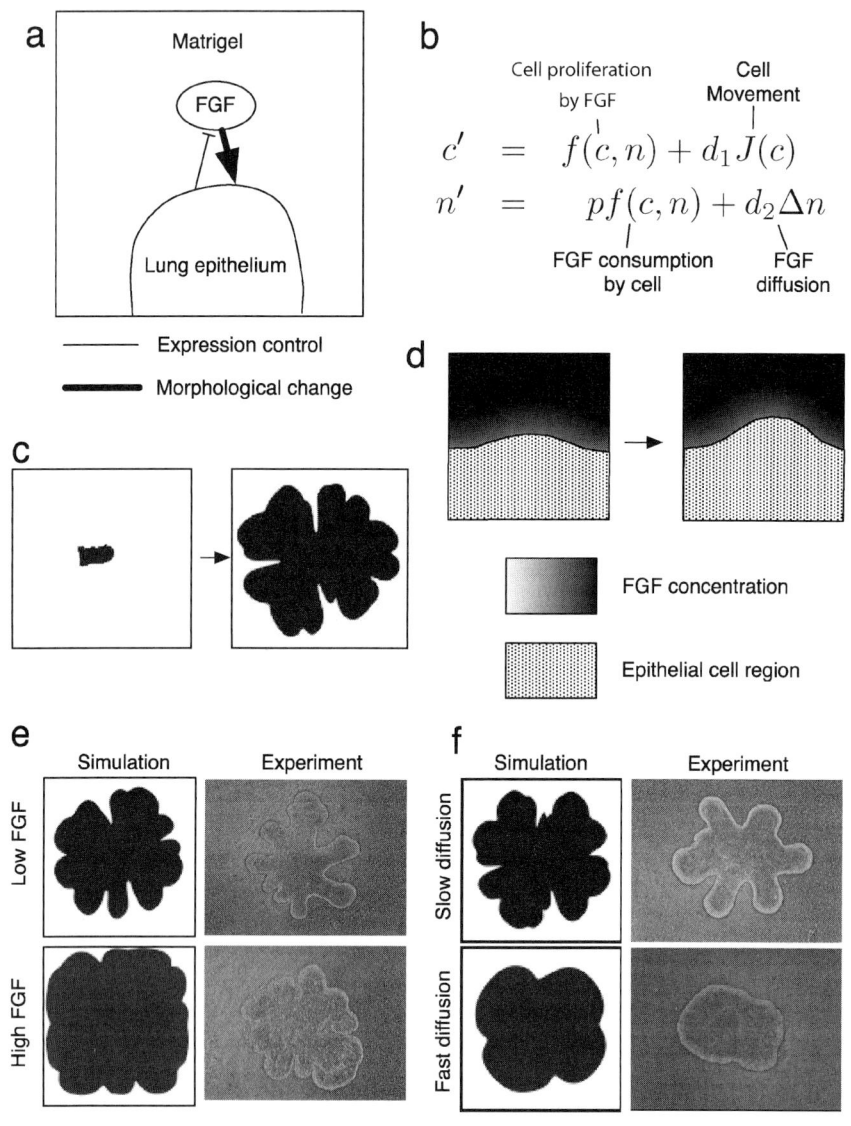

Figure 2 A model for *in vitro* lung branching morphogenesis. (a) Molecular interaction in *in vitro* lung branching system. (b) Governing equation of the model. (c) By running the simulation, branching structure is formed by the model. (d) Intuitive explanation of the model. Slightly protruded epithelial cell should be exposed to higher FGF concentration because it is distant from other FGF consuming tissues. As a result, slightly protruded area should grow further and results in interface instability. (e) Effect of high FGF concentration on branch morphology. (f) Effect of FGF diffusion coefficient on final pattern. From Miura and Shiota (2002).

C. Experimental Verification of the Model

After we model the phenomena, we can derive many experimentally testable predictions. First of all, FGF consumption by lung epithelium is the key factor to generate the "protrusion grows faster" tendency. Therefore, we predicted from the model that the FGF concentration should be lower around the lung explant because FGF should be taken up into the lung epithelium. We could actually observe that FGF concentration is lower around the epithelial explant by immunohistochemistry and fluorescently-labeled FGF (Miura and Shiota, 2002).

Next, because of the FGF consumption by lung epithelium, two neighboring lung epithelia should locally inhibit one another's branch formation. This phenomenon was observed in branching morphogenesis of bacterial colonies and called the *shielding effect* (Matsushita and Fujikawa, 1990). We could reproduce this phenomenon experimentally (Miura and Shiota, 2002). An intuitive explanation of this phenomenon is as follows: FGF concentration in the area between two epithelial explants should be low because FGF is consumed by both explants. Therefore, bud formation is suppressed between two explants. We predicted from numerical simulation that saturating amount of FGF can cancel this effect because enough FGF remains in the area between two explants even if FGF is consumed by both explants. Actual experimental results showed exactly the same behavior as numerical simulation, which supports the validity of the model (Miura and Shiota, 2002).

Another prediction we can make is that the branching pattern should depend on the initial FGF concentration. If FGF concentration is too low, the epithelial cells simply cannot grow and nothing happens. When FGF concentration is appropriate, the "protrusion grows faster" tendency appears by FGF depletion. If FGF levels are saturating, the growth speed of epithelium should be constant everywhere. Initial morphological perturbations are smeared out by this, resulting in a cystic pattern (Hartmann, 2007). We can actually observe these pattern changes experimentally (Fig. 2e), which further validates the model (Miura and Shiota, 2002; Hartmann and Miura, 2006a, 2006b).

The effect of type-I collagenase can be understood as a change in FGF diffusion coefficient. Because FGF diffusion is modulated by extracellular matrix components, degrading ECM by collagenase results in faster diffusion, which inhibits FGF to form a gradient appropriately steep so as to generate a regional difference. Numerical simulation with a higher diffusion coefficient results in a cyst-like pattern (Fig. 2f), and we could actually observe diffusion coefficient change by type I collagenase treatment using fluorescently-labeled FGF (Miura and Shiota, 2002).

Various morphological responses elicited by different FGF subtypes (Cardoso *et al.*, 1997) can also be elucidated using the model. From experiments described above, we can assume that cystic structure results from larger FGF diffusion coefficient. We predicted that matrix metalloproteinases, which degrade surrounding extracellular matrix, should increase by FGF7 treatment, which is indeed the case (Miura and Shiota, 2002). All these observations suggest that this simple model recapitulates essential aspects of branching morphogenesis *in vitro*.

III. Functional Modeling—Structure and Air Flow

We need multiscale modeling to understand both morphological and functional aspects of the bronchial tree. Although the model presented above is useful to understand mechanisms of branching morphogenesis, the model is not very useful from the functional point of view. We need additional kinds of model which efficiently depict functional aspects of this pattern. L-system based models are more appropriate because they contain information on connection between branches, diameter etc.

Such a model was formulated by Kitaoka *et al.* (1999). They derived a set of very simple branch bifurcation rules from previous morphometric studies and successfully generated a three-dimensional airway tree *in silico*. They used two principles: (1) the amount of fluid delivery through a branch is proportional to the volume of the region it supplies; and (2) the terminal branches are arranged homogeneously within the organ. Then they defined the contour of the lung and location of the main bronchi, and applied the principles iteratively to generate a branched structure. They also defined an index which represents functional efficiency of the tree structure, and modified the parameters to maximize the index. The resulting three-dimensional pattern is virtually indistinguishable from the real human airway tree. Although this model contains no information on the mechanism of branching, it should be very useful to study the structure–function relationship in the lung.

IV. Future Directions

As we have seen in previous sections, we know of numerous molecular players which are involved in lung branching morphogenesis from developmental biology. We also know of very abstract mechanism of branching morphogenesis from applied mathematics, and we also know some basic rules which the adult airway tree obeys from physiology. We combined the former two fields to elucidate the mechanism of initial branch formation under special *in vitro* condition. Another study combined latter two fields to generate branch structure which is useful for functional assays. However, there are still huge gaps between these fields.

The most important problem is how to model the *in vivo* situation—expression of the key morphological regulator FGF10 is restricted to mesenchymal tissue surrounding an epithelial tip, and how this expression site is regulated is still not understood. Moreover, the molecular interactions are unlikely to be so simple *in vivo*—we need to construct a huge computational model for which no mathematical analysis is available, or to construct a very abstract model for which relationship with actual experimental results is unclear. The geometry of lung mesenchyme is another problem. Mesenchymal tissue changes shape during branching morphogenesis, and morphogen diffusion should be limited to within this area. Such pattern forming mechanisms have been termed "morphodynamic" by Salazar-Ciudad *et al.* (2003) and their implementation is computationally very cumbersome.

The relationship between pattern formation and function is another important theme. Branching morphogenesis happens well before respiration begins, so the branching pattern is not closely related to function during development. However, the adult airway tree obeys certain rules, and as a result lung structure is optimized for its function (Kitaoka *et al.*, 1999). Therefore, it is highly probable that remodeling of branch structure takes place at later stages of development. How the air flow modulates airway structure after birth has not been studied in detail, and should be very important to experimentally verify the model.

Since information from developmental biology is mainly molecular, we started from modeling molecular interactions. However, since developing lung structure is also under the control of physical law, mechanical factors should also be involved (for review, see Davies, 2006). For example, Lubkin and Murray (1995) formulated a model in which purely mechanical interaction can generate branched patterns in lung branching morphogenesis (see also Lubkin and Li, 2002). Moore *et al.* (2005) studied the relationship between cell tension and branching morphogenesis and showed that modulating cell tension by the Rho signaling pathway influences the branch structure in organ culture. We also showed that decreasing cell tension by cytochalasin D treatment results in increased bud number in mesenchyme-free conditions (Hartmann and Miura, 2006b), which supports the role of mechanical factors during branching morphogenesis.

Species difference is another important theme. In fishes, the homologous organ for lung is swimbladder, which is a cyst-like structure without branches. In birds, the lung has both properties—the dorsal part consists of airway tree while the ventral side has cyst-like structures called *air sacs*. Sakiyama *et al.* (2000) has done tissue recombination experiments of chick lung and showed that the morphological difference is due to the surrounding mesenchyme. They also showed that Hoxb cluster genes are differently expressed between the dorsal and ventral part of chick lung at earlier stages, which may control the morphological difference. However, how the transcription factor modulates morphology has not been studied in detail, and theoretical models can provide working hypotheses for the targets of this transcription factor.

V. Numerical Simulations of Branching Morphogenesis Models

A. Programming Simulations with *Mathematica*

In this section, we present *Mathematica* simulation programs. There are several reasons to use this platform—it is reasonably fast, visualization of results is quite easy and we can make programs very short, which makes it easier to understand the main idea of the models. In the following subsections *Mathematica* code is described in **bold** typeface. Detailed instruction and many examples of pattern formation with *Mathe-*

matica can be found in (Gaylord and Wellin, 1995). Although all the executable codes are presented in this section, original source code is available on request.

B. Diffusion-Limited Aggregation

One of the simplest ways to implement branching morphogenesis is diffusion-limited aggregation (DLA). In this model, we define a set of simple rules on a two-dimensional grid, and iterative application of the rules finally results in branched structure which has noninteger fractal dimension. Detail of the model is described in Gaylord and Wellin (1995).

At first, we define size of two-dimensional grid.
gridNumber = 60;
Then we define a two-dimensional grid. We define an unoccupied grid site as 0 and an occupied grid site as 1.
grid = Table[Table[0, {gridNumber}], {gridNumber}];
We put initial "seed" at the center of the grid.
grid[[gridNumber/2, gridNumber/2]] = 1;
The result is a grid filled with 0 except at the center.
ListDensityPlot[grid];
Output of this command is shown in Fig. 3a.

Next, we define a list of vectors to generate random two-dimensional motion of a particle. We can generate random walk movement by choosing one of the four vectors randomly and adding it to the current coordinate.
motion = {{1, 0}, {0, 1}, {−1, 0}, {0, −1}};
We also define a function to judge whether a particle reaches the neighbor of preexisting cluster. If this function is not zero, that means the particle reaches one of the neighboring grids of the preexisting seed.
neighbour[{x_, y_}] :=
grid[[b[x + 1], y]] + grid[[b[x − 1], y]]+
grid[[x, b[y + 1]]] + grid[[x, b[y − 1]]];
Usually a particle which goes too far from preexisting cluster is canceled because the chance the particle reach the cluster becomes very small. Here we use a different rule by using periodic boundary—if a particle goes out of the grid, it appears again from the opposite side of the grid.
b[x_] := Mod[(x − 1), gridNumber] + 1;
We also define a function which sets the initial location of particle release. It comes from a random location on one of the four edges of the grid.
initialPoint :=
{{Random[Integer, {1, gridNumber}], 1},
{1, Random[Integer, {1, gridNumber}]},
{Random[Integer, {1, gridNumber}], gridNumber},
{gridNumber, Random[Integer, {1, gridNumber}]}}[[

10. Modeling Lung Branching Morphogenesis 303

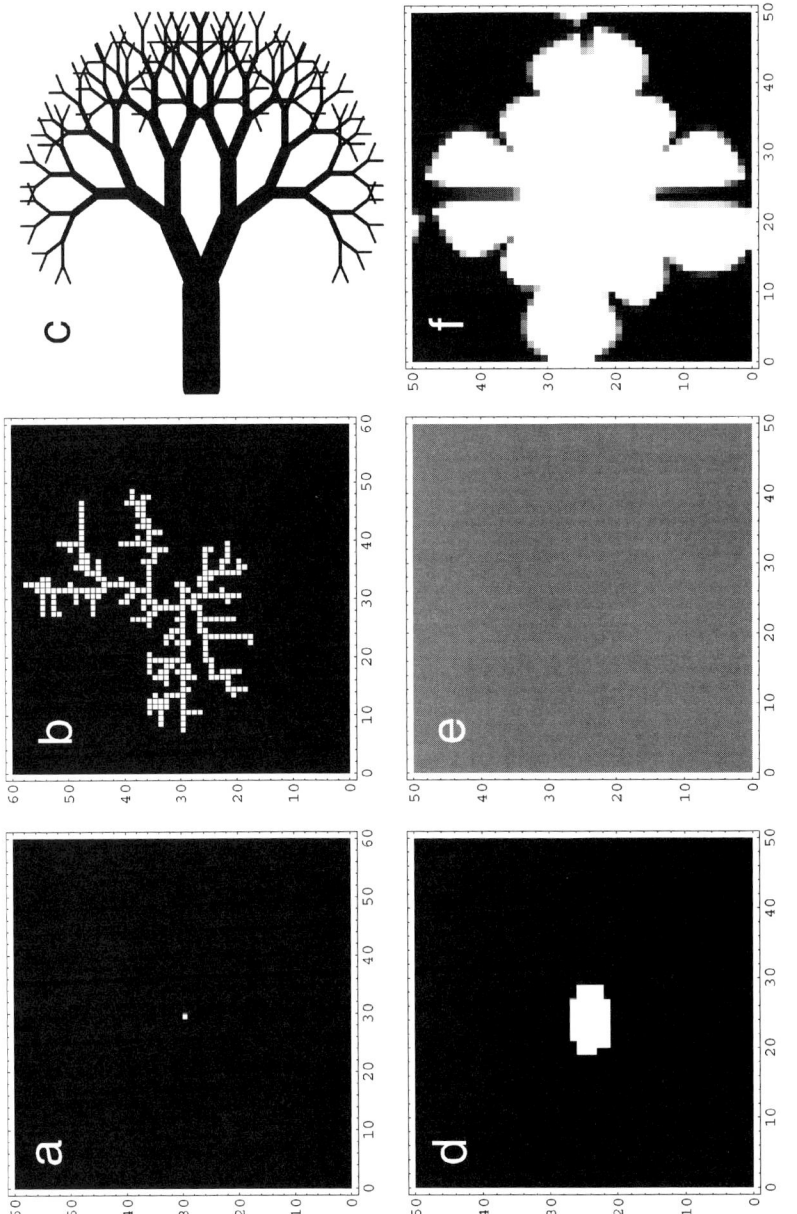

Figure 3 Graphic output of *Mathematica* code. (a–b) DLA program. (c) L-system program. (d–f) Reaction-diffusion based program.

Random[Integer, {1, 4}]]]]

Then we can run the main program—we release particles from the edge of the grid, let them move randomly in the grid, and when they reach the neighbor of preexisting grid make them part of the preexisting cluster. We iterate this process 300 times.

Do[*p* = initialPoint;
While[neighbour[*p*] == 0,
Module[{},
p = *b*[*p* + motion[[Random[Integer, {1, 4}]]]]];
grid[[*p*[[1]], *p*[[2]]]] = 1;
, {300}]//Timing
{45.6191Second, Null}
ListDensityPlot[grid];

Output of this command is shown in Fig. 3b. We can observe formation of a fine branched structure by this model.

C. L-System

The L-system is a way to describe branched structure with very simple rules. It defines the original root of the branch and a set of rules to describe how a single branch give rise to two branches; length, bifurcation angle and diameter change. This was originally developed by Lindenmayer (1975) to describe plant development.

At first, we define **l[p,v,t]** as a graphic object **Line**, which start from **p** and ends with **p + v** (**p** and **v** are vectors).

l[{*p*_, *v*_, *t*_}] :=
Graphics[{Thickness[*t*], Line[{*p*, *p* + *v*}]}]

Next, we define a rotation matrix.

$$\text{rotation}[v\_, \text{theta}\_] := \begin{pmatrix} \cos[\theta] & -\sin[\theta] \\ \sin[\theta] & \cos[\theta] \end{pmatrix} \cdot v$$

Then we define the position and thickness of the initial branch.

initialBranch = {{0, 0}, {1, 0}, .1};

Next, we define a rule how one branch bifurcates into two sister branches. In this case, sister branches rotate 60 and −60 degrees, and have 80% of the original length and 67% of the original thickness.

f[{p_, v_, t_}] :=
{{*p* + *v*, 0.8rotation[*v*, −Pi/6.], *t*/1.5},
{*p* + *v*, 0.8rotation[*v*, Pi/6.], *t*/1.5}}

Here we check how this function works on the original branch.

Map[*f*, {initialBranch}]
{{{{1, 0}, {0.69282, −0.4}, 0.0666667},
{{1, 0}, {0.69282, 0.4}, 0.0666667}}}

To avoid deep nested list, we use **Flatten** function. By using this, result of function **f** is a list of vectors which contain information on initial location, length and thickness of sister branches.

10. Modeling Lung Branching Morphogenesis

Flatten[%, 1]
{{{1, 0}, {0.69282, −0.4}, 0.0666667},
{{1, 0}, {0.69282, 0.4}, 0.0666667}}

Next, we define a function for one round of branching.

oneStep[l_] := Flatten[Map[f, l], 1]

Finally, we obtain a list of branches by applying the function repeatedly to the original branch.

branches =
Flatten[NestList[oneStep, {initialBranch}, 7], 1];

We visualize the tree structure by applying the list of coordinate data to **Line** function.

Show[Map[l, branches, {1}], AspectRatio → 1.0]

Output of this command is shown in Fig. 3c.

D. Reaction–Diffusion Model

At first, we define two variables—cell density c and FGF concentration n. Next, we set various simulation parameters.

domainSize = 50; dx = 1; dc = 0.3; dn = 0.5;
n0 = 0.5; c0 = 1; p = 1;
dt = 0.1;

Then we define initial distributions of cell density and FGF. We introduce random perturbation in the initial cell density distribution.

cInitial =
Table[
If[(x − domainSize/2 + Random[])^2 * 2+
(y − domainSize/2 + Random[])^2 <
(domainSize/10)^2, c0, 0],
{x, dx, domainSize, dx}, {y, dx, domainSize, dx}];
nInitial = Table[n0, {x, dx, domainSize, dx},
{y, dx, domainSize, dx}];

Initial distributions of cell density and FGF concentration are as follows:

ListDensityPlot[cInitial, Mesh → False]
ListDensityPlot[nInitial, Mesh → False]

Output of this command is shown in Figs. 3d–3e.

Reaction term is defined as follows:

f[c_, n_] := 5cn/(1 + 3n);

Then we define cell motion. Cell density is approximately the same in epithelial tissue, so we specify that cell moves to keep a certain constant value **c0**.

cM[l_] := (l − c0)(Sign[l − c0] + 1)/2;

diffusionC[l_] :=

dc*
(cM[RotateLeft[*l***]] + cM[RotateRight[***l***]]+**
cM[RotateLeft[*l***, {0, 1}]]+**
cM[RotateRight[*l***, {0, 1}]] − 4cM[***l***])/(dx * dx);**
diffusionN[l_] :=
dn*
(RotateLeft[*l***] + RotateRight[***l***] − 2***l*** +**
RotateLeft[*l***, {0, 1}] + RotateRight[***l***, {0, 1}]−**
2*l***)/(dx * dx);**

We define a function to calculate the value of cell density and FGF distribution after certain time.

cnAfterDt[{c_, n_}] :=
{
c + dt * (*f***[c, n] + diffusionC[c]),**
n + dt * (−*pf***[c, n] + diffusionN[n])**
};
cnAfter1Time[l_List] :=
Nest[cnAfterDt, *l*, Round[1/dt]];

Actual calculation result is stored in the "result" variable.

result = NestList[cnAfter1Time, {cInitial, nInitial},
500]; //Timing

Resulting pattern is visualized by *ListDensityPlot[ ]* function.

ListDensityPlot[result[[500, 1]], PlotRange → {0, 1},
Mesh → False]

Output of this command is shown in Fig. 3f. We can observe formation of branched structure.

References

Arman, E., Haffner-Krausz, R., Gorivodsky, M., and Lonai, P. (1999). Fgfr2 is required for limb outgrowth and lung-branching morphogenesis. *Proc. Natl. Acad. Sci. USA* **96**, 11895–11899.

Ball, P. (1999). "The Self-Made Tapestry." Oxford Univ. Press, Oxford.

Bellusci, S., Henderson, R., Winnier, G., Oikawa, T., and Hogan, B. L. (1996). Evidence from normal expression and targeted misexpression that bone morphogenetic protein (Bmp-4) plays a role in mouse embryonic lung morphogenesis. *Development* **122**, 1693–1702.

Bellusci, S., Furuta, Y., Rush, M. G., Henderson, R., Winnier, G., and Hogan, B. L. (1997a). Involvement of Sonic hedgehog (Shh) in mouse embryonic lung growth and morphogenesis. *Development* **124**, 53–63.

Bellusci, S., Grindley, J., Emoto, H., Itoh, N., and Hogan, B. L. (1997b). Fibroblast growth factor 10 (FGF10) and branching morphogenesis in the embryonic mouse lung. *Development* **124**, 4867–4878.

Ben-Jacob, E., and Levine, H. (2000). Cooperative self-organization of microorganisms. *Adv. Phys.* **49**, 395–554.

Bentley, W. A., and Humphreys, W. J. (1962). "Snow Chrystals." Dover, New York.

Bragg, A. D., Moses, H. L., and Serra, R. (2001). Signaling to the epithelium is not sufficient to mediate all of the effects of transforming growth factor beta and bone morphogenetic protein 4 on murine embryonic lung development. *Mech. Dev.* **109**, 13–26.

Cardoso, W. V., and Lue, J. (2006). Regulation of early lung morphogenesis: Questions, facts and controversies. *Development* **133**, 1611–1624.

Cardoso, W. V., Itoh, A., Nogawa, H., Mason, I., and Brody, J. S. (1997). FGF-1 and FGF-7 induce distinct patterns of growth and differentiation in embryonic lung epithelium. *Dev. Dyn.* **208**, 398–405.

Chen, C., Chen, H., Sun, J., Bringas, P., Chen, Y., Warburton, D., and Shi, W. (2005). Smad1 expression and function during mouse embryonic lung branching morphogenesis. *Am. J. Physiol. Lung. Cell Mol. Physiol.* **288**, L1033–L1039.

Chuang, P. T., and McMahon, A. P. (2003). Branching morphogenesis of the lung: New molecular insights into an old problem. *Trends Cell Biol.* **13**, 86–91.

Colvin, J. S., White, A. C., Pratt, S. J., and Ornitz, D. M. (2001). Lung hypoplasia and neonatal death in Fgf9-null mice identify this gene as an essential regulator of lung mesenchyme. *Development* **128**, 2095–2106.

Cross, M. C., and Hohenberg, P. C. (1993). Pattern formation outside equilibrium. *Rev. Mod. Phys.* **65**, 851–1112.

Davies, J. A. (2002). Do different branching epithelia use a conserved developmental mechanism? *BioEssays* **24**, 937–948.

Davies, J.A. (Ed.) 2006. "Branching Morphogenesis." Springer-Verlag, New York.

Dean, C. H., Miller, L. -A. D., Smith, A. N., Dufort, D., Lang, R. A., and Niswander, L. A. (2005). Canonical Wnt signaling negatively regulates branching morphogenesis of the lung and lacrimal gland. *Dev. Biol.* **286**, 270–286.

del Moral, P. -M., Langhe, S. P. D., Sala, F. G., Veltmaat, J. M., Tefft, D., Wang, K., Warburton, D., and Bellusci, S. (2006). Differential role of FGF9 on epithelium and mesenchyme in mouse embryonic lung. *Dev. Biol.* **293**, 77–89.

Desai, T. J., Malpel, S., Flentke, G. R., Smith, S. M., and Cardoso, W. V. (2004). Retinoic acid selectively regulates Fgf10 expression and maintains cell identity in the prospective lung field of the developing foregut. *Dev. Biol.* **273**, 402–415.

Desai, T. J., Chen, F. L. J., Qian, J., Niederreither, K., Doll, P., Chambon, P., and Cardoso, W. V. (2006). Distinct roles for retinoic acid receptors alpha and beta in early lung morphogenesis. *Dev. Biol.* **291**, 12–24.

Eblaghie, M. C., Reedy, M., Oliver, T., Mishina, Y., and Hogan, B. L. M. (2006). Evidence that autocrine signaling through Bmpr1a regulates the proliferation, survival and morphogenetic behavior of distal lung epithelial cells. *Dev. Biol.* **291**, 67–82.

Ganser, G. L., Stricklin, G. P., and Matrisian, L. M. (1991). EGF and TGF alpha influence in vitro lung development by the induction of matrix-degrading metalloproteinases. *Int. J. Dev. Biol.* **35**, 453–461.

Gaylord, R. J., and Wellin, P. R. (1995). "Computer Simulations With Mathematica: Explorations in Complex Physical and Biological Systems." Springer-Verlag, Berlin.

Gilbert, S. F. (2003). "Developmental Biology." Sinauer, Massachusettes.

Hartmann, D. (2004). Pattern formation in cultures of bacillus subtilis. *J. Biol. Syst.* **12**, 179–199.

Hartmann, D. (2007). Limit behavior of spatially growing cell cultures. *J. Theor. Biol.* **244**, 409–415.

Hartmann, D., and Miura, T. (2006a). Mathematical analysis of a free-boundary model for lung branching morphogenesis. *Math. Med. Biol.* **24**, 209–224.

Hartmann, D., and Miura, T. (2006b). Modeling in vitro lung branching morphogenesis during development. *J. Theor. Biol.* **242**, 862–872.

Heine, U. I., Munoz, E. F., Flanders, K. C., Roberts, A. B., and Sporn, M. B. (1990). Colocalization of TGF-beta 1 and collagen I and III, fibronectin and glycosaminoglycans during lung branching morphogenesis. *Development* **109**, 29–36.

Hogan, B. L. (1999). Morphogenesis. *Cell* **96**, 225–233.

Honda, H. (1971). Description of the form of trees by the parameters of the tree-like body: Effects of the branching angle and the branch length on the sample of the tree-like body. *J. Theor. Biol.* **31**, 331–338.

Honda, H., and Fisher, J. B. (1979). Ratio of tree branch lengths: The equitable distribution of leaf clusters on branches. *Proc. Natl. Acad. Sci. USA* **76**, 3875–3879.

Honda, H., Hatta, H., and Fisher, J. B. (1997). Branch geometry in Cornus kousa (Cornaceae): Computer simulations. *Am. J. Bot.* **84**, 745–755.

Hyatt, B. A., Shangguan, X., and Shannon, J. M. (2002). BMP4 modulates fibroblast growth factor-mediated induction of proximal and distal lung differentiation in mouse embryonic tracheal epithelium in mesenchyme-free culture. *Dev. Dyn.* **225**, 153–165.

Hyatt, B. A., Shangguan, X., and Shannon, J. M. (2004). FGF-10 induces SP-C and Bmp4 and regulates proximal-distal patterning in embryonic tracheal epithelium. *Am. J. Physiol. Lung Cell Mol. Physiol.* **287**, L1116–L1126.

Izvolsky, K. I., Shoykhet, D., Yang, Y., Yu, Q., Nugent, M. A., and Cardoso, W. V. (2003a). Heparan sulfate–FGF10 interactions during lung morphogenesis. *Dev. Biol.* **258**, 185–200.

Izvolsky, K. I., Zhong, L., Wei, L., Yu, Q., Nugent, M. A., and Cardoso, W. V. (2003b). Heparan sulfates expressed in the distal lung are required for Fgf10 binding to the epithelium and for airway branching. *Am. J. Physiol. Lung Cell Mol. Physiol.* **285**, L838–L846.

Kashimata, M., and Gresik, E. W. (1996). Contemporary approaches to the study of salivary gland morphogenesis. *Eur. J. Morphol.* **34**, 143–147.

Kawasaki, K., Mochizuki, A., Matsushita, M., Umeda, T., and Shigesada, N. (1997). Modeling spatiotemporal patterns generated by Bacillus subtilis. *J. Theor. Biol.* **188**, 177–185.

Kim, S. K., and MacDonald, R. J. (2002). Signaling and transcriptional control of pancreatic organogenesis. *Curr. Opin. Genet. Dev.* **12**, 540–547.

Kitaoka, H., Takaki, R., and Suki, B. (1999). A three-dimensional model of the human airway tree. *J. Appl. Physol.* **87**, 2207–2217.

Kumar, V. H., Lakshminrusimha, S., Abiad, M. T. E., Chess, P. R., and Ryan, R. M. (2005). Growth factors in lung development. *Adv. Clin. Chem.* **40**, 261–316.

Lebeche, D., Malpel, S., and Cardoso, W. V. (1999). Fibroblast growth factor interactions in the developing lung. *Mech. Dev.* **86**, 125–136.

Li, C., Hu, L., Xiao, J., Chen, H., Li, J. T., Bellusci, S., Delanghe, S., and Minoo, P. (2005). Wnt5a regulates Shh and Fgf10 signaling during lung development. *Dev. Biol.* **287**, 86–97.

Li, Y., Zhang, H., Choi, S. C., Litingtung, Y., and Chiang, C. (2004). Sonic hedgehog signaling regulates Gli3 processing, mesenchymal proliferation, and differentiation during mouse lung organogenesis. *Dev. Biol.* **270**, 214–231.

Lindenmayer, A. (1975). Developmental algorithms for multicellular organisms: A survey of L-systems. *J. Theor. Biol.* **54**, 3–22.

Lubkin, S. R., and Li, Z. (2002). Force and deformation on branching rudiments: Cleaving between hypotheses. *Biomech. Model Mechanobiol.* **1**, 5–16.

Lubkin, S. R., and Murray, J. D. (1995). A mechanism for early branching in lung morphogenesis. *J. Math. Biol.* **34**, 77–94.

Lue, J., Izvolsky, K. I., Qian, J., and Cardoso, W. V. (2005). Identification of FGF10 targets in the embryonic lung epithelium during bud morphogenesis. *J. Biol. Chem.* **280**, 4834–4841.

Mailleux, A. A., Tefft, D., Ndiaye, D., Itoh, N., Thiery, J. P., Warburton, D., and Bellusci, S. (2001). Evidence that SPROUTY2 functions as an inhibitor of mouse embryonic lung growth and morphogenesis. *Mech. Dev.* **102**, 81–94.

Makarenkova, H. P., Ito, M., Govindarajan, V., Faber, S. C., Sun, L., McMahon, G., Overbeek, P. A., and Lang, R. A. (2000). FGF10 is an inducer and Pax6 a competence factor for lacrimal gland development. *Development* **127**, 2563–2572.

Malpel, S., Mendelsohn, C., and Cardoso, W. V. (2000). Regulation of retinoic acid signaling during lung morphogenesis. *Development* **127**, 3057–3067.

Matsushita, M., and Fujikawa, H. (1990). Diffusion-limited growth in bacterial colony formation. *Physica A* **168**, 498–506.

Matsushita, M., Wakita, J., Itoh, H., Rafols, I., Matsuyama, T., Sakaguchi, H., and Mimura, M. (1998). Interface growth and pattern formation in bacterial colonies. *Physica A* **249**, 517–524.

Min, H., Danilenko, D. M., Scully, S. A., Bolon, B., Ring, B. D., Tarpley, J. E., DeRose, M., and Simonet, W. S. (1998). Fgf-10 is required for both limb and lung development and exhibits striking functional similarity to Drosophila branchless. *Genes Dev.* **12**, 3156–3161.

Miura, T., and Maini, P. K. (2004). Periodic pattern formation in reaction–diffusion systems: An introduction for numerical simulation. *Anat. Sci. Int.* **79**, 112–123.

Miura, T., and Shiota, K. (2000). Time-lapse observation of branching morphogenesis of the lung bud epithelium in mesenchyme-free culture and its relationship with the localization of actin filaments. *Int. J. Dev. Biol.* **44**, 899–902.

Miura, T., and Shiota, K. (2002). Depletion of FGF acts as a lateral inhibitory factor in lung branching morphogenesis in vitro. *Mech. Dev.* **116**, 29–38.

Moore, K. A., Polte, T., Huang, S., Shi, B., Alsberg, E., Sunday, M. E., and Ingber, D. E. (2005). Control of basement membrane remodeling and epithelial branching morphogenesis in embryonic lung by rho and cytoskeletal tension. *Dev. Dynam.* **232**, 268–281.

Nakanishi, Y., Sugiura, F., Kishi, J., and Hayakawa, T. (1986). Scanning electron microscopic observation of mouse embryonic submandibular glands during initial branching: Preferential localization of fibrillar structures at the mesenchymal ridges participating in cleft formation. *J. Embryol. Exp. Morphol.* **96**, 65–77.

Nelson, T. R., West, B. J., and Goldberger, A. L. (1990). The fractal lung: Universal and species-related scaling patterns. *Experientia* **46**, 251–254.

Nogawa, H., and Ito, T. (1995). Branching morphogenesis of embryonic mouse lung epithelium in mesenchyme-free culture. *Development* **121**, 1015–1022.

Ohuchi, H., Hori, Y., Yamasaki, M., Harada, H., Sekine, K., Kato, S., and Itoh, N. (2000). FGF10 acts as a major ligand for FGF receptor 2 IIIb in mouse multiorgan development. *Biochem. Biophys. Res. Commun.* **277**, 643–649.

Park, W. Y., Miranda, B., Lebeche, D., Hashimoto, G., and Cardoso, W. V. (1998). FGF-10 is a chemotactic factor for distal epithelial buds during lung development. *Dev. Biol.* **201**, 125–134.

Pepicelli, C. V., Lewis, P. M., and McMahon, A. P. (1998). Sonic hedgehog regulates branching morphogenesis in the mammalian lung. *Curr. Biol.* **8**, 1083–1086.

Perl, A. -K. T., Hokuto, I., Impagnatiello, M. -A., Christofori, G., and Whitsett, J. A. (2003). Temporal effects of Sprouty on lung morphogenesis. *Dev. Biol.* **258**, 154–168.

Pongracz, J. E., and Stockley, R. A. (2006). Wnt signaling in lung development and diseases. *Respir. Res.* **7**, 15.

Prusinkiewicz, P. (2004). Modeling plant growth and development. *Curr. Opin. Plant. Biol.* **7**, 79–83.

Sadler, T. W. (2004). "Langman's Medical Embryology," 9th ed. Lippincott/Williams&Wilkins, Maryland.

Sakiyama, J., Yokouchi, Y., and Kuroiwa, A. (2000). Coordinated expression of Hoxb genes and signaling molecules during development of the chick respiratory tract. *Dev. Biol.* **227**, 12–27.

Salazar-Ciudad, I., Jernvall, J., and Newman, S. A. (2003). Mechanisms of pattern formation in development and evolution. *Development* **130**, 2027–2037.

Sekine, K., Ohuchi, H., Fujiwara, M., Yamasaki, M., Yoshizawa, T., Sato, T., Yagishita, N., Matsui, D., Koga, Y., Itoh, N., and Kato, S. (1999). Fgf10 is essential for limb and lung formation. *Nat. Genet.* **21**, 138–141.

Shah, M. M., Sampogna, R. V., Sakurai, H., Bush, K. T., and Nigam, S. K. (2004). Branching morphogenesis and kidney disease. *Development* **131**, 1449–1462.

Shannon, J. M., and Hyatt, B. A. (2004). Epithelial–mesenchymal interactions in the developing lung. *Annu. Rev. Physiol.* **66**, 625–645.

Shu, W., Guttentag, S., Wang, Z., Andl, T., Ballard, P., Lu, M. M., Piccolo, S., Birchmeier, W., Whitsett, J. A., Millar, S. E., and Morrisey, E. E. (2005). Wnt/beta-catenin signaling acts upstream of N-myc, BMP4, and FGF signaling to regulate proximal–distal patterning in the lung. *Dev. Biol.* **283**, 226–239.

Takaki, R. (2005). Can morphogenesis be understood in terms of physical rules? *J. Biosci.* **30**, 87–92.

Tefft, D., Lee, M., Smith, S., Crowe, D. L., Bellusci, S., and Warburton, D. (2002). mSprouty2 inhibits FGF10-activated MAP kinase by differentially binding to upstream target proteins. *Am. J. Physiol. Lung Cell Mol. Physiol.* **283**, L700–L706.

Turing, A. M. (1952). The chemical basis of morphogenesis. *Philos. Trans. R. Soc. B* **237**, 37–72.

Urase, K., Mukasa, T., Igarashi, H., Ishii, Y., Yasugi, S., Momoi, M. Y., and Momoi, T. (1996). Spatial expression of Sonic hedgehog in the lung epithelium during branching morphogenesis. *Biochem. Biophys. Res. Commun.* **225**, 161–166.

Warburton, D., Zhao, J., Berberich, M. A., and Bernfield, M. (1999). Molecular embryology of the lung: Then, now, and in the future. *Am. J. Physiol.* **276**, L697–L704.

Warburton, D., Schwarz, M., Tefft, D., Flores-Delgado, G., Anderson, K. D., and Cardoso, W. V. (2000). The molecular basis of lung morphogenesis. *Mech. Dev.* **92**, 55–81.

Weaver, M., Dunn, N. R., and Hogan, B. L. (2000). Bmp4 and Fgf10 play opposing roles during lung bud morphogenesis. *Development* **127**, 2695–2704.

White, A. C., Xu, J., Yin, Y., Smith, C., Schmid, G., and Ornitz, D. M. (2006). FGF9 and SHH signaling coordinate lung growth and development through regulation of distinct mesenchymal domains. *Development* **133**, 1507–1517.

11

Multiscale Models for Vertebrate Limb Development

Stuart A. Newman, Scott Christley,[†,‡] Tilmann Glimm,[§]
H. G. E. Hentschel,[¶] Bogdan Kazmierczak,[∥] Yong-Tao Zhang,[‡,**]
Jianfeng Zhu,[**] and Mark Alber[‡,**]*

*Department of Cell Biology and Anatomy, New York Medical College, Valhalla, New York 10595
[†]Department of Computer Science and Engineering, University of Notre Dame, Notre Dame, Indiana 46556
[‡]Interdisciplinary Center for the Study of Biocomplexity, University of Notre Dame, Notre Dame, Indiana 46556
[§]Department of Mathematics, Western Washington University, Bellingham, Washington 98225
[¶]Department of Physics, Emory University, Atlanta, Georgia 30322
[∥]Polish Academy of Sciences, Institute of Fundamental Technological Research, 00-049 Warszawa, Poland
[**]Department of Mathematics, University of Notre Dame, Notre Dame, Indiana 46556

I. Introduction
II. Tissue Interactions and Gene Networks of Limb Development
III. Models for Chondrogenic Pattern Formation
   A. Limb Mesenchyme as a "Reactor–Diffusion" System
   B. The Core Patterning Network in a Geometric Setting
   C. "Bare-Bones" System of Reactor–Diffusion Equations
   D. Morphogen Dynamics in the Morphostatic Limit
IV. Simulations of Chondrogenic Pattern Formation
   A. Biological Questions Addressed by the Simulations
   B. Discrete Stochastic Models
   C. Continuum Models
V. Discussion and Future Directions
   Acknowledgments
   References

Dynamical systems in which geometrically extended model cells produce and interact with diffusible (morphogen) and nondiffusible (extracellular matrix) chemical fields have proved very useful as models for developmental processes. The embryonic vertebrate limb is an apt system for such mathematical and computational modeling since it has been the subject of hundreds of experimental studies, and its normal and variant morphologies and spatiotemporal organization of expressed genes are well known. Because of its stereotypical proximodistally generated increase in the number of parallel skeletal elements, the limb lends itself to being modeled by Turing-type systems which are capable of producing periodic, or quasiperiodic, arrangements of spot- and stripe-like elements. This chapter describes several such models, including, (i) a system of partial differential equations in which changing cell density enters into the dynamics explicitly, (ii) a model for morphogen dynamics alone, derived from the latter system in the "morphostatic limit" where cell movement relaxes on a much slower time-scale than cell differentiation, (iii) a discrete stochastic model for the simplified pattern formation that occurs when limb cells are placed in planar culture, and (iv) several hybrid models in which continuum morphogen systems interact with cells represented as energy-minimizing mesoscopic entities. Progress in devising computational methods for handling 3D, multiscale, multimodel simulations of organogenesis is discussed, as well as for simulating reaction–diffusion dynamics in domains of irregular shape. © 2008, Elsevier Inc.

I. Introduction

The vertebrate limb, an array of jointed skeletal elements and associated tissues that arose in fish-like ancestors nearly 400 million years ago, has long held central importance in the fields of developmental and evolutionary biology (reviewed in Newman and Müller, 2005). The developing limb is relatively easy to manipulate surgically in the embryos of avian species such as the chicken. In mammals, such as the human and mouse, it is subject to mutations of large effect that do not otherwise prove fatal to the organism. In fish and amphibians the paired limbs, or related structures, exist with variant anatomical characteristics and regenerative properties. In addition, limb mesenchymal cells can be grown in culture, where they undergo differentiation and pattern formation with a time-course and on a spatial scale similar to that in the embryo. The limb is therefore an ideal system for studying developmental dynamics, genetics, origination and plasticity of multicellular form. Over the last 60 years, knowledge of the tissue, cellular, and molecular interactions involved in generating a vertebrate limb has accumulated dramatically based on the incisive application of new technologies to all of these systems.

The products of scores of genes have been found to participate in limb development (reviewed in Tickle, 2003) and many of these are impaired by mutation or exogenous substances. But genes and their interactions are neither an exclusive nor exhaustive explanatory level for developmental change (Newman, 2002). The mesoscopic physics of viscoelastic materials and excitable media must also enter into the molding and patterning of tissues (Forgacs and Newman, 2005). In particular, they will contribute to the set of dynamic processes by which the interactions of limb bud cells with their various intra- and extracellular molecular components result in a series of articulated, well-arranged rods and nodules of cartilage, and later, bone (Newman and Frisch, 1979).

As with many complex, multiscale, phenomena in biology, insight into emergent organizational properties can be gained by, and indeed require, mathematical and computational modeling. Such modeling does not replace analysis at the cellular and molecular levels, but complements it. Mathematics and computational analysis are the best means we have for describing and understanding the spatiotemporal behaviors of systems containing many components, operating on multiple scales.

Developing organs have both discrete and continuous aspects; they may undergo changes according to deterministic or stochastic rules. Some embryonic tissues are planar and can be approximated as 2D sheets, whereas other tissues are space-filling and inherently 3D. Some developmental processes are synchronized over a spatial domain whereas others sweep across a region over time. Some changes occur autonomously within a given tissue type, while others only proceed by unidirectional or reciprocal interactions between pairs of tissues. In some cases, one developmental process will relax much faster or much slower than another, so that the two can be treated essentially independently of one another. In other cases, the only accurate representation is to treat the processes as mutually determinative and conditioning. Each of these possibilities presents a distinct problem for the modeler, and it is becoming increasingly clear that a fully satisfactory model for the development of any living organ must embody all of them. That is, it will be inescapably hybrid, mathematically and computationally.

This article presents several approaches taken by ourselves and others to modeling skeletal pattern formation during development of the vertebrate limb. Because of the constraints and technical difficulties of producing a multiscale, 3D, hybrid model, as well as the incompleteness of our knowledge of the relevant molecules and the topology, relative strengths and rates of their interactions, the models presented are partial and provisional. Nonetheless, the shortcomings of each of the component models and modeling attempts are usually quite obvious, and we also report on work in progress to remedy them in pursuit of increasingly realistic explanatory accounts.

II. Tissue Interactions and Gene Networks of Limb Development

The limb buds of vertebrates emerge from the body wall, or flank, at four discrete sites—two for the forelimbs and two for the hindlimbs. The paddle-shaped limb bud mesoblast, which gives rise to the skeleton and muscles, is surrounded by a layer of simple epithelium, the ectoderm. The skeletons of most vertebrate limbs develop as a series of precartilage primordia in a *proximodistal* fashion: that is, the structures closest to the body form first, followed, successively, by structures more and more distant from the body. For the forelimb of the chicken, for example, this means the humerus of the upper arm is generated first, followed by the radius and ulna of the mid-arm, the wrist bones, and finally the digits (Saunders, 1948; reviewed in Newman, 1988) (Fig. 1). Urodele salamanders appear to be an exception to this proximodistal progression (Franssen *et al.*, 2005). Cartilage is mostly replaced by bone in species with bony skeletons.

Before the cartilages of the limb skeleton form, the mesenchymal cells of the mesoblast are dispersed in a hydrated ECM, rich in the glycosaminoglycan hyaluro-

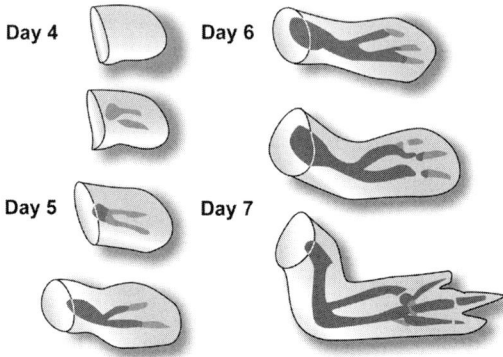

Figure 1 Progress of limb skeletal development in chicken forelimb (wing) between 3 and 7 days of embryogenesis. Gray represents precartilage condensation and black represents definitive cartilage. The limb bud, is paddle-shaped, being narrower in the dorsoventral dimension than in anteroposterior or the shoulder-to-finger tips direction in which it mainly grows. (Adapted, with changes, from Forgacs and Newman, 2005.)

nan. The first morphological evidence that cartilage will differentiate at a particular site in the mesoblast is the emergence of precartilage mesenchymal condensations. The cells at these sites then progress to fully differentiated cartilage elements by switching their transcriptional capabilities. Condensation involves the transient aggregation of cells within a mesenchymal tissue. This process is mediated first by the local production and secretion of ECM glycoproteins, in particular, fibronectin, which act to alter the movement of the cells and trap them in specific places. The aggregates are then consolidated by direct cell–cell adhesion. For this to occur the condensing cells need to express, at least temporarily, adhesion molecules such as N-CAM, N-cadherin, and possibly cadherin-11 (reviewed in Hall and Miyake, 1995, 2000; Forgacs and Newman, 2005).

Because all the precartilage cells of the limb mesoblast are capable of producing fibronectin and cadherins, but only those at sites destined to form skeletal elements do so, there clearly must be communication among the cells to divide the labor in this respect. This is mediated in part by secreted, diffusible factors of the TGF-β family of growth factors, which promote the production of fibronectin (Leonard *et al.*, 1991). Limb bud mesenchyme also shares with many other connective tissues the autoregulatory capability of producing more TGF-β upon stimulation with this factor (Miura and Shiota, 2000a; Van Obberghen-Schilling *et al.*, 1988).

The limb bud ectoderm performs several important functions. First, it is a source of fibroblast growth factors (FGFs) (Martin, 1998). Although the entire limb ectoderm produces FGFs, the particular mixture produced by the apical ectodermal ridge (AER), a narrow band of specialized ectodermal cells running in the anteroposterior direction along the tip of the growing limb bud in birds, mammals and reptiles, is essential to limb outgrowth and pattern formation. FGF8 is the most important of these (Mariani

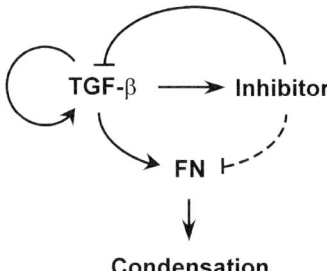

Figure 2 Core set of cell–gene product interactions leading to limb precartilage mesenchymal condensation. The molecular identity of the lateral inhibitor of condensation is unknown, but depends on interaction of ectodermal FGFs with mesenchymal FGF receptor 2 (Moftah *et al.*, 2002) as well as the Notch signaling pathway (Fujimaki *et al.*, 2006). This inhibitor may act at the level of TGF-β synthesis or activity (solid inhibitory vector), fibronectin synthesis (dashed inhibitory vector), or at some earlier stage. For the purposes of our computational model, we assume it acts on the activator (i.e., TGF-β). (Based on Hentschel *et al.*, 2004.)

and Martin, 2003). It is expressed mainly in the AER in amniotes (birds, mammals and reptiles), but is also expressed in limb bud mesenchyme in urodeles (Han *et al.*, 2001). The AER affects cell survival (Dudley *et al.*, 2002) and keeps the precondensed mesenchyme of the "apical zone" in a labile state (Kosher *et al.*, 1979). Its removal leads to terminal truncations of the skeleton (Saunders, 1948).

The FGFs produced by the ectoderm affect the developing limb tissues through one of three distinct FGF receptors. The apical zone is the only region of the mesoblast-containing cells that expresses FGF receptor 1 (FGFR1) (Peters *et al.*, 1992; Szebenyi *et al.*, 1995). In the developing chicken limb, cells begin to condense at a distance of approximately 0.3 mm from the AER. In this zone FGFR1 is downregulated and cells that express FGFR2 appear at the sites of incipient condensation (Peters *et al.*, 1992; Szebenyi *et al.*, 1995; Moftah *et al.*, 2002). Activation of these FGFR2-expressing cells by FGFs releases a laterally-acting (that is, peripheral to the condensations) inhibitor of cartilage differentiation (Moftah *et al.*, 2002). Although the molecular identity of this inhibitor is unknown, its behavior is consistent with that of a diffusible molecule, or one whose signaling effects propagate laterally in an analogous fashion. Recent work suggests that Notch signaling also plays a part in this lateral inhibitory effect (Fujimaki *et al.*, 2006). The roles of TGF-β, the putative lateral inhibitor, and fibronectin in mediating precartilage condensation in the limb bud mesenchyme can be schematized in the form of a "core" cell–molecular-genetic network (Fig. 2).

Finally, differentiated cartilage in the more mature region, proximal to the condensing cells, expresses FGFR3, which is involved in the growth control of this tissue (Ornitz and Marie, 2002). The ectoderm, by virtue of the FGFs it produces, thus regulates growth and differentiation of the mesenchyme and cartilage.

The limb ectoderm is also involved in shaping the limb bud. By itself, the mesenchyme, being an isotropic tissue with liquid-like properties, would tend to round up (Forgacs and Newman, 2005). Ensheathed by the ectoderm, however, it assumes a paddle shape. This appears to be due to the biomechanical influence of the epithelium, the underlying basal lamina of which is organized differently beneath the dorsoventral surfaces and the anteroposteriorly arranged AER (Newman et al., 1981). There is no entirely adequate biomechanical explanation for the control of limb bud shape by the ectoderm [but see Dillon and Othmer (1999) and Borkhvardt (2000) for suggestions].

III. Models for Chondrogenic Pattern Formation

A. Limb Mesenchyme as a "Reactor–Diffusion" System

Reaction–diffusion systems, networks of molecular species in which positive and negative feedbacks in their production and consumption, and disparate diffusion rates, cause compositionally uniform configurations to be potentially unstable, have attracted interest as biological pattern-forming mechanisms ever since their well known proposal as the "chemical basis of morphogenesis" by A. M. Turing half a century ago (Turing, 1952). Experimentally motivated reaction–diffusion-type models (not all of them conforming to Turing's precise scheme) have been gaining prominence in many areas of developmental biology (Forgacs and Newman, 2005; Maini et al., 2006), including the patterning of the pigmentation of animal skin (Yamaguchi et al., 2007), feather germs (Jiang et al., 2004), hair follicles (Sick et al., 2006), and teeth (Salazar-Ciudad and Jernvall, 2002). As we have seen, patterning of the limb skeleton is dependent on molecules of the TGF-β and FGF classes, which are demonstrably diffusible morphogens (Lander et al., 2002; Williams et al., 2004; Filion and Poper, 2004). Like many reaction–diffusion systems, moreover, the developing limb has self-organizing pattern-forming capability. It is well known, for example, that many transcription factors (e.g., the Hox proteins) and extracellular factors (e.g., Sonic hedgehog protein, retinoic acid) are present in spatiotemporal patterns during limb development, and disrupting their distributions leads to skeletal anomalies (Tickle, 2003). Nevertheless, randomized limb mesenchymal cells with disrupted gradients of Hox proteins, Shh, etc., give rise to digit-like structures *in vivo* (Ros et al., 1994) and discrete, regularly spaced cartilage nodules *in vitro* (Downie and Newman, 1994; Kiskowski et al., 2004). Moreover, simultaneous knockout of Shh and its inhibitory regulator Gli3 in mice yields limbs with numerous extra digits (Litingtung et al., 2002). If anything, therefore, such gradients limit and refine the self-organizing capacity of limb mesenchyme to produce skeletal elements rather than being required for it.

Beyond this, the following experimental findings count in favor of the relevance of a reaction–diffusion-type mechanism for limb pattern formation: (i) the pattern of precartilage condensations in limb mesenchyme *in vitro* changes in a fashion consistent with a reaction–diffusion mechanism (and not with an alternative mechanochemical mechanism) when the density of the surrounding matrix is varied (Miura and Shiota,

2000b); (ii) exogenous FGF perturbs the kinetics of condensation formation by limb precartilage mesenchymal cells *in vitro* in a fashion consistent with a role for this factor in regulating inhibitor production in a reaction–diffusion model (Miura and Maini, 2004); (iii) the "thick-thin" pattern of digits in the *Doublefoot* mouse mutant can be accounted for by the assumption that the normal pattern is governed by a reaction–diffusion process the parameters of which are modified by the mutation (Miura *et al.*, 2006).

The scale-dependence of reaction–diffusion systems (i.e., the addition or loss of pattern elements when the tissue primordium has variable size), sometimes considered to count against such mechanisms for developmental processes, may actually represent the biological reality in the developing limb. Experiments show, for example, that the number of digits that arise is sensitive to the anteroposterior (thumb-to-little finger breadth) of the developing limb bud, and will increase (Cooke and Summerbell, 1981) or decrease (Alberch and Gale, 1983) over typical values if the limb is broadened or narrowed.

Turing's notion of a reaction–diffusion system, though directed towards understanding biological pattern formation, used the formalism of chemical reaction systems (Turing, 1952). Given the nature of the core mechanism for chondrogenic pattern formation (see Section II) and the fact that a living cell's reactivity to its microenvironment and capacity to product new molecules in response to it is a complex function of many biosynthetic processes (which can be made increasingly explicit as more information accrues), we use the term "reactor–diffusion" system for the morphogenetic descriptions and models described below.

B. The Core Patterning Network in a Geometric Setting

The developing limb has a smooth, but nonstandard, geometric shape that changes over time. Moreover, different processes take place in different parts of the developing structure. As is the case with somitogenesis along the body axis (Pourquié, 2003; Schnell *et al.* and Baker *et al.*, this volume), the distal end of the pattern-forming system (the tail tip and the AER) produces FGFs that keep a zone of tissue within the high end of the gradient in an immature, unpatterned state.

To simplify the presentation of our basic limb development model, we made the following geometric idealization (Newman and Frisch, 1979; Hentschel *et al.*, 2004): the limb bud is considered as a parallelepiped of time-dependent proximodistal length, $L(t)$, taken along the x-axis, and fixed length, l_y, along the anteroposterior (thumb to little finger) direction (y-axis). The dorsoventral (back to front) width (z-axis) was collapsed to zero in this schematic representation (Fig. 3). The developing limb bud is considered, based on classic observations, to consist of three regions: an "apical zone" of size $l_{\text{apical}}(t)$, at the distal tip of the bud, consisting of noncondensing mesenchymal cells, an "active zone," proximal to the apical zone, of length $l_x(t)$, which contains differentiating and condensing cell types, and a "frozen zone"

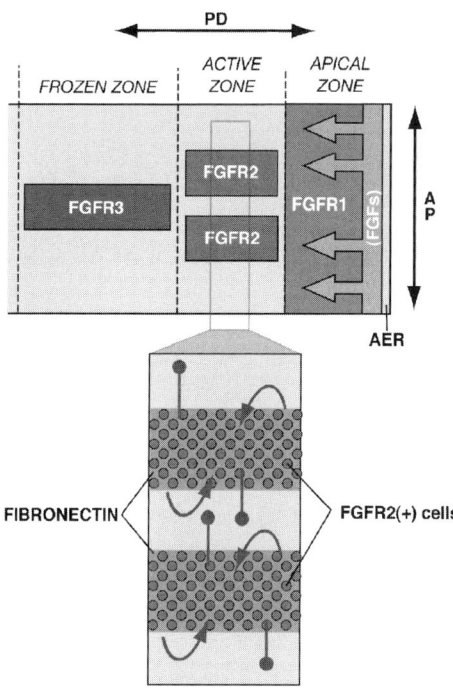

Figure 3 The interactions of the core mechanism are superimposed on a two-dimensional schematic limb bud organized into zones defined by experimentally-determined expression patterns of FGF receptors 1, 2 and 3. In the apical zone cell rearrangement is suppressed by the FGFs emanating from the AER. The active zone, a detailed view of which is shown below) is the site of spatiotemporal regulation of mesenchymal cell condensation (i.e., pattern formation). When cells leave the proximal end of the active zone and enter the frozen zone they differentiate into cartilage and their spatiotemporal pattern becomes fixed. The length of the dorsoventral axis is collapsed to zero in this simplified model. PD: proximodistal; AP: anteroposterior. In lower panel, curved arrows: positively autoregulatory activator; lines ending in circles: lateral inhibitor. (See Hentschel et al., 2004; figure based on Forgacs and Newman, 2005, with changes.) See color insert.

of length $l_{\text{frozen}}(t)$, containing differentiated cartilage cells, proximal to the active zone. The lengths of these zones add up to that of the entire limb bud: $L(t) = l_{\text{apical}}(t) + l_x(t) + l_{\text{frozen}}(t)$.

The dynamic model we present in the following section resides within, but is not uniquely tied to, the schematic shown in Fig. 3. While the division into apical, active and frozen zones is experimentally motivated and is inherent to our conception of the spatiotemporal organization of limb development, the 2D rectilinear template of Fig. 3 is presented for didactic purposes. Our goal, partly implemented in subsequent sections, is to model the cellular and molecular dynamics in the 3D space of a naturally contoured limb bud.

C. "Bare-Bones" System of Reactor–Diffusion Equations

We refer to the dynamic model for limb development presented here as "bare-bones," because while it incorporates the core mesenchymal cell–morphogen–ECM network summarized in Fig. 2, it omits the spatiotemporal modulatory factors such as Hox protein gradients, Shh, and so on, that cause the various skeletal elements (e.g., the different digits, the radius, and ulna) to appear different from one another. As a first pass, we attempt only to model the proximodistal temporal progression of skeletogenesis and the generally increasing number of elements along the proximodistal axis.

As described in earlier sections, limb skeletal patterning involves cycles of cell state changes and local cell movement: mesenchymal cells upregulate their production of fibronectin at particular tissue sites, leading to precartilage condensation. This is followed, in turn, by chondrogenesis (cartilage differentiation). Finally, the spatiotemporal control of these differentiative and morphogenetic changes in the mesenchyme is entirely dependent on products of the surrounding epithelium.

We begin with the hypothesis that the division of the distal portion of the limb into an apical and active zone reflects the activity of the AER in suppressing differentiation of the mesenchyme subjacent to it (Kosher et al., 1979). The spatial relationship between the apical and active zones results from the graded distribution of FGFs, the presumed AER-produced suppressive factors. The active zone, therefore, is where the mesenchyme cells no longer experience high levels of FGFs and therefore become responsive to the activator, TGF-β, and the factors that mediate lateral inhibition. The dynamic interactions of cells and morphogens in the active zone give rise to spatial patterns of condensations. The length of the active zone, $l_x(t)$ or alternatively, the active zone plus the apical zone, serves as a "control parameter" that influences the number of parallel stripes of cell condensation that form.

Cell proliferation enters into this scheme in the following fashion: cells are recruited into the active zone from the proximal end of the apical zone, as dividing cells move away from the influence of the AER. (This is similar to the role of the caudal FGF gradient in somitogenesis; Dubrulle et al., 2001). The active zone loses cells, in turn, to the proximal frozen zone, the region where cartilage differentiation has occurred and a portion of the definitive pattern has become set.

Four main types of mesenchymal cells are involved in chick limb skeletal pattern formation. These are represented in the model by their spatially and temporarily varying densities. The cell types are characterized by their expression of one of the three FGF receptors found in the developing limb. The cells expressing FGFR1, FGFR2 and FGFR3 are denoted, respectively, by R_1; $R_2 + R_2'$; and R_3. The apical zone consists of R_1 cells, and those of the frozen zone R_3 cells (reviewed in Ornitz and Marie, 2002). The active zone contains R_2 cells and the direct products of their differentiation, R_2' cells. These latter cells secrete enhanced levels of fibronectin. The R_1, R_2, and R_2' cells are mobile, while the R_3 (cartilage) cells are immobile.

According to our model (Hentschel et al., 2004), transitions and association between the different cell types are regulated by the gene products of the core mechanism

(Fig. 2): c, c_a, c_i, and ρ denote, respectively, the spatially and temporarily varying concentrations of FGFs (produced mainly by the ectoderm), TGF-β (produced throughout the mesenchyme), a diffusible inhibitor of chondrogenesis produced by R_2 cells, and fibronectin, produced mainly by R_2' cells. The model thus comprises eight variables, with an equation for the behavior of each of them [Eqs. (1)–(8)]. These eight variables correspond to the core set of interactions necessary to describe the development of a basic, "bare bones" skeletal pattern (see Fig. 2).

$$\partial c/\partial t = D\nabla^2 c - kc + J(x,t), \tag{1}$$

$$\partial c_a/\partial t = D_a\nabla^2 c_a - k_a c_i c_a + J_a^1 R_1 + J_a(c_a) R_2, \tag{2}$$

$$\partial c_i/\partial t = D_i\nabla^2 c_i - k_a c_i c_a + J_i(c_a) R_2, \tag{3}$$

$$\partial R_1/\partial t = D_{\text{cell}}\nabla^2 R_1 - \chi\nabla \cdot (R_1\nabla\rho) + rR_1(R_{\text{eq}} - R)$$
$$+ k_{21}R_2 - k_{12}(c, c_a)R_1, \tag{4}$$

$$\partial R_2/\partial t = D_{\text{cell}}\nabla^2 R_2 - \chi\nabla \cdot (R_2\nabla\rho) + rR_2(R_{\text{eq}} - R) + k_{12}R_1$$
$$- k_{21}R_2 - k_{22}R_2, \tag{5}$$

$$\partial R_2'/\partial t = D_{\text{cell}}\nabla^2 R_2' - \chi\nabla \cdot (R_2'\nabla\rho) + rR_2'(R_{\text{eq}} - R) + k_{22}R_2 - k_{23}R_2', \tag{6}$$

$$\partial R_3/\partial t = r_3 R_3(R_{\text{eq}} - R_3) + k_{23}R_2', \tag{7}$$

$$\partial\rho/\partial t = k_b(R_1 + R_2) + k_b' R_2' - k_c\rho. \tag{8}$$

In this set of equations, R_1, R_2, R_2', and R_3 are densities of the different kinds of cells and $R = R_1 + R_2 + R_2'$ is the overall density of the mobile cells. In addition, $J(x,t)$ is the flux of FGF, which we assume is predominantly released at the limb boundary by the AER. The term J_a^1 represents a low level flux of the activator (TGF-β) released by the R_1 cells, and $J_a(c_a)$ is the flux of activator released by the R_2 cells at a rate dependent on the local concentration of activator in the ECM. In general, we expect $J_a^1 \ll J_a(c_a)$.

As written, the equations are for a fixed domain. Allowing the domain to grow will add to each equation a convective term representing the local velocity field created by the domain growth. However, rough estimates indicate that the Péclet number, LV/D, where L is the characteristic length scale of the limb, V the characteristic velocity of the flow and D the morphogen diffusion coefficient, is relatively small, meaning that diffusion is more important than convection.

We have considered the system's behavior in a domain of two or three dimensions with a smooth boundary (i.e., the mathematical conditions pertaining to the space in which a cell layer or tissue mass resides), which is assumed to be fixed in space and time, under the additional assumptions that $\tau_m \ll \tau_g$, $\tau_d \ll \tau_g$, where τ_m, τ_d, and τ_g are the characteristic times of morphogen evolution, cell differentiation, and limb growth, respectively. In terms of the model parameters, $\tau_m \sim L^2/D$, where L is the characteristic length scale, as above, and D is the morphogen diffusivity, $\tau_d \sim 1/k$, where k is the rate-limiting kinetic term for morphogen production, and $\tau_g \sim L/V$ where V is the typical convective time scale for the viscoelastic tissue.

All of the functions are subject to no-flux boundary conditions

$$\frac{\partial c}{\partial n} = \frac{\partial c_a}{\partial n} = \frac{\partial c_i}{\partial n} = \frac{\partial \rho}{\partial n} = \frac{\partial R_1}{\partial n} = \frac{\partial R_2}{\partial n} = \frac{\partial R_2'}{\partial n} = \frac{\partial R_3}{\partial n} = 0. \tag{9}$$

Here n is the outward normal to the spatial domain, so the cells and secreted molecules have zero flux through the boundary.

The variables are intimately interconnected and interacting: diffusing morphogen-type growth factors (i.e., FGFs and TGF-β) and extracellular matrix molecules (i.e., fibronectin) secreted by some cells represent signals for others to move, differentiate or to produce other or more of the same molecules. The interactions among these variables constitute a reactor–diffusion system with pattern forming capability.

Simulations using the full system (1)–(8) are computationally formidable, but we have shown analytically that for a large class of initial conditions smooth solutions exist for it (Alber et al., 2005a). For computational purposes, however, we have sought simpler approximations that are justified both biologically and mathematically. In the initial description of the model we used a biologically motivated separation of time scales to reduce the system to four equations. We then used linear stability analysis to compute solutions on the plane for active zones of progressively decreasing width, arriving at a simulation that accurately portrayed the proximodistal order of appearance and increase in number of skeletal elements in a 2D projection (Hentschel et al., 2004). In a separate study we have shown the existence of the steady state solutions to the four-equation system in the form of stripes and spots in the 2D case and analogous extensions in a 3D case and demonstrated that they cannot be stable at the same time (Alber et al., 2005b).

Subsequently, using a multiscale, multimodel simulation environment (Izaguirre et al., 2004; Cickovski et al., 2007), we modeled cell rearrangement as an individual-based module in the presence of morphogen fields and cell-state transition rules based on simplifications of system (1)–(8). Although this strategy permitted us to generate 3D simulations with authentic developmental properties (Chaturvedi et al., 2005; Cickovski et al., 2005), the equations for the morphogen dynamics abstracted from the full system (where cell and morphogen changes are interconnected) were fairly ad hoc.

In the next section we describe one attempt to remedy this deficit by deriving an analytically rigorous reduction of the morphogen dynamics to a two-equation system (Alber et al., in press). To do this we have treated system (1)–(8) in the biologically plausible "morphostatic limit" (Salazar-Ciudad et al., 2003), i.e., under the assumption that cell differentiation occurs on a faster scale than cell rearrangement.

D. Morphogen Dynamics in the Morphostatic Limit

In this section we consider the bare-bones system for skeletal pattern formation (Eqs. (1)–(8)) in the limiting case in which the dynamics of cell differentiation is faster than the evolution of the overall cell density. This assumption is consistent with existing experimental evidence, though is not uniquely determined by it. An additional assumption is that the spatial variations in the densities of the various cell types involved are small and can be replaced by a constant density for the analysis of the evolution of the morphogen concentrations.

Rates of limb development vary widely across phylogenetic distances, and it is highly unlikely that the characteristic time scales of the various component processes of development all scale proportionally. For this reason, the limit we consider may pertain to limb formation in some species and not others, or only at some stages of limb evolution.

With the above assumptions, system (1)–(8) reduces to two evolution equations for the morphogens:

$$\frac{\partial c_a}{\partial t} = D_a \nabla^2 c_a + U(c_a) - k_a c_a c_i, \qquad (10)$$

$$\frac{\partial c_i}{\partial t} = D_i \nabla^2 c_i + V(c_a) - k_a c_a c_i. \qquad (11)$$

The equations are subject to no-flux boundary conditions. The terms $D_a \nabla^2 c_a$ and $D_i \nabla^2 c_i$ describe diffusion of the morphogens, the term $-k_a c_a c_i$ represents their decay or inactivation, and the terms $U(c_a)$ and $V(c_a)$ describe their production by cells. Specifically, $U(c_a)$ and $V(c_a)$ are given by

$$U(c_a) = \left(J_a^1 \alpha(c, c_a) + J_a(c_a)\beta(c, c_a)\right) R_{\text{eq}}; \quad V(c_a) = J_i(c_a)\beta(c, c_a) R_{\text{eq}}. \quad (12)$$

In these expressions, R_{eq} is the cell density (assumed to be approximately constant, see above), and the terms describing secretion of activator and inhibitor are denoted by J with appropriate subscripts. Details on the exact form of these terms are given in Alber et al., in press.

In the classification of developmental pattern forming mechanisms proposed by Salazar-Ciudad and coworkers (Salazar-Ciudad et al., 2003), a "morphostatic" mechanism is one which occurs in two distinct phases: cell interaction that leads to alteration of cell states, and hence changes in pattern, is followed by the action of one of various mechanisms (termed "morphogenetic") which causes spatial rearrangement of cells without changing their states. In contrast, "morphodynamic" mechanisms are ones in which cell state changes and cell rearrangements happen simultaneously. For the limb model we consider here, the morphogenetic process is precartilage mesenchymal condensation.

With reference to the classification of Salazar-Ciudad et al. (2003), then, the system (10)–(11) obtained from the assumption that differentiation dynamics relaxes faster than mesenchymal condensation can be considered the "morphostatic limit" of the full (morphodynamic) model of Hentschel et al. (2004). Because the relative rates of cell differentiation and cell movement are likely to be subject to natural selection, evolution of morphostatic mechanisms may represent a successful evolutionary strategy in certain cases.

The task of finding parameter ranges under which the system can give rise to patterns [what we refer to as the "Turing space" of the system (Alber *et al.*, in press)] is much more tractable in the reduced system (10)–(11) than in the full system (1)–(8). This is due to the smaller number of variables and parameters. In addition, the resulting reduced reaction–diffusion system could then be used in a variety of computational models in which geometries, additional morphogens and individual model cells (insofar as they behave according to the assumptions under which the morphogen subsystem was defined) could be introduced in a controlled fashion. In particular, the fact that the reduced equations are derived in a mathematically rigorous fashion from explicit biological assumptions makes them much more appropriate than the ad hoc reaction kinetics used in earlier hybrid continuum-discrete models of the limb (Chaturvedi *et al.*, 2005; Cickovski *et al.*, 2005), and adds to the authenticity of these models.

We have considered a broad class of Michaelis–Menten-type kinetics models for the rate terms in (10)–(11). The necessary conditions for the Turing instability are fulfilled across a wide range of parameter values, suggesting that a precise choice of these coefficients is not required for pattern formation. In this sense the system is robust.

For the system (10)–(11) to exhibit a Turing-type instability in the morphostatic limit several constraints on morphogen dynamics must be met. In particular, the results indicate the following qualitative predictions:

1) The maximum production rate of the inhibitor (in fact, by R_2 cells) exceeds their rate of production of the activator TGF-β.

2) The threshold levels of local TGF-β concentration which elicit maximal production rates by R_2 cells of TGF-β, and of inhibitor, must be of roughly the same order of magnitude. (See Alber *et al.*, in press.)

IV. Simulations of Chondrogenic Pattern Formation

A. Biological Questions Addressed by the Simulations

In this section we introduce several methods to address questions raised by the foregoing biological considerations. While our ultimate objective is to produce a simulation of a vertebrate limb in 3D based solely on biologically authentic cell behaviors, each component of this target hybrid model presents its own computational difficulties, and in most cases neither all of the biological variables nor the parameters of their interactions are available. Below we present a proof-of-principle that biologically authentic chondrogenic patterns can be generated using the postulated cell behaviors and molecular components in a multiscale model. While the system in which this is demonstrated is not the full, growing limb bud, but rather disk-like micromass cultures, which have been studied extensively as an *in vitro* model for skeletogenesis, it is a system for which detailed quantitative measurements are available. For our quasi-3D individual-based representation of these cultures we use a discrete form of the reaction–diffusion system for the morphogen dynamics.

Following this, we present the results of two attempts at modeling the growing 3D limb using different implementations of the Cellular Potts Model, based on energy-minimization of cell–ECM interactions. Unlike the case with the micromass cultures, the geometry of these 3D simulations is necessarily highly idealized relative to the living system. Moreover, in each of these simulations we use slightly different simplifications of the mechanism of Hentschel *et al.* (2004) (Eqs. (1)–(8)), based on different ad hoc separations of the morphogen dynamics from cell movement.

Finally, we have used a novel finite element-based computational strategy to explore the pattern-forming potential of Eqs. (10)–(11), the morphogen dynamics in the morphostatic limit of system (1)–(8), on irregular spatial domains. Although we have only accomplished this in 2D thus far, and the distal undifferentiated zone (i.e., apical plus active zones) of the developing limb (see Fig. 2) constitute 3D domains of changing shape at successive stages of development, this new result and simulation protocol should eventually enable us to return to the 3D framework and implement simulations with more natural shapes and authentic morphogen dynamics.

B. Discrete Stochastic Models

Stochastic discrete models are used in a variety of problems dealing with biological complexity. One motivation for this approach is the enormous range of length scales of typical biological phenomena. Treating cells as simplified interacting agents, one can simulate the interactions of tens of thousands to millions of cells and still have within reach the smaller-scale structures of tissues and organs that would be ignored in continuum (e.g., partial differential equation) approaches. At the same time, discrete stochastic models can be made sophisticated enough to reproduce almost all commonly observed types of cell behavior (Chaturvedi *et al.*, 2003, 2004; Casal *et al.*, 2004, 2005; Alber *et al.*, 2003, 2004; Sozinova *et al.*, 2006a, 2006b).

1. Proof of Principle: Pattern Formation in Mesenchymal Micromass Cultures

We have used a multiscale, stochastic, discrete approach to model chondrogenic pattern formation *in vitro* in the high-density limb bud mesenchyme micromass culture system (Kiskowski *et al.*, 2004; Christley *et al.*, 2007). The most recent of these models is multiscale (i.e., cell and molecular dynamics occur on distinct spatial and temporal scales) with cells represented as spatially extended objects that can change their shape (Fig. 4). It has been calibrated using experimental data and simulation

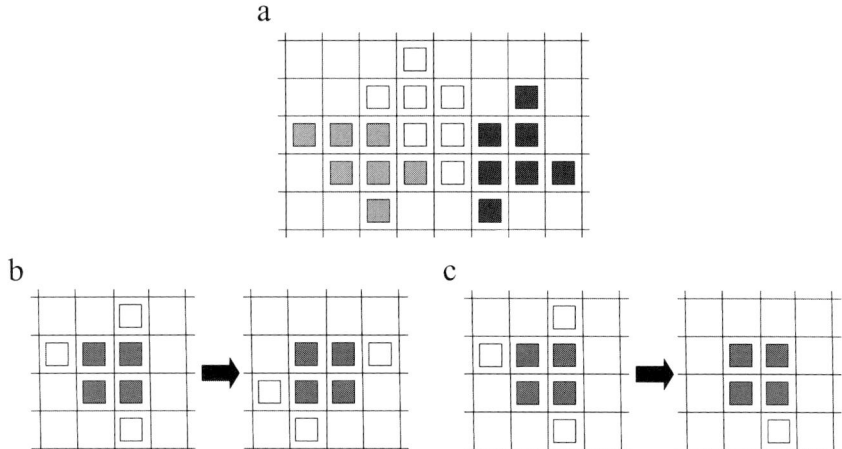

Figure 4 Multipixel spatial representation of limb mesenchymal cells. (A) Three cells on the spatial grid each occupying seven pixels. (B) Cell changes shape. The region of the cell that contains the nucleus, indicated by the four gray pixels, is structurally maintained; two border pixels move to new locations, and one border pixel (top right) displaces a nucleus pixel which gets shifted to the right. (C) Cell rounding-up on fibronectin. The surface area in the presence of suprathreshold amounts of fibronectin is reduced with two border pixels moving into a quasi third dimension above the cell. (From Christley *et al.*, 2007.)

results indicate that cells can form condensation patterns by undergoing small displacements of less than a cell diameter, packing more closely by changing their shapes, while maintaining a relatively uniform cell density across the entire spatial domain.

The simulations have disclosed two distinct dynamical regimes for pattern self-organization involving transient or stationary inductive patterns of morphogens (Fig. 5). The transient regime involves the appearance of morphogen concentrations in a spatial pattern for a brief period of time after which the morphogens are lost from the system by degradation or inactivation. This transient regime is oscillatory in behavior with the periodic reappearance and degradation of morphogens. The stationary regime involves a spatial pattern of morphogen concentrations that remains stable indefinitely over time. Sensitivity to changes in key parameters has been studied indicating robustness in pattern formation behavior but with variation in the morphological outcomes (Fig. 6). Formation of both spots and stripes of precartilage condensation can be produced by the model under slightly different parameter choices. This corresponds well to experimental results where either morphotype may be generated under similar initial conditions, and it supports the applicability of the core molecular-genetic mechanism we have used to the understanding of both *in vitro* and *in vivo* chondrogenic pattern formation.

Tuning of a core molecular-genetic mechanism (e.g., by natural selection or epigenetic factors) can provide both multiple dynamical pathways to the same phenotypic outcome and multiple phenotypes produced from the same dynamical pathway.

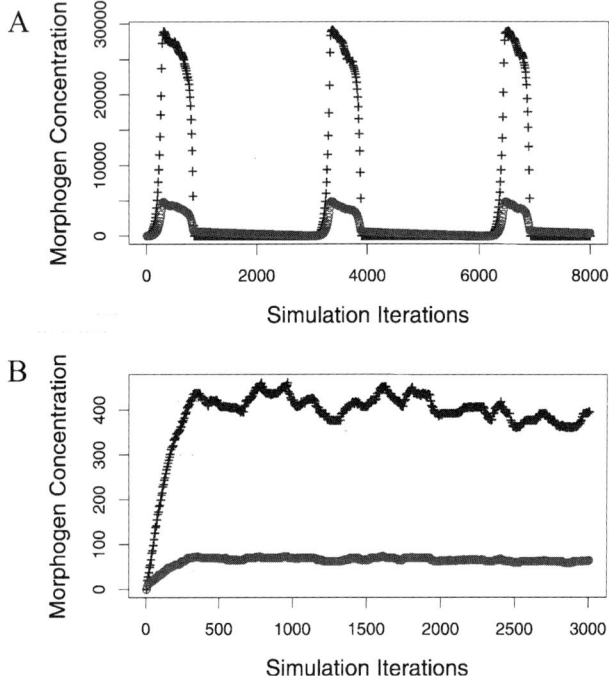

Figure 5 Dynamics of oscillatory and stationary regimes. (A) Oscillatory regime produces transient patterns that repeat over time but are spatially stochastic. (B) Stationary regime produces stable patterns with minor stochastic fluctuations around an equilibrium concentration. Graphs show the maximum concentration value for a single pixel across the entire molecular grid for activator (black) and inhibitor (blue) morphogens (Christley et al., 2007). See color insert.

An important implication is that limb development does not require a strict progression from one stable dynamical regime to another, but can occur by a succession of transient dynamical regimes or combination of stable and transient regimes to achieve a particular morphological outcome.

2. Simulations of Chondrogenic Patterning in 3D

a. Energetics of Cell Interactions and the Cellular Potts Model The mesoscopic Cellular Potts Model (CPM), first introduced by Graner and Glazier (1992) and Glazier and Graner (1993), is a cell-level, energy-minimization-based, lattice model that uses an effective energy H coupled to external fields, e.g., the local concentrations of diffusing chemicals, to describe cell–cell interactions, cell adhesion, motion, differentiation, division, and apoptosis. The effective energy mixes true energies, such as cell–cell adhesion, with terms that mimic energies, e.g., the response of a cell to a chemotactic or haptotactic gradient. Since the cells' environment is extremely viscous,

11. Multiscale Models for Vertebrate Limb Development 327

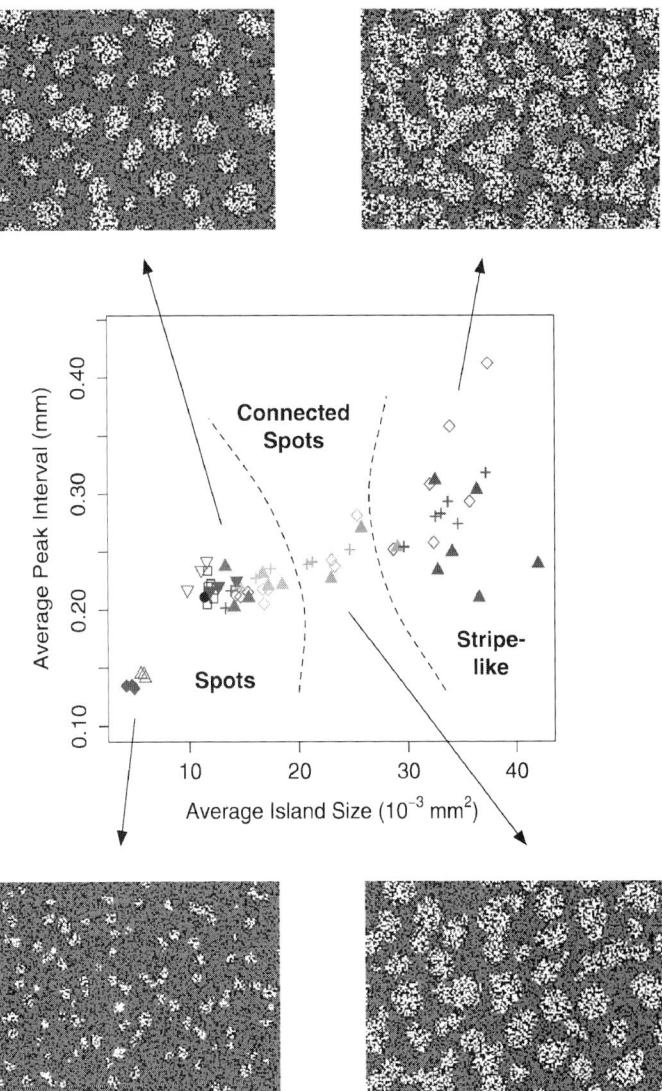

Figure 6 Variation in some of the key parameters induces morphological changes in the resultant spatial patterns from distinct spots to connected spots to stripe-like patterns. Average peak interval versus average island size for variations in the some of the key parameters are shown: +5% (diamond) and −5% (filled diamond) for activator self-regulation, +5% (triangle) and −5% (filled triangle) for activator regulation of inhibitor, +5% (down triangle) and −5% (filled down triangle) for inhibitor regulation of activator, +5% (plus) for inhibitor decay. The colored points are a gradient of variations: 1% (red), 2% (orange), 3% (green), 4% (blue), 5% (violet). Also shown are the five simulations (square) using the standard parameter values and the mean for twelve experiments (circle) (Christley et al., 2007). All simulations were run for 3000 iterations with periodic boundary conditions. (From Christley et al., 2007.) See color insert.

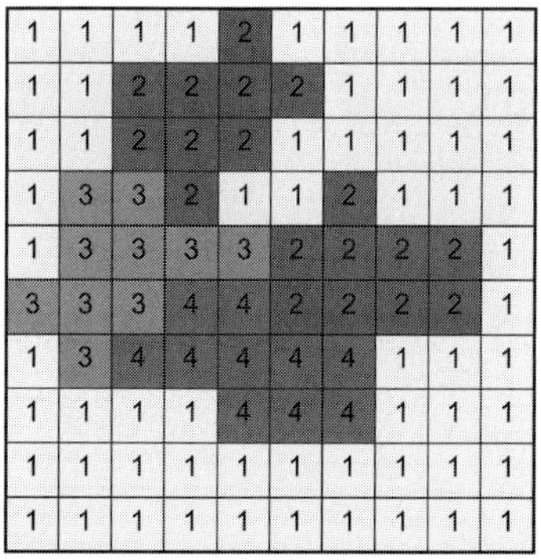

Figure 7 A typical 2D CPM configuration. The numerals indicate indices at lattice sites. The colors indicate cell type. A cell is collection of connected lattice points with same index. The number of lattice points in a cell is its volume and its number of links with other indices is its surface area. We represent ECM as a generalized cell with index 1. See color insert.

their motion is entirely dissipative, minimizing their total effective energy consistent with constraints and boundary conditions.

A CPM-based tissue morphogenesis model consists of a list of biological cells, a list of generalized cells, a set of chemical diffusants and a description of the biological and physical behaviors and interactions embodied in the effective energy, along with auxiliary equations to describe secretion, transport and absorption of diffusants and other extracellular materials, state changes within the cell, mitosis and cell death.

In the CPM, *generalized cells* can model noncellular materials such as ECM and fluid media. Such an approach is much simpler and faster than finite-element modeling, has better spatial fidelity than modeling cells as point particles and can be formally translated into a force-based description of cell behaviors. The CPM discretizes space into a 2D (as in Fig. 7) or 3D lattice. Each lattice point contains an integer index that identifies the cell, ECM element or other object to which it belongs. Separate lattices contain concentrations of diffusants, which evolve under partial differential equations (PDEs), while transitions between different cell states and types are governed by sets of state maps and ordinary differential equations (ODEs).

Each cell is represented by a cluster of pixels with the same index. Pixels evolve according to the classical Metropolis algorithm based on Boltzmann statistics and the effective energy E (Newman and Barkema, 1999). Namely, if a proposed change in a lattice configuration results in energy change ΔE_i, it is accepted with probability

$$\Phi(\Delta E_i) = \begin{cases} 1, & \Delta E_i \leqslant 0, \\ \dfrac{e^{-\frac{\Delta E_i}{T}}}{\sum_{j=1}^{N} e^{-\frac{\Delta E_j}{T}}}, & \Delta E > 0, \end{cases} \qquad (13)$$

where T represents an effective boundary fluctuation amplitude of model cells in units of energy and N is the total number of possible configuration changes. Thus the pattern evolves (and cells move) to minimize the total effective energy.

b. 3D CPM-Based Limb Simulations Chaturvedi *et al.* (2005) and Cickovski *et al.* (2005, 2007) have presented a unified, object-oriented, three-dimensional biomodeling environment. This framework allows the integration of multiple submodels at scales from subcellular to those of tissues and organs. The implementations in each case combined a modified CPM with a continuum reaction–diffusion model and a state automaton with well-defined conditions for cell differentiation transitions to model genetic regulation. This environment allows one to rapidly and compactly create computational models of a class of complex-developmental phenomena. Cickovski *et al.* (2005) describes in detail a computational package, CompuCell3D,[1] based on the hybrid modeling approach.

CPM-based simulations of vertebrate limb development in this multiscale, multimodel simulation environment are presented in both Chaturvedi *et al.* (2005) and Cickovski *et al.* (2005). The biological basis of each set of simulations was that described above: the "bare-bones" mechanism of Hentschel *et al.* (2004), projected onto the geometry of Fig. 3, but with the third, dorsoventral, dimension made explicit. Both studies separated the morphogen dynamics from cell rearrangement (intrinsically connected to each other in system (1)–(8)) using different ad hoc simplifications to obtain a two-equation PDE system for activator–inhibitor interactions.

Beyond this, each of the analyses used a different set of additional simplifications to make the 3D simulations tractable. Chaturvedi *et al.* (2005), for example, used the full range of cell types, R_1, R_2, R'_2, and R_3, in their cell-state transition map, whereas Cickovski *et al.* (2005) modeled only "noncondensing" and "condensing" cells. Chaturvedi *et al.* (2005) induced transitions between patterns in the different proximodistal domains of the developing limb by changing the values of the morphogen diffusion coefficients, which is formally equivalent to the biologically-justified alteration of the aspect ratio of the active zone (Newman and Frisch, 1979; Hentschel *et al.*, 2004). Successive stationary patterns were then computed. In Cickovski *et al.* (2005), the width of the active zone changed in an "automatic," or self-organizing, fashion by incorporating into the model the movement of proliferating cells away from the high point of FGF at the AER. Additional details can be found in the respective papers.

Examples of 3D simulations of TGF-β, fibronectin and condensation patterns from these two studies are shown in (Fig. 8). The similarities in the results, and the rough fidelity to actual development (e.g., proximodistal emergence of increasing numbers of elements) despite the varying simplifications used, are encouraging with regard to the validity of the common underlying assumptions. It should also be noted that other

---

[1] https://simtk.org/home/compucell3d.

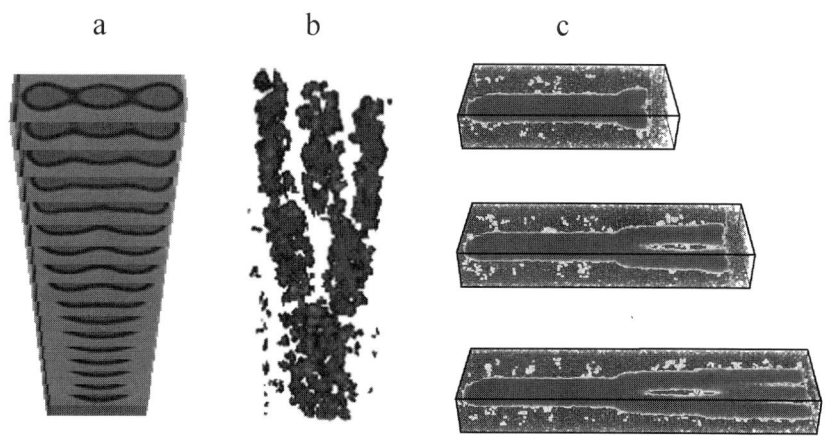

Figure 8 (a) CompuCell3D simulation of distribution of TGF-β (activator) over progressive developmental stages (time increasing from bottom to top); (b) CompuCell3D simulation of cell condensation into humerus (lower), ulna and radius (middle), and digits (top) after 1040 Monte Carlo steps. Visualization was produced using volume rendering; (c) CompuCell3D simulation successive stages (top to bottom) of proximodistal development of limb skeleton. Panels (a) and (b) from Chaturvedi et al. (2005); panel (e) from Forgacs and Newman (2005), based on Cickovski et al. (2005), courtesy of Trevor Cickovski. See color insert.

pattern forming mechanisms (e.g., chemotaxis) have generic features in common with reactor–diffusion models that enable the production of limb-like patterns (cp. Fig. 8a with Fig. 11 of Myerscough et al., 1998). The choice among competing models must be ultimately decided by experimental evaluation of aspects in which they differ.

C. Continuum Models

In continuum models some entities that are inherently discrete and exist in finite numbers are modeled instead by a continuous variable. This can be justified if the entities are present in large numbers, and the scale on which they are observed is much larger than the scale of entities themselves. Continuum models use families of differential or integrodifferential equations to describe "fields" of interaction. Discrete models describe individual (microscopic) behaviors. They are often applied to microscale events where a small number of elements can have a large (and stochastic) impact on a system. When we model organogenesis, such as vertebrate limb development, morphogen production and diffusion are on the molecular level, which has a much smaller scale than the cell level. Since our observations are on the cell and tissue levels, continuum models will work well for limb bud shaping and morphogen evolution, components of our multiscale model. We use reaction–diffusion PDEs to model diffusing morphogen molecules and their biochemical reactions. Efficient numerical methods for reaction–diffusion equations can be used to simulate the reaction–diffusion models. In general, continuum models are more computationally efficient than discrete models for the case of large number of molecules when the assumption of continuity does not alter the system behavior.

1. Morphogen Dynamics on an Irregular Domain

Patterns in reaction–diffusion systems depend sensitively on domain size and shape (Lyons and Harrison, 1992; Crampin *et al.*, 2002; Zykov and Engel, 2004). Since the natural shape of a limb bud, and its subdomains such as the active zone (see Fig. 2) have nonstandard geometries, we developed a mathematical formalism based on the Galerkin method (Johnson, 1987), to handle the complicated geometries and to solve the morphostatic reaction–diffusion system (10)–(11) numerically. The Galerkin method is a means for converting an ordinary or partial differential equation system into a problem represented by a system of algebraic equations in a more restricted space than that of the original system. Since it is a "variational" method, it employs "test functions" to approximate the system's behavior on the restricted space.

a. The Discontinuous Galerkin Finite Element Method. Our formalism, the *Discontinuous Galerkin finite element* (DGFE) method, is termed "discontinuous" due to the usage of completely discontinuous piecewise polynomial space for the numerical solution and the test functions. Major advances in the use of this approach were presented in a series of papers by Cockburn and Shu (1989, 1991, 1998a, 1998b, 1998), and Cockburn *et al.* (1989, 1990).

DG methods have several advantages that make them attractive for biological applications. These include their ability to easily handle complicated geometry and boundary conditions (an advantage shared by all finite element methods), their flexibility in using polynomials with different orders on neighboring elements to approximate the solution of the differential equation, and their permitting the use of very irregular computational meshes. These properties of DGFE methods facilitate the incorporation of adaptivity techniques, an important means for reducing computational cost. DG methods are also compact and hence efficient in parallel implementation. They can also be easily coordinated with finite volume techniques for application to problems with discontinuous or sharp gradient solutions.

Recently, Cheng and Shu (2007) developed a new DG method for solving time dependent PDEs with higher order spatial derivatives. The scheme is formulated by repeated integration by parts of the original equation and careful treatment of the discontinuity of the numerical solutions on the interface of the neighboring elements, which is important for the stability of the DG method. It is easier to formulate and implement, and needs less storage and CPU cost, than the usual DG method for PDEs with higher order spatial derivatives.

We adopted the discontinuous Galerkin finite element numerical approach of Cheng and Shu (2007) and implemented it on both 2D rectangular and triangular meshes to solve the reaction–diffusion system (10)–(11). The spatial discretization by the DG method transforms the reaction–diffusion PDEs to a system of ODEs. Mathematical characteristics of our reaction–diffusion system (i.e., a property known as "stiffness") would make it computationally expensive to solve by regular methods. We therefore used "operator splitting" (Strang, 1968) and nonstandard discretization techniques (Chou *et al.*, 2007) to achieve more efficient simulations.

Patterns in reaction–diffusion systems are sensitive to the domain size and geometrical shape. The shape of the developing limb bud undergoes continuous changes. The DG finite element approach can handle the irregular shapes easily by using triangular meshes to fit the domain. Both spot-like and stripe-like patterns are observed in simulations of the steady state of the reaction–diffusion system (10)–(11), derived in the morphostatic limit. To simulate the pattern of realistic shapes of morphogenetically active zones in the limb bud, we randomly perturbed the rectangular boundary of the active zone, and used triangular mesh to fit the irregular boundary. In Fig. 9A, the triangular meshes are presented to show the fit of the computational mesh to the irregular boundary. The vertical length of the domain is roughly 0.65, and the horizontal length is 0.15. A flood contour plot of the steady-state of the concentration of the activator c_a is shown. Different colors in the domain represent different values of c_a. We achieved stripe-like patterns as shown in Fig. 9A. To examine the pattern dependence on the ratio of horizontal length and vertical length of the active zone, we fixed the vertical length to be 1, but changed the horizontal length successively. The steady-state patterns are shown in Fig. 9B, and we can observe the changing of spot-like patterns to stripe-like patterns when the horizontal length is decreased. (See Alber *et al.*, in press, for parameter values and other details.)

This exercise demonstrates that the system describing the limb's morphogen dynamics in the morphostatic limit (Eqs. (10)–(11)) can form skeletal element-like patterns in domains of realistic proportions. Moreover, as required by the "bare-bones" model, the system undergoes transitions to different numbers of elements as a result of changes in the shape of the undifferentiated domain.

V. Discussion and Future Directions

Turing-type systems are mathematical models which have established the principle that coupling of reaction with diffusion is capable of generating spatial structure. The variables in Turing's original formulation (Turing, 1952), though meant to represent determinants of embryonic development, did not correspond to experimentally defined biological entities, and indeed took the form of a system of chemical reactions. The biosynthetic and secretory capabilities of living cells are unlike standard

11. Multiscale Models for Vertebrate Limb Development 333

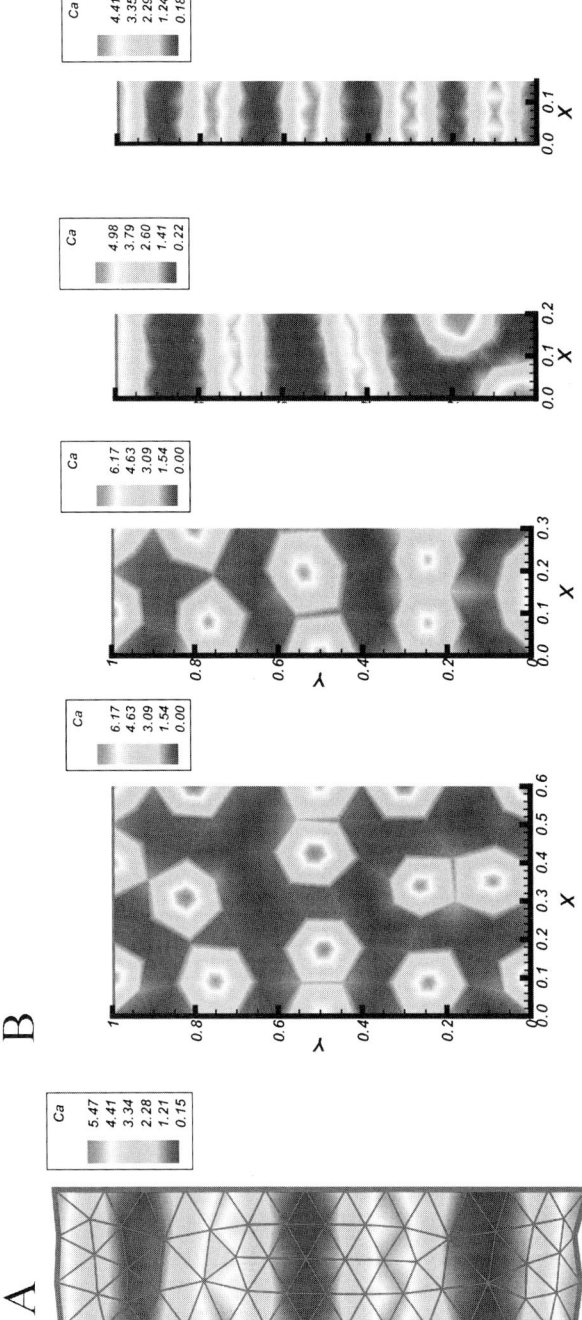

Figure 9 (A) Simulations on the active zone with an irregular shape. Contour plots of the steady state of the concentration of the activator. A triangular mesh is used to fit the irregular boundary of the domain. The vertical length of the domain is roughly 0.65, and the horizontal length is 0.15, arbitrary units. (B) Simulations on active zones with changing sizes in the horizontal direction, and the vertical size fixed at 1. Contour plots of the steady states of concentrations of the activator for different domain size. Triangular meshes are used. (Adapted from Alber et al., in press.) See color insert.

chemical reactions, and the associated diffusion-like processes are also more complex than their purely physical counterpart (Lander, 2007). While a generalized framework like the one Turing described is indeed contributing to the understanding of many aspects of biological development (reviewed in Forgacs and Newman, 2005; Maini *et al.*, 2006), valid models must inevitably have features that are different from, or go beyond, Turing's original proposal. These must include the ability of cells to produce, release and take up molecules, and to change the repertoire of substances they process by changing their differentiated state. Cells, acting thus as "reactors," also replicate and change their positions in response to their microenvironments, activities with no counterparts in the world of classic Turing systems.

While the reactor–diffusion mechanism for vertebrate limb development has yet to be confirmed in an unequivocal fashion, or more generally, a "local autoactivation-lateral inhibition" (LALI) system (Meinhardt and Gierer, 2000; Newman and Bhat, in press), it is the only available model that accounts for the spatiotemporal increase skeletal elements seen in all amniote species (reptiles, birds, mammals) in a natural fashion. We have found that under slightly different assumptions from our "barebones" mechanism (Hentschel *et al.*, 2004; Eqs. (1)–(8), above), which take account of the differences in tissue sources of key FGFs in urodele amphibians (Han *et al.*, 2001), our model can even account for the deviations from the proximodistal order of skeletogenesis in those species (Franssen *et al.*, 2005; Glimm and Newman, in preparation).

The reactor–diffusion model for limb development has clear affinities to the increasingly well-established "clock and wavefront" model for somitogenesis (Cooke and Zeeman, 1976; Pourquié, 2003; Baker *et al.*, Glazier *et al.*, this volume). Specifically, in both models mesenchymal cells become susceptible to a periodic patterning signal by escaping the suppressive activity of a gradient of FGF. In somitogenesis the periodic signal, which sweeps along the length of the presomitic mesoderm, is a Notch and Wnt-based molecular oscillator with both temporal and spatial aspects. In the limb, the molecular mechanism presumed to generate the condensation-inducing chemical standing waves is more elusive, not least because the nature of the lateral inhibitor remains unknown (but see Moftah *et al.*, 2002 and Fujimaki *et al.*, 2006). Measurement of morphogen diffusion rates in living tissues is notoriously difficult (Lander, 2007) and the reactor–diffusion mechanism may thus be difficult to subject to definitive tests for some time. A promising route to a partial test of this model is suggested by experimental evidence for temporal periodicities in the response of micromass cultures to perturbation by TGF-β (Leonard *et al.*, 1991). As noted above, discrete stochastic simulations of this culture system disclose a temporally periodic dynamical regime (Christley *et al.*, 2007; Fig. 5A) and analysis of the PDE system (1)–(8) under conditions in which diffusion is suppressed shows it to have strong oscillatory modes (Hentschel and Newman, in preparation). Recent observations, moreover, indicate that Notch-related signaling components undergo temporal oscillations in limb mesenchyme (Bhat and Newman, unpublished).

11. Multiscale Models for Vertebrate Limb Development

While the morphostatic limit for the system (1)–(8) was introduced to make analysis of morphogen patterns more tractable, it may also represent a reality of development for most or some tetrapod embryos. The establishment of the limb skeletal pattern in chickens occurs over 4 days, while the same process in humans occurs over 4 weeks. Since the spatial scales of limb development in the two species are similar, one or more dynamical processes—morphogen evolution, cell differentiation, cell mobility—must differ substantially. This indicates that the parameters determining rates of development, including, but not confined to those in the "bare-bones" formulation, have been subject to natural selection. Transformation of an inherently morphodynamic system into a morphostatic one by, for example, slowing the rate of cell movement, is a plausible evolutionary scenario for the limb and for other developmental systems, making them more resistant to radical changes in morphology when key genes are mutated (Salazar-Ciudad *et al.*, 2003; Salazar-Ciudad, 2006).

Simulating morphogen dynamics in irregularly shaped domains, as indicated above, is a necessary step to generating an authentic limb model. But rather than the domain shapes being chosen arbitrarily, ultimately they must arise from the material properties of the limb bud. Two major modifications need to be made to the 3D modeling strategy described in section IVB2 before it has a real possibility of capturing the essence of the developing limb. First, the growing limb bud needs to be modeled as a naturally shaped object with a curvilinear profile in all three dimensions. Thus far our 2D and 3D simulations have all used rectangular shapes (Hentschel *et al.*, 2004) and parallelepipeds (Chaturvedi *et al.*, 2005; Cickovski *et al.*, 2005). Modeling of shape change in developing systems has attracted much attention in recent years, with applications ranging over growth of tumors (Araujo and McElwain, 2004) and hydroid polyps (Beloussov and Grabovsky, 2006) to the developing human brain (Czanner *et al.*, 2001). The limb mesenchyme is a viscous material ensheathed by the ectoderm, a viscoelastic sheet. Murea and Hentschel (2007) have presented a finite element strategy for modeling limb bud outgrowth as a problem in creeping viscous flow under nonuniform surface tension in a (2D) system with a free-moving boundary. This continuously reshaped form, particularly in a 3D extension of this analysis, would constitute a realistic supporting medium and set of boundaries for both the morphogen dynamics (e.g., Eqs. (10)–(11)), and the model cells of the CPM or its equivalent, in a hybrid model that includes morphogenesis as well as skeletal pattern formation.

Secondly, the limb and its skeleton are not symmetrical. Although little is known about how they function, the various gradients in the limb bud, e.g., Sonic hedgehog emanating from the posterior Zone of Polarizing Activity (ZPA), various Hox proteins distributed along the proximodistal and anteroposterior axes, and Wnt7A distributed along the dorsoventral axis, influence the size and shapes of the various skeletal elements and endow the limb with polarity along these axes. Computational models such as those described here afford a way of testing hypotheses for the action of these modulating factors (Yokouchi *et al.*, 1995; Newman, 1996). By default, under symmetric conditions, simulations with the models described above lead to skeletons with no distinction among the different elements, other than they increase in number along the proximodistal axis as development proceeds (Fig. 8). By making the reasonable assumption that the graded modulators act on the efficiency and strength of components of the core mechanism (e.g., how much fibronectin is produced per unit of TGF-β; how efficient the inhibitor is in suppressing TGF-β) it should be possible to introduce such uniformities into the various models and observe the effects on the simulated limb morphologies. As these models becomes richer in the genetic interactions they represent and potentially more realistic, they can help explain the outcomes of experimental manipulations and effects of medically relevant mutations in the vertebrate limb, and suggest hypotheses for its evolutionary origination and diversification.

Acknowledgments

We thank Rajiv Chaturvedi, Trevor Cickovski, James Glazier, Jesús Izaguirre and Maria Kiskowski for their contributions to early stages of the work reviewed here. We acknowledge support from the National Science Foundation in the form of Grants Nos. IBN-034467 to S.A.N. and M.A., and EF-0526854, from the Frontiers in Integrative Biological Research program, to S.A.N. In addition, M.A. received support from Grant No. GM076692-01/NSF 04.6071 from the Interagency Opportunities in Multiscale Modeling in Biomedical, Biological and Behavioral Systems program, and B.K. from Grant No. 1P03A01230 from the Polish Ministry of Science and Higher Education.

References

Alber, M. S., Kiskowski, M. A., Glazier, J. A., and Jiang, Y. (2003). On cellular automaton approaches to modeling biological cells. *In* "Mathematical Systems Theory in Biology, Communication, and Finance." (J. Rosenthal and D. S. Gilliam, Eds.). *In* "IMA," Vol. 134, Springer-Verlag, New York, pp. 1–39.

Alber, M., Kiskowski, M., Jiang, Y., and Newman, S. A. (2004). Biological lattice gas models. *In* "Dynamics and Bifurcation of Patterns in Dissipative Systems." (G. Dangelmayr and I. Oprea, Eds.). *In* "World Scientific Series on Nonlinear Science," Vol. 12, World Scientific, Singapore, pp. 274–291.

Alber, M., Hentschel, H. G. E., Kazmierczak, B., and Newman, S. A. (2005a). Existence of solutions to a new model of biological pattern formation. *J. Math. Anal. Appl.* **308**, 175–194.

Alber, M., Glimm, T., Hentschel, H. G. E., Kazmierczak, B., and Newman, S. A. (2005b). Stability of n-dimensional patterns in a generalized Turing system: Implications for biological pattern formation. *Nonlinearity* **18**, 125–138.

Alber, M., Glimm, T., Hentschel, H. G. E., Kazmierczak, B., Zhang, Y. T., Zhu, J., and Newman, S. A., The morphostatic limit for a model of skeletal pattern formation in the vertebrate limb. *Bull. Math. Biol.* (in press).

Alberch, P., and Gale, E. A. (1983). Size dependence during the development of the amphibian foot. Colchicine-induced digital loss and reduction. *J. Embryol. Exp. Morphol.* **76**, 177–197.

Araujo, R., and McElwain, D. (2004). A history of the study of solid tumor growth: The contribution of mathematical modeling. *Bull. Math. Biol.* **66**, 1039–1091.

Borkhvardt, V. G. (2000). The growth and form development of the limb buds in vertebrate animals. *Ontogenez* **31**, 192–200.

Beloussov, L. V., and Grabovsky, V. I. (2006). Morphomechanics: Goals, basic experiments and models. *Int. J. Dev. Biol.* **50**, 81–92.
Casal, A., Sumen, C., Reddy, T., Alber, M., and Lee, P. (2004). "A Cellular Automata Model of Early T Cell Recognition." "Lecture Notes in Computer Science," Vol. 3305. Springer-Verlag, New York. pp. 553–560.
Casal, A., Sumen, C., Reddy, T., Alber, M., and Lee, P. (2005). Agent-based modeling of the context dependency in T cell recognition. *J. Theor. Biol.* **236**, 376–391.
Chaturvedi, R., Izaguirre, J. A., Huang, C., Cickovski, T., Virtue, P., Thomas, G., Forgacs, G., Alber, M., Hentschel, G., Newman, S. A., and Glazier, J. A. (2003). "Multimodel simulations of chicken limb morphogenesis." "Lecture Notes in Computer Science," Vol. 2659. Springer-Verlag, New York. pp. 39–49.
Chaturvedi, R., Huang, C., Izaguirre, J. A., Newman, S. A., Glazier, J. A., and Alber, M. (2004). "A hybrid discrete-continuum model for 3-D skeletogenesis of vertebrate limb." Lecture Notes in Computer Science, Vol. 3305. Springer-Verlag, New York. pp. 543–552.
Chaturvedi, R., Huang, C., Kazmierczak, B., Schneider, T., Izaguirre, J. A., Glimm, T., Hentschel, H. G., Glazier, J. A., Newman, S. A., and Alber, M. S. (2005). On multiscale approaches to 3-dimensional modeling of morphogenesis. *J. R. Soc. Interface* **2**, 237–253.
Cheng, Y., and Shu, C.-W., 2007. A discontinuous Galerkin finite element method for time dependent partial differential equations with higher order derivatives. *Math. Comput.* (in press).
Chou, C. -S., Zhang, Y. -T., Zhao, R., and Nie, Q. (2007). Numerical methods for stiff reaction–diffusion systems. *Discrete Contin. Dynam. Syst. Ser. B* **7**, 515–525.
Christley, S., Alber, M. S., and Newman, S. A. (2007). Patterns of mesenchymal condensation in a multiscale, discrete stochastic model. *PLoS Comput. Biol.* **3**(e76), 0743–0753.
Cickovski, T., Huang, C., Chaturvedi, R., Glimm, T., Hentschel, H. G. E., Alber, M., Glazier, J. A., Newman, S. A., and Izaguirre, J. A. (2005). A framework for three-dimensional simulation of morphogenesis. *IEEE/ACM Trans. Comput. Biol. Bioinform.* **2**, 273–288.
Cickovski, T., Kedar, A., Swat, M., Merks, R. M. H., Glimm, T., Hentschel, H. G. E., Alber, M. S., Glazier, J. A., Newman, S. A., and Izaguirre, J. A. (2007). From genes to organisms via the cell: A problem solving environment for multicellular development. *Comput. Sci. Eng.* **9**(July–August), 50–60.
Cockburn, B., and Shu, C. -W. (1989). TVB Runge–Kutta local projection discontinuous Galerkin finite element method for conservation laws. II. General framework. *Math. Comput.* **52**, 411–435.
Cockburn, B., and Shu, C. -W. (1991). The Runge–Kutta local projection P1-discontinuous Galerkin finite element method for scalar conservation laws. *Math. Model. Numer. Anal.* **25**, 337–361.
Cockburn, B., and Shu, C. -W. (1998a). The local discontinuous Galerkin finite element method for convection–diffusion system. *SIAM J. Numer. Anal.* **35**, 2440–2463.
Cockburn, B., and Shu, C. -W. (1998b). The Runge–Kutta discontinuous Galerkin method for conservation laws. V. Multidimensional systems. *J. Comput. Phys.* **141**, 199–224.
Cockburn, B., and Shu, C. -W. (1998). The local discontinuous Galerkin method for time-dependent convection–diffusion systems. *SIAM J. Numer. Anal.* **35**, 2440–2463.
Cockburn, B., Lin, S. -Y., and Shu, C. -W. (1989). TVB Runge–Kutta local projection discontinuous Galerkin finite element method for conservation laws. III. One dimensional systems. *J. Comput. Phys.* **84**, 90–113.
Cockburn, B., Hou, S., and Shu, C. -W. (1990). The Runge–Kutta local projection discontinuous Galerkin finite element method for conservation laws. IV. The multidimensional case. *Math. Comput.* **54**, 545–581.
Cooke, J., and Summerbell, D. (1981). Control of growth related to pattern specification in chick wing-bud mesenchyme. *J. Embryol. Exp. Morphol.* **65 Suppl**, 169–185.
Cooke, J., and Zeeman, E. C. (1976). A clock and waterfront model for the control of repeated structures during animal morphogenesis. *J. Theor. Biol.* **58**, 455–476.
Crampin, E. J., Hackborn, W. W., and Maini, P. K. (2002). Pattern formation in reaction–diffusion models with nonuniform domain growth. *Bull. Math. Biol.* **64**, 747–769.

Czanner, S., Durikovic, R., and Inoue, H., 2001. Growth simulation of human embryo brain. *In* "17th Spring Conference on Computer Graphics (SCCG '01)," p. 0139.

Dillon, R., and Othmer, H. G. (1999). A mathematical model for outgrowth and spatial patterning of the vertebrate limb bud. *J. Theor. Biol.* **197**, 295–330.

Downie, S. A., and Newman, S. A. (1994). Morphogenetic differences between fore and hind limb precartilage mesenchyme: Relation to mechanisms of skeletal pattern formation. *Dev. Biol.* **162**, 195–208.

Dubrulle, J., McGrew, M. J., and Pourquié, O. (2001). FGF signaling controls somite boundary position and regulates segmentation clock control of spatiotemporal Hox gene activation. *Cell* **106**, 219–232.

Dudley, A. T., Ros, M. A., and Tabin, C. J. (2002). A reexamination of proximodistal patterning during vertebrate limb development. *Nature* **418**, 539–544.

Filion, R. J., and Popel, A. S. (2004). A reaction–diffusion model of basic fibroblast growth factor interactions with cell surface receptors. *Ann. Biomed. Eng.* **32**, 645–663.

Forgacs, G., and Newman, S. A. (2005). "Biological Physics of the Developing Embryo." Cambridge Univ. Press, Cambridge.

Franssen, R. A., Marks, S., Wake, D., and Shubin, N. (2005). Limb chondrogenesis of the seepage salamander, *Desmognathus aeneus* (Amphibia: Plethodontidae). *J. Morphol.* **265**, 87–101.

Fujimaki, R., Toyama, Y., Hozumi., N., and Tezuka, K. (2006). Involvement of Notch signaling in initiation of prechondrogenic condensation and nodule formation in limb bud micromass cultures. *J. Bone Miner. Metab.* **24**, 191–198.

Glazier, J. A., and Graner, F. (1993). A simulation of the differential adhesion driven rearrangement of biological cells. *Phys. Rev. E Stat. Nonlin. Soft Matter Phys.* **47**, 2128–2154.

Graner, F., and Glazier, J. A. (1992). Simulation of biological cell sorting using a two-dimensional extended Potts model. *Phys. Rev. Lett.* **69**, 2013–2016.

Hall, B. K., and Miyake, T. (1995). Divide, accumulate, differentiate: Cell condensation in skeletal development revisited. *Int. J. Dev. Biol.* **39**, 881–893.

Hall, B. K., and Miyake, T. (2000). All for one and one for all: Condensations and the initiation of skeletal development. *BioEssays* **22**, 138–147.

Han, M. J., An, J. Y., and Kim, W. S. (2001). Expression patterns of FGF-8 during development and limb regeneration of the axolotl. *Dev. Dynam.* **220**, 40–48.

Hentschel, H. G. E., Glimm, T., Glazier, J. A., and Newman, S. A. (2004). *Proc. Roy. Soc. London B* **271**, 1713–1722.

Izaguirre, J. A., Chaturvedi, R., Huang, C., Cickovski, T., Coffland, J., Thomas, G., Forgacs, G., Alber, M., Newman, S. A., and Glazier, J. A. (2004). CompuCell, a multimodel framework for simulation of morphogenesis. *Bioinformatics* **20**, 1129–1137.

Jiang, T. X., Widelitz, R. B., Shen, W. M., Will, P., Wu, D. Y., Lin, C. M., Jung, H. S., and Chuong, C. -M. (2004). Integument pattern formation involves genetic and epigenetic controls: Feather arrays simulated by digital hormone models. *Int. J. Dev. Biol.* **48**, 117–135.

Johnson, C. (1987). "Numerical solution of partial differential equations by the finite element method." Cambridge Univ. Press, Cambridge/New York.

Kiskowski, M., Alber, M. S., Thomas, G. L., Glazier, J. A., Bronstein, N., Pu, J., and Newman, S. A. (2004). Interplay between activator–inhibitor coupling and cell–matrix adhesion in a cellular automaton model for chondrogenic patterning. *Dev. Biol.* **271**, 372–387.

Kosher, R. A., Savage, M. P., and Chan, S. C. (1979). In vitro studies on the morphogenesis and differentiation of the mesoderm subjacent to the apical ectodermal ridge of the embryonic chick limb-bud. *J. Embryol. Exp. Morphol.* **50**, 75–97.

Lander, A. D. (2007). Morpheus unbound: Reimagining the morphogen gradient. *Cell* **128**, 245–256.

Lander, A. D., Nie, Q., and Wan, F. Y. (2002). Do morphogen gradients arise by diffusion? *Dev. Cell* **2**, 785–796.

Leonard, C. M., Fuld, H. M., Frenz, D. A., Downie, S. A., Massague, J., and Newman, S. A. (1991). Role of transforming growth factor-β in chondrogenic pattern formation in the embryonic limb: Stimulation of mesenchymal condensation and fibronectin gene expression by exogenous TGF-β and evidence for endogenous TGF-β-like activity. *Dev. Biol.* **145**, 99–109.

Litingtung, Y., Dahn, R. D., Li, Y., Fallon, J. F., and Chiang, C. (2002). Shh and Gli3 are dispensable for limb skeleton formation but regulate digit number and identity. *Nature* **418**, 979–983.

Lyons, M. J., and Harrison, L. G. (1992). Stripe selection: An intrinsic property of some pattern-forming models with nonlinear dynamics. *Dev. Dynam.* **195**, 201–215.

Maini, P. K., Baker, R. E., and Chuong, C. M. (2006). Developmental biology: The Turing model comes of molecular age. *Science* **314**, 1397–1398.

Mariani, F. V., and Martin, G. R. (2003). Deciphering skeletal patterning: Clues from the limb. *Nature* **423**, 319–325.

Martin, G. R. (1998). The roles of FGFs in the early development of vertebrate limbs. *Genes Dev.* **12**, 1571–1586.

Meinhardt, H., and Gierer, A. (2000). Pattern formation by local self-activation and lateral inhibition. *BioEssays* **22**, 753–760.

Miura, T., and Maini, P. K. (2004). Speed of pattern appearance in reaction–diffusion models: Implications in the pattern formation of limb bud mesenchyme cells. *Bull. Math. Biol.* **66**, 627–649.

Miura, T., and Shiota, K. (2000a). Extracellular matrix environment influences chondrogenic pattern formation in limb bud micromass culture: Experimental verification of theoretical models. *Anat. Rec.* **258**, 100–107.

Miura, T., and Shiota, K. (2000b). TGFβ2 acts as an "activator" molecule in reaction–diffusion model and is involved in cell sorting phenomenon in mouse limb micromass culture. *Dev. Dynam.* **217**, 241–249.

Miura, T., Shiota, K., Moriss-Kay, G., and Maini, P. K. (2006). Mixed-mode pattern in Doublefoot mutant mouse limb—Turing reaction–diffusion model on a growing domain during limb development. *J. Theor. Biol.* **240**, 562–573.

Moftah, M. Z., Downie, S. A., Bronstein, N. B., Mezentseva, N., Pu, J., Maher, P. A., and Newman, S. A. (2002). Ectodermal FGFs induce perinodular inhibition of limb chondrogenesis in vitro and in vivo via FGF receptor 2. *Dev. Biol.* **249**, 270–282.

Murea, C. M., and Hentschel, H. G. E. (2007). A finite element method for growth in biological development. *Math. Biosci. Eng.* **4**, 339–353.

Myerscough, M. R., Maini, P. K., and Painter, K. J. (1998). Pattern formation in a generalized chemotactic model. *Bull. Math. Biol.* **60**, 1–26.

Newman, M. E. J., and Barkema, G. T. (1999). "Monte Carlo Methods in Statistical Physics." Oxford Univ. Press, Oxford.

Newman, S. A. (1988). Lineage and pattern in the developing vertebrate limb. *Trends Genet.* **4**, 329–332.

Newman, S. A. (1996). Sticky fingers: Hox genes and cell adhesion in vertebrate limb development. *BioEssays* **18**, 171–174.

Newman, S. A. (2002). Developmental mechanisms: Putting genes in their place. *J. Biosci.* **27**, 97–104.

Newman, S.A., and Bhat, R. Activator–inhibitor mechanisms of vertebrate limb pattern formation. *Birth Defects Res. C Embryo Today* (in press).

Newman, S. A., and Frisch, H. (1979). Dynamics of skeletal pattern formation in developing chick limb. *Science* **205**, 662–668.

Newman, S. A., Frisch, H. L., Perle, M. A., and Tomasek, J. J. (1981). Limb development: Aspects of differentiation, pattern formation and morphogenesis. *In* "Morphogenesis and Pattern Formation." (T. G. Connolly, L. L. Brinkley, and B. M. Carlson, Eds.), Raven Press, New York, pp. 163–178.

Newman, S. A., and Muller, G. B. (2005). Origination and innovation in the vertebrate limb skeleton: An epigenetic perspective. *J. Exp. Zool. B Mol. Dev. Evol.* **304**, 593–609.

Ornitz, D. M., and Marie, P. J. (2002). FGF signaling pathways in endochondral and intramembranous bone development and human genetic disease. *Genes Dev.* **16**, 1446–1465.

Peters, K. G., Werner, S., Chen, G., and Williams, L. T. (1992). Two FGF receptor genes are differentially expressed in epithelial and mesenchymal tissues during limb formation and organogenesis in the mouse. *Development* **114**, 233–243.

Pourquié, O. (2003). The segmentation clock: Converting embryonic time into spatial pattern. *Science* **301**, 328–330.

Ros, M. A., Lyons, G. E., Kosher, R. A., Upholt, W. B., Coelho, C. N., and Fallon, J. F. (1994). Recombinant limbs as a model to study homeobox gene regulation during limb development. *Dev. Biol.* **166**, 59–72.

Salazar-Ciudad, I. (2006). On the origins of morphological disparity and its diverse developmental bases. *BioEssays* **28**, 1112–1122.

Salazar-Ciudad, I., and Jernvall, J. (2002). A gene network model accounting for development and evolution of mammalian teeth. *Proc. Natl. Acad. Sci. USA* **99**, 8116–8120.

Salazar-Ciudad, I., Jernvall, J., and Newman, S. A. (2003). Mechanisms of pattern formation in development and evolution. *Development* **130**, 2027–2037.

Saunders Jr., J. W. (1948). The proximodistal sequence of origin of the parts of the chick wing and the role of the ectoderm. *J. Exp. Zool.* **108**, 363–402.

Sick, S., Reinker, S., Timmer, J., and Schlake, T. (2006). WNT and DKK determine hair follicle spacing through a reaction–diffusion mechanism. *Science* **314**, 1447–1450.

Sozinova, O., Jiang, Y., Kaiser, D., and Alber, M. (2006a). A three-dimensional model of Myxobacterial aggregation by contact-mediated interactions. *Proc. Natl. Acad. Sci. USA* **102**, 11308–11312.

Sozinova, O., Jiang, Y., Kaiser, D., and Alber, M. (2006b). A three-dimensional model of fruiting body formation. *Proc. Natl. Acad. Sci. USA* **103**, 17255–17259.

Strang, G. (1968). On the construction and comparison of difference schemes. *SIAM J. Numer. Anal.* **5**, 506–517.

Szebenyi, G., Savage, M. P., Olwin, B. B., and Fallon, J. F. (1995). Changes in the expression of fibroblast growth factor receptors mark distinct stages of chondrogenesis in vitro and during chick limb skeletal patterning. *Dev. Dynam.* **204**, 446–456.

Tickle, C. (2003). Patterning systems—from one end of the limb to the other. *Dev. Cell* **4**, 449–458.

Turing, A. M. (1952). The chemical basis of morphogenesis. *Philos. Trans. R. Soc. London B* **237**, 37–72.

Van Obberghen-Schilling, E., Roche, N. S., Flanders, K. C., Sporn, M. B., and Roberts, A. (1988). Transforming growth factor β-1 positively regulates its own expression in normal and transformed cells. *J. Biol. Chem.* **263**, 7741–7746.

Williams, P. H., Hagemann, A., Gonzalez-Gaitan, M., and Smith, J. C. (2004). Visualizing long-range movement of the morphogen Xnr2 in the Xenopus embryo. *Curr. Biol.* **14**, 1916–1923.

Yamaguchi, M., Yoshimoto, E., and Kondo, S. (2007). Pattern regulation in the stripe of zebrafish suggests an underlying dynamic and autonomous mechanism. *Proc. Natl. Acad. Sci. USA* **104**, 4790–4793.

Yokouchi, Y., Nakazato, S., Yamamoto, M., Goto, Y., Kameda, T., Iba, H., and Kuroiwa, A. (1995). Misexpression of Hoxa-13 induces cartilage homeotic transformation and changes cell adhesiveness in chick limb buds. *Genes Dev.* **9**, 2509–2522.

Zykov, V., and Engel, H. (2004). Dynamics of spiral waves under global feedback in excitable domains of different shapes. *Phys. Rev. E Stat. Nonlin. Soft Matter. Phys.* **70**, 016201.

12

Tooth Morphogenesis *in vivo*, *in vitro*, and *in silico*

Isaac Salazar-Ciudad

Developmental Biology Program, Institute of Biotechnology, P.O. Box 56, FIN-00014, University of Helsinki, Helsinki, Finland

---

    I. Introduction
   II. The Use of Mammalian Tooth for Developmental and Evolutionary Biology
  III. Morphological Changes During Tooth Development
  IV. Gene Networks in Tooth Development
   V. The Formation of the Cusps
  VI. Spacing Between Cusps
 VII. Morphodynamic Model 1
VIII. Model 1 and Tooth Dynamics
  IX. Morphodynamic Model 2
   X. What Do Model Dynamics Reveal About Developmental Dynamics
      A. The Formation of the Cusps
      B. The Positioning of Knots
      C. Cervical Loop Growth
      D. Morphodynamics
      E. Developmental Dynamics and Tooth Diversity
  XI. Tooth Model in Comparison to Other Models of Organ Development
 XII. Concluding Remarks
      Acknowledgments
      References

    One of the aims of developmental biology is to understand how a single egg cell gives rise to the complex spatial distributions of cell types and extracellular components of the adult phenotype. This review discusses the main genetic and epigenetic interactions known to play a role in tooth development and how they can be integrated into coherent models. Along the same lines, several hypotheses about aspects of tooth development that are currently not well understood are evaluated. This is done from their morphological consequences from the model and how these fit known morphological variation and change during tooth development. Thus the aim of this review is two-fold. On one hand the model and its comparison with experimental evidence will be used to outline our current understanding about tooth morphogenesis. On the other hand these same comparisons will be used to introduce a computational model that makes accurate predictions on three-dimensional morphology and patterns of gene expression by implementing cell signaling, proliferation and mechanical interactions between cells. In comparison with many other models of development this model includes reaction–diffusion-like dynamics confined to a diffusion chamber (the devel-

oping tooth) that changes in shape in three-dimensions over time. These changes are due to mechanical interactions between cells triggered by the proliferation enhancing effect of the reactants (growth factors). In general, tooth morphogenesis can be understood from the indirect cross-regulation between extracellular signals, the local regulation of proliferation and differentiation rates by these signals and the effect of intermediate developing morphology on the diffusion, dilution, and spatial distribution of these signals. Overall, this review should be interesting to either readers interested in the mechanistic bases of tooth morphogenesis, without necessarily being interested in modeling *per se*, and readers interested in development modeling in general.
© 2008, Elsevier Inc.

I. Introduction

One of the aims of developmental biology is to understand how a single egg cell gives rise to the complex spatial distributions of cell types and extracellular components of the adult phenotype. Ideally this problem can be separated into the problems of cell differentiation, or how cells become who they are, and the problem of how cells become arranged in specific spatial patterns, that is, how cells end up where they are. This review considers this second problem for the case of teeth, that is: how a tooth develops its morphology. This review discusses the main genetic and epigenetic interactions known to play a role in tooth development and how they can be integrated into coherent models. Computational models provide a method to explore the logical implications of hypotheses about developmental dynamics. It will be discussed whether existing models in tooth development provide precise predictions about how three-dimensional morphologies and patterns of gene expression change during development. By precise it is meant that computational models can give quantitative descriptions of morphology by specifying the three-dimensional positions of cells, which cells they have as neighbors and what are their levels of expression of several genes. More importantly computational models relate these morphological changes to a specific set of genetic and epigenetic interactions and to their specific organization of those. Then the morphological consequences of a genetic or epigenetic manipulation can be explored by making the same manipulations on model "genetic" and "epigenetic" interactions and seeing the *in silico* morphologies appearing from the model. This provides an objective way to see if the results of experimental manipulations are consistent or not with a specific hypothesis about morphogenesis. The studies and models here reviewed provide developmental explanations for morphological changes that are gradual and complex. The models presented do not focus, as many studies in development do, on gross and discrete morphological alterations of morphology so as to identify the genes involved in development. Instead the focus, and the model predictions, are on complex multivariate gradual changes and on how small genetic variation regulates these morphological changes and developmental dynamics. Along the same lines several hypotheses about aspects of tooth development that are currently not well understood are evaluated. This is done from their morphological consequences in the model and how these fit known morphological variation and change during tooth development. This review aims, thus, to provide mechanistic insights about overall tooth developmental dynamics and about specific aspects of it. In that sense tooth computational models are not an end result but a way to summarize understanding of the development of an organ and an analytical tool to allow precise comparison of hypothesis with results. Thus the aim of this review is two-fold. On one hand the model and its comparison with experimental evidence will be used to explain current understanding about tooth morphogenesis. On the other hand these same comparisons will be used to introduce a computational model that makes accurate predictions on three-dimensional morphology and patterns of gene expression by implementing cell signaling, proliferation and mechanical interactions between cells. The model includes a reaction–diffusion-like part and a biomechanical part that changes the three-dimensional shape in which diffusion and reaction is taking place. These changes are due to mechanical interactions between cells triggered by the proliferation enhancing effect of the reactants (growth factors). Overall, this review should be interesting to either readers interested in the mechanistic bases of tooth morphogenesis, without necessarily being interested in modeling *per se*, and readers interested in development modeling in general.

II. The Use of Mammalian Tooth for Developmental and Evolutionary Biology

There are several characteristics of teeth that are very convenient for studies in development and evolution. The tooth organ develops without too many influences from other parts of the body. In fact, teeth can develop *in vitro* to give rise to nearly normal morphologies. *In vitro* teeth, however, do not fully mineralize. In addition they exhibit more morphological variability than *in vivo* teeth. Tooth development is finished by the time teeth erupt. After that point tooth morphology changes only because of mechanical wear or accident. Tooth morphology is thus relatively independent of environmental effects. In that sense tooth morphology has been shown to exhibit high heritability (Matsumoto *et al.*, 1990), although more accurate measurement of morphology indicates that such heritabilities may not be as high as previously thought (Townsend *et al.*, 2003). Overall, teeth, in contrast to many other mammalian organs, can be studied in relative isolation.

Teeth are probably the mammalian organs that are most often preserved in the fossil record. In fact, many extinct mammalian species are known only from tooth remains. The fossil record, thus, provides a relatively good description of the morphological transitions that occurred in tooth evolution (Butler, 1983, 1995). There is also substantial knowledge about the patterns of variation in morphology within populations of several species (Colyer, 1936; Wolsan, 1989; Szuma, 2002). Overall, the tooth literature provides a solid knowledge of the morphological diversity that can arise from changes in tooth development and, thus, the tooth is a good system for studies in development and morphological evolution.

The developmental systems that are currently best understood are often characterized by a relative decoupling between cell signaling and cell movement. Thus early anteroposterior and dorsoventral patterning in *Drosophila* blastoderm, wing imaginal disc patterning in *Drosophila*, hair and feather primordia spacing in vertebrate skin, can all be basically understood from the signaling between cells that do not move or move little. In most cases these patterning events occur, or can be understood, in a two-dimensional field. These kinds of systems do not necessarily represent the whole of animal development. In general cells signal to each other most of the time and animal development involves numerous events of active or passive cell movements. Thus, it is often the case that cells are signaling to each other while their relative locations are changing. In that situation the spatial distribution of the cells receiving a signal changes depending on how the relative distances, shapes and orientations of those cells and the cells sending the signal change over time. In those development systems, called morphodynamic (Salazar-Ciudad *et al.*, 2003), it may be more complex and more difficult to understand how different cell types get arranged in specific spatial locations. Then common models of pattern formation and the insights acquired from more static developmental systems cannot directly be applied to morphodynamic systems (Salazar-Ciudad *et al.*, 2003). Teeth provide a relatively simple system in which cells change their relative locations while signaling. They are also eminently three-dimensional morphologies. In that sense teeth provide an advantageous entry point for the study of more complex and representative developmental dynamics and morphologies.

III. Morphological Changes During Tooth Development

This section describes the main morphological changes occurring during tooth development. This description is merely phenomenological. The genetic and epigenetic bases of such changes are covered in later sections. This section, and most of the article, is focused on the mouse first molar because it is the best described tooth but most of the discussion is valid for many other teeth. In that sense the description, in this section, of the morphological changes during tooth development should be seen as the description of the problem that the model and the research review in this article tries to understand.

Teeth start from an invagination of the oral epithelium. In this initial bud stage, the invagination deepens in the underlying mesenchyme and extends in an anterior and posterior direction from its starting point. By embryonic day E13 it is deep where it started and shallow in its anterior and posterior borders (Viriot *et al.*, 1997). The driving force for such invagination is unknown but at that stage most

12. Tooth Morphogenesis

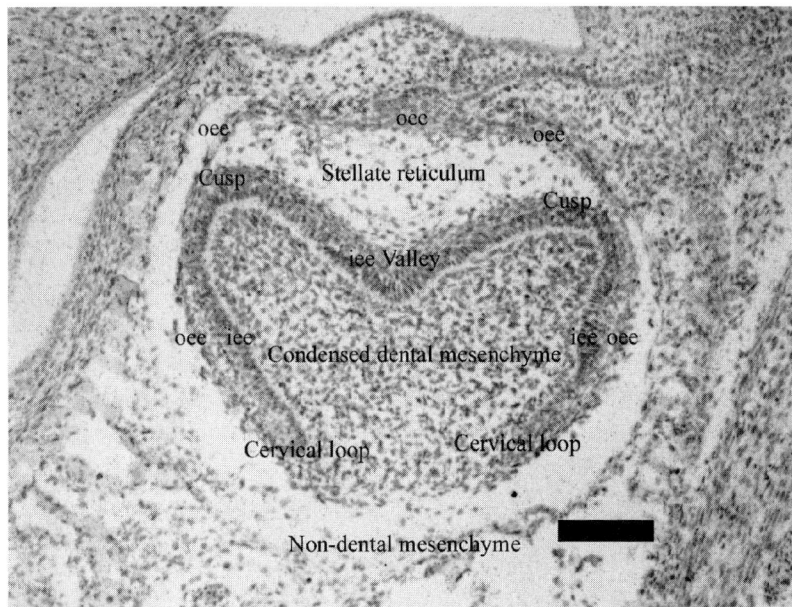

Figure 1 Frontal section of a first mouse molar at embryonic day 17 (E17) showing the basic anatomy of a developing tooth. Scale bar indicates 100 μm. Buccal is to the left and lingual to the right. Abbreviations: iee—inner enamel epithelium; oee—outer enamel epithelium.

dental epithelial cells are proliferating (Lesot *et al.*, 1996; Jernvall *et al.*, 1994; Coin *et al.*, 1999). The apical side of the invagination (the side in contact with the forming oral cavity) is very narrow and has cells that, over time, become vacuolized and form a pressurized structure called the stellate reticulum. The maturation of this structure also seems to proceed into an anterior and posterior direction and seems to be more advanced in parts that invaginate earlier, but no systematic study of that has been performed. The mesenchyme underlying the deepest parts of the invagination (see Figs. 1 and 2) starts to condense at this point. Soon after that, in the cap stage, the part of the epithelium just above the condensating mesenchyme stops growing and its cells become more packed (Jernvall *et al.*, 1994). This structure is called the enamel knot or, here, simply the knot. Next, the epithelium grows more extensively near the knot, but not in the knot itself, leading to the formation of two epithelial loops (called the cervical loops; one on the buccal side and one on the lingual) that invaginate deeper in the underlying mesenchyme. These loops also form in an anterior and posterior direction (see Fig. 2) and proceed at a distance from the knot. At E15, the beginning of the so-called bell stage, another knot forms buccally from the first one. Over time, the two knots end up in the tip of two bumps in the epithelium while the epithelium in between continues to proliferate and deepens into the underlying mesenchyme (Fig. 2). As a result, a valley forms in the epithelium between these bumps. By E16 a new

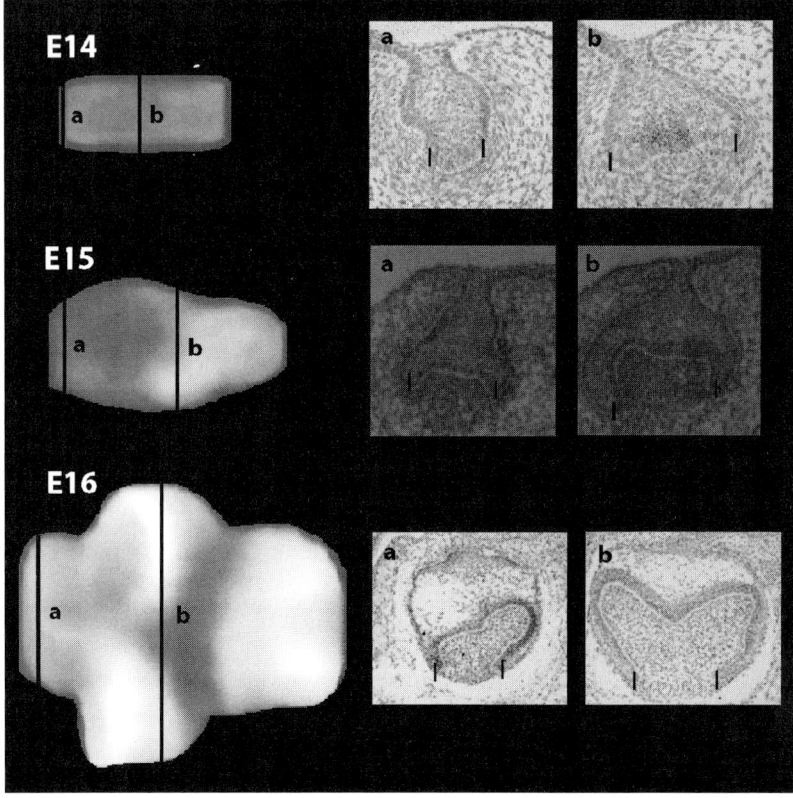

Figure 2 In the left side a three-dimensional reconstruction of the inner enamel epithelium surface for embryonic stages E14, E15, and late E16. For the three-dimensional reconstructions anterior is to the left and posterior to the right. The images in the right are frontal sections for each stage. The levels at which these sections were performed are indicated on the three-dimensional reconstructions. In each section two vertical black lines indicate the boundary between inner and outer enamel epithelium (the three-dimensional reconstructions include only the inner enamel epithelium).

knot forms, posteriorly from the first one, and, soon afterwards, a fourth one forms lingually from it. These also end up in the tips of some bumps in the epithelium while the epithelium around them continues to grow. Progressively, the epithelium that is not enclosed by the cervical loops (called the outer enamel epithelium in contrast with the inner enamel epithelium) stops proliferating and its cells flatten. Meanwhile the cervical loops continue to grow and extend in the anterior and posterior extremes of the tooth. They also change from growing laterally and downwards to grow mainly downwards. Progressively growth becomes confined to the tips of the cervical loops while the epithelium near the knots starts to become columnar and eventually starts mineralization. This process of differentiation proceeds as a wave from the knots to the cervical loops. Ultimately the inner enamel epithelium outlines the surface of the formed tooth. The cells that formed the knot end up at the tips of the tooth cusps (see Fig. 2) while the cervical loops form the lateral walls of the tooth. The cervical loops continue to develop to give rise, in mouse but not in all other rodents, to the tooth root, by a process that is beyond the scope of this review. The mechanisms by which teeth erupt is also outside the scope of this review.

IV. Gene Networks in Tooth Development

There is over 300 genes known to be expressed in teeth (Thesleff *et al.*, 2001; see http://bite-it.helsinki.fi/ for a database on those patterns). It is not the purpose of this review to consider all, or even a significant proportion, of them. There are already some excellent reviews on this (Jernvall and Thesleff, 2000; Thesleff *et al.*, 2001). Here I only consider genes that, when mutated, produce a specific change in tooth morphology (other than the nonformation of teeth). Until recently there were relatively few mutants with a clear specific effect on tooth morphology (Pispa *et al.*, 1999; Biben *et al.*, 2002; Wang *et al.*, 2004a, 2004b; Kangas *et al.*, 2004; Kassai *et al.*, 2005; Järvinen *et al.*, 2006). Many other genes are involved in the terminal differentiation and mineralization of teeth and are not considered in here. In general terms, most patterns of gene expression can be categorized in terms of few types. Many genes are expressed only in the main tooth tissues (or in combinations of them) such as in the epithelia (inner, outer or both), or in the mesenchyme or in the stellate reticulum. Many other genes seem to be expressed in concentric patterns around the knots (either in the epithelium or mesenchyme or both), or everywhere but in the knots. Some few genes are expressed in the anterior and/or posterior borders of the tooth primordia. So far no single gene has been shown to be expressed in only one knot (except for the first knot). This relative simplicity suggests that many insights about how gene interactions regulate the production of morphology can be acquired without a detailed description of all the genes involved.

Gene regulation ultimately acts on development by affecting a limited number of cell behaviors (proliferation, apoptosis, matrix and signal secretion, differentiation, adhesion and consequent changes in cell shape and motility). Some understanding of morphogenesis is achievable without having to consider the details of intracellular signaling but by simply considering in which ways those behaviors change when cells receive specific growth factors. In this review the account of developmental mechanisms is based on considering how cells respond to specific growth factors by secreting other growth factors, proliferating, differentiating, expressing some receptors and changing their adhesive properties. What happens inside cells is then largely simplified to facilitate the understanding of the collective behavior of enamel epithelium and dental mesenchyme during morphogenesis. As I will try to show, the rise of the whole tooth morphology can be understood from these relative simple developmental rules and some few epigenetic constraints.

V. The Formation of the Cusps

At E12, when a starting invagination is morphologically visible, intense expression of several signals appears in a subset of invaginating cells. This transient, early epithelial signaling center expresses signals in all four major morphogen families (BMPs, Shh, Wnt, and FGFs) as well as other genes associated with signaling such as p21, Msx2, and Lef1 (Dassule and McMahon, 1998; Keränen *et al.*, 1998). Many of these signals remain expressed during the bud stage, especially at the tip of the invagination, and in the first and subsequent knots (Vaahtokari *et al.*, 1996). It is not clear how the knots form but since the first knot appears in a location where all these signals are expressed intensively my hypothesis in current and previous models has been that epithelial cells differentiate into knot cells after receiving a threshold concentration of some signal. Which signal that may be is not totally clear but recent studies (Kassai *et al.*, 2005) show that Bmp4 added to *in vitro* tooth cultures accelerates the formation of knots and tooth differentiation. That seems to be especially the case in mutants lacking ectodin, a BMP sequesterer, which has been suggested to buffer the levels of diffusing Bmp4. Moreover, Bmp4 has been shown to induce p21 (Jernvall *et al.*, 1998), a gene known to arrest the cell cycle at the G1/S transition (Steinman *et al.*, 2001). p21 is one of the earliest markers of knot formation. Activin, another TGFß superfamily member, has also been shown to induce p21 and is expressed in the mesenchyme just under the knots (Wang *et al.*, 2004a). Fgf-2, -4, -8, and -9 have been shown to induce, *in vitro*, the proliferation of dental epithelium and mesenchyme (Jernvall *et al.*, 1994; Kettunen *et al.*, 1998) while Fgf-10 acts similarly only in the dental epithelium (Kettunen, 2000). These FGFs are expressed in the knots and several FGF receptors are expressed throughout the dental primordia (except, significantly in the knot itself). Shh is also expressed in the knot and has been suggested to promote proliferation in the dental epithelium (Cobourne *et al.*, 2001; Gritli-Linde *et al.*, 2002). Bmp2 and Bmp4 are also expressed in the knots but, as mentioned, they promote differentiation of the surrounding tissues. The spreading of differentiation from the knots downwards is correlated with a gradual spreading of the expression of many knot genes from the knots downwards to the cervical loops.

The proliferation that knots promote around them combined with their lack of proliferation has been proposed (Butler, 1956; Butler and Ramadan, 1962; Osman and Ruch, 1976; Ruch, 1990; Jernvall *et al.*, 1994; Jernvall and Thesleff, 2000) to produce the folding of the epithelium around the knots and the subsequent formation of cusps. The relative sharpness or bluntness of cusps has been proposed to arise from either the relative rates of growth of the epithelium (by itself or in proportion to mesenchymal growth) near the knots (Jernvall and Thesleff, 2000; Salazar-Ciudad and Jernvall, 2002) and/or the relative efficiency of Bmp2 and 4 in producing the differentiation of cells near the knot (Salazar-Ciudad and Jernvall, 2002). Although these hypotheses can explain the formation of single cusps (as in canine teeth) it does not explain how different cusps, and the spacing between them, arise nor how whole tooth morphology arises.

VI. Spacing Between Cusps

In mammals, cusps in a tooth appear at a distance from each other. Grossly the shape of a tooth can be described by the relative positions and heights of cusps and the border of the tooth. If, as suggested, knots appear after exposure of dental epithelium to Bmp4 there has to be some mechanism to restrict the expansion of these knots and allow the formation of different discrete knots (instead of a single large one). Knots have been suggested to secrete some signal that would inhibit neighboring cells from becoming knots (Jernvall and Thesleff, 2000). Since this hypothesis was proposed, our understanding on the signaling taking place during tooth development has greatly advanced. It seems that this later inhibition may involve several signals. Activin, that as mentioned is expressed in the mesenchyme under the knots, has been shown to promote the expression of follistatin (Heikinheimo *et al.*, 1998; Wang *et al.*, 2004a). Follistatin is a diffusible gene product that sequesters both activin and Bmp2 and Bmp4 (Nakamura *et al.*, 1990; Iemura *et al.*, 1998) and thus acts as its inhibitor. However, follistatin knock out mice seems to be able to form several individual knots and cusps (Wang *et al.*, 2004a). Thus follistatin cannot be the signal mediating lateral inhibition, at least not alone. Activin also promotes Edar expression, which in turn seems to promote Shh expression in the knots (Laurikkala *et al.*, 2001; Pummila *et al.*, 2007). It is possible that the proliferative effect of Shh or FGFs on dental epithelium may prevent it from differentiating and thus from forming a knot.

More generally it has been suggested that the spacing between knots in the tooth would arise by a reaction–diffusion or reaction–diffusion-like mechanism (Jernvall and Thesleff, 2000). In that respect, Bmp4 has been shown to, indirectly through Msx1, activate its own expression (Vainio *et al.*, 1993; Bei and Maas, 1998). Since it also seems to promote knot formation Bmp4 is a perfect candidate to be an activator in a reaction–diffusion mechanism, although activin could also be involved. Although the exact molecular mechanisms may not yet be clear, it is likely that Shh, FGFs and follistatin may be, or may mediate, the inhibitory molecule in this reaction–diffusion-like mechanism. In contrast with reaction–diffusion models, the inhibitor is not expressed by all cells that receive activator but only in knot cells. This is drawn from the previous experimental evidence and produces, as I will show in the model, dynamics that are slightly different from those of pure reaction–diffusion mechanisms.

Reaction–diffusion mechanisms produce symmetric regular patterns in which a number of repeated elements, either stripes or spots, are homogeneously spaced through a (cellular) field. However, when looking at the morphological disparity of mammalian molars it is clear that molars tend to be nonsymmetric and not all cusps are of the same size and height. In addition, not all cusps are equally spaced nor, always, regularly arranged in the tooth. In fact, the disparity of molar shape can be described as variation in these aspects of cusps. Therefore, a simple reaction–diffusion-like mechanism alone cannot account for the development of the morphologies of molars. As briefly explained in the introduction, growth and its interaction with signaling is the additional factor that needs to be considered. This growth signaling interdependence happens, as I will explain in the next section, at several levels. The components of the reaction–diffusion-like mechanism, at least BMP and Shh, affect proliferation in the epithelium and mesenchyme. This affects local growth that in turn affects the amount and shape of the space in which these molecules can diffuse. This regulates the distances at which Bmp4 concentration reaches the threshold to form new knots and indirectly regulates the heights and positions at which new knots form. Thus these morphodynamic mechanisms cannot be understood as the mere addition of growth to reaction–diffusion mechanisms, but, as I will show, represent a distinct class of dynamics that allows a repertoire of morphologies larger than that of reaction diffusion or growth alone (Salazar-Ciudad *et al.*, 2003). Even with growth, reaction–diffusion mechanisms cannot produce teeth that are asymmetric in the anteroposterior or bucolingual axis. This is achieved, as I will detail, because of asymmetries in Bmp4 and other molecular expression in the mesenchyme surrounding teeth.

The next sections describe these morphodynamic dynamics in detail. This is done by introducing the models and how they can be used to explore the implications of several alternative hypotheses about the interaction of signaling with specific aspects of growth. Thus, much relevant experimental evidence is introduced along with the models to make explicit the biological assumptions upon which the models are built. Two computational models of tooth morphogenesis will be reviewed here (Salazar-Ciudad and Jernvall, 2002, 2007). The models differ in the accuracy of their predictions and the detail and realism of the cellular behaviors and gene networks involved in tooth formation.

VII. Morphodynamic Model 1

The model includes four cell behaviors: cells can secrete signaling molecules; cells can receive signaling molecules (and change their behaviors in consequence); cells can also divide and differentiate. The model includes a network of gene products that regulates these behaviors and interact between them. Model 1 includes an activator which, as Bmp4 does indirectly, activates itself and, after a concentration threshold, knot formation. The inhibitor is produced from the knot at a rate proportional to the activator concentration in the knot and it inhibits activator production by epithelial cells.

The model starts with four epithelial cells distributed in a regular rectangular grid. Three layers of mesenchymal cells lie under these epithelial cells. All epithelial cells secrete activator at an intrinsic rate (k_3) and also in response to the local activator concentration. Over time, in areas where the local activator concentration exceeds a set threshold, the epithelial cells differentiate irreversibly into nondividing knot cells. These knot cells also secrete inhibitor at a rate equal to the local activator concentration. This inhibitor counteracts activator secretion and, in addition, enhances growth of the mesenchyme. The mesenchyme is mainly a three-dimensional space where diffusion and growth take place.

12. Tooth Morphogenesis

Diffusion takes place inside the three-dimensional space (subdivided into a three-dimensional grid of boxes) of the growing tooth. The system has zero-flux boundary conditions in the epithelium (diffusion is not allowed in their apical side) and open boundary conditions in the mesenchyme (molecules exit the system through the borders). The mesenchyme is surrounded by the epithelium (where diffusion is allowed), except in the ventral border where lies the nondental mesenchyme (where the activator and inhibitor can diffuse out of the system). The rate of activator secretion in nonknot epithelial cells is:

$$\frac{\partial A}{\partial t} = \frac{k_1[A]}{k_2[I]+1} + k_3 + D_A \nabla^2 [A], \tag{1}$$

where $D_A \nabla^2[A]$ is the diffusion term and D_A is the diffusion coefficient of the activator. The k_1 and k_2 constants can be related to biochemical aspects as the affinity of each molecule for its receptor or to the signal amplification produced by its chain of signal transduction. The rate of inhibitor secretion by knot cells is:

$$\frac{\partial I}{\partial t} = [A] + D_I \nabla^2 [I], \tag{2}$$

where $D_I \nabla^2[I]$ is the diffusion term and D_I is the diffusion coefficient of the inhibitor.

In model 1 all cells are positioned in a three-dimensional grid made of regular cubes. In that sense the initial tooth primordium is made of 4 columns of cells (with epithelial cells in the top of each column). Epithelial growth is implemented by making epithelia increase its depth into the mesenchyme. This is made by displacing epithelial cells downwards as they growth. Epithelial growth rate is $R_e - [A]$ and at least zero. When a single epithelial cell shifts ventrally one cell length into the mesenchyme, it displaces ventrally all the underlying cells in that column (including the stellate reticulum space apical to the top epithelial cell). Since this growth rate depends of the concentration of the activator it can differ between different parts of the epithelium. As a result of these processes, part of the epithelium folds into the mesenchyme leaving the knots isolated in the tips of the forming cusps. At the same time mesenchymal growth produces localized lateral expansion affecting cusp sharpness. Any cell which, due to the relative displacement of columns, gets in contact with the stellate reticulum is considered, in model 1, an epithelial cell.

Mesenchymal growth occurs mainly in the direction offering least resistance (away from the space apical to the epithelium). This growth is proportional to the concentration of inhibitor in each mesenchymal cell. Thus the inhibitor in the model has two effects (although *in vivo* these effects may be mediated by distinct molecules coming from the knots). This propensity of movement towards less occupied space was calculated as the sum of the concentration of inhibitor in all the mesenchymal cells of a column multiplied by a constant (R_m) that reflects the sensitivity of cells to the inhibitor's growth effect. Specifically, the lateral pressure of cells in a column i is distributed into four nearest neighboring columns (the anterior, posterior, buccal and lingual columns) by the following rules: (i) pressure distribution can only occur to columns shorter than column i. (ii) The resistance ($1/Sj$) of each neighboring column shorter than column i is the total number of cells that all the columns in one direction have (for example, all the posterior columns next to the column i). This reads:

$$S_j = 1 \bigg/ \left(\sum_{k=0}^{k=n(i,j)} m(k) \right), \qquad (3)$$

where j can be any of the four directions possible, p, posterior, a, anterior, b, buccal, l, lingual. These correspond to the four axes of a tooth. $n(i, j)$ is the number of columns between column i and the border of the tooth in the direction j, and $m(k)$ is the number of cells in column k. Note that $n(i, j)$ and $m(k)$ depend on tooth shape at each time point and are not external functions or fixed parameters of the model but, at each iteration, the result of previous dynamics. (iii) The pressure of column i is distributed to its neighbors in inverse proportion to their resistance. This is defined as

$$R_j(i) = D_j R_m \sum_{k=0}^{k=m(i)} [I]_{ik}, \qquad (4)$$

where

$$D_j = S_j / (S_p + S_a + S_b + S_l) \quad \text{for } j[p, a, b, l].$$

$R_j(i)$ is the resulting lateral pressure exerted by a column i in direction j. $[I]_{ik}$ is the concentration of the inhibitor in cell k in column i and R_m is the rate constant of mesenchymal growth. S_j is the inverse of the resistance and $(S_p + S_a + S_b + S_l)$ is the overall inverse of the resistance in all directions.

The lateral expansion is mimicked by adding new cells to a column when it receives lateral pressure in a given direction that exceeds a unit corresponding to a cell size in a given direction. For a column that is not in the border of the tooth, the lateral force received from neighboring columns causes the whole column to displace upwards. Note that mesenchymal pressure only moves from tall to short columns and thus this upward movement is equivalent to the growth of a column (or dome) to the sides. When this displacement upwards exceeds one unit a new cell is added at the bottom of the column. Mesenchymal pressure from a column in the border of the tooth to the border leads to the addition of a new column in that direction (when accumulated pressure in that direction exceeds unity). A new column starts as a single epithelial cell. This way the tooth can grow in width (and this is due to the effect of signals emanating from the knots). Mesenchymal cells are also added at the bottom of the tooth at the level of the deepest column. This does not represent a real biological phenomenon, it simply indicates that more space is considered for the model calculation (essentially these additional mesenchymal cells were already there but were not considered for the calculations). Lateral growth can be biased in the anterior, posterior, buccal and/or lingual direction by increasing the lateral force on cells in the perimeters of the tooth in any of these four directions. There is a bias in the posterior (B_p), anterior (B_a), buccal (B_b) and lingual (B_l) direction. For cells in the border equation (4) becomes:

$$R_j(i) = D_j R_m \sum_{k=0}^{k=m(i)} [I]_{ik} + B_j \quad \text{for } j[p,a,b,l]. \tag{5}$$

These biases are B_p, B_a, B_b, and B_l. These biases are necessary because without them all teeth would be symmetrical. The teeth of mouse and other mammals are known to grow asymmetrically in these directions.

VIII. Model 1 and Tooth Dynamics

In spite of the overall coarseness by which the model implements cell mechanical interactions and growth, the model is able to reproduce, to a large extent, the three-dimensional changes in morphology and patterns of gene expression observed during development in mouse and vole first molars (see Fig. 3). The model reproduces these changes with a relative timing very comparable to that of real mouse and vole tooth development. These molars were chosen for testing the model because they are the ones for which changes in morphology and patterns in gene expression are best described (Jernvall *et al.*, 2000). In addition, the model is also able to reproduce the molar morphology of many other species (Salazar-Ciudad and Jernvall, 2002).

The model also has some obvious limitations that have motivated the construction of a second model. These have to do, mainly, with the way growth and mechanical interactions between cells are implemented. These are also the aspects of tooth development that are less well understood. Model 1 does not explain how epithelial cells grow down but it simply takes it for granted. Although it is possible that on average increased epithelial grow leads to sharper cusps, dividing epithelial cells tend to bud-off in the plane of the epithelium. Thus, epithelial growth does not necessarily move cells downward but moves cells in the plane of the epithelium (which changes locally and over time). Although, on average, the implementation of mesenchymal growth pressure may give rise to lateral growth towards the borders, it is unlikely that this pressure is the result of summing up the growth along a column of cells. Instead, it may be more realistic to assume that every mesenchymal cell grows and divides in any possible direction. In the model the cells can only move as cohesive columns in a fixed rectangular grid. Lateral growth in a column is mimicked by vertical growth in neighboring columns. This provides a rather discrete description of space, cell position and cell displacements. Morphodynamic model 2 (Salazar-Ciudad and Jernvall, 2007) provides a more spatially continuous implementation of these aspects as well as a better low level implementation of epithelial and mesenchymal growth and cell division.

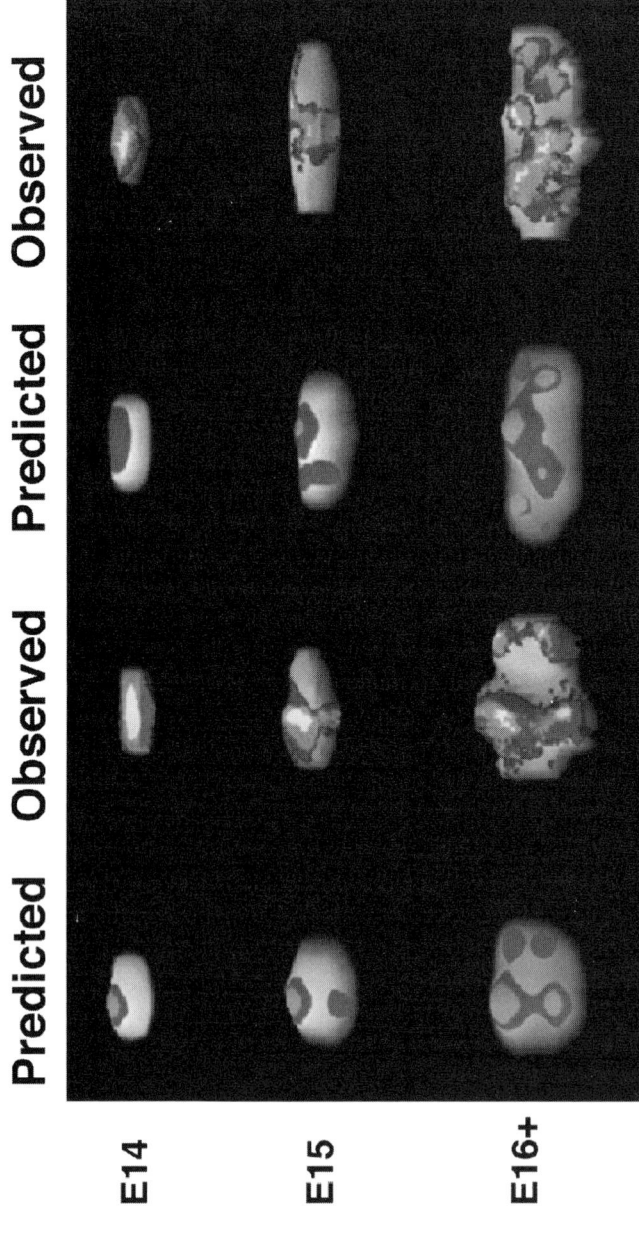

Figure 3 Model 1 predicts gene expression patterns in mouse (left) and vole (right). The activator and inhibitor concentration peaks predict the observed nested gene coexpression patterns among genes of different signaling families. On predicted shapes, the coexpression patterns of activator and inhibitor (in orange) inside the activator domain (in red) resemble the observed coexpression patterns where Fgf4, Shh, Lef1, and p21 (in yellow) mark the cores of the enamel knots, surrounded by areas lacking Fgf4 (in orange) and Fgf4ILef1 expressions (in red). Anterior side is toward the left and buccal side is toward the top; ages are in embryonic days. See color insert.

IX. Morphodynamic Model 2

Gene networks in model 2 are implemented in a similar way to model 1. Three more genes are included, *BMP2*, ectodin, and *FGF-4*. Ectodin is an extracellular sequesterer of several BMPs and in the model it acts by decreasing the concentration of free diffusible Bmp2 and 4. Fgf4 is also included. As in mouse teeth (Jernvall *et al.*, 1994), it is secreted from the knots and it promotes proliferation of the underlying mesenchyme. Also as in mouse teeth (Åberg *et al.*, 1997), Bmp2 is secreted from the knots and enhances differentiation. In the model, Bmp4 is the activator and Shh is again assumed to be the inhibitor. These genetic differences, however, have a rather mild effect on model dynamics. The more significant changes have been made at the level of growth dynamics, cell biomechanics and proliferation. In contrast to model 1, model 2 does not constrain cell position and displacement to a rectangular discrete grid. Instead, the cells in the epithelium form a grid that deforms and grows due to cell growth and division.

The model only considers tooth development from the later moments of the bud stage. The model's initial conditions consist of an initially flat epithelium. This epithelium represents the tip of the invagination before the first knot forms (see Fig. 2). The model starts with 19 hexagonal epithelial cells arranged in a hexagon (see Fig. 4B) with 20 layers of mesenchymal cells under them. Each cell has six neighbors and is situated at an arbitrary distance of 1 from them. Cells in the borders have only 3 or 4 neighbors.

Each cell is a three-dimensional volume that includes the cell itself and its immediate extracellular space. Molecular diffusion between two cells is proportional to the area of contact between those cells and their surroundings (finite volume method). This method is used because it allows accurate calculations even when cells change their shapes. The contour conditions for diffusion are the same than in model 1. The initial condition includes the expression of Bmp4 in the borders of the tooth at the level of the epithelium. The concentration of Bmp4 is kept constant and asymmetric in the borders all the time: Border epithelial cells that extend buccally from the midline have concentration set to a model parameter, *bib*, while cell extending lingually are set to *bil*. This asymmetry in Bmp4 expression in the borders of the tooth germ exists in mouse first molar through development (Åberg *et al.*, 1997).

Epithelial cells grow by pushing their neighbors in the plane of the epithelia. Each epithelial cell pushes each of its neighboring epithelial cells away from it. In other words, this pushing is, for each pair of cells, in the direction of a unit vector pointing from the center of one cell to the center of the other. For each cell the direction of push made by all neighboring cells is summed up and normalized to determine the direction of growth. The amount of growth by a cell is, in a similar way to model 1, proportional to R_e multiplied by 1 minus the amount of differentiation of that cell (cells differentiate at a rate proportional to Bmp2 concentration). When the distance between two original neighbor cells is larger than two, a new daughter cell appears at the midpoint in between them. The gene product concentrations of that daughter cell are the average of the two "mother" cells. This way the grid made by the positions of the epithelial cells grows by intercalation of new cells and the average size of cells remains relatively constant (as in many animal epithelia). New cells have as neighbors the two "mother" cells and a subset of their mother's neighbors (chosen in such a way that the lines uniting cells with their neighbors do not cross each other). As in model 1 knot cells do not grow nor divide.

With the above-mentioned rules the model's teeth would simply grow as flat epithelia. As explained above, once the first knot forms two epithelial folds form buccally and lingually from the knot. Sections at successive times show that the cervical loops tend to bend towards the midline as time progresses. In some molars that bending is so large that the buccal and lingual cervical loops get very close to each other at their tips (Fig. 1). Tooth biomechanics are currently not understood sufficiently well to explain this phenomenon. In fact it is not known why cervical loops grow downwards and why they bend medially. Several hypotheses can be advances to explain this. Here I present the one that is implemented in the current version of the model. Later sections describe how the other hypotheses have been tried in the model and how and why they fail to produce dynamics consistent with those found in mouse molars. My preferred hypothesis is that the cervical loops and their growth dynamics are a result of the epithelium having a strong binding affinity for the dental mesenchyme and the dental mesenchyme having a strong binding affinity for itself. This could happen either because dental and nondental mesenchyme express different adhesion molecules or simply because dental mesenchyme is more condensed and thus has more adhesion molecules per relative volume. That way the morphology and the medial bending of the cervical loops would be a result of the epithelium trying to engulf the dental mesenchyme (Steinberg, 1970). In that sense the cervical loops would expand in the interface between the dental mesenchyme and the normal mesenchyme of the jaw. The condensed dental mesenchyme extends as a spheroidal mass under the inner enamel epithelium. Morphologically it seems that the cervical loops do not surround mesenchyme that is not condensed. In that sense it is also possible that the contact with this dental mesenchyme is involved in making the inner epithelium different from the outer one. At the same time it is possible that signals emanating from the epithelium or from the knots are responsible for the condensation of the mesenchyme. This could happen early on in the bud or cap stage or continuously throughout tooth development. This differential affinity hypothesis is implemented in the model and produces growth dynamics consistent with those observed in the cervical loops of mouse molars.

Instead of simulating the whole dynamics of spreading and condensation of the dental mesenchyme the current version of the model takes a much simpler approach. The borders of the tooth in the model, analogous to the cervical loops, have a tendency to grow following the interface between the dental and jaw mesenchyme. The more condensed mesenchyme there is, the less steep is that interface and the more the epithelium grows laterally. In the model I assume that Fgf-4 enhances mesenchymal proliferation [as has been shown by bead experiments (Jernvall *et al.*, 1994)]. In practice this is implemented by altering the direction of displacement by adding an arbitrary downward vector and multiplying the x and y components of displacement by a factor proportional to mesenchymal growth (produced by Fgf-4). As in any other epithelial cell, however, the total amount of displacement is equal to $R_e(1 - d_i)$.

Mesenchymal growth is implemented as a pressure exerted in each epithelial cell in a direction normal to the apical surface of the epithelium. This force is proportional to the concentration of Fgf-4, to a model parameter specifying how strongly Fgf-4 promotes mesenchymal proliferation and to 1 minus the differentiation of the epithelial cell. This implementation assumes that proliferation in the mesenchyme produces local increases in its volume that exert a pressure in all directions. Pressures exerted downwards to the mesenchyme simply displace nondental mesenchyme and have no effect on the form of the epithelium. Then, only the pressures that are perpendicular to the epithelium surface need to be considered.

In addition to growth, cells interact by pushing each other if their centers become too close. This is meant to add physical realism to the model and can be visualized as the cells being united by an elastic spring. This simply simulates the inevitable physical resistance of cells to external pressures. A traction force that tends to decrease the distance between neighboring cells when they get very far away from each other is also implemented to provide physical realism.

Overall, all of these displacement equations are summed up in each iteration. It is assumed that the stellate reticulum exerts a pressure on the epithelia. Then, all displacements in its direction are discarded (this rarely happens however). In addition, knots are not allowed to move down because they are supposed to be attached to the stellate reticulum and they do not seem to move down in any species examined (Butler, 1956). Cells with odd shapes can produce, as in the case of finite element analysis, inaccuracy in the calculations, either on morphology or/and diffusion. The rules of division of large cells in model 2 ensure that there is a limited variation in cell size and shape and thus ensure the accuracy of calculations.

A special situation applies to the cells in the anterior and posterior borders. At these borders the cervical loops form very late while the oral epithelium is still extending anterioposteriorly. This seems to happen at different rates in the anterior and posterior border and this asymmetry seems to be responsible for the anteroposterior asymmetry of the molars. Thus, this asymmetry needs to be implemented in the model. However, the genetic or/and epigenetic interactions that give rise to this asymmetry are not known. The only thing that is known is that Bmp3 and follistatin, molecules that sequester Bmp2, Bmp4 and activin, are expressed in the anterior border of the first mouse molar at E14 (Åberg et al., 1997; Wang et al., 2004a). It is not clear how, or whether, that may affect the anteroposterior asymmetry. Due to this lack of specific understanding these biases are simply implemented as a factor, model parameters *bia* and *bip*, multiplying the displacement along the anteroposterior axis (this is the x coordinate) in the anterior, *bia*, and posterior, *bip*, borders of the tooth. Anterior border cells are those whose y coordinate is smaller (in absolute value) than a model parameter, *AP*, and whose x coordinate is larger than zero. Posterior cells are those whose y coordinate is smaller (in absolute value) than

AP and whose *x* coordinate is smaller than zero. In addition, the *z* coordinate (ventral direction) in both borders is multiplied by a factor, model parameter *bo*. This is used to phenomenologically describe the delayed formation of the cervical loops in the anterior and posterior borders and is assumed to be due to the existence of some incipient condensed dental mesenchyme through the anteroposterior axis along the invagination.

X. What Do Model Dynamics Reveal About Developmental Dynamics

Fig. 4 shows the changes in mouse *in silico* morphology over early tooth development from the initial conditions described. In spite of the simplicity with which the model deals with cell behaviors and gene interactions, the model provides a reasonable match with the three-dimensional changes in molar morphology and patterns of gene expression occurring during development. Moreover, these *in silico* morphological changes occur with a timing that is very close to the timing observed in mouse first molars. The model outcomes have not yet been compared with later stages because in mouse molars cusps exhibit, after E15, an anteroposterior bending that cannot be explained from present hypotheses.

Although the model can produce a tooth shape that reasonably resembles that of the adult mouse, the model does not currently include any aspects of the mineralization process. Thus, although the morphology of the epithelium gives a clear outline of the final adult morphology, the deposition of enamel may slightly modify that outline. This is specially the case in species which, unlike mouse, have substantially thick enamel. In general, more enamel deposition (Avishai *et al.*, 2004) makes cusps to appear relatively blunter.

Three points are worth considering in evaluating the validity of both models:

1. The models, implementations and parameters come from some understanding and hypotheses about the genetic and epigenetic interactions involved in tooth development. Thus the models are not statistical models that fit some parameters to a given target. Different morphologies can be produced by changing the parameters of the model but the model cannot be fit to any arbitrary morphology nor do the parameters keep their biological validity for all values. Both models are also able to produce tooth morphologies, and its change during development, of other species, for example: voles, seals and early tribosphenic mammals. Overall these are first (developmental) principles models where morphological changes are predicted from hypotheses and assumptions. More generally, however, these models provide a tool with which these hypotheses can be tested by comparing their morphological consequences in the model with real morphological variation arising from artificial manipulation of tooth development or arising in evolution. Thus the hypotheses of the models can be changed or revised according to how well they ex-

12. Tooth Morphogenesis

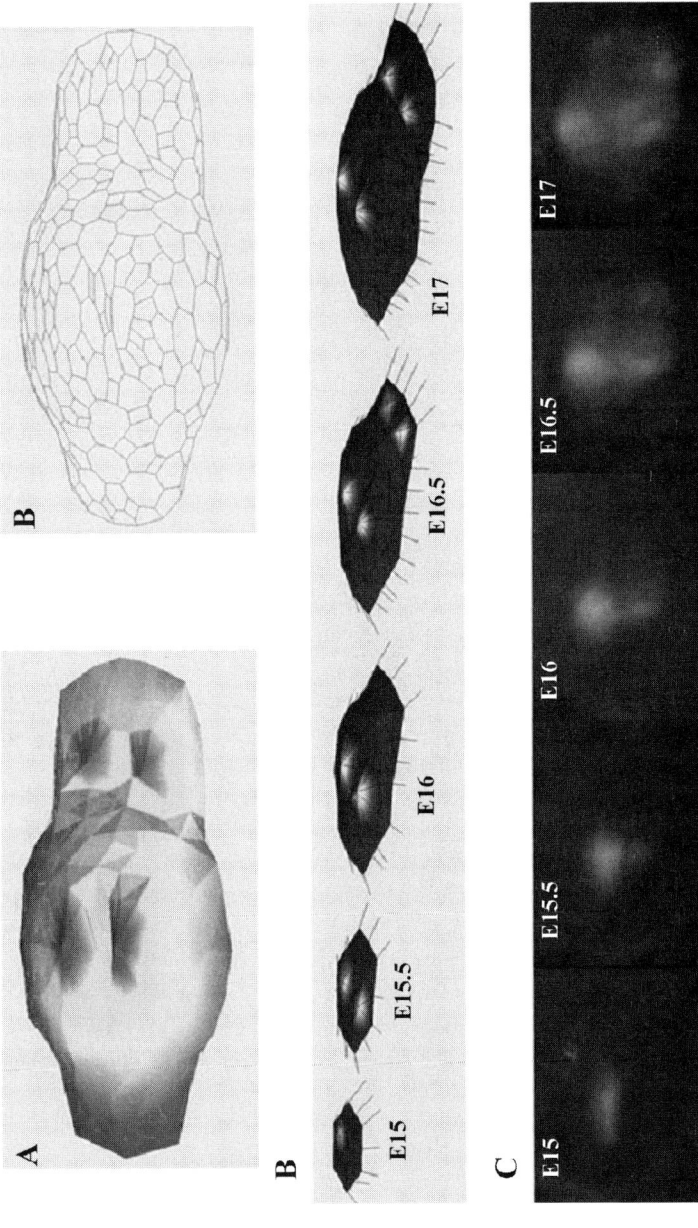

Figure 4 The figure shows in (A) an outcome of the model 2 for a middle stage in tooth development. In color it is plotted the concentration of Fgf-4 in the inner enamel epithelium (red is the highest concentration and blue is the lowest). In (B) the contours of inner enamel epithelium cells are plotted. In (C) several time slides of the outcomes of model 2 are plotted. In (D) it is shown the *in vitro* first molar development from embryonic day 15–17. The green color indicates the presence of fusion protein between GFP and Shh (in a heterozygous mouse) and thus indicates the spatial distribution that Shh may exhibit. See color insert.

plain new morphological variation. This has, in fact, been done to develop model 2 from model 1 (Salazar-Ciudad and Jernvall, 2002).
2. The morphological outcome of the models arises from the dynamics of the model. In other words, all the parameters and hypotheses of the models (except for the anteroposterior asymmetry) are at the low level of genetic and epigenetic interactions but explain high level phenomena such as the overall morphology of the tooth. This is performed without arguing for the existence of complex spatial prepatterns or genetic interpretations. In other words, nowhere in the models it is prespecified where knots are going to arise.
3. That the hypotheses of the models are consistent with mouse first molar morphology and development does not necessarily mean that the hypotheses of the models are correct. That would require the design and realization of further experiments on the bases of the models. In particular, other hypotheses could also explain the development of mouse first molar. In the next section model dynamics are discussed in detail and compared with tooth development to explain how specific hypotheses of model 2 produce specific outcomes and how alternative hypotheses can be discarded (or not).

A. The Formation of the Cusps

In model 2 Bmp2 signaling from the knot slows down proliferation rate in surrounding tissue by enhancing cell differentiation. This way cells around the knot are left behind by the rest of the epithelium that is dividing more intensively. This gives rise to a morphologically distinct cusp with slowly dividing cells near the top and the knot at the top. However, in the model, if Bmp2 secretion is totally inhibited cusps still form as long as there are nonproliferating knots. Without knots, proliferation in the epithelium, and the pushing between its cells, results, in the model, in the progressive flow of the epithelium towards the edges of the developing tooth (where the cervical loops are engulfing the dental mesenchyme). This leads to the formation of teeth that are nearly flat. This situation is also found in tooth tissue cultures in which a "tooth" without knots also forms (Iwatsuki *et al.*, 2006). The presence of a knot leads the epithelium to flow away from the knot in all directions. This mechanical effect is due to the nonproliferation of knots. This meant that the knot height does not change and that the knot is left behind as a morphologically distinct cusp. If two knots form close to each other, then the flow of epithelial cells from them leads to the formation of a valley in between. This is very similar to what happens in mouse molars and other teeth where valleys form between cusps. All these morphological changes occur without a prespecification, in the model, of where or how those cusps would form. The model simply specifies how individual cells push each other and how they respond to growth factors. In other words, the cusps simply emerge as a dynamic output of model 2. In that sense the model suggests a simple mechanistic basis for the folding of the epithelia.

12. Tooth Morphogenesis

Although the model suggests that Bmp2 (and its differentiation enhancing effect) may not be strictly required for cusp formation, the absence of Bmp2 leads to cusps that are much sharper than in mouse first molar. Bmp2 increases cusps bluntness by producing a gradual, rather than sharp, extension of the area around the knots where proliferation is decreased. A similar effect is produced by mesenchymal growth. These hypotheses about the effect of differentiation and mesenchymal growth on cusps sharpness could be further tested by specifically inhibiting Bmp2 (and other BMPs with possible redundant functions as Bmp4) expression in the knots. Similar experiments could be performed for Shh. These manipulations could be rather complex because these genes are extensively used through development. Conditional knockouts or/and overexpression transgenes would then be required. Ideally, manipulations that produce gradual changes in the expression of these genes from the knots would be much more informative.

In model 2 the formation of valleys between cusps also produces, indirectly, the movement of cusps away from one another (for the mouse mainly away from the midline buccally and lingually). For the stages considered in Fig. 4 this pushing is relatively small but increases at later stages. A similar displacement has been observed for mouse molars (Jernvall *et al.*, 2000). In the model this seems to occur by the simple traction of the knots by the epithelium flowing towards the cervical loop and by the pushing of the epithelium flowing towards the valley side (while the invagination of this valley is taking place). Then, again, the model suggests a simple explanation for complex morphogenetic phenomena on the bases of simple developmentally grounded cell to cell interactions.

The hypothesis that nondiving cells are responsible for the bending of the epithelium and the formation of cusps is technically more difficult to test because it is a biomechanical explanation that is not so readily approachable from the molecular techniques most commonly used in developmental biology. Ideally, this hypothesis could be tested by inhibiting proliferation in specific locations in the epithelia. However, the enamel epithelium is surrounded, even *in vitro*, by the vacuolated stellate reticulum and it is thus difficult to manipulate mechanically. Local administration of cell division inhibiting substances is, then, also difficult. This problem could be overcome by the design of transgenes in which a cell division inhibiting construct could be expressed in a specific area of the epithelium. However, as described, gene expression in the epithelium is either around the knots, in the whole epithelium or away from the knots. Thus it is difficult to target gene expression to arbitrary parts of the inner enamel epithelium for the stages considered in the model. A viable alternative would be a construct that is expressed in the enamel epithelium in random spots after some heat-shock or other signal is provided (as achieved by Zong *et al.*, 2005, for the cerebellum of mouse). That way proliferation could be inhibited in areas of the inner enamel epithelium where there are no knots. If the inhibition of proliferation is responsible for the formation of cusps these transgenes should produce a cusp under each nonproliferative zone in the enamel epithelium. The overall morphology and the positions of valleys and cusps should then be predictable from the model (for a given random distribution of nonproliferative zones). If that is not the case, then it is likely that something else (like some mechanical attachment of nonproliferative zones to the stellate reticulum or some signaling from the knot) is also required to explain cusp formation.

As mentioned above, bead experiments (Jernvall *et al.*, 1994) indicate that Fgf-4 secreted from the knots may induce proliferation in the enamel epithelium. However, if this proliferation-enhancing effect is concentration dependent, then the higher rates of proliferation should be encountered at medium distances from the knot. In other words, at short distances from a knot the differentiating effect of Bmp2 dominates while at large distances Fgf-4 concentration would be too low to induce strong proliferation. This spatial distribution of proliferation intensity (from the knots to the cervical loops) does not produce, when implemented in the models, cusp morphologies consistent with those found in most mammalian teeth. Nor does information about the distribution of proliferation rates (Jernvall *et al.*, 1994) support the existence of maximal proliferative rates at medium distances from the knots. This suggests that either the proliferative effect of Fgf-4 is not concentration dependent (cell proliferation would occur at a fixed rate after a Fgf-4 concentration threshold is reached or Fgf-4 concentration in the enamel epithelium would rapidly saturate its receptors in epithelial cells) or, as assumed in here, the proliferative effect of Fgf-4 is much stronger in the mesenchyme. If the model is changed to allow Fgf-4 to activate epithelium proliferation in a nonconcentration-dependent manner then the range of morphologies found is very similar to the ones found in the current version of the model (although the actual parameter combinations in which specific morphologies arise changes).

B. The Positioning of Knots

In mouse molars, as well as in many other rodent molars, several cusps have, roughly, the same height. Taking into account that knots form at a distance from each other and that the knots rapidly become higher than the surrounding proliferating epithelium it would seem difficult to explain how that happens. This may seem inconsistent with what is expected from a reaction diffusion model. In reaction–diffusion models the distances between peaks of activator concentration tend to be constant (over a field of cells) and depend on the several kinetic parameters (such as the diffusion rates and the inhibition of activator production). However, in model 2, the initial asymmetric expression of Bmp4 in the borders of the developing tooth results in the second knot forming very close to the first one. The second knots forms close to the side where Bmp4 is expressed less strongly (lingual side). This happens early, before too much growth occurs. In mouse first molar (see Fig. 4D) the second knot also forms on that side and before too much growth occurs. This predicts that in species where the second knot forms more anteriorly or posteriorly than the first knot, the Bmp4 bucolingual asymmetry would be smaller or/and the early bud stages in which the second knot forms would be narrower than in mouse. Both situations seem to apply to the vole (*Microtus rossameridionalis*), another species of rodent (Keränen *et al.*, 1998).

In the mouse first molar the third and fourth knots form at the same level in the anterior–posterior axis. This pattern seems to require some specific positioning of the Bmp-4 sources. As mentioned, Bmp4 is expressed in the borders of the tooth throughout tooth development. This means that new knots tend to form towards the borders of the tooth (and away from existing knots). Without this external source of Bmp-4 the third knot forms, in model 2, where there is more space (and away from existing knots). That is, the third knot forms in the midline (in the buccolingual axis) in the posterior part of the tooth. The fourth knot forms then much posteriorly and the resulting tooth morphology does not resemble that of the mouse first molar. The Bmp4 expressed in the borders results in the third knots forming close to the buccal border (Bmp4 border expression is larger: $bib > bil$) and a fourth knot can form close to the lingual border (and that same anterior–posterior level) as occurs in the mouse first molar.

In mouse first molar the distance from the first knot at which the second knot forms is much smaller than the distance from the first knot at which the third and fourth knots form. This seems to be due, as will be discussed in following sections, to the morphodynamic interrelationship between signaling and growth.

C. Cervical Loop Growth

Other than the hypothesis implemented in the current version of the model there could be other hypotheses explaining the bending and growth pattern of cervical loops. It could be that the cervical loops would bend due to differential growth rates between inner and outer enamel epithelium. It could also be that the dental mesenchyme exerts some mechanical pressure to the epithelium that is larger near the knots and produces, then, a tooth that is wider near the knots and narrower near the cervical loops. As a side effect of that the cervical loops would seem to bend towards the midline. With the model I have explored these possibilities. Differential growth between inner and outer epithelium is unlikely to be responsible for the morphological changes occurring in the cervical loops. Proliferation in the outer enamel epithelium seems to stop by E15. In addition, when the model included extensive outer enamel epithelium proliferation the tooth primordium developed into an inflated spherical balloon-like shape that did not resemble tooth primordia morphology at early or late stages. Mesenchymal growth is, as described, implemented in model 2. The vacuolization of the stellate reticulum seems to be stronger near old knots and away from the tips of the cervical loops. Increased vacuolization may imply higher hydrostatic pressure. This suggests that mesenchymal growth would more efficiently push the epithelium near the cervical loops (where the stellate reticulum is less developed). When that is implemented in the model the loops do not tend to bend inwards medially but outwards laterally. It is possible that the mesenchyme exerts some pressure on the epithelium but the previous arguments and simulations suggest that it may not be able to explain, by itself, the bending growth trajectory of the cervical loops. Thus, at present, the differential adhesion hypothesis implemented in the model is the only one that seems consistent with the cervical loop growth pattern observed mouse molars. This hypothesis could be further tested by experimentally manipulating the amount or adhesiveness of the dental mesenchyme. For example, specific inhibition of proliferation in the dental mesenchyme (locally) or experimental subtraction of part of it should lead to steeper cervical loops. These latter experiments are under way in our laboratory.

D. Morphodynamics

The fact that signaling and cell movement is happening at the same time has some important consequences for the dynamics of the model. In reaction–diffusion models the distance between knots (or activator peaks) depends on several kinetic parameters such as the diffusion rates and inhibitor and activator productions rates. The inclusion of growth and its interdependence with signaling means that knots can appear at a larger range of distances. Moreover, this interdependence affects where new knots form. First, new knots will appear where there is (epithelial) space for them. That depends on the growth biases (*bil* and *bib*) but also on the overall proliferation rates and relative rates of growth of the epithelium and mesenchyme. Since molecules are leaving the tooth from its borders the shape of those new areas, and not only their size, affects whether, and when, Bmp4 concentration can reach the threshold value to produce a knot. Thus, for example, slender epithelial contours with a high ratio between border perimeter and surface tend to produce more outflow of inhibitor and more inflow of activator from the tooth (and then knot formation).

In addition, the volume enclosed by tooth morphology also affects the concentration and dilution of growth factors. Thus, sharp cusps produce, because of smaller volume relative to blunter cusps, a smaller dilution of inhibitor and thus new knots form at larger distances from the knot in the tip of the cusp (although that also depends on the relative rates of diffusion of the inhibitor and activators). Blunter cusps produce the contrary effect and in general the local morphology of the tooth at each time point affects the patterns of diffusion of all the growth factors and thus subsequent morphological changes. Since growth factors also affect local growth in the epithelium and mesenchyme there is a strong interdependence between the effects of model parameters affecting signaling and model parameters affecting growth. Extensive model simulation has shown that this interdependence between growth and signaling allows for a larger spectrum of tooth morphologies compared with a situation where both things would act independently (Salazar-Ciudad and Jernvall, 2004). At the same time, however, this means that some small changes in one parameter can affect overall tooth morphology and that other parameter changes in model parameters may have a modest effect in morphology. Overall, morphodynamic mechanisms produce a complex relationship between genetic and morphological variation. This produces developmental dynamics, both in the model and in the organism, that are more difficult to understand in comparison with a nonmorphodynamic, morphostatic, situation (Salazar-Ciudad and Jernvall, 2004). In that sense mathematical modeling provides a useful tool to identify and study these complex dynamics.

E. Developmental Dynamics and Tooth Diversity

This review has explained how current experimental evidence in tooth development can be integrated into a predictive coherent model. The current and previous models have been shown to be able to produce the forms of mouse first molars and also many other teeth (Salazar-Ciudad and Jernvall, 2002, 2004, 2006). In that sense the model should be able to generate some hypotheses about which developmental changes are responsible for the generation of the morphological diversity of mammalian teeth. For example, as few as two changes in model parameters can transform a mouse first molar into a vole first molar (Salazar-Ciudad and Jernvall, 2002). In general by simply changing model parameters it is possible to produce a large diversity of mammalian-looking teeth. My hypothesis would then be that most mammalian tooth morphologies would be possible by tinkering with those parameters. Since most of these parameters have clear molecular bases the model can offer some hints about what genetic changes may be involved in the morphological differences between species or within populations. Of course, to the extent that the model is a necessary simplification of reality it may underestimate the number of parameters. In addition, our incomplete understanding of tooth developmental dynamics may conflate several parameters and changes in parameters with small changes in gene network topology. In that respect the present model should be considered as an identification of the basic possible developmental logic of tooth morphogenesis.

Studies in the previous version of the model have shown that the model is also able to predict the morphology of a mutant (Järvinen *et al.*, 2006). That mutant is a genetic construct in which β-catenin is under the regulation Keratin-14 promoter and thus canonical Wnt signaling is constitutively activated in the whole dental epithelium. Wnt is not included in the model but its effect has been suggested to be mediated through an increase in activator production. In that way the role of a gene not yet directly included in the model can be explored through its effect in some of the existing model parameters and its effect on the whole dynamics of tooth development.

Model 2 has also been shown to be able to produce the morphology of ringed seal (*Phoca hispida*) premolars (seals tend to be homodont), its change during development and the main patterns of morphological variation found in their populations (Jernvall and Salazar-Ciudad, 2007).

Most of the model parameters (all except *bip* and *bia*) are based on genetic or cellular features of tooth development and probably have a genetic basis. Variation in some of these parameters, like the diffusion rates, may arise from genetic variation in a single gene (or few genes). These could be genes such as the growth factor themselves (although chemical modification of these proteins may also affect their function and diffusion rates and then involves variation in other genes). Functional constraints on protein structure (like, for example, to be able to bind to its receptor) may result in numerical variation in these parameters being only possible in some ranges. Other parameters, like the efficiency of Fgf-4 in promoting dental mesenchyme proliferation R_m, could change due to variation in several genes (for example, in all the genes involved in the signal transduction of Fgf-4 in dental mesenchymal cells). Parameter changes, however, do not affect the topology of the gene network considered in the model (except for the trivial case that two genes stop interacting because their interaction strength is close to zero) nor which cell behaviors are regulated by growth factors, and how. These are aspects of development that can also change over the course of evolution (although they may require more or larger changes at the genetic level) and thus some tooth morphological transitions may be explainable by these kinds of changes. This view contrasts with claims that the diversity of teeth in a heterodont individual (like, for example, mouse) comes from the differential expression of transcriptional factors that precisely control the development of each tooth type (McCollum and Sharpe, 2001). This view is not very explicit about the mechanisms by which that may happen but it seems to favor a complex genetic hierarchic regulation ("control") of several unspecified morphogenetic events by each of these transcriptional factors. However, it may be easier to change, by random genetic variation, the strength by which a gene interacts with some others than the identity of the genes with which a gene interacts (Salazar-Ciudad, 2006). This suggests that morphological evolution may often proceed by changes in developmental parameter values (or to genetics changes leading to small topological changes) rather than by complex changes in network topologies (although that may of course also occur). In contrast, it would be simpler to suggest that the transcriptional factors regulating each tooth type act by simply regulating some gene products so as to change one or several of the developmental parameters (as in the model). That could involve, for example, subtle regulatory changes such as quantitative increases in the expression of a receptor, signal, signal transducer, or/and adhesive molecule.

XI. Tooth Model in Comparison to Other Models of Organ Development

With few exceptions (Goodwin and Trainor, 1985; Dillon *et al.*, 2003), mathematical and computational models of morphogenesis and pattern formation in animal development do not explicitly consider the interdependence between signaling and cell movements. There are, however, some other models that consider how signaling affects growth (Miura *et al.*, 2006) and also some models in which signaling and growth happen at the same time but signaling does not affect growth (Meinhardt, 1982, 2007; Harris *et al.*, 2005). There are also some few reaction–diffusion models on teeth. These consider, however, the relative positioning of teeth in the jaw in Alligator (Kulesa *et al.*, 1996a, 1996b; Murray and Kulesa, 1996) and not teeth morphology as in this review. Very few mathematical models of development give as an outcome complex three-dimensional morphologies and patterns of gene expression as the current and previous versions of the tooth morphodynamic model (Salazar-Ciudad and Jernvall, 2002). Some of the models that consider cell movements and signaling use a different simulation approach than the one presented in here. In Glazier's model (Glazier and Graner, 1993) cells are composed of elements that move in a prespecified rectangular grid. These elements move to neighbor grid positions depending on the energy change they would produce with the change. This mainly relates to the adhesive context on these grid positions. This approach has been successful in simulating several developmental systems (Chaturvedi *et al.*, 2005; Merks *et al.*, 2006) and is probably applicable to many more. Since each cell needs to consist of several of these elements, with increasing numbers of elements, cell behavior is more accurate and more computation is required per cell compared to my model. This model inevitably introduces a discretization of space while model 2 does not. More importantly model 2 may be more able to deal with mechanical changes occurring in tightly linked cells and only slightly deformable groups of cells (like in the invagination of epithelia or the traction of one tissue by another, as in this model). In contrast, my approach cannot deal, as these other approaches do, with phenomena in which cells dramatically change their neighborhood (like in adhesive cell sorting). In that sense this model provides a different approach to the simulation of organs development.

XII. Concluding Remarks

In summary this review has described a number of hypotheses about how tooth morphology develops from genetic and epigenetic interactions. By implementing these hypotheses into a mathematical developmental model the capacity of these hypotheses to produce the developmental and morphological transformation observed in mouse first molar have been evaluated. The studies and models here reviewed provide developmental accounts for morphological changes that are gradual and complex. In other words, these studies do not focus on gross and discrete morphological alterations of morphology to identify the genes involved in development. Instead, the focus, and the model predictions, is on complex multivariate gradual changes and on how small genetic variation regulates these morphological changes. The models have been used as a framework to compare different hypotheses in developmental biology. In this way I have reviewed our current understanding of tooth morphogenesis and outlined possible future lines of investigation in which this understanding can be improved. At the same time I have provided an example of how close collaboration between theoretical and experimental developmental biologists can give rise to operative models than can be used as reference for discussion and evaluation of hypotheses, experiments and computational implementations of organ development.

Acknowledgments

I thank Bruno Julia, Janne Hakanen, Alistair Evans, and Jukka Jernvall for comments and Juselius foundation (Helsinki, Finland) for funding.

References

Åberg, T., Wozney, J., and Thesleff, I. (1997). Expression patterns of bone morphogenetic proteins (Bmps) in the developing mouse tooth suggest roles in morphogenesis and cell differentiation. *Dev. Dynam.* **210**, 383–396.

Avishai, G., Muller, R., Gabet, Y., Bab, I., Zilberman, U., and , P. Smith (2004). New approach to quantifying developmental variation in the dentition using serial microtomographic imaging. *Microsc. Res. Techniq.* **65**, 263–269.

Bei, M., and Maas, R. (1998). FGFs and BMP4 induce both Msx1-independent and Msx1-dependent signaling pathways in early tooth development. *Development* **125**, 4325–4333.

Biben, C., Wang, C. C., and Harvey, R. P. (2002). NK-2 class homeobox genes and pharyngeal/oral patterning: Nkx2-3 is required for salivary gland and tooth morphogenesis. *Int. J. Dev. Biol.* **46**, 415–422.

Butler, P. M. (1956). The ontogeny of molar pattern. *Biol. Rev.* **31**, 30–70.

Butler, P. M. (1983). The evolution of mammalian dental morphology. *J. Biol. Buccale* **11**, 285–302.

Butler, P. M. (1995). Ontogenetic aspects of dental evolution. *Int. J. Dev. Biol.* **39**, 25–34.

Butler, P. M., and Ramadan, A. A. (1962). Distribution of mitosis in inner enamel epithelium of molar tooth germs of the mouse. *J. Dent. Res.* **41**, 1261–1262.

Chaturvedi, R., Huang, C., Kazmierczak, B., Schneider, T., Izaguirre, J. A., Glimm, T., Hentschel, H. G., Glazier, J. A., Newman, S. A., and Alber, M. S. (2005). On multiscale approaches to three-dimensional modeling of morphogenesis. *J. R. Soc. Interface* **2**, 237–253.

Cobourne, M. T., Hardcastle, Z., and Sharpe, P. T. (2001). Sonic hedgehog regulates epithelial proliferation and cell survival in the developing tooth germ. *J. Dent. Res.* **80**, 1974–1979.

Coin, R., Lesot, H., Vonesch, J. L., Haikel, Y., and Ruch, J. V. (1999). Aspects of cell proliferation kinetics of the inner dental epithelium during mouse molar and incisor morphogenesis: A reappraisal of the role of the enamel knot area. *Int. J. Dev. Biol.* **43**, 261–269.

Colyer, F. (1936). "Variations and Diseases of the Teeth of Animals." Bale, Sons, and Danielson, London.

Dassule, H., and McMahon, A. (1998). Analysis of epithelial–mesenchymal interactions in the initial morphogenesis of the mammalian tooth. *Dev. Biol.* **202**, 215–227.

Dillon, R., Gadgil, C., and Othmer, H. G. (2003). Short- and long-range effects of Sonic hedgehog in limb development. *Proc. Natl. Acad. Sci. USA* **100**, 10152–10157.

Glazier, J. A., and Graner, F. (1993). A simulation of the differential adhesion driven rearrangement of biological cells. *Phys. Rev. E* **47**, 2128–2154.

Goodwin, B. C., and Trainor, L. E. H. (1985). Tip and Whorl morphogenesis in acetabularia by calcium-regulated strain fields. *J. Theor. Biol.* **117**, 79–106.

Gritli-Linde, A., Bei, M., Maas, R., Zhang, X. M., Linde, A., and McMahon, A. P. (2002). Shh signaling within the dental epithelium is necessary for cell proliferation, growth and polarization. *Development* **129**, 5323–5337.

Harris, M. P., Williamson, S., Fallon, J. F., Meinhardt, H., and Prum, R. O. (2005). Molecular evidence for an activator–inhibitor mechanism in development of embryonic feather branching. *Proc. Natl. Acad. Sci. USA* **102**, 11734–11739.

Heikinheimo, K., Begue-Kirn, C., Ritvos, O., Tuuri, T., and Ruch, J. V. (1998). Activin and bone morphogenetic protein (BMP) signaling during tooth development. *Eur. J. Oral. Sci.* **1**, 167–173.

Iemura, S., Yamamoto, T. S., Takagi, C., Uchiyama, H., Natsume, T., Shimasaki, S., Sugino, H., and Ueno, N. (1998). Direct binding of follistatin to a complex of bone-morphogenetic protein and its receptor inhibits ventral and epidermal cell fates in early Xenopus embryo. *Proc. Natl. Acad. Sci. USA* **95**, 9337–9342.

Iwatsuki, S., Honda, M. J., Harada, H., and Ueda, M. (2006). Cell proliferation in teeth reconstructed from dispersed cells of embryonic tooth germs in a three-dimensional scaffold. *Eur. J. Oral Sci.* **114**, 310–317.

Järvinen, E., Salazar-Ciudad, I., Birchmeier, W., Taketo, M. M., Jernvall, J., and Thesleff, I. (2006). Continuous tooth generation in mouse is induced by activated epithelial Wnt/beta-catenin signaling. *Proc. Natl. Acad. Sci. USA* **103**, 18627–18632.

Jernvall, J., and Thesleff, I. (2000). Reiterative signaling and patterning during mammalian tooth morphogenesis. *Mech. Dev.* **92**, 19–29.

Jernvall, J., and Salazar-Ciudad, I., 2007. The economy of tinkering mammalian teeth. *In* "Tinkering: The Microevolution of Development" (G. Bock, J. Goodie Eds.), Novartis Foundation, vol. 284, pp. 207–216.

Jernvall, J., Kettunen, P., Karavanova, I., Martin, L. B., and Thesleff, I. (1994). Evidence for the role of the enamel knot as a control center in mammalian tooth cusp formation: Nondividing cells express growth stimulating Fgf-4 gene. *Int. J. Dev. Biol.* **38**, 463–469.

Jernvall, J., Aberg, T., *et al.* (1998). The life history of an embryonic signaling center: BMP-4 induces p21 and is associated with apoptosis in the mouse tooth enamel knot. *Development* **125**, 161–169.

Jernvall, J., Keränen, S. V., and Thesleff, I. (2000). Evolutionary modification of development in mammalian teeth: Quantifying gene expression patterns and topography. *Proc. Natl. Acad. Sci. USA* **97**, 14444–14448.

Kangas, A. T., Evans, A. R., Thesleff, I., and Jernvall, J. (2004). Nonindependence of mammalian dental characters. *Nature* **432**, 211–214.

Kassai, Y., Munne, P., Hotta, Y., Penttilä, E., Kavanagh, K., Ohbayashi, N., Takada, S., Thesleff, I., Jernvall, J., and Itoh, N. (2005). Regulation of mammalian tooth cusp patterning by ectodin. *Science* **309**, 2067–2070.

Kettunen, P., Karavanova, I., and Thesleff, I. (1998). Responsiveness of developing dental tissues to fibroblast growth factors: Expression of splicing alternatives of FGFR1, -2, -3, and of FGFR4; and stimulation of cell proliferation by FGF-2, -4, -8, and -9. *Dev. Genet.* **22**, 374–385.

Keränen, S. V. E., Åberg, T., Kettunen, P., Thesleff, I., and Jernvall, J. (1998). Association of developmental regulatory genes with the development of different molar tooth shapes in two species of rodents. *Dev. Genes Evol.* **208**, 477–486.

Kulesa, P. M., Cruywagen, G. C., Lubkin, S. R., Ferguson, M. W. J., and Murray, J. D. (1996a). Modeling the spatial patterning of teeth primordia in the alligator. *Acta Biotheor.* **44**, 153–164.

Kulesa, P. M., Cruywagen, G. C., Lubkin, S. R., Maini, P. K., Sneyd, J., Ferguson, M. W. J., and Murray, J. D. (1996b). On a model mechanism for the spatial patterning of teeth primordia in alligator. *J. Theor. Biol.* **180**, 287–296.

Laurikkala, J., Mikkola, M., Mustonen, T., Aberg, T., Koppinen, P., Pispa, J., Nieminen, P., Galceran, J., Grosschedl, R., and Thesleff, I. (2001). TNF signaling via the ligand–receptor pair ectodysplasin and edar controls the function of epithelial signaling centers and is regulated by Wnt and activin during tooth organogenesis. *Dev. Biol.* **229**, 443–455.

Lesot, H., Vonesch, J. L., Peterka, M., Tureckova, J., Peterkova, R., and Ruch, J. V. (1996). Mouse molar morphogenesis revisited by three-dimensional reconstruction. II. Spatial distribution of mitoses and apoptosis in cap to bell staged 1st and second upper molar teeth. *Int. J. Dev. Biol.* **40**, 1017–1031.

Matsumoto, T., Nonaka, K., and Nakata, M. (1990). Genetic study of the dentinal growth in mandibular first molars of mice. *J. Craniofac. Genet. Dev. Biol.* **10**, 373–389.

McCollum, M., and Sharpe, P. T. (2001). Evolution and development of teeth. *J. Anat.* **199**, 153–159.

Meinhardt, H. (1982). "Models of Biological Pattern Formation." Academic Press, London.
Meinhardt, H., 2007. Models of biological pattern formation: From elementary steps to the organization of embryonic axes. Curr. Top. Dev. Biol. (in press).
Merks, R. M., Brodsky, S. V., Goligorksy, M. S., Newman, S. A., and Glazier, J. A. (2006). Cell elongation is key to in silico replication of in vitro vasculogenesis and subsequent remodeling. *Dev. Biol.* **289**, 44–54.
Miura, T., Shiota, K., Morriss-Kay, G., and Maini, P. K. (2006). Mixed-mode pattern in Doublefoot mutant mouse limb—Turing reaction–diffusion model on a growing domain during limb development. *J. Theor. Biol.* **240**, 562–573.
Murray, J. D., and Kulesa, P. M. (1996). On a dynamic reaction–diffusion mechanism: The spatial patterning of teeth primordia in the alligator. *J. Chem. Soc. Faraday Trans.* **92**, 2927–2932.
Nakamura, T., Takio, K., *et al.* (1990). Activin-binding protein from rat ovary is follistatin. *Science* **247**, 836–838.
Osman, A., and Ruch, J. V. (1976). Répartition topographique des mitoses dans l'incisive et dans la 1ère molaire inférieures de l'embryon de souris. *J. Biol. Buccale* **4**, 331–348.
Pispa, J., Jung, H. S., *et al.* (1999). Cusp patterning defect in Tabby mouse teeth and its partial rescue by FGF. *Dev. Biol.* **15**, 521–534.
Pummila, M., Fliniaux, I., Jaatinen, R., James, M. J., Laurikkala, J., Schneider, P., Thesleff, I., and Mikkola, M. L. (2007). Ectodysplasin has a dual role in ectodermal organogenesis: Inhibition of Bmp activity and induction of Shh expression. *Development* **134**, 117–125.
Ruch, J. V. (1990). Patterned distribution of differentiating dental cells: Facts and hypotheses. *J. Biol. Buccale* **18**, 91–98.
Salazar-Ciudad, I. (2006). On the origins of morphological disparity and its diverse developmental bases. *BioEssays* **28**, 1112–1122.
Salazar-Ciudad, I., and Jernvall, J. (2002). A gene network model accounting for development and evolution of mammalian teeth. *Proc. Natl. Acad. Sci. USA* **99**, 8116–8120.
Salazar-Ciudad, I., and Jernvall, J. (2004). How different types of pattern formation mechanisms affect the evolution of form and development. *Evol. Dev.* **6**, 6–16.
Salazar-Ciudad, I., and Jernvall, J. (2006). Graduality and innovation in the evolution of complex phenotypes: Insights from development. *J. Exp. Zool. B Mol. Dev. Evol.* **304**, 619–631.
Salazar-Ciudad, I., and Jernvall, J., 2007. A genetic and cellular computational model of tooth development and morphogenesis (in preparation).
Salazar-Ciudad, I., Jernvall, J., and Newman, S. A. (2003). Mechanisms of pattern formation in development and evolution. *Development* **130**, 2027–2037.
Steinberg, M. S. (1970). Does differential adhesion govern self-assembly processes in histogenesis? Equilibrium configurations and the emergence of a hierarchy among populations of embryonic cells. *J. Exp. Zool.* **173**, 395–433.
Szuma, E. (2002). Dental polymorphism in a population of the red fox (*Vulpes vulpes*) from Poland. *J. Zool.* **256**, 243–253.
Steinman, R. A., Lu, Y., Yaroslavskiy, B., and Stehle, C. (2001). Cell cycle-independent upregulation of p27Kip1 by p21Waf1 in K562 cells. *Oncogene* **20**, 6524–6530.
Thesleff, I., Keranen, S., and Jernvall, J. (2001). Enamel knots as signaling centers linking tooth morphogenesis and odontoblast differentiation. *Adv. Dent. Res.* **15**, 14–18.
Townsend, G., Richards, L., Hughes, T., Pinkerton, S., and Schwerdt, W. (2003). Molar intercuspal dimensions: Genetic input to phenotypic variation. *J. Dent. Res.* **82**, 350–355.
Vaahtokari, A., Åberg, T., and Thesleff, I. (1996). Apoptosis in the developing tooth: Association with an embryonic signaling center and suppression by EGF and FGF-4. *Development* **122**, 121–126.
Vainio, S., Karavanova, I., Jowett, A., and Thesleff, I. (1993). Identification of BMP-4 as a signal mediating secondary induction between epithelial and mesenchymal tissues during early tooth development. *Cell* **75**, 45–58.

Viriot, L., Peterkova, R., Vonesch, J. L., Peterka, M., Ruch, J. V., and Lesot, H. (1997). Mouse molar morphogenesis revisited by three-dimensional reconstruction. III. Spatial distribution of mitoses and apoptoses up to bell-staged first lower molar teeth. *Int. J. Dev. Biol.* **41**, 679–690.

Wang, X. P., Suomalainen, M., Jorgez, C. J., Matzuk, M. M., Wankell, M., Werner, S., and Thesleff, I. (2004a). Modulation of activin/bone morphogenetic protein signaling by follistatin is required for the morphogenesis of mouse molar teeth. *Dev. Dynam.* **231**, 98–108.

Wang, X. P., Suomalainen, M., Jorgez, C. J., Matzuk, M. M., Werner, S., and Thesleff, I. (2004b). Follistatin regulates enamel patterning in mouse incisors by asymmetrically inhibiting BMP signaling and ameloblast differentiation. *Dev. Cell.* **7**, 719–730.

Wolsan, M. (1989). Dental polymorphism in the genus Martes (Carnivora: Mustelidae) and its evolutionary significance. *Acta Theriol.* **40**, 545–593.

Zong, H., Espinosa, J. S., Muzumdar, M. D., and Luo, L. (2005). Mosaic analysis with double markers in mice. *Cell* **121**, 479–492.

13

Delaunay-Object-Dynamics: Cell Mechanics with a 3D Kinetic and Dynamic Weighted Delaunay-Triangulation

Michael Meyer-Hermann

Frankfurt Institute for Advanced Studies (FIAS), Ruth-Moufang Str. 1,
60438 Frankfurt am Main, Germany

    I. Overview of Methods in Theoretical Biology
   II. Delaunay-Based Interaction
      A. Empty-Circumsphere-Criterion
      B. Empty-Orthosphere-Criterion
  III. Voronoi-Cells Approximate Real Cells
      A. Voronoi-Orthosphere-Relation
      B. Definition of Contact Area
  IV. Delaunay-Dynamics
      A. Radon's Theorem
   V. Equation of Motion for Vertices
      A. Drag Forces from the Environment
      B. Active Motility
      C. Passive Two-Cell Mechanics
      D. Compression
  VI. Mechanics Matters
      A. Avascular Tumor Growth
      B. Morphogenesis of Lymphoid Follicles
 VII. Conclusion
      Acknowledgments
      References

Mathematical methods in Biology are of increasing relevance for understanding the control and the dynamics of biological systems with medical relevance. In particular, agent-based methods turn more and more important because of fast increasing computational power which makes even large systems accessible. An overview of different mathematical methods used in Theoretical Biology is provided and a novel agent-based method for cell mechanics based on Delaunay-triangulations and Voronoi-tessellations is explained in more detail: The *Delaunay-Object-Dynamics* method. It is claimed that the model combines physically realistic cell mechanics with a reasonable computational load. The power of the approach is illustrated with two examples, avascular tumor growth and genesis of lymphoid tissue in a cell-flow equilibrium. © 2008, Elsevier Inc.

I. Overview of Methods in Theoretical Biology

Mathematical modeling of complex systems is a central focus of today's research. Complex systems can be found in any discipline and on any scale. However, the notion of *complex system* is employed in very different context. This chapter will focus on complex systems composed of interacting biological cells. Such systems are complex in the sense that (i) the number of constituents is large, (ii) the diversity of the constituents can be represented in a high-dimensional state space only, (iii) the number of constituents in each state may become very small, (iv) the constituents can be highly motile, (v) the shape of each agent is a dynamical quantity, (vi) the constituents undergo mechanical and chemical interactions with each other, and (vii) the constituents interact with soluble molecules and the extracellular matrix. A most relevant problem in today's Theoretical Biology is the definition of a methodological platform which is suitable for such a system. The small number of constituents per state counter-intuitively increases the complexity of the system. As will be argued below this property excludes a large number of modeling techniques. The second critical system property is the high motility of the constituents. In fact, a corresponding theoretical method does not exist yet, and it will be the goal of this chapter to present a novel method Delaunay-Object-Dynamics for biological complex systems that has the potential to provide such the methodological platform.

In the framework of Developmental Biology a number of suitable methods has been developed to describe and understand the morphogenesis of organs and organisms (Meinhardt, 1982). There is a long-standing tradition of modeling cellular systems with a set of (partial) differential equations (Murray, 1993). This method is appropriate if the system is characterized by large numbers of equal cells interacting with soluble chemicals or gradients of it. However, when turning to a system composed of either small numbers of cells or highly diverse cells, this approach may become inexact and in some cases might even fail to give qualitatively correct predictions (Bettelheim *et al.*, 2001). Small numbers of cells are found in the early stage of organogenesis or more generally in Developmental Biology, and also in the early phase of malignant tumor growth. Large diversity is found in many parts of the immune system. For example, the germinal center reaction, a site of antibody optimization (MacLennan, 1994), exhibits such a diversity. Cells are mutated and new cells occur in very small number in the beginning. These are then selected and multiplied. Such systems exhibit the necessity to be modeled by so-called agent-based (or individual-based) modeling techniques. It is the discreteness of the cell objects (Shnerb *et al.*, 2000), their individuality (Meyer-Hermann, 2002), and the small number of constituents per state that deserve a different representation.

Agent-based models differ from the differential equation approach in two aspects: First, the objects are discrete entities, second, every object can have dynamics of its own. This, in particular, implies that mechanical and chemical interactions mediated by pairs of molecules bound to the plasma membrane are possible with direct neighbors only (chemical interaction with solubles are treated separately). The cellular automaton (Lindenmayer and Rozenberg, 1976; Wolfram, 1986; Deutsch and Dormann, 2004) is a well-established agent-based model. It relies on a regular grid with every node representing a single cell. In a classical cellular automaton deterministic rules define the dynamics of the cells [e.g., Conway's game of life (Deutsch and Dormann, 2004)]. The rules include an effective representation of the mutual agent-interactions. Even though such automata seem to be rather rigid structures, the complexity of the emergent properties on the system level is appealing. A main limitation is the difficulty to link the dynamics to observable quantities. Thus, the parameters (rules) of the automaton may well represent our intuition but cannot be linked to quantities that can be measured in experiments.

A cellular automaton can be linked to experimentally accessible quantities by relating one step of a simulation update to a time scale. In a gauging procedure a well characterized process (for example, the division time of specific cells) is used to introduce the real world time into the system. Then all other quantities are defined relative to this time scale. The property vectors of each agent have either to be complemented by a kind of internal clock (Beyer *et al.*, 2002) or the processes are translated into reaction rates (Meyer-Hermann, 2002). In the latter case the deterministic cellular automaton is turned into a stochastic automaton. The reaction rates are measurable quantities. However, the interactions themselves are still very far from real chemical or mechanical observables. A school of "cellular automatists" claims that any physiological process can be translated into corresponding cellular automaton rules. This might be true. However, the impact of cell mechanics on the system has the potential to be derived from measurable quantities. While on the mechanical level the cellular automaton would only reproduce realistic system's behavior by introduction of additional rules, a theory of the cell mechanics would have predictive power in this respect.

There are several situations in which the rigid space of cellular automata and the effective rules are not appropriate for modeling. For example, cell division leads to the occupation of a new (previously empty) site. In reality a cell is first growing and then dividing while keeping the total volume constant during mitosis. Another difficulty appears for highly motile cells in dense tissue: The occupation of neighboring sites on the latticc leads to a considerable underestimation of the motility. Such situations might be solved using lattice-gas models (Deutsch and Dormann, 2004)—an extension of cellular automata to a lattice on which every node can carry more than a single ccll (in various velocity states). This mainly shifts the problem to another scale, and is effectively a representation of the system at a lower resolution.

The problem of poor representation of cell mechanics might be solved by including a subcellular level. The classical attempt in this direction is an extended Potts model modified for Biology (Graner and Glazier, 1992). This model mathematically goes back to the Ising model. A discrete variable called *spin* is used to define objects. The philosophy of the Potts model is to define each cell by a set of nodes with the same spin. A different spin is attributed to each cell. The interaction of different spins is defined by an interaction matrix in a Hamiltonian. The dynamics are the result of an energy minimization in a Monte-Carlo-like process (using a Boltzmann-weight for the probability of changing to a new configuration). The process involves spin-flips of neighboring nodes, thus a node is assumed to either keep its own spin or to adopt the spin of a neighbor node. Thus, a cell has to be represented by several nodes because a single node-cell would risk to be deleted from the system when it is displaced (thus flipped).

The Hamiltonian generally used represents a surface energy of cells in order to get a spherical shape for a *free* cell. The process of energy minimization then leads to the elimination of cells because the state of minimal energy would be a single cell in the whole system (which minimizes the surface energy). Thus, a volume conserving term is introduced into the Hamiltonian (Graner and Glazier, 1992). The system equilibrates between two conflicting goals, which might be associated with surface minimization and volume conservation. Both, the temperature at which the energy is minimized and the ratio of volume and interaction term are parameters of the model which cannot be easily related to physiological observables. This model allows for several additional interactions which might be incorporated into the Hamiltonian.

As in the Ising model Potts models run into a state of minimal energy, unless explicitly driven out of the equilibrium state. This nice property is very suitable for morphogenesis of organs or organisms. For example, the chemotaxis-based dynamics of *Dictyostelium discoideum* were successfully described with this method (Maree and Hogeweg, 2001). This property also delineates the limits of this method. The shape of cells is an inherent property of the energy minimization dynamics. It is not clear how these cell shapes relate to real cell behavior. In fact it is unclear whether energy minimization is a suitable starting point for the description of cell mechanics at all. Cells are actively producing energy all the time. They do not minimize energy but actively get out of energy minima. If one considers lymphocytes, for example, they are actively moving around in lymphoid tissue. In other words the Potts model does not explain how physically cell migration is actually achieved (Palsson, 2001).

Beside this conceptual question the Potts model does not inherently include a differentiation of the cell surface. Only two types of nodes build up a cell: Internal- and surface-nodes. The surface nodes interact with the neighbors according to the interaction matrix. One may add a polarity vector to the nodes with a specific spin, and the polarity might be related to external signals (Merks *et al.*, 2006). An attribution of local properties on the cell surface would go beyond the classical Potts model (unless every surface property would be associated with an extra spin-state). Thus, a differential expression of molecules on the cell surface as it occurs during cell polarizations has to be accounted for with additional techniques.

Alternative models on the subcellular level which do not use energy minimization and which allow for differentiated expression of molecules on the cell surface and even within the cell are hyphasma (Meyer-Hermann and Maini, 2005) and the subcellular element method (Newman, 2005). As in the Potts model one biological cell is represented by a set of nodes. While in the subcellular element method the volume elements are connected by springs, in hyphasma the cells are hold together by the surface

13. Delaunay-Object-Dynamics

tension associated with a network of actin filaments. Both methods allow for a differentiation of each node within each cell, however, this was not explicitly included yet. In contrast to the Potts model the interaction with the environment is defined by active forces that are exerted on the connective tissue or on interaction partners. Thereby the force balance is always respected, i.e., forces are exerted onto other cells or to the extracellular matrix. Energy minimization is not an issue in these models. On the contrary the cells remain far from energetic minima all the time. These cells deform during active displacement and in the case of hyphasma the cell shape is inherently connected to the cell motility. This allows for a realistic description of cell motility data as found in intravital imaging experiments (Meyer-Hermann and Maini, 2005). In principle, hyphasma and the subcellular element method both provide a structure that fulfills the aforementioned list of requirements. However, these methods are not yet in a stage of development that allows a detailed evaluation of their potential.

Hyphasma, the subcellular element method, as well as the Potts model are computationally rather expensive and allow only a limited number of biological cells (around 10,000) to be modeled. It is therefore necessary to think about methods that include cell shape properties, cell physics, and differentiated expression of surface properties while ignoring subcellular length scales.

Two very nice models reduce biological cells to visco-elastic objects in a lattice free space (Drasdo *et al.*, 1995; Palsson, 2001). The first is a two-dimensional Monte-Carlo simulation (Drasdo *et al.*, 1995) describing cell interactions with potentials. The visco-elastic cells have an incompressible core and a smoother rim which can adhesively or chemically interact with neighbors. This is combined with a statistical description of cell shapes. The model is focused on cell division and cell population growth.

In contrast the second approach is purely based on balanced Newtonian force equation in the over-damped approximation (Palsson, 2001). Cells are represented as ellipsoids. This model consistently respects physical properties of the cells (elasticity, compressibility), includes active motion, and adhesive or chemical interactions. Only a little anisotropy is introduced by the restriction that forces act only on the symmetry axes of the ellipsoids (Drasdo, 2003). While being a most realistic model with flexibility in applications there are two points of possible improvement: At first, the contact area is calculated as a virtual overlap of the ellipsoids. In rather dense tissue this overestimates the contact area between cells. Secondly, both implementations (Drasdo *et al.*, 1995; Palsson, 2001) need to localize the potential interaction partners at every time step, in particular, if the cells are highly motile. Using a grid for an estimate of the possible partners these codes scale with N (N is the number of cells). It is the factor in front of this scaling behavior, which depends on the size of the area scanned for neighbors, that limits the number of modeled cells.

The present chapter will describe the Delaunay-Object-Dynamics method, a model architecture which combines the advantages of agent- and force-based modeling with a lattice-free description of space and a linear scaling of the implementation. This is only possible on the basis of the aforementioned models. A weighted Delaunay-triangulation provides the neighborhood-topology of the system of interacting particles [as previously used for granular media (Ferrez, 2001)]. Every cell is attributed to a vertex. Vertices move according to balanced force equations. They can be created and deleted. This gives rise to the implementation of a dynamic and kinetic Delaunay-triangulation in three dimensions. As the number of cells is supposed to be large, a parallel update algorithm is needed for the Delaunay-triangulation (Beyer *et al.*, 2005).

The Delaunay-based interaction of visco-elastic spheres in wide parts corresponds to the model of Palsson (2001). The mentioned slight anisotropy was deleted at the price of replacing ellipsoids by spheres, which restricts the possible shape of cells. However, optionally, the dual Voronoi-tessellation can be used to describe the cell volume and the interaction faces to neighboring cells in dense tissue. It turns out that this is an improvement of cell shape representation in dense tissue. As an example application this modeling architecture will be used to describe tumor growth and lymphoid follicle morphogenesis with special emphasis on the relevance of a comprehensive treatment of mechanical properties of the cells.

II. Delaunay-Based Interaction

If representing visco-elastic objects like biological cells in a lattice-free space the interaction partners are not *a priori* defined. When updating a dynamic evolution of such a system, every object has to check with every other object of the system, what the potential interaction would be. Such a system scales badly with N^2 when N is the number of objects. If the number of objects that are treated with such a method becomes large (in the order of 100,000 and more), this turns to be an important problem of the implementation. It is desirable to find a solution that scales linearly with N: In the following the first kinetic and dynamic Delaunay-triangulation in three dimensions suitable for large sets of biological cells is presented.

The Delaunay-triangulation provides a neighborhood-topology of a set of vertices (object positions). If the vertices are in general position (Muecke, 1998) a unique triangulation is defined by the empty-circumsphere-criterion.

A. Empty-Circumsphere-Criterion

No vertex of the triangulation lies within the circumsphere of the simplices in the triangulation.

A simplex is defined as the convex hull of four vertices. In three dimensions the corresponding geometrical object is a tetrahedron. The circumsphere of a simplex is the sphere with a surface touching all vertices v_i (with $i = 1, \ldots, 4$) that build the simplex, thus, fulfilling

$$(\mathbf{m} - \mathbf{v}_i)^2 = R_m^2, \tag{1}$$

where **m** and R_m are the position and the radius of the circumsphere, respectively.

This natural definition of neighborhood is suitable for objects that are of the same size and was successfully applied to granular media (Ferrez, 2001). In the case of biological cells, the problem is more complicated. Cells are inherently of different size and, also, may change their size dynamically. In particular, cell growth before mitosis is a relevant process in any simulation of organo- or tumorogenesis. Thus, the objects attributed to the vertices shall be of different size. An extension of the Delaunay-triangulation to the weighted Delaunay-triangulation accounts for this.

Every vertex $X = (\mathbf{x}, \omega_X)$ is defined as a set of a position $\mathbf{x}$ and a weight ω_X. The weight alters the distance measure. The orthogonal distance (also called power distance) is defined as

$$\pi(X, Y) = (\mathbf{x} - \mathbf{y})^2 - \omega_X - \omega_Y. \tag{2}$$

Every vertex is complemented by a sphere with a radius corresponding to the square-root of the vertex weight. Two vertices are orthogonal if their orthogonal distance vanishes.

In analogy to the circumsphere an orthosphere can be considered as a quasi-vertex $(\mathbf{m}_\perp, R_{m_\perp}^2)$ with center $\mathbf{m}_\perp$ and weight $R_{m_\perp}^2$. Then the orthosphere is defined as the quasi-vertex being orthogonal to all weighted vertices $(v_i, R_i^2$ (with $i = 1, \ldots, 4$) building a simplex. Thus, the orthosphere $(\mathbf{m}_\perp, R_{m_\perp}^2)$ fulfills

$$(\mathbf{m}_\perp - v_i)^2 = R_{m_\perp}^2 + R_i^2. \tag{3}$$

If all vertices carry the same weight this reduces to the equation for the circumsphere. Now the empty-orthosphere-criterion can be formulated in full analogy to the simple Delaunay-triangulation (see Fig. 1).

B. Empty-Orthosphere-Criterion

The weighted Delaunay-triangulation is a triangulation where no vertex lies inside the orthospheres attributed to the simplices of the triangulation.

In two dimensions a vertex lies *inside* the orthosphere when the tangential line of the sphere (attributed to the vertex) which passes through the center of the orthosphere touches the sphere at a point nearer to the orthosphere center than its radius (see Fig. 1). In three dimensions this generalizes to a cone of tangential lines. The touching points with the vertex-sphere build a circle and either lie all inside or outside the orthosphere (see, for example, Ferrez, 2001). An effective way to calculate the empty-orthosphere-criterion is provided by the projection of vertices on a paraboloid in four dimensions (Muecke, 1998; Ferrez, 2001; Aurenhammer and Imai, 1988). The question whether a vertex lies inside an orthosphere is reduced to the orientation of simplices, thus, to the calculation of a determinant.

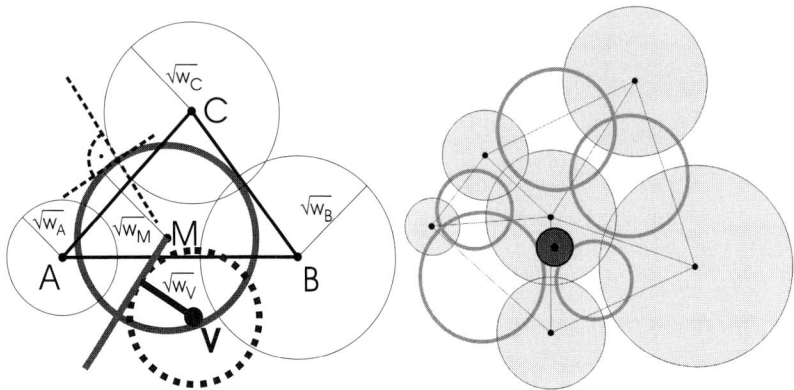

Figure 1 Weighted Delaunay-triangulation and orthosphere: (Left panel) A two-dimensional representation of an orthosphere attributed to the simplex for the vertices $A = (\mathbf{a}, \omega_A)$, $B = (\mathbf{b}, \omega_B)$, and $C = (\mathbf{c}, \omega_C)$. The orthosphere M (thick full line) with radius $\sqrt{\omega_M}$ intersects perpendicularly (thin dashed lines) with the spheres of all vertices A, B, C (thin full lines). The empty-orthosphere-criterion is violated by the vertex $V = (\mathbf{v}, \omega_V)$. It is not the position $\mathbf{v}$ which is essential. The criterion is violated when the tangent through the center of the orthosphere (thick full line) touches the sphere attributed to V (thick dotted line) at a point nearer to the orthosphere center than its radius. (Right panel) An example of a two-dimensional triangulation (thin lines) of a set of vertices (dots) with attributed spheres (gray disks). The orthospheres (thick circles) of the simplices (triangles) are shown. Note that one vertex (dark disk) is not connected to any other vertex because it is covered by the disk of a different vertex. This vertex is called redundant. Adapted from Beyer *et al.* (2005).

III. Voronoi-Cells Approximate Real Cells

The dual of the Delaunay-triangulation is the Voronoi-tessellation. It provides a subdivision of space into disjoint subspaces. Every Voronoi-cell is a polygonal object which can be used as an approximation for a real cell shape. In dense cellular tissue the Voronoi-cells are, indeed, very similar to real cells. However, in loose tissue or considering cells in solution the Voronoi-tessellation induces Voronoi-cells with a far too large volume. In the limit of low density tissue a different approach is necessary.

The vertices that are used to define a weighted Delaunay-triangulation can be considered as the generators for this spatial tessellation. The tessellation is defined to collect all points that are nearer to the generator $A = (\mathbf{a}, \omega_A)$ than to any other vertex:

$$V_A = \left\{ \mathbf{x} \in \mathbb{R}^3, \ X = (\mathbf{x}, \omega_X = 0) \colon \tilde{\pi}(X, A) < \tilde{\pi}(X, B) \ \forall B \neq A \right\}. \tag{4}$$

In a weighted Delaunay-triangulation, thus using the weighted distance measure Eq. (2), this property defines a weighted Voronoi-tessellation. It contains the unweighted tessellation as the special case of equal weights for all vertices (generators). The cells of these tessellations are disjoint and together cover the three-dimensional space. Such a Voronoi-tessellation is shown in Fig. 2.

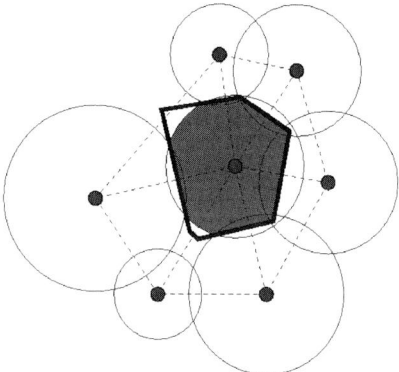

Figure 2 Voronoi-cells in a weighted Delaunay-triangulation: The points represent vertices of a two-dimensional triangulation (dashed line). The vertex weights are the squared radii of the associated disks (spheres in three dimensions). The polygonal structure is the Voronoi-cell of the central vertex. For equal weights the Voronoi-cell has faces at the center of the connections of two vertices. Note that the faces of the Voronoi-cell are sometimes larger and sometimes smaller than the corresponding virtual overlap of the disks. Previously published in Schaller and Meyer-Hermann (2005).

The definition Eq. (4) also defines the common face of two Voronoi-cells. The intersection of three such contact faces defines a corner point of the contact area. It can be proven that these points are related to the centers of the orthospheres of the triangulation.

A. Voronoi-Orthosphere-Relation

If the generators of a Voronoi-tessellation are the vertices of a weighted Delaunay-triangulation, the corners of the Voronoi-cells are the centers of the orthospheres associated with the simplices of the triangulation.

This identification simplifies the calculation of contact areas and cell volumes.

In order to also cover the case of low density cells the Voronoi-tessellation is supplemented by a second concept. The spheres attributed to the vertices serve as approximation for the cells in the low density limit. The forces acting by and on the vertices (they will be defined later) are related to the contact area between two cells connected in the triangulation. The following definition of the contact area is a suitable approximation for cell–cell-contact in the limits of low and high density tissue.

B. Definition of Contact Area

The contact area between two connected vertices in the weighted Delaunay-triangulation is the minimum of the Voronoi-face and the area of virtual overlap between the spheres associated to the vertices as weights.

This is also illustrated in Fig. 2. The position of the sphere overlap (if any) and the Voronoi-face always coincides, such that the contact area is well defined.

Note that a weighted Delaunay-triangulation allows for a natural definition of cell growth: Cell growth (or shrinkage) is represented as dynamic weight. For example, a cell in cell cycle grows during the G-phases. If this is thought to be a deterministic process of constant protein synthesis extending over the time τ, during which the cell volume doubles, the growth dynamics are described by a differential equation for the radius

$$\frac{dR}{dt} = \frac{1}{6\tau} \frac{R_m^3}{R^2}, \tag{5}$$

where R_m is the target radius, and the squared radius is associated with the vertex weight. The weighted Delaunay-concepts allows for the implementation of more sophisticated models of cell growth when the weights are subject to dynamics.

IV. Delaunay-Dynamics

Cellular systems are dynamic systems. The growth process mentioned in the previous section is part of these dynamics. A changed weight might easily invalidate a weighted Delaunay-triangulation in the sense that the empty-orthosphere-criterion is violated: The changed weight changes the basis of the calculation of the triangulation. There are other cellular processes that might have the same effect. One obvious example is cell division. Here a new vertex is inserted into the triangulation. If this happens within the convex hull of the triangulation the novel vertex will generally lie within one of the orthospheres. Also the counter process, cell death, will change the triangulation. Finally, cells are also motile. They exhibit passive and active movements depending on the type of cells considered. A moving vertex will frequently enter a different orthosphere.

The conclusion of the existence of all these dynamical processes is clear. The triangulation cannot be calculated once and remains then valid for the rest of the simulation. Either the triangulation has to be recalculated after each time step, or it has to be locally adapted. Naively retriangulation scales with N^2 (N is the number of vertices). A more efficient way of recalculating the triangulation is by the incremental insertion algorithm, which was proven to work for weighted Delaunay-triangulations in three dimensions (Edelsbrunner and Shah, 1996). This algorithm starts from one artificial simplex that is sufficiently large to cover the convex hull of all vertices of the triangulation. Then all vertices are inserted to the triangulation. The update of the triangulation might follow different algorithms: The Bowyer–Watson algorithm (Bowyer, 1981; Watson, 1981) localizes simplices with violated empty-circumsphere-criterion, deletes these simplices and retriangulates the induced cavity in the triangulation. The Green–Sibson algorithm (Boissonnat and Teillaud, 1993) makes a one to four flip, which is a topological transformation and recursively performs other topological transformations until the empty-orthosphere-criterion of all simplices is fulfilled.

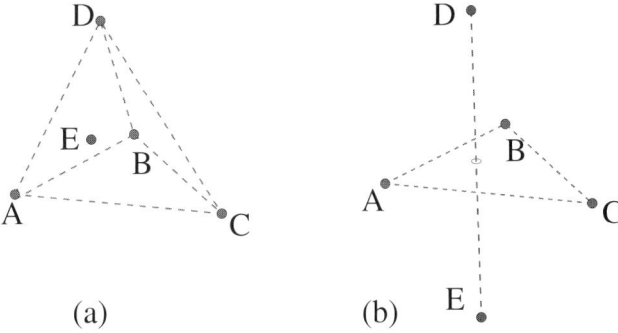

Figure 3 Radon's theorem: A set of five points in three dimensions can be divided into two sets of points such that their convex hulls intersect with each other.

In both cases the localization of the first invalidated simplex is necessary which is efficiently solved by the simplex visibility walk (Devillers *et al.*, 2001). It scales with the distance of an initial simplex to the target simplex. This adds to the otherwise linear scaling of the algorithm, thus, inferring a $N^{4/3}$-scaling. Retriangulation after every time step fails to exhibit linear scaling (Cignoni *et al.*, 1993) as required for the large number of vertices aimed to be treated with this method.

For a linear scaling of a Delaunay-based modeling architecture a local update of the three-dimensional weighted Delaunay-triangulation is needed. This can be achieved in most situations by flipping algorithms. The possibility of flip-algorithms goes back to Radon's theorem (see Fig. 3 and Goodman and O'Rourke, 1997).

A. Radon's Theorem

Consider a set X of five points in three-dimensional space. Then a partition $X = X_1 \cup X_2$ with $X_1 \cap X_2 = \emptyset$ exists such that $\mathrm{conv}(X_1) \cap \mathrm{conv}(X_2) \neq \emptyset$. If the points are in general position this partition is unique.

Both configurations in Fig. 3 correspond to different realizations of triangulations. The vertex E, which lies inside the convex hull defined by A, B, C, D in configuration (a), is either inserted or deleted giving rise to a flip between 1 or 4 simplices. Correspondingly, the points in the configuration (b) of Fig. 3 can be triangulated to 2 or 3 simplices, depending on the proximity of the points D and E. For moving points D and E the transition between the configurations with 2 or 3 simplices is a flip algorithm which can locally restore the empty-orthosphere-criterion (see Fig. 4).

In almost all cases, flips combined with an adaptive step-size control will prevent the necessity of a total retriangulation. Exceptions are when the five points are in a nonconvex configuration or when the vertex not only enters an orthosphere but also changes the orientation of incident simplices. Corresponding algorithm exist (Edelsbrunner and Shah, 1996; Schaller and Meyer-Hermann, 2004) or were recently

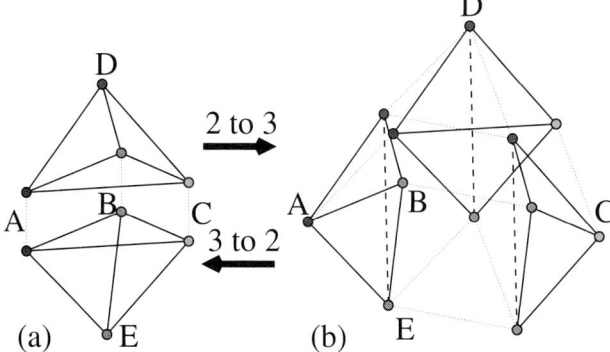

Figure 4 Flips restore Delaunay: A transition from a configuration of five vertices with two or three simplices can locally restore the Delaunay-triangulation. This might be used in an incremental insertion algorithm or after a vertex moving into the orthosphere of another simplex. Previously published in (Schaller and Meyer-Hermann, 2004).

extended for parallel kinetic and dynamic weighted Delaunay-triangulations (Beyer *et al.*, 2005; Beyer and Meyer-Hermann, 2006). They scale linearly in N as required for large systems.

Note that such flips change not only the neighborhood-topology but also the number of simplices associated with the same set of vertices. This property allows for a differentiated representation of cell–cell-contact in the simulation of biological cells. A typical representation of the distribution of the number of neighbors of a cell during the simulation of tumor growth (Schaller and Meyer-Hermann, 2005) is shown in Fig. 5. If compared to lattice-based models the variability of occurring neighborhood topologies is relatively large. The maximum number of neighbors in a rigid squared lattice in the nearest-neighbor approximation is 6. It is 26 if the next-to-nearest-neighbors are included. For an fcc-grid (dense packing of spheres) the number of direct neighbors is 12, which is quite near of the average value of 14.2 (see Fig. 5). 40 cells in contact to a single cell, as sometimes found in the present simulation of relatively densely packed cells of various sizes, can never be observed in a model architecture based on rigid grids with a one-node–one-cell philosophy.

V. Equation of Motion for Vertices

Every cell is subject to active and passive forces. Passive forces include adhesion between cells or between cells and the extracellular matrix. In addition, cells have elastic properties inducing a repelling force if another cell approaches. The induced motion is subject to friction with the extracellular matrix and other cells. Active forces are, in particular, important for highly motile cells of the immune system. These forces are exerted onto the extracellular matrix.

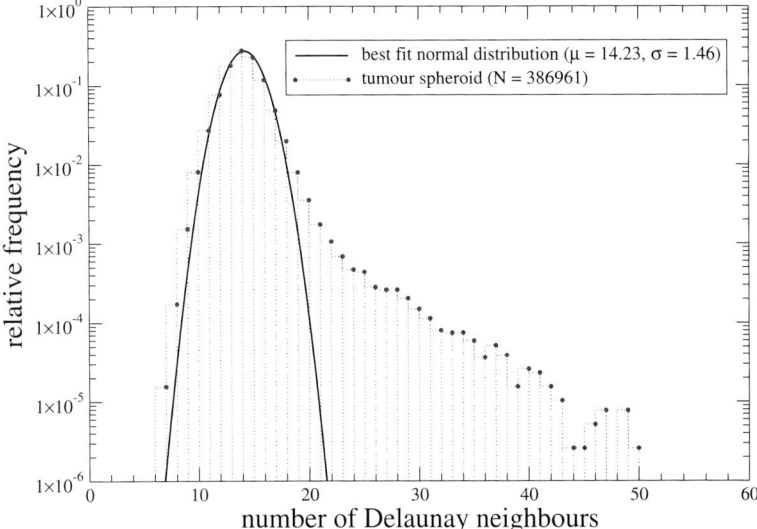

Figure 5 Complexity of the neighborhood topology: The frequency of occurrence of different numbers of neighbors during the simulation of tumor growth is shown. The tumor consists of roughly 400,000 cells (vertices). The diversity of different numbers of neighbors is relatively large and diverse if compared to lattice-based methods. The maximum of the distribution is at 14.2 neighbors. The full line shows a normal distribution around this peak. Note that large numbers of neighbors occur more frequently than expected by the normal distribution. Courtesy of Gernot Schaller (Schaller, 2005).

All these forces are assumed to act on the vertices of the weighted Delaunay-triangulation. The corresponding equation of motion reads in a general form

$$m\ddot{\mathbf{x}}_i = \mathbf{F}_{i,M}^{\text{drag}}(\dot{\mathbf{x}}_i) + \sum_{j \in \mathcal{N}_i} \mathbf{F}_{ij}^{\text{drag}}(\dot{\mathbf{x}}_j - \dot{\mathbf{x}}_i)$$
$$+ \mathbf{F}_i^{\text{active}}(\phi_i) + \sum_{j \in \mathcal{N}_i^{\text{Delaunay}}} \mathbf{F}_{ij}^{\text{active}}(\phi_i) - \sum_{j \in \mathcal{N}_i} \mathbf{F}_{ji}^{\text{active}}(\phi_j)$$
$$+ \sum_{j \in \mathcal{N}_i} \mathbf{F}_{ij}^{\text{passive}}(\mathbf{x}_j, \mathbf{x}_j) + \sum_{j \in \mathcal{N}_i} \mathbf{F}_{ij}^{\text{press}}(V_i, V_j)$$
$$\approx 0. \tag{6}$$

$\mathbf{x}_i$ denotes the position of the vertex i and the dots symbolize time derivatives. ϕ_i denotes internal properties of cell i which are themselves subject to dynamics. Two types of neighborhood are introduced: $\mathcal{N}_i^{\text{Delaunay}}$ denotes all vertices connected to the vertex i in the Delaunay-triangulation. $\mathcal{N}_i$ only denotes the subset of these that correspond to a real cell–cell-contact involving a contact area between the cells. The terms **F** and their interpretation are explained in the following subsections in the sequence of appearance on the right-hand side of Eq. (6). Generally speaking there are three classes of forces acting on cells: Active forces are associated with active motility of cells, passive and pressure forces are related to mechanic properties like elasticity and surface energy, and drag forces are related to the extracellular environment (including other cells) and its viscosity or adhesion properties.

The Newtonian equations of motion are solved in the overdamped approximation which is reliable for cells in tissue (Palsson, 2001; Schaller and Meyer-Hermann, 2005). This approximation assumes that there is no memory of the momentum in the system. Thus, the cell velocities are always a direct result of active forces and the forces exerted by the environment. This realistic assumption reduces the second order differential equations to a set of first order equations.

A torque acting on cells might be induced by friction forces between different cells. In a consistent overdamped approximation this gives rise to an additional equation for the angular momentum of the cell (Schaller and Meyer-Hermann, 2007). These alter the drag forces. It is assumed here that cell torque is a minor effect during cell motility. For actively moving lymphocytes the polarity of the cells is conserved during movement across a dense tissue of lymphocytes (Gunzer et al., 2004). For these cells the neglect of torque is justified. However, this might be reconsidered for cells that dominantly undergo passive forces without a clearly defined polarity.

A. Drag Forces from the Environment

When a cell moves in an environment of fluid and through an extracellular matrix, active forces are counteracted by a general drag force. A viscosity can obviously be attributed to the extracellular medium. The bending and deformation of the extracellular matrix may also be associated with a viscosity. Because of low Reynolds numbers for cellular flow (Howard, 2001; Palsson, 2001; Schaller and Meyer-Hermann, 2005), one can safely assume a linear dependence of the drag force on the cell velocity. In the coordinate system of the extracellular matrix the drag forces mediated by the medium and the extracellular matrix are directed against the direction of motion and are given by

$$\mathbf{F}_{i,M}^{\text{drag}}(\dot{\mathbf{x}}_i) = -\gamma_M \dot{\mathbf{x}}_i = -\eta_M r_i \left(1 - \frac{\sum_{j \in \mathcal{N}_i} a_{ij}}{A_i}\right) \dot{\mathbf{x}}_i \tag{7}$$

with γ_M being the corresponding friction coefficient. η_M is the viscosity of the medium, and r_i is the radius of the cell. The bracket calculates the area of the cell being in contact to the medium, thus, not being in contact to other cells, where A_i is the total cell surface and the sum adds up all contact faces a_{ij} to other cells j. The usage of a specific viscosity coefficient takes advantage of the clearly defined surface areas of the cells in the Delaunay–Voronoi-concept.

The medium drag force always acts against the direction of motion. Relative to the orientation of the surface in contact to the medium this implies that the drag force includes a normal and a tangential component. This is different for the drag forces stemming from other cells which are assumed to always act tangential to the contact area. In the case of cell–cell-contacts, the effective viscosity of the *medium cell* is determined by the properties of the cytosol and the cytoskeleton. These also mediate the elastic forces which, however, act normal to the contact area. A drag term acting normal to the contact area would correspond to a flow of the cytosol which can be safely neglected (Palsson, 2001; Dallon and Othmer, 2004; Schaller and Meyer-Hermann, 2005). Thus, the model explicitly distinguishes between passive forces mediated by the environment and elasticity. They correspond to the tangential and normal components, respectively. The drag force mediated by cell–cell-contact is then given by

$$\mathbf{F}_{ij}^{\text{drag}}(\dot{\mathbf{x}}_j - \dot{\mathbf{x}}_i) = (\eta_i r_j + \eta_j r_i)\frac{a_{ij}}{A_i}\big[(\dot{\mathbf{x}}_j - \dot{\mathbf{x}}_i) - \{\mathbf{e}_{ij} \cdot (\dot{\mathbf{x}}_j - \dot{\mathbf{x}}_i)\}\mathbf{e}_{ij}\big], \tag{8}$$

where $\mathbf{e}_{ij}$ denotes the unit vector normal to the contact area a_{ij} between the cells i and j. $\eta_{i,j}$ are the viscosities mediated by the cells.

B. Active Motility

Biological cells have the potential to actively displace themselves. In particular in the framework of the immune system active motility of cells is a most important process in order to efficiently fight against infections. The agents of the immune system have to reach the site of infection. They also transport infectious substances (antigens) to sites of antigen presentation (lymphoid tissue) and present the antigen to a diversity of cells constantly passing lymphoid tissue. As all these processes involve the exertion of forces either onto the connective tissue or onto other cells, such active forces have to be accounted for in the modeling framework.

Cell motility is characterized by the polarity of the cell. The polarity is described with a unit vector **p**. It is itself a dynamic quantity. A set of motility inducing soluble substances determines the direction of polarity (chemotaxis) and the speed of the displacement (chemokinesis). If the chemoattractants are membrane bound on other agents of the connective tissue the induced motility is associated with the term haptotaxis.

A basic requirement for any kind of motility induced by the presence of molecules is the expression of the corresponding receptors on the surface of the moving cell. The expression of the receptors is also highly dynamic and is determined by internal states of the cell as well as by external stimuli (Moser *et al.*, 2004; Cyster, 2005). For example, naive B cells in primary follicles are kept in the follicles by the interaction of the chemokine CXCL13 with the expressed receptor CXCR5. Upon antigenic stimulation, they upregulate the receptor CCR7, which is responsive to the soluble chemokine CCL21 present in the surrounding of the follicle. This induces a movement towards the border of the follicle where the activated B cells may get help by T cells, which are mainly absent in the follicle. Thus, motility responses are on one hand a result of a complex calculation of responses to various chemokines, and on the other hand relevant to home the cells to the right physiological compartment.

How are such active forces for cell displacement transmitted? One possibility is the exertion of forces onto the extracellular matrix. The expression of adhesion molecules allow the cells to anchor to the connective tissue and then to either walk along or across fibers. It is assumed in the following that the direction of movement is determined by the intrinsic polarity **p** of the cell and not by the direction of the fibers as sometimes suggested (Bajenoff et al., 2006). Then the force is simply given by

$$\mathbf{F}_i^{\text{active}}(\phi_i) = f_i(\phi_i)\mathbf{p}_i. \tag{9}$$

Note that both, the force f_i (which is directly related to the target speed of the cell) and the polarity are a result of internal properties and the environment. The concrete assumptions that determine forces and polarity are very different and depend on the considered cell type and environment.

For example, the polarity of lymphocytes is not constantly adapted to the gradients of the chemoattractants. In contrast, a persistence time was observed (Gunzer et al., 2004) during which the direction of the motility is conserved. It is suggested that the direction of movement is adapted to the gradients only between these time intervals which are in the order of 2 minutes. With this assumption, indeed, speed and displacement description could be easily explained (Meyer-Hermann and Maini, 2005).

1. Nonlocal Interactions

The derivation of motility parameters from chemokine distributions call for a hybrid architecture of the model. The chemokines are secreted by cells, thus the Delaunay-triangulation contains the sources of chemokines. As the cells bind and internalize the chemokines, the vertices of the triangulation are also sinks for the chemokines. The chemokines diffuse such that the whole interaction has to be modeled as a reaction–diffusion system. This is realized on an independent regular grid with appropriate resolution.

2. Active Forces on and by Neighbors: The Constriction Ring

The active force as defined by Eq. (9) tell us about the capability of a cell to exert forces onto the extracellular matrix in order to displace itself. However, nothing is said about the mechanism behind this. In contrast to other cell types it was shown for T lymphocytes that they move independent of a cleavage of fibers. So the extracellular matrix remains intact during T cell movements. At the front the T cell attaches a *constriction ring* to the extracellular matrix and then squeezes through this ring by retracting forces on the back side of the cell. The forces are a result of the reorganization of parts of the cytoskeleton. This concept is rather old (Lewis, 1931). The ring is clearly visible in recent experiments (Wolf et al., 2003).

The concept of a constriction ring is based on hydrostatic pressure in the cell. The retraction of the back of the cell induces a pressure in the cell. It is assumed that the pressure is equilibrated in the cell. The level of pressure P that can be induced by the reorganization of the cytoskeleton is a parameter of the model and can be adjusted to motility data. The pressure is transformed into a force which is not only directed perpendicular to the plane defined by the constriction ring but acts on all neighboring cells which are in contact. The two-cell interaction induced by this mode of motility is (Beyer and Meyer-Hermann, 2007a)

$$\mathbf{F}_{ij}^{\text{active}}(\phi_i) = a_{ij} P_i \, \text{sign}\left((\mathbf{x}_{ij} - \mathbf{x}_i^{\text{ring}}) \cdot \mathbf{p}_i\right) \frac{\mathbf{x}_{ij} - \mathbf{x}_i^{\text{ring}}}{\|\mathbf{x}_{ij} - \mathbf{x}_i^{\text{ring}}\|}, \tag{10}$$

where $\mathbf{x}_i^{\text{ring}}$ is the center of the constriction ring of the moving cell i, and $\mathbf{x}_{ij}$ is the center of the contact area a_{ij} between cells i and j. If the cell i does not stay in physical contact to its neighbor j, a virtual surface area a_{ij} of the cell i in contact to the extracellular matrix is inserted instead. This surface area is defined as the minimum of the Voronoi-face and the virtual overlap of cell i with its mirror image at the equilibrium distance of the Johnson–Kendall–Roberts force (see Section C). The direction of the force is determined by the sign of the scalar product with the polarity of the cell. Thus, on the back the force is pulling on neighbors while at the front it is pushing.

Corresponding hydrostatic pressure-derived forces of other cells will also act on the cell under consideration. Thus, the counter forces are defined by

$$\mathbf{F}_{ji}^{\text{active}}(\phi_j) = a_{ij} P_j \, \text{sign}\left((\mathbf{x}_{ij} - \mathbf{x}_j^{\text{ring}}) \cdot \mathbf{p}_j\right) \frac{\mathbf{x}_{ij} - \mathbf{x}_j^{\text{ring}}}{\|\mathbf{x}_{ij} - \mathbf{x}_j^{\text{ring}}\|}. \tag{11}$$

The exertion of these forces depends on the presence of neighbors. If the environment is asymmetrically populated with cells a drift of the moving cell with respect to the direction of its polarity is induced.

Note that the cell is also moving without any cellular neighbor which is in real contact to the moving cell. The contact areas with other cells are then replaced by virtual contact areas with the extracellular matrix. If the extracellular matrix would be modeled explicitly, it would take up the forces transmitted by the constriction ring and those defined in Eq. (9).

C. Passive Two-Cell Mechanics

Two cells when they are in physical contact mechanically interact with each other. The interaction is exerted by the cytoskeleton. For example, the actin-filaments build a network of fibers with elastic properties on the intracellular side of the plasma-membrane. Adhesion molecules (selectins, integrins, cadherins) mediate cell–cell binding. This not only allows to transmit the elastic forces of the cytoskeleton onto the interaction partner. Adhesion molecules also induce attracting forces between both cells.

An approximation of the elastic forces and adhesion is provided by the Johnson–Kendall–Roberts (JKR) model (Johnson et al., 1971) which was shown to be a good approximation for rubber (Brilliantov and Poeschel, 2004) but also for cells (Chu et al., 2005). The model is a generalization of the Hertz model (Hertz, 1882) which only describes the elastic properties of two spheres. The adhesive force is added as a surface energy of the cell. The surface energy can be assumed to depend on the expression level of the corresponding adhesion molecules on both cells:

$$\sigma_{ij} = \frac{\sigma_0}{2}(a_i l_j + l_i a_j), \tag{12}$$

where σ_0 denotes the surface energy per surface concentration of receptor–ligand pairs. a and l are the surface concentrations of the adhesion molecules and the ligands, respectively.

With ρ_{ij} the radius of the contact area of two elastic spheres, the effective radius R_{ij} defined by

$$\frac{1}{R_{ij}} = \frac{1}{r_i} + \frac{1}{r_j}, \tag{13}$$

and the elastic coefficient

$$K_{ij} = \frac{4}{3}\left[\frac{1-v_i^2}{E_i} + \frac{1-v_j^2}{E_j}\right]^{-1} \tag{14}$$

defined by the elastic modulus $E_{i,j}$ and the Poisson ratios $v_{i,j}$, the JKR-force is then given by

$$\mathbf{F}_{ij}^{\text{JKR}}(\mathbf{x}_j, \mathbf{x}_i) = \left(\frac{K_{ij}\rho_{ij}^3}{R_{ij}} - \sqrt{6\pi\sigma_{ij}K_{ij}\rho_{ij}^3}\right)\mathbf{e}_{ij} \tag{15}$$

where an homogeneous surface energy on the contact surface is assumed.

Note that this formulation based on contact areas is well defined for pairs of cells, however, multicell interactions might induce artifacts. In a dense configuration the contact area is given by the Voronoi-cell. Normally adhesion increases the cell surface. If multiple cells push on one cell the surface might shrink and the Voronoi-cell might turn too small. The adhesion term further reduces the Voronoi-faces and the vertex risks to become redundant (the vertex is fully covered by the spheres attributed to the neighbor vertices, see Fig. 1). Redundant vertices are not connected to the triangulation anymore. A volume conserving term might rescue the cell from becoming redundant (see Section D).

In some cases it might be favored to clearly separate two-cell from multiple-cell interactions. To this end, the passive force has to be reformulated in terms independent of the contact area: The radius of the contact area is approximated by

$$\rho_{ij} \approx \sqrt{h_{ij}R_{ij}} \quad \text{with } h_{ij} \stackrel{\text{def}}{=} r_i + r_j - \|\mathbf{x}_j - \mathbf{x}_i\| \tag{16}$$

denoting the virtual overlap of the undeformed spheres. This approximation corresponds to the indentation as induced by the Hertz model and corresponds to the limit of weak adhesion compared to the elastic forces. The passive force is then defined as the JKR-force in this approximation

$$\mathbf{F}_{ij}^{\text{passive}}(\mathbf{x}_j, \mathbf{x}_j) = \left(K_{ij} R_{ij}^2 \left(\frac{h_{ij}}{R_{ij}} \right)^{\frac{3}{2}} - \sqrt{6\pi \sigma_{ij} K_{ij} R_{ij}^3} \left(\frac{h_{ij}}{R_{ij}} \right)^{\frac{3}{4}} \right) \mathbf{e}_{ij} \qquad (17)$$

and relies solely on the positions and the weights of the two vertices (not on the contact surface), thus, defining a pure two-cell interaction (also interaction with the extracellular matrix of connective tissue is not included).

For vanishing surface energy the purely elastic Hertz model is recovered which corresponds to the first term. The second adhesive term implies a minimum in the corresponding interaction potential, thus, inducing an equilibrium distance between two cells. At equilibrium, adhesive and elastic forces cancel each other. The force acts only perpendicular to the contact face. The tangential part associated with adhesive friction is included in the friction model, Eq. (8).

An active reorganization of the cell which reinforces initial bindings by adhesion molecules and adapts the cytoskeleton to stress configurations, as well as the dynamics in the wrapped membrane which can have a larger surface than the one corresponding to the apparent contact surface (Evans, 1995), are not included in the JKR-force. Such effects can only be included in a subcellular modeling. However, on the whole cell scale, the present description allows for the measurement of the physical properties of the cells (which will differ for different cell types). Thus, the model gains predictive power because the parameters of the model are not free parameters but observable quantities. These quantities have to be measured for every system under consideration.

D. Compression

Considering pairs of cells the cell volume is well approximated by the minimum of the Voronoi-cell and the sphere attributed to the vertex. Only small deviations from the actual target volume of the cell are to be expected. However, if several cells interact with a cell under consideration, it is well possible that the volume of the Voronoi-cell becomes smaller than the target volume. The cell would stay in a highly compressed state for rather long times because the two-cell elastic forces do not account for a corresponding repulsive force.

In order to effectively include the impact of multicell interactions, a corresponding pressure of the cells is calculated

$$P_i = \frac{E_i}{3(1 - 2\nu_i)} \left(1 - \frac{V_i}{V_i^{\text{target}}} \right), \qquad (18)$$

where the factor describes the compressibility in terms of the elastic modulus and the Poisson-ratio. The current volume V_i is the minimum of Voronoi- and sphere-volume. The target volume might itself be a dynamical quantity if cell growth is considered.

On the basis of the contact surface, the pressure is then transformed into an additional force between pairs of cells

$$\mathbf{F}_{ij}^{\text{press}}(V_i, V_j) = a_{ij}(P_i - P_j)\mathbf{e}_{ij}, \tag{19}$$

which acts perpendicular to the contact area.

VI. Mechanics Matters

The Delaunay-Object-Dynamics model developed and outlined in the previous sections strongly emphasizes the mechanics of cells and cellular interactions. How important are mechanical properties of biological cells? In the following some examples for the relevance of cell mechanics are given. This shall support the modeling architecture which is considered to be well defined in terms of physical interactions. However, it is not excluded that the characteristics demonstrated in the following might also be achieved with a different modeling philosophy. In particular, modeling at a subcellular resolution generally allows to get similar results. Thus, the strong point of the presented method is to combine advanced cell mechanics with a one-node–one-cell philosophy.

A. Avascular Tumor Growth

Avascular tumor growth *in vitro* is characterized by an exponential growth in the beginning which soon turns into a power law growth. The dependency on the nutrient concentrations was investigated and quantified (Freyer and Sutherland, 1986). The role of contact inhibition in the turn from exponential to power law growth is not clear from the data. In principle it could by hypothesized that it is a pure result of lack of nutrients. However, as it is known that *in vivo* tumors are generally under pressure the cell mechanics might well be of general relevance for tumor growth.

Cells in the Delaunay-Object-Dynamics model grow in dependence on the local nutrient concentration. The nutrients follow a reaction–diffusion equation with Dirichlet boundary conditions. The diffusion constants in the tumor are corrected for the tissue constraints and differ from the diffusion constants in the solution. The data (Freyer and Sutherland, 1986) can only be reproduced with the assumption that the viability of the cells depends on the product of two considered nutrients (oxygen and glucose) (Schaller and Meyer-Hermann, 2005). Subcritical products lead to necrosis.

It is assumed that the cells enter the G_0 phase in dependence on the mechanical tension they experience. This explicitly introduces contact inhibition (Galle *et al.*, 2005)

Figure 6 Avascular tumor growth: A central slice of a growing tumor in a three-dimensional simulation is shown at day 23. From left to right the tumors are under different nutrient (oxygen, glucose) conditions (in mMol): (0.07, 0.8), (0.07, 16.5), (0.28, 0.8), (0.28, 16.5), respectively. These values correspond to those used in Freyer and Sutherland (1986). The largest tumor has a diameter of 750 μm. The upper row shows the state of the cells: Necrotic cells (black, gray), quiescent cells (light gray, yellow), mitotic cells (gray, red), cells in G_1-phase (gray, green), and cells in S/G_2-phase (dark gray, blue). The proliferation is restricted to the outer rim of the tumor. In the center a necrotic core develops. Necrotic cells are removed from the simulation (phagocytosis). The lower row depicts the cellular tension with free cells in dark/blue and strongly compressed cells in light/red. The removal of necrotic cells in the necrotic core leads to a relaxation of the cell tensions in the model. The light/red region in large tumors corresponds to the site of contact inhibition. Only at the outer rim the tension is sufficiently small to allow viable cells to divide. Previously published in Schaller and Meyer-Hermann (2005). The second colors refer to the colored version of the figure as electronically published. See color insert.

as a possible state into the model. It turns out that this state is frequently found during tumor growth provided the nutrient concentrations are sufficiently large (Schaller and Meyer-Hermann, 2005). For lower concentrations contact inhibition is not found (see Fig. 6). Thus, the turnover from exponential to power law growth is related to different reasons depending on the nutrient concentration.

Saturation of tumor growth is generally assumed to occur (Folkman and Hochberg, 1973; Freyer and Sutherland, 1986) but is not clearly visible from the data. Chemotaxis towards the necrotic core was hypothesized to induce saturation (Dormann and Deutsch, 2002). Even though there are good reasons to expect chemotactic signals in the necrotic core, one might think of an alternative hypothesis based on the cell mechanics. If adhesion between tumor cells is increased (in the Delaunay-based model) a flow equilibrium between proliferation at the outer rim of the tumor and an adhesion-driven attraction to the necrotic core is found (Schaller and Meyer-Hermann, 2005). Thus, knowing the adhesive properties of the cells from experiment would allow to draw conclusions about the role of mechanical versus chemical signals during tumor growth.

B. Morphogenesis of Lymphoid Follicles

A strong point of the presented Delaunay-Object-Dynamics modeling architecture is the applicability to fast moving agents. The immune system is composed of highly motile cells which constantly enter and leave tissue. For example, T- and B-lymphocytes enter lymph nodes via so-called high-endothelial venules (thus, coming from the blood), stay there for some hours and then leave the lymph nodes through the lymphatic vessel system.

Remarkably, despite the extreme motility of the agents a quasi-static organization of the tissue is induced. Follicles are egg-shaped structures situated near the cortex of the lymph node. They are largely dominated by B-lymphocytes and follicular dendritic cells (FDC, of unknown origin). The population of such follicles is composed of 10,000–100,000 B-lymphocytes and is exchanged on the scale of a day. Thus, a follicle is an example of a highly motile stable structure in a dynamic flow equilibrium. This differs from the characteristics of most pattern formation systems in Developmental Biology, where stable structures are more associated with little cell kinetics. Even the epidermis turns out to exhibit slow dynamics compared to the structures of the lymphoid system.

T- and B-lymphocytes both enter the lymph nodes at similar high rates. However, their path in the lymphoid tissue is different and guided by chemokines. A critical signal of B-lymphocytes induces the generation of FDC which secrete the chemokine (CXCL13) attracting further B-lymphocytes. In a positive feedback-loop, this promotes generation of the FDC-inducing signal. It turns out that the follicles generated by these assumptions, well-supported by experiment, are not stable. An unknown negative control of this positive feedback-loop is a requirement for the stability of the follicle. This is a prediction of the model (Beyer and Meyer-Hermann, 2007b).

The genesis of the follicle turns out to exhibit an interesting intermediate stage (see Fig. 7). It is characterized by a ring-zone of B-lymphocytes fully enclosing T-lymphocytes. Indeed, such ring-structures were also observed in mice (Fu *et al.*, 1997; Tumanov *et al.*, 2003) in sheep (Halleraker *et al.*, 1994) and in tertiary lymphoid tissue (Cupedo *et al.*, 2004). In fact it is the final state of lymphoid tissue organization in sharks representing a rather old adaptive immune system (Rumfelt *et al.*, 2002). This can be simulated by knocking out the FDC-inducing signal in the model. Then the genesis is stopped in the ring-phase.

What determines the ring-structure? It turns out that the ring is a result of cell mechanics. T-lymphocytes are faster than B-lymphocytes. Exchanging cell motility properties, while preserving all chemical interactions, inverts the morphology of primary follicles: Now the T-lymphocytes wrap around the B-lymphocytes (see Fig. 7). Here a clear connection of cell mechanics and tissue morphology is demonstrated.

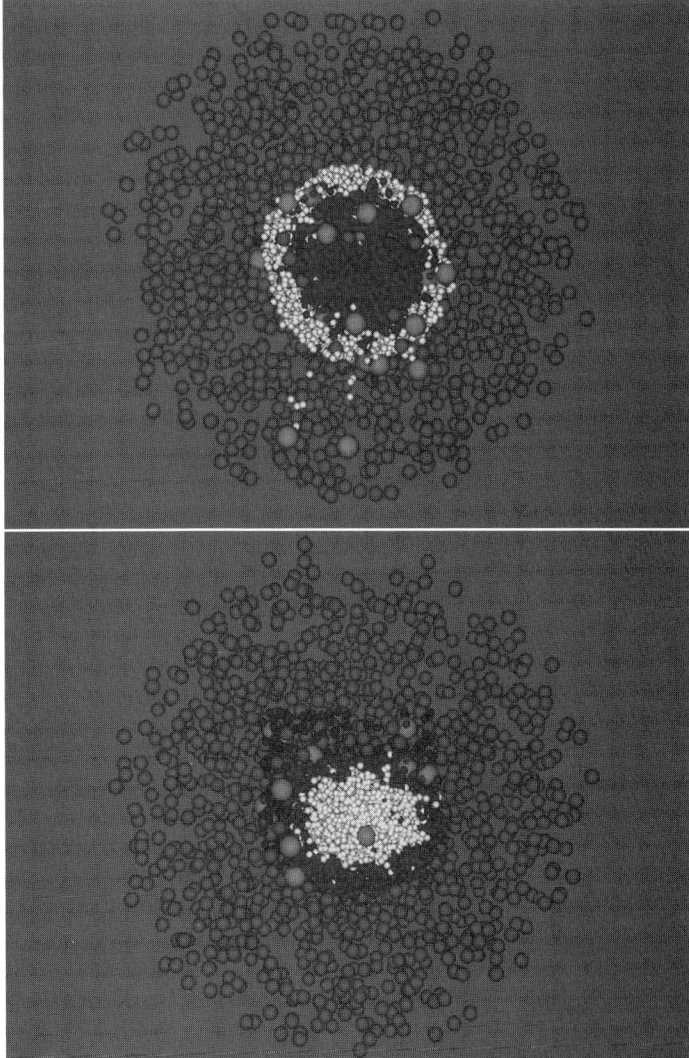

Figure 7 An intermediate stage of follicle morphogenesis: During genesis of lymphoid follicles B cells (small white, white) enter through high endothelial venules (large gray, red) and organize themselves into egg-shaped follicles before they leave via efferent lymphatic vessels (small dark gray, dark gray). A negatively regulated positive feedback with fibroblastic reticular cells (medium-sized gray, green) is an essential step in the morphogenesis. Here an intermediate stage is shown (upper panel) which is characterized by a ring of B cells around T cells (small black, dark blue). If the motility parameters of T cells and B cells are exchanged, also the ring structure is inverted (lower panel), while the final state of the follicle remains unaltered (not shown). The rings have a diameter of 200 μm. Courtesy of Tilo Beyer (Beyer, 2007). The second colors refer to the colored version of the figure as electronically published. See color insert.

VII. Conclusion

It is concluded that the construction of a physically well-defined modeling architecture for dynamic cellular systems is most important in order to gain predictive power. Only when the parameters of the model are observable quantities, the model acquires the potential to be falsified, which is a prerequisite of any scientific approach. The Delaunay-Object-Dynamics method exhibits such a physical basis and scales linearly with the number of constituents, thus, allowing the modeling of large systems. A refinement of the physical forces in this model architecture as well as the measurement of the corresponding mechanical properties of the cells are most important to further improve this modeling approach. Current restrictions come from the inherent geometrical shape of the Voronoi-cells or the attributed visco-elastic spheres. While providing a large flexibility in mechanical properties, these structures are always convex. Thus, nonconvex cells like dendritic cells or neurons cannot be appropriately represented in this model. However, nonconvex cells can be constructed as a compound of connected vertices.

The Delaunay-Object-Dynamics model architecture inherently connects different scales of modeling and, thus, provides a multiscale modeling platform (Schaller and Meyer-Hermann, 2005; Beyer and Meyer-Hermann, 2007b). Intracellular content has the potential to be differentially expressed on the cell surface and defines the interaction behavior of cells with each other or with the extracellular environment. These interactions give rise to specific self-organization of cellular tissue.

Multiscale modeling is a recent development in Theoretical Biology which connects intracellular molecular interactions to cell behavior and tissue organization (see Schaller and Meyer-Hermann, 2005; Jiang *et al.*, 2005; Ribba *et al.*, 2006; Beyer and Meyer-Hermann, 2007b; Schnell *et al.*, 2007). The large potential of multiscale modeling is related to the philosophy of a bottom-up approach: The properties of the tissue emerge from underlying interactions on a different time- or length-scale. It was exactly this deductive (instead of analytic) philosophy that made physics such successful in the 20th century. Also in Biology the predictive power will be enhanced by concisely built multiscale approaches. However, it is not yet clear whether it will have the same success as in Physics because of the inherent complexity of biological systems.

Acknowledgments

I thank Gernot Schaller and Tilo Beyer for the excellent implementation of this code in the framework of their Ph.D. theses, and for providing the figures of this book chapter. I thank Tilo Beyer, Marc Thilo Figge, and Ulrike Ruttmann for revising the manuscript. FIAS is supported by the ALTANA AG. Meyer-Hermann is supported by the EU-NEST Project MAMOCELL within FP6.

References

Aurenhammer, F., and Imai, H. (1988). Geometric relations among Voronoi diagrams. *Geometriae Dedicata* **27**, 65–75.
Bajenoff, M., Egen, J. G., Koo, L. Y., Laugier, J. P., Brau, F., Glaichenhaus, N., and Germain, R. N. (2006). Stromal cell networks regulate lymphocyte entry, migration, and territoriality in lymph nodes. *Immunity* **25**, 989–1001.
Bettelheim, E., Agam, O., and Shnerb, N. M. (2001). "Quantum phase transitions" in classical nonequilibrium processes. *Physica E* **9**, 600–608.
Beyer, T. (2007). Spatio-temporal dynamics of primary lymphoid follicles during organogenesis and lymphoneogenesis. Ph.D. thesis.
Beyer, T., and Meyer-Hermann, M. (2006). The treatment of non-flippable configurations in three-dimensional regular triangulations. *WSEAS Trans. Syst.* **5**, 1100–1107.
Beyer, T., and Meyer-Hermann, M. (2007a). Modeling emergent tissue organization involving high-speed migrating cells in a flow equilibrium. Phys. Rev. E, http://arXiv.org/, q-bio.TO/0611057 (in press).
Beyer, T., and Meyer-Hermann, M. (2007b). Mechanisms of organogenesis of primary lymphoid follicles. http://arXiv.org/, q-bio.TO/0611058.
Beyer, T., Meyer-Hermann, M., and Soff, G. (2002). A possible role of chemotaxis in germinal center formation. *Int. Immunol.* **14**, 1369–1381.
Beyer, T., Schaller, G., Deutsch, A., and Meyer-Hermann, M. (2005). Parallel dynamic and kinetic regular triangulation in three dimensions. *Comp. Phys. Commun.* **172**, 86–108.
Boissonnat, J. -D., and Teillaud, M. (1993). On the randomized construction of the Delaunay tree. *Theor. Comput. Sci.* **112**, 339–354.
Bowyer, A. (1981). Computing Dirichlet tessellations. *Comput. J.* **24**, 162–166.
Brilliantov, N. N., and Poeschel, T. (2004). Collision of adhesive viscoelastic particles. *In* "The Physics of Granular Media." (H. Hinrichsen and D. E. Wolf, Eds.), Wiley–VCH, Berlin.
Chu, Y. S., Dufour, S., Thiery, J. P., Perez, E., and Pincet, F. (2005). Johnson–Kendall–Roberts theory applied to living cells. *Phys. Rev. Lett.* **94**, 028102.
Cignoni, P., Montani, C., Perego, R., and Scopigno, R. (1993). Parallel 3D Delaunay-triangulation. *In* "Eurographics 1993." (R. J. Hubbold and R. Juan), Vol. 12. Blackwell Publishers, Cambridge, MA, pp. C129–C142.
Cupedo, T., Jansen, W., Kraal, G., and Mebius, R. E. (2004). Induction of secondary and tertiary lymphoid structures in the skin. *Immunity* **21**, 655–667.
Cyster, J. G. (2005). Chemokines, sphingosine-1-phosphate, and cell migration in secondary lymphoid tissue. *Annu. Rev. Immunol.* **23**, 127–159.
Dallon, J. C., and Othmer, H. G. (2004). How cellular movement determines the collective force generated by the *Dictyostelium discoideum* slug. *J. Theor. Biol.* **231**, 203–222.
Deutsch, A., and Dormann, S. (2004). "Cellular Automaton Modeling of Biological Pattern Formation." Birkhäuser, Basel.
Devillers, O., Pion, S., and Teillaud, M. (2001). Walking in a triangulation. *In*: Proceedings of the 17th Annual ACM Symposium on Computational Geometry, pp. 106–14.
Dormann, S., and Deutsch, A. (2002). Modeling of self-organized avascular tumor growth with a hybrid cellular automaton. *In Silico Biol.* **2**, 393–406.
Drasdo, D. (2003). On selected individual-based approaches to the dynamics in multicellular systems. *In* "Polymer and Cell Dynamics: Multiscale Modeling and Numerical Simulations." (W. Alt, M. Chaplain, M. Griebel, and J. Lenz, Eds.). Birkhäuser, pp. 169–204.
Drasdo, D., Kree, R., and McCaskill, J. S. (1995). Monte Carlo approach to tissue-cell populations. *Phys. Rev. E* **52**, 6635–6657.
Edelsbrunner, H., and Shah, N. R. (1996). Incremental topological flipping works for regular triangulations. *Algorithmica* **15**, 223–241.
Evans E., (1995). Physical actions in biological adhesion. *In* "Handbook of Biological Physics." (R. Lipowsky and E. Sackmann), Vol. I. Elsevier, Amsterdam, pp. 723–753.

Ferrez, J.-A. (2001). Dynamic triangulations for efficient 3d simulations of granular materials. EPFL thesis.
Folkman, J., and Hochberg, M. (1973). Self-regulation of growth in three dimensions. *J. Exp. Med.* **138**, 745–753.
Freyer, J. P., and Sutherland, R. M. (1986). Proliferative and clonogenic heterogeneity of cells from EMT6/Ro multicellular spheroids induced by the glucose and oxygen supply. *Cancer Res.* **46**, 3504–3512.
Fu, Y. X., Molina, H., Matsumoto, M., Huang, G., Min, J., and Chaplin, D. D. (1997). Lymphotoxin-alpha (LTalpha) supports development of splenic follicular structure that is required for IgG responses. *J. Exp. Med.* **185**, 2111–2120.
Galle, J., Loeffler, M., and Drasdo, D. (2005). Modeling the effect of deregulated proliferation and apoptosis on the growth dynamics of epithelial cell populations *in vitro*. *Biophys. J.* **88**, 62–75.
Goodman, J., and O'Rourke, J. (1997). "Handbook of Discrete and Computational Geometry." CRC Press, Boca Raton, FL.
Graner, F., and Glazier, J. A. (1992). Simulation of biological cell sorting using a two-dimensional extended Potts model. *Phys. Rev. Lett.* **69**, 2013–2016.
Gunzer, M., Weishaupt, C., Hillmer, A., Basoglu, Y., Friedl, P., Dittmar, K. E., Kolanus, W., Varga, G., and Grabbe, S. (2004). A spectrum of biophysical interaction modes between T cells and different antigen-presenting cells during priming in 3-D collagen and *in vivo*. *Blood* **104**, 2801–2809.
Halleraker, M., Press, C. M., and Landsverk, T. (1994). Development and cell phenotypes in primary follicles of foetal sheep lymph nodes. *Cell. Tissue Res.* **275**, 51–62.
Hertz, H. (1882). Über die Berührung fester elastischer Körper. *J. Reine Angew. Math.* **92**, 156–171.
Howard, J. (2001). "Mechanics of Motor Proteins and the Cytoskeleton." Sinauer Associates Inc.
Jiang, Y., Pjesivac-Grbovic, J., Cantrell, C., and Freyer, J. P. (2005). A multiscale model for avascular tumor growth. *Biophys. J.* **89**, 3884–3894.
Johnson, K. L., Kendall, K., and Roberts, A. D. (1971). Surface energy and the contact of elastic solids. *Proc. R. Soc. London Ser. A* **324**, 301–313.
Lewis, W. (1931). Locomotion of lymphocytes. *Bull. Johns Hopkins Hosp.* **49**, 29–36.
Lindenmayer, A., and Rozenberg, G. (1976). "Automata, Languages, Development." North-Holland, Amsterdam.
MacLennan, I. C. M. (1994). Germinal Centers. *Annu. Rev. Immunol.* **12**, 117–139.
Maree, A. F., and Hogeweg, P. (2001). How amoeboids self-organize into a fruiting body: Multicellular coordination in *Dictyostelium discoideum*. *Proc. Natl. Acad. Sci. USA* **98**, 3879–3883.
Meinhardt, H. (1982). "Models of Biological Pattern Formation." Academic Press.
Merks, R. M., Brodsky, S. V., Goligorksy, M. S., Newman, S. A., and Glazier, J. A. (2006). Cell elongation is key to in silico replication of *in vitro* vasculogenesis and subsequent remodeling. *Dev. Biol.* **289**, 44–54.
Meyer-Hermann, M. (2002). A mathematical model for the germinal center morphology and affinity maturation. *J. Theor. Biol.* **216**, 273–300.
Meyer-Hermann, M., and Maini, P. K. (2005). Interpreting two-photon imaging data of lymphocyte motility. *Phys. Rev. E* **71**. 061912-1–12.
Moser, B., Wolf, M., Walz, A., and Loetscher, P. (2004). Chemokines: Multiple levels of leukocyte migration control. *Trends Immunol.* **25**, 75–84.
Muecke, E. (1998). A robust implementation for three-dimensional Delaunay-triangulations. *Int. J. Comput. Geom. Appl.* **2**, 255–276.
Murray, J. D. (1993). "Mathematical Biology," 2 ed. Springer-Verlag, New York.
Newman, T. J. (2005). Modeling multicellular systems using subcellular elements. *Math. Biosci. Eng.* **2**, 611–622.
Palsson, E. (2001). A three-dimensional model of cell movement in multicellular systems. *Future Gener. Comput. Syst.* **17**, 835–852.
Ribba, B., Colin, T., and Schnell, S. (2006). A multiscale mathematical model of cancer, and its use in analyzing irradiation therapies. *Theor. Biol. Med. Mod.* **3**, 7.

Rumfelt, L. L., McKinney, E. C., Taylor, E., and Flajnik, M. F. (2002). The development of primary and secondary lymphoid tissues in the nurse shark *Ginglymostoma cirratum*: B-cell zones precede dendritic cell immigration and T-cell zone formation during ontogeny of the spleen. *Scand. J. Immunol.* **56**, 130–148.

Schaller, G. (2005). On selected numerical approaches to cellular tissue. Ph.D. thesis.

Schaller, G., and Meyer-Hermann, M. (2004). Kinetic and dynamic Delaunay tetrahedralizations in three dimensions. *Comput. Phys. Commun.* **162**, 9–23.

Schaller, G., and Meyer-Hermann, M. (2005). Multicellular tumor spheroid in an off-lattice Voronoi/Delaunay cell model. *Phys. Rev. E* **71**, 051910-1-16.

Schaller, G., and Meyer-Hermann, M. (2007). A modeling approach towards an epidermal homoeostasis control. *J. Theor. Biol.* **247**, 554–573.

Schnell, S., Grima, R., and Maini, P. K. (2007). Multiscale modeling in biology. *Am. Sci.* **95**, 134–142.

Shnerb, N. M., Louzon, Y., Bettelheim, E., and Solomon, S. (2000). The importance of being discrete: Life always wins on the surface. *Proc. Natl. Acad. Sci. USA* **97**, 10322–10324.

Tumanov, A. V., Grivennikov, S. I., Shakhov, A. N., Rybtsov, S. A., Koroleva, E. P., Takeda, J., Nedospasov, S. A., and Kuprash, D. V. (2003). Dissecting the role of lymphotoxin in lymphoid organs by conditional targeting. *Immunol. Rev.* **195**, 106–116.

Watson, D. F. (1981). Computing the n-dimensional Delaunay-tessellation with application to Voronoi polytopes. *Comput. J.* **24**, 167–172.

Wolf, K., Muller, R., Borgmann, S., Brocker, E. B., and Friedl, P. (2003). Amoeboid shape change and contact guidance: T-lymphocyte crawling through fibrillar collagen is independent of matrix remodeling by MMPs and other proteases. *Blood* **102**, 3262–3269.

Wolfram, S. (1986). "Theory and Application of Cellular Automata." World Publishing Co., Singapore.

14

Cellular Automata as Microscopic Models of Cell Migration in Heterogeneous Environments

Haralambos Hatzikirou and Andreas Deutsch
Center for Information Services and High-Performance Computing, Technische Universität Dresden, Nöthnitzerstr. 46, 01069 Dresden, Germany

I. Introduction
  A. Types of Cell Motion
  B. Mathematical Models of Cell Migration
  C. Overview of the Paper
II. Idea of the LGCA Modeling Approach
III. LGCA Models of Cell Motion in a Static Environment
  A. Model I
  B. Model II
IV. Analysis of the LGCA Models
  A. Model I
  B. Model II
V. Results and Discussion
  Acknowledgments
  Appendix A
  A.1. States in Lattice-Gas Cellular Automata
  A.2. Dynamics in Lattice-Gas Cellular Automata
  Appendix B
  Appendix C
  Appendix D
  References

Understanding the precise interplay of moving cells with their typically heterogeneous environment is crucial for central biological processes as embryonic morphogenesis, wound healing, immune reactions or tumor growth. Mathematical models allow for the analysis of cell migration strategies involving complex feedback mechanisms between the cells and their microenvironment. Here, we introduce a cellular automaton (especially lattice-gas cellular automaton—LGCA) as a microscopic model of cell migration together with a (mathematical) tensor characterization of different biological environments. Furthermore, we show how mathematical analysis of the LGCA model can yield an estimate for the cell dispersion speed within a given environment. Novel imaging techniques like diffusion tensor imaging (DTI) may provide tensor data of biological microenvironments. As an application, we present LGCA simulations of a proliferating cell population moving in an external field defined by clinical DTI data. This system can serve as a model of *in vivo* glioma cell invasion.

© 2008, Elsevier Inc.

I. Introduction

Alan Turing in his landmark paper of 1952 introduced the concept of self-organization to biology (Turing, 1952). He suggested "that a system of chemical substances, called morphogens, reacting together and diffusing through a tissue, is adequate to account for the main phenomena of morphogenesis. Such a system, although it may originally be quite homogeneous, may later develop a pattern or structure due to an instability of the homogeneous equilibrium, which is triggered off by random disturbances." Today, it is realized that, in addition to diffusible signals, the role of cells in morphogenesis is crucial. In particular, living cells possess migration strategies that go far beyond the merely random displacements characterizing nonliving molecules (diffusion). It has been shown that the microenvironment plays a an important role in the way that cells select their migration strategies (Friedl and Broecker, 2000). Moreover, the microenvironment provides the prototypic substrate for cell migration in embryonic morphogenesis, immune defense, wound repair or even tumor invasion.

The cellular microenvironment is a highly heterogeneous medium for cell motion including the extracellular matrix composed of fibrillar structures, collagen matrices, diffusible chemical signals as well as other mobile and immobile cells. Cells move within their environment by responding to variuos stimuli. In addition, cells change their environment locally by producing or absorbing chemicals and/or by degrading the neighboring tissue. This interplay establishes a dynamic relationship between individual cells and the surrounding substrate. In the following subsection, we provide more details about the different cell migration strategies in various environments. Environmental heterogeneity contributes to the complexity of the resulting cellular behaviors. Moreover, cell migration and interactions with the environment are taking place at different spatiotemporal scales. Mathematical modeling has proven extremely useful in getting insights into such multiscale systems. In this paper, we show how a suitable microscopical mathematical model (a cellular automaton) can contribute to understand the interplay of moving cells with their heterogeneous environment. A broad spectrum of challenging questions can be addressed, in particular:

- What kind of spatiotemporal patterns are formed by moving cells using different strategies?
- How does the moving cell population affect its environment and what is the feedback to its motion?
- What is the spreading speed of a cell population within a heterogeneous environment?

A. Types of Cell Motion

The cell migration type is strongly coupled to the kind of environment that hosts the cell population. A range of external cues impart information to the cells which regulate

14. Cellular Automata as Microscopic Models of Cell Migration

Table I Diversity in cell migration strategies (after Friedl et al., 2004)

Type/motion	Random Walk	Cell–Cell Adhesion	Cell–ECM Adhesion	Proteolysis
Amoeboid	++	−/+	−/+	+/−
Mesenchymal	−	−/+	+	+
Collective	−	++	++	+

In different tissue environments, different cell types exhibit either individual (amoeboid or mesenchymal) or collective migration mechanisms to overcome and integrate into tissue scaffolds (see text for explanations).

their movement, including long-range diffusible chemicals (e.g., chemoattractants), contact with membrane-bound molecules on neighboring cells (mediating cell–cell adhesion and contact inhibition) and contact with the extracellular matrix (ECM) surrounding the cells (contact guidance, haptotaxis). Accordingly, the environment can act on the cell motion in many different ways.

Recently, Friedl et al. (2000, 2004) have investigated in depth the kinds of observed cell movement in tissues. The main processes that influence cell motion are identified by: cell–ECM adhesion forces introducing integrin-induced motion and cell–cell adhesive forces leading to cadherin-induced motion. The different contributions of these two kinds of adhesive forces characterize the particular type of cell motion. Table I gives an overview of the possible types of cell migration in the ECM.

Amoeboid motion is the simplest kind of cell motion and can be characterized as random motion of cells without being affected by the integrin concentration of the underlying matrix. Amoeboidly migrating cells develop a dynamic leading edge rich in small pseudopodia, a roundish or ellipsoid main cell body and a trailing small uropod. Cells, like neutrophils, perceive the tissue as a porous medium, where their flexibility allows them to move through the tissue without significantly changing it. On the other hand, mesenchymal motion of cells (for instance, glioma cells) leads to alignment with the fibers of the ECM, since the cells are responding to environmental cues of nondiffusible molecules bound to the matrix and follow the underlying structure. Mesenchymal cells retain an adhesive, tissue-dependent phenotype and develop a spindle-shaped elongation in the ECM. In addition, the proteolytic activity (metalloproteinases production) of such cells allows for the remodeling of the matrix and establishes a dynamical environment. The final category is collective motion of cells (e.g., endothelial cells) that respond to cadherins and create cell–cell bounds. Clusters of cells can move through the adjacent connective tissue. Leading cells provide the migratory traction and, via cell–cell junctions, pull the following group forward.

One can think about two distinct ways of cells responding to environmental stimuli: either the cells are following a certain direction and/or the environment imposes an orientational preference leading to alignment. An example of the directed case is the graded spatial distribution of adhesion ligands along the ECM which is thought to influence the direction of cell migration (McCarthy and Furcht, 1984), a phenomenon

Table II In this table, we classify the environmental effect with respect to different cell migration strategies

	Static	Dynamic
Direction	Haptotaxis	Chemotaxis
Orientation	Amoeboid	Mesenchymal

One can distinguish static and dynamic environments. In addition, we differentiate environments that impart directional or only orientational information for migrating cells (see text for explanations).

known as haptotaxis (Carter, 1965). Chemotaxis mediated by diffusible chemotactic signals provides a further example of directed cell motion in a dynamically changing environment. On the other hand, alignment is observed in fibrous environments where amoeboid and mesenchymal cells change their orientation according to the fiber structure. Mesenchymal cells use additionally proteolysis to facilitate their movement and remodel the neighboring tissue (dynamic environment). Table II summarizes the above statements.

It has been shown that the basic strategies of cell migration are retained in tumor cells (Friedl and Wolf, 2003). However, it seems that tumor cells can adapt their strategy, i.e., the cancer cell's migration mechanisms can be reprogrammed, allowing it to maintain its invasive properties via morphological and functional dedifferentiation (Friedl and Wolf, 2003). Furthermore, it has been demonstrated that the microenvironment is crucial for cancer cell migration, e.g., fiber tracks in the brain's white matter facilitate glioma cell motion (Swanson *et al.*, 2002; Hatzikirou *et al.*, 2005). Therefore, a better understanding of cell migration strategies in heterogeneous environments is particularly crucial for designing new cancer therapies.

B. Mathematical Models of Cell Migration

The present paper focuses on the analysis of cell motion in heterogeneous environments. A large number of mathematical models has already been proposed to model various aspects of cell motion. Reaction–diffusion equations have been used to model the phenomenology of motion in various environments, like diffusible chemicals [Keller–Segel chemotaxis model (Keller and Segel, 1971), etc.] and mechanical ECM stresses (Murray *et al.*, 1983). Integrodifferential equations have been introduced to model fiber alignment in the work of Dallon, Sherratt, and Maini (Dallon *et al.*, 2001). Navier–Stokes equations and the theory of fluid dynamics provided insight of "flow" of cells within complex environments, for instance, modeling cell motion as flow in porous media (Byrne and Preziosi, 2003). However, the previous continuous models describe cell motion at a macroscopic level neglecting the microscopical cell–cell and cell–environment interactions. Kinetic equations (Dolak and Schmeiser, 2005;

Chauviere *et al.*, 2007) and especially transport equations (Othmer *et al.*, 1988; Dickinson and Tranquillo, 1993; Dickinson and Tranquillo, 1995; Hillen, 2006) have been proposed as models of cell motion along tissues, at a mesoscopic level of description (the equations describe the behavior of cells within a small partition of space). Microscopical experimental data and the need to analyze populations consisting of a low number of cells call for models that describe the phenomena at the level of cell–cell interactions, e.g., interacting particle systems (Liggett, 1985), cellular automata (CA) (Deutsch and Dormann, 2005), off-lattice Langevin methods (Galle *et al.*, 2006; Grima, 2007; Newman and Grima, 2004), active Brownian particles (Schweitzer, 2003; Peruani and Morelli, 2007), and other microscopic stochastic models (Othmer and Stevens, 1997; Okubo and Levin, 2002).

The focus of this study is to introduce CAs, and especially a subclass of them called lattice-gas cellular automata (LGCA), as a microscopic model of cell motion (Deutsch and Dormann, 2005). LGCA (and lattice Boltzmann equation (LBE) models) have been originally introduced as discrete models of fluid dynamics (Chopard and Droz, 1998). In contrast with other CA models, LGCA allow for modeling of migrating particles in a straightforward manner (see Section II). In the following, we show that LGCA can be extended to serve as models for migrating cell populations. Additionally, their discrete nature allows for the description of the cell–cell and cell–environment interactions at the microscopic level of single cells but at the same time enables us to observe the macroscopical evolution of the whole population.

C. Overview of the Paper

In this paper, we will explore the role of the environment (both in a directional and orientational sense) for cell movement. Moreover, we consider cells that lack metalloproteinase production (proteolytic proteins) and do not change the ECM structure, introducing static environments (additionally, we do not consider diffusible environments). Moreover, we will consider populations with a constant number of cells (no proliferation/death of cells) in time.

In Section II, we introduce the modeling framework of CA and particularly LGCA. In Section III, we define LGCA models of moving cells in different environments that impart directional and orientational information to the moving cells. Furthermore, we provide a tensor characterization for the environmental impact on migrating cell populations. As an example, we present simulations of a proliferative cell population in a tensor field defined by clinical DTI data. This system can serve as a model of *in vivo* glioma cell invasion. In Section IV, we show how mathematical analysis of the LGCA model can yield an estimate for the cell dispersion speed within a given environment. Finally, in Section V, we sum up the results, we critically argue on the advantages and disadvantages of using LGCA, and we discuss potential venues for analysis, extensions and applications.

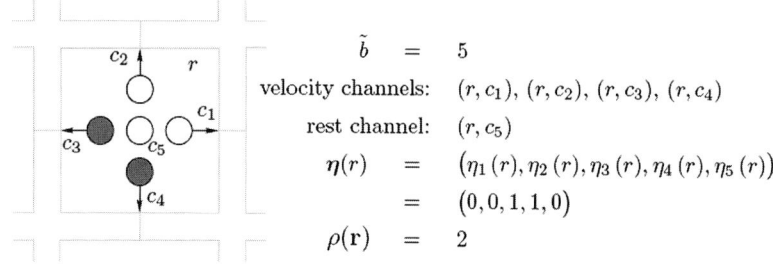

Figure 1 Node configuration: channels of node r in a two-dimensional square lattice ($b = 4$) with one rest channel ($\beta = 1$). Gray dots denote the presence of a particle in the respective channel.

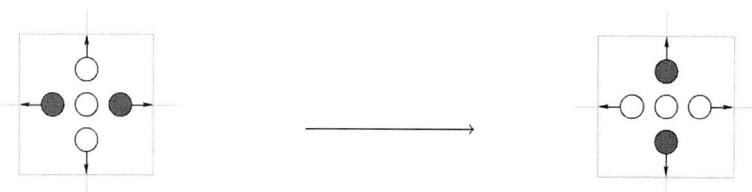

Figure 2 Example for interaction of particles at two-dimensional square lattice node r; gray dots denote the presence of a particle in the respective channel. No confusion should arise by the arrows indicating channel directions.

II. Idea of the LGCA Modeling Approach

The strength of the lattice-gas method lies in its potential to unravel effects of movement and interaction of individuals (e.g., cells). In traditional cellular automaton models implementing movement of individuals is not straightforward, as one node in the lattice can typically only contain one individual, and consequently movement of individuals can cause collisions when two individuals want to move into the same empty node. In a lattice-gas model this problem is avoided by having separate channels for each direction of movement. The channels specify the direction and magnitude of movement, which may include zero velocity (resting) states. For example, a square lattice has four nonzero velocity channels and an arbitrary number of rest channels (Fig. 1). Moreover, LGCA impose an exclusion principle on channel occupation, i.e. each channel may at most host one particle.

The transition rule of a LGCA can be decomposed into two steps. An *interaction* step updates the state of each channel at each lattice site. Particles may change their velocity state and appear or disappear (birth/death) as long as they do not violate the exclusion principle (Fig. 2). In the *propagation* step, cells move synchronously in the direction and by the distance specified by their velocity state (Fig. 3). The propagation step is deterministic and conserves mass and momentum. Synchronous transport prevents particle collisions which would violate the exclusion principle (other models

14. Cellular Automata as Microscopic Models of Cell Migration 407

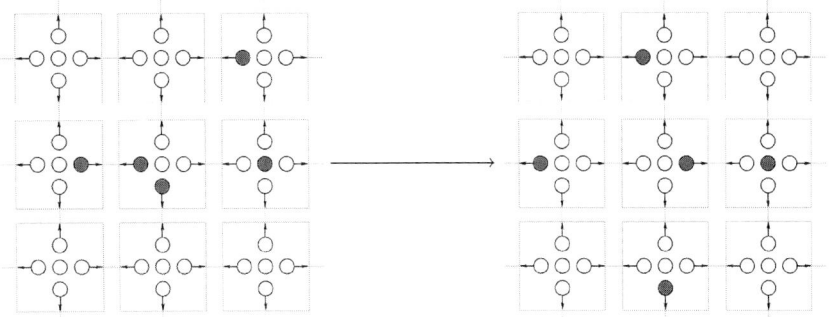

Figure 3 Propagation in a two-dimensional square lattice with speed $m = 1$; lattice configurations before and after the propagation step; gray dots denote the presence of a particle in the respective channel.

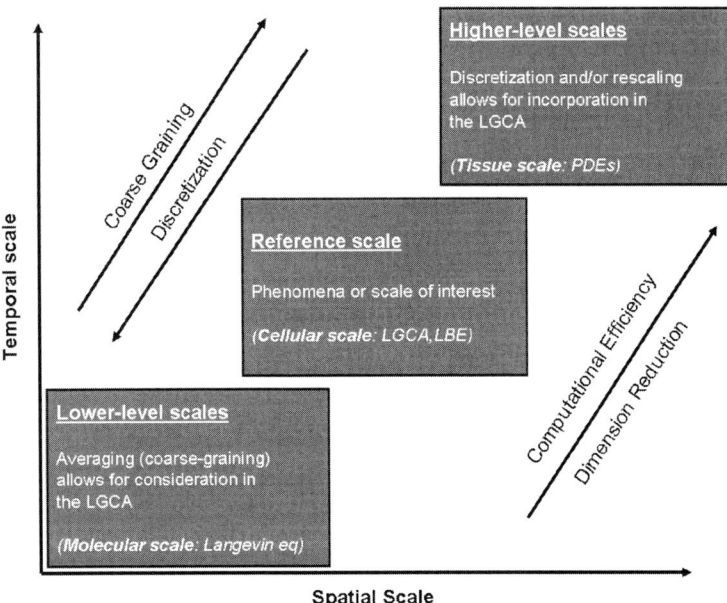

Figure 4 The sketch visualizes the hierarchy of scales relevant for the LGCA models introduced in this chapter (see text for details).

have to define a collision resolution algorithm). LGCA models allow parallel synchronous movement and updating of a large number of particles.

The basic idea of lattice-gas automaton models is to mimic complex dynamical system behavior by the repeated application of simple local migration and interaction rules. The reference scale (Fig. 4) of the LGCA models introduced in this chapter is that of a finite set of cells. The dynamics evolving at smaller scales (intracellular) are included in a coarse-grained manner, by introducing an appropriate stochastic interaction rule. This means that our lattice-gas automata impose a microscopic, though not truly molecular, view of the system by conducting fictive microdynamics on a regular lattice. Please note that the theoretically inclined reader can find, in Appendix A, a detailed definition of the LGCA mathematical nomenclature (after Deutsch and Dormann, 2005).

III. LGCA Models of Cell Motion in a Static Environment

In this section, we define two LGCA models that describe cell motion in static environments. We especially address two problems:

- How can we model the environment?
- How should the automaton rules be chosen to model cell motion in a specific environment?

As we already stated, we want to mathematically model biologically relevant environments. According to Table II, we distinguish a "directional" and an "orientational" environment, respectively. The mathematical entity that allows for the modeling of such environments is called a *tensor field*. A tensor field is a collection of different tensors which are distributed over a spatial domain. A *tensor* is (in an informal sense) a generalized linear 'quantity' or 'geometrical entity' that can be expressed as a multidimensional array relative to a choice of basis of the particular space on which the tensor is defined. The intuition underlying the tensor concept is inherently geometrical: a tensor is independent of any chosen frame of reference.

The rank of a particular tensor is the number of array indices required to describe it. For example, mass, temperature, and other scalar quantities are tensors of rank 0; while force, momentum, velocity and further vector-like quantities are tensors of rank 1. A linear transformation such as an anisotropic relationship between velocity vectors in different directions (diffusion tensors) is a tensor of rank 2. Thus, we can represent an environment with directional information as a vector (tensor of rank 1) field. The geometric intuition for a vector field corresponds to an 'arrow' attached to each point of a region, with variable length and direction (Fig. 5). The idea of a vector field on a curved space is illustrated by the example of a weather map showing wind velocity, at each point of the earth's surface. An environment that carries orientational information for each geometrical point can be modeled by a tensor field of rank 2. A geometrical visualization of a second order tensor field can be represented as a collection of ellipsoids, assigned to each geometrical point (Fig. 6). The ellipsoids represent the orientational information that is encoded into tensors.

To model cell motion in a given tensor field (environment)—of rank 1 or 2—we should modify the interaction rule of the LGCA. We use a special kind of interaction

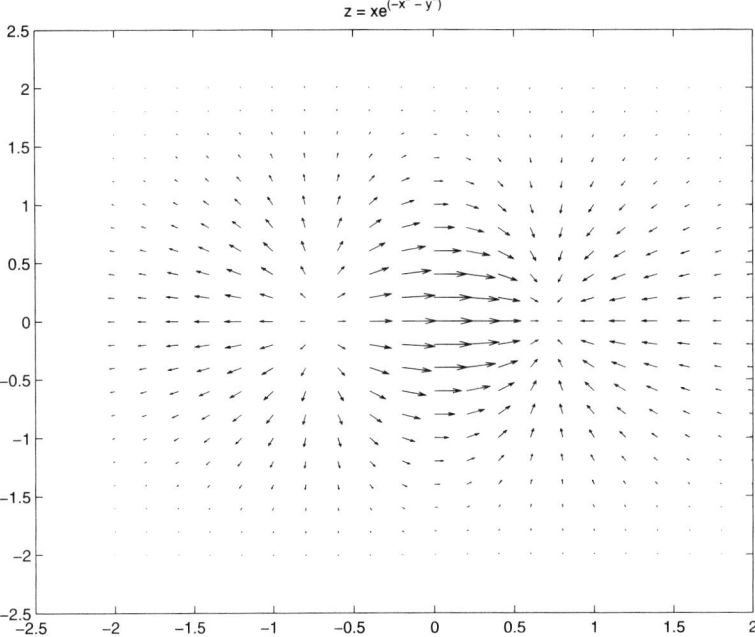

Figure 5 An example of a vector field (tensor field of rank 1). The vectors (e.g., integrin receptor density gradients) show the direction and the strength of the environmental drive.

rules for the LGCA dynamics, firstly introduced by Alexander *et al.* (1992). We consider biological cells as random walkers that are reoriented by maximizing a potential-like term. Assuming that the cell motion is affected by cell–cell and cell–environment interactions, we can define the potential as the sum of these two interactions:

$$G(r, k) = \sum_j G_j(r, k) = G_{cc}(r, k) + G_{ce}(r, k), \tag{1}$$

where $G_j(r, k)$, $j = $ cc, ce is the subpotential that is related to cell–cell and cell–environment interactions, respectively.

Interaction rules are formulated in such a way that cells preferably reorient into directions which maximize (or minimize) the potential, that is according to the gradients of the potential $G'(r, k) = \nabla G(r, k)$.

Consider a lattice-gas cellular automaton defined on a two-dimensional lattice with b velocity channels ($b = 4$ or $b = 6$). Let the number of particles at node r at time k be denoted by

$$\rho(r, k) = \sum_{i=1}^{\tilde{b}} \eta_i(r, k),$$

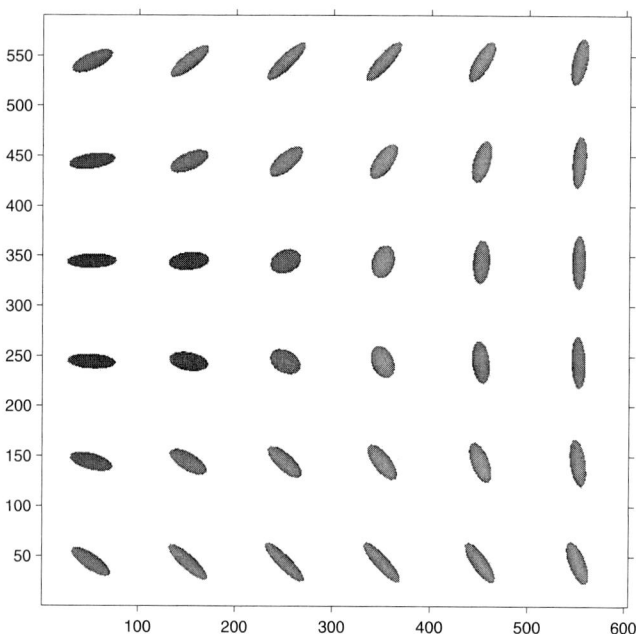

Figure 6 An example of a tensor field (tensor field of rank 2). We represent the local information of the tensor as ellipsoids. The ellipsoids can encode, e.g., the degree of alignment of a fibrillar tissue. The colors are denoting the orientation of the ellipsoids. See color insert.

and the *flux* be denoted by

$$\mathbf{J}(\eta(r,k)) = \sum_{i=1}^{b} \mathbf{c}_i \eta_i(r,k).$$

The probability that η^C is the outcome of an interaction at node r is defined by

$$P(\eta \to \eta^C | G(r,k)) = \frac{1}{Z} \exp[\alpha F(G'(r,k), \mathbf{J}(\eta^C))] \delta(\rho(r,k), \rho^C(r,k)), \quad (2)$$

where η is the preinteraction state at r and the Kronecker's δ assumes the mass conservation of this operator. The sensitivity is tuned by the parameter α. The normalization factor is given by

$$Z = Z(\eta(r,k)) = \sum_{\eta^C \in \mathcal{E}} \exp[\alpha F(G'(r,k), \mathbf{J}(\eta^C))] \delta(\rho(r,k), \rho^C(r,k)).$$

$F(\cdot)$ is a function that defines the effect of the G' gradients on the new configuration. A common choice of $F(\cdot)$ is the inner product $\langle,\rangle$, which favors (or penalizes) the configurations that tend to have the same (or inverse) direction of the gradient G'. Accordingly, the dynamics are fully specified by the following microdynamical equation

$$\eta_i(r + \mathbf{c}_i, k + 1) = \eta_i^C(r, k).$$

In the following, we present two stochastic potential-based interaction rules that correspond to the motion of cells in a vector field and a rank 2 tensor field, respectively. We exclude any other cell–cell interactions and we consider that the population has a fixed number of cells (mass conservation).

A. Model I

The first rule describes cell motion in a static environment that carries directional information expressed by a vector field $\mathbf{E}$. Biologically relevant examples are the motion of cells that respond to fixed integrin concentrations along the ECM (haptotaxis). The spatial concentration differences of integrin proteins constitute a gradient field that creates a kind of "drift" $\mathbf{E}$ (Dickinson and Tranquillo, 1993). We choose a two-dimensional LGCA without rest channels and the stochastic interaction rule of the automaton follows the definition of the potential-based rules [Eq. (1) with $\alpha = 1$]:

$$P(\eta \to \eta^C)(r, k) = \frac{1}{Z} \exp(\langle \mathbf{E}(r), \mathbf{J}(\eta^C(r, k)) \rangle) \delta(\rho(r, k), \rho^C(r, k)). \qquad (3)$$

We simulate our LGCA for spatially homogeneous fields $\mathbf{E}$ for various intensities and directions. In Fig. 7, we observe the time evolution of a cell cluster under the influence of a given field. We see that the cells collectively move in the gradient direction and they roughly keep the shape of the initial cluster. The simulations in Fig. 8 show the evolution of the system for different fields. It is evident that the "cells" follow the direction of the field and their speed responds positively to an increase of the field intensity.

B. Model II

We now focus on cell migration in environments that promote alignment (orientational changes). Examples of such motion are provided by neutrophil or leukocyte movement through the pores of the ECM, the motion of cells along fibrillar tissues or the motion of glioma cells along fiber track structures. As stated before, such an environment can be modeled by the use of a second rank tensor field that introduces a spatial anisotropy along the tissue. In each point, a tensor (i.e., a matrix) informs the cells about the local orientation and strength of the anisotropy and proposes a principle (local) axis of movement. For instance, the brain's fiber tracks impose a spatial anisotropy and their degree of alignment affects the strength of anisotropy.

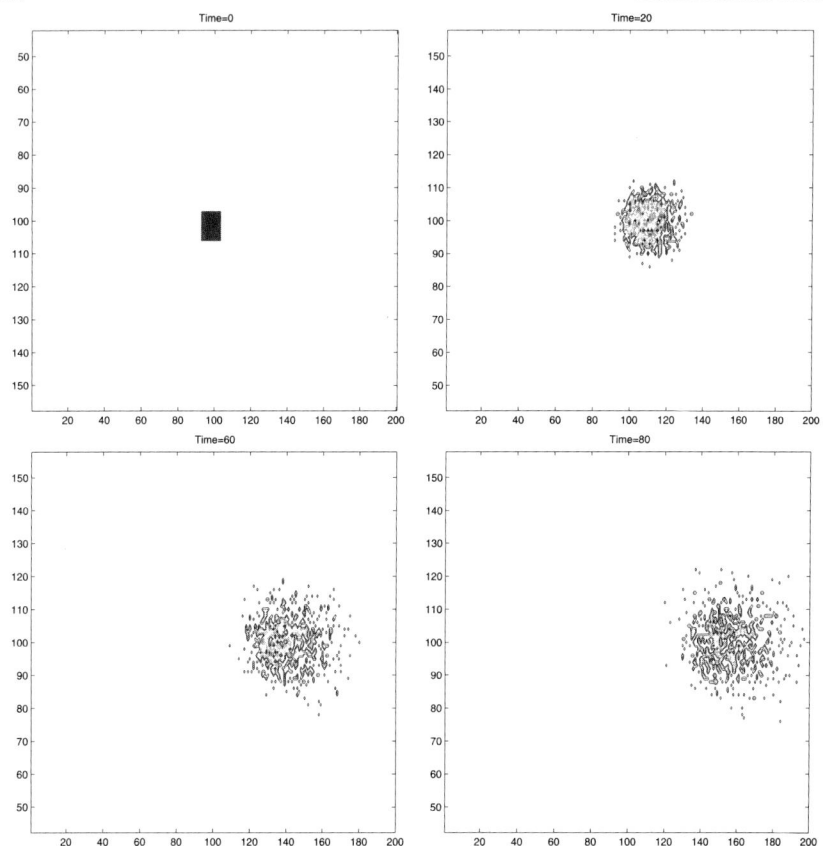

Figure 7 Time evolution of a cell population under the effect of a field $E = (1, 0)$. One can observe that the environmental drive moves all the cells of the cluster into the direction of the vector field. The blue color stands for low, the yellow for intermediate and red for high densities. See color insert.

Here, we use the information of the principal eigenvector of the diffusion tensor which defines the local principle axis of cell movement. Thus, we end up again with a vector field but in this case we exploit only the orientational information of the vector. The new rule for cell movement in an "oriented environment" is:

$$P(\eta \to \eta^C)(r, k) = \frac{1}{Z} \exp(|\langle \mathbf{E}(r), \mathbf{J}(\eta^C(r, k))\rangle|)\delta(\rho(r, k), \rho^C(r, k)). \qquad (4)$$

In Fig. 9, we show the time evolution of a simulation of model II for a given field. Fig. 10 shows the typical resulting patterns for different choices of tensor fields. We observe that the anisotropy leads to the creation of an ellipsoidal pattern, where the length of the main ellipsoid's axis correlates positively with the anisotropy strength.

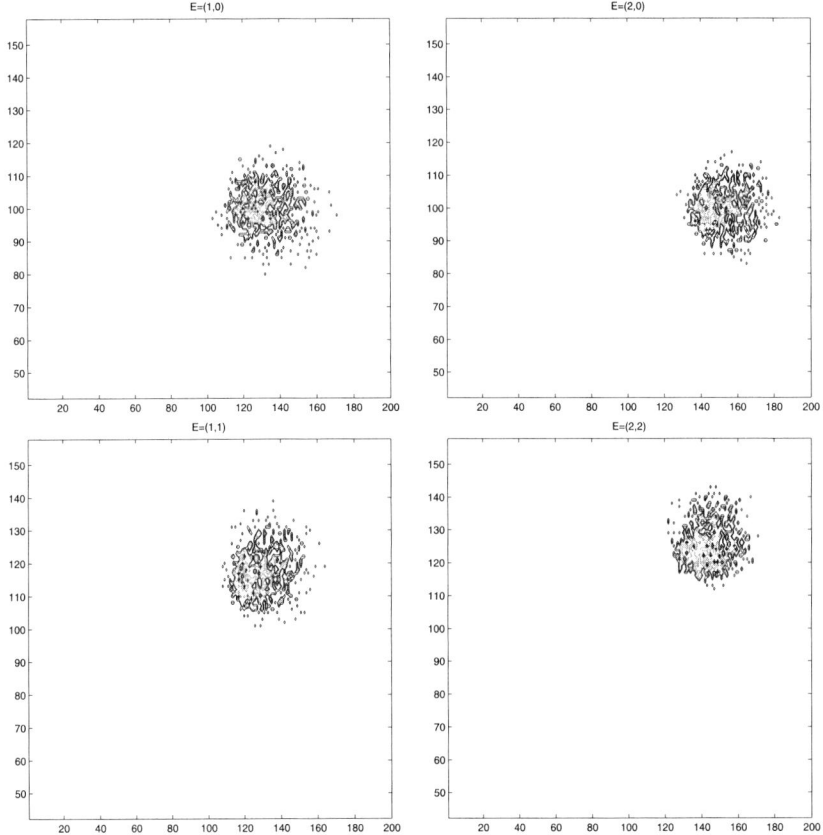

Figure 8 The figure shows the evolution of the cell population under the influence of different fields (100 time steps). Increasing the strength of the field, we observe that the cell cluster is moving faster in the direction of the field. This behavior is characteristic of a haptotactically moving cell population. The initial condition is a small cluster of cells in the center of the lattice. Colors denote different node densities (as in Fig. 7). See color insert.

This rule can, for example, be used to model the migration of glioma cells within the brain. Glioma cells tend to spread faster along fiber tracks. Diffusion Tensor Imaging (DTI) is a MRI-based method that provides the local anisotropy information in terms of diffusion tensors. High anisotropy points belong to the brain's white matter, which consists of fiber tracks. A preprocessing of the diffusion tensor field can lead to the principle eigenvectors' extraction of the diffusion tensors, that provides us with the local principle axis of motion. By considering a proliferative cell population, as in Hatzikirou *et al.* (2007), and using the resulting eigenvector field we can model and simulate glioma cell invasion. In Fig. 11, we simulate an example of

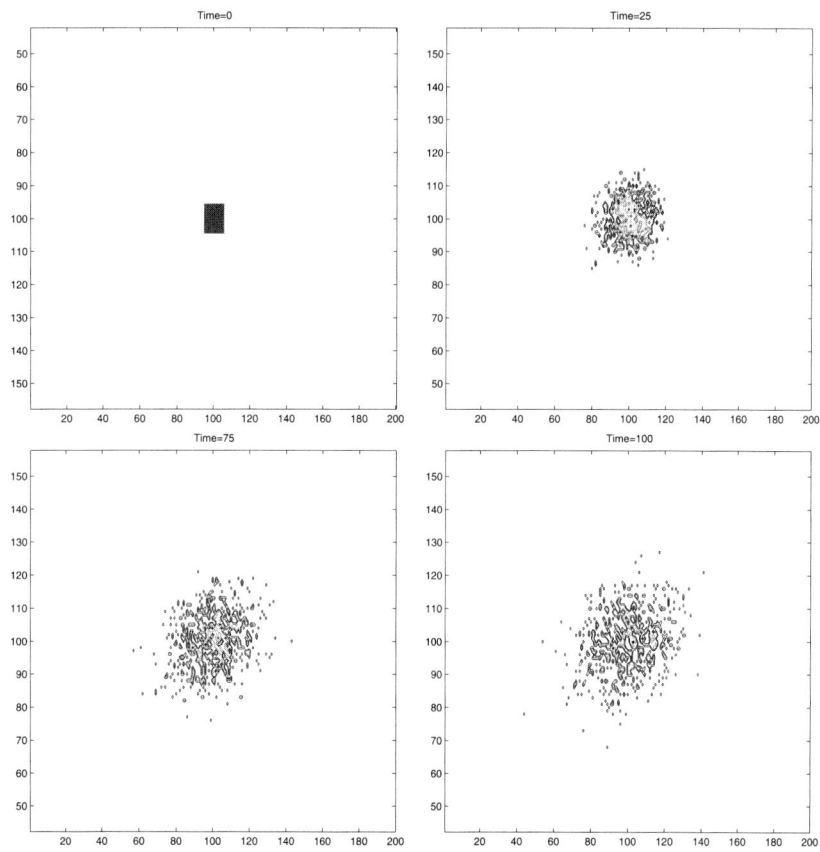

Figure 9 Time evolution of a cell population under the effect of a tensor field with principal eigenvector (principal orientation axis) $E = (2, 2)$. We observe cell alignment along the orientation of the axis defined by E, as time evolves. Moreover, the initial rectangular shape of the cell cluster is transformed into an ellipsoidal pattern with principal axis along the field E. Colors denote the node density (as in Fig. 7). See color insert.

glioma growth and show the effect of fiber tracks in tumor growth using the DTI information.

IV. Analysis of the LGCA Models

In this section, we provide a theoretical analysis of the proposed LGCA models. Our aim is to calculate the equilibrium cell distribution and to estimate the speed of cell dispersion under environmental variations. Finally, we compare our theoretical results with the simulations.

14. Cellular Automata as Microscopic Models of Cell Migration

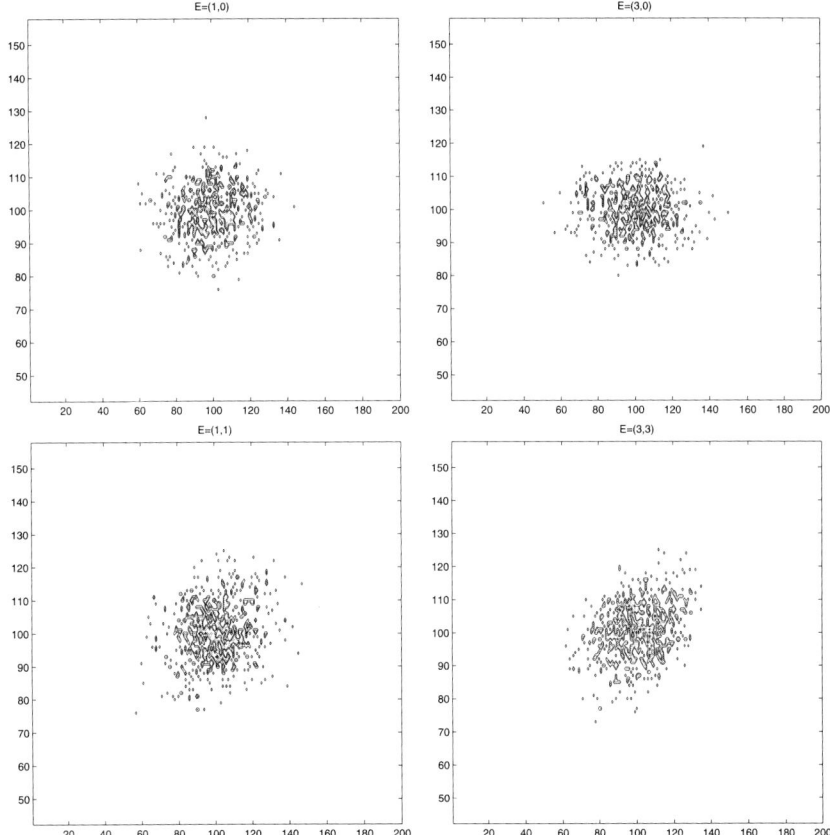

Figure 10 In this graph, we show the evolution of the pattern for four different tensor fields (100 time steps). We observe the elongation of the ellipsoidal cell cluster when the strength is increased. Above each figure the principal eigenvector of the tensor field is denoted. The initial conditions is always a small cluster of cells in the center of the lattice. The colors denote the density per node (as in Fig. 7). See color insert.

A. Model I

In this subsection, we analyze model I and we derive an estimate of the cell spreading speed in dependence of the environmental field strength. The first idea is to choose a macroscopically accessible observable that can be measured experimentally. A reasonable choice is the mean lattice flux $\langle \mathbf{J}(\eta^C) \rangle_\mathbf{E}$, which characterizes the mean motion of the cells, with respect to changes of the field's strength $|\mathbf{E}|$:

$$\langle \mathbf{J}(\eta^C) \rangle_\mathbf{E} = \sum_i \mathbf{c}_i f_i^{\text{eq}}, \tag{5}$$

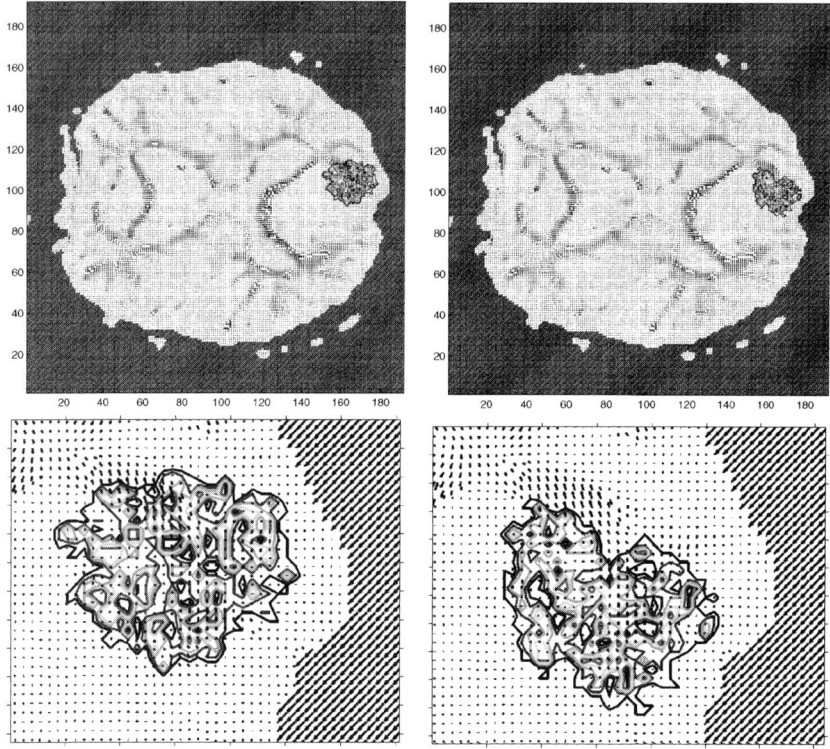

Figure 11 We show the brain's fiber track effect on glioma growth. We use a LGCA of a proliferating cancer cell population (for definition see Hatzikirou et al., 2007) moving in a tensor field provided by clinical DTI data, representing the brain's fiber tracks. (Top) The left figure is a simulation without any environmental information (only diffusion). In the top right figure the effect of the fiber tracks in the brain on the evolution of the glioma growth is obvious. (Bottom) The two figures display magnification of the tumor area in the simulations above. This is an example of how environmental heterogeneity affects cell migration (in this case tumor cell migration). See color insert.

where $f_i^{\text{eq}}, i = 1, \ldots, b$ is the equilibrium density distribution of each channel. Mathematically, this is the mean flux *response* to changes of the external vector field $\mathbf{E}$. The quantity that measures the linear response of the system to the environmental stimuli is called *susceptibility*:

$$\chi = \frac{\partial \langle \mathbf{J} \rangle_\mathbf{E}}{\partial \mathbf{E}}. \tag{6}$$

It appears if we expand the mean flux in terms of small fields as

$$\langle \mathbf{J} \rangle_\mathbf{E} = \langle \mathbf{J} \rangle_{\mathbf{E}=0} + \frac{\partial \langle \mathbf{J} \rangle_\mathbf{E}}{\partial \mathbf{E}} \mathbf{E} + O(\mathbf{E}^2). \tag{7}$$

For the zero-field case, the mean flux is zero since the cells are moving randomly within the medium (diffusion). Accordingly, for small fields $\mathbf{E} = \begin{pmatrix} e_1 \\ e_2 \end{pmatrix}$ the linear approximation reads

$$\langle \mathbf{J} \rangle_{\mathbf{E}} = \frac{\partial \langle \mathbf{J} \rangle_{\mathbf{E}}}{\partial \mathbf{E}} \mathbf{E}.$$

The *general linear response relation* is

$$\langle \mathbf{J}(\eta^C) \rangle_{\mathbf{E}} = \chi_{\alpha\beta} e_\beta = \chi e_\alpha, \tag{8}$$

where the second rank tensor $\chi_{\alpha\beta} = \chi \delta_{\alpha\beta}$ is assumed to be isotropic. In biological terms, we want to study the response of cell motion with respect to changes of the spatial distribution of the integrin concentration along the ECM, corresponding to changes in the resulting gradient field.

The aim is to estimate the stationary mean flux for fields $\mathbf{E}$. At first, we have to calculate the equilibrium distribution that depends on the external field. The external drive breaks down the detailed balance (DB) conditions[1] that would lead to a Gibbs equilibrium distribution. In the case of nonzero external field, the system is as out of equilibrium. The external field (environment) induces a breakdown of the spatial symmetry which leads to nontrivial equilibrium distributions depending on the details of the transition probabilities. The (Fermi) exclusion principle allows us to assume that the equilibrium distribution follows a kind of Fermi–Dirac distribution (Frisch *et al.*, 1987):

$$f_i^{eq} = \frac{1}{1 + e^{x(\mathbf{E})}}, \tag{9}$$

where $x(\mathbf{E})$ is a quantity that depends on the field $\mathbf{E}$ and the mass of the system (if the DB conditions were fulfilled, the argument of the exponential would depend only on the invariants of the system). Thus, one can write the following *ansatz*:

$$x(\mathbf{E}) = h_0 + h_1 \mathbf{c}_i \mathbf{E} + h_2 \mathbf{E}^2. \tag{10}$$

After some algebra (the details can be found in Appendix A), for small fields $\mathbf{E}$, one finds that the equilibrium distribution looks like:

$$f_i^{eq} = d + d(d-1)h_1 \mathbf{c}_i \mathbf{E} + \frac{1}{2}d(d-1)(2d-1)h_1^2 \sum_\alpha c_{i\alpha}^2 e_\alpha^2$$

$$+ d(d-1)h_2 \mathbf{E}^2, \tag{11}$$

---

[1] The detailed balance (DB) and the semidetailed balance (SDB) imposes the following condition for the microscopic transition probabilities: $P(\eta \to \eta^C) = P(\eta^C \to \eta)$ and $\forall \eta^C \in \mathcal{E} : \sum_\eta P(\eta \to \eta^C) = 1$. Intuitively, the DB condition means that the system jumps to a new microconfiguration and comes back to the old one with the same probability (mircoreversibility). The relaxed SDB does not imply this symmetry. However, SDB guarantees the existence of steady states and the sole dependence of the Gibbs steady state distribution on the invariants of the system (conserved quantities).

where $d = \rho/b$ and $\rho = \sum_{i=1}^{b} f_i^{eq}$ is the mean node density (which coincides with the macroscopic cell density) and the parameters h_1, h_2 have to be determined. Using the mass conservation condition, we find a relation between the two parameters (see Appendix A):

$$h_2 = \frac{1-2d}{4} h_1^2. \tag{12}$$

Finally, the equilibrium distribution can be explicitly calculated for small driving fields:

$$f_i^{eq} = d + d(d-1)h_1 \mathbf{c}_i \mathbf{E} + \frac{1}{2}d(d-1)(2d-1)h_1^2 Q_{\alpha\beta} e_\alpha e_\beta, \tag{13}$$

where $Q_{\alpha\beta} = c_{i\alpha} c_{i\beta} - \frac{1}{2}\delta_{\alpha\beta}$ is a second order tensor.

If we calculate the mean flux, using the equilibrium distribution up to first order terms of $\mathbf{E}$, we obtain from Eq. (5) the linear response relation:

$$\langle \mathbf{J}(\eta^C) \rangle = \sum_i c_{i\alpha} f_i^{eq} = \frac{b}{2} d(d-1) h_1 \mathbf{E}. \tag{14}$$

Thus, the susceptibility reads:

$$\chi = \frac{1}{2} bd(d-1)h_1 = -\frac{1}{2} b g_{eq} h_1, \tag{15}$$

where $g_{eq} = f_i^{eq}(1 - f_i^{eq})$ is the equilibrium single particle fluctuation. In Bussemaker (1996), the equilibrium distribution is directly calculated from the nonlinear lattice Boltzmann equation corresponding to a LGCA with the same rule for small external fields. In the same work, the corresponding susceptibility is determined and this result coincides with ours for $h_1 = -1$. Accordingly, we consider that $h_1 = -1$ in the following.

Our method allows us to proceed beyond the linear case, since we have explicitly calculated the equilibrium distribution of our LGCA:

$$f_i^{eq} = \frac{1}{1 + \exp\{\ln(\frac{1-d}{d}) - \mathbf{c}_i \mathbf{E} + \frac{1-2d}{4}\mathbf{E}^2\}} \longrightarrow 1 + \exp\{\ldots\}. \tag{16}$$

Using the definition of the mean lattice flux Eq. (5), we can obtain a good theoretical estimation for larger values of the field. Fig. 12 shows the behavior of the system's normalized flux obtained by simulations and a comparison with our theoretical findings. For small values of the field intensity $|\mathbf{E}|$ the linear approximation performs rather well and for larger values the agreement of our nonlinear estimate with the simulated values is more than satisfactory. One observes that the flux response to large fields saturates. This is a biologically justified result, since the speed of cells is finite and an infinite increase of the field intensity should not lead to infinite fluxes (the mean flux is proportional to the mean velocity). Experimental findings in systems of cell migration mediated by

14. Cellular Automata as Microscopic Models of Cell Migration

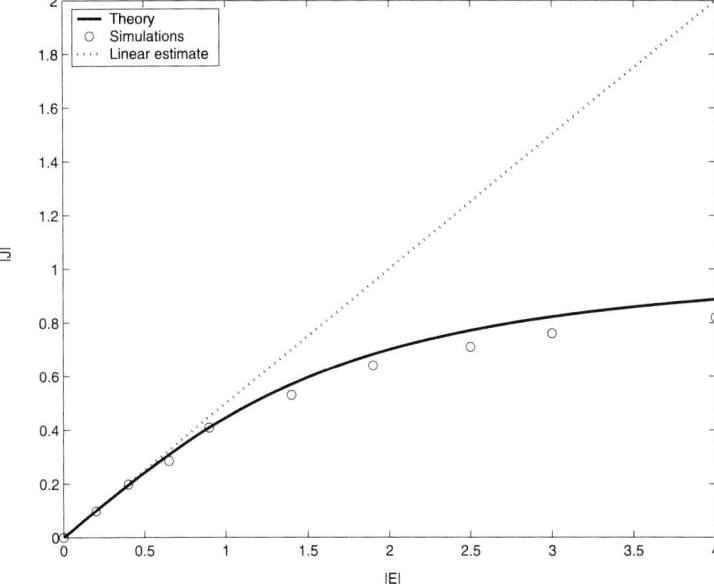

Figure 12 This figure shows the variation of the normalized measure of the total lattice flux $|\mathbf{J}|$ against the field intensity $|\mathbf{E}|$. We compare the simulated values with the theoretical calculations (for the linear and nonlinear theory). We observe that the linear theory predicts the flux strength for low field intensities. Using the full distribution, the theoretical flux is close to the simulated values also for larger field strengths.

adhesion receptors, such as ECM integrins, support the model's behavior (Palecek et al., 1997; Zaman et al., 2006). One could propose and analyze more observables related to the cell motion, e.g., mean square displacement, but this is beyond the scope of the paper. Here we aim to outline examples of typical analysis and not to reproduce the full repertoire of LGCA analysis for such problems.

B. Model II

In the following section, our analysis characterizes cell motion by a different measurable macroscopic variable and provides an estimate of the cell dispersion for model II. In this case, it is obvious that the average flux, defined in (5), is zero (due to the symmetry of the interaction rule). In order to measure the anisotropy, we introduce the flux difference between $\mathbf{v}_1$ and $\mathbf{v}_2$, where the $\mathbf{v}_i$'s are eigenvectors of the anisotropy matrix (they are linear combinations of $\mathbf{c}_i$'s). For simplicity of the calculations, we consider $b = 4$ and X–Y anisotropy. We define:

$$\left|\langle \mathbf{J}_{\mathbf{v}_1}\rangle - \langle \mathbf{J}_{\mathbf{v}_2}\rangle\right| = \left|\langle \mathbf{J}_{x^+}\rangle - \langle \mathbf{J}_{y^+}\rangle\right| = \left|c_{11} f_1^{\text{eq}} - c_{22} f_2^{\text{eq}}\right|. \tag{17}$$

As before, we expand the equilibrium distribution around the field $\mathbf{E}$ and we obtain Eq. (28) (see Appendix A). With similar arguments as for the previous model I, we can assume that the equilibrium distribution follows a kind of Fermi–Dirac distribution (compare with Eq. (9)). This time our *ansatz* has the following form,

$$x(\mathbf{E}) = h_0 + h_1|\mathbf{c}_i\mathbf{E}| + h_2\mathbf{E}^2, \qquad (18)$$

because the rule is symmetric under the rotation $\mathbf{c}_i \to -\mathbf{c}_i$. Conducting similar calculations (Appendix B) as in the previous subsection, one can derive the following expression for the equilibrium distribution:

$$f_i^{eq} = d + d(d-1)h_1|\mathbf{c}_i\mathbf{E}| + \frac{1}{2}d(d-1)(2d-1)h_1^2\sum_\alpha c_{i\alpha}^2 e_\alpha^2$$
$$+ d(d-1)(2d-1)h_1^2|c_{i\alpha}c_{i\beta}|e_\alpha e_\beta + d(d-1)h_2\mathbf{E}^2. \qquad (19)$$

In Appendix B, we identify a relation between h_1 and h_2 using the microscopic mass conservation law. To simplify the calculations we assume a square lattice (similar calculations can also be carried out for the hexagonal lattice case) and using $c_{11} = c_{22} = 1$, we derive the difference of fluxes along the X–Y axes (we restrict ourselves here to the linear approximation):

$$\left|f_1^{eq} - f_2^{eq}\right| = d(d-1)h_1\left|\sum_\alpha |c_{1\alpha}|e_\alpha - \sum_\alpha |c_{2\alpha}|e_\alpha\right| = d(d-1)h_1|e_1 - e_2|. \qquad (20)$$

We observe that the parameter h_1 is still free and we should find a way to calculate it. In Appendix C, we use a method similar to the work of Bussemaker (Bussemaker, 1996) and we find that $h_1 = -1/2$. Substituting this value into the last relation and comparing with simulations (Fig. 13), we observe again a very good agreement between the linear approximation and the simulations.

V. Results and Discussion

In this study, our first goal was to interpret in mathematical terms the environment related to cell migration. We have distinguished both static and dynamic environments, depending on the interactions with the cell populations. Mathematical entities called tensors enable us to extract local information about the local geometrical structure of the tissue. Technological advances, like DTI (diffusion tensor imaging), in image analysis allow us to identify the microstructure of *in vivo* tissues. The knowledge of the microenvironment gives us a detailed picture of the medium through which the cells move, at the cellular length scale. Microscopical models are able to exploit this microscaled information and capture the dynamics.

To study and analyze the effects of the microenvironment on cell migration, we have introduced a microscopical modeling method called LGCA. We have identified and modeled the two main effects of static environments on cell migration:

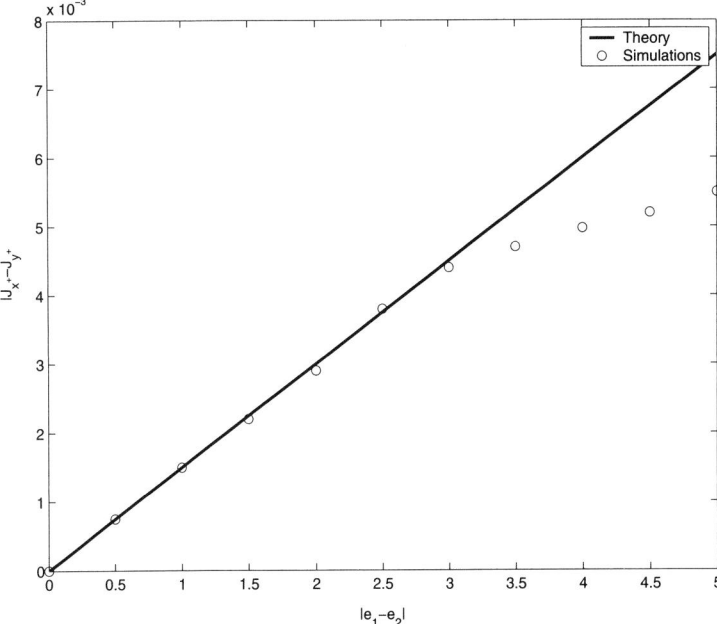

Figure 13 The figure shows the variation of the X–Y flux difference against the anisotropy strength (according to model II). We compare the simulated values with the linear theory and observe a good agreement for low field strength. The range of agreement, in the linear theory, is larger than in the case of model I.

- The first model addresses motion in an environment providing directional information. Such environments can be mediated by integrin density gradient fields or diffusible chemical signals leading to haptotactical or chemotactical movement, respectively. We have carried out simulations for different static fields, in order to understand the environmental effect on pattern formation. The main conclusion is that such an environment favors the collective motion of the cells in the direction of the gradients. Interestingly, we observe in Fig. 7 that the cell population coarsely keeps the shape of the initial cluster and moves towards the same direction. This suggests that collective motion is not necessary an alternative cell migration strategy, as described in (Friedl, 2004). According to our results, collective motion can be interpreted as emergent behavior in a population of amoeboidly moving cells in a directed environment. Finally, we have calculated theoretically an estimator of the cell spreading speed, i.e., the mean flux for variations of the gradient field strength. The results exhibit a positive response of the cell flux to increasing field strength. The saturation of the response for large stimulus emphasizes the biological relevance of the model.
- The second model describes cell migration in an environment that influences the orientation of the cells (e.g., alignment). Fibrillar ECMs induce cell alignment and

can be considered as an example of an environment that affects cell orientation. Simulations show that such motion produces alignment along a principal orientation (i.e., fiber) and the cells tend to disperse along it (Fig. 9). Like model I, we have calculated the cell response to variations of the field strength, in terms of the flux difference between the principal axis of motion and its perpendicular. This difference gives us an estimate of the speed and the direction of cell dispersion. Finally, we observe a similar saturation plateau for large fields, as in model I. Moreover, we gave an application of the second model for the case of brain tumor growth using DTI data (Fig. 11).

As we have shown, the environment can influence cell motion in different ways. An interesting observation is that directional movement favors collective motion in the direction imposed by the environment. In contrast, model II imposes diffusion of cells along a principal axis of anisotropy and leads to dispersion of the cells. For both models, the cells respond positively to an increase of the field strength and their response saturates for infinitely large drives.

In the following discussion we focus on a critical evaluation of the modeling potential of lattice-gas cellular automata suggested in this paper as models of a migrating cell population in heterogeneous environments. Firstly, we discuss the advantages of the method:

- The LGCA rules can mimic the microscopic processes at the cellular level (coarse-grained subcellular dynamics). Here we focused on the analysis of two selected microscopic interaction rules. Moreover, we showed that with the help of methods motivated by statistical mechanics, we can estimate the macroscopic behavior of the whole population (e.g., mean flux).
- The discrete nature of the LGCA can be extremely useful for deeper investigations of the boundary layer of the cell population. Recent research by Bru et al. (Bru et al., 2003) has shown that the fractal properties of tumor surfaces (calculated by means of fractal scaling analysis) can provide new insights for a deeper understanding of the cancer phenomenon. In a forthcoming paper by De Franciscis et al. (De Franciscis et al., 2007), we give an example of the surface dynamics analysis of a LGCA model for tumor growth.
- Motion through heterogeneous media involves phenomena at various spatial and temporal scales. These cannot be captured in a purely macroscopic modeling approach. In macroscopic models of heterogeneous media diffusion is treated by using powerful methods that homogenize the environment by the definition of an effective diffusion coefficient (the homogenization process can be perceived as an intelligent averaging of the environment in terms of diffusion coefficients). Continuous limits and effective descriptions require characteristic scales to be bounded and their validity lies far above these bounds (Lesne, 2007). In particular, it is found that in motion through heterogeneous media, anomalous diffusion (subdiffusion) describes the particles' movement over relevant experimental time scales, particularly if the environment is fractal (Saxton, 1994); present macroscopic continuum equations

14. Cellular Automata as Microscopic Models of Cell Migration

cannot describe such phenomena. On the other hand, discrete microscopic models, like LGCA, can capture different spatiotemporal scales and they are well-suited for simulating such phenomena.
- Moreover, the discrete structure of the LGCA facilitates the implementation of extremely complicated environments (in the form of tensor fields) without any of the computational problems characterizing continuous models.
- LGCA are perfect examples of parallelizable algorithms. This fact makes them extremely computationally efficient.

While defining a new LGCA model, the following points have to be treated with special caution:

- An important aspect of the LGCA modeling approach to multiscale phenomena is the correct choice of the spatiotemporal scales. In particular, cell migration in heterogeneous environments involves various processes at different spatiotemporal scales. The LGCA models considered in this paper focus at the cellular scale. Moreover, intracellular effects are incorporated in a coarse-grained manner. If one attempts to deal with smaller scales explicitly, extensions of the suggested models are required. In the LGCA (and LBE) literature one can find various solutions for multiscale phenomena, as extension of the state space or multigrid techniques (Succi, 2001).
- An arbitrary choice of a lattice can introduce artificial anisotropies in the evolution of some macroscopic quantities, e.g., in the square lattice the 4th rank tensor is anisotropic (for instance the macroscopic momentum and pressure tensor). These problems can be solved by the use of hexagonal lattices (Frisch *et al.*, 1987) in 2D and the FCHC (Face Centered HyperCubic) lattices (Succi, 2001).
- The simplified LGCA dynamics and the discreteness of space and time can introduce spurious staggered invariants, which can produce undesired artifacts like spurious conservation laws (Kadanoff *et al.*, 1989).

In this paper, we have analyzed LGCA models for nonproliferative cell motion in static heterogeneous environments. A straightforward extension of our models would be the explicit modeling of dynamic cell–environment interactions, i.e., proteolytic activity and ECM remodeling. The works of Chauviere *et al.* (2007) and Hillen (2006) have modeled and analyzed the interactions of cell populations and dynamic environment in terms of kinetic models and transport equations, respectively. The models have shown self-organization of a random environment into a network that facilitates cell motion. However, these models are mass-conserving (the cell population is not changing in time). Our goal is to model and analyze similar situations with LGCA for a cell population that interacts with its environment and changes its density along time.

Moreover, the introduction of proliferation allows the triggering of traveling fronts. In Hatzikirou *et al.* (2007) we have modeled tumor invasion as a diffusive, proliferative cell population (no explicit consideration of any environment) and we have calculated the speed of the traveling front in terms of microscopically accessible parameters. It would be interesting to analyze the environmental effect on the speed of the tumor expansion.

A further interesting aspect of LGCA is that the corresponding Lattice Boltzmann Equations (LBE) can be used as efficient numerical solvers of continuous macroscopic equations. In particular, LBEs which are LGCA with continuous state space, have been used as efficient solvers of Navier–Stokes equations for fluid dynamical problems (Succi, 2001). In an on-going project with K. Painter (Hatzikirou *et al.*, 2007), we try to use LBE models as efficient numerical solvers of kinetic models.

Note that apart from cell migration, the microenvironment plays an important role in the evolutionary dynamics (as a kind of selective pressure) of evolving cellular systems, like cancer (Anderson *et al.*, 2006; Basanta *et al.*, 2007). It is evident that a profound understanding of microenvironmental effects could help not only to understand developmental processes but also to design novel therapies for diseases such as cancer.

In summary, a module-oriented modeling approach, as demonstrated in this paper, hopefully contributes to an understanding of migration strategies which contribute to the astonishing phenomena of embryonic morphogenesis, immune defense, wound repair and cancer evolution.

Acknowledgments

The authors thank their colleagues Fernando Peruani, Edward Flach, and David Basanta (Dresden) for their valuable comments and corrections. Moreover, us thank Thomas Hillen (Edmonton) and Kevin Painter (Edinburgh) for fruitful discussions. We thank the Marie-Curie Training Network "Modeling, Mathematical Methods and Computer Simulation of Tumor Growth and Therapy" for financial support (through Grant EU-RTD-IST-2001-38923). We gratefully acknowledge support by the systems biology network HepatoSys of the German Ministry for Education and Research through Grant 0313082C. Andreas Deutsch is a member of the DFG-Center for Regenerative Therapies Dresden—Cluster of Excellence—and gratefully acknowledges support by the Center.

Appendix A

A lattice-gas cellular automaton is a cellular automaton with a particular state space and dynamics. Therefore, we start with the introduction of cellular automata which are defined as a class of spatially and temporally discrete dynamical systems based on local interactions. In particular, a cellular automaton is a 4-tuple $(\mathcal{L}, \mathcal{E}, \mathcal{N}, \mathcal{R})$, where

- $\mathcal{L}$ is an infinite regular lattice of nodes (discrete space),
- $\mathcal{E}$ is a finite set of states (discrete state space); each node $r \in \mathcal{L}$ is assigned a state $s \in \mathcal{E}$,

- $\mathcal{N}$ is a finite set of neighbors, indicating the position of one node relative to another node on the lattice $\mathcal{L}$ (neighborhood); Moore and von Neumann neighborhoods are typical neighborhoods on the square lattice,
- $\mathcal{R}$ is a deterministic or probabilistic map

$$\mathcal{R}: \mathcal{E}^{|\mathcal{N}|} \to \mathcal{E},$$
$$\{s_i\}_{i \in \mathcal{N}} \mapsto s,$$

which assigns a new state to a node depending on the states of all its neighbors indicated by $\mathcal{N}$ (local rule).

The temporal evolution of a cellular automaton is defined by applying the function $\mathcal{R}$ synchronously to all nodes of the lattice $\mathcal{L}$ (homogeneity in space and time).

A.1. States in Lattice-Gas Cellular Automata

In lattice-gas cellular automata, velocity channels $(r, \mathbf{c}_i)$, $\mathbf{c}_i \in \mathcal{N}_b(r)$, $i = 1, \ldots, b$, are associated with each node r of the lattice. Here, b is the *coordination number* (for details see below). In addition, a variable number $\beta \in \mathbb{N}_0$ of rest channels (zero-velocity channels), $(r, \mathbf{c}_i)$, $b < i \leq b + \beta$, with $\mathbf{c}_i = \{0\}^\beta$ may be introduced. Furthermore, an exclusion principle is imposed. This requires, that not more than one particle can be at the same node within the same channel. As a consequence, each node r can host up to $\tilde{b} = b + \beta$ particles, which are distributed in different channels $(r, \mathbf{c}_i)$ with at most one particle per channel. Therefore, state $s(r)$ is given by

$$s(r) = \big(\eta_1(r), \ldots, \eta_{\tilde{b}}(r)\big) =: \boldsymbol{\eta}(r),$$

where $\boldsymbol{\eta}(r)$ is called node configuration and $\eta_i(r) \in \{0, 1\}$, $i = 1, \ldots, \tilde{b}$ are called occupation numbers which are Boolean variables that indicate the presence ($\eta_i(r) = 1$) or absence ($\eta_i(r) = 0$) of a particle in the respective channel $(r, \mathbf{c}_i)$. Therefore, the set of elementary states $\mathcal{E}$ of a single node is given by

$$\mathcal{E} = \{0, 1\}^{\tilde{b}}.$$

For any node $r \in \mathcal{L}$, the nearest lattice neighborhood $\mathcal{N}_b(r)$ is a finite list of neighboring nodes and is defined as

$$\mathcal{N}_b(r) := \{r + \mathbf{c}_i : \mathbf{c}_i \in \mathcal{N}_b, \ i = 1, \ldots, b\},$$

where b is the *coordination number*, i.e., the number of nearest neighbors on the lattice.

Fig. 1 gives an example of the representation of a node on a two-dimensional lattice with $b = 4$ and $\beta = 1$, i.e., $\tilde{b} = 5$.

In multicomponent LGCA, ς different types (σ) of particles reside on separate lattices ($\mathcal{L}_\sigma$) and the exclusion principle is applied independently to each lattice. The state variable is given by

$$s(r) = \eta(r) = \left(\eta_\sigma(r)\right)_{\sigma=1}^\varsigma = \left(\eta_{\sigma,1}(r), \ldots, \eta_{\sigma,\tilde{b}}(r)\right)_{\sigma=1}^\varsigma \in \mathcal{E} = \{0, 1\}^{\tilde{b}\varsigma}.$$

A.2. Dynamics in Lattice-Gas Cellular Automata

The dynamics of a LGCA arises from repetitive application of superpositions of local (probabilistic) *interaction* and deterministic *propagation* (migration) steps applied simultaneously at all lattice nodes at each discrete time step. The definitions of these steps have to satisfy the exclusion principle, i.e., two or more particles are not allowed to occupy the same channel.

According to a model-specific *interaction* rule ($\mathcal{R}^C$), particles can change channels (see Fig. 2) and/or are created or destroyed. The temporal evolution of a state $s(r, k) = \eta(r, k) \in \{0, 1\}^{\tilde{b}}$ in a LGCA is determined by the temporal evolution of the occupation numbers $\eta_i(r, k)$ for each $i \in \{1, \ldots, \tilde{b}\}$ at node r and time k. Accordingly, the preinteraction state $\eta_i(r, k)$ is replaced by the postinteraction state $\eta_i^C(r, k)$ determined by

$$\eta_i^C(r, k) = \mathcal{R}_i^C\left(\boldsymbol{\eta}_{\mathcal{N}(r)}(k)\right), \tag{21}$$

$$\mathcal{R}^C\left(\boldsymbol{\eta}_{\mathcal{N}(r)}(k)\right) = \left(\mathcal{R}_i^C\left(\boldsymbol{\eta}_{\mathcal{N}(r)}(k)\right)\right)_{i=1}^{\tilde{b}} = z \text{ with probability } P(\boldsymbol{\eta}_{\mathcal{N}(r)}(k) \to z)$$

with $z \in (0, 1)^{\tilde{b}}$ and the time-independent transition probability P.

In the deterministic *propagation* or streaming step (P), all particles are moved simultaneously to nodes in the direction of their velocity, i.e., a particle residing in channel $(r, \mathbf{c}_i)$ at time k is moved to channel $(r+m\mathbf{c}_i, \mathbf{c}_i)$ during one time step (Fig. 3). Here, $m \in \mathbb{N}$ determines the *speed* and $m\mathbf{c}_i$ the *translocation* of the particle. Because all particles residing at velocity channels move the same number m of lattice units, the exclusion principle is maintained. Particles occupying rest channels do not move since they have "zero velocity." In terms of occupation numbers, the state of channel $(r + m\mathbf{c}_i, \mathbf{c}_i)$ after propagation is given by

$$\eta_i^P(r + m\mathbf{c}_i, k + 1) = \eta_i(r, k), \quad \mathbf{c}_i \in \mathcal{N}_b. \tag{22}$$

Hence, if only the propagation step would be applied then particles would simply move along straight lines in directions corresponding to particle velocities.

Combining interactive dynamics with propagation, Eqs. (21) and (22) imply that

$$\eta_i^{CP}(r + m\mathbf{c}_i, k) = \eta_i(r + m\mathbf{c}_i, k + 1) = \eta_i^C(r, k). \tag{23}$$

This can be rewritten as the *microdynamical difference equations*

$$\begin{aligned}\mathcal{R}_i^C\left(\boldsymbol{\eta}_{\mathcal{N}(r)}(k)\right) - \eta_i(r, k) &= \eta_i^C(r, k) - \eta_i(r, k) \\ &= \eta_i(r + m\mathbf{c}_i, k + 1) - \eta_i(r, k) \\ &=: \mathcal{C}_i\left(\boldsymbol{\eta}_{\mathcal{N}(r)}(k)\right), \quad i = 1, \ldots, \tilde{b},\end{aligned} \tag{24}$$

14. Cellular Automata as Microscopic Models of Cell Migration 427

where the *change in the occupation numbers* due to interaction is given by

$$C_i(\eta_{\mathcal{N}(r)}(k)) = \begin{cases} 1, & \text{creation of a particle in channel } (r, \mathbf{c}_i), \\ 0, & \text{no change in channel } (r, \mathbf{c}_i), \\ -1, & \text{annihilation of a particle in channel } (r, \mathbf{c}_i). \end{cases} \quad (25)$$

In a multicomponent system with $\sigma = 1, \ldots, \varsigma$ components, Eq. (24) becomes

$$\eta_{\sigma,i}^C(r, k) - \eta_{\sigma,i}(r, k) = \eta_{\sigma,i}(r + m_\sigma \mathbf{c}_i, k+1) - \eta_{\sigma,i}(r, k)$$
$$= C_{\sigma,i}(\eta_{\mathcal{N}(r)}(k)), \quad (26)$$

for $i = 1, \ldots, \tilde{b}$, with speeds $m_\sigma \in \mathbb{N}$ for each component $\sigma = 1, \ldots, \varsigma$. Here, the change in the occupation numbers due to interaction is given by

$$C_{\sigma,i}(\eta_{\mathcal{N}(r)}(k)) = \begin{cases} 1, & \text{creation of a particle in channel } (r, \mathbf{c}_i)_\sigma, \\ 0, & \text{no change in channel } (r, \mathbf{c}_i)_\sigma, \\ -1, & \text{annihilation of a particle in channel } (r, \mathbf{c}_i)_\sigma, \end{cases} \quad (27)$$

where $(r, \mathbf{c}_i)_\sigma$ specifies the ith channel associated with node r of the lattice $\mathcal{L}_\sigma$.

Appendix B

In this appendix, we calculate in detail the equilibrium distribution for model I. For the zero-field case, we know that the equilibrium distribution is $f_i^{eq} = \rho/b = d$. Thus, we can easily find that $h_0 = \ln(\frac{1-d}{d})$. For simplicity of the notation we use f_i instead of f_i^{eq}.

The next step is to expand the equilibrium distribution around $\mathbf{E} = 0$ and we obtain:

$$f_i = f_i(\mathbf{E} = 0) + \nabla_\mathbf{E} f_i \mathbf{E} + \frac{1}{2} \mathbf{E}^T \nabla_\mathbf{E}^2 f_i \mathbf{E}. \quad (28)$$

In the following, we present the detailed calculations. The chain rule gives:

$$\frac{\partial f_i}{\partial e_\alpha} = \frac{\partial f_i}{\partial x} \frac{\partial x}{\partial e_\alpha}. \quad (29)$$

Then using Eqs. (9) and (10):

$$\frac{\partial f_i}{\partial x} = -\frac{e^x}{(1+e^x)^2} \to d(d-1), \quad (30)$$

$$\frac{\partial x}{\partial e_\alpha} = \frac{\partial}{\partial e_\alpha}(h_0 + h_1 \mathbf{c}_i \mathbf{E} + h_2 \mathbf{E}^2) = h_1 c_{i\alpha} + 2h_2 e_\alpha. \quad (31)$$

For $\mathbf{E} = 0$ we set:

$$\frac{\partial f_i}{\partial e_\alpha} = d(d-1)h_1 c_{i\alpha}, \quad (32)$$

where $\alpha = 1, 2$. Then, we calculate the second-order derivatives:

$$\frac{\partial^2 f_i}{\partial e_\alpha^2} = \frac{\partial}{\partial e_\alpha}\left(\frac{\partial f_i}{\partial x}\frac{\partial x}{\partial e_\alpha}\right) = \frac{\partial^2 f_i}{\partial x \partial e_\alpha}\frac{\partial x}{\partial e_\alpha} + \frac{\partial f_i}{\partial x}\frac{\partial^2 x}{\partial^2 e_\alpha}$$

$$= \frac{\partial^2 f_i}{\partial x^2}\left(\frac{\partial x}{\partial e_\alpha}\right)^2 + \frac{\partial f_i}{\partial x}\frac{\partial^2 x}{\partial e_\alpha^2}. \tag{33}$$

Especially:

$$\frac{\partial^2 f_i}{\partial x^2} = \frac{e^x(e^x - 1)}{(1 + e^x)^3} = d(d-1)(2d-1), \tag{34}$$

$$\frac{\partial^2 x}{\partial e_\alpha^2} = 2h_2. \tag{35}$$

Thus, relation (33) reads:

$$\frac{\partial^2 f_i}{\partial e_\alpha^2} = d(d-1)(2d-1)h_1^2 c_{i\alpha} + d(d-1)2h_2. \tag{36}$$

For the case $\alpha \neq \beta$ ($\alpha, \beta = 1, 2$), we have:

$$\frac{\partial^2 f_i}{\partial e_\alpha \partial e_\beta} = \frac{\partial}{\partial e_\beta}\left(\frac{\partial f_i}{\partial x}\frac{\partial x}{\partial e_\alpha}\right) = \frac{\partial^2 f_i}{\partial x \partial e_\beta}\frac{\partial x}{\partial e_\alpha} + \frac{\partial f_i}{\partial x}\frac{\partial^2 x}{\partial e_\alpha \partial e_\beta}$$

$$= \frac{\partial^2 f_i}{\partial x^2}\frac{\partial x}{\partial e_\alpha}\frac{\partial x}{\partial e_\beta} + \frac{\partial f_i}{\partial x}\frac{\partial^2 x}{\partial e_\alpha \partial e_\beta}. \tag{37}$$

We can easily derive:

$$\frac{\partial^2 x}{\partial e_\alpha \partial e_\beta} = 0. \tag{38}$$

Thus, Eq. (37) becomes:

$$\frac{\partial^2 f_i}{\partial e_\alpha \partial e_\beta} = d(d-1)(2d-1)h_1^2 c_{i\alpha} c_{i\beta}. \tag{39}$$

Finally, the equilibrium distribution is:

$$f_i = d + d(d-1)h_1 \mathbf{c}_i \mathbf{E} + \frac{1}{2}d(d-1)(2d-1)h_1^2 \sum_\alpha c_{i\alpha}^2 e_\alpha^2 + d(d-1)h_2 \mathbf{E}^2. \tag{40}$$

In the last relation, we have to determine the free parameters h_1, h_2. Using the mass conservation law, we can find a relation between h_1 and h_2:

$$\rho = \sum_{i=1}^{b} f_i$$

$$= \underbrace{\sum_i d}_{\rho} + d(d-1)h_1 \underbrace{\sum_i \mathbf{c}_i \mathbf{E}}_{0} + \frac{1}{2}d(d-1)(2d-1)h_1^2 \underbrace{\sum_i \sum_\alpha c_{i\alpha}^2 e_\alpha^2}_{\frac{b}{2}\mathbf{E}^2}$$

$$+ d(d-1)h_2 \sum_i h_2 \mathbf{E}^2. \tag{41}$$

For any choice of the lattice, we find:

$$h_2 = \frac{1-2d}{4}h_1^2. \tag{42}$$

Finally, the equilibrium distribution can be explicitly calculated for small driving fields:

$$f_i = d + d(d-1)h_1 \mathbf{c}_i \mathbf{E} + \frac{1}{2}d(d-1)(2d-1)h_1^2 Q_{\alpha\beta} e_\alpha e_\beta, \tag{43}$$

where $Q_{\alpha\beta} = c_{i\alpha}c_{i\beta} - \frac{1}{2}\delta_{\alpha\beta}$ is a second-order tensor.

We now calculate the mean flux, in order to obtain the linear response relation:

$$\langle \mathbf{J}(\eta^C) \rangle = \sum_i c_{i\alpha} f_i = \frac{b}{2}d(d-1)h_1 \mathbf{E}. \tag{44}$$

Thus, the susceptibility reads:

$$\chi = \frac{1}{2}bd(d-1)h_1 = -\frac{1}{2}b g_{\text{eq}} h_1. \tag{45}$$

Appendix C

In this appendix, we present details of the calculation of the equilibrium distribution for model II. To simplify the calculations, we restrict ourselves to the case of the square lattice ($b = 4$).

Using the mass conservation law, allows us to calculate the relation between h_1, h_2.

$$\rho = \sum_{i=1}^{b} f_i$$

$$= \underbrace{\sum_i d}_{\rho} + d(d-1)h_1 \sum_i |\mathbf{c}_i| \mathbf{E} + \frac{1}{2}d(d-1)(2d-1)h_1^2 \underbrace{\sum_i \sum_\alpha c_{i\alpha}^2 e_\alpha^2}_{\frac{b}{2}\mathbf{E}^2}$$

$$+ d(d-1)(2d-1)h_1^2 \underbrace{\sum_i |c_{i\alpha}c_{i\beta}|e_\alpha e_\beta}_{\frac{b}{2}\delta_{\alpha\beta}e_\alpha e_\beta} + d(d-1)h_2 \sum_i \mathbf{E}^2. \qquad (46)$$

Finally, the previous equation becomes:

$$2d(d-1)h_1 \sum_\alpha e_\alpha + d(d-1)(2d-1)h_1^2\mathbf{E}^2 + 4d(d-1)h_2\mathbf{E}^2 = 0, \qquad (47)$$

and we find:

$$h_2 = \frac{1-2d}{4}h_1^2 - \frac{1}{2}\frac{e_1+e_2}{e_1^2+e_2^2}h_1. \qquad (48)$$

Appendix D

In this appendix, we estimate the free parameter h_1 for model II.

The field $\mathbf{E}$ induces a spatially homogeneous deviation from the field-free equilibrium state $f_i(\mathbf{r}|\mathbf{E}=\mathbf{0}) = f_{\text{eq}}$ of the form:

$$f_i(\mathbf{r}|\mathbf{E}) = f_{\text{eq}} + \delta f_i(\mathbf{E}). \qquad (49)$$

We denote the transition probability as $P(\boldsymbol{\eta} \to \boldsymbol{\eta}^C) = A_{\eta\eta^C}$. The average flux is given by:

$$\langle \mathbf{J}(\boldsymbol{\eta}^C)\rangle = \sum_{i=1}^b \mathbf{c}_i \delta f_i(\mathbf{E}). \qquad (50)$$

For small $\mathbf{E}$ we expand Eq. (4) as:

$$A_{\eta\eta^C}(\mathbf{E}) \simeq A_{\eta\eta^C}(0)\{1 + [|\mathbf{J}(\boldsymbol{\eta}^C)| - |\overline{\mathbf{J}(\boldsymbol{\eta}^C)}|]\mathbf{E}\}, \qquad (51)$$

where we have defined the expectation value of $\mathbf{J}(\boldsymbol{\eta}^C)$ averaged over all possible outcomes η^C of a collision taking place in a field-free situation:

$$|\overline{\mathbf{J}(\boldsymbol{\eta}^C)}| = \sum_{\eta^C} |\mathbf{J}(\boldsymbol{\eta}^C)| A_{\eta\eta^C}(0). \qquad (52)$$

In the mean-field approximation the deviations $\delta f(\mathbf{E})$ are implicitly defined as stationary solutions of the nonlinear Boltzmann equation for a given $\mathbf{E}$, i.e.

$$\Omega_i^{10}[f_{\text{eq}} + \delta f_i(\mathbf{E})] = 0. \qquad (53)$$

Here the nonlinear Boltzmann operator is defined by:

14. Cellular Automata as Microscopic Models of Cell Migration

$$\Omega_i^{10}(\mathbf{r}, t) = \langle \eta_i^C(\mathbf{r}, t) - \eta_i(\mathbf{r}, t) \rangle_{MF}$$
$$= \sum_{\eta^C} \sum_{\eta} [\eta_i^C(\mathbf{r}, t) - \eta_i(\mathbf{r}, t)] A_{\eta\eta^C}(\mathbf{E}) F(\eta, \mathbf{r}, t), \quad (54)$$

where the factorized single node distribution is defined as:

$$F(\eta, \mathbf{r}, t) = \prod_i [f_i(\mathbf{r}, t)]^{\eta_i} [1 - f_i(\mathbf{r}, t)]^{1-\eta_i}. \quad (55)$$

Linearizing around the equilibrium distribution yields:

$$\Omega_i^{10}[f_{eq} + \delta f_i(\mathbf{E})] = \Omega_i^{10}(f_i) + \sum_j \Omega_{ij}^{11}(f_{eq}) \delta f_j(\mathbf{E}), \quad (56)$$

where $\Omega_{ij}^{11} = \frac{\partial \Omega_i^{10}}{\partial f_j}$. Moreover:

$$\Omega_i^{10}(f_{eq}) = \sum_{\eta^C, \eta} (\eta_i^C - \eta_i)\{1 + [|\mathbf{J}(\eta^C)| - \overline{|\mathbf{J}(\eta^C)|}]\mathbf{E}\} A_{\eta\eta^C}(0) F(\eta). \quad (57)$$

Using the relations $\sum (\eta_i^C - \eta_i) A_{\eta\eta^C}(0) F(\eta) = 0$ and $\sum (\eta_i^C - \eta_i) \overline{|\mathbf{J}(\eta^C)|} A_{\eta\eta^C}(0) \times F(\eta) = 0$, we obtain:

$$\Omega_i^{10}(f_{eq}) = \langle (\eta_i^C - \eta_i) |\mathbf{J}(\eta^C)| \rangle \mathbf{E}. \quad (58)$$

Around $\mathbf{E} = \mathbf{0}$ we set:

$$\sum_j \Omega_{ij}^{11}(f_{eq}) \delta f_j(\mathbf{E}) + \langle (\eta_i^C - \eta_i) |\mathbf{J}(\eta^C)| \rangle_{MF} \mathbf{E} = 0. \quad (59)$$

Solving the above equation involves the inversion of the symmetric matrix $\Omega_{ij}^{11} = 1/b - \delta_{ij}$. It can be proven that the linearized Boltzmann operator looks like:

$$\Omega_{ij}^{11} = \left\langle (\delta\eta_i^C - \delta\eta_i) \frac{\delta\eta_j}{g_{eq}} \right\rangle = \frac{1}{g_{eq}} (\langle \delta\eta_i^C, \delta\eta_j \rangle - \langle \delta\eta_i, \delta\eta_j \rangle), \quad (60)$$

where $\delta\eta_i = \eta_i - f_{eq}$ and the single particle fluctuation $g_{eq} = f_{eq}(1 - f_{eq})$. For the second term of the last part of Eq. (60), we have $\langle \delta\eta_i, \delta\eta_j \rangle = \delta_{ij} g_{eq}$. To evaluate the first term, we note that the outcome of the collision rule only depends on $\eta(\mathbf{r})$ through $\rho(\mathbf{r})$, so that the first quantity does not depend on the i and j and

$$\langle \delta\eta_i^C, \delta\eta_j \rangle = \frac{1}{b^2} \langle [\delta\rho(\mathbf{r})]^2 \rangle = \frac{1}{b} g_{eq}, \quad (61)$$

where we have used $\rho(\eta) = \rho(\eta^C)$. Thus Eq. (60) takes the value $(1/b - \delta_{ij})$.

Returning to the calculation of the generalized inverse of Ω^{11}, we observe that its null space is spanned by the vector $\overbrace{(1, \ldots, 1)}^{b}$, which corresponds to the conservation of particles

$$\sum_i \delta f_i(\mathbf{E}) = 0. \tag{62}$$

The relation satisfies the solvability condition of the Fredholm alternative for Eq. (59), which enables us to invert the matrix within the orthogonal complement of the null space. With some linear algebra, we can prove that the generalized inverse $[\Omega^{11}]^{-1}$ has the same eigenvectors but inverse eigenvalues as the original matrix Ω^{11}. In particular, it can be verified that since $\mathbf{c}_{\alpha i}$, $\alpha = 1, 2$ (where 1, 2 stands for x- and y-axis, respectively) are eigenvectors of Ω^{11} with eigenvalue -1, we have

$$\sum_j [\Omega_{ij}^{11}]^{-1} c_{aj} = -c_{ai}. \tag{63}$$

Now we can calculate the flux of particles for one direction:

$$\langle J_{a^+}(\eta^C) \rangle = -\sum_j c_{aj} [\Omega_{ij}^{11}]^{-1} \langle (\eta_i^C - \eta_i) | \mathbf{J}_\beta(\eta^C) | \rangle e_a$$
$$= c_{ai} \langle (\eta_i^C - \eta_i) | \mathbf{J}_\beta(\eta^C) | \rangle e_a. \tag{64}$$

Calculating in detail the last relation:

$$c_{ai} \langle (\eta_i^C - \eta_i) | \mathbf{J}_\beta(\eta^C) | \rangle = \tag{65}$$
$$= c_{ai} \sum_j |c_{\beta j}| \langle (\delta \eta_i^C - \delta \eta_i) \delta \eta_j^C \rangle \tag{66}$$
$$= \frac{1}{2} g_{\text{eq}} c_{ai}. \tag{67}$$

The observable quantity that we want to calculate for the second rule is:

$$|\langle \mathbf{J}_{x^+}(\eta^C) \rangle - \langle \mathbf{J}_{y^+}(\eta^C) \rangle| = \frac{1}{2} g_{\text{eq}} |e_1 - e_2|, \tag{68}$$

since $c_{11} = c_{22} = 1$.

References

Alexander, F. J., Edrei, I., Garrido, P. L., and Lebowitz, J. L. (1992). Phase transitions in a probabilistic cellular automaton: Growth kinetics and critical properties. *J. Stat. Phys.* **68**(3/4), 497–514.

Anderson, A. R., Weaver, A. M., Cummings, P. T., and Quaranta, V. (2006). Tumor morphology and phenotypic evolution driven by selective pressure from the microenvironment. *Cell* **127**(5), 905–915.

Basanta, D., Simon, M., Hatzikirou, H., and Deutsch, A. (2007). Evolutionary game theory elucidates the role of glycolysis in glioma progression and invasion. *Cancer Res.* submitted for publication.

Bru, A., Albertos, S., Subiza, J. L., Lopez Garcia-Asenjo, J., and Bru, I. (2003). The universal dynamics of tumor growth. *Biophys. J.* **85**, 2948–2961.

Bussemaker, H. (1996). Analysis of a pattern forming lattice gas automaton: Mean field theory and beyond. *Phys. Rev. E* **53**(2), 1644–1661.

14. Cellular Automata as Microscopic Models of Cell Migration

Byrne, H., and Preziosi, H. (2003). Modeling solid tumor growth using the theory of mixtures. *Math. Med. Biol.* **20**(4), 341–366.

Carter, S. B. (1965). Principles of cell motility: The direction of cell movement and cancer invasion. *Nature* **208**(5016), 1183–1187.

Chauviere, A., Hillen, T., and Preziosi L. (2007). Modeling the motion of a cell population in the extracellular matrix. *Discr. Cont. Dyn. Syst.* (to appear).

Chopard, B., and Droz, M. (1998). "Cellular Automata Modeling of Physical Systems." Cambridge Univ. Press, Cambridge.

Dallon, J. C., Sherratt, J. A., and Maini, P. K. (2001). Modeling the effects of transforming growth factor on extracellular alignment in dermal wound repair. *Wound Rep. Reg.* **9**, 278–286.

De Franciscis, S., Hatzikirou, H., and Deutsch, A. (2007). Evaluation of discrete models of avascular tumor growth by means of fractal scaling analysis (preprint).

Deutsch, A., and Dormann, S. (2005). "Cellular Automaton Modeling of Biological Pattern Formation." Birkhäuser, Boston.

Dickinson, R. B., and Tranquillo, R. T. (1993). A stochastic model for cell random motility and haptotaxis based on adhesion receptor fluctuations. *J. Math. Biol.* **31**, 563–600.

Dickinson, R. B., and Tranquillo, R. T. (1995). Transport equations and cell movement indices based on single cell properties. *SIAM J. Appl. Math.* **55**(5), 1419–1454.

Dolak, Y., and Schmeiser, C. (2005). Kinetic models for chemotaxis: Hydrodynamic limits and spatiotemporal mechanics. *J. Math. Biol.* **51**, 595–615.

Friedl, P. (2004). Prespecification and plasticity: Shifting mechanisms of cell migration. *Curr. Opin. Cell. Biol.* **16**(1), 14–23.

Friedl, P., and Broecker, E. B. (2000). The biology of cell locomotion within a three dimensional extracellular matrix. *Cell Motil. Life Sci.* **57**, 41–64.

Friedl, P., and Wolf, K. (2003). Tumor-cell invasion and migration: Diversity and escape mechanisms. *Nat. Rev.* **3**, 362–374.

Frisch, U., d'Humieres, D., Hasslacher, B., Lallemand, P., Pomeau, Y., and Rivet, J. -P. (1987). Lattice gas hydrodynamics in two and three dimensions. *Compl. Syst.* **1**, 649–707.

Galle, J., Aust, G., Schaller, G., Beyer, T., and Drasdo, D. (2006). Individual cell-based models of the spatial-temporal organization of multicellular systems—achievements and limitations. *Cytom. Part A* **69A**(7), 704–710.

Grima, R. (2007). Directed cell migration in the presence of obstacles. *Theor. Biol. Med. Model.* **4**, 2.

Hatzikirou, H., Brusch, L., Schaller, C., Simon, M., and Deutsch, A. (2007). Characterization of traveling front behavior in a lattice gas cellular automaton model of glioma invasion. *Math. Comp. Mod.* (in print).

Hatzikirou, H., Deutsch, A., Schaller, C., Simon, M., and Swanson, K. (2005). Mathematical modeling of glioblastoma tumor development: A review. *Math. Mod. Meth. Appl. Sci.* **15**(11), 1779–1794.

Hatzikirou, H., Painter, K., and Deutsch, A. (2007). Numerical solvers of transport equations modeling individual cell motion. *J. Math. Biol.* (in preparation).

Hillen, T. (2006). (M5) Mesoscopic and macroscopic models for mesenchymal motion. *J. Math. Biol.* **53**, 585–616.

Kadanoff, L. P., McNamara, G. R., and Zanetti, G. (1989). From automata to fluid flow: Comparisons of simulation and theory. *Phys. Rev. A* **40**, 4527–4541.

Keller, E. F., and Segel, L. A. (1971). Traveling bands of chemotactic bacteria: A theoretical analysis. *J. Theor. Biol.* **30**, 235–248.

Lesne, A. (2007). Discrete vs continuous controversy in physics. *Math. Struct. Comp. Sci.* (in print).

Liggett, T. M. (1985). "Interacting Particle Systems." Springer-Verlag, Berlin.

McCarthy, J. B., and Furcht, L. T. (1984). Laminin and fibronectin promote the haptotactic migration of B16 mouse melanoma cells. *J. Cell Biol.* **98**(4), 1474–1480.

Murray, J. D., Oster, G. F., and Harris, A. K. (1983). A mechanical model for mesenchymal morphogenesis. *J. Math. Biol.* **17**, 125–129.

Newman, T. J., and Grima, R. (2004). Many-body theory of chemotactic cell–cell interactions. *Phys. Rev. E* **70**, 051916.

Okubo, A., and Levin, S. A. (2002). "Diffusion and Ecological Problems: Modern Perspectives." Springer-Verlag, New York.

Othmer, H. G., and Stevens, A. (1997). Aggregation, blowup and collapse: The ABC's of taxis in reinforced random walks. *SIAM J. Appl. Math.* **57**, 1044–1081.

Othmer, H. G., Dunbar, S. R., and Alt, W. (1988). Models of dispersal in biological systems. *J. Math. Biol.* **26**, 263–298.

Palecek, S. P., Loftus, J. C., Ginsberg, M. H., Lauffenburger, D. A., and Horwitz, A. F. (1997). Integrin-ligand binding governs cell-substratum adhesiveness. *Nature* **388**(6638), 210.

Peruani, F., and Morelli, L. (2007). Self-propelled particles with fluctuating speed. *Phys. Rev. Lett.* **99**, 010602.

Saxton, M. (1994). Anomalous diffusion due to obstacles: A Monte Carlo study. *Biophys. J.* **66**, 394–401.

Schweitzer, F. (2003). "Brownian Agents and Active Particles." Springer-Verlag, Berlin.

Succi, S. (2001). "The Lattice Boltzmann Equation: For Fluid Dynamics and Beyond." "Numerical Mathematics and Scientific Computation." Oxford Univ. Press, Oxford, NY.

Swanson, K. R., Alvord Jr., E. C., and Murray, J. D. (2002). Virtual brain tumors (gliomas) enhance the reality of medical imaging and highlights inadequacies of current therapy. *Brit. J. Canc.* **86**, 14–18.

Turing, A. M. (1952). The chemical basis of morphogenesis. *Philos. Trans. R. Soc. London B* **237**, 37–72.

Zaman, M. H., Matsudaira, P., and Lauffenburger, D. A. (2006). Understanding effects of matrix protease and matrix organization on directional persistence and translational speed in three-dimensional cell migration. *Ann. Biomed. Eng.* **35**(1), 91–100.

15

Multiscale Modeling of Biological Pattern Formation

Ramon Grima
Institute for Mathematical Sciences, Imperial College, London SW7 2PG, United Kingdom

I. Introduction
II. Quantitative Modeling
III. Building Cellular and Tissue-Level Models for a Simple Biological System
   A. Tissue-Level Modeling
   B. Cell-Level Modeling
IV. Mean-Field Theory and the Interrelationship of Models at Different Spatial Scales
   A. Coarse Graining
   B. Mean-Field Theory
V. Multiple Scale Analysis
   A. Weak Intercellular Interactions
   B. Strong Intercellular Interactions
VI. Discussion
   References

In the past few decades, it has become increasingly popular and important to utilize mathematical models to understand how microscopic intercellular interactions lead to the macroscopic pattern formation ubiquitous in the biological world. Modeling methodologies come in a large variety and presently it is unclear what is their interrelationship and the assumptions implicit in their use. They can be broadly divided into three categories according to the spatial scale they purport to describe: the molecular, the cellular and the tissue scales. Most models address dynamics at the tissue-scale, few address the cellular scale and very few address the molecular scale. Of course there would be no dissent between models or at least the underlying assumptions would be known if they were all rigorously derived from a molecular level model, in which case the laws of physics and chemistry are very well known. However in practice this is not possible due to the immense complexity of the problem. A simpler approach is to derive models at a coarse scale from an intermediate scale model which has the special property of being based on biology and physics which are experimentally well studied. In this article we use such an approach to understand the assumptions inherent in the use of the most popular models, the tissue-level ones. Such models are found to invariably rely on the hidden assumption that statistical correlations between cells can be neglected. This often means that the predictions of these models are qualitatively correct but may fail in spatial regions where cell concentration is small, particularly if there are strong long-range correlations in cell movement. Such behavior can only be properly taken into account by cellu-

lar models. However such models unlike the tissue-level models are frequently not easily amenable to analysis, except when the number of interacting cells is small or when the interactions are weak, and thus are rather more suited for simulation. Hence it is our conclusion that the simultaneous theoretical and numerical analysis of models of the same biological system at different spatial scales provides a more robust method of understanding biological systems than the utilization of a single scale model. In particular this enables one to clearly separate nonphysical predictions stemming from model artifacts from those due to genuine physiological behavior.
© 2008, Elsevier Inc.

I. Introduction

Cell migration is at the basis of much of pattern formation in nature. Cell movement particularly that directed by extracellular chemical signals, e.g., chemotaxis and haptotaxis, plays a fundamental role in a wide range of developmental processes, a few examples being: primordial germ cell migration in the chick, mouse and zebrafish embryos (Reichman-Fried *et al.*, 2004; Weidinger *et al.*, 2002); gastrulation and limb development in the chick embryo (Li and Muneoka, 1999; Yang *et al.*, 2002); attraction of sperm to the ova or oviducts in a diverse range of invertebrates and vertebrates (Ralt *et al.*, 1994; Xiang *et al.*, 2004; Yoshida *et al.*, 1993); the wiring of the developing nervous system (Ming *et al.*, 2002). Its significance is however not restricted to the developing embryo, but is also of crucial importance in disease (Ridley *et al.*, 2003), e.g., cancer and arthritis, and in biofilm dynamics (Hall-Stoodley *et al.*, 2004), a subject of current technological importance.

The study of cell movement has a long history, starting with Anthony van Leeuwenhoek who in 1674 was the first to observe swimming cells in a drop of water using a primitive microscope. Since then, advances in microscope technology have enabled us to study in detail the mechanics of individual cell motion. This has shed light on how the coupling of intracellular processes such as actin polymerization (in amoeboid cells) and receptor activation are responsible for how cells alter their movement in response to external stimuli (Bray, 2001). Having made substantial progress in understanding the basic units of life, the primary challenge ahead lies in understanding how these units, the cells, interact with each other to produce the large-scale organization typical of tissues. Studies which directly address such problems can be experimental or theoretical, the latter only coming into fashion in the last half of the twentieth century. In this article we will discuss the multiscale modeling of tissue, in particular focusing on how the choice of the mathematical approach to the problem depends on the questions of interest and on the spatial and temporal length scales inherent to the biological processes under investigation.

In the last fifty years, the role of mathematical modeling in biology has vastly increased in scope and application, and nowadays it is considered an important tool

15. Multiscale Modeling of Biological Systems

to interpret experimental data and also to test hypotheses and generate new predictions (Mogilner *et al.*, 2006; Schnell *et al.*, 2007). One may ask why there is the need at all of such modeling when the basic details about the fundamental units, the cells, and their interactions are well known. The answer is that a system of discrete entities (these can represent molecules, cells or whole organisms) interacting with each other via very simple rules can produce incredibly complex and intricate behavior which generally is difficult or impossible to predict by "simply thinking about it." An example of this is Conway's Game of Life (Gardner, 1970). This is a two-dimensional computer simulation (a cellular automaton) in which two types of discrete entities, here referred to as A and B, are placed on a square grid and interact via four rules: (i) if an A has less than two A neighbors then it changes to B, (ii) if an A has more than three A neighbors it changes to B, (iii) if A has two or three A neighbors then it remains unchanged, (iv) if a B has exactly three A neighbors then it changes to A. These rules are very simple but when they are iterated it is found that the system exhibits a large number of different life-like behaviors such as the appearance of still and oscillating ordered patterns from initially random ones and arrangements of entities which glide over the grid. These nontrivial patterns can be qualitatively predicted in many instances by theory (Deutsch and Dormann, 2005). Simple cellular automaton models such as the above illustrate the need of computer simulation and mathematical modeling in understanding how cell–cell interactions at the microscopic level lead to the macroscopic form and order which is ubiquitous in nature.

The remainder of this article is organized as follows. In Section II we discuss how mathematical models of biological systems can be constructed at various spatial scales of interest and how these models can, separately or combined, provide insight in the system under study. In Section III we illustrate the process of model building at two different scales for a rudimentary chemotactic intercellular communication system which incidentally is a popular abstraction of a number of different biological systems. In Section IV we discuss the relationship of models at different scales to each other in the framework of mean-field theory. This sheds light on the models' range of validity and applicability. We conclude in Section V by showing how multiple scale analysis of the same biological system is more robust then single scale analysis. Such an analysis enables one to weed out nonphysical predictions due to model artifacts and hence to build a coherent picture of the biological dynamics.

II. Quantitative Modeling

Mathematical models can be constructed at various scales of interest. Three common spatial scales which are of experimental relevance are the molecular, the cellular and the supercellular or tissue scale. In principle, given complete knowledge of the molecular-level processes occurring inside a cell, one can construct a model which is correct and valid for all spatial scales of interest. However, in practice, because this knowledge is incomplete and also because of the immense number of equations (one needs equations for the position and velocity of each individual molecule in a cell!) needed to encode such detail, it is not feasible to build and analyze such detailed microscopic models of cell movement. We shall not treat such models further but instead focus on cellular and tissue models.

By cellular models, we mean those which precisely describe the movement of each individual cell. In these models, the natural spatial scale is set by the size of the cell and so chemical molecules which are much smaller than this scale are modeled by means of a continuum concentration field, not as discrete entities. In this description, a colony of N cells interacting via M different chemicals would be modeled by $3N$ equations of motion for the cells in terms of the position and velocities of their center of mass and M equations describing how the chemical concentrations vary with space and time. These models can be based on cellular automata (Deutsch and Dormann, 2005) or lattice-free Monte Carlo simulations (Drasdo et al., 1995; Grima, 2007; Newman and Grima, 2004; Grima and Schnell, 2007).

Tissue-level models are even coarser than cellular models. The spatial scale is here much larger than that of a cell, meaning that one cannot distinguish between individual cells or molecules. These are population-type models in which one is interested in the collective dynamics of the colony rather than in its individual components. Hence in this case, both cells and chemicals are modeled by means of continuum concentration fields. A colony of N cells interacting via M different chemicals would be described by just $1 + M$ equations, a substantial reduction in complexity compared to the cellular model. Tissue-level models are by far the most common in the scientific literature mainly owing to the fact that they are based on sets of coupled ordinary or partial differential equations for which many methods of analysis are available (Murray, 2002; Murray, 2003). The differences between cell and tissue-level modeling are illustrated in Fig. 1.

It is to be emphasized that models can be built at other scales, for example, one could construct sub-cellular models in which the relevant spatial scale is a fraction of the cell diameter. Examples of such models are the Cellular Potts model (Graner and Glazier, 1992), Hyphasma (Meyer-Hermann and Maini, 2005), the SubCellular Element Model (Newman, 2005), and continuum models based on a viscoelastic description of the intracellular environment (Gracheva and Othmer, 2004). These models are currently for the most part more phenomenological than cellular models due to the large detail involved in describing the intracellular environment.

Models at different scales can also be combined in a module-like fashion to form a supermodel which describes in an approximate way a biological system over a wide range of spatial and temporal scales. These models are not usually amenable to mathematical analysis but are suitable for studying, via computer simulation, very complex processes such as those involved in cancer (Alarcon et al., 2004).

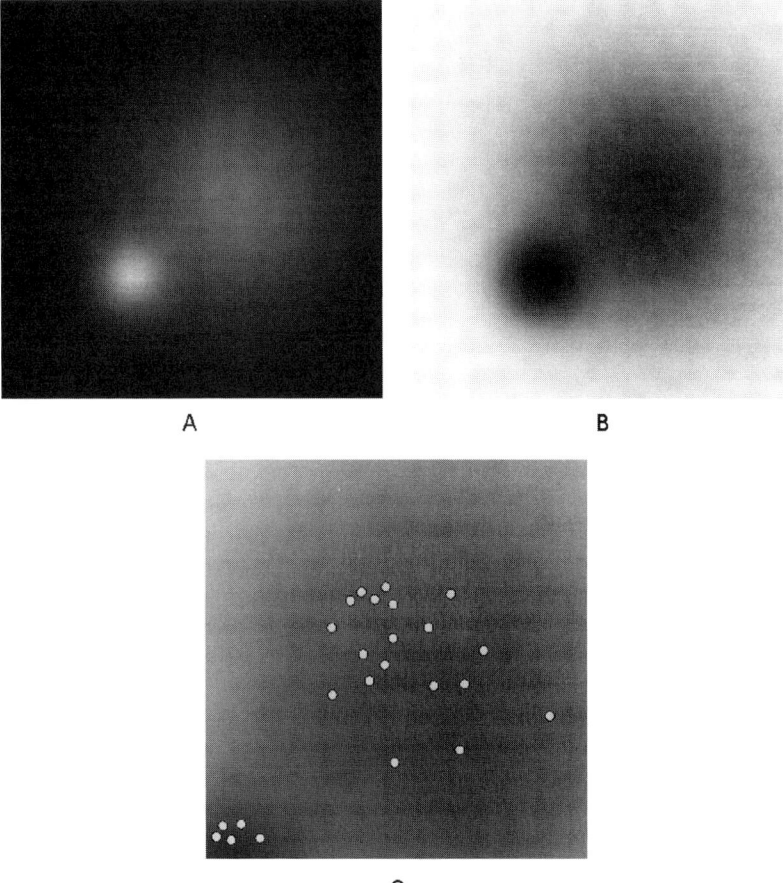

Figure 1 Schematic illustrating tissue (A and B) and cellular models (C). Tissue-level models are at a scale much larger than that of an individual cell and so the only variables are cell concentration (shown by the intensity of the color green in A) and chemical concentration (shown by the gray scale in B where white denotes regions of high chemical concentration). Note that in this example, chemical is absorbed by cells and thus regions of large cell concentration correspond to regions of low chemical concentration. However generally the relationship between the two is much more complex. Cellular-level models are at a much finer spatial scale such that individual cells (solid green circles in C) can be distinguished; molecules are much smaller than cells and hence only a chemical concentration can be defined at this level (represented by gray-scale background in C). See color insert.

III. Building Cellular and Tissue-Level Models for a Simple Biological System

In this section we illustrate the process of model building at two different spatial scales, namely at the cellular and tissue scales. To make our discussions concrete,

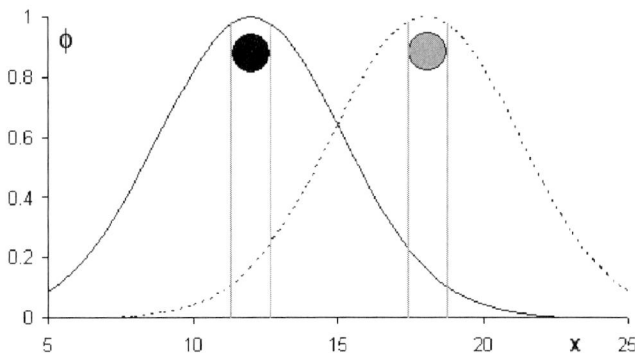

Figure 2 Two chemotactic cells interacting via chemical fields. The cells continuously secrete chemical which then diffuses and decays at a later time. Thus at a snapshot in time, t, the chemical concentration ϕ around a cell due to its own secretion is a Gaussian (solid and dashed curves). Each cell measures the difference in chemical concentration across its body; the black cell senses a chemical gradient pointing towards the gray cell (difference in concentration indicated by points where the solid vertical gray lines intersect with dotted and vertical Gaussians). The same is valid for the gray cell. Hence cells will move towards or away from each other, depending on whether they are positively or negatively chemotactic.

we consider a simple biological system composed of cells which continuously secrete a chemical at a constant rate and which chemotactically respond to it by moving up (positive chemotaxis) or down (negative chemotaxis) the local chemical gradient present across the cell. The chemical diffuses and degrades in solution after some time. This is amongst the simplest examples of a cell–cell communication system. If the cells are strongly chemotactic then the expected net result is aggregation if the cells are positively chemotactic and dispersion if they are negatively chemotactic (Fig. 2).

This simple system was originally postulated as a crude model for the aggregation stage of the life-cycle of the slime mold (Keller and Segel, 1970). In that case, chemotaxis is strictly positive and the secreted chemical is cAMP. By adding more processes to the core model described above, one finds that it describes a wide range of biological systems. It can be applied to model the preaggregation stage of the slime mold where the amoeboid cells secrete a chemorepellent, i.e., a chemical to which they do negative chemotaxis to, while simultaneously performing positive chemotaxis to folic acid trails left by their food source, bacteria. Other examples in which the model finds use are: (a) the formation of plaques in Alzheimer's disease, which form by aggregation of microglia (Edelstein-Keshet and Spiros, 2002), (b) the formation of regular striking patterns of bacterial colonies of *Escherichia coli* and *Salmonella typhimurium* feeding on intermediates of the tricarboxylic acid cycle (Murray, 2003), (c) the dynamics of blood vessel formation (Gamba *et al.*, 2003).

A. Tissue-Level Modeling

The tissue-level model for this simple biological system was first written down by Keller and Segel (1970) in the context of slime mold aggregation. It has two main variables: the chemical concentration Φ and the cell number density or concentration ρ. These two completely define the state of the system at a given time and point in space. The equations defining the model are:

$$\frac{\partial \rho}{\partial t} = D_0 \nabla^2 \rho - \alpha \nabla \cdot \rho \nabla \Phi, \tag{1}$$

$$\frac{\partial \Phi}{\partial t} = D_1 \nabla^2 \Phi - \lambda \Phi + \beta \rho. \tag{2}$$

The second equation follows directly from well-known physical principles. This is a reaction–diffusion equation modeling the diffusion of chemical with diffusion coefficient D_1, its decay or degradation in solution at a rate λ and chemical secretion by cells at a rate β.

The first equation was written down as an approximate phenomenological description of the average movement of a colony of cells with time. This was done by analogy with two physical laws, namely those governing diffusion and heat transfer (Keller and Segel, 1971). The first term on the right of Eq. (1) represents cell motion in the absence of chemotaxis while the second term takes into account chemotaxis. In the absence of a chemical gradient, $\nabla \Phi = 0$, the equation becomes identical to the diffusion equation. This models the fact that in the absence of chemotaxis, the random motion of cells appears to be similar to that of molecules. In this case D_0 is a measure of the space spanned by motile cells per unit time, otherwise called the cell diffusion coefficient. The second term which describes chemotaxis can be deduced by analogy with Fourier's law of heat conduction which states that the heat flux through a material is proportional to the temperature gradient. In a similar way one can hypothesize that cell flux due to chemotaxis is proportional to the chemical gradient. The parameter α is the chemotactic sensitivity which is analogous to conductivity in heat transfer problems; it is positive for positive chemotaxis and negative for negative chemotaxis. We note that this model has not been derived from a finer-scale model or from first principles but rather by analogy with unrelated though qualitatively similar physical phenomena. This is a general common problem to many tissue-models which means that frequently the assumptions implicit in such models are not exactly known, a topic we address in detail in Section IV.

B. Cell-Level Modeling

We now consider a cellular model for the above biological system. As we have previously remarked this approach consists of writing down equations describing the movement of each individual cell rather than a population-type description as for the tissue-level model. If we set the spatial scale of interest to be about an order of magnitude larger than the cell diameter then small scale details such as the actual cell shape can be ignored meaning that the dynamics of the cell's center of mass is the only relevant process for modeling purposes. As we now show, the model can be rigorously built using experimental data of single cell motion in the absence and presence of chemotaxis.

Experiments show that in the absence of chemical or adhesion gradients, cell velocities are correlated for some short time τ (which varies with the cell type) and uncorrelated for times longer than τ (Mombach and Glazier, 1996; Rieu et al., 2000). Hence provided τ is sufficiently small, cells can be considered to perform random walks and their movement statistics to be analogous to the statistics of the Brownian motion of molecules in solution. Thus by analogy cells have a diffusion coefficient D_0 characterizing their diffusion-like motion. Note that the origin of the random walk behavior of cells has physically nothing to do with that of molecules; the two simply share a common mathematical underpinning. For cells, D_0 simply reflects the incessant fluctuations in the actin polymerization dynamics of their cytoskeleton while for solute molecules it reflects the rate of collision of the smaller solvent molecules with them (Rieu et al., 2000). Typically the cell diffusion coefficient is one to two orders of magnitude smaller than the chemical diffusion coefficient of free molecules in dilute solutions.

It thus follows from experiment that a cell can be modeled as a random walker. Using Newton's second law one can then immediately deduce that the equation of motion for the center of mass coordinates, $\vec{x}_i(t) = \{x_i(t), y_i(t), z_i(t)\}$, of a single cell, labeled i, is given by:

$$m \frac{\partial^2 \vec{x}_i(t)}{\partial t^2} = -k \frac{\partial \vec{x}_i(t)}{\partial t} + \vec{\eta}_i(t) + F_c(\vec{x}_i(t), t). \qquad (3)$$

The forces acting on the cell are the viscous drag force, $k \frac{\partial \vec{x}_i(t)}{\partial t}$, due to the surrounding fluid, a stochastic force, $\vec{\eta}_i(t)$, which takes into account the cell's apparent random motion and a force F_c which directs the cell towards or away from the source of the chemotactic stimulus. The stochastic force term is usually modeled as white noise which has two main statistical properties, namely that both the mean force and the correlation time τ are zero. This is appropriate for modeling cells with small τ. A more general description involves using colored noise instead of white noise which has a nonzero correlation time.

Cells and microorganisms live in an environment with different hydrodynamic properties than the one we experience in our every day life. Viscous forces dominate their movement whereas inertia is an insignificant factor for cell motility (Berg, 1993; Bray, 2001; Purcell, 1977). Practically speaking this means that water jet propulsion is not efficient for cells and that streamlined shapes are also of no help to their movement. More importantly for our discussion this means that the acceleration term on the left hand side of Eq. (3) is negligible compared to the viscous drag force term and so it can be neglected, leading to a simplified model equation:

$$\frac{\partial \vec{x}_i(t)}{\partial t} = \vec{\eta}_i(t) + F_c(\vec{x}_i(t), t). \tag{4}$$

The last task in completing the mathematical description of individual cell movement involves specifying the chemotactic force, F_c. It is experimentally found that in the presence of a chemical gradient, the diffusion-like behavior persists but the mean-velocity of chemotactic cells, which was zero in the absence of chemotaxis, is now a function of the magnitude of the local chemical gradient and of the absolute value of the local chemical concentration (Lewus and Ford, 2001; Tranquillo et al., 1988). If the chemical concentration is not too large, it is found that this mean velocity is directly proportional to the chemical gradient but independent of the absolute value of the chemical concentration. From Eq. (4) it follows that the mean cell velocity is proportional to the chemotactic force and thus to comply with experiment, we set $F_c = \alpha \nabla_i \phi$, where the subscript i denotes that the gradient is evaluated at the current position of cell i. Hence the equations of motion for a cell colony composed of N cells are:

$$\frac{\partial \vec{x}_i(t)}{\partial t} = \vec{\eta}_i(t) + \alpha \nabla_i \phi, \quad i = 1, \ldots, N. \tag{5}$$

Note that each cell in the colony has associated with it a distinguishing index i. Note also that in the tissue-level model the concentration field was denoted by Φ rather than ϕ as in Eq. (5). The reason for the use of this notation will become clear in the next section. The chemotactic sensitivity α is a measure of an individual cell's response to an external chemical stimulus.

To complete the cellular model one needs to specify an equation for how the chemical concentration varies in time and space due to the three processes of diffusion, decay and secretion by the cells. This follows straightforwardly from well-known physical principles:

$$\frac{\partial \phi}{\partial t} = D_1 \nabla^2 \phi - \lambda \phi + \beta \sum_{i=1}^{N} \delta(\vec{x} - \vec{x}_i(t)). \tag{6}$$

Note that chemical is actually secreted by cells on their surface but since we are assuming that the spatial scale is an order of magnitude or so greater than cell size, this is effectively the same as saying that secretion occurs at the current position of their center of mass coordinates. This is the reasoning behind the delta function term on the right-hand side of the above equation.

The cellular model is now complete. It is fully described by the $3N + 1$ equations given by Eqs. (5) and (6). Note that there is a one-to-one correspondence between all the parameters in the cellular and tissue models. The cellular model was introduced and first studied by Newman and Grima (2004).

IV. Mean-Field Theory and the Interrelationship of Models at Different Spatial Scales

As we saw in Sections II and III there are many varied modeling methodologies for studying biological systems. Each modeling approach has its strengths and limitations and naturally is only valid provided certain conditions are satisfied. Presently much work is being done to understand the assumptions implicit in such approaches, how different models are related to each other and how to separate results due to model artifacts from those due to genuine physiological processes. For example, model results can in some instances be highly sensitive to the topology of an artificial spatial grid on which simulation of molecular or cellular movement occurs (Deutsch and Dormann, 2005; Grima and Schnell, 2006). Of course there would be no dissent between models at different scales if they were all rigorously derived from a molecular level model, in which case the laws of physics and chemistry are very well understood. However in practice this is not possible. A simpler approach is to derive models at a coarse scale from an intermediate scale model which has the special property of being based on biology and physics which are experimentally well studied. In this section we use such an approach to understand the assumptions inherent in the use and application of the tissue-level model of our simple biological system. This will involve the derivation of this phenomenological model from the experimentally based and finer scale cell-level model introduced in the previous section.

A. Coarse Graining

The process of deriving models at a certain scale from those at a finer model is called coarse-graining. This usually implies performing some form of averaging of the dynamics of interaction over a certain length scale. Examples from statistical mechanics would be the derivation of the macroscopic ideal-gas laws and the stress-strain constitutive relations (Goldhirsch and Goldenberg, 2002) starting from the microscopic equations of motion for individual molecular motion. The process of going from the cellular model to the tissue-level model involves going from a description in terms of the $3N$ positions and velocities of individual cells to one in terms of a single parameter, the average number density (or concentration) of cells ρ. The definition of the number density is based on the idea of an ensemble average from statistical mechanics. Consider M separate biological systems (or experiments) which initially at $t = 0$ are all identical, meaning that they have the same exact spatial distribution of cells. Each cell's movement is described by an equation of motion as in our cellular model. Then as time progresses, the spatial distribution of cells in each of the systems is not anymore the same due to the inherent stochasticity in each cell's movement. The average cell concentration at a given point in space and time, $\rho(\vec{x}, t)$, can then be obtained by counting the total number of cells at time t having center of mass coordinates in a small volume ΔV centered about $\vec{x}$ in all the biological systems and dividing by

15. Multiscale Modeling of Biological Systems

$M \Delta V$. This is the basic idea behind ensemble averaging, which is mathematically encapsulated by the equation:

$$\rho(\vec{x}, t) = \left\langle \sum_{i=1}^{N} \delta(\vec{x} - \vec{x}_i(t)) \right\rangle. \tag{7}$$

Note that the ensemble average is denoted by the angular brackets. Taking the time derivative of Eq. (7) one obtains an equation for the time evolution of the cell concentration as a function of the positions and velocities of all cells:

$$\frac{\partial \rho}{\partial t} = -\sum_{i=1}^{N} \nabla \cdot \left\langle \frac{\partial \vec{x}_i(t)}{\partial t} \delta(\vec{x} - \vec{x}_i(t)) \right\rangle. \tag{8}$$

This equation provides the link between the tissue-level parameter, ρ, and the cell-level parameters which are the cell position and velocity. By substituting the equations defining our cellular model, Eqs. (5) and (6) in Eq. (8), one obtains the coarse-grained version of the cellular model (a detailed derivation can be found in Newman and Grima, 2004). We shall call this derived model, Model A, as opposed to the tissue-model written down directly using phenomenological arguments, defined by Eqs. (1) and (2), which we refer to as Model B.

The question of interest here is whether Models A and B are the same. The answer is no, except under certain conditions. The general, nonintuitive conclusion of the coarse-graining procedure is that generally it is not possible to derive a tissue-level model, i.e., a model in terms of only the concentrations of chemicals and cells. In particular it is found that the exact determination of how the cell concentration varies with time requires full knowledge of the statistical correlations between cells, i.e., the two-cell joint probability distribution, the three-cell joint probability distribution functions, etc. This fact suggests that Model B is an approximate description which results from Model A by making some assumptions regarding the nature of cell–cell correlations.

Statistical correlations between cells depend sensitively on a number of factors. Each cell's motion has two components; one is stochastic which models the incessant actin polymerization dynamics of their cytoskeleton (cell diffusion) while the other component leads to directed movement up or down chemical gradients (chemotaxis). Since cells themselves produce the chemotactic chemical, it follows that the second component of their motion leads to correlations between different cells. On the contrary the first component of their motion dilutes intercell correlations. Hence the strength of statistical correlations between cells depends on the relative strengths of cell diffusion and chemotaxis. Similarly one expects that if cells have a large space in which to roam then their correlations are bound to be less important than if they were constrained to move in a smaller space. If the number of cells is particularly large then it is plausible that the correlations between any two cells i and j is small. This is since the chemical field sensed by cell i will be due to the sum total of chemical secreted by all other cells at all previous times, which effectively swamps the chemical specifically secreted by cell j. However it must be emphasized that since chemotaxis is a long-range force, the magnitude of intercell correlations in our biological system is always appreciable particularly when compared to other biological systems in which cell interaction occurs only via contact-mediated forces such as adhesion. Note that in the latter case there are only correlations between nearest neighboring cells rather than between all cells as in the case of chemotaxis.

It can be shown that by assuming that cell–cell statistical correlations are very weak or negligible, one recovers Model B from Model A (similar conclusions are also borne out by a different derivation using a different model due to Stevens, 2000). Without this assumption, it is not generally possible to derive a tissue-level model from a finer scale one. This conclusion is true not only for our simple biological chemotactic system but is generally valid. It is a statement following from many-body theory, which is a general mathematical framework for studying systems consisting of many interacting discrete entities (Mattuck, 1992). The weak correlation assumption inherent in the tissue-model implies that such models are only valid for modeling biological systems where there are large numbers of cells and where chemotaxis and other-long range forces are weak compared to local forces due to neighboring cells. Another situation where the use of such models is warranted is when a chemical gradient is set up by some artificial means in a laboratory, e.g., Boyden chamber experiments. In such a case the chemical providing the stimulus is not produced by the cells and thus there cannot be any cell–cell correlations due to chemotaxis. The use of tissue-level models is suspicious for understanding the biological dynamics in spatial regions where cell concentration is low, for example, at the edges of a cell colony or regions where the cell death rate is high or situations where interactions are limited to few hundreds of cells. The use of such models is also problematic in cases where (i) the effective dimensionality of the environment in which cells move is less than three dimensions, e.g., haptotactic cell motion, which restricts cell movement to following adhesive gradients on surfaces, (ii) the extracellular chemical signals secreted by cells decay slowly compared to the typical timescale of cell movement. In such cases it would be wise to compare the predictions of tissue-level models with those of a finer scale model such as a cellular model before making any final conclusions.

B. Mean-Field Theory

Models which are obtained by neglecting correlations between interacting entities are generally called mean-field theories (or self-consistent field theories) in the physical sciences. Though this idea is new to biology and mathematical biology, mean-field theory (MFT) has been extensively used and its implications investigated in various fields of physics particularly in statistical mechanics (Kadanoff, 2001). The idea behind MFT is to make the approximation that each entity behaves as if it was an independent entity sitting in a mean field produced by all other entities. Note that the mean referred to in this case is the average field computed by: (i) calculating the total field due to all other entities in a set of N independent realizations at a time t which leads to the quantities $\phi_1(t), \phi_2(t), \ldots, \phi_N(t)$; (ii) calculating the average of the latter quantities leading to $\Phi(t)$. Since realizations are independent from each other, an entity is uncorrelated with the mean field $\Phi(t)$ produced by all the other entities.

MFT is expected to be correct when fluctuations in the field, i.e., the standard deviation of the set of values $\phi_1(t), \phi_2(t), \ldots, \phi_N(t)$, are much smaller than its mean value, $\Phi(t)$. This condition is usually satisfied provided the number of interacting entities is particularly large. Thus it is generally found that MFT provides a qualitatively correct picture of the dynamics of fluids provided they are not near their critical points. At these points, phase transitions occur and due to the strong fluctuations in mass density, the predictions of MFT are incorrect in both two and three dimensions. Indeed this is the underlying reason why the Van der Waals theory of fluids fails to describe the experimental data of fluid behavior near the critical points (Kadanoff, 2001).

The main difference between MFT applied to cellular models to understand tissue dynamics and that applied to molecular models of fluids is that the latter involves Avogadro number of interacting molecules whereas the former frequently involves few hundreds or thousands of cells. This suggests that fluctuations in the chemical fields through which cells interact may play an important role in determining the overall behavior of a multicellular system and that mean-field theories such as the tissue model previously introduced may not be able to effectively capture such behavior. The chemical field sensed by cells in the tissue-models is an ensemble averaged field, Φ, and hence is an artificially smooth field devoid of any local spatial fluctuations whereas the chemical field in the cellular model, ϕ, contains such fluctuations. Note that the cellular model itself is the result of the application of the mean-field assumption on a molecular-level model and hence in reality the tissue-model represented by the Keller–Segel equations, Eqs. (1) and (2), encapsulates two consecutive and separate mean-field assumptions, as illustrated in Fig. 3.

Unfortunately it is frequently difficult to develop a tractable analytical theory for strongly correlated behavior and indeed to date most attempts at understanding such behavior have been via direct computer simulation. The mathematical tools to understanding such systems include diagrammatic perturbation methods (Mattuck, 1992), field theory and renormalization group theory (Cardy, 1997) and small noise perturbation theories (Gardiner, 2004). In the next section we discuss some of the effects that fluctuations bear on model predictions.

V. Multiple Scale Analysis

In this section we discuss the effects of cell correlations in the simple biological example introduced in previous sections. In particular we investigate the validity of the mean-field assumption inherent in tissue-level models. Depending on whether

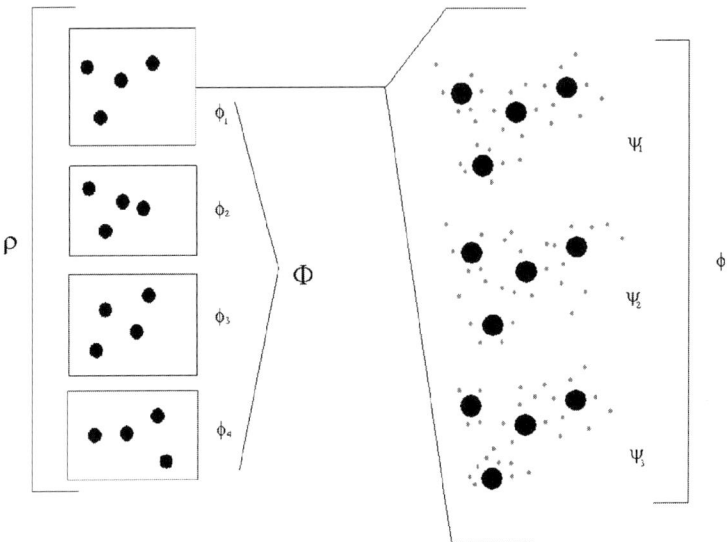

Figure 3 Schematic showing a pictorial representation of a tissue-level model in the framework of mean-field theory. The black circles are cells and the smaller red circles are chemical molecules. We show a system of four interacting cells. At each instant in time, the variables of the tissue-level model (ρ, Φ) are derived by averaging an ensemble of similar cellular-scale models each one having slightly different positions for the four cells. Each of the cellular models has associated with it a mesoscopic chemical field ϕ_i which is dependent on the spatial configuration of cells. The ensemble average over this field gives the macroscopic field Φ while the density field ρ results from the ensemble average of the spatial configuration of cells. In a similar manner, the cellular models are related to the microscopic models: the ensemble average over the microscopic field ψ_j produced by the different spatial configurations of the chemical molecules in a molecular-level model leads to the mesoscopic fields. See color insert.

the chemotaxis is positive or negative, cells either aggregate or disperse. Aggregation leads to progressively higher cell concentrations suggesting that the Keller–Segel tissue-model (MFT) is at least overall qualitatively correct in predicting the dynamics. The converse is true of chemotaxis-induced dispersion; in such a case it is not immediately clear as to how trustworthy is MFT.

A. Weak Intercellular Interactions

We first consider the case of weak chemotactic signaling between the cells. Consider an experimental setup in which initially cells are confined to a small volume. Given this setup, what is the qualitative and quantitative temporal dynamics of cell motility? The cells will experience two forces: a cohesive (dispersive) force due to positive (negative) chemotaxis and a dispersive force due to cell diffusion.

In the absence of chemotaxis, it is clear that only cell diffusion is present and thus the outer edge of the cell aggregate will necessarily increase as the square root of time elapsed, at a rate determined by the cell diffusion coefficient. When chemotaxis is turned on, albeit very weakly, it is found that this behavior is qualitatively unchanged. This is predicted by both cellular and tissue-level models (Newman and Grima, 2004). However tissue-level models predict that the rate at which the outer edge of the aggregate outwardly expands is also unchanged whereas cellular models predict a smaller rate when chemotaxis is positive (cohesive forces act against cell diffusion) and larger when chemotaxis is negative (dispersive forces enhance cell diffusion). These results are illustrated in Fig. 4. The discrepancies between the two models can be shown to arise due to the fact that for very weak chemotaxis the cells primarily interact with their own secreted chemical rather than with that of other cells. The mean-field approach underestimates effects due to self-interaction since its main underlying assumption, namely that it is valid in the limit of a very large number of interacting cells, also implies that the self-interaction is negligible compared to that due to all other cells. The above results are obtained via perturbation analysis in the strength of cell–cell coupling. By the term "coupling" here we mean how strongly cells feel each other through chemical fields or other forces. For example, the coupling strength in our simple biological system would be proportional to the product of the chemotactic sensitivity and the chemical secretion rate, $\varepsilon = \alpha\beta$.

Thus in the case of weak intercellular interactions the MFT assumption leads to a qualitatively correct picture but quantitatively incorrect predictions. This we expect to generally be the case when modeling cells which are interacting solely through local forces such as adhesion or through weak long-range interactions. This statement is particularly true when the strength of these cell–cell interaction forces are small compared to those stemming from the incessant fluctuations in the actin polymerization dynamics of the cell's cytoskeleton. Another instance when the MFT assumption holds is when cells are only chemotactic for periodic short spans of time, in between of which cell motion is random and completely unaffected by external chemical stimuli (pure cell diffusion). These conjectures have been directly or implicitly verified for models of tissue growth (Deutsch and Dormann, 2005) and of in vitro monolayer cultures (Drasdo, 2005).

B. Strong Intercellular Interactions

Perturbation analysis is typically not useful for strongly-coupled many-body systems which in general elude analysis and their exploration is thus presently for the most part confined to computer simulation. We shall describe two complimentary approximate analytical methods which can afford insight into such problems. The first assumes that there is a very high average cell concentration in the system such that MFT can be directly applied whilst the second assumes that the cell concentration is so low that a single-cell model captures the dynamics. These two approaches are extremes and

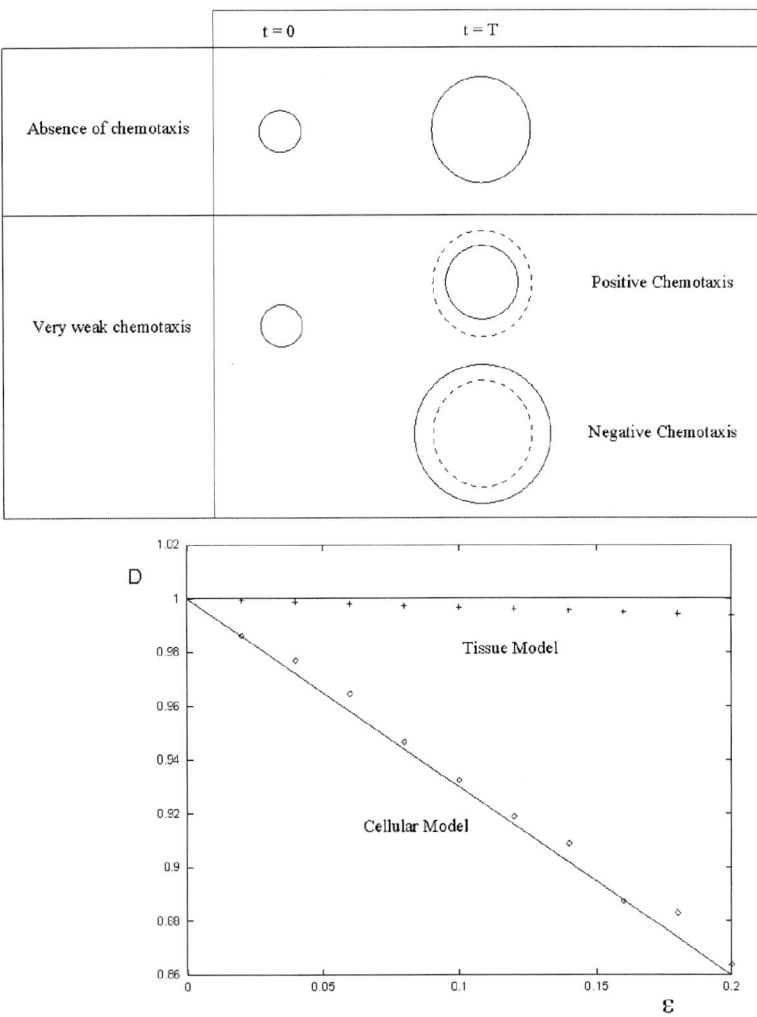

Figure 4 Differences in prediction between cellular and tissue-level models for a simple biological system consisting of chemotactic cells which are constantly secreting a chemoattractant (positive chemotaxis) or a chemorepellent (negative chemotaxis). The cells at $t = 0$ occupy a small compact volume. In the absence of chemotaxis, as time progresses the cells diffuse and the cell aggregate becomes correspondingly larger; the size of the predicted aggregate in this case is the same for both models at any point in time (shown for $t = T$). However when very weak chemotactic forces are present, the predictions of the two models at later times disagree (solid circles represent the prediction of the cellular model whereas dashed circles represent those of the tissue model). The graph at the bottom of the figure compares the predictions for the effective cell diffusion coefficient D using both theory (solid curves) and computer simulation (data points) for the case of positive chemotaxis. The parameters are $D_0 = D_1 = 1$, $\lambda = 0.05$ and the number of realizations is 10^5. The models disagree because the tissue model incorrectly ignores cell–cell correlations induced by chemotactic signaling.

hence in reality the dynamics of the tissue will be exactly described by neither but will be approximated by both, in a manner dependent on the actual cell concentration. The single cell approach can sometimes properly take into account cell–cell correlations even when strong but its predictions may quantitatively (and maybe qualitatively as well) fail when there are large numbers of interacting cells. The MFT approach cannot take into account cell–cell correlations but since effects due to the latter are diluted for large cell concentrations it may offer a picture of the dynamics in such a limit. The two approaches are thus complimentary in many ways.

1. Mean-Field Analysis

Analysis of MFT is frequently a necessary first step in shedding light on the complex dynamics. This is since many methods from the general theory of ordinary and partial differential equations, the building blocks of mean-field models, can be brought to bear on the problem. The most common and popular technique is that of stability analysis (phase-plane analysis) about the equilibrium points of the system (Deutsch and Dormann, 2005; Murray, 2002). For our simple biological system, the equilibrium point corresponds to the case when the cells and the chemical are uniformly distributed all over space such that the rate of chemical secretion by the cells equals exactly the rate of chemical degradation. In such a case the system neither aggregates nor disperses but if fluctuations in cell density occur then it is plausible that the system will move away from this equilibrium state. Stability analysis indicates that such fluctuations will cause positively chemotactic cells to aggregate provided the chemotactic coupling between them is sufficiently large. In particular, for cells constrained to move in a very thin capillary, i.e., a one-dimensional space, aggregation occurs only if the chemotactic sensitivity, α, satisfies the condition:

$$\alpha > \frac{D_0}{\beta \rho_0} \left(\frac{4 D_1^2 \pi^2}{\Lambda^2} + \lambda \right), \tag{9}$$

where Λ is the spatial wavelength of the fluctuations in the cell concentration. If this condition is not satisfied then any fluctuations in cell concentration will die away by cell diffusion thus reverting the system back to the equilibrium state. Of course if the cells are negatively chemotactic then the system always goes back to its equilibrium state as the chemotaxis-induced dispersion enhances the effects of cell diffusion. It must be borne in mind that these conclusions are correct for our simple biological system initially set up such that cells and chemical are uniformly distributed all over space in equilibrium. A more general realistic initial condition is one in which cells occupy a certain finite spatial region of size L. In such a case and for positive chemotaxis, aggregation occurs if Eq. (9) with $\Lambda = L$ is approximately satisfied. For negative chemotaxis, the cells will invariably disperse themselves over a larger region than L at a rate larger than that solely due to cell diffusion.

Whether these predictions hold or not for the strongly-coupled many-cell case can be ascertained by simultaneous computer simulation of the cellular and tissue

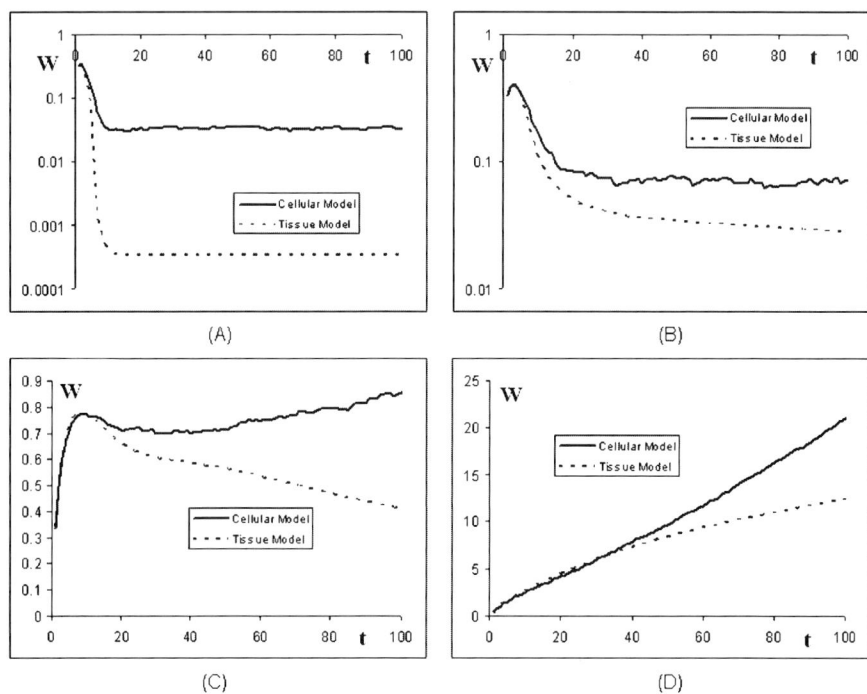

Figure 5 Comparison of cellular and tissue-model predictions for the temporal variation of the size W of a strongly coupled aggregate of twenty positively chemotactic cells. The cells are initially uniformly distributed over the spatial region $[-1, 1]$. W is the standard deviation of the cell positions. The parameters are $D_1 = \alpha = \beta = \lambda = 1$ and the cell diffusion coefficient is $D_0 = 0.05$ (A), $D_0 = 0.5$ (B), $D_0 = 1.0$ (C) and $D_0 = 2$ (D). The data for the tissue-model is obtained by numerical integration of Eqs. (1) and (2), while the data from the stochastic cellular model defined by Eqs. (5) and (6) is obtained by averaging simulation data for 1000 independent realizations. Proper convergence of results with decreasing size of time step, δt, has been verified by repeating simulations using various values of δt; data shown is using $\delta t = 0.005$ or 0.01. There is qualitative agreement between the two models except when the parameters are such that the multicellular system is close to the transition from aggregative behavior to dispersive behavior (C).

models for the same set of parameters. Such a comparison is shown in Fig. 5 for the case of twenty positively chemotactic cells which are initially uniformly distributed in a region of length two spatial units in a one-dimensional space. Note that in the cellular model the cells aggregate when the cell diffusion coefficient, D_0, is less than some critical value between 0.5 and 1, and disperse otherwise. Given the parameters $\alpha = \beta = D_1 = \lambda = 1$ used in the simulation, the stability analysis on the mean-field (tissue) model, i.e., Eq. (9), predicts aggregation if D_0 is less than 0.9. Hence in this case the mean-field analysis has fared remarkably well even though the system consists of a small number of strongly-interacting cells.

Note that qualitative differences between cellular and tissue-models are present only when the parameters force the system to be close to the transition between aggregative and dispersive behavior (Fig. 5C), a phenomenon which indeed parallels the breakdown of mean-field models of fluid phase transitions, i.e., near their critical points. In all other cases, there is qualitative but not necessarily quantitative agreement. Generally it is found that tissue-level models predict artificially small cell aggregate sizes, particularly when aggregation is strong, e.g., the final aggregate size for the tissue model in Fig. 5A is approximately 100 times smaller than that predicted by the cellular model. This can be explained by the following thought experiment. Consider the case where positively chemotactic cells at some point in time are exactly distributed according to a Gaussian distribution; this we call Case A. This case is not really achievable in practice. The real case, Case B, would have small cell concentration fluctuations about the Gaussian distribution of Case A. Indeed Case A is the ensemble average of a large number of Case B systems. Since the chemotactic chemical is produced by the cells, then the chemical concentration will have an identical spatial distribution as the one for cell concentration. Hence in Case A, cells on both sides of the central peak of the Gaussian distribution will invariably sense a chemical gradient directed towards this peak. This is not the situation in Case B, where cells will sense chemical gradients both towards and away from the central peak according to the local fluctuations (the local minima and maxima) in chemical concentration. This naturally implies that in the real case, Case B, cells will not be as strongly attracted to the global cell concentration peak as in Case A, which explains the observed differences in final aggregate width in Fig. 5.

There is another reason supporting such observations. A cell, due to its own constant secretion and positive chemotaxis to its own chemical, will tend to stay in the region of space which it presently occupies; this self-interaction opposes the aggregation of the multicellular system. However this behavior is counteracted by the response to the chemical secreted by other cells which favors aggregation. The mean-field approach is strictly valid in the limit of a very large number of interacting cells in which case the self-interaction is negligible compared to that due to all other cells. Hence MFT has to predict artificially stronger aggregation than cellular models. This shows one of the implications of assuming cell sensitivity to a mean chemical field (tissue-level model) as opposed to sensitivity to the actual field (cellular model).

2. The Single-Cell Approach

The single-cell system is not always amenable to analysis, but in cases where it is, it can provide useful insight into the strongly-coupled dynamics of cell movement. This approach takes into account the self-interaction but ignores interaction with other cells and hence is the polar opposite of the mean-field approach. The two methods are thus complimentary and together they give a more complete picture of the dynamics than that obtained solely from one approach.

For a single noninteracting cell, the model reduces to that of a pure random walker. For a cell interacting via positive or negative chemotaxis to its own secreted chemical, it reduces to a consideration of self-attracting or self-repelling random walks, the study of which is indeed a subject in its own right. One may rightly wonder what is the use of studying the dynamics of a self-interacting cell since such a model appears at first sight far fetched from reality. Besides the obvious fact that understanding a one cell, strongly-coupled system paves the way for understanding strongly-coupled many-body effects, such a simple model finds direct application in certain types of autocrine signaling (Alberts *et al.*, 1994) as well as modeling the movement of a single cell in a spatial region characterized by a low cell concentration. In the latter context, cells respond to chemical gradients which are primarily generated by themselves while gradients generated by other cells must have less important, secondary effects. The critical cell concentration below which such circumstances occur can be estimated as follows. Cells typically diffuse an order of magnitude or more slower than chemicals in solution. Thus the chemical concentration due to a single cell at position x at any given time t is to a good approximation given by considering the cell to be fixed at this position while secreted chemical diffuses at a rate determined by its diffusion coefficient D_1 and decays in solution at a rate λ. This implies that the typical distance traveled by the chemical away from the cell, before it decays, is of the order $L_c = \sqrt{D_1/\lambda}$ (this is called the Kuramoto length; see, for example, Togashi and Kaneko, 2004). Hence cell–cell interactions via chemotaxis will dominate over self-interaction if the average intercellular distance is smaller than L_c. If the intercell distance is larger than L_c then cells will primarily interact with their self-generated chemical, in which case the single-cell approach will approximate the dynamics of cell movement better than the mean-field approach.

It turns out that even the single self-interacting cell problem is nontrivial. The cell constantly modifies its physical environment through its continuous chemical secretion and simultaneously reacts to its environment via chemotactic sensing and directed motion. This feedback gives the cell memory of its past movements since the chemical field sensed at a particular time is due to secretion at all previous spatial locations which the cell visited. This nonMarkovian behavior (Gardner, 2004) is generally not possible to analyze unless some further assumptions are made about the problem. One reasonable assumption is that the cell diffusion coefficient is much smaller than its chemical counterpart. In such a case one can make progress by use of small noise perturbation methods (Grima, 2005) which leads to the following two main results: (i) for positive chemotaxis, a cell's motion is always (in the presence or absence of chemotaxis) similar to diffusion, i.e., its mean square displacement is proportional to time; (ii) for negative chemotaxis, the cell's motion is akin to that of diffusion when the coupling is less than a certain threshold (weak interaction) and ballistic, i.e., its mean square displacement is proportional to time squared, when the threshold is exceeded (strong interaction). Thus if Λ is the power that time is raised to in the expression for the mean-square displacement (*msd*), i.e., $msd \propto t^{\Lambda}$, then the single-cell approach predicts a transition from $\Lambda = 1$ to $\Lambda = 2$ as the coupling strength is increased. This

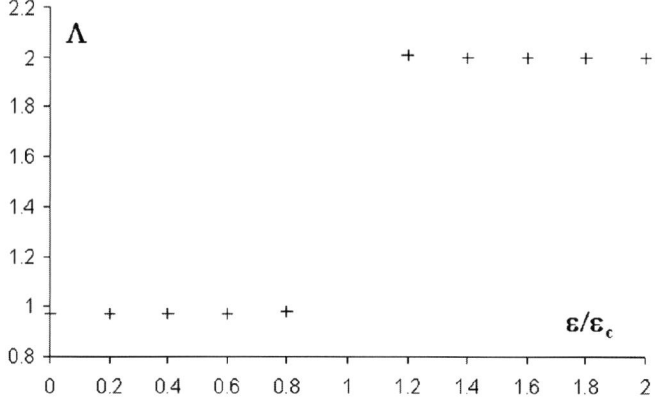

Figure 6 Graph showing the variation of the exponent Λ with the relative strength of coupling ϵ/ϵ_c for a single cell in one dimension (data from numerical simulations). Λ is the power that time is raised to in the expression for the mean-square displacement (*msd*), i.e., $msd \propto t^{\Lambda}$. The critical coupling strength at which the transition is predicted to occur by theory is denoted by ϵ_c. The parameters are $D_0 = 0.001$, $D_1 = 1$ and $\lambda = 0.1$. The *msd* is obtained by averaging the square of the single cell displacement from 1000 independent numerical simulations of the cellular model; Λ is then extracted from the slope of log–log plots of *msd* versus time for data in the time range: $t \in [500, 50000]$. The cell behavior is similar to diffusion for low coupling strength ($\epsilon < \epsilon_c$) and ballistic otherwise.

conjecture for negative chemotaxis is verified by simulation (Fig. 6). Note that this behavior is not transient. The cell's motion is ballistic after a time of the order of $1/\lambda$ has elapsed since chemical secretion started.

We note that transitions of this kind cannot be predicted by stability analysis of mean-field models. These transitions occur in all dimensions. It is found that the sharpness of the transition apparent in Fig. 6 occurs only for cell diffusion coefficients which are approximately two orders of magnitude smaller than the chemical diffusion coefficient. For larger ratios of the cell to chemical diffusion coefficients, the transition occurs but is not sharp, meaning that for intermediate coupling the cell's movement is neither diffusion nor ballistic but rather superdiffusive, $1 < \Lambda < 2$. The physical and biological implications of this transition can be appreciated by plotting typical cell trajectories for different coupling strengths (Figs. 7A–7C). Note that the cell direction changes very frequently for low coupling strength but for high coupling strength the cell moves in a certain randomly chosen direction for a significant amount of time before changing direction. This shows that the statistics of cell movement, in particular the correlation time T_c, is sensitively dependent on the feedback between cytoskeletal dynamics and surface receptor activation by chemorepellents (Fig. 7D).

As we now show, the possible biological relevance of this phenomenon lies in enabling cells to optimize their search strategy for catching moving prey or for finding target areas. Simulations (Viswanathan *et al.*, 1999; Bartumeus *et al.*, 2002) show that in general for searchers faster than their targets, the optimal searching strategy is for

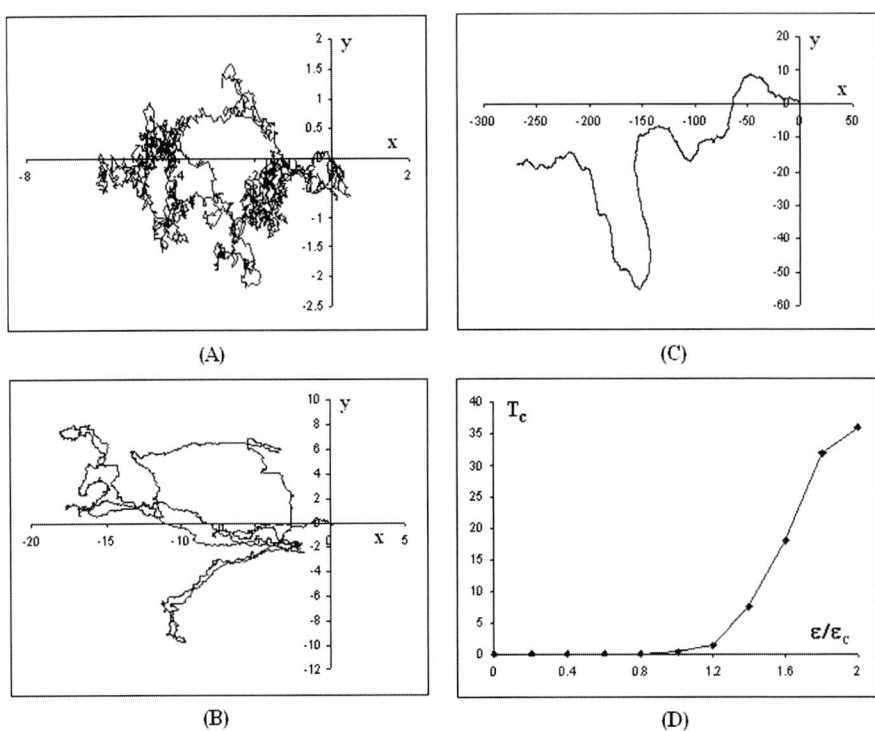

Figure 7 Sample trajectory of a cell self-interacting through negative chemotaxis in two dimensions in the time period: $t \in [0, 300]$. The parameters are $D_0 = 0.01$, $D_1 = 1$, and $\lambda = 0.1$. We show the cell behavior for (A) $\epsilon = 0.5$, (B) $\epsilon = \epsilon_c$, (C) $\epsilon = 2\epsilon_c$. Note how as the transition from diffusive to ballistic behavior occurs, the cell tends to move for longer times in one direction before switching to a new random direction. This phenomenon is quantified by plotting the correlation time T_c versus the coupling strength (D). T_c is extracted from the velocity autocorrelation data.

searchers to have a mean square displacement which scales quadratic in time (ballistic motion, also called Levy walks with $\mu = 2$); if the searchers are slower than their targets then searching is optimal if the searcher's mean square displacement scales linearly with time (Brownian motion). Note that here by the word "searchers" we do not specifically mean a chemotactic cell but rather any entity which through a random walk of some sort aspires to find its target or prey. In the light of these results, the previous prediction of a transition in cell motility acquires biological meaning. Cells dispersing through negative chemotaxis (such as the slime mold in its preaggregative life stage) can fine tune their optimal strategy to find their prey by controlling the rate of chemorepellent secretion: low secretion rate is effective for finding fast moving targets and high secretion rate for finding slow moving targets.

Our results are based on the stochastic cellular model defined by Eqs. (5) and (6) which assume that the mean chemotactic velocity of a cell is directly proportional to the local chemical gradient sensed by the cell. As previously remarked this "assumption" agrees with experiment provided the absolute value of the local chemical concentration is not too large. For large concentrations, the chemotactic response saturates and may, for example, become logarithmic (this is referred to as the Weber–Fechner law; see, for example, Brown and Berg, 1974). For such cases, the transition in cell motility for negative chemotaxis still holds (Grima, 2006).

We note that for the case of negative chemotaxis, independent of the initial cell concentration, the repulsive effects between cells will invariably lead to a gradual increase in the intercell distance suggesting that the transition predicted by the single-cell approach is as well valid for this many-cell situation. Such can explain why in-vitro experiments investigating the negative chemotaxis phase of an initially compact aggregate of *Dictyostelium discoideum*, show that the edge of the aggregate grows proportional to time (ballistic behavior) and not to the square root of time (diffusive behavior) as normal nonchemotactic cells do (Keating and Bonner, 1977).

We finish this section by noting that mean-field models cannot generally easily capture the qualitative and quantitative behavior predicted by the single-cell approach. This is since at the low-cell concentrations at which the latter approach is valid, the chemical concentration sensed by the cells will exhibit wild spatial fluctuations not the smooth behavior implicit in the application of mean-field modeling. The discrepancy between mean-field and discrete stochastic models is not limited to models of tissue but also appears at the molecular scale (Togashi and Kaneko, 2004) in reaction–diffusion phenomena and also in the large scale population dynamics of predator–prey systems (Durrett and Levin, 1994; McKane and Newman, 2004).

VI. Discussion

In this article, motivated by the complexity inherent in the biological world, we have discussed how mathematical models can be constructed at different spatial scales to provide insight into the fundamental biological processes at the heart of pattern formation and also to assist experimental data interpretation. In particular we have illustrated model construction at the cell and tissue-level scales for a simple multicellular chemotactic system which finds broad application by itself or in conjunction with other processes in understanding a wide range of phenomena. The crucial feature which makes this system have broad appeal is the fact that it is amongst the simplest biological systems in which the individual cell dynamics are regulated via interaction with other cells; such feedback is at the heart of the large scale self-organization of tissue.

The process of model building at a particular relevant spatial scale should in principle result from a rigorous derivation starting from a molecular level model. However in practice this is not possible due to the huge complexity of the processes involved. This has led to model development based on phenomenology, namely biological processes are modeled by similarity with other well understood physical processes. There are many ways in which this can be done and thus it is not always clear if the prediction of a model is genuine or if it is due to artifacts introduced by the particular phenomenological approach. One approach to solving this problem is to derive models at a coarse scale from an intermediate scale model which has the special property of being based on biology and physics which are experimentally well studied.

In this article starting from a cellular model we have used such an approach to understand the assumptions implicit in using tissue-level models, which are the dominant modeling methodology in biology and mathematical biology. Such models are found to be correct only when statistical correlations between cells are ignored which limits their application to weakly-interacting multicellular systems. Hence tissue-models based on sets of coupled partial differential equations for the cell and chemical concentrations are akin to mean-field theories in the physical sciences. This suggests that tissue-level models break down when the fluctuations in cell and chemical concentrations become comparable to their average, a situation which would be expected to occur for biological systems with few hundreds of strongly-interacting cells. Nevertheless for our simple biological system it is found that tissue-level models do generally correctly capture some of the major qualitative features though they may fare badly in predicting the correct quantitative details. Theoretical and numerical analysis of the corresponding cellular models provides the missing quantitative information. This suggests that the simultaneous use of multiple modeling methodologies at different spatial scales is useful to weed out model artifacts and to ensure both qualitative and quantitative accuracy in model prediction.

References

Alarcon, T., Byrne, H. M., and Maini, P. K. (2004). Towards whole-organ modeling of tumor growth. *Prog. Biophys. Mol. Biol.* **85**, 451–472.

Alberts, B., Bray, D., Lewis, J., Raff, M., Roberts, K., and Watson, J. D. (1994). "Molecular Biology of the Cell," 3rd ed. Garland Publishing, New York.

Bartumeus, F., Catalan, J., Fulco, U. L., Lyra, M. L., and Viswanathan, G. M. (2002). Optimizing the encounter rate in biological interactions: Levy versus Brownian strategies. *Phys. Rev. Lett.* **88**, 097901.

Berg, H. C. (1993). "Random Walks in Biology." Princeton Univ. Press, New Jersey.

Bray, D. (2001). "Cell Movements: From Molecules to Motility," 2nd ed. Garland Publishing, New York.

Brown, D. A., and Berg, H. C. (1974). Temporal stimulation of chemotaxis in *Escherichia coli*. *Proc. Natl. Acad. Sci.* **71**, 1388–1392.

Cardy, J. L. (1997). *In* "The mathematical beauty of physics." (J. M. Drouffe and J. B. Zuber, Eds.), World Scientific, Singapore.

Deutsch, A., and Dormann, S. (2005). "Cellular Automaton Modeling of Biological Pattern Formation: Characterization, Applications and Analysis." Birkhäuser, Boston.

Drasdo, D. (2005). Coarse graining in simulated cell populations. *Adv. Complex Syst.* **8**, 319–363.

Drasdo, D., Kree, R., and McCaskill, J. S. (1995). Monte-Carlo approach to tissue-cell populations. *Phys. Rev. E* **52**, 6635–6657.

Durrett, R., and Levin, S. A. (1994). Stochastic spatial models—a users guide to ecological applications. *Philos. Trans. R. Soc. London B* **343**, 329–350.

Edelstein-Keshet, L., and Spiros, A. (2002). Exploring the formation of Alzheimer's disease: Senile plaques in silico. *J. Theor. Biol.* **216**, 301–326.

Gamba, A., Ambrosi, D., Coniglio, A., de Candia, A., Di Talia, S., Giraudo, E., Serini, G., Preziosi, L., and Bussolino, F. (2003). Percolation, morphogenesis, and Burgers dynamics in blood vessels. *Phys. Rev. Lett.* **66**, 118101.

Gardiner, C. W. (2004). "Handbook of Stochastic Methods for Physics, Chemistry, and the Natural Sciences," 3rd ed. Springer-Verlag, Berlin.

Gardner, M. (1970). Fantastic combinations of John Conway's new solitaire game of life. *Sci. Am.* **223**, 120–123.

Goldhirsch, I., and Goldenberg, C. (2002). On the microscopic foundations of elasticity. *Eur. Phys. J. E* **9**, 245–251.

Gracheva, M. E., and Othmer, H. G. (2004). A continuum model of motility in ameboid cells. *Bull. Math. Biol.* **66**, 167–193.

Graner, F., and Glazier, J. A. (1992). Simulation of biological cell sorting using a 2-dimensional extended potts-model. *Phys. Rev. Lett.* **69**, 2013–2016.

Grima, R. (2005). Strong-coupling dynamics of a multicellular chemotactic system. *Phys. Rev. Lett.* **95**, 128103.

Grima, R. (2006). Phase transitions and superuniversality in the dynamics of a self-driven particle. *Phys. Rev. E* **74**, 011125.

Grima, R. (2007). Directed cell migration in the presence of obstacles. *Theor. Biol. Med. Model.* **4**, 2.

Grima, R., and Schnell, S. (2006). A systematic investigation of the rate laws valid in intracellular environments. *Biophys. Chem.* **124**, 1–10.

Grima, R., and Schnell, S. (2007). Can tissue surface tension drive somite formation? *Dev. Biol.* **307**, 248–257.

Hall-Stoodley, L., Costerton, J. W., and Stoodley, P. (2004). Bacterial biofilms: From the natural environment to infectious diseases. *Nat. Rev. Microbiol.* **2**, 95–108.

Kadanoff, L. P. (2001). "Statistical Physics: Statics, Dynamics and Renormalization." World Scientific, New Jersey.

Keating, M. T., and Bonner, J. T. (1977). Negative chemotaxis in cellular slime-molds. *J. Bacteriol.* **130**, 144–147.

Keller, E. F., and Segel, L. A. (1970). Initiation of slime mold aggregation viewed as an instability. *J. Theor. Biol.* **26**, 399–415.

Keller, E. F., and Segel, L. A. (1971). Traveling bands of chemotactic bacteria—a theoretical analysis. *J. Theor. Biol.* **30**, 235–248.

Lewus, P., and Ford, R. M. (2001). Quantification of random motility and chemotaxis bacterial transport coefficients using individual-cell and population-scale assays. *Biotechnol. Bioeng.* **75**, 292–304.

Li, S. G., and Muneoka, K. (1999). Cell migration and chick limb development: Chemotactic action of FGF-4 and the AER. *Dev. Biol.* **211**, 335–347.

Mattuck, R. D. (1992). "A guide to Feynman diagrams in the many-body problem," 2nd ed. Dover, New York.

McKane, A. J., and Newman, T. J. (2004). Stochastic models of population dynamics and their deterministic analogs. *Phys. Rev. E* **70**, 041902.

Meyer-Hermann, M. E., and Maini, P. K. (2005). Interpreting two-photon imaging data of lymphocyte motility. *Phys. Rev. E* **71**, 061912.

Ming, G. L., Wong, S. T., Henley, J., Yuan, X. B., Song, H. J., Spitzer, N. C., and Poo, M. M. (2002). Adaptation in the chemotactic guidance of nerve growth cones. *Nature* **417**, 411–418.

Mogilner, A., Wollman, R., and Marshall, W. F. (2006). Quantitative modeling in cell biology: What is it good for? *Dev. Cell* **11**, 279–287.

Mombach, J. C. M., and Glazier, J. A. (1996). Single cell motion in aggregates of embryonic cells. *Phys. Rev. Lett.* **76**, 3032–3035.

Murray, J. D. (2002). "Mathematical Biology, Vol. I. An Introduction," 3rd ed. Springer-Verlag, New York.

Murray, J. D. (2003). "Mathematical Biology, Vol. II. Spatial Models and Biomedical applications," 3rd ed. Springer-Verlag, New York.

Newman, T. J. (2005). Modeling multicellular systems using subcellular elements. *Math. Biosci. Eng.* **2**, 611–622.

Newman, T. J., and Grima, R. (2004). Many-body theory of chemotactic cell–cell interactions. *Phys. Rev. E* **70**, 051916.

Purcell, E. M. (1977). Life at low Reynolds-number. *Am. J. Phys.* **45**, 3–11.

Ralt, D., Manor, M., Cohendayag, A., Turkaspa, I., Benshlomo, I., Makler, A., Yuli, I., Dor, J., Blumberg, S., Mashiach, S., and Eisenbach, M. (1994). Chemotaxis and chemokinesis of human spermatozoa to follicular factors. *Biol. Rep.* **50**, 774–785.

Reichman-Fried, M., Minina, S., and Raz, E. (2004). Autonomous modes of behavior in primordial germ cell migration. *Dev. Cell* **6**, 107–114.

Ridley, A. J., Schwartz, M. A., Burridge, K., Firtel, R. A., Ginsberg, M. H., Borisy, G., Parsons, J. T., and Horwitz, A. R. (2003). Cell migration: Integrating signals from front to back. *Science* **302**, 1704–1709.

Rieu, J. P., Upadhyaya, A., Glazier, J. A., Ouchi, N. B., and Sawada, Y. (2000). Diffusion and deformations of single hydra cells in cellular aggregates. *Biophys. J.* **79**, 1903–1914.

Schnell, S., Grima, R., and Maini, P. K. (2007). Multiscale modeling in biology. *Am. Sci.* **95**, 134–142.

Stevens, A. (2000). The derivation of chemotaxis equations as limit dynamics of moderately interacting stochastic many-particle systems. *Siam. J. Appl. Math.* **61**, 183–212.

Togashi, Y., and Kaneko, K. (2004). Molecular discreteness in reaction–diffusion systems yields steady states not seen in the continuum limit. *Phys. Rev. E* **70**, 020901.

Tranquillo, R. T., Zigmond, S. H., and Lauffenburger, D. A. (1988). Measurement of the chemotaxis coefficient for human neutrophils in the under-agarose migration assay. *Cell. Motil. Cytoskel.* **11**, 1–15.

Viswanathan, G. M., Buldyrev, S. V., Havlin, S., da Luz, M. G. E., Raposo, E. P., and Stanley, H. E. (1999). Optimizing the success of random searches. *Nature* **401**, 911–914.

Weidinger, G., Wolke, U., Koprunner, M., Thisse, C., Thisse, B., and Raz, E. (2002). Regulation of zebrafish primordial germ cell migration by attraction towards an intermediate target. *Development* **129**, 25–36.

Xiang, X. Y., Burnett, L., Rawls, A., Bieber, A., and Chandler, D. (2004). The sperm chemoattractant "allurin" is expressed and secreted from the Xenopus oviduct in a hormone-regulated manner. *Dev. Biol.* **275**, 343–355.

Yang, X. S., Dormann, D., Munsterberg, A. E., and Weijer, C. J. (2002). Cell movement patterns during gastrulation in the chick are controlled by chemotaxis mediated by positive and negative FGF4 and FGF8. *Dev. Cell* **3**, 425–437.

Yoshida, M., Inaba, K., and Morisawa, M. (1993). Sperm chemotaxis during the process of fertilization in the ascidians *Ciona savignyi* and *Ciona intestinalis*. *Dev. Biol.* **157**, 497–506.

16

Relating Biophysical Properties Across Scales

Elijah Flenner,* Francoise Marga,* Adrian Neagu,*,† Ioan Kosztin,* and Gabor Forgacs*,‡

*Department of Physics and Astronomy, University of Missouri-Columbia, Columbia, Missouri 65211
†University of Medicine and Pharmacy Timisoara, 300041 Timisoara, Romania
‡Department of Biological Sciences, University of Missouri-Columbia, Columbia, Missouri 65211

I. Introduction
II. Theory and Computer Modeling
   A. Tissue Liquidity
   B. The Monte Carlo Method
   C. Cellular Particle Dynamics Method
   D. Tissue Fusion
III. Results
   A. Monte Carlo Simulations
   B. Cellular Particle Dynamics (CPD) Simulations
   C. Comparison of Experiments, Theory and Computer Simulations
IV. Conclusions
   Acknowledgments
   References

A distinguishing feature of a multicellular living system is that it operates at various scales, from the intracellular to organismal. Genes and molecules set up the conditions for the physical processes to act, in particular to shape the embryo. As development continues the changes brought about by the physical processes lead to changes in gene expression. It is this coordinated interplay between genetic and generic (i.e., physical and chemical) processes that constitutes the modern understanding of early morphogenesis. It is natural to assume that in this multiscale process the smaller defines the larger. In case of biophysical properties, in particular, those at the subcellular level are expected to give rise to those at the tissue level and beyond. Indeed, the physical properties of tissues vary greatly from the liquid to solid. Very little is known at present on how tissue level properties are related to cell and subcellular properties. Modern measurement techniques provide quantitative results at both the intracellular and tissue level, but not on the connection between these. In the present work we outline a framework to address this connection. We specifically concentrate on the morphogenetic process of tissue fusion, by following the coalescence of two contiguous multicellular aggregates. The time evolution of this process can accurately be described by the theory of viscous liquids. We also study fusion by Monte Carlo simulations and a novel Cellular Particle Dynamics (CPD) model, which is similar to the earlier introduced Subcellular Element Model (SEM; Newman, 2005). Using the combination of experiments, theory and modeling we are able to relate the measured tissue level biophysical quantities to subcellular parameters. Our approach has validity beyond the particular morphogenetic process considered here and provides a general way to relate biophysical properties across scales. © 2008, Elsevier Inc.

I. Introduction

In most organisms, embryonic development starts with a more or less spherical zygote. The end product of a series of morphogenetic transformations is anything but spherical. Moreover, the internal structure of the adult, composed of a multitude of organs with varying shape, interwoven by tubes and held together by the extracellular matrix, in itself is miraculously complex.

Morphogenetic shape changes require the coordinated movement of cells, which in turn requires physical forces. Movement and forces can be characterized in terms of well-defined concepts and physical parameters, such as diffusion, elasticity, viscosity, friction, velocity, etc. (Forgacs and Newman, 2005). Thus, cells also must possess physical characteristics and these, indeed, have been measured for various cell types. On the other hand cells carry out their functions according to instructions generated by their genome, a purely biochemical entity. It is the genes that set up cellular physical properties through their control of protein synthesis, the major "workers" within the cell. Changes in the number, variety and structure of proteins result in vastly differing cellular properties. An example is the actin cytoskeleton of a locomoting cell. In order to translocate, the cell needs to build a motile apparatus by the continuous polymerization/depolymerization of its filamentous actin (i.e., F actin) network. The organization of the actin cytoskeleton is thus quite different in a moving, as opposed to resting cell and this difference is manifest in distinct intracellular viscoelastic properties.

As development proceeds individual cells organize into tissues, which themselves greatly differ in physical properties: blood is liquid, bone is solid. In between these extremes lie most of the organs and tissues with typically intermediate viscoelastic properties. However, a blood cell is not the same as a liquid drop, a bone-forming cell (i.e., osteoblast) itself is not a solid and a single cardyomiocyte has totally different physical properties than the mature heart. How do intracellular and cellular physical properties determine the physical attributes of tissues, structures composed of a large number of individual cells? From the physical point of view the question may seem innocent: provided the short range interactions between cells are known, macroscopic collective tissue properties follow from the application of statistical mechanics. However, the situation is considerably more complicated when the true biological nature of cells is taken into account. Cells can produce "stuff" and thus change their interactions, just to mention one complication. Nevertheless it is intuitively obvious that ultimately the biophysical properties of tissues must derive from intracellular and cellular properties.

This article aims at elucidating how cell and subcellular physical properties may give rise to tissue level properties. We begin by developing a theoretical and computational framework to address this question. Next we apply the formalism to the fusion of tissue fragments, a morphogenetic process analogous to the coalescence of liquid drops. Specifically, we relate surface tension and viscosity, parameters characterizing fusion, to those that describe physical processes inside the cell and between cells.

II. Theory and Computer Modeling

A. Tissue Liquidity

A careful scrutiny of early embryonic development led Steinberg to formulate the differential adhesion hypothesis (DAH) (Steinberg, 1963), which attributes morphogenetic events to differences in the cell adhesion apparatus of the different cell types. DAH implies that early morphogenesis is a self-assembly process (Whitesides and Boncheva, 2002), whereby mobile and interacting subunits spontaneously give rise to structure (Foty and Steinberg, 2005; Gonzalez-Reyes and St. Johnston, 1998; Perez-Pomares and Foty, 2006; Steinberg, 1970). In light of DAH, embryonic tissues mimic the behavior of highly viscous, incompressible liquids (Steinberg and Poole, 1982). Their liquid-like behavior is indeed manifest in the rounding-up of initially irregular tissue fragments, the fusion of two or more contiguous tissue droplets into a single cellular spheroid (Gordon *et al.*, 1972), the engulfment via spreading of one tissue type over the surface of another (Steinberg and Takeichi, 1994), and the segregation or sorting of various cell types in heterotypic cell mixtures (Foty *et al.*, 1994; Technau and Holstein, 1992). All these phenomena can be interpreted using the theory of ordinary liquids. For example, in the absence of external forces a liquid droplet assumes a spherical shape (just as an originally irregular tissue fragment) because its constituent molecules attract each other and adopt positions that maximize their contact and minimize the overall surface area. Immiscible liquids, initially randomly intermixed, separate (similarly to the sorting of heterotypic cells) into a configuration with the more cohesive liquid being surrounded by the less cohesive one (e.g., water drop surrounded by an oil drop). It is however, important to keep in mind that true liquid molecules move due to thermal agitation, whereas cellular motion is powered by metabolic energy. Thus, the properties of embryonic tissues are analogous, but not identical to those of liquids. The physical properties that best characterize liquids are surface or interfacial tension (γ) and viscosity (η). Such quantities can effectively be attributed to embryonic tissues through measurement techniques employed in case of liquids (Forgacs *et al.*, 1998; Foty *et al.*, 1994, 1996; Gordon *et al.*, 1972). Indeed, numerous embryonic tissues have been characterized in terms of effective tissue surface tension and the measured tensions were found to be consistent with the mutual sorting behavior of these tissues (Foty *et al.*, 1996), as predicted by DAH. The latter provides the molecular basis of tissue surface tension by postulating a connection between this macroscopic quantity and the strength of adhesion between cells constituting the tissue. On theoretical grounds it was established that tissue surface tension has to be proportional to the number of cell surface adhesion molecules (Forgacs *et al.*, 1998), a conclusion confirmed later experimentally (Foty and Steinberg, 2005). These findings imply a connection between tissue and cellular level quantities. The implications of DAH have been confirmed not only *in vitro* and *in silico* (Graner and Glazier, 1992; Glazier and Graner, 1993), but also *in vivo* (Godt and Tepass, 1998; Gonzalez-Reyes and St. Johnston, 1998).

DAH provides useful predictions on equilibrium tissue configurations, but cannot address the question of how these configurations are arrived at in time. Since embryonic development is all about changing shapes, the more complete quantitative understanding of early morphogenetic processes necessitates a dynamical approach. It is an intriguing question whether the liquid analogy can be extended beyond equilibrium. Liquid molecules and cells both interact via short-range forces that can be characterized by their strength (ε) and range (δ). Dynamical behavior is governed by Newton's second law, which in case of cells [due to the irrelevance of inertial forces (Odell *et al.*, 1981)] simplifies to frictional (i.e., Langevin) dynamics. The macroscopic properties of liquids, such as γ and η can be determined from the intermolecular forces in terms of ε and δ (Israelachvili, 1992). Thus, if tissue liquidity remains a useful concept in dynamical developmental processes then measurable tissue level physical parameters (tissue surface tension and viscosity) could possibly be related to the strength and range of interaction between cell adhesion molecules (CAMs), such as cadherins (Takeichi, 1990). Since CAMs typically are transmembrane proteins with attachment to intracellular organelles [e.g., cadherins associate with the actin cytoskeleton (Gumbiner, 1996)], the liquid analogy, in principle, could open the possibility to establish connection between intracellular molecular entities and tissue level macroscopic physical parameters. In what follows we develop two theoretical/computational approaches to carry out this program.

B. The Monte Carlo Method

The DAH has inspired several theoretical models of living tissues. According to DAH morphogenesis results from the movement of cells seeking positions that lead to a minimum of the total energy of adhesion (Steinberg, 1963, 1996). Early lattice models based on DAH yielded important insight into cellular pattern formation (Leith and Goel, 1971). Computer simulations based on these models were limited by the available computer power and the implementation of deterministic motility rules (Goel and Rogers, 1978; Rogers and Goel, 1978). In contrast, the Metropolis algorithm (Metropolis *et al.*, 1953), based on purely stochastic motility rules, turned out to be a convenient and fast method for finding conformations emerging by energy minimization. Its effectiveness in describing tissue liquidity has been shown in Monte Carlo simulations based on the Potts model, a widely used model in statistical physics (Graner and Glazier, 1992; Glazier and Graner, 1993). In this model the tissue is represented on a lattice. Each cell is composed of several contiguous lattice sites (i.e., subcellular elements) that are labeled by an identification number (the same for each subcellular element) and a cell-type index. The average number of sites per cell is maintained around a target value. Deviations from this target value are constrained by an elastic term in the overall energy assigned to the system. Cells interact with their close neighbors. Evolution, described by the Metropolis algorithm, accounts for cell migration and shape changes resulting from the movement of subcellular elements (Graner and Glazier, 1992; Glazier and Graner, 1993). This approach was successfully applied to cell sorting and the mutual engulfment of adjacent tissue fragments. Interestingly, it was shown that cell motility could be associated with an effective temperature-like parameter (Mombach et al., 1995).

Inspired by the approach of Glazier and Graner, we have constructed a three-dimensional lattice model that enabled us to simulate tissue liquidity in systems composed of large numbers (10^5–10^6) of interacting cells [i.e., model tissue; (Jakab et al., 2004; Neagu et al., 2005)]. In this model cells and similar-sized volume elements of medium or extracellular matrix (ECM) are represented as particles on sites of a cubic lattice. The type of the particle on a given site r is specified by an integer $\sigma_r \in \{1, 2, \ldots, T\}$. The contact interaction energy $J(\sigma_r, \sigma_{r'})$ between two particles, located at neighboring sites r and r', of type $\sigma_r = i$ and $\sigma_{r'} = j$, respectively, is given by $J(\sigma_r, \sigma_{r'}) = -\varepsilon_{ij}$, where ε_{ij} is the mechanical work needed to break the bond between them. For $i \neq j$ ($i = j$) the energy ε_{ij} is referred to as the work of adhesion (cohesion). The total interaction energy of the model tissue is written as

$$E = \sum_{\langle r, r' \rangle} J(\sigma_r, \sigma_{r'}), \tag{1}$$

where $\langle r, r' \rangle$ indicates that the summation involves only close neighbors. Specifically, we consider interactions between nearest, next-nearest and second-nearest neighbors. Thus, the total number of neighbors interacting with a given particle is $n = 26$. For simplicity we assume that neighbors of the same type interact with the same strength, irrespective whether they are nearest, next-nearest and second-nearest neighbors. Thus, in the case of a homocellular tissue fragment (particles of type 2) in cell culture medium (particles of type 1), $J(\sigma_r, \sigma_{r'})$ may take either of the values $J(1, 1) = -\varepsilon_{11}$, $J(2, 2) = -\varepsilon_{22}$, and $J(1, 2) = J(2, 1) = -\varepsilon_{12}$. Note that in this case $\varepsilon_{11} \ll \varepsilon_{22}$ and $\varepsilon_{12} \ll \varepsilon_{22}$.

By separating interfacial terms in the sum, the adhesive energy of a system made of T types of particles, up to an irrelevant additive constant, becomes (Neagu et al., 2006)

$$E = \sum_{\substack{i,j=1 \\ i<j}}^{T} \gamma_{ij} N_{ij}, \tag{2}$$

provided that particles do not change their type. Here N_{ij} denotes the total number of bonds between particles of type i and type j ($i \neq j$). It is noteworthy that the total interaction energy depends only on the number of heterotypic bonds, N_{ij} and the interfacial tension parameters (Jakab et al., 2004)

$$\gamma_{ij} = \frac{1}{2}(\varepsilon_{ii} + \varepsilon_{jj}) - \varepsilon_{ij}. \tag{3}$$

The model can readily be extended to include cell differentiation, proliferation or death, as well as the remodeling of the ECM by cells. In such situations the interaction energy in Eq. (1) depends explicitly on the "works of cohesion"

$$E = \sum_{\substack{i,j=1 \\ i<j}}^{T} \gamma_{ij} N_{ij} - \frac{1}{2} n \sum_{i=1}^{T} N_i \varepsilon_{ii}. \tag{4}$$

Now N_i (number of type i particles in the system) may vary during the simulation.

We have applied the Metropolis Monte Carlo method (Amar, 2006; Metropolis et al., 1953) to the above model to investigate the evolution of a cellular system that displays random motility and is consistent with DAH. Specifically, the employed computational algorithm consisted of the following steps.

1. The initial state of the system is constructed by assigning to each lattice site a cell type index that specifies occupancy.
2. Particles that are in contact with other type of particles, and thus represent interfacial cells, are identified.
3. Each interfacial cell is given the chance to move by exchanging its position with a randomly selected neighbor of a different type. The corresponding energy change ΔE is computed after each move and the new configuration is accepted with a probability $\min(1, \exp(-\Delta E/E_T))$. Thus an energy-lowering movement is accepted with probability one. Changes in position that lead to an increase in energy are also allowed, albeit with smaller probability given by the Boltzmann factor.
4. The configuration, the interaction energy, and the values of interfacial areas are written into output files.
5. Interfacial cells are identified in the new configuration and the process is repeated from step 3 until the desired number of steps is completed.

The energy E_T in the expression for the acceptance probability is an effective measure of cell motility. It is the biological analogue of $k_B T$, the thermal fluctuation energy (k_B—Boltzmann constant, T—absolute temperature). Being related to cytoskeleton-driven cell membrane ruffling (Mombach et al., 1995), E_T is referred to as the biological fluctuation energy. It has experimentally been assessed for certain embryonic cell types (Beysens et al., 2000). The interaction energies in Eqs. (1)–(4) are expressed in units of E_T.

A Monte Carlo step (MCS) is defined as the set of operations during which each interfacial cell has been given the chance to experience a change. The simulations were implemented with fixed boundary conditions by confining cell movement to the volume of the simulated region. The concern that finite-size effects may interfere with the energetically-driven rearrangement of cells is justified only when a significant fraction of the cells migrates away from the model tissue construct and reaches the system boundary. This is usually not the case in our simulations.

C. Cellular Particle Dynamics Method

A more realistic approach to simulate the motion and self-assembly of large cell aggregates is to model the individual cells as a set of interacting *cellular particles* (CPs), and follow their 3D spatial trajectories in time by integrating numerically the corresponding equations of motion in the spirit of Newman's Subcellular Element Model (SEM; Newman, 2005). CPs result from a coarse-graining description of the cells. The number and type of the CPs that constitute the cell are determined by the chosen level of coarse-graining, thus making the model flexible. The motion of a cell, as well as dynamical changes in its shape, are determined by the collective motion of its constituent CPs.

At a given time the location of a CP is determined by its position vector, $r_{i,n}(t)$, where the index $i = 1, \ldots, N_{CP}$ labels the CPs within the cell. Here we consider a system of N cells labeled by the index n ($n = 1, \ldots, N$). The precise form of the interaction between CPs is dictated by the detailed biology of the cell, whose complexity makes the determination of this interaction potential practically impossible. However, it is quite reasonable to assume that CPs interact through local biomechanical forces which (similarly to liquid molecules) can conveniently be modeled by the Lennard–Jones (LJ) potential energy

$$V_{LJ}(r; \varepsilon, \sigma) = 4\varepsilon \left[\left(\frac{\sigma}{r}\right)^{12} - \left(\frac{\sigma}{r}\right)^{6} \right], \tag{5}$$

where r is the distance between the CPs, ε is the energy required to separate the CPs and σ gives the size (diameter) of a CP. The *intercellular* interaction potential between two CPs that belong to two different cells is $V_2(r) = V_{LJ}(r; \varepsilon_2, \sigma_2)$. If the two CPs belong to the same cell, the corresponding *intracellular* potential has the form $V_1(r) = V_{LJ}(r; \varepsilon_1, \sigma_1) + V_c(r; k, \alpha)$. Here V_c represents a *confining* potential energy of the form

$$V_c(r; k, \alpha) = \begin{cases} (k/2)(r - \alpha)^2, & \text{for } r > \alpha, \\ 0, & \text{for } r < \alpha, \end{cases} \tag{6}$$

which assures that CPs for a given cell remain inside the cell (Flenner et al., 2007). Thus V_c preserves the integrity of the cell. Indeed, whenever the separation between two CPs within a given cell exceeds a predefined distance α, an elastic force $F_c = -\partial V_c/\partial r = -k(r - \alpha)$ is turned on to prevent the two CPs to move further apart.

Thus by tuning the values of k and α in the confining potential one can control, respectively, the stiffness and the size of a cell. While the LJ parameters corresponding to the inter- and intracellular interactions may have different values, here, for simplicity, we shall assume that these are equal, i.e., $\varepsilon_1 = \varepsilon_2 = \varepsilon$ and $\sigma_1 = \sigma_2 = \sigma$. In addition, each CP interacts with its highly viscous and stochastic environment (i.e., cytosol). These interactions can be described by a friction force $\boldsymbol{F}_f = -\mu \boldsymbol{v}$ (μ is the friction coefficient and $\boldsymbol{v} = \dot{\boldsymbol{r}}$ is the velocity of the CP), and a stochastic force $\boldsymbol{\xi}(t)$, modeled as a Gaussian white noise with zero mean and variance $\langle \xi_i(t)\xi_j(0) \rangle = 2D\mu^2 \delta(t)\delta_{ij}$ (D is the self-diffusion coefficient of a CP). Although living cellular systems are not in thermodynamic equilibrium, we assume that the "Einstein relation" $D\mu = E_T$ still holds, where E_T is the biological fluctuation energy introduced earlier. Finally, assuming that the motion of the CPs in the highly viscous medium is overdamped (i.e., inertia can be neglected), the corresponding Langevin equation of motion for the ith CP in the nth cell can be written as

$$\mu \dot{\boldsymbol{r}}_{i,n} = \boldsymbol{\xi}_{i,n}(t) - \underbrace{\sum_{j \neq i} \partial V_1(|\boldsymbol{r}_{i,n} - \boldsymbol{r}_{j,n}|)/\partial \boldsymbol{r}_{i,n}}_{F_{1i,n}} - \underbrace{\sum_{j,m \neq n} \partial V_2(|\boldsymbol{r}_{i,n} - \boldsymbol{r}_{j,m}|)/\partial \boldsymbol{r}_{i,n}}_{F_{2i,n}},$$
$$i = 1, \ldots, N_{cp}, \qquad n, m = 1, \ldots, N, \tag{7}$$

where $\boldsymbol{F}_{1i,n}$ ($\boldsymbol{F}_{2i,n}$) represents the net intra- (inter-)cellular interaction force exerted by the other CPs.

Similarly to SEM (Newman, 2005), the CPD approach consists in numerically integrating the above equations in order to determine the individual trajectories of all CPs. This can be achieved most efficiently by properly implementing the intra- and intercellular interaction forces and a Langevin dynamics integrator in one of the freely available massively parallel molecular dynamics (MD) packages (Flenner et al., 2007). We have successfully implemented CPD into two such MD programs, NAMD (Phillips et al., 2005) and LAMMPS (Plimpton, 1995). We used both implementations to simulate the fusion process of two identical spherical cell aggregates as described in Section III. The obtained results were insensitive to which MD package was used.

D. Tissue Fusion

Tissue fusion is a ubiquitous morphogenetic process (Perez-Pomares and Foty, 2006). A specific example in early development is the fusion of embryonic cushions, a process leading to the formation of the 4-chambered heart, i.e., septation (Wessels and Sedmera, 2003). We have investigated tissue fusion experimentally using roundup fragments of several cell types. When two embryonic tissue spheroids (in tissue culture medium) were arranged contiguously by gravitational forcing in hanging drop configuration (on a lid of an inverted Petri dish) they fused into a single aggregate.

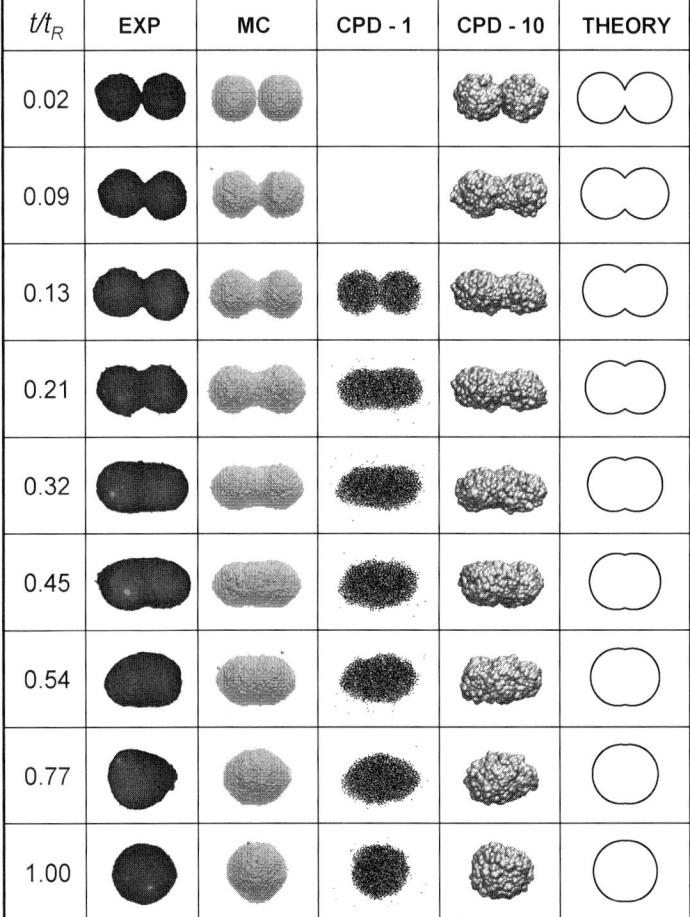

Figure 1 Snapshots of the fusion of spherical cell aggregates obtained from experiment, MC and CPD simulations (CPD-1 and CPD-10), and from theoretical modeling. The snapshots in the horizontal rows were taken at well defined instants of time (expressed in terms of the rounding, or total fusion time t_R) listed in the first column, except the MC, results which were taken after 0, 3, 12, 50, 150, 300, 500, 800, and 1200 thousands of MCS. In the experiment the rounding time was 214 hours.

Fig. 1 shows snapshots of fusion in the case of aggregates of smooth muscle cells. (For the preparation of such aggregates, see, for example, Foty *et al.* (1996), Hegedus *et al.* (2006). To quantify the fusion process, we followed the time evolution of the interfacial area of contact between the fusing tissue drops (Fig. 1). The process was then simulated using both the Monte Carlo and the CPD methods, the latter with either one (CPD-1) or ten (CPD-10) CPs per cell (Fig. 1).

As will be shown, the fusion of two cellular aggregates is similar to the coalescence of two viscous liquid drops. The process is driven by surface tension γ and is resisted by viscosity η. In principle, the precise theoretical description of shape evolution during fusion can be given using the laws of hydrodynamics. Under some reasonable approximations applicable to highly viscous incompressible liquids (Frenkel, 1945), we have found an exact solution to the problem, which is in excellent agreement with both our experimental and computer simulation results (Flenner et al., 2007). In our analytical model, the fusing drops are spherical caps whose radii increase as their centers move closer together, eventually forming a single spherical drop with radius $R_f = 2^{1/3} R_0$, determined from volume conservation. Here R_0 is the initial radius of the two identical spherical drops. Accordingly, the interfacial contact between the fusing drops is a circle of radius r_0, which increases from 0 to R_f. While the mathematical expression of the time dependence of $(r_0/R_0)^2$ (i.e., the area of interfacial contact πr_0^2 expressed in units of πR_0^2), is rather complicated, the solution is almost indistinguishable from the simple formula

$$(r_0/R_0)^2 = 2^{2/3}\left(1 - e^{-t/1.35\tau_0}\right), \quad \tau_0 = \eta R_0/\gamma. \tag{8}$$

Because $(r_0/R_0)^2$ approaches exponentially its limiting value $(R_f/R_0)^2 = 2^{2/3}$, with a time constant $\tau = 1.35\tau_0$, the definition of the total fusion or rounding time t_R is somewhat arbitrary. We define t_R as the instant of time when r_0 reaches 96% of the final radius R_f, i.e., $r_0(t_R) = 0.96 R_f$. At this time, for all practical purposes, the fused aggregates are completely rounded up (see also Fig. 1). From Eq. (8) it follows that t_R is related to the characteristic time τ_0 through the formula $t_R = 3.45\tau_0 = 3.45\eta R_0/\gamma$. Furthermore, as discussed later, Eq. (8) allows relating tissue level quantities to molecular, cell and subcellular quantities.

III. Results

A. Monte Carlo Simulations

The Monte Carlo method was used to study a number of quantities characterizing the fusion of two cellular aggregates, such as the surface energy and the degree of mixing of cells. Snapshots of simulated structural changes in the model system are shown in Fig. 1.

The Monte Carlo simulations presented in this article were performed with the "works of cohesion" $\varepsilon_{cc} = 1$ and $\varepsilon_{mm} = 0$, describing respectively cell–cell interactions and medium–medium interactions and $\varepsilon_{cm} = 0$, the associated cell–medium interaction (all interactions are expressed in units of the cellular fluctuation energy, E_T). This choice of parameters corresponds to a cellular aggregate placed in cell culture medium (with which its interaction is negligible) and results in a cell-aggregate–medium interfacial tension $\gamma_{cm} = 0.5$ [see Eq. (3)].

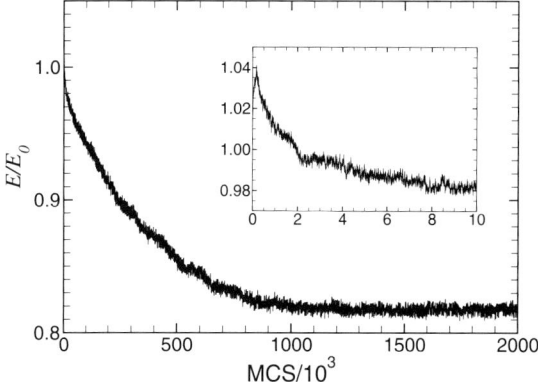

Figure 2 Simulated evolution of the surface energy E of the model tissue, expressed in units of its initial E_0 value. Here MCS stands for the number of elapsed Monte Carlo steps. The inset depicts the same dependency for the first 10^4 MCS. The tissue–medium interfacial tension parameter was $\gamma_{cm} = 0.5$.

Fig. 2 illustrates how the interfacial energy, and thereby the surface area, of the model tissue decreases during fusion. The lowering of the interfacial energy continues until it levels off, showing that the fusion and complete rounding of the initial spheroids occurs in about 1.2×10^6 MCS. As shown in the inset of Fig. 2, fluctuation-driven deviations from the spherical shape of the initial aggregates cause a slight increase in surface area right after the start. The interfacial energy begins to decrease as soon as the two spheroids make contact and start to coalesce. Fig. 2 demonstrates the efficiency of the Metropolis algorithm in selecting energetically favorable rearrangements of motile, interacting particles (thus mimicking liquidity despite the fact that no real time evolution is being considered). An important feature of our implementation of the Metropolis algorithm is that it only includes conformational changes that result from the movement of cells. This is in contrast with Monte Carlo simulations based on completely random changes in particle arrangement.

According to the first two columns in Fig. 1, Metropolis Monte Carlo (MC) simulations qualitatively reproduce the experimental sequence of intermediate stages observed during fusion. We also investigated if this similarity pertains as well to the movement of individual cells within the fusing aggregates. To this end, we studied how cells originating from one of the spheroids diffuse into the other one. The simulations started with two contiguous aggregates made of model cells of the same type, but labeled and colored differently for bookkeeping (Fig. 3). The system was "sectioned" normally to the longitudinal axis of symmetry of the initial configuration. The position of each transversal plane was specified by the coordinate x of its intersection with the longitudinal axis. The fractions f_L and f_R of cells initially located in the left and right aggregates were respectively defined by

$$f_\beta(x) = \frac{n_\beta(x)}{n_L(x) + n_R(x)}, \quad \beta = L, R, \tag{9}$$

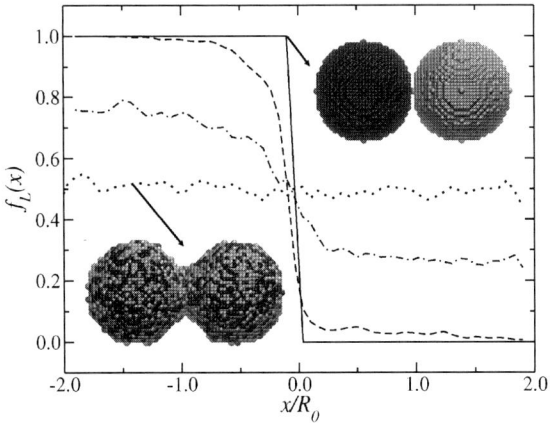

Figure 3 The mixing of cells during aggregate fusion simulated by the Metropolis Monte Carlo algorithm. The starting system is shown in the upper right corner: two aggregates made of the same type of cells, colored differently in order to keep track of mixing. The curves plot the fraction vs. position for cells originating from the left aggregate. Cells are counted in cross-sections taken normally to the longitudinal axis of symmetry of the system (Ox). The coordinate of the intersection of the axis with the plane of cross-section is expressed in units of the initial aggregate radius R_0; the origin corresponds to the center of symmetry of the initial configuration. Each curve refers to a different stage of the simulation as follows: initial state—step function, and intermediate states at 300 MCS (dashed), 1000 MCS (dashed–dotted), and 3000 MCS (dotted) gradually converging to the horizontal line corresponding to complete mixing (fraction of cells from either aggregate in each cross-section is 0.5). The model tissue conformation obtained in 3000 MCS is depicted in the lower left corner. The interfacial tension parameter was $\gamma_{cm} = 0.5$.

where $n_L(x)$ ($n_R(x)$) is the number of cells originating from the left (right) aggregate and located in the cross-section of coordinate x. In our lattice formulation, the coordinate x only takes on the discrete values x_s, with $s = 1, 2, \ldots, S$, with S being the total number of those cross-sections that contain model cells.

As a global measure of particle mixing in a given configuration, we define the degree of mixing d_m by the relation

$$d_m = \frac{1}{N} \int f_L(x) f_R(x)\, dx = \frac{1}{N} \sum_{s=1}^{S} f_L(x_s) f_R(x_s). \tag{10}$$

The normalization factor N is chosen such that $d_m = 1$ for complete mixing, when $f_L(x) = f_R(x) = 0.5$ in each cross-section that contains cells. In the discrete formulation this condition yields $N = S/4$.

The curves in Fig. 3 represent the variation of f_L along the symmetry axis of the system at different instants of time. In the initial state, $f_L(x) = 1$ for $x < 0$ and $f_L(x) = 0$ for $x > 0$. As the fusion evolves, cells mix and the graph of $f_L(x)$ flattens, approaching the value 0.5 associated with a degree of mixing $d_m = 1$. Fig. 3 shows that within 3×10^3 MCS the cells starting from different initial aggregates mix

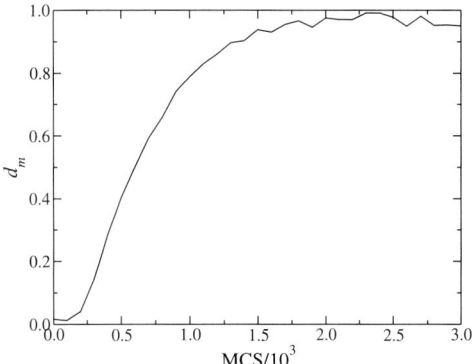

Figure 4 Degree of mixing (Eq. (6)) vs elapsed MCS.

completely, though the fusion is still in an incipient stage (see the snapshot of the corresponding state in the lower left corner of Fig. 3). The plot of the degree of mixing versus the number of elapsed MCS (Fig. 4) also supports the conclusion that over 90% mixing occurs within about 0.1% of the number of MCS needed for complete rounding. This finding is at variance with experiments (i.e., confocal microscopic analysis of the series in Fig. 1) showing negligible cell mixing until the aggregates coalesce forming a cylindrical shape with hemispherical caps.

During simulations of fusion we monitored the radius of the circular contact area of the two spherical aggregates. Using the method illustrated in Fig. 5, we calculated the surface density of cells in this plane for intermediate states of the simulation as a function of the distance r from the center of symmetry of the initial system. It was found that this density is well described by the function

$$\rho(r) = \frac{1}{2}\left[1 - \tanh\left(\frac{2(r - r_0)}{b}\right)\right], \tag{11}$$

which is also used in analyses of simulations of spherical liquid drops (Thompson et al., 1984). We identify the parameter r_0 in the fit as the radius of the contact area. The second fitting parameter, b, is related to the steepness of the density profile in the vicinity of the surface.

In Fig. 6 the evolution of $(r_0/R_0)^2$ is plotted as a function of the elapsed MCS. The data points in Fig. 6 represent the mean values of $(r_0/R_0)^2$ obtained by averaging the results of 30 simulations started with different seeds of the random number generator. After about 1.2×10^6 MCS complete rounding takes place and $(r_0/R_0)^2$ reaches its limiting value $(R_f/R_0)^2 = 2^{2/3} \approx 1.59$, in agreement with volume conservation. Furthermore, according to Fig. 2, at this stage of the simulation the surface area and interfacial energy approach their minimum.

We conclude this section by noting that Monte Carlo simulations based on a lattice representation of a living tissue qualitatively describe the liquid-like behavior

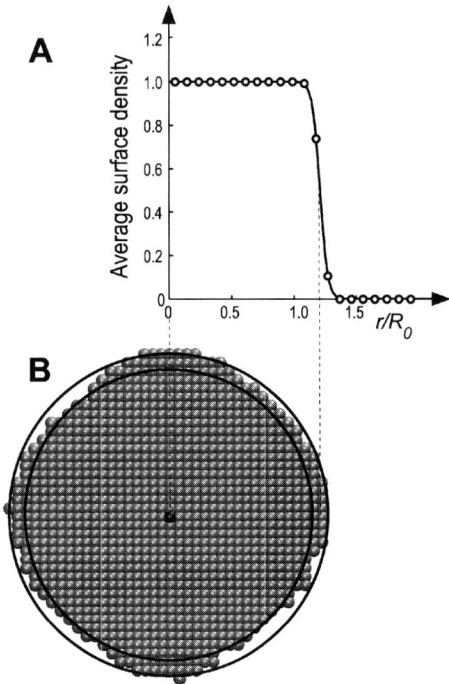

Figure 5 (A) Definition of r_0, the radius of the interfacial contact. The contact surface between the fusing aggregates is approximated by a circle. The surface density of model cells in this surface is plotted vs the distance from the center of the circle. The density is fitted by the function given in Eq. (11). The disk radius is defined as the radial coordinate at which this function drops to half of its maximum. (B) The contact disk obtained in 7×10^5 MCS.

of embryonic tissues. However, discrepancy between the experimental and simulated mixing of cells during aggregate fusion shows that the Metropolis algorithm is inappropriate for describing individual cell motility in a three-dimensional tissue. Another shortcoming of the Metropolis Monte Carlo method is that *a priori* there is no relationship between real time and the number of performed MCS.

B. Cellular Particle Dynamics (CPD) Simulations

Here we present results of two CPD simulations for the fusion of two cellular aggregates. In the first simulation, referred to as CPD-1, each cell was formed by a single CP, whereas in the second simulation, referred to as CPD-10, each cell contained ten CPs. While both models lead to liquid-like behavior, each model has advantages and disadvantages. We first describe the simulations in detail. Then, based on the analysis

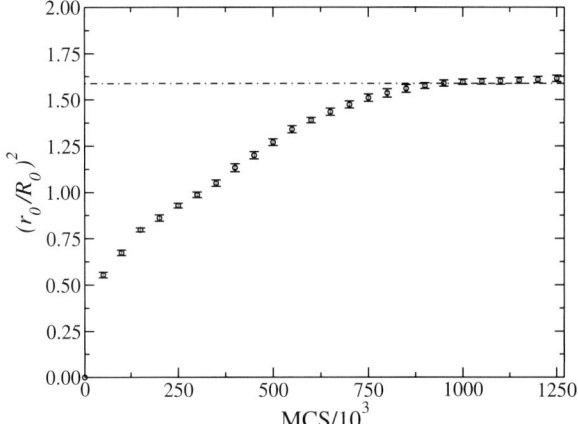

Figure 6 Simulated evolution of $(r_0/R_0)^2$. The open circles represent mean values obtained from 30 MC simulations with interfacial tension parameter $\gamma_{cm} = 0.5$ (the simulations differed only by the seed of the random number generator). The dash–dot line shows the target value derived from volume conservation. Error bars are based on standard deviation around the mean.

of the obtained CPD trajectories, we calculate the same quantities as in the MC simulations, i.e., the degree of mixing of the cells and the time evolution of the contact area between the drops. For convenience, we employ as length, energy and time units respectively σ, E_T, and $\sigma^2\mu/E_T$. In terms of these units the CPD equations of motion (Eq. (7)) are equivalent to setting $\sigma = E_T = \mu = 1$ in them. Thus, we only need to specify the dimensionless values for ε, k, and α (the latter two enter only in CPD-10).

In CPD-1 the LJ energy parameter was set to $\varepsilon = 2.5$. This relatively large energy value was motivated by the need to maintain the integrity of the cell aggregates against the escape of cells located at their surface. To make the LJ interaction between cells truly short ranged it was truncated at two length units. The starting configuration of the spheroidal cell aggregates was generated as follows. First, from a compact cubic lattice of cells a spherical aggregate (centered about the origin of the lattice) of radius $R_0 = 11$ was created, by removing all the cells situated outside this sphere. Next, two copies of this spherical aggregate were placed such that the distance between their centers was $2R_0$. Initially, each of the aggregates contained 2108 cells, and the two aggregates had a small but finite contact region, which facilitated the fusion process right at the beginning of the CPD simulation. We used periodic boundary conditions with a simulation box of side 300σ. During the simulation only a few cells crossed the simulation box. The Langevin CPD equations were integrated using the Euler algorithm with a time step $\Delta t = 10^{-5}$.

In the CPD-10 simulation the LJ energy parameter was set to $\varepsilon = 1$, which is smaller than in the CPD-1 case. The integrity of the individual cells was enforced by the confining potential with $\alpha = 2.5$ and $k = 5$. Similarly to CPD-1, a spherical aggregate of 200 cells was constructed on a lattice and then equilibrated through a

CPD run until a steady state, with small energy fluctuations, was reached. The initial configuration of the system was built from two contiguous equilibrated aggregates.

A sequence of snapshots of the fusion process for both CPD simulations is shown in Fig. 1. In agreement with our theoretical prediction of the time evolution of the fusion process, the shapes of the fusing aggregates are similar for both experiment and simulations. The rounding time was found by examining the time evolution of r_0 [defined in Eq. (11)]. Note that in CPD-1 r_0 had a finite value at the start of the simulation. The snapshots show that both CPD models reproduce the experimentally observed fusion process. However, a notable difference between the two simulations is that while in CPD-1 by the end of the fusion process approximately 10% of the cells escaped from the two aggregates, in case of CPD-10 none of the cells detached. Since maintaining the integrity of the system in case of CPD-1 is problematic, the general conclusion is that in CPD simulations cells must be built from at least a few CPs. The actual choice of the number of CPs depends on the problem at hand.

Another important difference between the two CPD simulations concerns the degree of mixing of the (otherwise identical) cells from different aggregates, which is noticeably larger in case of CPD-1. To examine the degree of mixing we have used the continuum version of Eq. (9),

$$f_\beta(x,t) = \frac{\sum_{i=1}^{N_\beta} \delta(x - x_i(t))}{\sum_{i=1}^{N} \delta(x - x_i(t))}, \quad \beta = L, R, \tag{12}$$

where $x_i(t)$ is the position of particle i at time t along the x-axis (as defined earlier). In the numerator for $\beta = L(R)$ the sum is over particles that originate from the left (right) aggregate, whereas in the denominator the sum is over all particles. Here $\delta(x)$ stands for the Dirac delta function (i.e., $\delta(x) = 0$ except $x = 0$, and $\int_{-\infty}^{\infty} \delta(x)\,dx = 1$). In Fig. 7 we plot $f_L(x,t)$ for the two CPD simulations. The different curves correspond to three different snapshots, $t = 0, 0.5t_R$ and t_R. Recall that in CPD-1 the fusion process has already started at $t = 0$, meaning that the degree of mixing in this case is underestimated. However, contrary to the Monte Carlo simulations, complete mixing does not occur by the end of the fusion process even in the case of CPD-1.

The degree of mixing $d_m(t)$, given by Eq. (10), was calculated for both simulations. As the results in Fig. 8 show, the rate of mixing appears to be constant throughout both simulations. The rate of mixing, however, is noticeably smaller for CPD-10 than CPD-1.

From the CPD trajectories, the time evolution of the relative contact area between the two fusing aggregates $(r_0/R_0)^2$ was also determined, as shown in Fig. 9. The algorithm for identifying the contact area and its radius r_0 was similar to the one employed in the case of MC simulations. In the CPD simulations, for the (dimensionless) rounding time we used $t_R = N_{t_R}\Delta t$ as the smallest time t when $r_0(t) = R_f$, the average final (stationary) radius of the fused system.

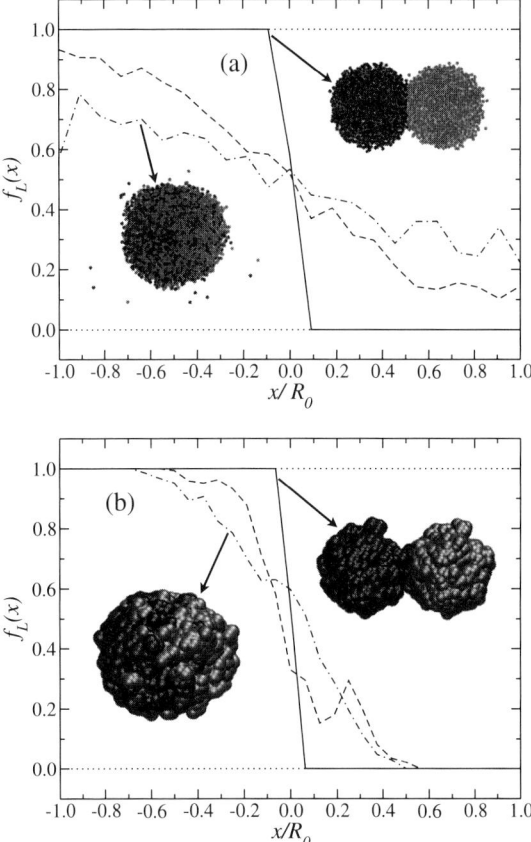

Figure 7 The fraction $f_L(x)$ in the two CPD simulations that originate from the left aggregate for $t = 0$ (solid line), $t = 0.5t_R$ (dashed line), and $t = t_R$ (dashed dotted line) in case of simulations (a) CPD-1 and (b) CPD-10. Snapshots in both panels are shown in the initial state at $t = 0$ (upper right) and $t = t_R$ (lower left).

C. Comparison of Experiments, Theory and Computer Simulations

Fig. 1 shows the comparison of shape evolution of the fusing aggregates as obtained in the experiments, the various simulations and by the theoretical analysis of the process. In order to quantitatively compare the various methods described we use the time dependence of $(r_0/R_0)^2$. Experimental data on this quantity was obtained by following the fusion of two aggregates composed of embryonic smooth muscle cells. Results of the comparison are shown in Fig. 10.

Equation (8) was used to fit the data. This provided the ratio η/γ for the tissue, but, more importantly, also allowed connecting the macroscopic, tissue level quanti-

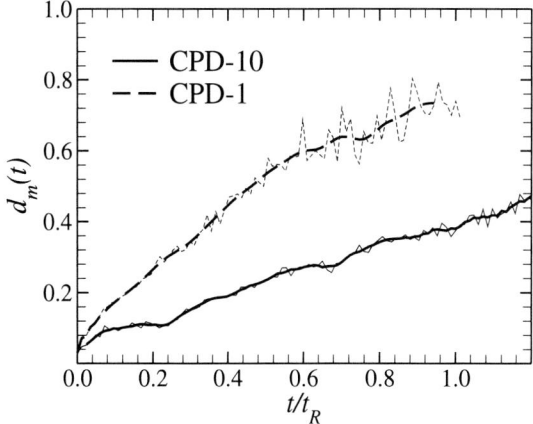

Figure 8 Mixing parameter d_m for the CPD-10 (solid line) and CPD-1 (dashed line) simulations. The fluctuations (noise) in the simulation data (thin curves) were reduced by an averaging procedure ("moving window averaging"), leading to the smooth thick line.

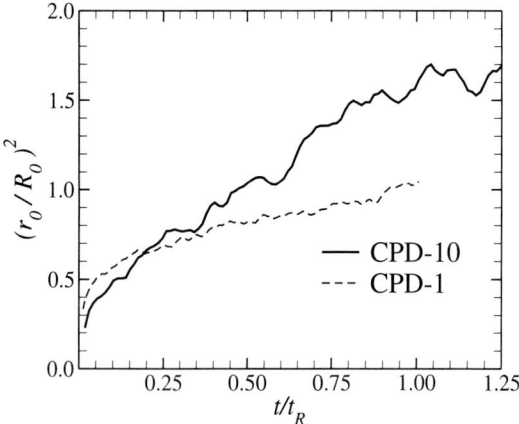

Figure 9 Time evolution of the contact area between the two fusing cell aggregates obtained in simulations CPD-10 (solid line) and CPD-1 (dashed line).

ties (γ, η) to the molecular, cell and subcellular quantities that characterize the CPD model (i.e., ε, σ—parameters in the LJ potential, k, α—parameters in the confining potential, and μ, E_T—cytosolic, "environmental" parameters). A practical way to implement this connection is as follows. As in any particle dynamics simulation, we use dimensionless quantities (as defined earlier). Furthermore, it is natural to choose the range of the confining potential to be the linear size of the cell d, i.e., $\alpha \sim d \sim N_{CP}^{1/3} \sigma$, where N_{CP} is the number of CPs in a cell. In our simulation $N_{CP} = 10$ and, in di-

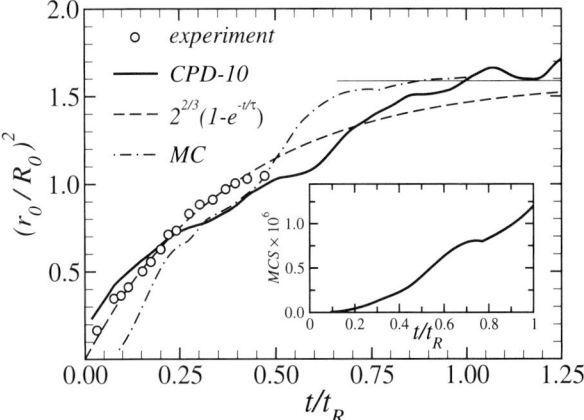

Figure 10 Comparison of the time evolutions of the interfacial area between two fusing aggregates obtained from experiment (open circles), CPD-10 simulation (solid line), MC simulation (dot-dashed line) and theoretical prediction (dashed line). The horizontal line indicates the limiting value $2^{2/3}$ of the contact area. Inset: Correlation between the MCS and t/t_R determined from shape correspondence in Fig. 1.

mensionless units, we set $\alpha = 2.5$. Thus, we only need to specify the values of ε and k. Although, in principle, these can be considered as adjustable parameters in the CPD simulation, we set $\varepsilon = 1$ and $k = 5$. Finally, choosing a suitable integration time step Δt (typically $\sim 10^{-4}$–10^{-5}), by starting from the initial configuration of two contiguous spherical cellular aggregates (connected through a point like region), we integrated the CPD equations of motion for N_t time steps using (the modified) LAMMPS code. By visualizing the obtained CPD trajectory [e.g., by using the program VMD (Humphrey et al., 1996)] we followed the time evolution of the fusion process (showed in Fig. 1). From the CPD trajectory, one can easily determine the rounding time $t_R = N_{t_R} \Delta t$ (in dimensionless units), where N_{t_R} is the corresponding number of simulation steps. In case of CPD-10, the radius of the final fused spheroid R_f was found to be equal (within the errors due to the fluctuations of the system) to the theoretical value $2^{1/3} \approx 1.26 P_0$ dictated by volume conservation. However, in case of CPD-1, where a noticeable fraction of the CPs escaped the aggregates during their fusion, R_f was only ~ 1, a value consistent with the number of lost CPs. As shown in Fig. 10, the contact area vs time in case of CPD-10 compares rather well both with the experimental data and the corresponding theoretical curve. The difference in the results, as well as the irregular shape of the CPD-10 curve, are due to the small system size, i.e., only 200 cells per aggregate. For a much larger system containing 2000 cells per aggregate the corresponding CPD-10 curve (data not shown) matches very well the theoretical result. We emphasize that the $(r_0/R_0)^2$ vs t/t_R curve is universal (i.e., free of any fitting parameter) as long as the fusing drops are incompressible and very viscous. Thus, once the fusion time t_R is determined, the data points $(r_0/R_0)^2$ obtained from experiment or simulations should lie on the theoretical curve given by Eq. (8).

The CPD trajectory provides $(r_0/R_0)^2$, as a function of $t/t_R = N_t/N_{t_R}$, whose graphical representation can readily be compared with both experimental and theoretical results (see Fig. 10). As already mentioned, in terms of τ_0 the rounding time $t_R \approx 3.45\tau_0 = 3.45\eta R_0/\gamma$. On the other hand, in physical units, the CPD rounding time is $t_R = N_{t_R}\Delta t\sigma^2\mu/E_T$. Equating these two expressions allows to determine μ, a cellular level quantity (that characterizes the viscous properties of the cell's interior, i.e., cytosol) in terms of tissue level properties that are accessible by biophysical measurements. As already mentioned, σ is determined by the known size of the cell and the number of CPs in the simulation (in our case 1 or 10).

Since the surface tension γ is an equilibrium quantity, one expects that it is related only to the LJ parameters in the CPD model. Indeed, by calculating the energy per unit area of the CPs in the surface layer of an equilibrated spherical aggregate, one finds that $\gamma = c\varepsilon/\sigma^2$ (where c is a coefficient of order unity that is only slightly model dependent), leading to yet another relationship between measurable tissue parameters and molecular quantities. Furthermore, this result implies that for two CPD models (having the same number of CPs) describing two distinct tissues,

$$\frac{\varepsilon_2}{\varepsilon_1} = \frac{\gamma_2}{\gamma_1}\left(\frac{\sigma_1}{\sigma_2}\right)^2 = \frac{\gamma_2}{\gamma_1}\left(\frac{d_1}{d_2}\right)^2. \tag{13}$$

Here d_1 and d_2 are the linear sizes at the two cell types. This equation relates the LJ energy parameters of one cell type to those of another. The above approach to calculate γ can also be used in the MC simulations. Such a calculation makes it possible to relate the model parameters in MC to those in CPD and thus to experimentally measurable equilibrium quantities.

Finally, we discuss the possibility of comparing the MC and CPD simulations results on the time evolution of the contact area. In general, there is no meaningful association between MCS and time. However, in the case of fusion of two aggregates, as shown in Fig. 1, during the course of the MC simulation the system seems to go through similar intermediate configurations between the initial and final states as in the CPD simulations. This allows establishing a nonlinear relationship between the MCS and time in CPD based on the shape similarity. The result is shown in the inset to Fig. 10. With this correspondence it is possible to represent on the same plot in Fig. 10 the "pseudo-time" dependence of the contact area as obtained from MC simulations.

IV. Conclusions

Biological systems are extremely complex, primarily because they extend over vastly differing scales, from the molecular to the organismal. How the genetic level information stored in the sequence of nucleotides with the size of nanometers is processed and eventually implemented at the size of meters is an unresolved puzzle. It is clear that problems of such magnitude will not be possible to solve with the traditional methods of the individual scientific disciplines such as physics, biology and chemistry. Furthermore, the vast number of degrees of freedom characterizing living systems necessitates the use of powerful computational methods. Contrary to nonliving physical systems, where reduction in the degrees of freedom is often possible (depending on the specific question asked), in living systems almost everything matters. Simplifying the system through the elimination of degrees of freedom may result in "throwing out the baby with the bathwater."

In the present work, combining experimental tools with theory and computer simulations we outlined the initial steps of a program to tackle the problem of multiple scales in living systems. On one hand we measured tissue level biophysical properties, such as surface tension and viscosity (in fact here we only concentrated on the ratio of these quantities). Such macroscopic material properties of living tissues must originate from cell and subcellular processes. No matter how evident this statement might be, no systematic method exists today to address the question of how cell and subcellular processes and properties (in particular mechanical ones, studied here) give rise to those at the level of tissues. To make progress, we considered a specific early developmental process, tissue fusion. Relying on the postulates of Steinberg's Differential Adhesion Hypothesis on the liquid-like nature of embryonic tissues (amply supported by *in vitro* and *in vivo* experiments), we modeled tissue fusion both theoretically and through computer simulations. The theoretical approach employed the basic laws of hydrodynamics. Computer simulations were carried out using both Monte Carlo and Cellular Particle Dynamics methods. The latter approach is similar to SEM (Newman, 2005).

The essence of our approach is that biological complexity at the molecular scale, where accurate quantitative measurements are difficult and scarce, is treated with specifically designed computational methods that connect cellular level quantities to tissue level quantities. Specifically, in our CPD model we used the Lennard–Jones potential that accurately describes intermolecular interactions in liquids to represent intra- and intercellular interactions. Theoretical models, like the one employed here, have no problem bridging the gap between molecular and macroscopic scales. The laws of hydrodynamics provide a coarse-grained description of liquids (in terms of densities and associated velocities), thus apply to the mesoscopic/macroscopic scale. However, they have their solid molecular foundation. We employed hydrodynamics to solve the problem of fusion of two viscous liquids and provide analytical expressions as a convenient way to compare the results of CPD and MC simulations with experimental results generated at the tissue level. The analytical expressions allowed to directly relate quantitative parameters at the cell and subcellular (i.e., molecular) level with those characterizing multicellular tissue. We believe our approach has utility beyond the specific example of tissue fusion considered here and provides a framework for further investigation on the connection of biological processes across scales.

Acknowledgments

This work was supported by the National Science Foundation under Grant FIBR-0526854. We gratefully acknowledge the computational resources provided by the University of Missouri Bioinformatics Consortium.

References

Amar, J. G. (2006). The Monte Carlo method in science and engineering. *Comput. Sci. Eng.* **8**, 9–19.
Beysens, D. A., Forgacs, G., and Glazier, J. A. (2000). Cell sorting is analogous to phase ordering in fluids. *Proc. Natl. Acad. Sci. USA* **97**, 9467–9471.
Flenner, E., Janosi, L., Bogdan, B., Neagu, A., Forgacs, G., and Kosztin, I., 2007. Theoretical and computer modeling of the time evolution of the dynamics of cellular aggregates (to be published).
Forgacs, G., and Newman, S. A. (2005). "Biological physics of the developing embryo." Cambridge Univ. Press, Cambridge.
Forgacs, G., Foty, R. A., Shafrir, Y., and Steinberg, M. S. (1998). Viscoelastic properties of living embryonic tissues: A quantitative study. *Biophys. J.* **74**, 2227–2234.
Foty, R. A., and Steinberg, M. S. (2005). The differential adhesion hypothesis: A direct evaluation. *Dev. Biol.* **278**, 255–263.
Foty, R. A., Forgacs, G., Pfleger, C. M., and Steinberg, M. S. (1994). Liquid properties of embryonic-tissues—measurement of interfacial-tensions. *Phys. Rev. Lett.* **72**, 2298–2301.
Foty, R. A., Pfleger, C. M., Forgacs, G., and Steinberg, M. S. (1996). Surface tensions of embryonic tissues predict their mutual envelopment behavior. *Development* **122**, 1611–1620.
Frenkel, J. (1945). Viscous flow of crystalline bodies under the action of surface tension. *J. Phys. (USSR)* **9**, 385–391.
Glazier, J. A., and Graner, F. (1993). Simulation of the differential adhesion driven rearrangement of biological cells. *Phys. Rev. E* **47**, 2128–2154.
Godt, D., and Tepass, U. (1998). Drosophila oocyte localization is mediated by differential cadherin-based adhesion. *Nature* **395**, 387–391.
Goel, N. S., and Rogers, G. (1978). Computer simulation of engulfment and other movements of embryonic tissues. *J. Theor. Biol.* **71**, 103–140.
Gonzalez-Reyes, A., and St. Johnston, D. (1998). The *Drosophila* AP axis is polarized by the cadherin-mediated positioning of the oocyte. *Development* **125**, 3635–3644.
Gordon, R., Goel, N. S., Steinberg, M. S., and Wiseman, L. L. (1972). A rheological mechanism sufficient to explain the kinetics of cell sorting. *J. Theor. Biol.* **37**, 43–73.
Graner, F., and Glazier, J. A. (1992). Simulation of biological cell sorting using a 2-dimensional extended potts-model. *Phys. Rev. Lett.* **69**, 2013–2016.
Gumbiner, B. M. (1996). Cell adhesion: The molecular basis of tissue architecture and morphogenesis. *Cell* **84**, 345–357.
Hegedus, B., Marga, F., Jakab, K., Sharpe-Timms, K. L., and Forgacs, G. (2006). The interplay of cell–cell and cell–matrix interactions in the invasive properties of brain tumors. *Biophys. J.* **91**, 2708–2716.
Humphrey, W., Dalke, A., and Schulten, K. (1996). VMD: Visual molecular dynamics. *J. Mol. Graphics* **14**, 33–38.
Israelachvili, J. (1992). "Intermolecular and surface forces." Academic Press, London.
Jakab, K., Neagu, A., Mironov, V., Markwald, R. R., and Forgacs, G. (2004). Engineering biological structures of prescribed shape using self-assembling multicellular systems. *Proc. Nat. Acad. Sci. USA* **101**, 2864–2869.
Leith, A. G., and Goel, N. S. (1971). Simulation of movement of cells during self-sorting. *J. Theor. Biol.* **33**, 171–188.

Metropolis, N., Rosenbluth, A. W., Rosenbluth, M. N., and Teller, A. H. (1953). Equation of state calculations by fast computing machines. *J. Chem. Phys.* **21**, 1087–1097.

Mombach, J. C., Glazier, J. A., Raphael, R. C., and Zajac, M. (1995). Quantitative comparison between differential adhesion models and cell sorting in the presence and absence of fluctuations. *Phys. Rev. Lett.* **75**, 2244–2247.

Neagu, A., Jakab, K., Jamison, R., and Forgacs, G. (2005). Role of physical mechanisms in biological self-organization. *Phys. Rev. Lett.* **95**. 178104.

Neagu, A., Kosztin, I., Jakab, K., Barz, B., Neagu, M., Jamison, R., and Forgacs, G. (2006). Computational modeling of tissue self-assembly. *Mod. Phys. Lett. B* **20**, 1217–1231.

Newman, T. J. (2005). Modeling multicellular systems using subcellular elements. *Math. Biosci. Eng.* **2**, 613–624.

Odell, G. M., Oster, G., Alberch, P., and Burnside, B. (1981). The mechanical basis of morphogenesis. I. Epithelial folding and invagination. *Dev. Biol.* **85**, 446–462.

Perez-Pomares, J. M., and Foty, R. A. (2006). Tissue fusion and cell sorting in embryonic development and disease: Biomedical implications. *BioEssays* **28**, 809–821.

Phillips, J. C., Braun, R., Wang, W., Gumbart, J., Tajkhorshid, E., Villa, E., Chipot, C., Skeel, R. D., Kale, L., and Schulten, K. (2005). Scalable molecular dynamics with NAMD. *J. Comput. Chem.* **26**, 1781–1802.

Plimpton, S. (1995). Fast parallel algorithms for short-range molecular-dynamics. *J. Comput. Phys.* **117**, 1–19.

Rogers, G., and Goel, N. S. (1978). Computer simulation of cellular movements: Cell-sorting, cellular migration through a mass of cells and contact inhibition. *J. Theor. Biol.* **71**, 141–166.

Steinberg, M. S. (1963). Reconstruction of tissues by dissociated cells: Some morphogenetic tissue movements and the sorting out of embryonic cells may have a common explanation. *Science* **141**, 401–408.

Steinberg, M. S. (1970). Does differential adhesion govern self-assembly processes in histogenesis? Equilibrium configurations and the emergence of a hierarchy among populations of embryonic cells. *J. Exp. Zool.* **173**, 395–434.

Steinberg, M. S. (1996). Adhesion in development: An historical overview. *Dev. Biol.* **180**, 377–388.

Steinberg, M. S., and Poole, T. J. (1982). Liquid behavior of embryonic tissues. *In* "Cell Behavior." (Bellairs R., Curtis A. S. G., and Dunn G., Eds.). Cambridge Univ. Press, Cambridge, pp. 583–607.

Steinberg, M. S., and Takeichi, M. (1994). Experimental specification of cell sorting, tissue spreading, and specific spatial patterning by quantitative differences in cadherin expression. *Proc. Natl. Acad. Sci. USA* **91**, 206–209.

Takeichi, M. (1990). Cadherins: A molecular family important in selective cell–cell adhesion. *Annu. Rev. Biochem.* **59**, 237–252.

Technau, U., and Holstein, T. W. (1992). Cell sorting during the regeneration of Hydra from reaggregated cells. *Dev. Biol.* **151**, 117–127.

Thompson, S. M., Gubbins, K. E., Walton, J. P. R. B., Chantry, R. A. R., and Rowlinson, J. S. (1984). A molecular dynamics study of liquid drops. *J. Chem. Phys.* **81**, 530–542.

Wessels, A., and Sedmera, D. (2003). Developmental anatomy of the heart: A tale of mice and man. *Physiol. Genomics* **15**, 165–176.

Whitesides, G. M., and Boncheva, M. (2002). Supramolecular chemistry and self-assembly special feature: Beyond molecules: Self-assembly of mesoscopic and macroscopic components. *Proc. Natl. Acad. Sci.* **99**, 4769–4774.

17

Complex Multicellular Systems and Immune Competition: New Paradigms Looking for a Mathematical Theory*

Nicola Bellomo* and Guido Forni[†]
*Department of Mathematics, Politecnico, Turin, Italy
[†]Department of Clinical and Biological Sciences, University of Turin, Turin, Italy

I. Introduction
II. Conceptual Lines Towards a Mathematical Biological Theory
III. From Hartwell's Theory of Modules to Mathematical Structures
IV. A Simple Application and Perspectives
V. What Is Still Missing for a Biological Mathematical Theory
   References

    This chapter deals with the modeling and simulation of large systems of interacting entities whose microscopic state includes not only geometrical and mechanical variables (typically position and velocity), but also biological functions or specific activities. The main issue looks at the development of a biological mathematical theory for multicellular systems. The first part is devoted to the derivation of mathematical structures to be properly used to model a variety of biological phenomena with special focus on immune competition. Then, some specific applications are proposed referring to the competition between neoplastic and immune cells. Finally, the last part is devoted to research perspectives towards the objective of developing a mathematical–biological theory. A critique is presented of what has already been achieved towards the above target and what is still missing with special focus on multiscale systems.
© 2008, Elsevier Inc.

I. Introduction

This chapter deals with the challenging objective of developing a mathematical theory of biological multicellular systems, with special focus on the competition between the immune system and cancer cells. The overall project does not simply deal with the design of a mathematical model, but it will lead to a robust mathematical description

---

* Partially supported by Marie Curie Network MRTN-CT-2004-503661—Modelling, Mathematical Methods, and Computer Simulations of Tumor Growth and Therapy.

of biological activities. This is a fascinating perspective that is worth pursuing despite its enormous conceptual difficulties. Towards this goal, the heuristic experimental approach, which is the traditional method of investigation in biological sciences, should be gradually integrated with new methods and paradigms generated by progressive interaction with the mathematical sciences.

A mathematical description of living matter is conceptually different from the description of inert matter. As observed by May (2004), mathematical theory and experimental investigation have always marched together in the physical sciences. Mathematics has been less intrusive in the life sciences, possibly because they have been until recently descriptive, lacking the invariance principles and fundamental natural constants of physics.

Nevertheless, several hints to interdisciplinary approaches have been offered. Hartwell *et al.* (1999) point to the fact that although living systems obey the laws of physics and chemistry, the notion of function and purpose differentiates biology from other natural sciences. Reproduction, competition, cell cycle, ability to communicate with other entities are features absent in classic Newtonian mechanics. Furthermore, the macroscopic features of a system constituted by millions of cells shows only the output of cooperative behavior and not the activities of the various cells. Remarkably similar concepts are proposed by Reed (2004) that consider the issue from the view point of applied mathematics.

Hartwell's suggestions are not limited to general speculations, but provide a theory of functional modules that actually contributes to the development of a mathematical theory for biological systems and that will be used in the present study. A functional module is defined as a discrete entity whose function is separable from those of other modules.

The present approach will use the mathematical kinetic theory for living particles in order to describe complex multicellular systems dealing with cell expansions, cell death and immune supervision. The analysis will be developed at the cellular scale, as an intermediate between the subcellular and macroscopic scales. Our kinetic theory for living systems is based on functional modules and will be implemented using game theory to describe cell interactions leading to selective expansions and death.

II. Conceptual Lines Towards a Mathematical Biological Theory

In the present chapter we use the mathematical kinetic theory for living particles in order to describe complex multicellular systems dealing with cell expansions, cell death and immune surveillance. The modeling is developed at the cellular scale, as an intermediate between the subcellular and macroscopic scales.

The traditional approach of the mathematical sciences to biology generates mathematical models which we separate from a mathematical theory. Models may even occasionally reproduce specific aspects of biological phenomena, while they are

17. Complex Multicellular Systems and Immune Competition

rarely able to capture their essential features. On the other hand, a mathematical theory provides structures rigorously deduced on the basis of mathematical assumptions whose validity is consistent with a large variety of physical systems.

Our kinetic theory attempts to describe the statistical evolution of large systems of interacting particles whose microscopic state includes activity, a variable related to the expression of biological function. In addition to geometrical and mechanical variables, it is the kinetic theory for active particles. Admittedly, additional work is needed to develop a comprehensive mathematical–biological theory.

In the framework of mathematical kinetic theory we will evaluate the statistical distribution of the activity (defined as state) of the cells forming the different populations considered in the model. The third step involves the derivation of mathematical equations describing the evolution of the cell state. This will be pursued through conservation equations describing the numerical modulation of cells denoted by a defined state.

In the physical sciences the analog of a multicellular system is a gas mixture of particles. Usually this involves a small number of components, whereas a biological system is commonly denoted by a *game* with a large number of cell populations. Interactions among particles obey the laws of classical Newtonian mechanics without amplification and death events. By contrast, the latter are the hallmarks of cellular systems. Detailed information on how cell interactions are regulated by signals emitted or perceived by the cells and transduced to the nucleus, where they participate in the modulation of cell activities, is provided by biological studies. As a consequence of these interactions the cell acquires a new state.

Mathematical models describe these cell activities with ordinary differential equations or Boolean networks, while multicellular systems are modeled by nonlinear integrodifferential equations similar to those of nonlinear kinetic theory (the Boltzmann equation) or by individual-based models which give rise to a large set of discrete reactions. On the other hand, applied mathematics attempts to deal with the biological complexity of cell behavior by isolating a few activities and studying them through differential equations. However, it is not generally possible to isolate specific biological activities from the whole system under consideration. Interactions among cell subpopulations may dramatically modulate the resulting cell activity, and thus the model cannot be limited to one specific activity isolated from its cellular microenvironment.

Moreover, a mathematical theory developed at the cellular scale should retain suitable information from the lower molecular scale, while it should allow the derivation of macroscopic equations by suitable asymptotic limits related to condensation and fragmentation events. This matter is even more relevant (Bellomo and Maini, 2006) considering that one of the major problems in modeling biological systems is the multiscale nature of most biological systems.

III. From Hartwell's Theory of Modules to Mathematical Structures

Our mathematical kinetic theory deals with the analysis of the evolution of multicellular systems, far from equilibrium. In dealing with the enormous difficulty in deriving a mathematical framework, reference to physical sciences provides a useful background: the state of each microscopic entity, the particle, is identified by geometrical and mechanical variables: position and orientation, as well as velocity and rotation. In our theory, the state of a cell includes specific activities which may differ among cell populations as well as common activities such as cell expansion and death.

To deal with the biological complexity, Hartwell's (1999) theory of functional modules can be exploited. The behavior of cell populations engaged in a defined activity should be considered as a collective whole.

Consider a large system of n interacting cell populations homogeneously distributed in space. The microscopic state of cells is identified by a scalar variable $u \in D_u$ which represents the relevant biological function expressed by cells of each population.

The biological function differs from population to population, while the overall statistical representation of the system is described by the distribution functions:

$$f_i = f_i(t, u) : [0, T] \times D_u \to \mathbb{R}_+, \quad i = 1, \ldots, n, \qquad (1)$$

where the subscript i labels the ith population, and, by definition, $dN_i = f_i(t, u) \, du$ denotes the number of cells, regarded as active particles, which, at time t, are in the element $[u, u + du]$ of the space of the microscopic states.

A mathematical model should describe the evolution in time of the distribution functions f_i. When these functions are obtained by solution of suitable mathematical problems, then gross averaged quantities can be computed. For instance, the local *number density* of cells is computed, under suitable integrability assumptions on f_i, as follows:

$$n_i(t) = \int_{D_u} f_i(t, u) \, du, \qquad (2)$$

while the *activation* and the *activation densities* are, respectively, given by

$$a_i = a[f_i](t) = \int_{D_u} u f_i(t, u) \, du, \qquad (3)$$

and

$$A_i = A[f_i](t) = \frac{a[f_i](t)}{n_i(t)}. \qquad (4)$$

Analogous calculations can be developed for higher order moments.

Considering the kinetic theory for active particles Bellouquid and Delitala (2006) applied to modeling multicellular systems, the derivation of a mathematical framework suitable to describe the evolution of the distribution functions $f_i = f_i(t, u)$

17. Complex Multicellular Systems and Immune Competition

appears appropriate to describe the evolution of the system under consideration. Such a framework acts as a general paradigm for the derivation of specific models generated by a detailed modeling of cellular interactions. It needs to be stressed that only a few activities that change the state of the cells are taken into account, such as:

- Stochastic modification of the microscopic state of cells due to binary interactions with other cells of the same or of different populations. These interactions are called *conservative* as they do not modify the number density of the various populations
- Genetic alteration of cells which may either increase the progression of tumor cells or even generate, by clonal selection, new cells in a new population of cancer cells with higher level of malignancy
- Proliferation or destruction of cells due to interaction with other cells of the same or of different populations. Proliferation refers to both tumor and immune cells.

The formal equation, which describes the evolution of f_i, is obtained by the balance of particles in the elementary volume of the space of the microscopic states. The technical derivation proposed in Bellouquid and Delitala (2005, 2006) is as follows:

$$\frac{\partial}{\partial t} f_i(t, u) = J_i[f](t, u) = C_i[f](t, u) + P_i[f](t, u) + D_i[f](t, u), \tag{5}$$

where the right-hand side term models the flow, at the time t, into the elementary volume $[u, u + du]$ of the state space of the ith population due to transport and interactions. In detail:

- $C_i[f](t, u)$ models the flow, at time t, into the elementary volume of the state space of the ith population due to conservative interactions:

$$C_i[f](t, u) = \sum_{j=1}^{n} \eta_{ij} \int\int_{D_u D_u} \mathcal{B}_{ij}(u_*, u^*; u) f_i(t, u_*) f_j(t, u^*) \, du_* \, du^*$$

$$- f_i(t, u) \sum_{j=1}^{n} \eta_{ij} \int_{D_u} f_j(t, u^*) \, du^*, \tag{6}$$

where η_{ij} is the encounter rate of a *candidate particle*, with state u_* in the ith population, and a *field particle*, with state u^* in the jth population. $\mathcal{B}_{ij}(u_*, u^*; u)$ denotes the probability density that the candidate particles fall into the state u remaining in the same populations. Conservative equations modify the microscopic state, but not capacity to produce a clonal expansion.

- $P_i[f](t, u)$ models the flow, at time t, into the elementary volume of the state space of the ith population due to proliferating interactions with transition of population:

$$P_i[f](t, u) = \sum_{h=1}^{n}\sum_{k=1}^{n} \eta_{hk} \int\int_{D_u D_u} \mu_{hk}^i(u_*, u^*; u) f_h(t, u_*)$$

$$\times f_k(t, u^*) \, du_* \, du^*, \tag{7}$$

where $\mu^i_{hk}(u_*, u^*; u)$ models the net proliferation into the ith population, due to interactions, which occur with rate η_{hk}, of the *candidate particle*, with state u_*, of the hth population and the *field particle*, with state u^*, of the kth population.

- $D_i[f](t, u)$ models the net flow, at time t, into the elementary volume of the state space of the ith population due to proliferative and destructive interactions without transition of population:

$$D_i[f](t, u) = \sum_{j=1}^{n} \eta_{ij} \int_{D_u} \mu_{ij}(u_*, u^*, u) f_i(t, u_*) f_j(t, u^*) \, du_* \, du^*, \qquad (8)$$

where $\mu_{ij}(u, u^*)$ models net flux within the same population due to interactions, which occur with rate η_{ij}, of the *test particle*, with state u, of the ith population and the *field particle*, with state u^*, of the jth population.

Substituting the above expression into (5), yields:

$$\frac{\partial}{\partial t} f_i(t, u) = \sum_{j=1}^{n} \eta_{ij} \int_{D_u} \int_{D_u} \mathcal{B}_{ij}(u_*, u^*; u) f_i(t, u_*) f_j(t, u^*) \, du_* \, du^*$$

$$- f_i(t, u) \sum_{j=1}^{n} \eta_{ij} \int_{D_u} f_j(t, u^*) \, du^*$$

$$+ \sum_{h=1}^{n} \sum_{k=1}^{n} \eta_{hk} \int_{D_u} \int_{D_u} \mu^i_{hk}(u_*, u^*; u) f_h(t, u_*) f_k(t, u^*) \, du_* \, du^*$$

$$+ \sum_{j=1}^{n} \eta_{ij} \int_{D_u} \mu_{ij}(u_*, u^*, u) f_i(t, u_*) f_j(t, u^*) \, du_* \, du^*. \qquad (9)$$

The above structure acts as a paradigm for the derivation of specific models, to be obtained by a detailed modeling of microscopic interactions. The modeling needs to be finalized to generate well defined expressions of the terms η, $\mathcal{B}$, and μ. However, following the idea of reducing complexity, each population is identified only by the ability of performing one activity. This reduction of complexity fits well with the suggestion, due to Hartwell, to regard a population as a modulus with the ability of expressing a well defined activity. A simpler structure is obtained in the absence of generation of particles in a population different from that of the interacting pairs with proliferation and destruction in the state of the test particle. In this case, the mathematical structure is as follows:

$$\frac{\partial}{\partial t} f_i(t, u) = \sum_{j=1}^{n} \eta_{ij} \int_{D_u} \int_{D_u} \mathcal{B}_{ij}(u_*, u^*; u) f_i(t, u_*) f_j(t, u^*) \, du_* \, du^*$$

$$- f_i(t, u) \sum_{j=1}^{n} \eta_{ij} \int_{D_u} f_j(t, u^*) \, du^*$$

$$+ f_i(t, u_*) \sum_{j=1}^{n} \eta_{ij} \int_{D_u} \mu_{ij}(u, u^*) f_j(t, u^*) \, du_* \, du^*. \tag{10}$$

As already mentioned, the above reduction of complexity corresponds to an interpretation of the theory of functional modules. Hence, each population is regarded as a module with the ability of expressing a well defined biological function.

IV. A Simple Application and Perspectives

The literature on the modeling of cellular systems by the methods of mathematical kinetic theory for active particles was first proposed by us, Bellomo and Forni (1994), and then developed by various authors as documented by, for example, Kolev (2003, 2005), Bellouquid and Delitala (2004, 2005, 2006), De Angelis and Jabin (2003), Derbel (2004). Models are related to biological theories of cell competition, see Dunn et al. (2002), Blankenstein (2005), Friedl et al. (2005).

The above literature shows how models describe the output of the competition according to an appropriate choice the parameters which can be possibly modified by suitable therapeutical actions. The aforementioned parameters have a well defined biological meaning and can be identified by appropriate experiments.

Specifically, the models describe, depending on the above mentioned parameters, progression and heterogeneity phenomena (Greller et al., 1996; Nowell, 2002), as well as various aspects of immune competition. They show either the growth and blow-up of tumor cells after progressive inhibition of the immune cells, or the destruction of tumor cell density by the immune system which remains active. The output depends, not only on the proliferation ability of progressing cells, but also on the ability of immune cells to target antigens and then destroy tumor cells.

Some models include the description of therapeutical actions by adding new populations which have the ability to modify the output of the competition, e.g., Kolev et al. (2005), Brazzoli and Chauviere (2006), De Angelis and Jabin (2005). In some cases immune activation, e.g., d'Onofrio (2006), and multiple therapies have to be taken into account, e.g., De Pillis et al. (2006). These papers show the flexibility of this theory as applied to different models of cell expansion, differentiation and surveillance. It also permits us to assess the consequences of therapeutical interventions. In a few models, the inclusion of additional cell populations changes the final result, while in others the effects of particular immune stimulations or multiple therapies have been taken into account.

Before dealing with an illustrative application of various aspects of the aforementioned competition, it is worth developing a critical analysis of the existing literature in the field in view of further research perspectives. Specifically, let us stress that all models cited above are characterized by the following:

i) The number of interacting cell populations is small compared with the number of cell populations and subpopulations that probably play a role *in vivo*. For instance, in the modeling of tumor–host interactions only normal, neoplastic, and immune cell populations are considered.

ii) The derivation of all models needs a detailed description of interactions at the microscopic level. These specific models are obtained by a phenomenological interpretation, based on causality effects, of biological reality.

iii) Therapeutical actions are modeled by addition of new populations which have the ability to activate the immune response of weakening the progression of tumor cells.

A more accurate modeling of the onset of cancer, the inclusion into the model of additional cell populations denoted by specialized cell functions, will improve the predictive ability of the model as shown by Pappalardo *et al.* (2005) and Lollini *et al.* (2006). The significance of a model is markedly increased by its ability not only to provide a quantitative simulation of what can be experimentally observed, but also to focus on events which are not experimentally evident. Subsequent experiments could be planned to specifically assess and validate model predictions.

This section analyzes how various phenomena of interest in biology can be described by models related to the mathematical structures (9) and (10), where the various terms characterizing cell interactions are obtained by a phenomenological interpretation of physical reality. Simple models can be improved to include the description of additional events. Specifically, a relevant biological characteristic, remarked in the paper by Greller *et al.* (1996), is the *heterogeneity* over the activity variable, which means that biological functions are not the same for all cells, but are statistically distributed (for progression) as already discussed in the previous section.

The simple model below, referring to the mathematical structure (9), is described as the technical interpretation of the contents of the above cited book by Bellouquid and Delitala (2006). Then some developments are analyzed taking advantage of several interesting directions recently proposed by Delitala and Forni (2007) who present a model which describes sequential genetic mutations of a neoplastic cell with increasing level of malignancy.

Consider a system constituted by two populations whose microscopic state $u \in (-\infty, \infty)$ has a different meaning for each population:

$i = 1$: *Environmental cells*—The state u refers to *natural* state (normal stromal cells) for negative values of u; to abnormal state, i.e., cells which have lost their differentiated state and become progressing cells, for positive values of u, with the additional ability to inhibit immune cells.

$i = 2$: *Immune cells*—Negative values of u correspond to nonactivity or *inhibition*; positive values of u to *activation* and hence their ability to *contrast* the growth of tumor cells.

Before providing a detailed description of microscopic interactions, some preliminary assumptions, which reduce the complexity of the system, are stated:

$H.1$. The number and distribution of cells of the first population is denoted by $f_1(0, u) = f_{10}(u)$ and is supposed to be constant in time for $u \in (-\infty, 0]$, while the distribution functions of all cell populations are normalized with respect to the initial number density of that population:

$$n_{10} = \int_{-\infty}^{0} f_{10}(u)\,du. \tag{11}$$

H.2. The distribution over the velocity variable is constant in time, therefore, the encounter rate is constant for all interacting pairs. For simplicity it will be assumed that $\eta_{ij} = 1$, for all i, j.

H.3. The term $\mathcal{B}_{ij}$, related to the transition probability density, is assumed to be defined by a given delta distribution identified by the most probable output $m_{ij}(u_*, u^*)$, which depends on the microscopic state of the interacting pairs:

$$\mathcal{B}_{ij}(u_*, u^*; u) = \delta(u - m_{ij}(u_*, u^*)). \tag{12}$$

Then, after the above preliminary assumptions, a detailed description of the nontrivial interactions, assumed to play a role in the evolution of the system, is given according to the phenomenological models described in what follows, while interactions which do not affect the evolution of the system are not reported.

- *Conservative Interactions*

    C.1. Interactions between cells of the first population generate a continuous trend in this population towards progressing states identified by the most probable output:

    $$m_{11} = u_* + \alpha_{11},$$

    where α_{11} is a parameter related to the inner tendency of both a normal and mutated cell to degenerate.

    C.2. The most probable output of the interaction between an active immune cell with a progressing cell is given as follows:

    $$u_* \geqslant 0, u^* \geqslant 0: \quad m_{21} = u_* - \alpha_{21},$$

    where α_{21} is a parameter which indicates the ability of mutated cells to inhibit immune cells.

- *Proliferating-Destructive Interactions*

    P.-D.1. Progressing cells undergo uncontrolled mitosis stimulated by encounters with nonprogressing cells due to their angiogenic ability:

    $$\mu_{11}(u, u^*) = \beta_{11} U_{[0,\infty)}(u) U_{(-\infty,0]}(u^*),$$

    where β_{11} is a parameter which characterizes the proliferating ability of mutated cells.

*P.-D.*2. Active immune cells proliferate due to encounters with progressing cells, although some of them may be inhibited:

$$\mu_{21}(v, w) = \beta_{21} U_{[0,\infty)}(v) U_{[0,\infty)}(w),$$

where β_{21} is a parameter which characterizes the proliferating ability of immune cells.

*P.-D.*3. Progressing cells are partially destroyed due to encounters with active immune cells:

$$\mu_{12}(u, u^*) = -\beta_{12} U_{[0,\infty)}(u) U_{[0,\infty)}(u^*),$$

where β_{12} is a parameter which characterizes the destructive ability of active immune cells.

Based on the above modeling of cell interactions, the evolution equation, Eq. (10) generates the following model:

$$\frac{\partial f_1}{\partial t}(t, u) = \left[f_1(t, u - \alpha_{11}) - f_1(t, u) \right] \int_{-\infty}^{0} f_1(t, u) \, du$$

$$+ f_1(t, u) \left[\beta_{11} \int_{-\infty}^{0} -\beta_{12} \int_{0}^{\infty} f_2(t, u) \, du \right] U_{[0,\infty)}(u), \quad (13)$$

$$\frac{\partial f_2}{\partial t}(t, u) = \int_{0}^{\infty} f_1(t, u) \, du \, f_2(t, u + \alpha_{21}) U_{[0,\infty)}(u + \alpha_{21})$$

$$+ (\beta_{21} - 1) \int_{0}^{\infty} f_1(t, u) \, du \, f_2(t, u) U_{[0,\infty)}(u), \quad (14)$$

where the stepwise function, $U_{[a,b]}(z)$ is such that: $U_{[a,b]}(z) = 1$, if $z \in [a, b]$ and $U_{[a,b]}(z) = 0$, if $z \notin [a, b]$.

The above model is characterized by five positive phenomenological parameters, which are small with respect to one: α_{11} corresponds to the tendency of cells to acquire progression through mutation, α_{21} corresponds to the ability of mutated cells to inhibit the active immune cells, β_{11} corresponds to the proliferation rate of mutated cells, β_{12} corresponds to the ability of immune cells to destroy mutated cells, β_{21} corresponds to the proliferation rate of immune cells.

The α-parameters are related to conservative encounters, while the β-parameters are related to proliferation and destruction.

In general, it is interesting to analyze the influence of the parameters of the model and of the mathematical problem over the following two different behaviors:

i) Blow-up of progressing cells which are not sufficiently controlled by immune cells due both to the fast progression of tumor cells and to the weak proliferation of immune cells.
ii) Destruction of progressing cells due to the action of the immune system which has a sufficient proliferation rate.

Simulations may be used to analyze the role of the parameters in determining the outcome of the competition. For instance, to show how different progression rates may lead to different outputs of competition.

The above two different behaviors are visualized in Figs. 1, 2, according to the solution of the initial value problem for Eqs. (13)–(14) corresponding to different selections of the parameter values. In particular, Fig. 1 shows how the distribution function of tumor cells moves towards greater values of progression. Correspondingly immune cells are progressively inhibited. Simulations are obtained for the following values of the parameters: $\alpha_{11} = \alpha_{21} = 0.1$, $\beta_{11} = 0.5$, and $\beta_{12} = \beta_{21} = 0.05$.

The opposite behavior is shown in Fig. 2, where the parameters related to the activities of immune cells have been selected corresponding to a higher defence ability: $\beta_{12} = 0.5$ and $\beta_{21} = 0.9$. In this case the activation of the immune system reduces progression of tumor cells and finally suppresses them.

The mathematical model described above has been derived within the framework defined by Eq. (10) which does not include the modeling of progressive genetic instability of the tumor. It follows that the model can describe biological phenomena on a short timescale, while for longer time genetic mutations play an important role in the evolution of the system. Moreover, the model has been derived for a small number of populations (regarded as modules expressing a well defined biological function), so that their mathematical structure is not too complex for a computational analysis.

Definitely, the above modeling approach has to be regarded as an approximation of physical reality considering that biological functions are the output of the collective behavior of several cell populations. A relatively more refined interpretation of the theory may identify the function expressed by each population, so that their number is consistently increased. For instance, the modeling of immune cells can be developed by looking at the role of each specific population rather than looking at the whole as one module only. Moreover, genetic instability in tumors favors the onset of a finite number of new populations with increasing degrees of malignancy related to subsequent genetic transitions. The downside of the above approach is that the number of parameters also increases and their identification may become very difficult or even impossible.

The above developments can be referred to Hartwell's theory of modules, as they correspond to assuming that genetic modifications are identified at different stages by discrete variables and that each population is regarded as a module which expresses the same type of biological function.

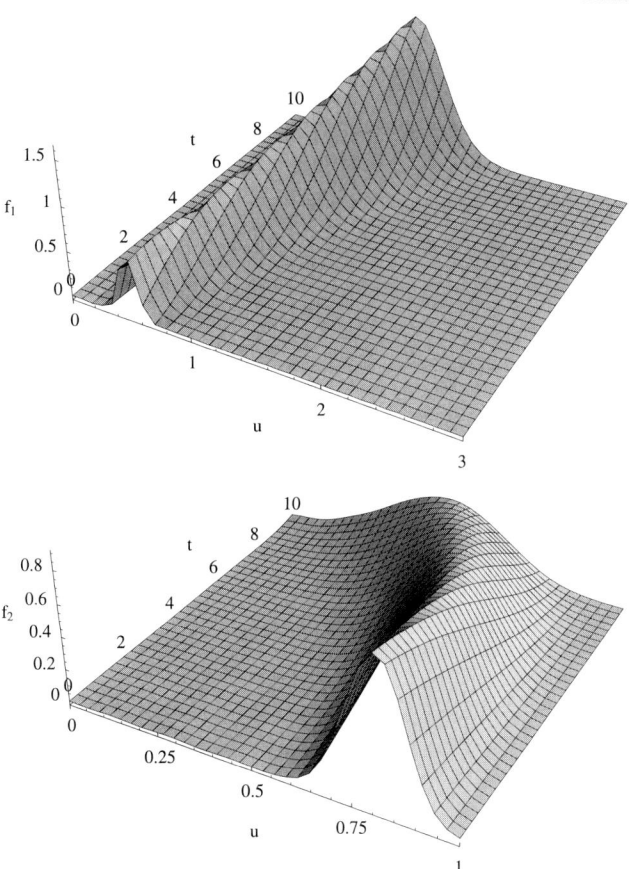

Figure 1 Heterogeneity and progression of tumor cells which increase their progression and proliferate due to their ability to inhibit immune cells (top) and progressive inhibition of immune cells (bottom). Simulations are obtained by solution of Eqs. (13)–(14) in the case of large values of α_{21}.

V. What Is Still Missing for a Biological Mathematical Theory

This final section goes back to the main issue of this chapter, the assessment of additional work needed to develop a proper biological–mathematical theory along the conceptual lines already defined. Some useful guidelines can be extracted again from Hartwell *et al.* (1999). Referring to the contents of the preceding sections, three items acquire a critical importance:

I) The notion of function or purpose differentiates living systems from those of inert matter. Biological functions have the ability to modify the conservation laws of classical mechanics and, in addition, can generate destructive and/or proliferating events.

17. Complex Multicellular Systems and Immune Competition 497

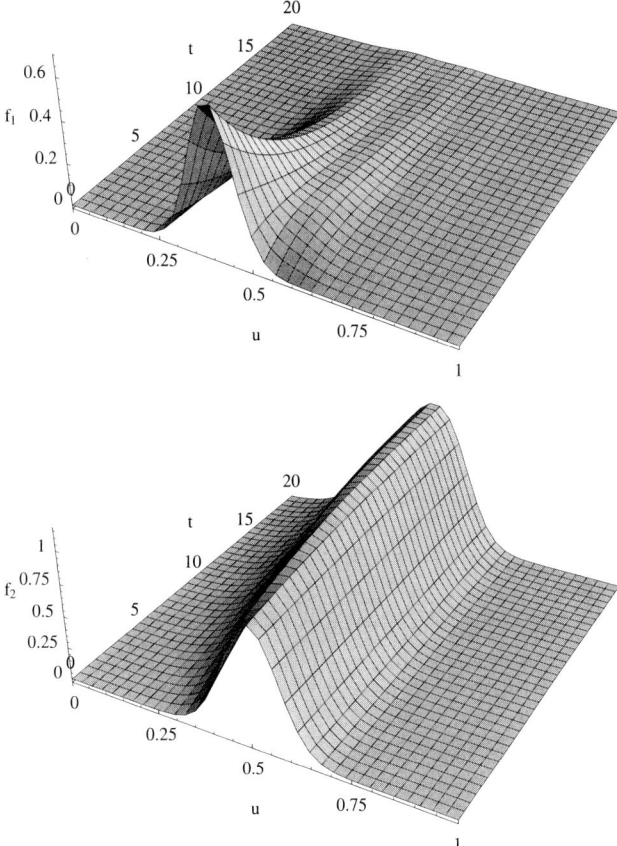

Figure 2 Heterogeneity and progressive destruction of tumor cells Eqs. (13)–(14) which decrease their progression and are progressively suppressed by immune cells (top), while (bottom) immune cells remain sufficiently active. Simulations are obtained by solution of Eqs. (13)–(14) in the case of small values of α_{21}.

II) Biological cells contain a large number of copies, each characterized by specific functions.

III) Systems in biology cannot be simply observed and interpreted at a macroscopic level. A system constituted by millions of cells shows at the macroscopic level only the output of cooperative and organized behaviors which may not, or are not, singularly observed.

Developing a research activity towards the perspective we have defined means that the paradigms of the traditional approach should be replaced by new ones, while applied mathematics should not attempt to describe complex biological systems by simple paradigms and equations. This approach has often generated various unsuccessful attempts and ultimately constitutes even an effective obstacle.

Therefore, we need to verify that the mathematical structures we propose are effectively consistent, according to the above three critical items. The answer is definitely positive for the first two items, considering that biological functions are represented by the activity variables and that the application of the theory of modules can reduce the number of cell populations which effectively "*play the game*." A greater (or smaller) number of populations can possibly provide a more (or less) accurate description of the system under consideration. The price to pay, when the number of populations is increased, is that additional work has to be done to assess all parameters of the model.

Moreover, both the mathematical structure and the model we propose have shown that biological events, such as proliferative and destructive interactions or generation of new populations related to progressive genetic mutations, can be described by mathematical equations.

On the other hand, while our mathematical structure is a candidate to derive specific models, it is not yet a biological–mathematical theory. Possibly, the aforementioned structure refers to a mathematical theory, Bellomo and Forni (2006), considering that a new class of equations has been generated. Indeed, a rigorous framework is given for the derivation of models, when a mathematical description of cell interactions can be derived, by phenomenological interpretation, from empirical data. On the other hand, only when the above interactions are determined by a theoretical interpretation given within the framework of biological sciences, may we talk about a *biological–mathematical theory*.

In more detail, biological sciences should provide, by a robust theory, the various terms which characterize the class of equations (9): the encounter rate η_{hk}; the transition probability density $\mathcal{B}_{hk}$; the population transition terms μ_{hk}^i; the proliferating/destructive terms μ_{hk}; where, in general, the above quantities may depend on the microscopic states.

An analogy can be given with the physics of classical particles whose dynamics is ruled by particle interaction models described by attractive–repulsive potentials. Newtonian mechanics provides the necessary mathematical background to describe particle interactions by attraction–repulsion potentials of the interacting particles, or by mechanical collisions which preserve mass, momentum and energy. It is worth stressing that a deep analysis of the inner structure of atoms or molecules is not necessary, but simply a theoretical description of the interaction potentials which govern pair interactions between particles will suffice. In the case of multicellular systems, the cell state includes, in addition to the mechanical state, biological functions which have the ability of modifying their mechanical behavior. In our case, biology should contribute, by experiments and theoretical interpretations, to describe the outcome of cellular interactions. Specifically, the above four terms should be elucidated at the molecular level, i.e., at the lower scale.

Although at present such a theory is not yet available, it is well understood that the objective can be achieved only through a detailed analysis of gene expression related to biological functions at the molecular scale. This research topic has been dealt with by, among others, Nowak and Sigmund (2004), Komarova (2006), Gatenby *et al.*

(2005), related to specific theories in the field of biological sciences, e.g., Hanahan and Weinberg (2000), Baylin and Ohm (2006), Merlo *et al.* (2006), Anderson *et al.* (2006). The above cited papers give evidence the evolutionary and ecological aspect of cancer onset and evolution referred to gene expression.

The various theoretical approaches known in the literature postulate probabilistic models of gene expression, while gene interactions among themselves and with the external environment should be taken into account. Considering that a robust theory is not yet available, a conjecture is proposed here to develop at the molecular scale some ideas already exploited at the cellular scale. In other words, we conjecture that the structures (9) or (10) can be used to describe the dynamics at the molecular scale to derive, out of this dynamics, the above main cellular interaction terms.

Finally, let us deal with the third critical issue. As we have seen, although the modeling and analysis has been developed at the cellular scale, looking at the lower scale appears to be necessary to recover the above interaction terms. The underlying microscopic description should provide the macroscopic description, which can be possibly observed as the output of the collective behavior of multicellular systems. The mathematical theory, developed at a well defined observation and representation scale, needs to be consistent with the whole set of scales which represent the system.

The analysis of this issue, with reference to the greater scale, generates, quite naturally, the following questions:

i) Which type of macroscopic phenomena can be accurately described by models at the multicellular scale?
ii) Supposing that the above problems are technically solved, is it sufficient to describe the overall system, or, is it necessary to consider the problem as composed of a series of interacting subsystems, each operating at a specific scale?
iii) Is the selection of one scale only sufficient to model the behavior of each subsystem or, even at this level, it is necessary to consider more than one scale?

Several authors have proposed various models of the macroscopic behavior of cancer tissues. A survey of models is presented in the review paper by Bellomo *et al.* (2003). The issue of moving boundary problems has been addressed by, among others, Bertuzzi *et al.* (2004, 2005), Cui and Friedman (2003), Friedman and Lolas (2005), Tao *et al.* (2004), Tao and Zhang (2007). In some cases different scales are taken into account in the model, e.g., Chaplain and Lolas (2005), Levine *et al.* (2001), Owen and Sherratt (2000), Bru *et al.* (1998), Bru and Herrero (2006). Finally various models have been recently proposed to describe networks of interconnected systems of several interacting subsystems, see for example, Alarcón *et al.* (2004), Byrne *et al.* (2006), Anderson (2005).

However complex is the multiscale problem, one has to tackle the problem of selecting the structure of mathematical equations to be used. Asymptotic methods can be used to derive macroscopic equations from the underlying microscopic description. Various papers have been recently proposed towards the above aim by presenting models where spatial phenomena are modeled by adding to the spatially homogeneous equations a stochastic perturbation corresponding to a velocity jump process (Othmer and Hillen, 2002; Stevens, 2002; Lachowicz, 2005; Chalub *et al.*, 2006).

It is worth stressing again that the analysis needs to be implemented by a theoretical input from biology. Specifically, the analysis of Bellomo and Bellouquid (2006) shows how, under a proper scaling, parabolic diffusion equations with a source term are obtained, where the presence of the source term appears if the proliferation rate is sufficiently large.

Therefore, we simply remark that the structure of the equations of tissues depends again on the predominance of one of the three aspects of the biological dynamics, i.e., encounter rate between cells, mutations, and proliferating/destructive events, with respect to the other two. Moreover, the structure of the mathematical equations modeling tissues may evolve in time due to the aforementioned dynamics. So far, the formal approach provides a scenario of structurally different macroscopic equations to be identified by a characterization by biological sciences.

References

Alarcón, T., Byrne, H. M., and Maini, P. K. (2004). A mathematical model of the effects of hypoxia on the cell-cycle of normal cancer cells. *J. Theor. Biol.* **229**, 395–411.

Anderson, A. R. A. (2005). A hybrid mathematical model of solid tumor invasion: The importance of cell adhesion. *Math. Med. Biol.* **22**, 163–186.

Anderson, A. R. A., Weaver, A. M., Cummings, P. T., and Quaranta, V. (2006). Tumor morphology and phenotypic evolution driven by selective pressure from the microenvironment. *Cell* **127**, 905–915.

Baylin, S. B., and Ohm, J. E. (2006). Epigenetic gene silencing in cancer—a mechanism for early oncogenic pathway addition. *Nat. Rev. Cancer* **6**, 107–116.

Bellomo, N., and Forni, G. (1994). Dynamics of tumor interaction with the host immune system. *Math. Comput. Model.* **20**, 107–122.

Bellomo, N., and Bellouquid, A. (2006). On the onset of nonlinearity for diffusion models of binary mixtures of biological materials by asymptotic analysis. *Int. J. Nonlinear Mech.* **41**, 281–293.

Bellomo, N., and Forni, G. (2006). Looking for new paradigms towards a biological–mathematical theory of complex multicellular systems. *Math. Mod. Meth. Appl. Sci.* **16**, 1001–1029.

Bellomo, N., and Maini, P. K. (2006). Preface. *Math. Mod. Meth. Appl. Sci.* **16**(7b), iii–vii.

Bellomo, N., De Angelis, E., and Preziosi, L. (2003). Multiscale modeling and mathematical problems related to tumor evolution and medical therapy. *J. Theor. Med.* **5**, 111–136.

Bertuzzi, A., Fasano, A., and Gandolfi, A. (2004). A free boundary problem with unilateral constraints describing the evolution of a tumor cord under the influence of cell killing agents. *SIAM J. Math. Anal.* **36**, 882–915.

Bertuzzi, A., Fasano, A., and Gandolfi, A. (2005). A mathematical model for tumor cords incorporating the flow of interstitial fluids. *Math. Mod. Meth. Appl. Sci.* **15**, 1735–1777.

Bellouquid, A., and Delitala, M. (2004). Kinetic (cellular) models of cell progression and competition with the immune system. *Z. Angew. Math. Phys.* **55**, 295–317.

Bellouquid, A., and Delitala, M. (2005). Mathematical methods and tools of kinetic theory towards modeling complex biological systems. *Math. Mod. Meth. Appl. Sci.* **15**, 1639–1666.

Bellouquid, A., and Delitala, M. (2006). "Modelling Complex Multicellular Systems—A Kinetic Theory Approach." Birkäuser, Boston.

Blankenstein, T. (2005). The role of tumor stroma in the interaction between tumor and immune system. *Curr. Opin. Immunol.* **17**, 180–186.

Brazzoli, I., and Chauviere, A. (2006). On the discrete kinetic theory for active particles: Modelling the immune competition. *Comput. Math. Meth. Med.* **7**, 142–1158.

Bru, A., and Herrero, M. A. (2006). From the physical laws of tumor growth to modeling cancer process. *Math. Mod. Meth. Appl. Sci.* **16**, 1199–1218.

Bru, A., Pastor, J. M., Fernaud, I., Bru, I., Melle, S., and Berenguer, C. (1998). Superrough dynamics on tumor growth. *Phys. Rev. Lett.* **81**, 4008–4011.

Byrne, H. M., Alarcon, T. A., Murphy, J., and Maini, P. K. (2006). Modelling the response of vascular tumors to chemotherapy: A multiscale approach. *Math. Mod. Meth. Appl. Sci.* **16**, 1219–1241.

Chalub, F., Dolak-Struss, Y., Markowich, P., Oeltz, D., Schmeiser, C., and Soref, A. (2006). Model hierarchies for cell aggregation by chemotaxis. *Math. Mod. Meth. Appl. Sci.* **16**, 1173–1198.

Chaplain, M. A. J., and Lolas, G. (2005). Spatiotemporal heterogeneity arising in a mathematical model of cancer invasion of tissue. *Math. Mod. Meth. Appl. Sci.* **15**, 1685–1734.

Cui, S., and Friedman, A. (2003). Hyperbolic free boundary problem modeling tumor growth. *Interfaces Free Bound.* **5**, 159–181.

De Angelis, E., and Jabin, P. E. (2003). Qualitative analysis of a mean field model of tumor–immune system competition. *Math. Mod. Meth. Appl. Sci.* **13**, 187–206.

De Angelis, E., and Jabin, P. E. (2005). Qualitative analysis of a mean field model of tumor–immune system competition. *Math. Meth. Appl. Sci.* **28**, 2061–2083.

Delitala, M., and Forni, G. (2007). From the mathematical kinetic theory of active particles to modeling genetic mutations and immune competition. Internal Report. Dept. Mathematics, Politecnico, Torino.

De Pillis, L. G., Gu, W., and Radunskaya, A. E. (2006). Mixed immunotherapy and chemotherapy of tumors: Modeling, applications, and biological interpretations. *J. Theor. Biol.* **238**, 841–862.

Derbel, L. (2004). Analysis of a new model for tumor–immune system competition including long time scale effects. *Math. Mod. Meth. Appl. Sci.* **14**, 1657–1681.

d'Onofrio, A. (2006). Tumor–immune system interaction: Modeling the tumor-stimulated proliferation of effectors and immunotherapy. *Math. Mod. Meth. Appl. Sci.* **16**, 1375–1401.

Dunn, G. P., Bruce, A. T., Ikeda, H., Old, L. J., and Schreiber, R. D. (2002). Cancer immunoediting: From immunosurveillance to tumor escape. *Nat. Immunol.* **3**, 991–998.

Friedl, P., de Boer, A. T., and Gunzer, M. (2005). Tuning immune responses: Diversity and adaptation of the immunological synapse. *Nat. Rev. Immunol.* **5**, 532–545.

Gatenby, R. A., Vincent, T. L., and Gillies, R. J. (2005). Evolutionary dynamics in carcinogenesis. *Math. Mod. Meth. Appl. Sci.* **15**, 1619–1638.

Greller, L., Tobin, F., and Poste, G. (1996). Tumor heterogeneity and progression: Conceptual foundation for modeling. *Invas. Metast.* **16**, 177–208.

Friedman, A., and Lolas, G. (2005). Analysis of a mathematical model of tumor lymphangiogenesis. *Math. Mod. Meth. Appl. Sci.* **15**, 95–107.

Hanahan, D., and Weinberg, R. A. (2000). The Hallmarks of cancer. *Cell* **100**, 57–70.

Hartwell, H. L., Hopfield, J. J., Leibner, S., and Murray, A. W. (1999). From molecular to modular cell biology. *Nature* **402**, c47–c52.

Kolev, M. (2003). Mathematical modeling of the competition between acquired immunity and cancer. *Appl. Math. Comput. Sci.* **13**, 289–297.

Kolev, M. (2005). A mathematical model of cellular immune response to leukemia. *Math. Comput. Model.* **41**, 1071–1082.

Kolev, M., Kozlowska, E., and Lachowicz, M. (2005). Mathematical model of tumor invasion along linear or tubular structures. *Math. Comput. Model.* **41**, 1083–1096.

Komarova, N. (2006). Stochastic modeling of drug resistance in cancer. *J. Theor. Biol.* **239**, 351–366.

Lachowicz, M. (2005). Micro and meso scales of description corresponding to a model of tissue invasion by solid tumors. *Math. Mod. Meth. Appl. Sci.* **15**, 1667–1684.

Lollini, P. L., Motta, S., and Pappalardo, F. (2006). Modelling tumor immunology. *Math. Mod. Meth. Appl. Sci.* **16**, 1091–1125.

Levine, H., Pamuk, S., Sleeman, B., and Nilsen-Hamilton, S. (2001). Mathematical modeling of capillary formation and development in tumor angiogenesis: Penetration into the stroma. *Bull. Math. Biol.* **63**, 801–863.

May, R. M. (2004). Uses and abuses of mathematics in biology. *Science* **303**, 790–793.

Merlo, L. M. F., Pepper, J. W., Reid, B. J., and Maley, C. C. (2006). Cancer as an evolutionary and ecological process. *Nat. Rev. Cancer* **6**, 924–935.

Nowak, M. A., and Sigmund, K. (2004). Evolutionary dynamics of biological games. *Science* **303**, 793–799.

Nowell, P. C. (2002). Tumor progression: A brief historical perspective. *Semin. Cancer Biol.* **12**, 261–266.

Othmer, H. G., and Hillen, T. (2002). The diffusion limit of transport equations. II. Chemotaxis equations. *SIAM J. Appl. Math.* **62**, 1222–1250.

Owen, T., and Sherrat, J. (2000). Pattern formation and spatiotemporal irregularity in a model for macrophage–tumor interactions. *J. Teor. Biol.* **189**, 63–80.

Pappalardo, F., Lollini, P. L., Castiglione, F., and Motta, S. (2005). Modelling and simulation of cancer immunoprevention vaccine. *Bioinformatics* **21**, 2891–2897.

Reed, R. (2004). Why is mathematical biology so hard? *Not. Am. Math. Soc.* **51**, 338–342.

Stevens, A. (2002). The derivation of chemotaxis equations as limit dynamics of moderately interacting stochastic many-particles systems. *SIAM J. Appl. Math.* **61**, 183–212.

Tao, Y., and Zhang, H. (2007). A parabolic–hyperbolic free boundary problem modeling tumor treatment with virus. *Math. Mod. Meth. Appl. Sci.* **17**, 63–80.

Tao, Y., Yoshida, N., and Guo, Q. (2004). Nonlinear analysis of a model of vascular tumor growth and treatment. *Nonlinearity* **17**, 867–895.

Index

A

actin cytoskeleton 143, 209, 210, 462
activation 1, 8, 11, 13, 16, 18, 29, 32, 35–37, 39, 41, 42, 44, 45, 47, 79, 83, 84, 87
activator 7–14, 16–18, 20, 21, 29, 33, 34, 41–43, 50, 52, 319, 320, 322, 323, 326, 327, 329, 330, 333, 349–351, 354, 355, 364
  concentration, local 350
  maximum 11, 15, 18, 20, 29
  production 12–14, 18, 362, 365
  transcriptional 84
activin 67, 348, 349, 357
adhesion 113, 127, 149, 167, 205, 206, 209, 211, 214, 221, 222, 224–226, 249, 255, 258, 269, 347, 390
  cell-matrix 121
  energies 219, 221, 226, 230
  molecules 140, 149, 150, 258, 314, 356, 388–391
adhesiveness 149, 364
adult phenotype 341, 342
agent-based models 374, 375
aggregates 314, 451, 469, 472, 475–477, 480
  fusing 471, 474, 476, 477, 479
aggregation 14, 16, 147, 148, 158, 440, 448, 451, 453
amniotes 136, 157, 164, 315
animal pole 115, 116, 119, 123, 126
anisotropy 279, 283, 284, 377, 378, 411, 412, 419, 422
  local 279, 280, 283
antagonist 8, 9, 14, 16, 20, 21
antagonistic reaction 1, 2, 6, 15, 16, 20, 28
anterior
  axis 171, 172
  compartments 43, 51, 209, 213–215, 241
  direction 135, 137, 139–141, 172, 185
  region 218
anteroposterior
  asymmetry 357, 360
  axis 22, 25, 26, 30, 46, 49, 78, 185–187, 189, 190, 194, 196–199, 207, 217, 335, 357, 358
  patterning 22, 24, 77–79, 107

anti-dorsalizing morphogenetic protein (ADMP) 1, 27, 28, 30
apical
  ectodermal ridge (AER) 48, 314, 315, 317–320, 329
  zone 315, 317–319, 324
apoptosis 249, 251, 253, 326, 347
area
  opaca 135, 138, 141, 142, 150, 151, 166, 167, 169, 171, 177
    cells 169, 171
  pellucida 135, 141, 166–169, 172, 176, 177, 270
asymmetry 15, 27, 31, 33, 350, 355, 357
attractants 147–149
attraction 85, 86, 149, 150, 283, 393, 436, 498
autocatalysis 8, 12–14, 18, 20, 24, 29
avascular tumor growth 373, 392, 393
avian embryos 270, 272
axis formation 29, 31

B

basal lamina 148, 151, 249, 251, 252, 256, 264, 316
basement membrane 151, 296
Bicoid distribution 65, 78, 79, 82
biological
  activities 292, 486, 487
  cells 328, 376–379, 384, 387, 392, 409, 497
  functions 485, 487, 488, 492, 495, 496, 498
  defined 491, 495
  pattern formation 1, 9, 54, 317, 435
  processes 9, 107, 162, 199, 401, 436, 457, 481
  systems 72, 373, 396, 436–441, 443–447, 449, 451, 453, 455, 457, 458, 480, 486, 487
    complex 497
    simple 435, 439–441, 444, 449–451, 458
  theory 485, 487, 491
biology 88, 163–165, 207, 221, 312, 373, 375, 396, 402, 435, 436, 444, 446, 458, 467, 486, 497, 498
  molecular 183
  theoretical 373, 374, 396

503

biophysical properties 461, 462
birds 11, 157, 164, 301, 314, 315, 334
BMP (Bone Morphogenetic Protein) 6, 21–22, 24–26, 27–28, 30, 53, 88–90, 91–108, 141–143, 149–150, 167, 291–294, 296–297, 348–350, 355, 357, 360–364
body
  axes, main 1, 3–5, 21, 22, 33
  column 4, 38, 39
  pattern 2, 6, 7, 42
Boolean
  functions 84, 87
  model 83, 84, 86, 87
borders 2, 7, 25, 26, 35, 41, 43, 44, 47–50, 80, 349, 351–353, 355–358, 362–364, 387
boundaries 10, 54, 67, 72, 74, 75, 78, 79, 96, 171, 200, 210, 211, 213, 226, 229, 321, 335, 346
  irregular 332, 333
brain 6, 22, 26, 30, 50, 53, 404, 413, 416
branched
  ducts 249, 250
    morphogenesis of 250, 251
  systems 250, 251
branches 52, 252, 259, 261, 294–296, 300, 301, 304, 305
branching 52, 249, 250, 252, 253, 257–260, 262, 300, 305
  models of 249, 256, 258
  morphogenesis 249, 251–263, 265, 292, 294–297, 299–302
  mechanism of 291, 295
bud formation 5, 39, 40, 253, 299

C

cadherins 145, 222, 236, 277, 314, 389, 403
cancer 424, 436, 438, 492
  cells 485, 489
cartilage 312–315, 318, 319
cell adhesion molecules (CAMs) 151, 184, 186, 196, 197, 205, 208, 210, 214
cells
  adhesion 121, 123, 124, 128, 150, 151, 161, 207, 210, 221, 269, 275, 314, 326, 403
  adjacent 10, 14, 37, 43, 120, 127, 272, 279
  aggregations 148, 149, 196, 197
  apposing 210, 214, 221
  area pellucida 167, 169, 171
  biology 113, 128, 162

chemotactic 20, 21, 440, 443, 450–453
clusters 162, 165, 210, 403, 411, 413, 414
concentration 435, 439, 445, 446, 448, 449, 451, 453, 454
condensing 209, 314, 315, 329
density 196, 197, 256, 272, 275, 279–281, 285, 295, 297, 305, 322, 451
diameters 90, 159, 168, 169, 257, 325, 438, 442
differentiated 50, 52, 270
diffusion 196, 422, 445, 448, 449
ellipsoidal 161, 175
epiblast 148, 149, 151, 167
expansions 486, 488, 491
growth 162, 163, 297, 355, 379, 382, 392
identity 115
intercalation 127, 135, 144, 146
  mediolateral 113, 123, 124, 127, 128
isolated 13, 42, 137
labeled 125, 144, 277
mechanics 373, 375, 376, 392–394
migration 54, 251, 376, 401–405, 407, 409, 411, 413, 415–421, 423–425, 427, 429, 431, 436, 465
  strategies 401–404, 421
misdifferentiated 206, 229
motility 122, 127, 128, 216, 223, 229, 282, 377, 386, 387, 442, 448, 457, 465, 466, 474
motion 157, 163, 167, 169, 170, 175, 252, 305, 401–405, 408, 409, 411, 419, 423, 436, 441, 449
movement 26, 114, 115, 119, 121, 137, 138, 143, 148, 151, 169, 174, 323, 324, 366, 367, 412, 435, 436, 446, 471
polarity 48, 115, 157, 169, 172, 180
populations 159, 402, 412–414, 420–423, 487, 488, 491–493, 495, 498
rearrangements 128, 186, 199, 206, 207, 249, 254, 255, 321, 322, 329, 467
single 12, 16, 26, 52, 53, 113, 114, 145, 162, 192, 194, 236, 239, 240, 280, 375, 376, 384, 405, 454
sorting 128, 142, 229, 465
states 43–47, 322, 328, 487, 498
types 12, 49, 65, 79, 160, 217–219, 221, 226, 230, 233, 234, 283, 319, 328, 329, 388, 462, 463, 480
velocities 159, 168, 169, 174, 177, 180, 386, 442, 443

Index 505

cellular
  automata 158, 375, 401–403, 405, 407, 409, 411, 413, 415, 417, 419, 421–425, 427, 429, 431, 437, 438
  automaton models 437
  interactions 392, 468, 489, 498
  models 438, 439, 441, 443–447, 449, 450, 452, 453, 456, 458
Cellular Particle Dynamics (CPD)
  methods 461, 467, 481
  models 476, 481
  simulations 469, 474–477, 480
cellular particles (CPs) 467–469, 474, 476
Cellular Potts Model 215, 324, 326
cellularization 78, 80
cervical loops 345–348, 356, 358, 360–364
channels 406, 411, 416, 426, 427
chemical
  concentration 295, 438–441, 443, 445, 453, 454, 457, 458
  gradients 169, 441, 443, 445, 446, 453, 454
  interactions 374, 377, 394
chemicals 66, 438, 445, 454
chemoattraction 147, 148
chemokines 388, 394
chemotaxis 135, 147, 159, 180, 199, 270, 287, 296, 330, 387, 404, 436, 440–443, 445, 446, 448–450
  negative 440, 450, 454–457
chick 18, 30, 46, 47, 136, 164, 176, 179, 185, 205–207, 211, 436
  embryo 45, 46, 48, 135–137, 139, 141–145, 147, 149, 151, 153, 157, 158, 163, 164, 168, 174, 207, 209, 436
    embryogenesis 174
    somitogenesis 206, 207, 209
chondrogenesis 319, 320
chondrogenic pattern formation 316, 317, 323
cleft
  formation 119, 249, 253, 263
  regions 252, 255
clefting 257, 260–262
Clock and Wavefront Model 183, 186, 194, 196
clusters, preexisting 302, 304
colony 438, 443
complex systems 28, 374
complexity, biological 160, 324, 481, 487, 488
computational models 80, 199, 295, 300, 315, 323, 329, 335, 341–343, 350, 366
computer
  modeling 461, 463

simulations 2, 54, 135, 157, 162, 167, 169, 170, 206, 207, 437, 438, 449, 450, 461, 464, 477, 481
condensations 210, 276, 314, 315, 319, 356, 487
connective tissue 314, 377, 387, 388, 391
contact
  area 373, 377, 381, 382, 387, 389, 390, 392, 473, 475, 476, 478–480
  inhibition 210, 281, 392, 393, 403
  surface 390–392, 474
continuum models 159, 166, 330, 331, 438
contractility 249, 257, 263
contraction 96, 106, 108, 117, 127, 145–147
control species 74, 75
convergent extension 113, 119, 120, 123, 124, 127, 145
convex hull 378, 382, 383
coordinate system, near-Cartesian 22, 23, 25
core mechanism 317–319, 336
cues 121, 127, 143, 169
culture medium 259, 281, 296
cusps 348, 349, 351, 358, 360–362, 364
  formation 361, 362
cytoplasm 14, 16, 82, 193
cytoskeleton 157, 162, 210, 387–389, 391, 445

D

deformations 257, 259–261, 264, 276, 386
Delaunay-based interaction 373, 378
Delaunay-Object-Dynamics
  method 373, 377, 396
  model 392
Delaunay-triangulation 373, 378–380, 384
  weighted 379–382
dental
  epithelium 348, 349, 365
  mesenchyme 347, 356, 360, 363, 364
developmental
  biology 1, 65, 113, 135, 157, 183, 200, 205, 249, 265, 269, 291, 300, 301, 311, 341, 342, 373, 374
  dynamics 312, 342, 358, 365
differential adhesion hypothesis (DAH) 364, 463, 464, 466, 481
diffusion
  coefficients 72, 73, 75, 79, 81, 90, 98, 101, 107, 279, 280, 422
  mechanisms 152, 316, 334, 349, 350

models 207, 305, 317, 329–331, 334, 349, 362, 364
systems 6, 316, 317, 321, 323, 324, 331, 332, 388
tensors 408, 412, 413
discrete
  entities 374, 437, 438, 486
  models 159, 324, 330, 331, 405
distribution
  functions 488, 493
  homogeneous 7, 8
dorsal
  aortae 275, 276
  ectoderm 88, 89, 96, 98
  midline 26, 89, 90, 96, 98, 106, 118
dorsoventral (DV)
  axes 5, 6, 24, 48, 94
  border 25, 26, 48, 51
  patterning 65, 87, 91, 107
Drosophila 36, 41, 43–45, 65, 67, 68, 79, 80, 87, 145, 147, 168, 344
  embryo 98, 147, 168, 178
dynamic
  environments 404, 420, 423
  microenvironments 113, 121
  systems 20, 65, 70, 311, 382
    discrete 84, 424
dynamics
  biological 437, 446, 500
  molecular 318, 324, 468

E

ectoderm 86, 114, 117–120, 122, 127, 136, 313, 315, 316, 320, 335
egg 3, 15, 22, 68, 78, 87, 114, 136, 142
elongation, sequential 24–26
embryo 3, 33, 34, 48, 49, 78–80, 88–90, 94–96, 98, 99, 101, 102, 113–125, 128, 135–138, 140, 141, 143, 146, 151, 271, 272
  bird 269, 270
  complete 3, 17, 140
  developing 19, 69, 73, 124, 127, 436
  early 22, 24, 115, 135, 136, 143, 150, 152
  large 28, 29
  quail 273, 276, 287
  spider 44
embryogenesis 13, 314
embryonic
  axes 1, 17, 22, 29, 115, 120, 276
  lung 261, 262

morphogenesis 402, 424
tissues 258, 259, 261, 313, 463, 474, 481
empty-orthosphere-criterion 373, 379, 382, 383
enamel epithelium 347, 361, 362
  inner 345, 346, 356, 359, 361
  outer 345, 346, 363
endoderm 114–117, 120, 123, 136
endothelial
  cells 269–273, 275–277, 281, 403
    vascular 277, 282
  progenitors 270, 271
energy 169, 222, 234, 235, 326, 329, 376, 465–467, 475, 498
  effective 216, 219, 326, 328
  interfacial 471
  minimization 376, 377, 464
environment
  interactions 404, 405, 409, 423
  static 401, 405, 408, 411, 420
ephrins 210, 218, 222
epiblast 135–143, 145–151, 157, 164–168, 170–174, 176, 177, 179, 180
epigenetic interactions 341, 342, 358, 360, 367
epithelial
  cells 117, 118, 210, 215, 217, 218, 221, 225, 255, 258, 297, 299, 348, 350, 351, 353, 355–357, 360, 362
  explants 297, 299
  surface 255–257, 262
epithelialization 209, 210, 215
epithelium 251–256, 259–262, 264, 265, 292, 293, 297, 299, 313, 316, 345–348, 350, 351, 353, 355–358, 360, 361, 363, 364, 367
  finger of 251, 252
  growing 292, 293
equations of motion 159, 170, 179, 180, 438
equilibrium distribution 417, 418, 420, 427–429, 431
evolution 4, 54, 84, 94, 96, 98, 106, 322, 323, 343, 358, 366, 411, 415, 416, 465, 466, 487–489, 493
evolutionary biology 312, 343
exclusion principle 406, 417, 426
explants 124, 297, 299
expression 5, 7, 11, 24, 26, 46, 47, 69, 78–80, 84, 87–89, 140–142, 225, 226, 293, 294, 348, 349, 361, 387
  analytical 481
  differential 214, 366, 376

extracellular matrix (ECM) 121, 122, 124, 127, 128, 216–222, 226, 251, 252, 254–256, 269, 272, 273, 275, 328, 384, 386, 388, 389, 402, 403, 465, 466
  fibers 255–257, 262

F

feedback loops, negative 3, 187, 194, 195
FGFs (Fibroblast Growth Factors) 46–48, 53, 56, 141, 149, 151, 167, 185–191, 194–196, 198–199, 206, 211–213, 253, 291–294, 296–300, 305–306, 314–321, 329, 334, 348–349, 354–359, 362, 366
fiber tracks 404, 413, 414, 416
fibers 269, 388, 389, 403, 422
fibrillar extracellular matrix 124, 127, 421
fibrils 122–124
fibronectin 122, 123, 256, 258, 314, 315, 319–321, 325, 329, 336
field
  chemical 160, 311, 440, 445, 447, 449, 453, 454
  external 326, 401, 417, 418
  intensity 411, 418, 419
  small 31, 416, 417
  strengths 419, 422
  tensor 405, 408, 410–412, 414–416, 423
  vector 167, 168, 408, 409, 411, 412
fluids, viscous 179, 261, 294, 295
follicles 387, 394, 395
follistatin 349, 357
forces
  active 377, 384, 386–388
  drag 373, 386
  elastic 387, 389–391
  passive 384, 386, 387
French flag model 73, 78
functional modules 486
  theory of 486, 488, 491
fusion 461, 463, 468–474, 477, 480, 481
  process 468, 469, 475, 476
  vascular 275, 276

G

Galerkin method 331
gastrulation 88, 113, 114, 116–121, 135, 136, 147, 151, 152, 164, 167, 176, 206, 254, 272, 436

gene
  activation 35, 37, 44–47, 49
  copy number 90, 91
  expression 7, 35, 65, 68, 83, 87, 115, 176, 196, 198, 205, 207, 208, 341, 343, 353, 498, 499
    control target 79
    prepattern of 184, 208
genetic
  information 8, 10
  prepattern 198, 199, 205, 271
Glazier–Graner–Hogeweg (GGH)
  model 205, 215, 233
  simulation 216, 219, 230
Glazier's model 367
glioma cells 403, 411, 413
growth
  dynamics 355, 356
  factors 115, 121, 148, 259, 314, 342, 343, 347, 360, 364, 366
  glioma 414, 416
  lateral 352, 353
  mesenchymal 348, 351–353, 357, 361, 363

H

Hertz model 390, 391
heterodimer-based model 95, 99, 102–104
heterodimers 92, 99, 107
  formation 91–93
heterogeneous environments 194, 401, 402, 404, 422, 423
higher organisms 1, 2, 4–7, 14, 33, 41, 50, 67, 147
Hill functions 83
Hox genes 30, 43, 45
hydra 4–7, 9, 17, 19, 29, 54
  organizer 2, 9, 22, 23
  tentacles 1, 38, 40
  tissue 4
hyphasma 376, 377, 438
hypoblast 135, 137, 140, 141, 143, 148
  secondary 137, 140, 141
hypostome 4, 7, 9, 38, 40
hypotheses 129, 135, 169, 184, 190, 211, 271, 287, 335, 336, 341–343, 348–350, 356, 358, 360, 361, 363–365, 367, 368

I

immune
  cells 485, 489, 491, 492, 495–497
    active 493, 494
    competition 485, 487, 489, 491, 493, 495, 497, 499
    system 374, 384, 387, 394, 485, 491, 495
  individual cells 4, 17, 66, 137, 143, 146, 159, 170, 179, 207, 211, 282, 402, 438, 439, 441, 462
    migration of 147, 272
  inhibition 1, 12, 14, 15, 19–21, 25, 28–30, 35, 41, 52, 78, 84, 141, 143, 257, 349, 362
    competitive 41
    long-range 19, 27, 35, 53
  inhibitor 7–9, 11–14, 16–19, 28, 29, 32, 33, 41, 42, 83, 84, 90, 91, 107, 140, 141, 315, 322, 323, 326, 327, 349–352, 354, 355, 364
    concentration 9, 12, 29, 351
    systems 8, 12, 14, 16, 20, 21, 34, 42, 50
insects 2, 6, 11, 13, 24–26, 31, 33–35, 50, 53
integrins 122, 123, 226, 277, 389
interacting cells 436, 448, 451, 465
interaction
  energy 221, 466
  rules 407–409, 419, 426
intercalation 128, 143–145, 151, 356
intrasomitic notch 225, 226
invagination 344, 345, 348, 355, 358, 361, 367
involution 116, 118, 123–125

K

kidney 53, 114, 250, 252, 257, 258, 292
kinetic
  interactions 72, 73, 99, 107
  parameters 71, 73, 362, 364
  theory 486–488
knot cells 348–350, 356
knots 345–352, 355–357, 360–364
Koller's sickle 3, 137, 138, 142, 144, 146, 147, 149, 150, 166–169, 171, 174, 176, 177, 180
  region 172, 177, 179

L

Langevin equations 157, 159, 160, 282
lateral inhibition, reduced 24

lattice 181, 216, 219, 328, 375, 377, 406, 413, 415, 423, 425, 429, 465, 475
Lattice Boltzmann Equation (LBE) 405, 423, 424
lattice-gas cellular automaton (LGCA) 401, 405, 406, 408, 411, 416, 418, 420, 422–424, 426
  models 401, 405, 407, 408, 414, 422, 423
limb 48, 49, 311, 312, 316, 317, 319, 320, 323, 324, 333–335
  bud 313–315, 317, 318, 331, 332, 335
  development 48, 293, 312, 313, 316, 318, 319, 322, 326, 334, 335, 436
  field 48, 49
  mesenchyme 316, 334, 335
local
  dynamics 71, 91, 93, 105, 107
  exclusion 42, 43
  self-enhancement 2
lung 50, 250, 252, 253, 257, 258, 261, 262, 291, 292, 294, 300, 301
  branching morphogenesis 291, 292, 295, 300, 301
  developing 261, 296, 297
  development 291
  epithelium 259, 292, 293, 299
lymph nodes 394
lymphocytes 376, 386, 388

M

macroscopic equations 487, 499, 500
mammalian teeth 362, 365
marginal zone 1, 2, 22–24, 26, 28, 29, 116–119, 123, 124, 137, 139–141, 143, 166, 167, 169, 170, 177, 178
mathematical
  analysis 197, 198, 297, 300, 401, 405, 438
  biology 446, 458
  equations 487, 498, 499
  formulation 3, 186, 194, 198
  framework 446, 488
  kinetic theory 486–488, 491
  models 67, 71, 89, 90, 105, 129, 183, 186–190, 194, 196, 198, 207, 272, 332, 367, 437, 485–488
  structures 485, 488, 492, 495, 498
  theory 485–487, 496, 498, 499
    biological 485, 496
matrix 151, 347, 403, 411, 432
  fibrillar 122, 124
mean-field theory (MFT) 435, 444, 446–449, 451, 453, 458

Index 509

mechanical interactions 160, 161, 179, 301,
    341–343, 353
mechanisms
    biological 208
    counting 45, 46
    epithelial 264
    generic 250
    morphostatic 322, 323
    of pattern formation 295
Menten-type kinetics models 323
mesenchymal cells 137, 206, 209, 217, 218,
    225, 254, 255, 262, 313, 319, 351–353,
    403, 404
    layers of 350, 355
mesenchyme 251–256, 258–262, 264, 265,
    293, 294, 297, 315, 319, 320, 345,
    347–351, 356, 357, 362, 364
mesoderm 9, 31, 46, 114–116, 120, 123, 124,
    127, 135, 136, 142, 144, 270
    cells 26, 122, 135, 138, 141, 142, 147
    formation 141, 142
    induction 27, 135, 140, 152
methods
    finite element 331
    subcellular element 376, 377
Metropolis algorithm 464, 465, 471, 474
Metropolis Monte Carlo method 466, 471, 472,
    474
microenvironments 115, 118, 122, 124, 125,
    128, 317, 334, 401, 402, 404, 420, 424,
    432
microscopic models 401, 405, 438
microsurgery 114, 115
midline
    formation 1, 2, 9, 14, 22, 24–26, 33, 34
    signal 33, 34
    system 33
migrating cells 403, 404
misdifferentiation 229, 240, 241
mitoses 251, 253, 255, 328, 375, 379
mixing, degree of 470, 473, 475, 476
modeling 12, 37, 39, 53, 54, 135, 151–153,
    158, 159, 174, 175, 199, 200, 296, 297,
    312, 313, 374, 375, 396, 485–487,
    489–492, 494, 495
    architecture 378, 392
    branching morphogenesis 254
    cells 328, 449
    computational 311, 312
    explicit 190, 423
    framework 387, 405

mathematical 152, 184, 200, 263, 265, 365,
    402, 436, 437
multicellular systems 157, 162, 488
multiscale 200, 300, 396, 436
tissue-level 435, 438, 441
models
    agent-based 374, 375
    architecture 377, 384, 396
    artifacts 436, 437, 444, 458
    biologically-accurate 200
    boundary 48, 51
    building 437, 439, 457
    calculations 147, 150, 352
    cells 282, 323, 329, 335, 465, 471, 472, 474
    classic 164
    clock-and-wavefront 206, 207
    coherent 341, 342
    complex biological 218
    conceptual 256, 262, 263
    continuous 152, 404, 423
    diffusion-type 316
    dynamic 318, 319
    grid-free 158, 162, 167, 174, 176
    heterodimer 95, 99, 101, 104
    hybrid 312, 313, 323, 335
    kinetic 423, 424
    lattice-gas 375, 406
    macroscopic 422
    mean-field 451, 453, 455, 457
    mechanical 258
    mechanochemical 271, 275, 281
    molecular 435, 444, 447, 457
    multiscale 323, 330
    parameters 171, 180, 195, 197, 199, 296,
        355, 357, 364, 365
    phenomenological 444, 493
    population-type 438
    predictions 99, 342, 367, 447, 458
    simplified 30, 197, 318
    simulation 364, 412
    system 28, 29, 54, 65, 68, 113, 114, 164, 470
    tissue 465, 467, 471
    two-step 87
modules 91, 105, 106, 170, 231, 304, 485, 486,
    488, 491, 495
molars 344, 349, 353, 355–363, 365, 367
molecular mechanisms 114, 115, 135, 149, 334
Monte Carlo
    simulations 461, 464, 470, 471, 473, 475,
        476, 479, 480

steps 216, 217, 219, 222–225, 227, 229, 230, 236, 238, 240, 242, 330, 466, 469, 471–474, 479, 480
morphogen 35, 37, 49, 51, 66, 67, 72, 74, 75, 78, 80, 101, 105, 107, 196, 197, 319, 322, 323, 325, 326
   dynamics 312, 321, 323, 324, 329, 331, 335
   gradient 1, 36, 49, 171
morphogenesis 107, 113–115, 120, 127, 128, 249, 251, 253, 254, 264, 297, 335, 342, 347, 359, 366, 395, 402
morphogenetic mechanisms 253
morphological
   changes 198, 207, 208, 327, 342, 344, 358, 360, 363, 364, 367
   consequences 341–343, 358
   differences 259, 301, 365
   variation 341, 343, 360, 365
morphology 225, 260–263, 292, 296, 301, 335, 342–344, 347, 349, 350, 353, 356, 358, 360–362, 364, 365, 394
morphostatic limit 312, 321–324, 332, 335
motility 225, 256, 270, 347, 375, 387–389
motion
   collective 179, 403, 421, 422, 467
   vortex 163, 168, 174, 177
mouse 29, 136, 184, 199, 211, 292, 312, 344, 347, 353–355, 358, 360–363, 365–367, 436
   molars 345, 356, 357, 360–362
   teeth 353, 355
moving cells 387, 389, 402, 405, 421
   interplay of 401, 402
multicell morphogenesis modeling 216
multicellular systems 157, 158, 160, 447, 452, 453, 485, 487, 488, 498, 499
   complex 485–487, 489, 491, 493, 495, 497, 499
multimodel simulation environment 321, 329
mutant
   embryos 98, 99, 194
   profile 101, 102
mutants 26, 85, 86, 98, 141, 348, 365

N

N-cadherin 205–207, 209–211, 213–215, 217–219, 221–223, 225–227, 229, 231, 233, 235, 237, 239, 241, 243, 314
   levels 209, 214, 215, 217

neighboring cells 37, 127, 142, 144, 146, 149, 152, 161, 181, 211, 355, 357, 378, 389, 403, 446
neoblasts 39
network
   interconnected 285–287
   polygonal 270, 271, 285
   segment polarity 65, 80, 84, 86, 87
nomenclature 205, 208
Notch pathway 47, 53, 87, 211, 213, 315, 334
nuclei 8, 75, 80, 81, 87, 89, 104, 325, 487
numerical
   simulations 95, 183, 186, 192, 195, 197, 295, 299
   solution 189, 191, 193, 195, 197, 331
nutrient 113, 160, 162, 295, 392, 393
   concentrations 162, 392, 393

O

ordinary differential equations 70, 94, 328, 331, 348, 451, 487
organizer 1, 4, 11, 17–19, 22–32, 38, 150
   formation 5, 19, 27–29
   hydra-type 1, 22, 26
   primary 2, 4, 18, 38, 39
organizing regions 1–4, 10, 17, 19, 22, 24, 32, 48, 49, 54
   secondary 18, 19, 28, 32, 40
organogenesis 114, 115, 121, 312, 374
organs 35, 53, 250, 252, 261, 263, 270, 292, 300, 324, 329, 343, 374, 376, 462
   branched 249, 251, 253, 259, 260, 263
orientation 3, 5, 17, 21, 22, 26, 50, 115, 120, 166, 256, 344, 379, 383, 386, 404, 421, 422
orthospheres 379–384
oscillations 16, 17, 30, 44, 46, 47, 185, 192–194, 198, 199, 211

P

pair-rule genes 78, 80
parameters
   free 391, 428, 430
   key 222, 325, 327
parasegments 41, 43, 78, 80, 85, 86
partial differential equations 152, 188, 324, 328, 331, 374, 438, 451, 458

Index 511

pattern
  formation 1, 4, 6–8, 10, 14–16, 27, 28, 30, 32, 35, 54, 66, 67, 72, 257, 263, 301, 312
  periodic 1, 4, 11, 17, 18, 44, 47, 207
  regulation 3, 11, 17, 19
patterning 2, 4, 8, 21, 30, 66–68, 72, 73, 86, 87, 89, 90, 98, 104–106, 206, 219, 269, 271, 316
  dorsal surface 88, 91, 105, 108
  extracellular 91, 94, 99, 101
  models 90, 91
periphery 11, 80, 150, 171, 209, 214, 217, 218, 295
perturbations 3, 67–70, 93, 101, 104, 106, 187, 190, 197, 198, 225, 334
phenomena, biological 20, 324, 485, 486, 495
Phenomenological Cell-Based Model 157, 170
physical
  processes 457, 461, 463
  properties 265, 377, 391, 461–463
  sciences 446, 458, 486–488
physics 157, 256, 263, 294, 311, 396, 435, 444, 446, 458, 481, 486, 498
planar polarity 157, 166–170, 173, 175, 178, 180, 181
planarians 1, 4, 24–26, 39
polarity 17, 41–43, 48, 115, 168–172, 179, 335, 376, 386–389
  field, planar 168–170, 181
populations 39, 181, 343, 365, 394, 405, 411, 421, 422, 487–489, 491–493, 495, 498
positional information 22, 37, 49, 50, 66, 67, 74, 121, 199
positive
  chemotaxis 440, 450, 453, 454
  feedback 2, 8, 18, 20, 65, 91, 105–108
posterior
  borders 207, 344, 347, 357, 358
  cells 44, 85, 211, 357
  compartments 41, 43, 51, 208, 209, 213, 215, 226
  direction 137, 140, 143, 185, 344, 345
  half-somites 46, 47
  pole 25, 26, 44–46, 78, 95, 135, 137, 140, 141, 173
  somite compartments 225, 226
posteriorization 29, 31
Potts model 375–377, 464
precondition 5, 24, 25, 27, 28, 38, 43, 51

predictions 13, 43, 46, 47, 84, 87, 190, 194, 198, 275, 299, 341–343, 350, 394, 435, 437, 450, 451
prepattern 183, 184, 196, 199, 207, 235, 238
presomitic mesoderm (PSM) 185, 187, 188, 190, 191, 196, 197, 199, 206–211, 213, 215, 216, 218, 222, 226, 228, 235, 237, 239, 242
  boundary 206–208
  cells 185, 187, 196, 197, 217, 226
pressure
  lateral 351, 352
  mesenchymal 352
primitive streak 135, 136, 141, 144, 147, 164–166, 174, 177, 179
  formation 157, 159, 161, 163, 165–167, 169, 171, 173, 175–177, 179, 181
primordial endothelial cells 269, 278
progressing cells 491–495
proliferation 21, 143, 341–343, 347, 348, 350, 355, 357, 360–363, 393, 423, 466, 489
  mesenchymal 356, 357
proteins 82–84, 86, 88, 101, 105, 114, 128, 145, 192–194, 366
protrusions 53, 127, 144, 275, 282, 292, 295
Python scripts 230, 234

Q

quantitative modeling 435, 437

R

radial intercalation 113, 116, 122, 123, 127
rate of protein synthesis 192, 193
reaction–diffusion model 6, 71, 72, 152, 158, 207, 295, 303, 305, 312, 316–317, 323, 324, 329, 330–332, 341, 343, 349–350, 362, 364, 366, 388, 392, 404, 441, 457
reactor 316, 317, 319, 321, 330, 334
rearrangements, structural 275
receptors 8, 28, 53, 66, 75, 77, 79, 81, 87, 89, 90, 105, 106, 108, 148, 149, 214, 366, 387
recombination experiments 258–260, 262
regeneration 4, 8, 11, 17, 18, 28, 39, 40, 42, 256
regions
  caudal 167
  confined 2, 35
  dorsal 89, 95–98, 102
  pacemaker 16, 20
  somite-wide 46

remodeling 121, 122, 301, 403, 466
repellents 140, 147–149
repulsion 25, 149, 205, 210, 211, 221, 222, 224–226, 282
   energy 225
robustness 65, 67–71, 86, 87, 90–93, 99, 101–105, 107, 108, 170, 195, 222, 262, 325

S

scale-invariance 66, 73, 107
scales
   cellular 423, 435, 486, 487, 499
   spatiotemporal 402, 423
scaling 79–82, 162, 500
   linear 170, 377, 383
segment polarity
   genes 65, 78, 80, 82, 84–86, 107
   network 65, 80, 84, 86, 87
segmentation 1, 41, 43, 44, 46, 78, 184, 185, 205, 209, 211, 213–215, 223, 225, 226, 229
   clock 183, 185–188, 190–192, 194, 198, 199
   intrasomitic 224
self-enhancing reaction 14–16, 18, 21, 27
self-inhibition 41, 44
self-interaction 449, 453
sequential pattern 1, 41, 44, 45
Shh (Sonic hedgehog) 21, 48, 53, 253, 291–294, 316, 319, 335, 348–350, 354–355, 359, 361
signal
   concentration 37, 188
   generation 50, 52, 53
   local 50, 52, 53
signaling 68, 89, 105, 107, 151, 152, 199, 344, 348–350, 362–364, 366
   centers 2, 17, 22, 24, 35, 167
   model 187, 196
   molecules 35, 148, 151, 186–189, 191, 296, 350
   systems 6, 151, 152
signatures 82, 84, 86, 107
simplex 378–384
simulations, multimodel 312
simulator 231, 232, 237, 242, 243
skeletal pattern formation 321, 335
skeletons 313, 315, 335, 336

somite 24, 41, 45–47, 183–187, 190, 196, 197, 206–211, 214, 216, 217, 221, 223–227, 275, 276
   anterior compartments of 217, 218
   coherent 187, 188, 199, 207, 223
   formation 1, 16, 44–46, 183, 185–187, 189–191, 193, 195–200, 206–209, 211, 226
   morphology 206, 223, 224
   physical 196, 198
   posterior compartments of 218, 227
   presumptive 196, 207, 208, 213
   segmentation 194, 209, 210, 215
      normal 211, 225, 229
somitic factor 186, 188, 189
   concentration 188
somitogenesis 183, 184, 198, 199, 205–208, 210, 211, 214, 220, 317, 319, 334
   simulations 205, 230, 236
spatial
   patterns 47, 66, 69, 73, 90, 197, 206, 319, 325, 327, 342
   profiles 96, 97
   regions 435, 446, 452, 454
   scales 73, 312, 335, 435–439, 442–444, 457, 458
Spemann-type organizer 1, 22, 24, 26–28
sprouts 269, 273, 275, 277, 279, 281
   formation of 269, 281, 284, 287
stable patterns 7, 16, 19, 20, 326
stages
   early 3, 4, 27, 136, 137, 142, 145, 149, 164, 167, 199, 374
   intermediate 394, 395, 471
stationary regimes 325, 326
steady states 17, 30, 66, 76, 85–87, 96, 332, 333, 417, 476
stellate reticulum 345, 347, 351, 357, 362, 363
steps
   elementary 1–4
   propagation 406, 407, 426
stochastic models 194
   discrete 312, 324, 457
streak 135, 137, 138, 140–144, 146–149, 152, 157, 164–168, 179
   formation 135–138, 140–145, 147–150, 152, 157, 168, 170, 174
   inhibition of 141, 147
stripes 12–14, 24–26, 31, 78, 86, 321, 325, 349
structural stability 70

Index 513

structures
  adjacent 38, 39, 41, 42
  anisotropic 283–285
  branched 52, 259, 291, 294, 297, 300, 302, 304
  elongated 135, 269, 270, 277, 281, 282, 287
  multilayered 140
  periodic 11, 30, 46
  repetitive 41
  stable 394
  terminal 4, 25, 42
Subcellular Element Model (SEM) 152, 157, 161–163, 165, 170, 175, 462, 467, 468, 481
subcellular elements 161–163, 465
subpatterns 1, 33
surface
  area 114, 117, 261, 325, 328, 389, 463, 471
  energy 242, 376, 386, 390, 470, 471
  tension 258, 260, 261, 263, 264, 463, 470, 480, 481
susceptibility 416, 418, 429
symmetry 24, 417, 419, 472, 473

T

tensors 408, 410, 411, 420
tentacle 4, 6, 39, 40
  activation 38, 40
  formation 4, 39, 40
TGF (Transforming Growth Factor)-β 253, 314–315, 316, 319–321, 323, 329–330, 334, 336
three-dimensional
  morphologies 341–344
  reconstructions 346
time
  evolution 229, 411, 412, 414, 445, 461, 469, 475, 476, 478–480
  steps 84, 85, 87, 162, 192, 195, 377, 382, 383, 413, 415, 426
timing 116, 213, 304, 358
tissue 38, 39, 50, 51, 66, 67, 115, 116, 119–122, 124, 250, 251, 258, 259, 270, 312, 313, 394, 402, 403, 436, 451, 461, 462, 464, 465
  architecture 120, 121, 124
  components 121, 273, 275
  dense 375, 377, 378, 386
  explants 115, 124
  extra-embryonic 136

  fragments, irregular 463
  fusion 461, 468, 481
  liquidity 461, 463, 464
  models 445–447, 453, 458
  movements 115, 120, 129, 272, 277, 287
  scales 435, 437, 439
  separation 116, 119, 120
  shear 124
  stiffness 120, 121
  structure 272
  types 68, 69, 88, 249, 251, 313, 463
tissue-level models 435, 436, 438, 439, 441–450, 453, 458
toolkit molecules 292, 293
tooth 316, 342–344, 346, 347, 349, 350, 352, 353, 355–357, 359, 360, 363–367
  development 341–344, 347, 349, 353, 355, 356, 358–360, 363, 365
  morphogenesis 341–343, 345, 347, 349–351, 353, 355, 357, 361, 363, 365, 367
  morphology 343, 347, 348, 358, 363, 364, 367
topological transformations 114, 382
traction 113, 120, 124, 127, 148, 249, 256, 262, 263, 270, 357, 361, 367
transcription 83, 84, 193, 198
  factors 68, 78, 80, 81, 84, 89, 142, 191, 192, 213, 301, 316, 366
  rate 83
transient
  evolution 90, 95, 96, 98
  regimes 325, 326
translation 82, 84, 193, 198, 213, 263
transport 28, 67, 98, 102, 107, 198, 250, 263, 328, 387
  mechanisms 90, 249, 250
traveling waves 1, 16, 20, 21, 46, 195
triangulation 378, 379, 381–383, 388, 390
trunk 1, 2, 7, 22, 25, 26, 30, 41
  formation 30, 45
tumor 270, 335, 385, 392, 393, 489, 492, 495
  cells 270, 393, 404, 491, 492, 495
    progression of 489, 492, 495, 496
  growth 335, 378, 392, 393, 401, 414, 422, 432, 485
  simulation of 384, 385
Turing model 72, 73

U

unspecific induction 1, 29, 32

V

values
  absolute 102, 357, 443, 457
  simulated 418, 419, 421
vascular structures 272, 277
vasculature 114, 273, 276, 287
vasculogenesis 263, 269–271, 281
vasculogenic sprouts 269, 271, 273, 275, 277, 279, 281, 283, 285, 287
vegetal cells 27, 117
veins 50, 52, 53
ventral side 2, 21, 25, 301
vertebrate 1, 6, 7, 21, 22, 24–27, 30, 31, 33, 37, 40, 41, 45, 50, 149, 206, 269, 313, 333, 436
  embryos 114, 184, 186
  limb 48, 312, 313, 323, 336
    development 311, 313, 315, 317, 319, 321, 323, 325, 327, 329–331, 334, 335
vertex 275, 373, 378–385, 388, 390, 391
viscosity 249, 258–261, 263, 386, 462, 463, 481
  mesenchymal 259, 260
viscous fingering 294, 295
volume
  elementary 489, 490
  increased 217, 218
  target 222, 391, 392
Voronoi-cells 380, 381, 390, 391, 396
Voronoi-tessellation 380, 381
vortices 145, 157, 165–167, 169–175, 177, 179

W

wavefront 187, 189, 191
  model 183, 186, 189, 191, 194, 196, 198
waves 16, 20, 21, 46, 47, 195, 196, 346
Wnt pathway 5–6, 9, 17, 22, 27, 28–29, 41, 47–48, 72, 138, 141–145, 148–149, 151, 199, 206, 211–213, 294, 334, 335, 348, 365

X

Xenopus 18, 24, 27, 29, 30, 32, 67, 120, 124, 168
  embryos 12, 206
  laevis 113–116

Z

zebrafish 12, 18, 114, 184, 225
zones
  active 317–319, 321, 324, 329, 331–333
  frozen 317–319
  nonproliferative 361, 362

Contents of Previous Volumes

Volume 47

1 Early Events of Somitogenesis in Higher Vertebrates: Allocation of Precursor Cells during Gastrulation and the Organization of a Moristic Pattern in the Paraxial Mesoderm

   Patrick P. L. Tarn, Devorah Goldman, Anne Camus, and Gary C. Shoenwolf

2 Retrospective Tracing of the Developmental Lineage of the Mouse Myotome

   Sophie Eloy-Trinquet, Luc Mathis, and Jean-Francois Nicolas

3 Segmentation of the Paraxial Mesoderm and Vertebrate Somitogenesis

   Olivier Pourqulé

4 Segmentation: A View from the Border

   Claudio D. Stern and Daniel Vasiliauskas

5 Genetic Regulation of Somite Formation

   Alan Rawls, Jeanne Wilson-Rawls, and Eric N. Olsen

6 Hox Genes and the Global Patterning of the Semitic Mesoderm

   Ann Campbell Burke

7 The Origin and Morphogenesis of Amphibian Somites

   Ray Keller

8 Somitogenesis in Zebrafish

   Scoff A. Halley and Christiana Nüsslain-Volhard

9 Rostrocaudal Differences within the Somites Confer Segmental Pattern to Trunk Neural Crest Migration

   Marianne Bronner-Fraser

Volume 48

1. **Evolution and Development of Distinct Cell Lineages Derived from Somites**
   Beate Brand-Saberi and Bodo Christ
2. **Duality of Molecular Signaling Involved in Vertebral Chondrogenesis**
   Anne-Hélène Monsoro-Burq and Nicole Le Douarin
3. **Sclerotome Induction and Differentiation**
   Jennifer L. Docker
4. **Genetics of Muscle Determination and Development**
   Hans-Henning Arnold and Thomas Braun
5. **Multiple Tissue Interactions and Signal Transduction Pathways Control Somite Myogenesis**
   Anne-Gaëlle Borycki and Charles P. Emerson, Jr.
6. **The Birth of Muscle Progenitor Cells in the Mouse: Spatiotemporal Considerations**
   Shahragim Tajbakhsh and Margaret Buckingham
7. **Mouse-Chick Chimera: An Experimental System for Study of Somite Development**
   Josiane Fontaine-Pérus
8. **Transcriptional Regulation during Somitogenesis**
   Dennis Summerbell and Peter W. J. Rigby
9. **Determination and Morphogenesis in Myogenic Progenitor Cells: An Experimental Embryological Approach**
   Charles P. Ordahl, Brian A. Williams, and Wilfred Denetclaw

Volume 49

1. **The Centrosome and Parthenogenesis**
   Thomas Küntziger and Michel Bornens
2. **γ-Tubulin**
   Berl R. Oakley

3 γ-Tubulin Complexes and Their Role in Microtubule Nucleation

   Ruwanthi N. Gunawardane, Sofia B. Lizarraga, Christiane Wiese, Andrew Wilde, and Yixian Zheng

4 γ-Tubulin of Budding Yeast

   Jackie Vogel and Michael Snyder

5 The Spindle Pole Body of *Saccharomyces cerevisiae*: Architecture and Assembly of the Core Components

   Susan E. Francis and Trisha N. Davis

6 The Microtubule Organizing Centers of *Schizosaccharomyces pombe*

   Iain M. Hagan and Janni Petersen

7 Comparative Structural, Molecular, and Functional Aspects of the *Dictyostelium discoideum* Centrosome

   Ralph Gräf, Nicole Brusis, Christine Daunderer, Ursula Euteneuer, Andrea Hestermann, Manfred Schliwa, and Masahiro Ueda

8 Are There Nucleic Acids in the Centrosome?

   Wallace F. Marshall and Joel L. Rosenbaum

9 Basal Bodies and Centrioles: Their Function and Structure

   Andrea M. Preble, Thomas M. Giddings, Jr., and Susan K. Dutcher

10 Centriole Duplication and Maturation in Animal Cells

   B. M. H. Lange, A. I. Faragher, P. March, and K. Gull

11 Centrosome Replication in Somatic Cells: The Significance of the G_1 Phase

   Ron Balczon

12 The Coordination of Centrosome Reproduction with Nuclear Events during the Cell Cycle

   Greenfield Sluder and Edward H. Hinchcliffe

13 Regulating Centrosomes by Protein Phosphorylation

   Andrew M. Fry, Thibault Mayor, and Erich A. Nigg

14 The Role of the Centrosome in the Development of Malignant Tumors

   Wilma L. Lingle and Jeffrey L. Salisbury

15 The Centrosome-Associated Aurora/Ipl-like Kinase Family

   T. M. Goepfert and B. R. Brinkley

16 Centrosome Reduction during Mammalian Spermiogenesis

   G. Manandhar, C. Simerly, and G. Schatten

17 The Centrosome of the Early *C. elegans* Embryo: Inheritance, Assembly, Replication, and Developmental Roles

   Kevin F. O'Connell

18 The Centrosome in *Drosophila* Oocyte Development

   Timothy L. Megraw and Thomas C. Kaufman

19 The Centrosome in Early *Drosophila* Embryogenesis

   W. F. Rothwell and W. Sullivan

20 Centrosome Maturation

   Robert E. Palazzo, Jacalyn M. Vogel, Bradley J. Schnackenberg, Dawn R. Hull, and Xingyong Wu

Volume 50

1 Patterning the Early Sea Urchin Embryo

   Charles A. Ettensohn and Hyla C. Sweet

2 Turning Mesoderm into Blood: The Formation of Hematopoietic Stem Cells during Embryogenesis

   Alan J. Davidson and Leonard I. Zon

3 Mechanisms of Plant Embryo Development

   Shunong Bai, Lingjing Chen, Mary Alice Yund, and Zinmay Rence Sung

4 Sperm-Mediated Gene Transfer

   Anthony W. S. Chan, C. Marc Luetjens, and Gerald P. Schatten

5 Gonocyte-Sertoli Cell Interactions during Development of the Neonatal Rodent Testis

   Joanne M. Orth, William F. Jester, Ling-Hong Li, and Andrew L. Laslett

6 Attributes and Dynamics of the Endoplasmic Reticulum in Mammalian Eggs

   Douglas Kline

7 Germ Plasm and Molecular Determinants of Germ Cell Fate

   Douglas W. Houston and Mary Lou King

Contents of Previous Volumes

Volume 51

1 **Patterning and Lineage Specification in the Amphibian Embryo**
   Agnes P. Chan and Laurence D. Etkin
2 **Transcriptional Programs Regulating Vascular Smooth Muscle Cell Development and Differentiation**
   Michael S. Parmacek
3 **Myofibroblasts: Molecular Crossdressers**
   Gennyne A. Walker, Ivan A. Guerrero, and Leslie A. Leinwand
4 **Checkpoint and DNA-Repair Proteins Are Associated with the Cores of Mammalian Meiotic Chromosomes**
   Madalena Tarsounas and Peter B. Moens
5 **Cytoskeletal and Ca^{2+} Regulation of Hyphal Tip Growth and Initiation**
   author
   Torralba and I. Brent Heath
6 **Pattern Formation during C. elegans Vulval Induction**
   Minqin Wang and Paul W. Sternberg
7 **A Molecular Clock Involved in Somite Segmentation**
   Miguel Maroto and Olivier Pourquie

Volume 52

1 **Mechanism and Control of Meiotic Recombination Initiation**
   Scoff Keeney
2 **Osmoregulation and Cell Volume Regulation in the Preimplantation Embryo**
   Jay M. Baltz
3 **Cell–Cell Interactions in Vascular Development**
   Diane C. Darland and Patricia A. D'Amore
4 **Genetic Regulation of Preimplantation Embryo Survival**
   Carol M. Warner and Carol A. Brenner

Volume 53

1. **Developmental Roles and Clinical Significance of Hedgehog Signaling**
   Andrew P. McMahon, Philip W. Ingham, and Clifford J. Tabin
2. **Genomic Imprinting: Could the Chromatin Structure Be the Driving Force?**
   Andras Paldi
3. **Ontogeny of Hematopoiesis: Examining the Emergence of Hematopoietic Cells in the Vertebrate Embryo**
   Jenna L. Galloway and Leonard I. Zon
4. **Patterning the Sea Urchin Embryo: Gene Regulatory Networks, Signaling Pathways, and Cellular Interactions**
   Lynne M. Angerer and Robert C. Angerer

Volume 54

1. **Membrane Type-Matrix Metal loproteinases (MT-MMP)**
   Stanley Zucker, Duanqing Pei, Jian Cao, and Carlos Lopez-Otin
2. **Surface Association of Secreted Matrix Metalloproteinases**
   Rafael Fridman
3. **Biochemical Properties and Functions of Membrane-Anchored Metalloprotease-Disintegrin Proteins (ADAMs)**
   J. David Becherer and Carl P. Blobel
4. **Shedding of Plasma Membrane Proteins**
   Joaquín Arribas and Anna Merlos-Suárez
5. **Expression of Meprins in Health and Disease**
   Lourdes P. Norman, Gail L. Matters, Jacqueline M. Crisman, and Judith S. Bond
6. **Type II Transmembrane Serine Proteases**
   Qingyu Wu
7. **DPPIV, Seprase, and Related Serine Peptidases in Multiple Cellular Functions**
   Wen-Tien Chen, Thomas Kelly, and Giulio Ghersi

8 The Secretases of Alzheimer's Disease

   Michael S. Wolfe

9 Plasminogen Activation at the Cell Surface

   Vincent Ellis

10 Cell-Surface Cathepsin B: Understanding Its Functional Significance

   Dora Cavallo-Medved and Bonnie F. Sloane

11 Protease-Activated Receptors

   Wadie F. Bahou

12 Emmprin (CD147), a Cell Surface Regulator of Matrix Metalloproteinase Production and Function

   Bryan P. Toole

13 The Evolving Roles of Cell Surface Proteases in Health and Disease: Implications for Developmental, Adaptive, Inflammatory, and Neoplastic Processes

   Joseph A. Madri

14 Shed Membrane Vesicles and Clustering of Membrane-Bound Proteolytic Enzymes

   M. Letizia Vittorelli

Volume 55

1 The Dynamics of Chromosome Replication in Yeast

   Isabelle A. Lucas and M. K. Raghuraman

2 Micromechanical Studies of Mitotic Chromosomes

   M. G. Poirier and John F. Marko

3 Patterning of the Zebrafish Embryo by Nodal Signals

   Jennifer O. Liang and Amy L. Rubinstein

4 Folding Chromosomes in Bacteria: Examining the Role of Csp Proteins and Other Small Nucleic Acid-Binding Proteins

   Nancy Trun and Danielle Johnston

Volume 56

1. **Selfishness in Moderation: Evolutionary Success of the Yeast Plasmid**
   Soundarapandian Velmurugan, Shwetal Mehta, and Makkuni Jayaram
2. **Nongenomic Actions of Androgen in Sertoli Cells**
   William H. Walker
3. **Regulation of Chromatin Structure and Gene Activity by Poly(ADP-Ribose) Polymerases**
   Alexei Tulin, Yurli Chinenov, and Allan Spradling
4. **Centrosomes and Kinetochores, Who needs 'Em? The Role of Noncentromeric Chromatin in Spindle Assembly**
   Priya Prakash Budde and Rebecca Heald
5. **Modeling Cardiogenesis: The Challenges and Promises of 3D Reconstruction**
   Jeffrey O. Penetcost, Claudio Silva, Maurice Pesticelli, Jr., and Kent L. Thornburg
6. **Plasmid and Chromosome Traffic Control: How ParA and ParB Drive Partition**
   Jennifer A. Surtees and Barbara E. Funnell

Volume 57

1. **Molecular Conservation and Novelties in Vertebrate Ear Development**
   B. Fritzsch and K. W. Beisel
2. **Use of Mouse Genetics for Studying Inner Ear Development**
   Elizabeth Quint and Karen P. Steel
3. **Formation of the Outer and Middle Ear, Molecular Mechanisms**
   Moisés Mallo
4. **Molecular Basis of Inner Ear Induction**
   Stephen T. Brown, Kareen Martin, and Andrew K. Groves
5. **Molecular Basis of Otic Commitment and Morphogenesis: A Role for Homeodomain-Containing Transcription Factors and Signaling Molecules**
   Eva Sober, Silke Rinkwitz, and Heike Herbrand

6 Growth Factors and Early Development of Otic Neurons: Interactions between Intrinsic and Extrinsic Signals

Berta Alsina, Fernando Giraldez, and Isabel Varela-Nieto

7 Neurotrophic Factors during Inner Ear Development

Ulla Pirvola and Jukka Ylikoski

8 FGF Signaling in Ear Development and Innervation

Tracy J. Wright and Suzanne L. Mansour

9 The Roles of Retinoic Acid during Inner Ear Development

Raymond Romand

10 Hair Cell Development in Higher Vertebrates

Wei-Qiang Gao

11 Cell Adhesion Molecules during Inner Ear and Hair Cell Development, Including Notch and Its Ligands

Matthew W. Kelley

12 Genes Controlling the Development of the Zebrafish Inner Ear and Hair Cells

Bruce B. Riley

13 Functional Development of Hair Cells

Ruth Anne Eatock and Karen M. Hurley

14 The Cell Cycle and the Development and Regeneration of Hair Cells

Allen F. Ryan

Volume 58

1 A Role for Endogenous Electric Fields in Wound Healing

Richard Nuccitelli

2 The Role of Mitotic Checkpoint in Maintaining Genomic Stability

Song-Tao Liu, Jan M. van Deursen, and Tim J. Yen

3 The Regulation of Oocyte Maturation

Ekaterina Voronina and Gary M. Wessel

4 Stem Cells: A Promising Source of Pancreatic Islets for Transplantation in Type 1 Diabetes

Cale N. Street, Ray V. Rajotte, and Gregory S. Korbutt

5 Differentiation Potential of Adipose Derived Adult Stem (ADAS) Cells

Jeffrey M. Gimble and Farshid Guilak

Volume 59

1 The Balbiani Body and Germ Cell Determinants: 150 Years Later

Malgorzata Kloc, Szczepan Bilinski, and Laurence D. Etkin

2 Fetal-Maternal Interactions: Prenatal Psychobiological Precursors to Adaptive Infant Development

Matthew F. S. X. Novak

3 Paradoxical Role of Methyl-CpG-Binding Protein 2 in Rett Syndrome

Janine M. LaSalle

4 Genetic Approaches to Analyzing Mitochondrial Outer Membrane Permeability

Brett H. Graham and William J. Craigen

5 Mitochondrial Dynamics in Mammals

Hsiuchen Chen and David C. Chan

6 Histone Modification in Corepressor Functions

Judith K. Davie and Sharon Y. R. Dent

7 Death by Abl: A Matter of Location

Jiangyu Zhu and Jean Y. J. Wang

Volume 60

1 Therapeutic Cloning and Tissue Engineering

Chester J. Koh and Anthony Atala

2 α-Synuclein: Normal Function and Role in Neurodegenerative Diseases

Erin H. Norris, Benoit I. Giasson, and Virginia M.-Y. Lee

3 Structure and Function of Eukaryotic DMA Methyltransferases

Taiping Chen and En Li

4 Mechanical Signals as Regulators of Stem Cell Fate

Bradley T. Estes, Jeffrey M. Gimble, and Farshid Guilak

5 Origins of Mammalian Hematopoiesis: *In Vivo* Paradigms and *In Vitro* Models

   M. William Lensch and George Q. Daley

6 Regulation of Gene Activity and Repression: A Consideration of Unifying Themes

   Anne C. Ferguson-Smith, Shau-Ping Lin, and Neil Youngson

7 Molecular Basis for the Chloride Channel Activity of Cystic Fibrosis Transmembrane Conductance Regulator and the Consequences of Disease-Causing Mutations

   Jackie F. Kidd, Iiana Kogan, and Christine E. Bear

Volume 61

1 Hepatic Oval Cells: Helping Redefine a Paradigm in Stem Cell Biology

   P. N. Newsome, M. A. Hussain, and N. D. Theise

2 Meiotic DMA Replication

   Randy Strich

3 Pollen Tube Guidance: The Role of Adhesion and Chemotropic Molecules

   Sunran Kim, Juan Dong, and Elizabeth M. Lord

4 The Biology and Diagnostic Applications of Fetal DMA and RNA in Maternal Plasma

   Rossa W. K. Chiu and Y. M. Dennis Lo

5 Advances in Tissue Engineering

   Shulamit Levenberg and Robert Langer

6 Directions in Cell Migration Along the Rostral Migratory Stream: The Pathway for Migration in the Brain

   Shin-ichi Murase and Alan F. Horwitz

7 Retinoids in Lung Development and Regeneration

   Malcolm Maden

8 Structural Organization and Functions of the Nucleus in Development, Aging, and Disease

   Leslie Mounkes and Colin L. Stewart

Volume 62

1. **Blood Vessel Signals During Development and Beyond**
   Ondine Cleaver
2. **HIFs, Hypoxia, and Vascular Development**
   Kelly L. Covello and M. Celeste Simon
3. **Blood Vessel Patterning at the Embryonic Midline**
   Kelly A. Hogan and Victoria L. Bautch
4. **Wiring the Vascular Circuitry: From Growth Factors to Guidance Cues**
   Lisa D. Urness and Dean Y. Li
5. **Vascular Endothelial Growth Factor and Its Receptors in Embryonic Zebrafish Blood Vessel Development**
   Katsutoshi Goishi and Michael Klagsbrun
6. **Vascular Extracellular Matrix and Aortic Development**
   Cassandra M. Kelleher, Sean E. McLean, and Robert P. Mecham
7. **Genetics in Zebrafish, Mice, and Humans to Dissect Congenital Heart Disease: Insights in the Role of VEGF**
   Diether Lambrechts and Peter Carmeliet
8. **Development of Coronary Vessels**
   Mark W. Majesky
9. **Identifying Early Vascular Genes Through Gene Trapping in Mouse Embryonic Stem Cells**
   Frank Kuhnert and Heidi Stuhlmann

Volume 63

1. **Early Events in the DMA Damage Response**
   Irene Ward and Junjie Chen
2. **Afrotherian Origins and Interrelationships: New Views and Future Prospects**
   Terence J. Robinson and Erik R. Seiffert
3. **The Role of Antisense Transcription in the Regulation of X-Inactivation**
   Claire Rougeulle and Philip Avner

4 The Genetics of Hiding the Corpse: Engulfment and Degradation of Apoptotic Cells in *C. elegans* and *D. melanogaster*

   Zheng Zhou, Paolo M. Mangahas, and Xiaomeng Yu

5 Beginning and Ending an Actin Filament: Control at the Barbed End

   Sally H. Zigmond

6 Life Extension in the Dwarf Mouse

   Andrzej Bartke and Holly Brown-Borg

Volume 64

1 Stem/Progenitor Cells in Lung Morphogenesis, Repair, and Regeneration

   David Warburton, Mary Anne Berberich, and Barbara Driscoll

2 Lessons from a Canine Model of Compensatory Lung Growth

   Connie C. W. Hsia

3 Airway Glandular Development and Stem Cells

   Xiaoming Liu, Ryan R. Driskell, and John F. Engelhardt

4 Gene Expression Studies in Lung Development and Lung Stem Cell Biology

   Thomas J. Mariani and Naftali Kaminski

5 Mechanisms and Regulation of Lung Vascular Development

   Michelle Haynes Pauling and Thiennu H. Vu

6 The Engineering of Tissues Using Progenitor Cells

   Nancy L. Parenteau, Lawrence Rosenberg, and Janet Hardin-Young

7 Adult Bone Marrow-Derived Hemangioblasts, Endothelial Cell Progenitors, and EPCs

   Gina C. Schatteman

8 Synthetic Extracellular Matrices for Tissue Engineering and Regeneration

   Eduardo A. Silva and David J. Mooney

9 Integrins and Angiogenesis

   D. G. Stupack and D. A. Cheresh

Volume 65

1. **Tales of Cannibalism, Suicide, and Murder: Programmed Cell Death in *C. elegans***

   Jason M. Kinchen and Michael O. Hengartner

2. **From Guts to Brains: Using Zebrafish Genetics to Understand the Innards of Organogenesis**

   Carsten Stuckenholz, Paul E. Ulanch, and Nathan Bahary

3. **Synaptic Vesicle Docking: A Putative Role for the Munc18/Sec1 Protein Family**

   Robby M. Weimer and Janet E. Richmond

4. **ATP-Dependent Chromatin Remodeling**

   Corey L. Smith and Craig L. Peterson

5. **Self-Destruct Programs in the Processes of Developing Neurons**

   David Shepherd and V. Hugh Perry

6. **Multiple Roles of Vascular Endothelial Growth Factor (VEGF) in Skeletal Development, Growth, and Repair**

   Elazar Zelzer and Bjorn R. Olsen

7. **G-Protein Coupled Receptors and Calcium Signaling in Development**

   Geoffrey E. Woodard and Juan A. Rosado

8. **Differential Functions of 14-3-3 Isoforms in Vertebrate Development**

   Anthony J. Muslin and Jeffrey M. C. Lau

9. **Zebrafish Notochordal Basement Membrane: Signaling and Structure**

   Annabelle Scott and Derek L. Stemple

10. **Sonic Hedgehog Signaling and the Developing Tooth**

    Martyn T. Cobourne and Paul T. Sharpe

Volume 66

1. **Stepwise Commitment from Embryonic Stem to Hematopoietic and Endothelial Cells**

   Changwon Park, Jesse J. Lugus, and Kyunghee Choi

2 **Fibroblast Growth Factor Signaling and the Function and Assembly of Basement Membranes**

   Peter Lonai

3 **TGF-β Superfamily and Mouse Craniofacial Development: Interplay of Morphogenetic Proteins and Receptor Signaling Controls Normal Formation of the Face**

   Marek Dudas and Vesa Kaartinen

4 **The Colors of Autumn Leaves as Symptoms of Cellular Recycling and Defenses Against Environmental Stresses**

   Helen J. Ougham, Phillip Morris, and Howard Thomas

5 **Extracellular Proteases: Biological and Behavioral Roles in the Mammalian Central Nervous System**

   Van Zhang, Kostas Pothakos, and Styliana-Anna (Stella) Tsirka

6 **The Genetic Architecture of House Fly Mating Behavior**

   Lisa M. Meffert and Kara L. Hagenbuch

7 **Phototropins, Other Photoreceptors, and Associated Signaling: The Lead and Supporting Cast in the Control of Plant Movement Responses**

   Bethany B. Stone, C. Alex Esmon, and Emmanuel Liscum

8 **Evolving Concepts in Bone Tissue Engineering**

   Catherine M. Cowan, Chia Soo, Kang Ting, and Benjamin Wu

9 **Cranial Suture Biology**

   Kelly A Lenton, Randall P. Nacamuli, Derrick C. Wan, Jill A. Helms, and Michael T. Longaker

Volume 67

1 **Deer Antlers as a Model of Mammalian Regeneration**

   Joanna Price, Corrine Faucheux, and Steve Allen

2 **The Molecular and Genetic Control of Leaf Senescence and Longevity in *Arabidopsis***

   Pyung Ok Lim and Hong Gil Nam

3 **Cripto-1: An Oncofetal Gene with Many Faces**

   Caterina Bianco, Luigi Strizzi, Nicola Normanno, Nadia Khan, and David S. Salomon

4 Programmed Cell Death in Plant Embryogenesis

Peter V. Bozhkov, Lada H. Filonova, and Maria F. Suarez

5 Physiological Roles of Aquaporins in the Choroid Plexus

Daniela Boassa and Andrea J. Yool

6 Control of Food Intake Through Regulation of cAMP

Allan Z. Zhao

7 Factors Affecting Male Song Evolution in *Drosophila montana*

Anneli Hoikkala, Kirsten Klappert, and Dominique Mazzi

8 Prostanoids and Phosphodiesterase Inhibitors in Experimental Pulmonary Hypertension

Ralph Theo Schermuly, Hossein Ardeschir Ghofrani, and Norbert Weissmann

9 14-3-3 Protein Signaling in Development and Growth Factor Responses

Daniel Thomas, Mark Guthridge, Jo Woodcock, and Angel Lopez

10 Skeletal Stem Cells in Regenerative Medicine

Wataru Sonoyama, Carolyn Coppe, Stan Gronthos, and Songtao Shi

Volume 68

1 Prolactin and Growth Hormone Signaling

Beverly Chilton and Aveline Hewetson

2 Alterations in cAMP-Mediated Signaling and Their Role in the Pathophysiology of Dilated Cardiomyopathy

Matthew A. Movsesian and Michael R. Bristow

3 Corpus Luteum Development: Lessons from Genetic Models in Mice

Anne Bachelot and Nadine Binart

4 Comparative Developmental Biology of the Mammalian Uterus

Thomas E. Spencer, Kanako Hayashi, Jianbo Hu, and Karen D. Carpenter

5 Sarcopenia of Aging and Its Metabolic Impact

Helen Karakelides and K. Sreekumaran Nair

6 Chemokine Receptor CXCR3: An Unexpected Enigma

Liping Liu, Melissa K. Callahan, DeRen Huang, and Richard M. Ransohoff

7 Assembly and Signaling of Adhesion Complexes

Jorge L. Sepulveda, Vasiliki Gkretsi, and Chuanyue Wu

8 Signaling Mechanisms of Higher Plant Photoreceptors: A Structure–Function Perspective

Haiyang Wang

9 Initial Failure in Myoblast Transplantation Therapy Has Led the Way Toward the Isolation of Muscle Stem Cells: Potential for Tissue Regeneration

Kenneth Urish, Yasunari Kanda, and Johnny Huard

10 Role of 14-3-3 Proteins in Eukaryotic Signaling and Development

Dawn L. Darling, Jessica Yingling, and Anthony Wynshaw-Boris

Volume 69

1 Flipping Coins in the Fly Retina

Tamara Mikeladze-Dvali, Claude Desplan, and Daniela Pistillo

2 Unraveling the Molecular Pathways That Regulate Early Telencephalon Development

Jean M. Hébert

3 Glia-Neuron Interactions in Nervous System Function and Development

Shai Shaham

4 The Novel Roles of Glial Cells Revisited: The Contribution of Radial Glia and Astrocytes to Neurogenesis

Tetsuji Mori, Annalisa Buffo, and Magdalena Gotz

5 Classical Embryological Studies and Modern Genetic Analysis of Midbrain and Cerebellum Development

Mark Zervas, Sandra Blaess, and Alexandra L. Joyner

6 Brain Development and Susceptibility to Damage; Ion Levels and Movements

Maria Erecinska, Shobha Cherian, and Ian A. Silver

7 Thinking about Visual Behavior; Learning about Photoreceptor Function

Kwang-Min Choe and Thomas R. Clandinin

8 Critical Period Mechanisms in Developing Visual Cortex

Takao K. Hensch

9 Brawn for Brains: The Role of MEF2 Proteins in the Developing Nervous System

Aryaman K. Shalizi and Azad Bonni

10 Mechanisms of Axon Guidance in the Developing Nervous System

Céline Plachez and Linda J. Richards

Volume 70

1 Magnetic Resonance Imaging: Utility as a Molecular Imaging Modality

James P. Basilion, Susan Yeon, and René Botnar

2 Magnetic Resonance Imaging Contrast Agents in the Study of Development

Angelique Louie

3 $^1H/^{19}F$ Magnetic Resonance Molecular Imaging with Perfluorocarbon Nanoparticles

Gregory M. Lanza, Patrick M. Winter, Anne M. Neubauer, Sheelton D. Caruthers, Franklin D. Hockett, and Samuel A. Wickline

4 Loss of Cell Ion Homeostasis and Cell Viability in the Brain: What Sodium MRI Can Tell Us

Fernando E. Boada, George La Verde, Charles jungreis, Edwin Nemoto, Costin Tanase, and Ileana Hancu

5 Quantum Dot Surfaces for Use *In Vivo* and *In Vitro*

Byron Ballou

6 *In Vivo* Cell Biology of Cancer Cells Visualized with Fluorescent Proteins

Robert M. Hoffman

7 Modulation of Tracer Accumulation in Malignant Tumors: Gene Expression, Gene Transfer, and Phage Display

Uwe Haberkorn

8 Amyloid Imaging: From Benchtop to Bedside

Chungying Wu, Victor W. Pike, and Yanming Wang

9 *In Vivo* Imaging of Autoimmune Disease in Model Systems

Eric T. Ahrens and Penelope A. Morel

Contents of Previous Volumes

Volume 71

1 **The Choroid Plexus–Cerebrospinal Fluid System: From Development to Aging**

   Zoran B. Redzic, Jane E. Preston, John A. Duncan, Adam Chodobski, and Joanna Szmydynger-Chodobska

2 **Zebrafish Genetics and Formation of Embryonic Vasculature**

   Tao P. Zhong

3 **Leaf Senescence: Signals, Execution, and Regulation**

   Yongfeng Guo and Susheng Can

4 **Muscle Stem Cells and Regenerative Myogenesis**

   Iain W. McKinnell, Gianni Parise, and Michael A. Rudnicki

5 **Gene Regulation in Spermatogenesis**

   James A. Maclean II and Miles F. Wilkinson

6 **Modeling Age-Related Diseases in *Drosophila*: Can this Fly?**

   Kinga Michno, Diana van de Hoef, Hong Wu, and Gabrielle L. Boulianne

7 **Cell Death and Organ Development in Plants**

   Hilary J. Rogers

8 **The Blood-Testis Barrier: Its Biology, Regulation, and Physiological Role in Spermatogenesis**

   Ching-Hang Wong and C. Van Cheng

9 **Angiogenic Factors in the Pathogenesis of Preeclampsia**

   Hai-Tao Yuan, David Haig, and S. Ananth Karumanchi

Volume 72

1 **Defending the Zygote: Search for the Ancestral Animal Block to Polyspermy**

   Julian L. Wong and Gary M. Wessel

2 **Dishevelled: A Mobile Scaffold Catalyzing Development**

   Craig C. Malbon and Hsien-yu Wang

3 **Sensory Organs: Making and Breaking the Pre-Placodal Region**

   Andrew P. Bailey and Andrea Streit

4 Regulation of Hepatocyte Cell Cycle Progression and Differentiation by Type I Collagen Structure
Linda K. Hansen, Joshua Wilhelm, and John T. Fassett

5 Engineering Stem Cells into Organs: Topobiological Transformations Demonstrated by Beak, Feather, and Other Ectodermal Organ Morphogenesis
Cheng-Ming Chuong, Ping Wu, Maksim Plikus, Ting-Xin Jiang, and Randall Bruce Widelitz

6 Fur Seal Adaptations to Lactation: Insights into Mammary Gland Function
Julie A. Sharp, Kylie N. Cane, Christophe Lefevre, John P. Y. Arnould, and Kevin R. Nicholas

Volume 73

1 The Molecular Origins of Species-Specific Facial Pattern
Samantha A. Brugmann, Minal D. Tapadia, and Jill A. Helms

2 Molecular Bases of the Regulation of Bone Remodeling by the Canonical Wnt Signaling Pathway
Donald A. Glass II and Gerard Karsenty

3 Calcium Sensing Receptors and Calcium Oscillations: Calcium as a First Messenger
Gerda E. Breitwieser

4 Signal Relay During the Life Cycle of *Dictyostelium*
Dana C. Mahadeo and Carole A. Parent

5 Biological Principles for *Ex Vivo* Adult Stem Cell Expansion
Jean-François Paré and James L. Sherley

6 Histone Deacetylation as a Target for Radiosensitization
David Cerna, Kevin Camphausen, and Philip J. Tofilon

7 Chaperone-Mediated Autophagy in Aging and Disease
Ashish C. Massey, Cong Zhang, and Ana Maria Cuervo

8 Extracellular Matrix Macroassembly Dynamics in Early Vertebrate Embryos
Andras Czirok, Evan A. Zamir, Michael B. Filla, Charles D. Little, and Brenda J. Rongish

Volume 74

1 **Membrane Origin for Autophagy**
   Fulvio Reggiori

2 **Chromatin Assembly with H3 Histones: Full Throttle Down Multiple Pathways**
   Brian E. Schwartz and Kami Ahmad

3 **Protein-Protein Interactions of the Developing Enamel Matrix**
   John D. Bartlett, Bernhard Ganss, Michel Goldberg, Janet Moradian-Oldak, Michael L. Paine, Malcolm L. Snead, Xin Wen, Shane N. White, and Van L. Zhou

4 **Stem and Progenitor Cells in the Formation of the Pulmonary Vasculature**
   Kimberly A. Fisher and Ross S. Summer

5 **Mechanisms of Disordered Granulopoiesis in Congenital Neutropenia**
   David S. Grenda and Daniel C. Link

6 **Social Dominance and Serotonin Receptor Genes in Crayfish**
   Donald H. Edwards and Nadja Spitzer

7 **Transplantation of Undifferentiated, Bone Marrow-Derived Stem Cells**
   Karen Ann Pauwelyn and Catherine M. Verfaillie

8 **The Development and Evolution of Division of Labor and Foraging Specialization in a Social Insect (*Apis mellifera* L.)**
   Robert E. Page Jr., Ricarda Scheiner, Joachim Erber, and Gro V. Amdam

Volume 75

1 **Dynamics of Assembly and Reorganization of Extracellular Matrix Proteins**
   Sarah L. Dallas, Qian Chen, and Pitchumani Sivakumar

2 **Selective Neuronal Degeneration in Huntington's Disease**
   Catherine M. Cowan and Lynn A. Raymond

3 **RNAi Therapy for Neurodegenerative Diseases**
   Ryan L. Boudreau and Beverly L. Davidson

4 **Fibrillins: From Biogenesis of Microfibrils to Signaling Functions**
   Dirk Hubmacher, Kerstin Tiedemann, and Dieter P. Reinhardt

5 Proteasomes from Structure to Function: Perspectives from Archaea
Julie A. Maupin-Furlow, Matthew A. Humbard, P. Aaron Kirkland, Wei Li, Christopher J. Reuter, Amy J. Wright, and G. Zhou

6 The Cytomatrix as a Cooperative System of Macromolecular and Water Networks
V. A. Shepherd

7 Intracellular Targeting of Phosphodiesterase-4 Underpins Compartmentalized cAMP Signaling
Martin J. Lynch, Elaine V. Hill, and Miles D. Houslay

Volume 76

1 BMP Signaling in the Cartilage Growth Plate
Robert Pogue and Karen Lyons

2 The CLIP-170 Orthologue Bik1 p and Positioning the Mitotic Spindle in Yeast
Rita K. Miller, Sonia D'Silva, Jeffrey K. Moore, and Holly V. Goodson

3 Aggregate-Prone Proteins Are Cleared from the Cytosol by Autophagy: Therapeutic Implications
Andrea Williams, Luca Jahreiss, Sovan Sarkar, Shinji Saiki, Fiona M. Menzies, Brinda Ravikumar, and David C. Rubinsztein

4 Wnt Signaling: A Key Regulator of Bone Mass
Roland Baron, Georges Rawadi, and Sergio Roman-Roman

5 Eukaryotic DMA Replication in a Chromatin Context
Angel P. Tabancay, Jr. and Susan L. Forsburg

6 The Regulatory Network Controlling the Proliferation-Meiotic Entry Decision in the *Caenorhabditis elegans* Germ Line
Dave Hansen and Tim Schedl

7 Regulation of Angiogenesis by Hypoxia and Hypoxia-Inducible Factors
Michele M. Hickey and M. Celeste Simon

Volume 77

1 The Role of the Mitochondrion in Sperm Function: Is There a Place for Oxidative Phosphorylation or Is this a Purely Glycolytic Process?
Eduardo Ruiz-Pesini, Carmen Díez-Sánchez, Manuel José López-Pérez, and José Antonio Enríquez

2 The Role of Mitochondrial Function in the Oocyte and Embryo

Rémi Dumollard, Michael Duchen, and John Carroll

3 Mitochondrial DMA in the Oocyte and the Developing Embryo

Pascale May-Panloup, Marie-Françoise Chretien, Yves Malthiery, and Pascal Reynier

4 Mitochondrial DMA and the Mammalian Oocyte

Eric A. Shoubridge and Timothy Wai

5 Mitochondrial Disease—Its Impact, Etiology, and Pathology

R. McFarland, R. W. Taylor, and D. M. Turnbull

6 Cybrid Models of mtDNA Disease and Transmission, from Cells to Mice

Ian A. Trounce and Carl A. Pinkert

7 The Use of Micromanipulation Methods as a Tool to Prevention of Transmission of Mutated Mitochondrial DMA

Helena Fulka and Josef Fulka, Jr.

8 Difficulties and Possible Solutions in the Genetic Management of mtDNA Disease in the Preimplantation Embryo

J. Poulton, P. Oakeshott, and S. Kennedy

9 Impact of Assisted Reproductive Techniques: A Mitochondrial Perspective from the Cytoplasmic Transplantation

A. J. Harvey, T. C. Gibson, T. M. Quebedeaux, and C. A. Brenner

10 Nuclear Transfer: Preservation of a Nuclear Genome at the Expense of Its Associated mtDNA Genome(s)

Emma J. Bowles, Keith H. S. Campbell, and Justin C. St. John

Volume 78

1 Contribution of Membrane Mucins to Tumor Progression Through Modulation of Cellular Growth Signaling Pathways

Kermit L. Carraway III, Melanie Funes, Heather C. Workman, and Colleen Sweeney

2 Regulation of the Epithelial Na^+ Channel by Peptidases

Carole Planes and George H. Caughey

3 Advances in Defining Regulators of Cementum Development and Periodontal Regeneration
   Brian L. Foster, Tracy E. Popowics, Hanson K. Fong, and Martha J. Somerman
4 Anabolic Agents and the Bone Morphogenetic Protein Pathway
   I. R. Garrett
5 The Role of Mammalian Circadian Proteins in Normal Physiology and Genotoxic Stress Responses
   Roman V. Kondratov, Victoria Y. Gorbacheva, and Marina P. Antoch
6 Autophagy and Cell Death
   Devrim Gozuacik and Adi Kimchi

Volume 79

1 The Development of Synovial Joints
   I. M. Khan, S. N. Redman, R. Williams, G. P. Dowthwaite, S. F. Oldfield, and C. W. Archer
2 Development of a Sexually Differentiated Behavior and Its Underlying CMS Arousal Functions
   Lee-Ming Kow, Cristina Florea, Marlene Schwanzel-Fukuda, Nino Devidze, Hosein Kami Kia, Anna Lee, Jin Zhou, David MacLaughlin, Patricia Donahoe, and Donald Pfaff
3 Phosphodiesterases Regulate Airway Smooth Muscle Function in Health and Disease
   Vera P. Krymskaya and Reynold A. Panettieri, Jr.
4 Role of Astrocytes in Matching Blood Flow to Neuronal Activity
   Danica Jakovcevic and David R. Harder
5 Elastin-Elastases and Inflamm-Aging
   Frank Antonicelli, Georges Bellon, Laurent Debelle, and William Hornebeck
6 A Phylogenetic Approach to Mapping Cell Fate
   Stephen J. Salipante and Marshall S. Horwitz

Volume 80

1 Similarities Between Angiogenesis and Neural Development: What Small Animal Models Can Tell Us
   Serena Zacchigna, Carmen Ruiz de Almodovar, and Peter Carmeliet

2 **Junction Restructuring and Spermatogenesis: The Biology, Regulation, and Implication in Male Contraceptive Development**
   Helen H. N. Van, Dolores D. Mruk, and C. Van Cheng

3 **Substrates of the Methionine Sulfoxide Reductase System and Their Physiological Relevance**
   Derek B. Oien and Jackob Moskovitz

4 **Organic Anion-Transporting Polypeptides at the Blood–Brain and Blood–Cerebrospinal Fluid Barriers**
   Daniel E. Westholm, Jon N. Rumbley, David R. Salo, Timothy P. Rich, and Grant W. Anderson

5 **Mechanisms and Evolution of Environmental Responses in *Caenorhabditis elegans***
   Christian Braendle, Josselin Milloz, and Marie-Anne Felix

6 **Molluscan Shell Proteins: Primary Structure, Origin, and Evolution**
   Frederic Marin, Cities Luquet, Benjamin Marie, and Davorin Medakovic

7 **Pathophysiology of the Blood–Brain Barrier: Animal Models and Methods**
   Brian T. Hawkins and Richard D. Egleton

8 **Genetic Manipulation of Megakaryocytes to Study Platelet Function**
   Jun Liu, Jan DeNofrio, Weiping Yuan, Zhengyan Wang, Andrew W. McFadden, and Leslie V. Pa rise

9 **Genetics and Epigenetics of the Multifunctional Protein CTCF**
   Galina N. Filippova

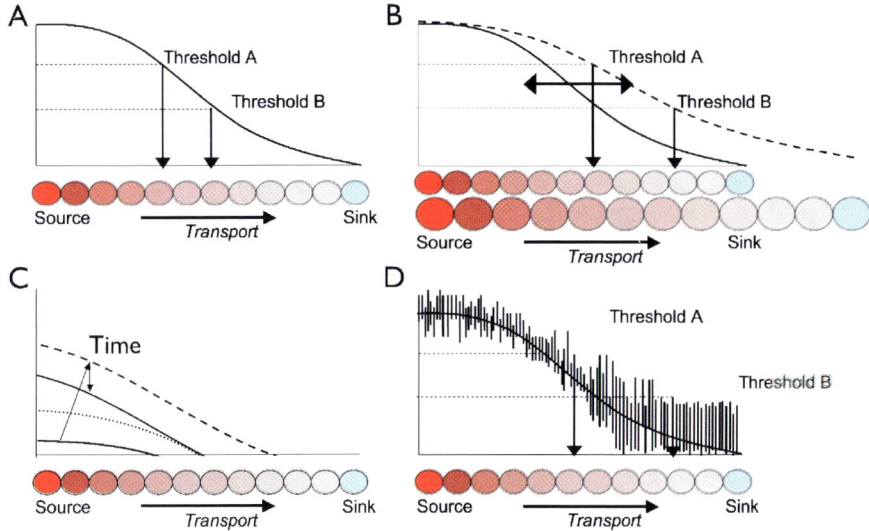

Chapter 2, Figure 1.

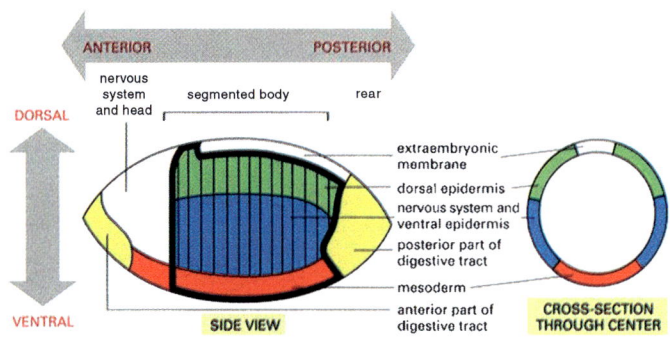

Chapter 2, Figure 2.

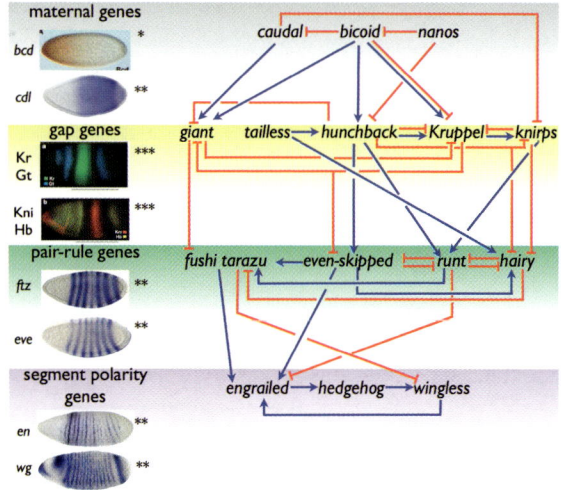

Chapter 2, Figure 3.

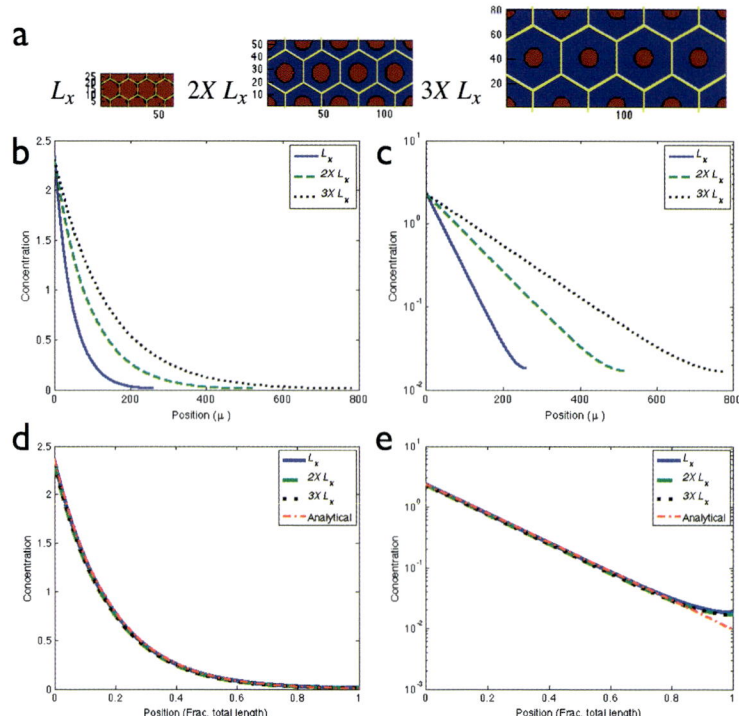

Chapter 2, Figure 5.

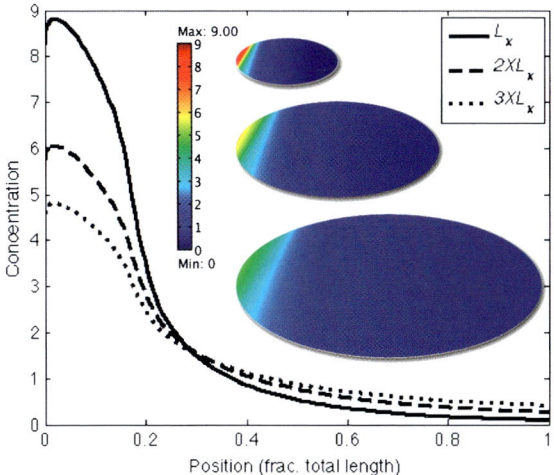

Chapter 2, Figure 6.

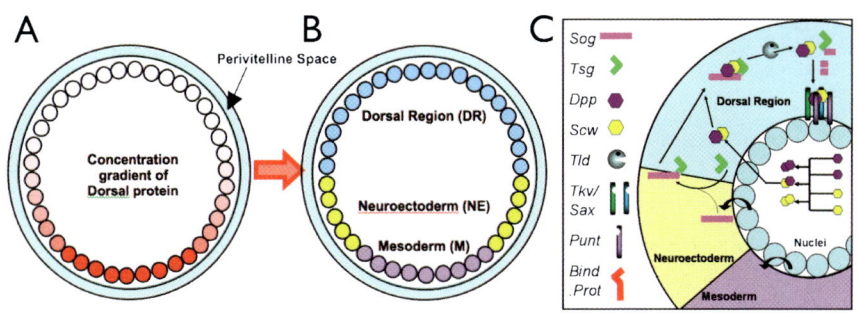

Chapter 2, Figure 9.

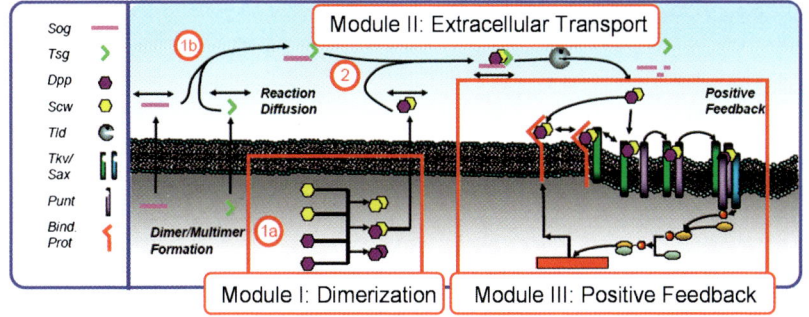

Chapter 2, Figure 10.

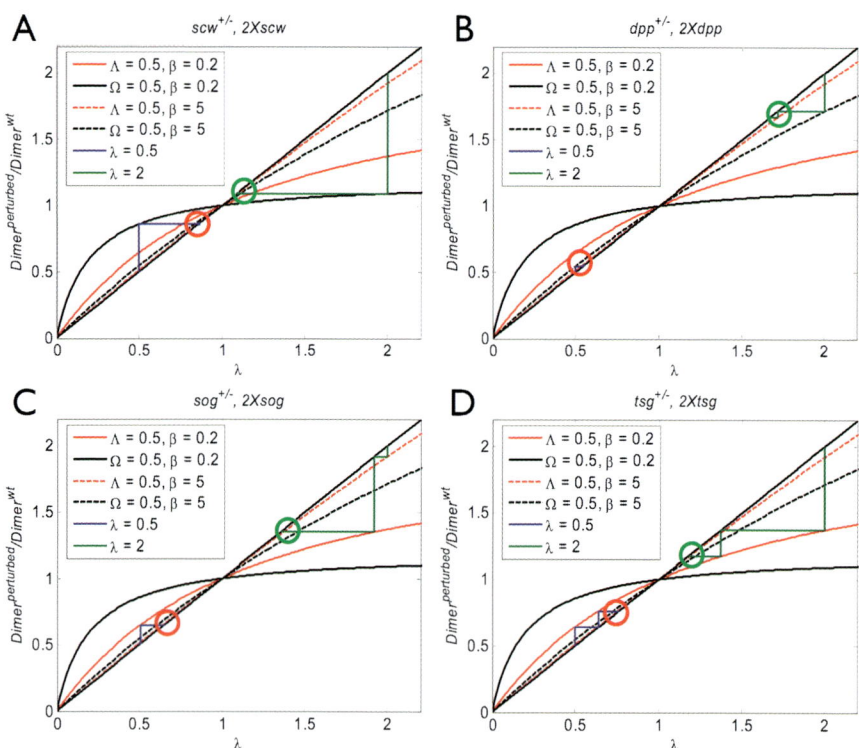

Chapter 2, Figure 11.

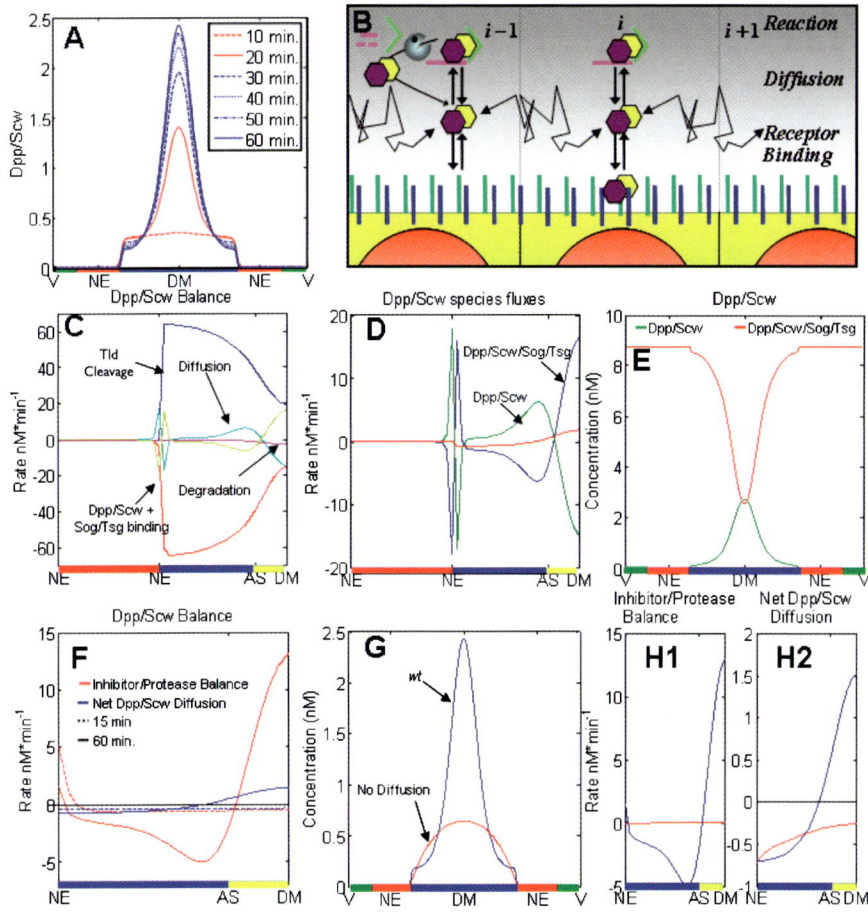

Chapter 2, Figure 12.

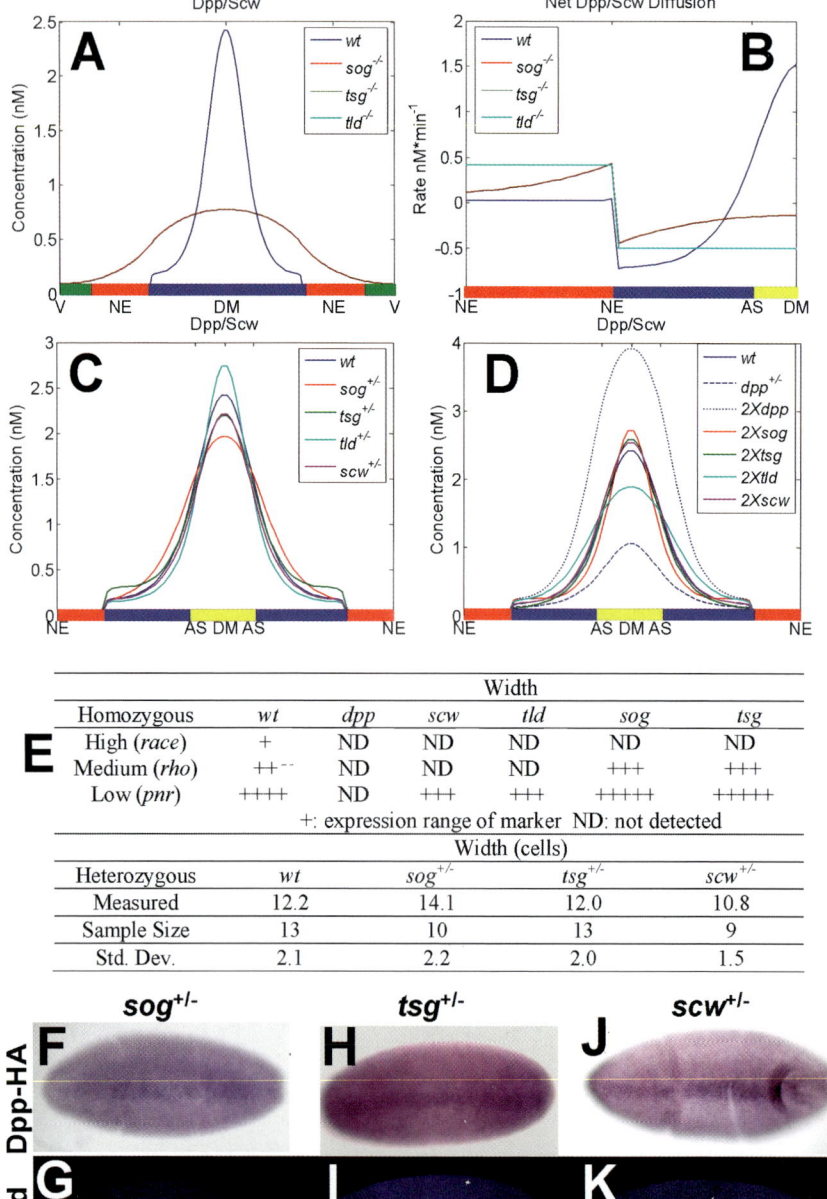

Chapter 2, Figure 13.

A

Summary of Computations	Heterodimer Model		Model I	
Total Simulations	99,186		99,399	
Solutions that vary 2-fold in dorsal 20%	3,385	(3.4% of total)	2,452	(2.5% of total)
Sog-Tsg Compensation for $sog^{+/-}$, $tsg^{+/-}$				
Number (%) passed sog+/- for 0.2-0.8	2,403	(71%)	1081	(44%)
Number (%) passed tsg+/- for 0.2-0.8	3,122	(92%)	-	
Number (%) passed sog+/- and tsg+/-	2,195	(65%)	1081	(44%)
$tld^{+/-}$, $scw^{+/-}$ Compensation				
Number (%) passed tld+/- for 0.2-0.8	2,823	(83%)	2102	(86%)
Number (%) passed scw+/- for 0.2-0.8	600	(18%)	11	(0.5%)
Threshold Dependence				
Threshold at 0.2 (% of 2-fold change)	1,925	(57%)	252	(10%)
Threshold at 0.3	2,138	(63%)	175	(7%)
Threshold at 0.4	1,874	(55%)	107	(4%)
Threshold at 0.5	1,399	(41%)	60	(2%)
Threshold at 0.6	937	(28%)	22	(0.9%)
Threshold at 0.7	575	(17%)	7	(0.3%)
Threshold at 0.8	94	(3%)	2	(0.1%)
Average 0.4-0.6	1,282	(38%)	32	(1.3%)
Average 0.3-0.7	938	(28%)	7	(0.3%)
Average 0.2-0.8	227	(7%)	3	(0.1%)

Chapter 2, Figure 15.

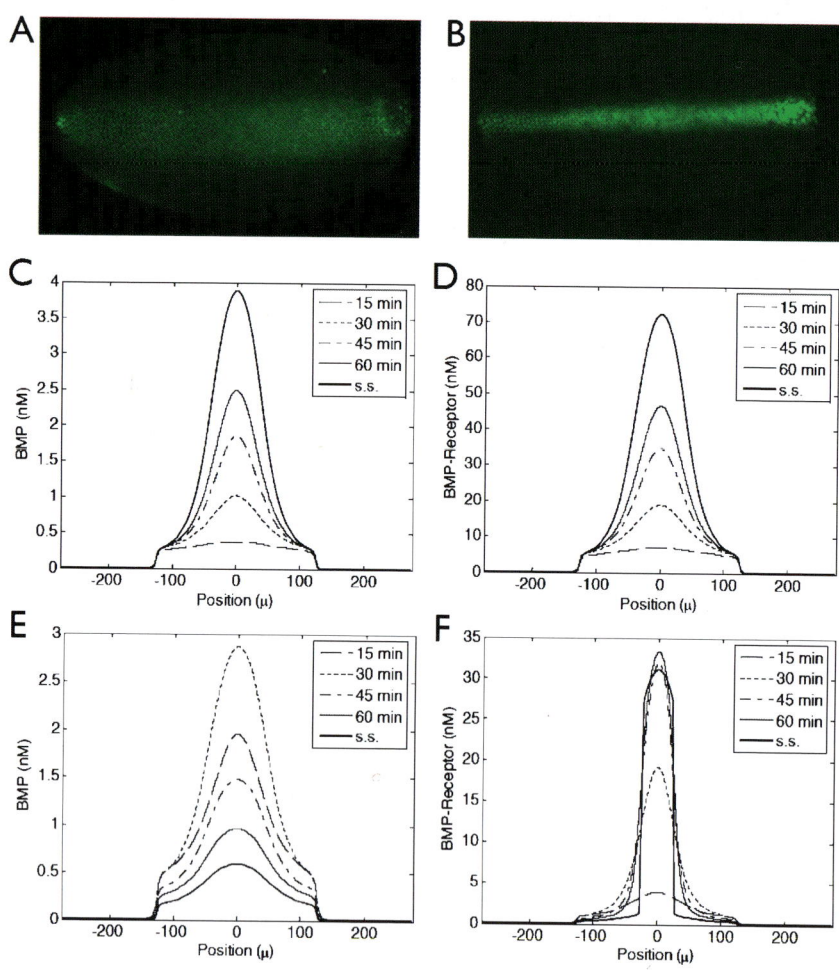

Chapter 2, Figure 16.

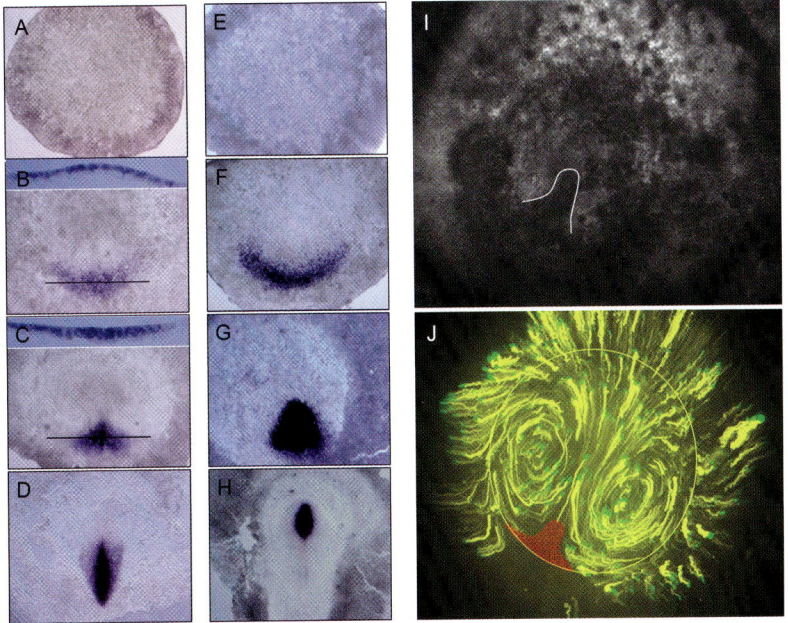

Chapter 4, Figure 1.

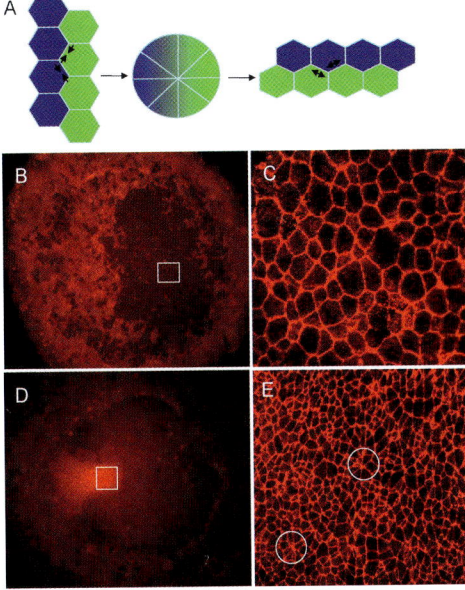

Chapter 4, Figure 2.

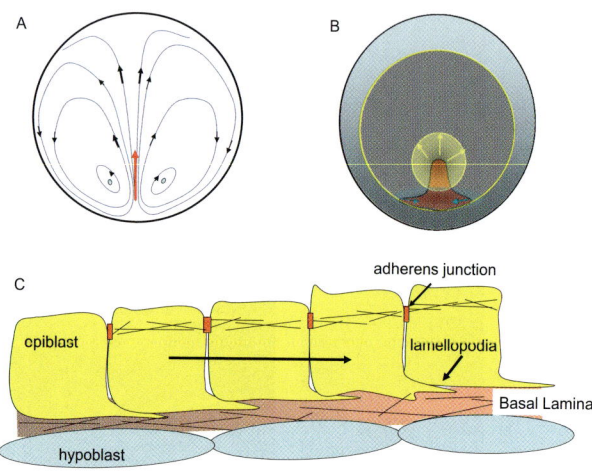

Chapter 4, Figure 3.

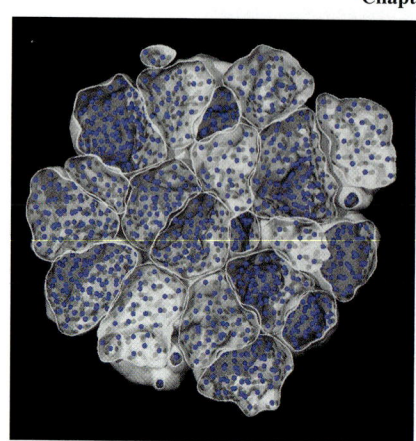

Chapter 5, Figure 1.

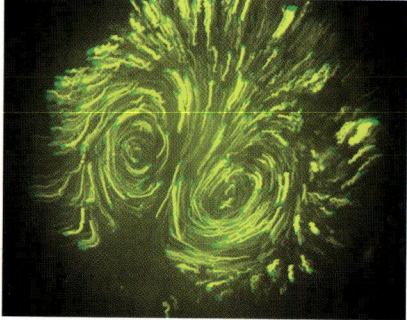

Chapter 5, Figure 5. **Chapter 5, Figure 6.**

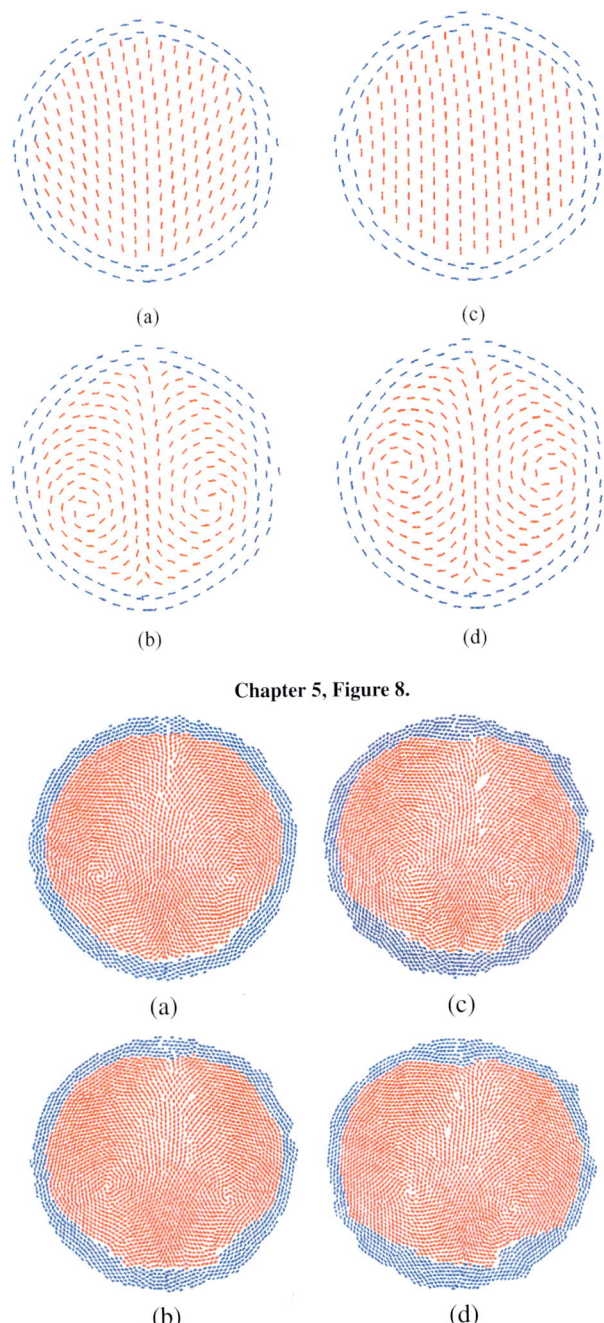

(a)

(c)

(b)

(d)

Chapter 5, Figure 8.

(a)

(c)

(b)

(d)

Chapter 5, Figure 9.

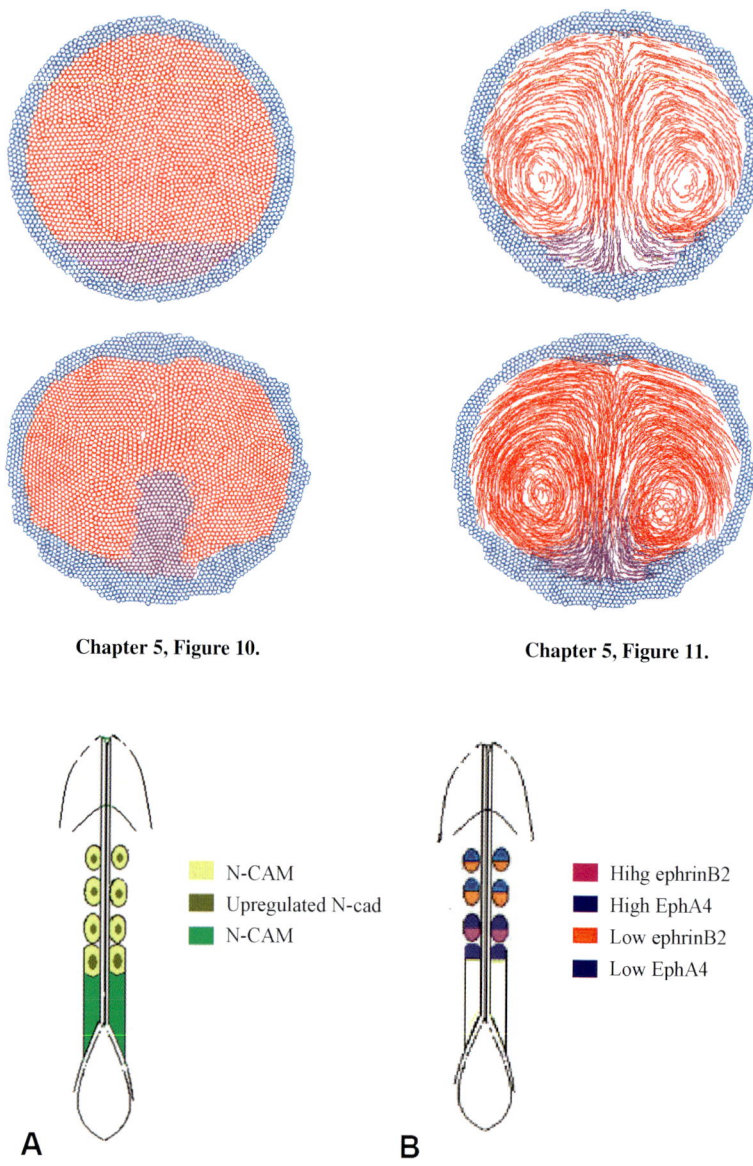

Chapter 5, Figure 10.

Chapter 5, Figure 11.

Chapter 7, Figure 1.

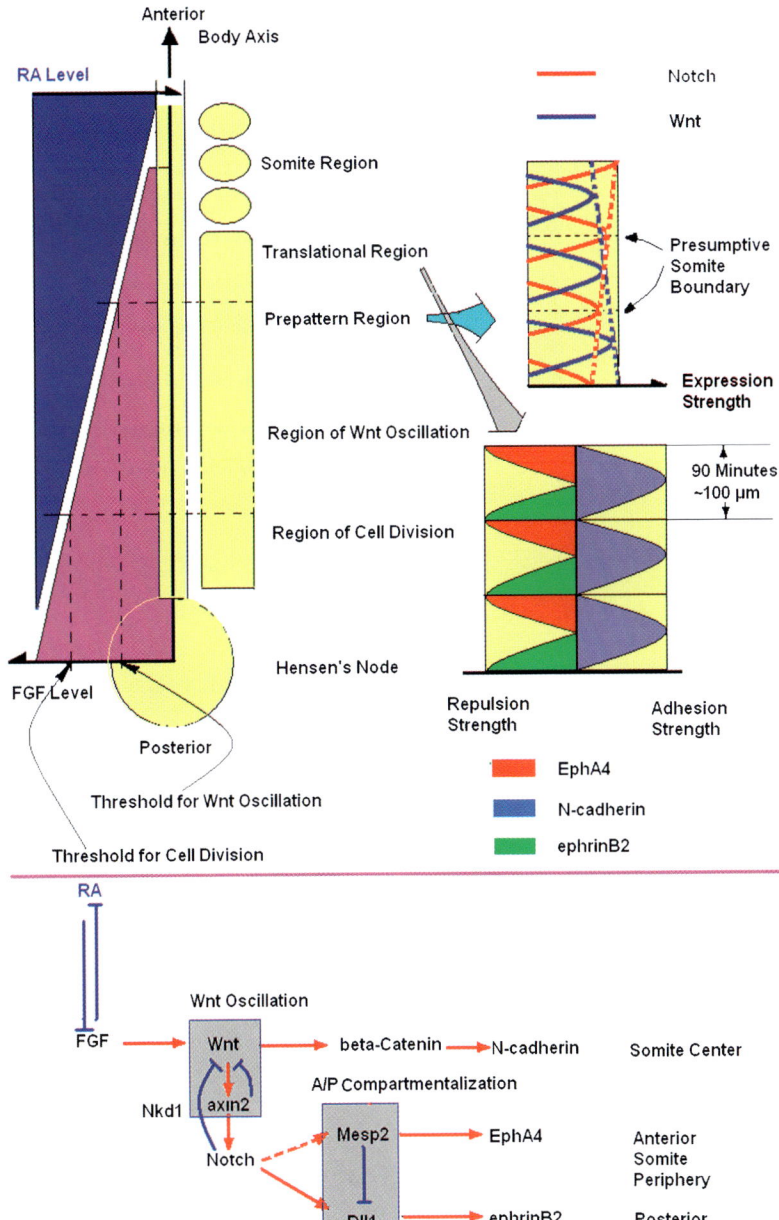

Chapter 7, Figure 2.

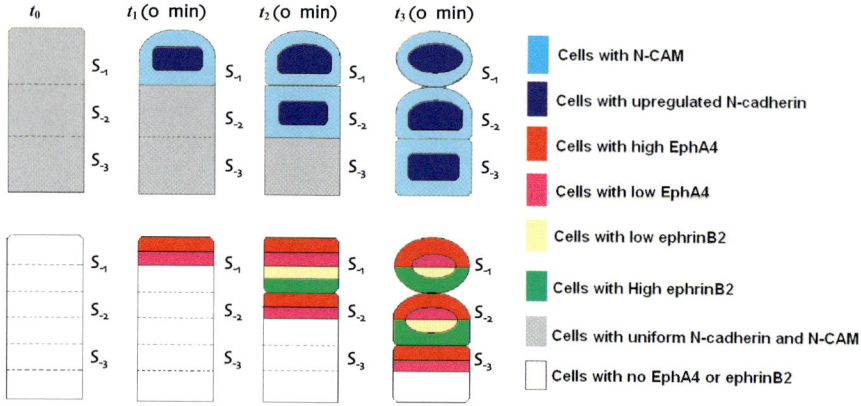

Chapter 7, Figure 3.

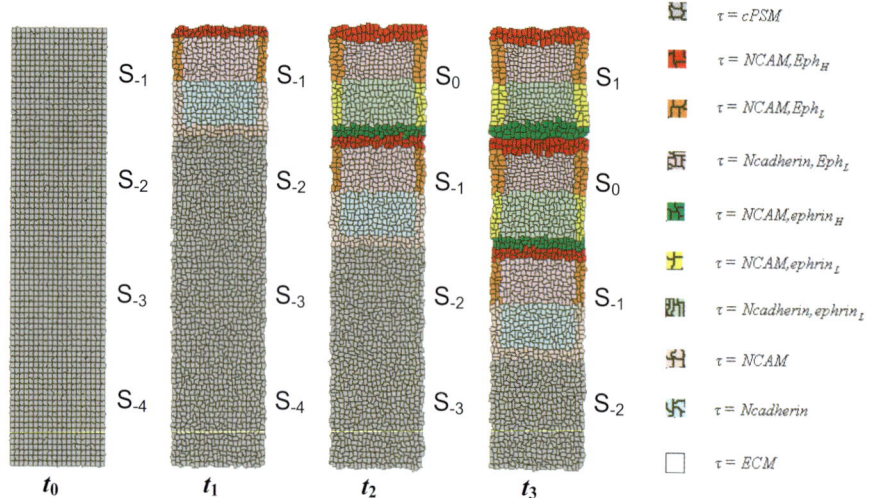

Chapter 7, Figure 4.

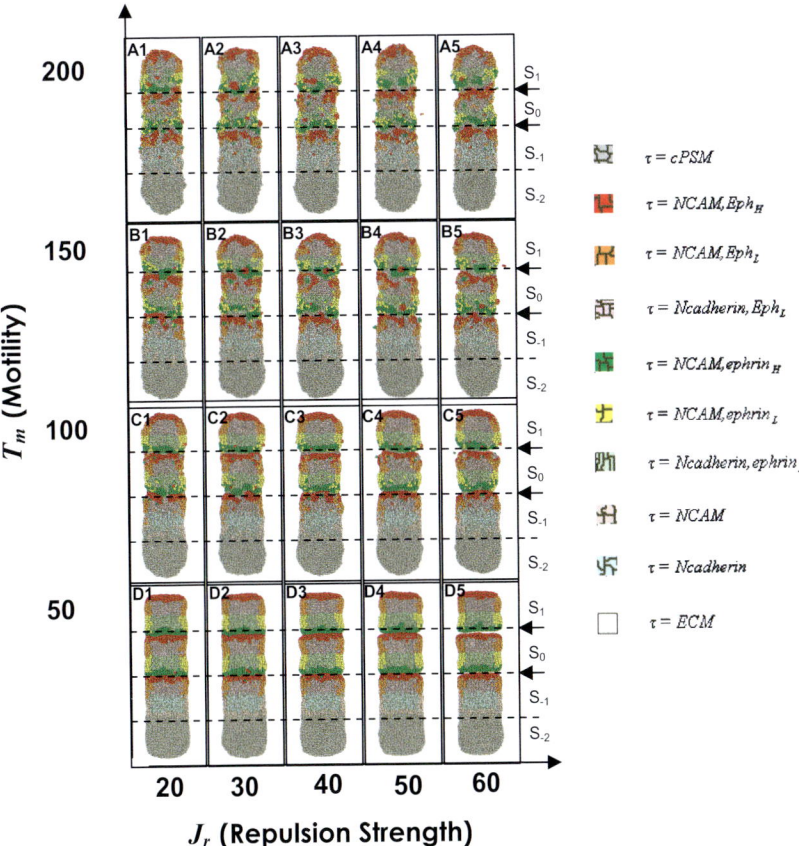

Chapter 7, Figure 5.

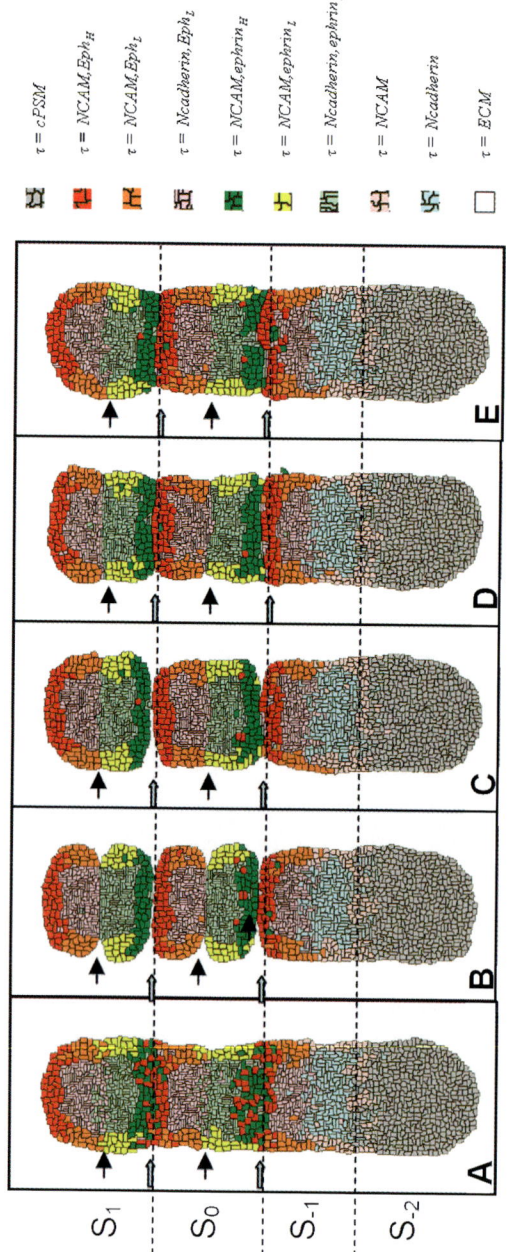

Chapter 7, Figure 6.

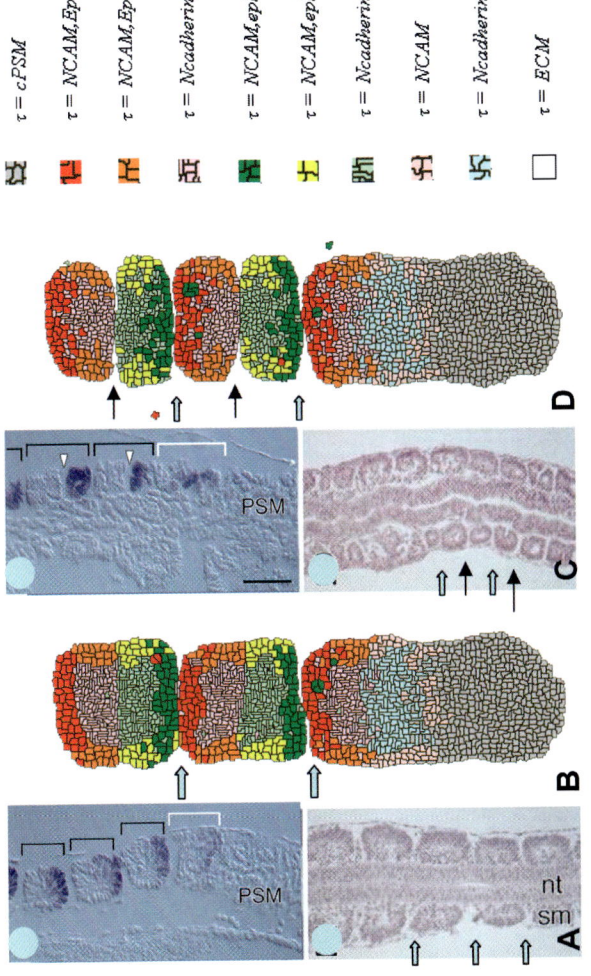

Chapter 7, Figure 7.

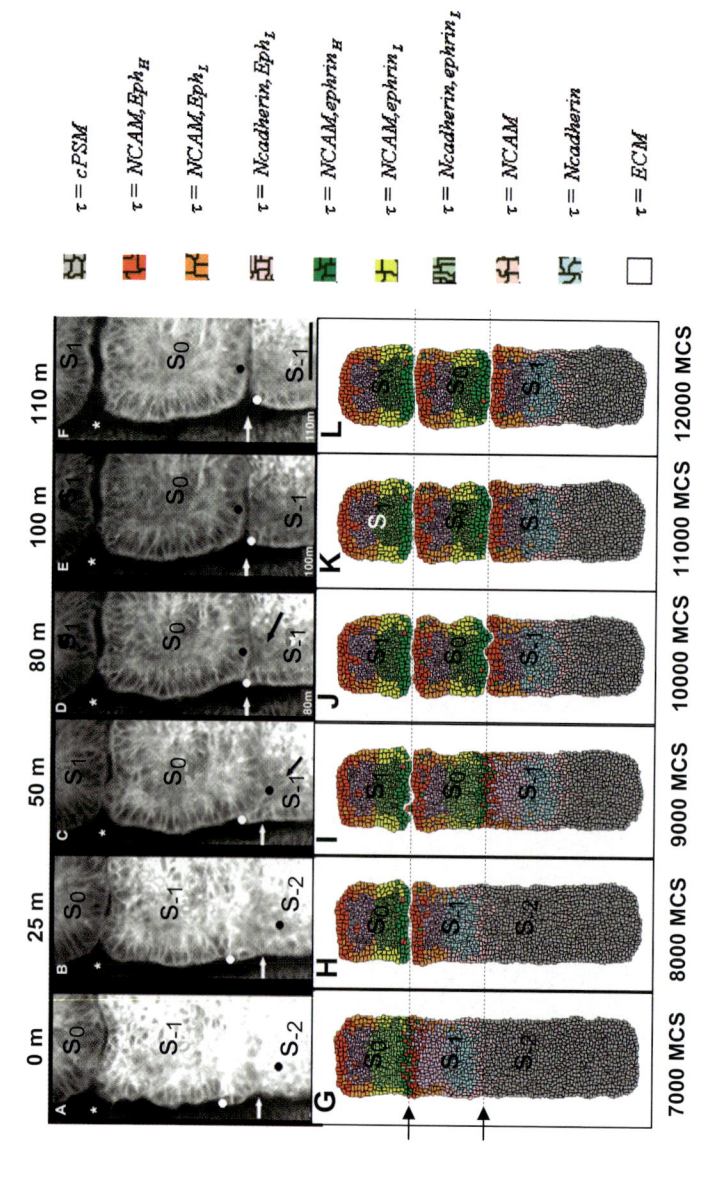

Chapter 7, Figure 8.

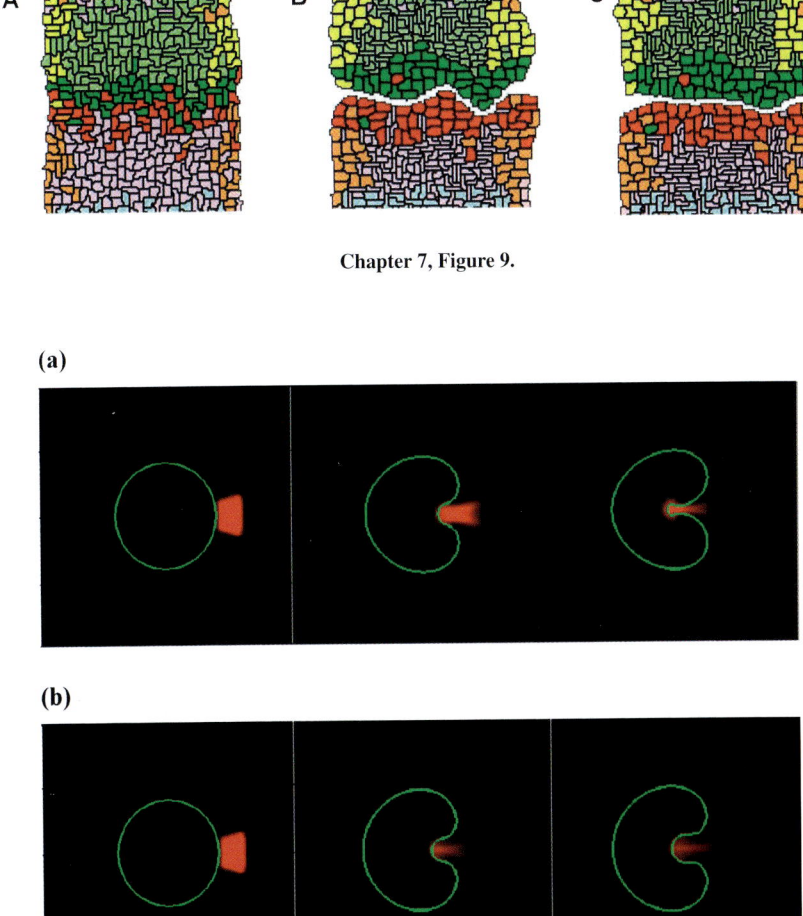

Chapter 7, Figure 9.

Chapter 8, Figure 11.

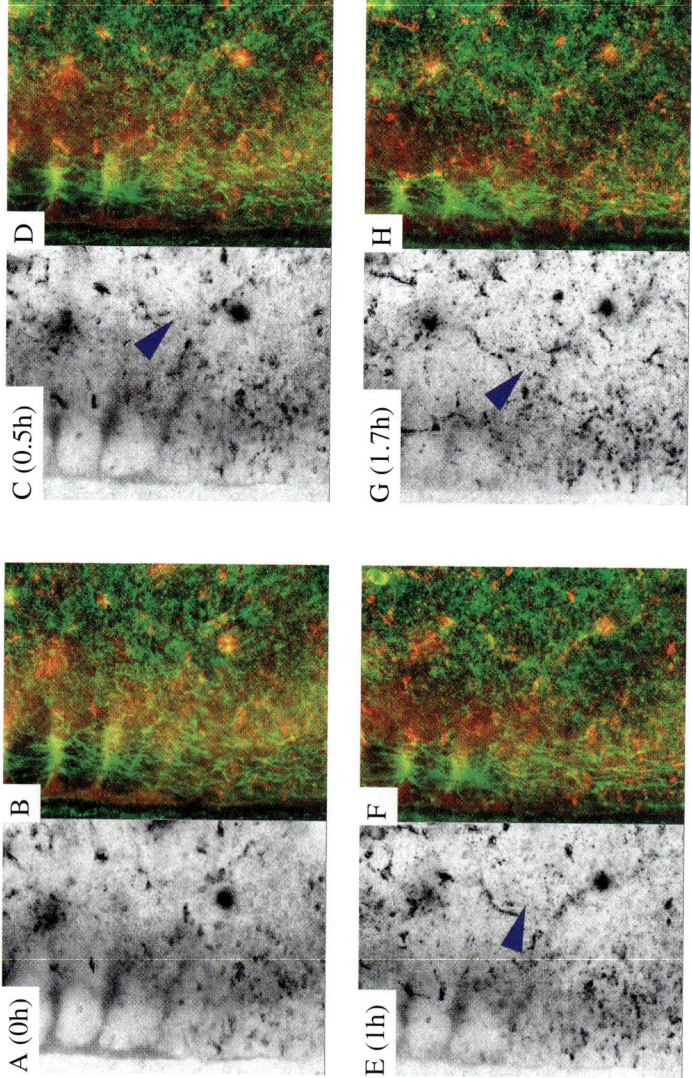

Chapter 9, Figure 1.

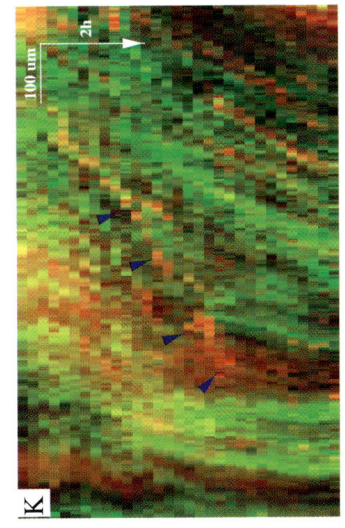
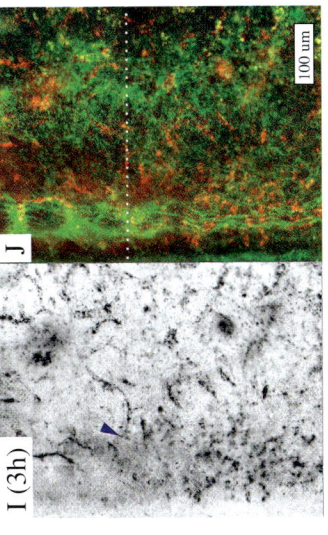

Chapter 9, Figure 1 (*continued*).

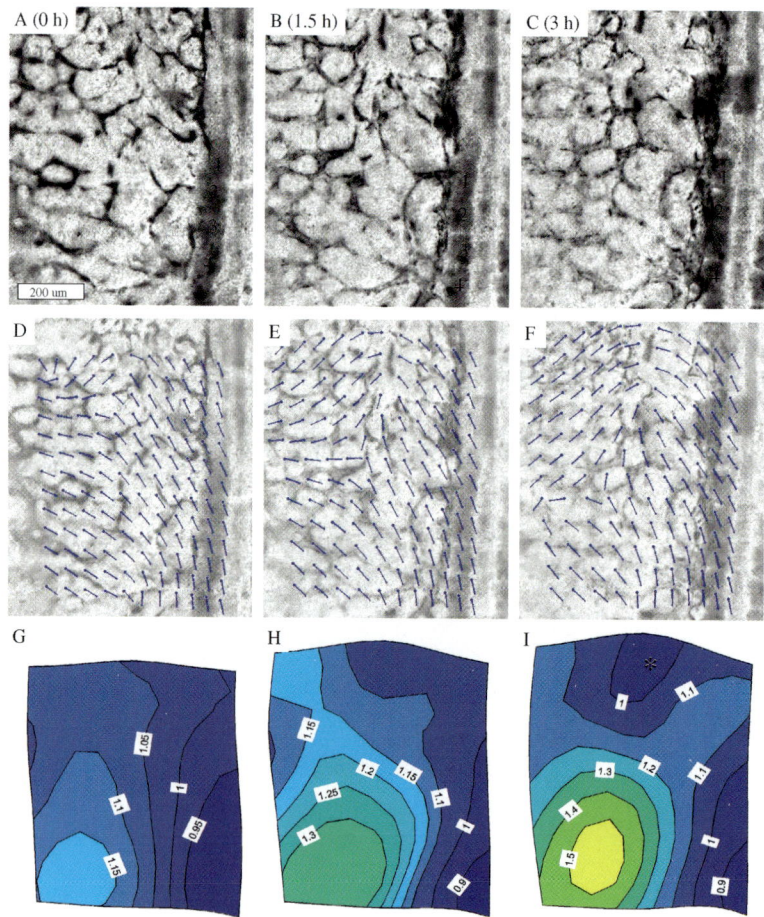

Chapter 9, Figure 2.

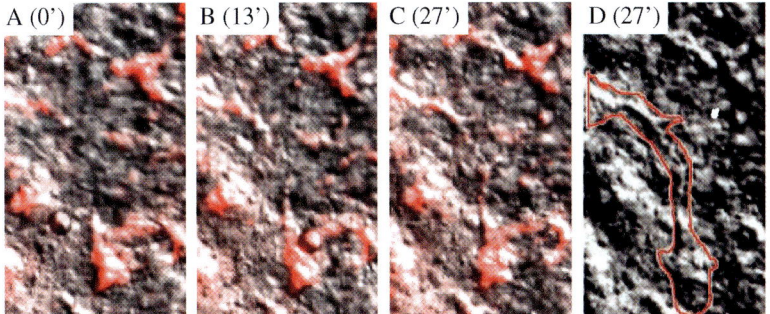

Chapter 9, Figure 3.

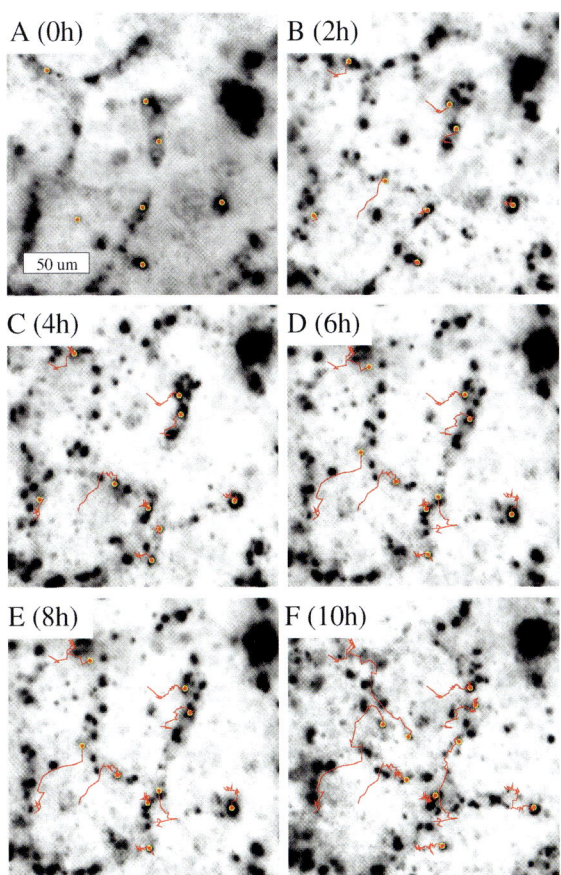

Chapter 9, Figure 4.

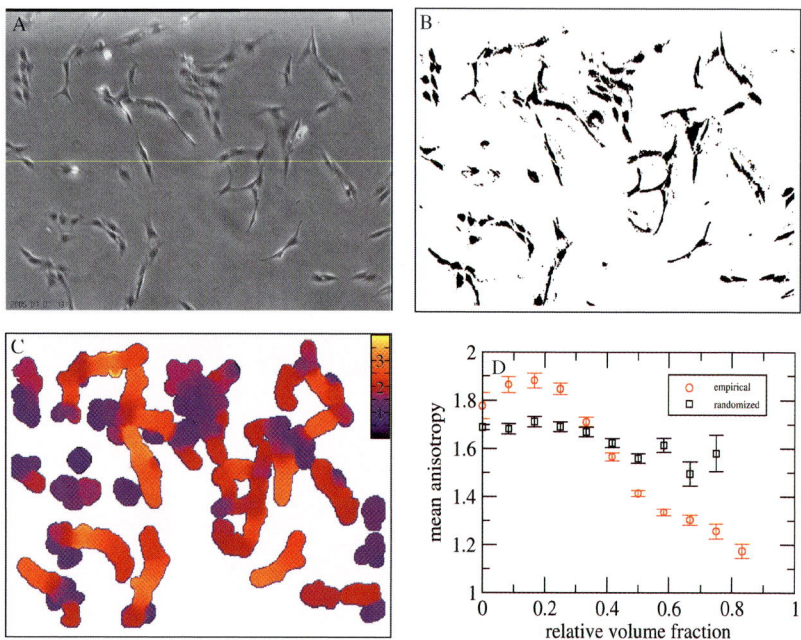

Chapter 9, Figure 5.

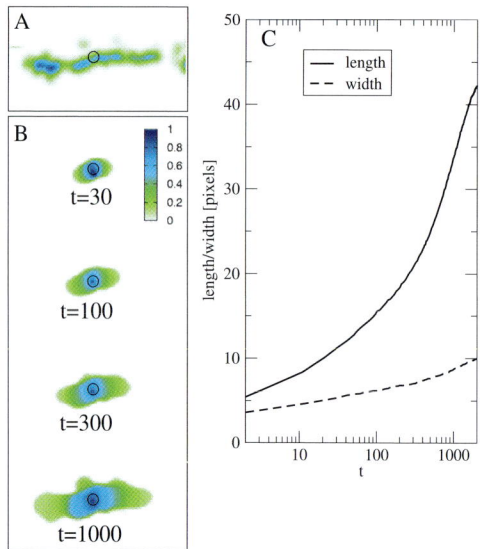

Chapter 9, Figure 6.

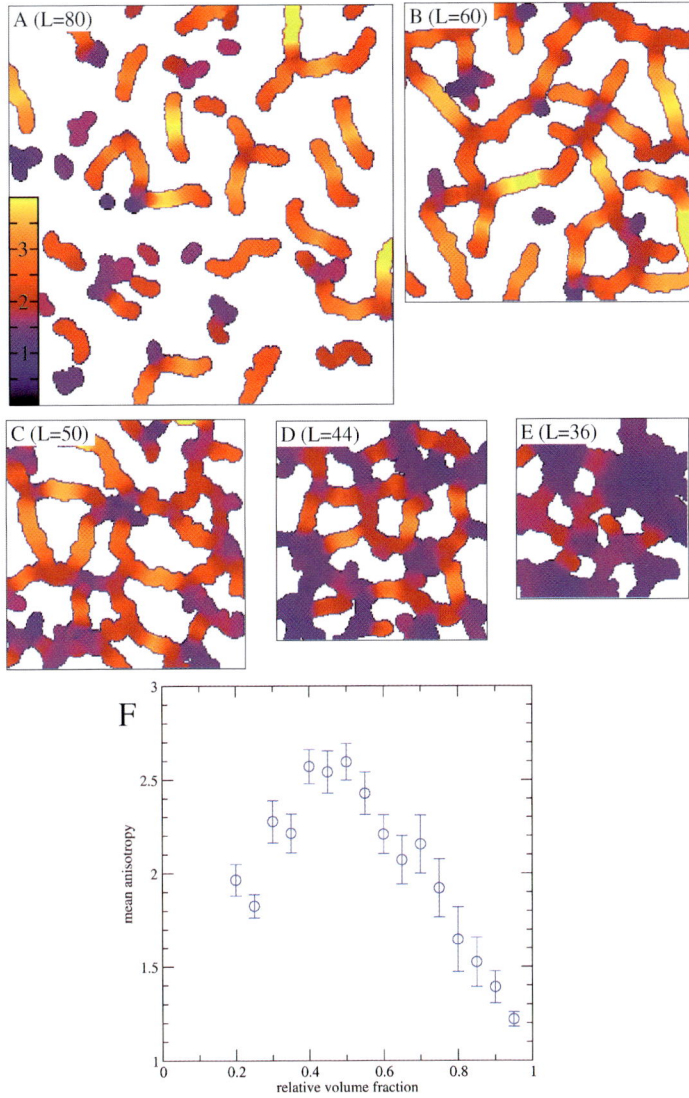

Chapter 9, Figure 10.

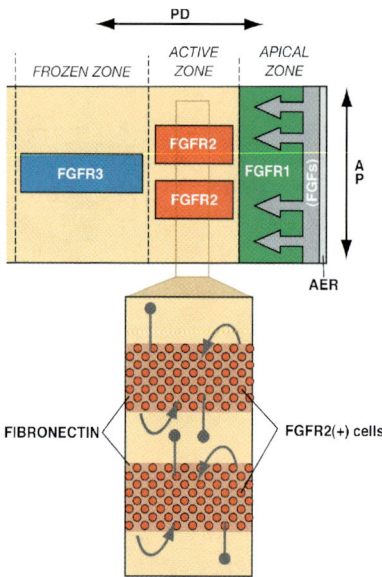

Chapter 11, Figure 3.

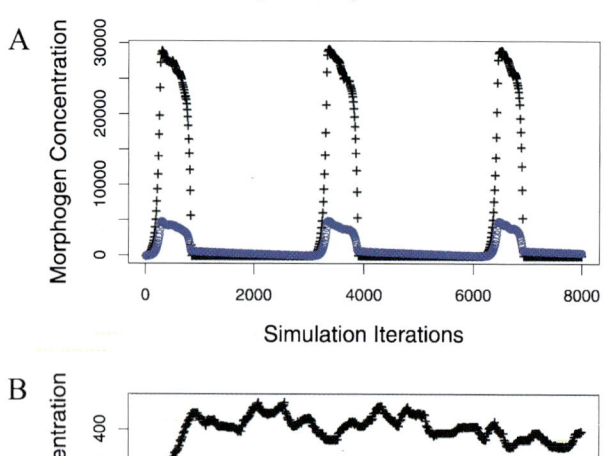

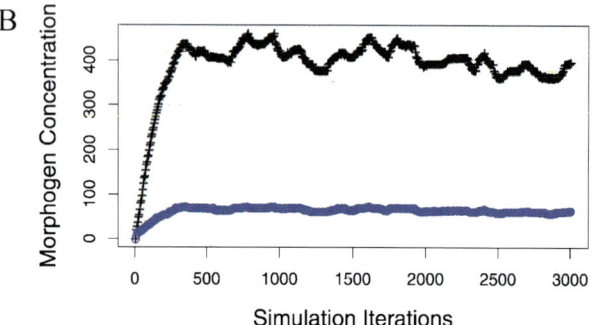

Chapter 11, Figure 5.

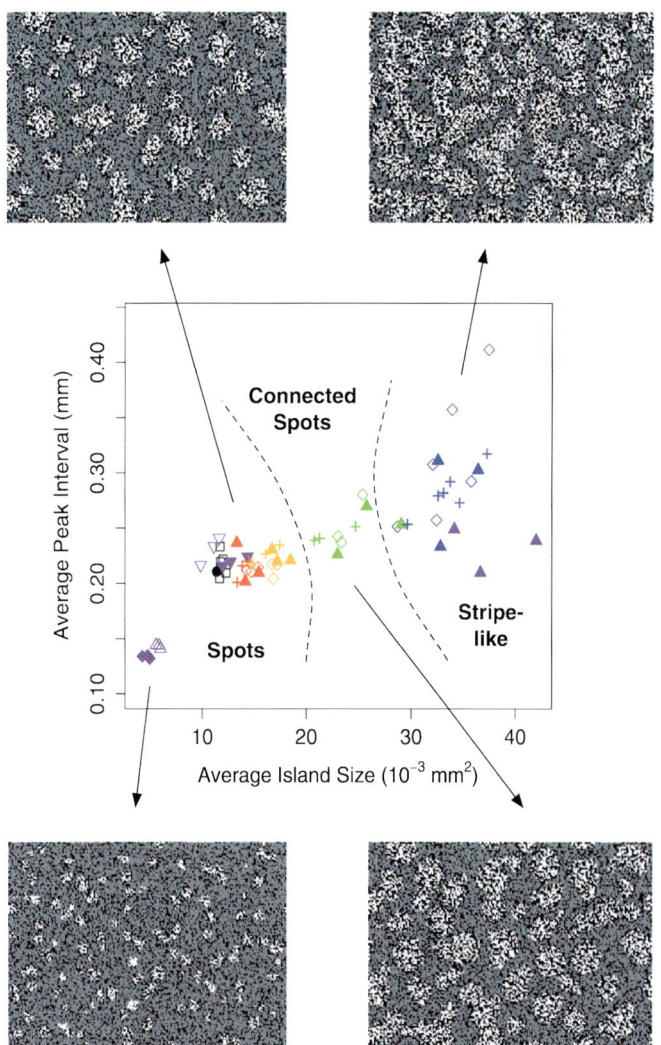

Chapter 11, Figure 6.

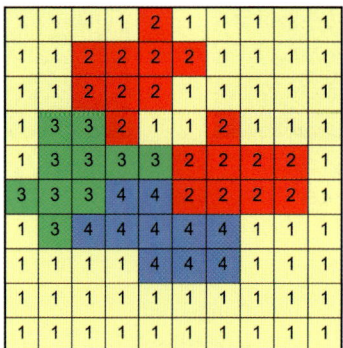

Chapter 11, Figure 7.

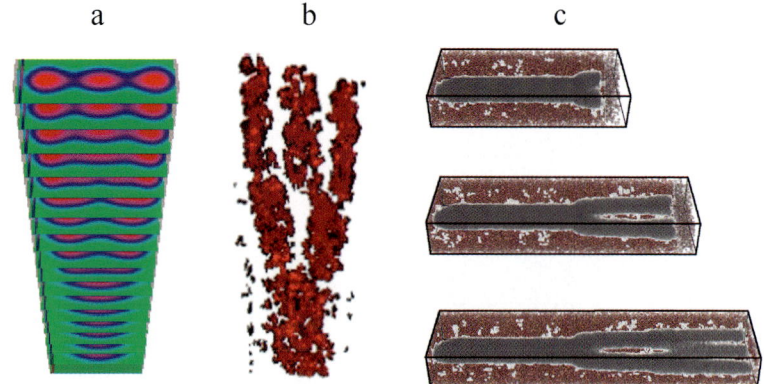

Chapter 11, Figure 8.

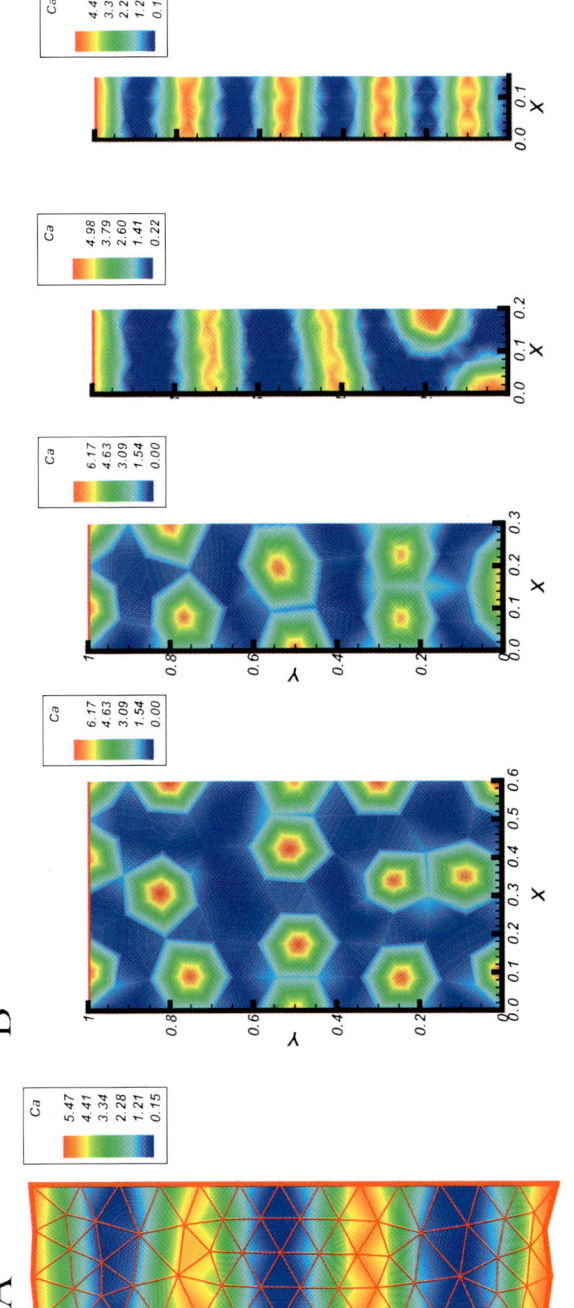

Chapter 11, Figure 9.

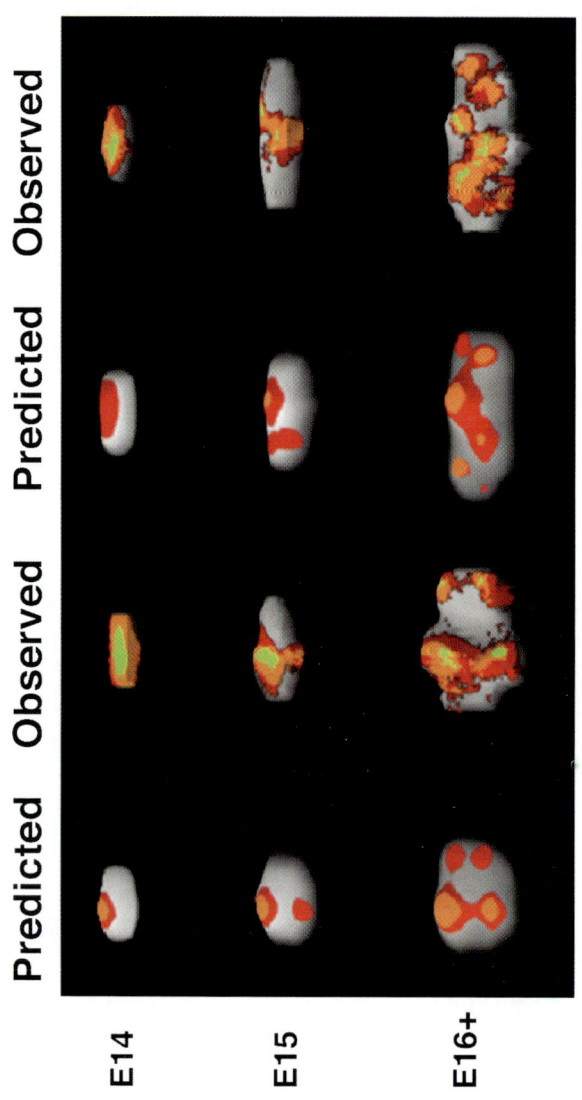

Chapter 12, Figure 3.

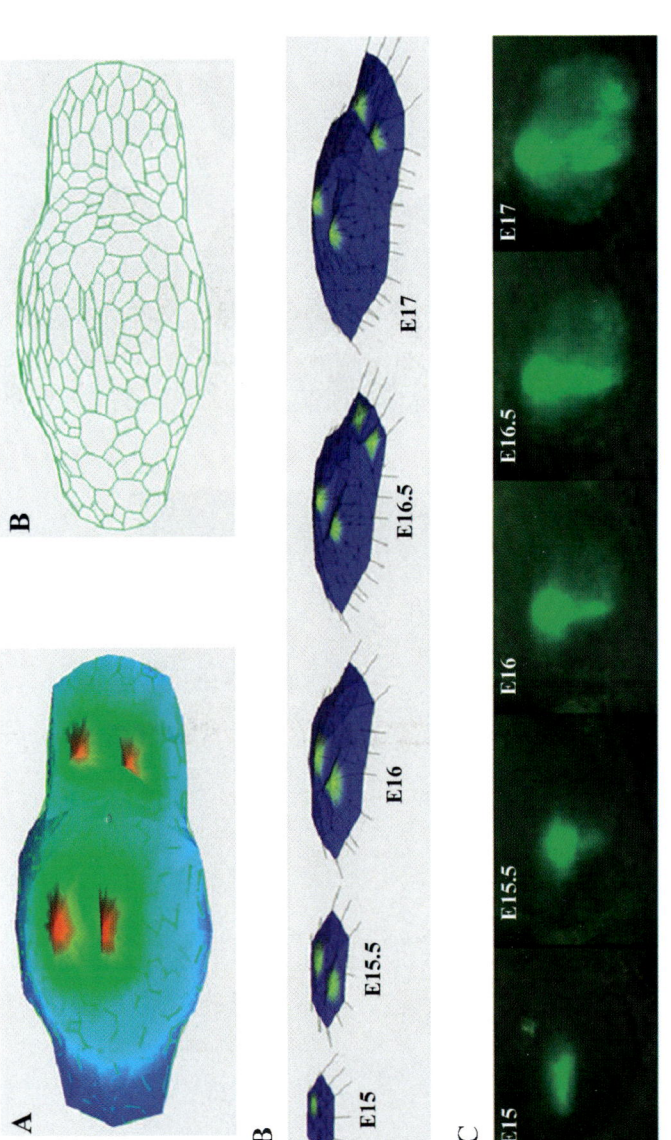

Chapter 12, Figure 4.

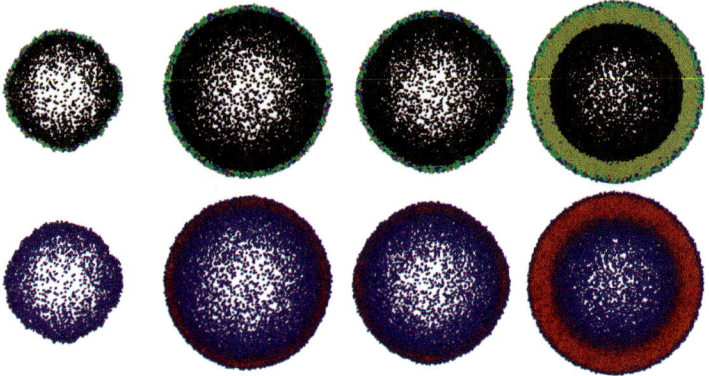

Chapter 13, Figure 6.

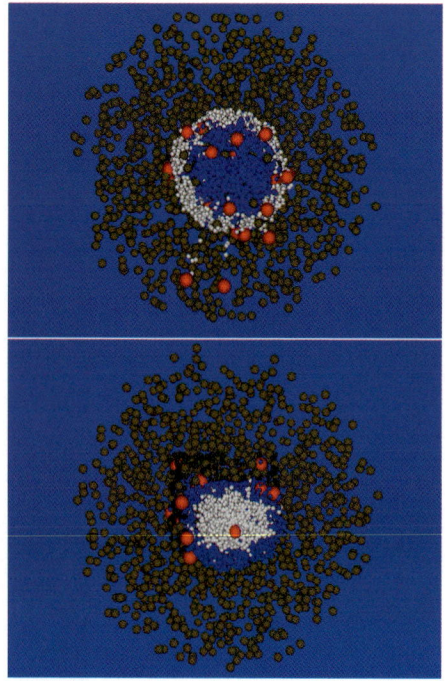

Chapter 13, Figure 7.

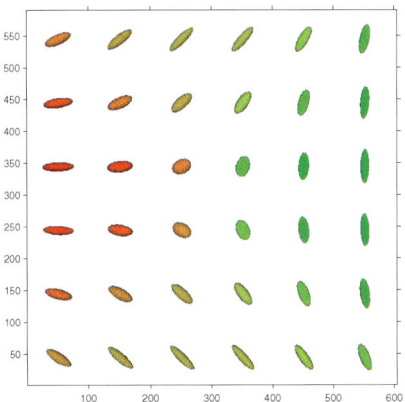

Chapter 14, Figure 6.

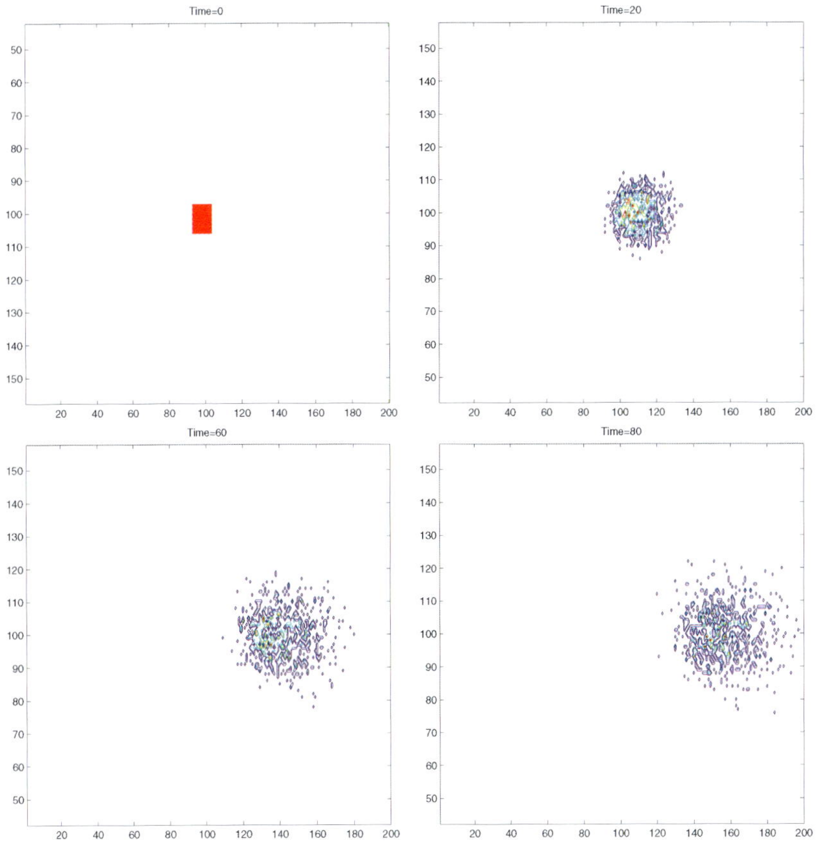

Chapter 14, Figure 7.

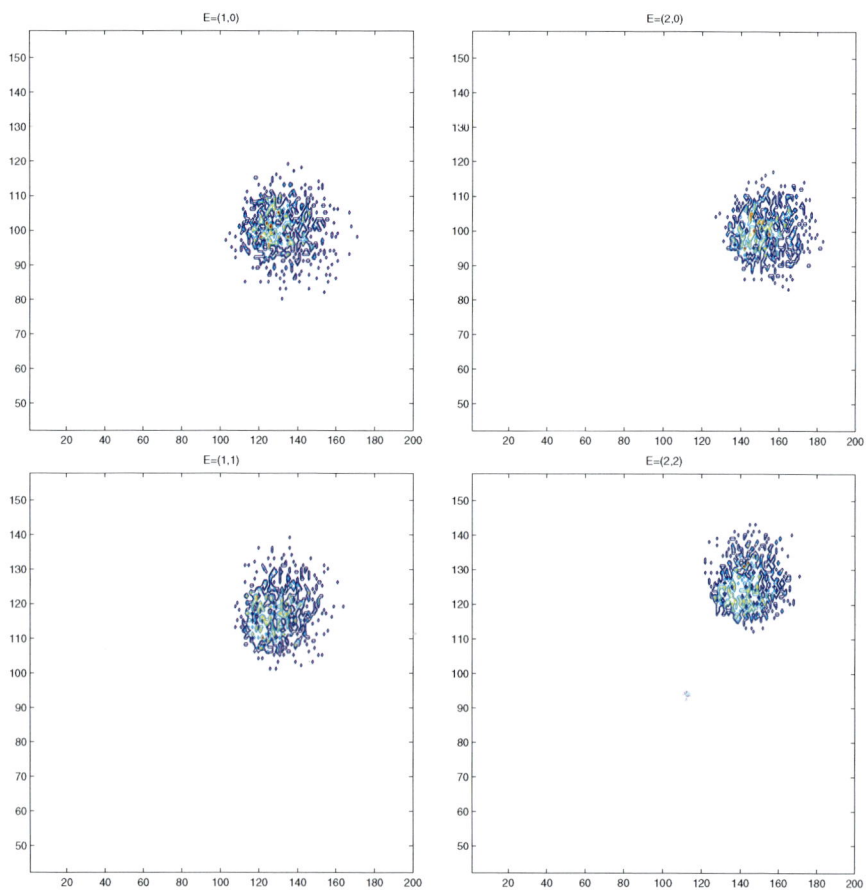

Chapter 14, Figure 8.

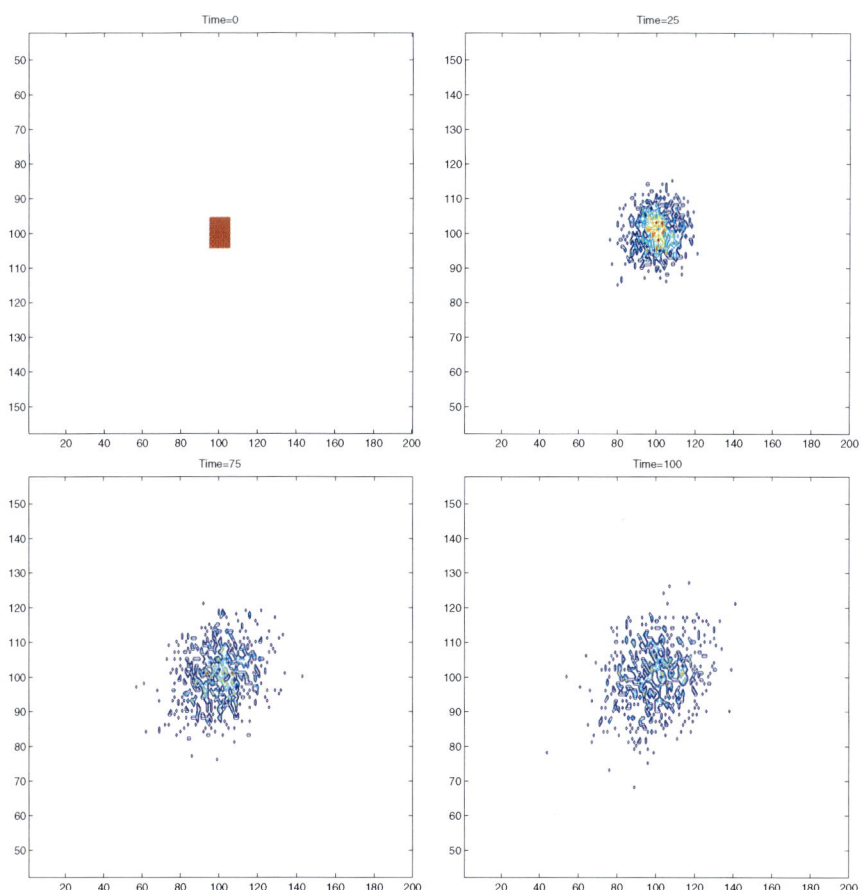

Chapter 14, Figure 9.

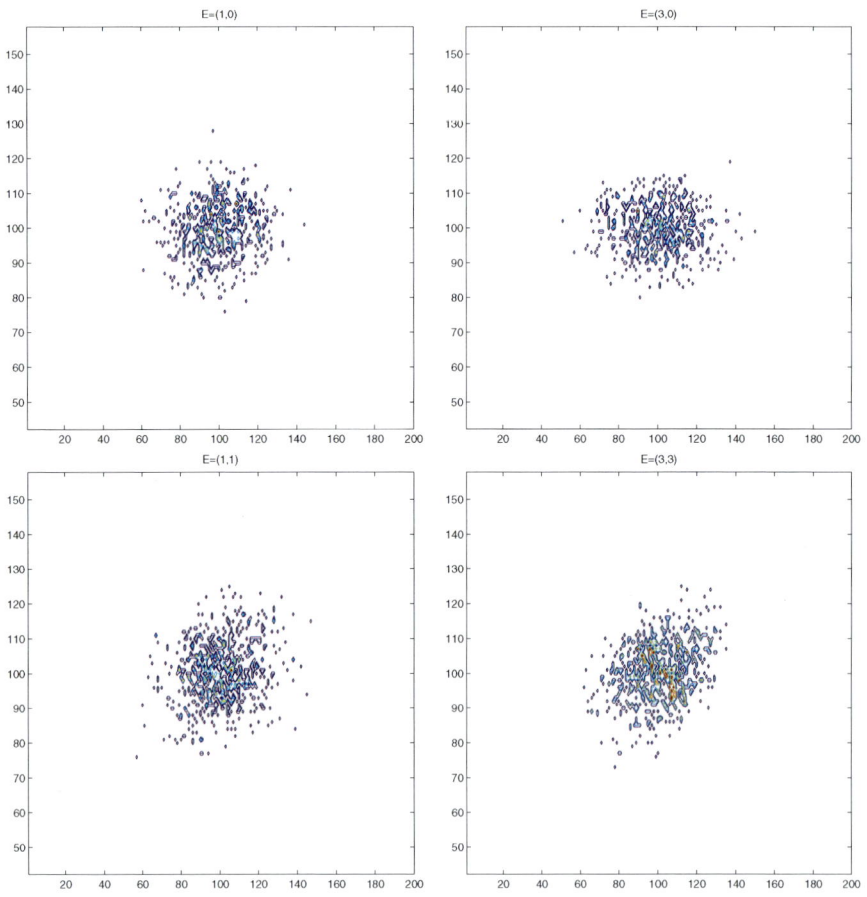

Chapter 14, Figure 10.

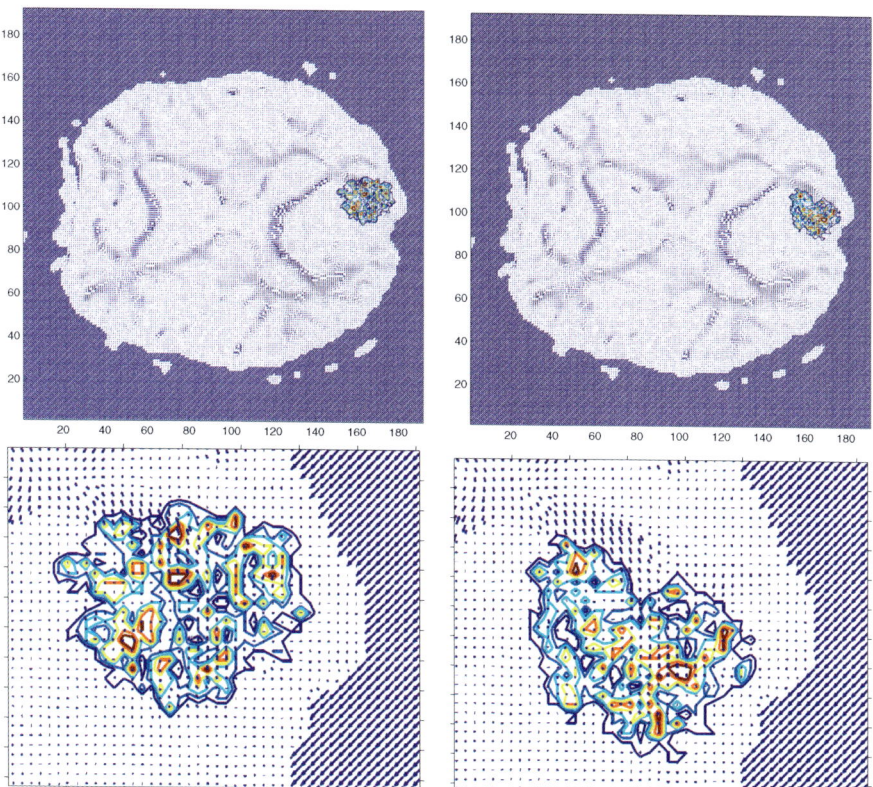

Chapter 14, Figure 11.

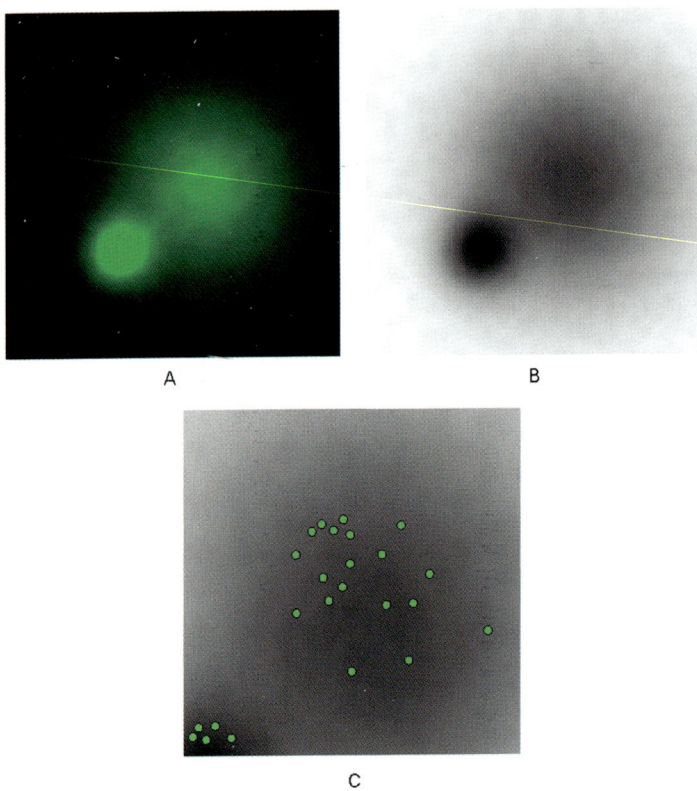

Chapter 15, Figure 1.

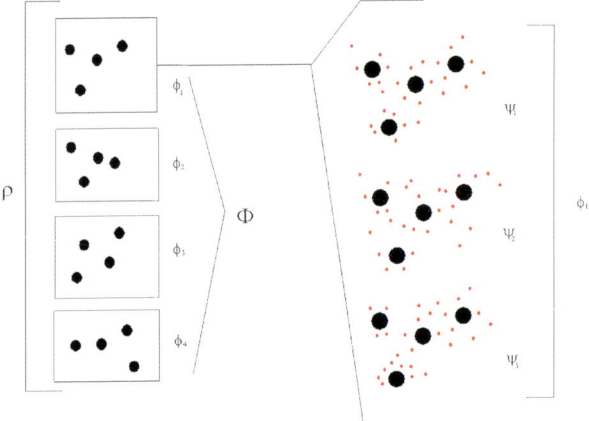

Chapter 15, Figure 3.